SPRINGER STUDY EDITION

Siegfried Flügge

Practical Quantum Mechanics

With 78 Figures

Springer-Verlag Berlin Heidelberg New York 1974

Prof. Dr. Siegfried Flügge
Fakultät für Physik der Universität Freiburg i. Br.

The first English edition of this book has been published in two volumes as:
Die Grundlehren der mathematischen Wissenschaften,
Band 177 and 178

AMS Subject Classification (1970)
Primary 81-01, 81-02, 81 A 06, 81 A 10, 81 A 63, 81 A 69, 81 A 75, 81 A 81
Secondary 34 E 20, 35 J 10

ISBN 3-540-07050-8 Springer-Verlag Berlin Heidelberg New York
ISBN 0-387-07050-8 Springer-Verlag New York Heidelberg Berlin

 Library of Congress Catalog Card Number 70-140508. Printed in Germany. Monophoto typesetting and offset printing: Zechnersche Buchdruckerei, Speyer. Bookbinding: Konrad Triltsch, Graphischer Betrieb, Würzburg
2141/3140-54321

Preface

This work was first published in 1947 in German under the title "Rechenmethoden der Quantentheorie". It was meant to serve a double purpose: to help both, the student when first confronted with quantum mechanics and the experimental scientist, who has never before used it as a tool, to learn how to apply the general theory to practical problems of atomic physics. Since that early date, many excellent books have been written introducing into the general framework of the theory and thus indispensable to a deeper understanding. It seems, however, that the more practical side has been somewhat neglected, except, of course, for the flood of special monographs going into broad detail on rather restricted topics. In other words, an all-round introduction to the practical use of quantum mechanics seems, so far, not to exist and may still be helpful.

It was in the hope of filling this gap that the author has fallen in with the publishers' wish to bring the earlier German editions up to date and to make the work more useful to the worldwide community of science students and scientists by writing the new edition in English.

From the beginning there could be no doubt that the work had to be much enlarged. New approximation methods and other developments, especially in the field of scattering, had to be added. It seemed necessary to include relativistic quantum mechanics and to offer, at least, a glimpse of radiation theory as an example of wave field quantization. The choice of the problems, included in the old days in a somewhat happy-go-lucky way, had now to be carefully reconsidered.

Thus a total of about twice as many problems as in the last German edition has resulted. Not one of the original problems has been simply translated; not more than about fifty have only undergone reshaping from the earlier text; the bulk, however, is going to be presented here for the first time. Nevertheless, the general character has remained the same, with perhaps a slight tendency to arrive at even more applicable results and numerical values at the end of each problem.

The more elementary problems, such as square-well potentials, have not been omitted but somewhat abridged. The general introduction to the German edition, some twenty odd pages surveying the basic equations and their meaning, has been discarded. Any student using the problems will be sufficiently well acquainted with the general framework to justify that omission. On the other hand, the extensive use of special functions made throughout the work seemed to make a mathematical appendix useful in which such formulae as occur in the problems have been collected and, in part, derived.

With considerable hesitation but giving way to the publishers' practical arguments the author has consented to having this edition divided into two separate volumes, hoping that no serious damage has thus been done to the intrinsic structure and continuity of the work. To facilitate its use, the complete index for both parts has been printed twice and will appear at the end of each volume. The numbers, therefore, refer to the problems in question, not to pages.

Hinterzarten, March 1971

The Author

Preface to the Paperback Edition

More than three years have elapsed, since this work appeared as a two-volume cloth edition. Author and publisher have been much gratified to learn how well it was received by scientists in many countries and how useful it proved to students of physics. To serve as a regular supplement to text books, however, it was hampered by its necessarily high price. Hence, the author has gratefully accepted the publisher's suggestion of a much less expensive one-volume Springer Study Edition; the more so, since it has always been his wish, as already expressed in the former edition's preface, to see the total work re-united in one volume. To save every conceivable additional cost, no corrections or alterations have been made; an errata sheet listing trivial errors, however, has been prepared, see p. XVI. Furthermore, the original pagination has been kept unchanged which, we trust, should cause very little inconvenience.

Author and publisher very much hope that a more widespread distribution of the book will thereby result and that it will meet especially the needs of the student.

Hinterzarten, September 1974

The Author

Contents Volume I

I. General Concepts

II. One-Body Problems without Spin

Contents Volume II

III. Particles with Spin

IV. Many-Body Problems

V. Non-Stationary Problems

VI. The Relativistic Dirac Equation

VII. Radiation Theory

Mathematical Appendix

Errata

Volume I

p. 51. In Fig. 8f ($c^2=225$) an odd parity level just below the well edge has been omitted.
p. 82. In Eq. (35.13) read ε_2 in the two last terms instead of ε_1.
p. 85. In Eq. (36.4) read 105 instead of 15.
p. 102. In Eq. (40.8) read $\xi^{3/2}$ instead of $\xi^{2/3}$.
p. 131, line 3, read Eq. (48.1) instead of Eq. (47.8).
p. 155. The fourth formula on this page should run

$$Q_{zz}=\langle r^2\rangle\, 3\,[(a_{l,m}^2+a_{l-1,m}^2)-1]$$

and the sixth formula

$$Q_{zz}=\tfrac{4}{5}\,\langle r^2\rangle\,(1-\tfrac{3}{2}\,m^2).$$

p. 299. Omit the factor $1/\pi$ in Eq. (113.3) in front of the integral.
p. 300, line 2. The residues are $\dfrac{1}{\pi}(-1)^{l+1}$.
p. 329. In Eq. (128.7) and in the equation immediately before it, the factor $\frac{1}{6}$ should be replaced by $\frac{1}{3}$.

Volume II

p. 11. In the formula for $\boldsymbol{J}^2\psi$ following Eq. (133.7d) the second factor under the radical in the third line should read $(l-m_j+\frac{1}{2})$.
p. 56. In Eq. (152.6) omit the summation sign.
p. 95. The right-hand side of Eq. (166.5) should run $W_\nu \boldsymbol{u}_\nu(\boldsymbol{r}_2)$.
p. 232. In the formula for W (line 12 from above) read $\omega\omega'/c^2$ instead of $\omega\omega'/c$. On line 18 from above read ω^2/c^2 instead of ω^2/c.
p. 265. Eq. (9a) should read $P_l(\pm 1)=(\pm 1)^l$. On line 9 from bottom replace $1/l$ by $1/\sqrt{l}$.
p. 269. On the first line of Eq. (14) both spherical harmonics on the right-hand side should have $m\pm 1$ for second subscripts. The right-hand side of the second line of this equation should begin $-l b_{l,m}$ etc.
p. 274. In Eq. (1) a factor x in front of the first term should be added.

I. General Concepts

Problem 1. Law of probability conservation

If the normalization relation

$$\int d^3x\,\psi^*\psi = 1 \tag{1.1}$$

is interpreted in the sense of probability theory, so that $d^3x\,\psi^*\psi$ is the probability of finding the particle under consideration in the volume element d^3x, then there must be a conservation law. This is to be derived. How may it be interpreted classically?

Solution. The conservation law sought must have the form of an equation of continuity,

$$\operatorname{div}\mathbf{s} + \frac{\partial\rho}{\partial t} = 0 \tag{1.2}$$

with

$$\rho = \psi^*\psi \tag{1.3}$$

the probability density, and $\mathbf{s}$ the probability current density. As ρ is a bilinear form of ψ and its complex conjugate, Eq. (1.2) can be constructed only by a combination of the two Schrödinger equations

$$H\psi = -\frac{\hbar}{i}\frac{\partial\psi}{\partial t}; \qquad H\psi^* = \frac{\hbar}{i}\frac{\partial\psi^*}{\partial t} \tag{1.4}$$

with the hamiltonian

$$H = -\frac{\hbar^2}{2m}\nabla^2 + V(\mathbf{r}) \tag{1.5}$$

the same for both equations. Thus we find

$$\psi^* H\psi - \psi H\psi^* = -\frac{\hbar}{i}\frac{\partial\rho}{\partial t}.$$

According to (1.2) it ought to be possible to write the left-hand side in the form of a divergence. Indeed we have

$$\psi^* H\psi - \psi H\psi^* = -\frac{\hbar^2}{2m}(\psi^*\nabla^2\psi - \psi\nabla^2\psi^*) = -\frac{\hbar^2}{2m}\operatorname{div}(\psi^*\nabla\psi - \psi\nabla\psi^*)$$

so that we may identify

$$s = \frac{\hbar}{2mi}(\psi^*\nabla\psi - \psi\nabla\psi^*). \tag{1.6}$$

Classical interpretations may be arrived at as follows. If the quantities ρ and s are both multiplied by m, the mass of the particle, we obtain mass density ρ_m and momentum density $\boldsymbol{g}$:

$$\rho_m = m\rho\,; \qquad \boldsymbol{g} = m\,\boldsymbol{s}, \tag{1.7}$$

and the equation of continuity may be interpreted as the law of mass conservation. In the same way, multiplication by the particle charge, e, yields charge density ρ_e and electric current density $\boldsymbol{j}$:

$$\rho_e = e\rho\,; \qquad \boldsymbol{j} = e\,\boldsymbol{s}, \tag{1.8}$$

and (1.2) becomes the law of charge conservation.

It is remarkable that the conservation laws of both mass and charge are essentially identical. This derives from the fact that one particle by its convection current causes both.

The expression for the total momentum of the Schrödinger field, derived from (1.6) and (1.7),

$$\boldsymbol{p} = \int d^3x\,\boldsymbol{g} = \frac{\hbar}{2i}\int d^3x(\psi^*\nabla\psi - \psi\nabla\psi^*),$$

may by partial integration in the second term be reduced to

$$\boldsymbol{p} = \int d^3x\,\psi^*\left(\frac{\hbar}{i}\nabla\right)\psi \tag{1.9}$$

corresponding to its explanation as the expectation value of the momentum operator $(\hbar/i)\,\nabla$ in the quantum state ψ (cf. Problem 3).

Problem 2. Variational principle of Schrödinger

To replace the Schrödinger equation

$$-\frac{\hbar^2}{2m}\nabla^2\psi + V(\boldsymbol{r})\psi = E\psi \tag{2.1}$$

by a variational principle for the energy.

Solution. Since the constraint

$$\int d^3x\,\psi^*\psi = 1 \tag{2.2}$$

holds for any solution ψ of the differential equation (2.1), the energy will be found by multiplying (2.1) with ψ^* and integrating over the whole space:

$$E = \int d^3x\,\psi^* \left\{-\frac{\hbar^2}{2m}\nabla^2\psi + V(\mathbf{r})\psi\right\}. \tag{2.3}$$

A partial integration in the first term yields, according to Green's law,

$$\int d^3x\,\psi^*\nabla^2\psi = \oint d\mathbf{f}\cdot\psi^*\nabla\psi - \int d^3x\,\nabla\psi^*\cdot\nabla\psi\,. \tag{2.4}$$

Now, the normalization integral (2.2) exists only if, at large distances r, the solution ψ vanishes at least as

$$\psi \propto r^{-\frac{3}{2}-\varepsilon}; \qquad \varepsilon > 0\,.$$

Under this condition, however, the surface integral in (2.4) vanishes when taken over an infinitely remote sphere so that (2.3) may be written

$$E = \int d^3x \left\{\frac{\hbar^2}{2m}\nabla\psi^*\cdot\nabla\psi + \psi^* V(\mathbf{r})\psi\right\}. \tag{2.5}$$

This equation is completely symmetrical in the functions ψ and ψ^*, as is the normalization (2.2), so that it might equally well have been derived from the complex conjugate of Eq. (2.1),

$$-\frac{\hbar^2}{2m}\nabla^2\psi^* + V(\mathbf{r})\psi^* = E\psi^*\,. \tag{2.1*}$$

It would not be difficult to show that (2.1) and (2.1*) are the Euler equations of the variational problem to extremize the integral (2.5) with the constraint (2.2). We shall, however, make no use of the apparatus of variational theory and prefer a direct proof, instead.

Let ψ_λ be a solution of (2.1) that belongs to its eigenvalue E_λ. It will give the integral (2.5) the value E_λ. Let us then replace ψ_λ by a neighbouring function $\psi_\lambda+\delta\psi$ with $|\delta\psi|$ being small but arbitrary, except for (2.2) still to hold for $\psi_\lambda+\delta\psi$ as well as for ψ_λ:

$$\int d^3x(\psi_\lambda^*+\delta\psi^*)(\psi_\lambda+\delta\psi) = 1$$

and therefore

$$\int d^3x(\psi_\lambda\delta\psi^*+\psi_\lambda^*\delta\psi) + \int d^3x\,\delta\psi^*\delta\psi = 0\,. \tag{2.6}$$

Setting $\psi_\lambda+\delta\psi$ into the energy integral (2.5), the energy becomes $E_\lambda+\delta E_\lambda$ with

$$\delta E_\lambda=\int d^3x\left\{\frac{\hbar^2}{2m}(\nabla\psi_\lambda^*\cdot\nabla\delta\psi+\nabla\psi_\lambda\cdot\nabla\delta\psi^*)+V(\psi_\lambda\delta\psi^*+\psi_\lambda^*\delta\psi)\right\}$$
$$+\int d^3x\left\{\frac{\hbar^2}{2m}\nabla\delta\psi^*\cdot\nabla\delta\psi+V\delta\psi^*\delta\psi\right\}. \tag{2.7}$$

Here the first-order changes stand in the first, and the second-order changes in the second line. By partial integration in the sense opposite to the one above we fall back, in the first line, on $\delta\psi\nabla^2\psi_\lambda^*$ and $\delta\psi^*\nabla^2\psi_\lambda$ where (2.1) and (2.1*) may be used to eliminate the derivatives. E.g. we then have

$$\int d^3x\left\{\frac{\hbar^2}{2m}\nabla\psi_\lambda^*\cdot\nabla\delta\psi+V\psi_\lambda^*\delta\psi\right\}=E_\lambda\int d^3x\,\delta\psi\,\psi_\lambda^*,$$

so that with the help of Eq. (2.6), the first line of (2.7) may finally be reduced to second-order contributions only:

$$\delta E_\lambda=\int d^3x\left\{\frac{\hbar^2}{2m}|\nabla\delta\psi|^2+(V-E_\lambda)|\delta\psi|^2\right\}. \tag{2.8}$$

Since no linear contribution in $\delta\psi$ or $\delta\psi^*$ remains, E_λ clearly is a maximum or a minimum for $\delta\psi=0$, i.e. for ψ_λ being a solution of the Schrödinger equation. Whether we get a maximum or a minimum will be decided by the sign of (2.8).

To get some insight into this last question we make use of the set $\{\psi_\nu\}$ of solutions to (2.1) to form a complete orthogonal system of functions,

$$\int d^3x\,\psi_\mu^*\psi_\nu=\delta_{\mu\nu}. \tag{2.9}$$

We then expand $\delta\psi$ with respect to this system:

$$\delta\psi=\sum_\nu c_\nu\psi_\nu. \tag{2.10}$$

Eq. (2.8) then renders

$$\delta E_\lambda=\sum_\mu\sum_\nu c_\mu^*c_\nu\int d^3x\left\{\frac{\hbar^2}{2m}\nabla\psi_\mu^*\cdot\nabla\psi_\nu+(V-E_\lambda)\psi_\mu^*\psi_\nu\right\}$$
$$=\sum_\mu\sum_\nu c_\mu^*c_\nu\int d^3x\,\psi_\mu^*\left\{-\frac{\hbar^2}{2m}\nabla^2\psi_\nu+(V-E_\lambda)\psi_\nu\right\}$$

or, using (2.1) and (2.9):

$$\delta E_\lambda=\sum_\mu|c_\mu|^2(E_\mu-E_\lambda). \tag{2.11}$$

If E_λ is the ground state, we have $E_\mu \geqslant E_\lambda$ for all states μ, so that the sum (2.11) is positive. The variational principle therefore makes E_λ a minimum. No such general rule can be established for excited states where the sum (2.11) consists of positive and negative terms.

Problem 3. Classical mechanics for space averages

To show that Newton's fundamental equation of classical dynamics,

$$\frac{d\boldsymbol{p}}{dt} = \boldsymbol{F}\,, \tag{3.1}$$

with $\boldsymbol{p}$ the momentum of, and $\boldsymbol{F}$ the force acting upon the particle, still holds for the space averages (expectation values) of the corresponding operators in quantum mechanics.

Solution. If the force $\boldsymbol{F}$ derives from a potential, $\boldsymbol{F} = -\nabla V$, and momentum is replaced by the operator $(\hbar/i)\nabla$, then the two space averages in Eq. (3.1) are defined by

$$\boldsymbol{p} = \frac{\hbar}{i}\int d^3x\, \psi^* \nabla \psi\,; \tag{3.2}$$

$$\boldsymbol{F} = -\int d^3x\, \psi^* (\nabla V)\psi\,. \tag{3.3}$$

Our task then is to prove that (3.1) is valid for the integrals (3.2) and (3.3) if ψ and ψ^* satisfy the Schrödinger equations

$$\left.\begin{aligned} -\frac{\hbar}{i}\frac{\partial\psi}{\partial t} &= -\frac{\hbar^2}{2m}\nabla^2\psi + V\psi\,; \\ +\frac{\hbar}{i}\frac{\partial\psi^*}{\partial t} &= -\frac{\hbar^2}{2m}\nabla^2\psi^* + V\psi^*\,. \end{aligned}\right\} \tag{3.4}$$

We start our proof with the time derivative of Eq. (3.2):

$$\dot{\boldsymbol{p}} = \frac{\hbar}{i}\int d^3x(\dot\psi^*\nabla\psi + \psi^*\nabla\dot\psi) = \frac{\hbar}{i}\int d^3x(\dot\psi^*\nabla\psi - \dot\psi\nabla\psi^*)$$

where the surface contribution of the partial integration in the last term vanishes and has been omitted. Replacing $\dot\psi^*$ and $\dot\psi$ according to (3.4), we may proceed to

$$\dot{\boldsymbol{p}} = -\frac{\hbar^2}{2m}\int d^3x(\nabla^2\psi^*\nabla\psi + \nabla^2\psi\nabla\psi^*)$$
$$+\int d^3x(\psi^* V\nabla\psi + V\psi\nabla\psi^*)\,. \tag{3.5}$$

A partial integration,

$$\int d^3x\,\nabla^2\psi^*\nabla\psi = -\int d^3x\,\nabla\psi^*\nabla^2\psi$$

shows that the two terms of the first integral cancel each other out. In the other integral of Eq. (3.5) we perform a partial integration in the last term,

$$\dot{\boldsymbol{p}} = \int d^3x\,\psi^*\{V\nabla\psi - \nabla(V\psi)\}\,.$$

Making use of

$$\nabla(V\psi) = V\nabla\psi + \psi\nabla V,$$

we finally arrive at

$$\dot{\boldsymbol{p}} = -\int d^3x\,\psi^*(\nabla V)\psi = \boldsymbol{F}\,,$$

as was to be proved.

Problem 4. Classical laws for angular motion

To show that the classical relation between angular momentum $\boldsymbol{L} = \boldsymbol{r}\times\boldsymbol{p}$ and torque $\boldsymbol{T} = \boldsymbol{r}\times\boldsymbol{F}$ (where $\boldsymbol{p}$ stands for linear momentum and $\boldsymbol{F}$ for force),

$$\frac{d\boldsymbol{L}}{dt} = \boldsymbol{T}\,, \tag{4.1}$$

still holds for the space averages in quantum mechanics.

Solution. As in the preceding problem, we start by constructing the space averages

$$\boldsymbol{L} = \frac{\hbar}{i}\int d^3x\,\psi^*(\boldsymbol{r}\times\nabla)\psi \tag{4.2}$$

and

$$\boldsymbol{T} = -\int d^3x\,\psi^*(\boldsymbol{r}\times\nabla V)\psi\,. \tag{4.3}$$

The wave functions, ψ and ψ^*, are again supposed to satisfy the Schrödinger equations (3.4).

We begin the proof of Eq. (4.1) by differentiating $\boldsymbol{L}$, Eq. (4.2):

$$\dot{\boldsymbol{L}} = \frac{\hbar}{i} \int d^3x \{\dot{\psi}^*(\boldsymbol{r} \times \nabla\psi) + \psi^*(\boldsymbol{r} \times \nabla\dot{\psi})\} .$$

In the second term we use the identity

$$\psi^* \nabla \dot{\psi} = \nabla(\psi^* \dot{\psi}) - \dot{\psi} \nabla \psi^* ,$$

to the first term of which we apply the general vector rule

$$\int d^3x\, \boldsymbol{r} \times \nabla f = 0 \tag{4.4}$$

with $f = \psi^* \dot{\psi}$. Thus we arrive at

$$\dot{\boldsymbol{L}} = \frac{\hbar}{i} \int d^3x\, \boldsymbol{r} \times (\dot{\psi}^* \nabla \psi - \dot{\psi} \nabla \psi^*)$$

where we replace the time derivatives $\dot{\psi}^*$ and $\dot{\psi}$ according to (3.4):

$$\begin{aligned} \dot{\boldsymbol{L}} = &-\frac{\hbar^2}{2m} \int d^3x\, \boldsymbol{r} \times (\nabla^2 \psi^* \nabla \psi + \nabla^2 \psi \nabla \psi^*) \\ &+ \int d^3x\, V \boldsymbol{r} \times (\psi^* \nabla \psi + \psi \nabla \psi^*) . \end{aligned} \tag{4.5}$$

Now, in the first integral, the bracket

$$\nabla^2 \psi^* \nabla \psi + \nabla^2 \psi \nabla \psi^* = \nabla(\nabla \psi^* \cdot \nabla \psi)$$

is the gradient of the scalar function $f = \nabla \psi^* \cdot \nabla \psi$ so that, according to (4.4), this integral vanishes. In the second integral, the bracket is equal to $\nabla(\psi^* \psi)$. We then use the identity

$$V \nabla(\psi^* \psi) = \nabla(V \psi^* \psi) - \psi^* \psi \nabla V$$

and for $f = V \psi^* \psi$ again the vector rule (4.4). Then the integral becomes finally

$$\int d^3x\, \boldsymbol{r} \times [V \nabla(\psi^* \psi)] = -\int d^3x\, \boldsymbol{r} \times (\psi^* \psi \nabla V) ,$$

i.e. it becomes identical with the torque average (4.3), as was to be proved.

Problem 5. Energy conservation law

If the energy content of a Schrödinger wave field is described by the space integral (2.5) of problem 2, the law of energy conservation should be of the form

$$\frac{\partial W}{\partial t} + \operatorname{div} \boldsymbol{S} = 0 \tag{5.1}$$

with W the energy density and $\boldsymbol{S}$ the energy flux vector. This law shall by derived by constructing the flux vector $\boldsymbol{S}$.

Solution. We have found in (2.5) that

$$E = \int d^3x\, W \tag{5.2}$$

with

$$W = \frac{\hbar^2}{2m} \nabla\psi^* \cdot \nabla\psi + \psi^* V \psi\,, \tag{5.3}$$

where the first term is the kinetic, the second the potential energy density. According to Eq. (5.1) we need the derivative

$$\dot{W} = \frac{\hbar^2}{2m}(\nabla\dot{\psi}^* \cdot \nabla\psi + \nabla\psi^* \cdot \nabla\dot{\psi}) + V(\dot{\psi}^*\psi + \psi^*\dot{\psi})\,. \tag{5.4}$$

Since

$$\nabla\dot{\psi}^* \nabla\psi = \nabla(\dot{\psi}^* \nabla\psi) - \dot{\psi}^* \nabla^2\psi$$

and

$$\nabla\psi^* \nabla\dot{\psi} = \nabla(\dot{\psi} \nabla\psi^*) - \dot{\psi} \nabla^2\psi^*$$

we can reshape the kinetic energy part of (5.4) and write

$$\dot{W} = \nabla\left\{\frac{\hbar^2}{2m}(\dot{\psi}^* \nabla\psi + \dot{\psi}\nabla\psi^*)\right\} - \frac{\hbar^2}{2m}\dot{\psi}^*\nabla^2\psi - \frac{\hbar^2}{2m}\dot{\psi}\nabla^2\psi^*$$
$$+ \dot{\psi}^* V\psi \qquad + \dot{\psi} V \psi^*\,. \tag{5.5}$$

In the last terms, use of the Schrödinger equations (3.4) permits us to replace space derivatives and potential by time derivatives. The resulting terms

$$\dot{\psi}^*\left(-\frac{\hbar}{i}\dot{\psi}\right) + \dot{\psi}\left(\frac{\hbar}{i}\dot{\psi}^*\right) = 0$$

exactly cancel so that Eq. (5.5) indeed is of the form (5.1) to be proved with

$$\boldsymbol{S} = -\frac{\hbar^2}{2m}(\dot{\psi}^* \nabla\psi + \dot{\psi}\nabla\psi^*) \tag{5.6}$$

the energy flux vector.

Problem 6. Hermitian conjugate

The hermitian conjugate $\Omega^\dagger$ of an operator Ω is defined by

$$\int d\tau (\Omega\psi)^* \varphi = \int d\tau \psi^* \Omega^\dagger \varphi \tag{6.1 a}$$

or, in Hilbert space notation,

$$\langle \Omega\psi | \varphi \rangle = \langle \psi | \Omega^\dagger \varphi \rangle \tag{6.1 b}$$

with ψ and φ any two functions normalized according to

$$\langle \psi | \psi \rangle = 1; \quad \langle \varphi | \varphi \rangle = 1 . \tag{6.2}$$

This definition shall be translated into a matrix relation. What follows for the eigenvalues of a hermitian operator defined by $\Omega = \Omega^\dagger$?

Solution. The matrix of an operator is defined with respect to a complete set of orthonormal functions, $\{u_\nu\}$:

$$\langle u_\nu | u_\mu \rangle = \delta_{\mu\nu} . \tag{6.3}$$

The arbitrary, but normalized functions ψ and φ then may be expanded,

$$\psi = \sum_\nu a_\nu u_\nu , \quad \varphi = \sum_\mu b_\mu u_\mu . \tag{6.4}$$

Putting (6.4) into (6.1), we get

$$\sum_\mu \sum_\nu a_\nu^* b_\mu \langle \Omega u_\nu | u_\mu \rangle = \sum_\mu \sum_\nu a_\nu^* b_\mu \langle u_\nu | \Omega^\dagger u_\mu \rangle$$

and, since this is supposed to hold for any pair ψ, φ, it must hold for each term separately,

$$\langle \Omega u_\nu | u_\mu \rangle = \langle u_\nu | \Omega^\dagger u_\mu \rangle . \tag{6.5}$$

We now use the set $\{u_\nu\}$ for matrix definition, writing the right-hand side of (6.5)

$$\langle u_\nu | \Omega^\dagger | u_\mu \rangle = (\Omega^\dagger)_{\mu\nu} .$$

The left-hand side may be reshaped as follows

$$\langle \Omega u_\nu | u_\mu \rangle = \int d\tau (\Omega u_\nu)^* u_\mu = \{ \int d\tau u_\mu^* (\Omega u_\nu) \}^* = \langle u_\mu | \Omega u_\nu \rangle^* = \Omega_{\nu\mu}^* .$$

Hence, it follows from (6.5) for the matrix elements of Ω and $\Omega^\dagger$ that

$$(\Omega^\dagger)_{\mu\nu} = \Omega_{\nu\mu}^* , \tag{6.6}$$

i.e. the elements of the hermitian conjugate (or adjoint) matrix are obtained by transposing ($\mu \rightleftharpoons \nu$), and taking the complex conjugate of, the elements of Ω.

It may be noted that from (6.6) we get immediately $\Omega^{\dagger\dagger} = \Omega$.

For a hermitian (or self-adjoint) matrix Ω with $\Omega = \Omega^\dagger$, according to (6.6), we have

$$\Omega_{\mu\nu} = \Omega^*_{\nu\mu}\,. \tag{6.7}$$

With $\mu = \nu$ this shows the diagonal elements to be real. That holds for any orthonormal set (i.e. for any coordinate system in Hilbert space), thus also if the set $\{u_\nu\}$ is so chosen as to make

$$\Omega_{\mu\nu} = \omega_\mu \delta_{\mu\nu} \tag{6.8}$$

a diagonal matrix with the eigenvalues ω_μ. It follows that a hermitian matrix has real eigenvalues.

NB. The last result is the reason for all operators corresponding to physical quantities (observables) being hermitian operators.

Problem 7. Construction of an hermitian operator

To translate the classical quantity xp_x into a quantum mechanical operator.

Solution. a) Since the operators x and p_x do not commute but satisfy the commutation relation

$$p_x x - x p_x = \frac{\hbar}{i}\,, \tag{7.1}$$

which is easily checked in the Schrödinger representation

$$p_x = \frac{\hbar}{i}\frac{\partial}{\partial x}\,, \tag{7.2}$$

we expect every operator

$$\Omega = (1-\alpha)x p_x + \alpha p_x x \tag{7.3}$$

to correspond to the classical quantity xp_x. Let us first suppose the constant α to be real. It must now be determined in such a way that the expectation value of Ω in any quantum state described by a wave function ψ is a real number:

$$\langle\Omega\rangle = \int d^3x\,\psi^*\,\Omega\psi = \text{real}. \tag{7.4}$$

Using (7.2) for p_x, Eq. (7.4) in more detail becomes

$$\langle\Omega\rangle = \frac{\hbar}{i}\int d^3x\,\psi^*\left\{(1-\alpha)x\frac{\partial\psi}{\partial x} + \alpha\frac{\partial}{\partial x}(x\psi)\right\}$$

$$= \frac{\hbar}{i}\int d^3x\,\psi^*\left(x\frac{\partial\psi}{\partial x} + \alpha\psi\right).$$

Breaking up

$$\psi = f + ig$$

into its real and imaginary parts, this leads to

$$\langle\Omega\rangle = \frac{\hbar}{i}\int d^3x\left\{x\left(f\frac{\partial f}{\partial x} + g\frac{\partial g}{\partial x}\right) + \alpha(f^2+g^2)\right\}$$

$$+\hbar\int d^3x\,x\left(f\frac{\partial g}{\partial x} - g\frac{\partial f}{\partial x}\right).$$

The second integral is real. The first being purely imaginary must vanish in order to satisfy (7.4). Since

$$\int d^3x(f^2+g^2) = 1$$

this condition may be written

$$\frac{1}{2}\int d^3x\,x\frac{\partial}{\partial x}(f^2+g^2) = -\alpha$$

or, by partial integration,

$$\alpha = \tfrac{1}{2} \tag{7.5}$$

so that the symmetrical combination

$$\Omega = \tfrac{1}{2}(xp_x + p_x x) \tag{7.6}$$

is the correct, because hermitian, operator.

If we admit complex values of α, any value

$$\alpha = \tfrac{1}{2} + i\beta \tag{7.7}$$

with arbitrary real β will do, because it leads to

$$\Omega = \tfrac{1}{2}(xp_x + p_x x) + i\beta(p_x x - xp_x) \tag{7.8}$$

where we have just shown the first part to have a real expectation value, and where the second part will in consequence of the commutation relation (7.1) become the real constant $\beta\hbar$, independent of the special quantum state. This term therefore has no physical significance and may be omitted.

b) Hermiticity of an operator may equally well be defined by the relation

$$\langle u|\Omega v\rangle = \langle\Omega u|v\rangle \tag{7.9a}$$

or, in more detail,

$$\int d^3x\,u^*\,\Omega v = \int d^3x(\Omega u)^*\,v \tag{7.9b}$$

with two arbitrary functions u and v, both complex and only chosen so that the integrals exist. With Ω the operator (7.3) and α real, this yields

$$\frac{\hbar}{i}\int d^3x\,u^*\left[(1-\alpha)x\frac{\partial v}{\partial x}+\alpha\frac{\partial(xv)}{\partial x}\right]=-\frac{\hbar}{i}\int d^3x\left[(1-\alpha)x\frac{\partial u^*}{\partial x}+\alpha\frac{\partial(xu^*)}{\partial x}\right]v$$

or

$$\int d^3x\,u^*\left[x\frac{\partial v}{\partial x}+\alpha v\right]=-\int d^3x\left[x\frac{\partial u^*}{\partial x}+\alpha u^*\right]v\,.$$

Reordering leads on to

$$\int d^3x\,x\frac{\partial}{\partial x}(u^*v)=-2\alpha\int d^3x\,u^*v$$

or, by partial integration on the left-hand side, to

$$-\int d^3x\,u^*v=-2\alpha\int d^3x\,u^*v$$

which, of course, yields the result of $\alpha=\frac{1}{2}$, Eq. (7.5), again.

Problem 8. Derivatives of an operator

Let $f(p,x)$ be an integer function of the operators p_k,x_k. Then the general relations

$$\frac{\partial f}{\partial x_k}=-[f,p_k] \tag{8.1}$$

and

$$\frac{\partial f}{\partial p_k}=[f,x_k] \tag{8.2}$$

with the abbreviation

$$[f,g]=\frac{i}{\hbar}(fg-gf)$$

shall be derived from the commutation rules.

Solution. The commutation rules are

$$[p_k,p_l]=0;\quad [x_k,x_l]=0;\quad [p_k,x_l]=\delta_{kl}\,. \tag{8.3}$$

From these we construct (8.1) and (8.2) in four consecutive steps:

1. Let $f=p_l$, then we have $\partial f/\partial x_k=0$ and $\partial f/\partial p_k=\delta_{kl}$. Hence, Eqs. (8.1) and (8.2) become $[p_k,p_l]=0$ and $[p_l,x_k]=\delta_{kl}$; i.e. they are satisfied according to (8.3). In the same way, for $f=x_l$, $\partial f/\partial x_k=\delta_{kl}$, $\partial f/\partial p_k=0$, they may be proved to hold.

2. Let (8.1) and (8.2) hold for two functions f and g. Then they hold as well for any linear combination $c_1 f + c_2 g$ with complex numbers c_1 and c_2, in consequence of their linearity.

3. With f and g, they hold for the product fg. For (8.1) this is easily checked by direct computation:

$$\frac{\partial}{\partial x_k}(fg) = f\frac{\partial g}{\partial x_k} + \frac{\partial f}{\partial x_k} g = -\{f[g,p_k] + [f,p_k]g\}$$

$$= -\frac{i}{\hbar}\{fgp_k - fp_k g + fp_k g - p_k fg\} = -[fg, p_k]$$

with the two central terms cancelling. An analogous computation is easily performed to prove (8.2).

4. It then follows that the relations hold for any linear combination of products consisting of an arbitrary number of p_k's and x_k's, i.e. for any integer function in these variables, as had to be proved.

Problem 9. Time rate of an expectation value

Let $\langle A \rangle$ be the expectation value of an operator not explicitly dependent upon time in a time dependent state ψ. How does $\langle A \rangle$ change with time? What follows for $\langle x_k \rangle$ and $\langle p_k \rangle$?

Solution. The expectation value

$$\langle A \rangle = \langle \psi | A | \psi \rangle = \int d\tau \, \psi^*(t) A \psi(t) \tag{9.1}$$

has the time rate

$$\frac{d}{dt}\langle A \rangle = \int d\tau \{\dot{\psi}^* A \psi + \psi^* A \dot{\psi}\}. \tag{9.2}$$

The time derivatives of the two wave functions ψ and ψ^* satisfy the Schrödinger equations

$$-\frac{\hbar}{i}\dot{\psi} = H\psi; \quad \frac{\hbar}{i}\dot{\psi}^* = H^\dagger \psi^* \tag{9.3}$$

with the hamiltonian operator H being hermitian: $H = H^\dagger$. Putting (9.3) into (9.2), we find

$$\frac{d}{dt}\langle A \rangle = \frac{i}{\hbar}\int d\tau \{(H^\dagger \psi^*) A \psi - \psi^* A H \psi\}$$

or, in Hilbert space notation,

$$\frac{d}{dt}\langle A \rangle = \frac{i}{\hbar}\{\langle H\psi | A\psi \rangle - \langle \psi | A H \psi \rangle\}. \tag{9.4}$$

In the first term of (9.4) we use the identity (cf. Eq. (6.1 b))

$$\langle \Omega\psi|\varphi\rangle = \langle\psi|\Omega^\dagger\varphi\rangle$$

or

$$\langle H\psi|A\psi\rangle = \langle\psi|H^\dagger A\psi\rangle = \langle\psi|HA\psi\rangle .$$

Eq. (9.4) then becomes

$$\frac{d}{dt}\langle A\rangle = \frac{i}{\hbar}\langle\psi|HA - AH|\psi\rangle \tag{9.5}$$

or, briefly,

$$\frac{d}{dt}\langle A\rangle = \langle[H,A]\rangle . \tag{9.6}$$

Applying (9.6) to the special operators $A = x_k$ and $A = p_k$ and using Eqs. (8.1) and (8.2), we find the canonical equations of classical mechanics to hold for the expectation values:

$$\frac{d}{dt}\langle x_k\rangle = \left\langle\frac{\partial H}{\partial p_k}\right\rangle; \quad \frac{d}{dt}\langle p_k\rangle = -\left\langle\frac{\partial H}{\partial x_k}\right\rangle . \tag{9.7}$$

Problem 10. Schrödinger and Heisenberg representations

In the Hilbert space coordinate system chosen in the so-called *Schrödinger representation*, the operators p_k, q_k do not depend upon time, but the state vector ψ does and satisfies the Schrödinger equation

$$-\frac{\hbar}{i}\dot{\psi} = H\psi \tag{10.1}$$

with $H(p_k, q_k)$ the hamiltonian of the system. Let H not depend explicitly upon time. It shall then be shown that a "rotating" Hilbert coordinate system with the state vector ψ at rest *(Heisenberg representation)* follows by an unitary transformation $U(t)$ and that the operators p_k, q_k in that frame depend on time according to the canonical equations

$$\dot{p}_k = -\frac{\partial H}{\partial q_k}; \quad \dot{q}_k = \frac{\partial H}{\partial p_k} . \tag{10.2}$$

Solution. Let all time-independent operators be denoted by a superscript 0 and let the time, when by "rotation" of the Heisenberg frame the coordinate systems begin to differ, be denoted by $t=0$. Then any Heisenberg operator $\Omega(t)$ coincides with its Schrödinger form at $t=0$, $\Omega(0) = \Omega^\circ$.

The state vectors in the Schrödinger frame, $\psi(t)$, and in the Heisenberg frame, ψ°, then are connected by the relation

$$\psi(t) = U(t)\,\psi^\circ . \tag{10.3}$$

It should be noted that, within normalization, the choice of $\psi(0) = \psi^\circ$ is still free. The normalization conditions,

$$\langle\psi|\psi\rangle = 1 \quad \text{and} \quad \langle\psi^\circ|\psi^\circ\rangle = 1 \tag{10.4}$$

then cause U to be a unitary operator satisfying

$$U^\dagger U = 1 . \tag{10.5}$$

This can easily be seen:

$$\langle\psi|\psi\rangle = \langle U\psi^\circ|U\psi^\circ\rangle = \langle\psi^\circ|U^\dagger U|\psi^\circ\rangle$$

and that equals $\langle\psi^\circ|\psi^\circ\rangle$ if, and only if, (10.5) is satisfied.

The operator U may be determined by putting (10.3) into (10.1),

$$-\frac{\hbar}{i}\dot{U} = H^\circ U \tag{10.6}$$

where we have now cautiously written H° instead of H, because in the Schrödinger frame an operator not depending upon time is used in (10.1). This differential equation, with the initial condition $U(0) = 1$, is solved by

$$U(t) = \mathrm{e}^{-\frac{i}{\hbar}H^\circ t} \tag{10.7}$$

where the exponential is explained by its power expansion.

Any operator, known in the Schrödinger frame, has to be constructed in the Heisenberg representation in such a way as to make its expectation value for any state ψ of the system independent of the special frame, i.e.

$$\langle\psi(t)|\Omega^\circ|\psi(t)\rangle = \langle\psi^\circ|\Omega(t)|\psi^\circ\rangle . \tag{10.8}$$

This equation is obviously satisfied for, and only for,

$$\Omega(t) = U^\dagger(t)\,\Omega^\circ\, U(t) \tag{10.9}$$

because

$$\langle\psi|\Omega^\circ|\psi\rangle = \langle U\psi^\circ|\Omega^\circ|U\psi^\circ\rangle = \langle\psi^\circ|U^\dagger\Omega^\circ U|\psi^\circ\rangle .$$

It should be noted that the operator H° commutes with U, according to (10.7), so that

$$H(t) = U^\dagger H^\circ U = U^\dagger U H^\circ = H^\circ$$

is invariant under this transformation. It therefore is not necessary to use different notations, H° and H, justifying the notation in (10.1) a posteriori.

In order now to prove the canonical equations (10.2) let us construct the time derivative of (10.9),

$$\dot{\Omega} = \dot{U}^\dagger \Omega^\circ U + U^\dagger \Omega^\circ \dot{U}. \tag{10.10}$$

From the equation adjoint to (10.1),

$$+\frac{\hbar}{i}\dot{\psi}^\dagger = H\psi^\dagger \tag{10.1a}$$

(holding because $H = H^\dagger$ is a hermitian operator), we get, in analogy to (10.6),

$$+\frac{\hbar}{i}\dot{U}^\dagger = H U^\dagger \tag{10.6a}$$

and may thus replace the derivatives in (10.10):

$$\dot{\Omega} = \frac{i}{\hbar}\{H U^\dagger \Omega^\circ U - U^\dagger \Omega^\circ H U\} = \frac{i}{\hbar}(H\Omega - \Omega H). \tag{10.11}$$

This is an essential law: The time derivative of an operator Ω, not explicitly depending on time, is its commutator with the hamiltonian:

$$\left.\begin{aligned} [H,\Omega] &= \frac{i}{\hbar}(H\Omega - \Omega H); \\ \dot{\Omega} &= [H,\Omega]. \end{aligned}\right\} \tag{10.12}$$

The law (10.12), when applied to the operators p_k and q_k directly yields

$$\dot{p}_k = [H,p_k]; \quad \dot{q}_k = [H,q_k]. \tag{10.13}$$

These algebraic relations may be reshaped using the relations derived in problem 8 from the basic commutation relations,

$$[H,p_k] = -\frac{\partial H}{\partial q_k}; \quad [H,q_k] = \frac{\partial H}{\partial p_k}, \tag{10.14}$$

thus directly leading to the canonical equations (10.2).

NB. The hamiltonian H as well as the operators Ω have been supposed not to depend upon time in the Schrödinger frame. If they do, additional partial derivatives with respect to time will occur. The whole formalism then will become more involved, e.g. $U(t)$ will no longer be a simple exponential because in the power expansion H will occur at different times.

Problem 11. Time dependent hamiltonian

Let H in the preceding problem depend explicitly on time (e.g. by the action of an alternating electric field). The unitary operator $U(t)$ shall be determined in this more general case.

Solution. Even with a hamiltonian $H(p_k, q_k; t)$ we still have the Schrödinger equation

$$-\frac{\hbar}{i}\dot{\psi} = H(t)\psi(t) \tag{11.1}$$

where now the transition to Heisenberg coordinates is performed by an unitary operator $U(t)$,

$$\psi(t) = U(t)\psi^{\circ} \tag{11.2}$$

satisfying the differential equation

$$-\frac{\hbar}{i}\dot{U} = H(t)U(t) \tag{11.3}$$

with the initial condition

$$U(0) = 1. \tag{11.4}$$

Let us write for the sake of brevity

$$-\frac{i}{\hbar} = \kappa \tag{11.5}$$

then

$$\dot{U} = \kappa H U$$

is solved with condition (11.4) by the infinite series

$$U(t) = 1 + \kappa \int_0^t dt' H(t') + \kappa^2 \int_0^t dt' H(t') \int_0^{t'} dt'' H(t'')$$
$$+ \kappa^3 \int_0^t dt' H(t') \int_0^{t'} dt'' H(t'') \int_0^{t''} dt''' H(t''') + \cdots. \tag{11.6}$$

This can easily be seen by differentation of the solution:

$$\dot{U}(t) = \kappa H(t) + \kappa^2 H(t) \int_0^t dt'' H(t'') + \kappa^3 H(t) \int_0^t dt'' H(t'') \int_0^{t''} dt''' H(t''') \ldots$$
$$= \kappa H(t) \left\{ 1 + \kappa \int_0^t dt'' H(t'') + \kappa^2 \int_0^t dt'' H(t'') \int_0^{t''} dt''' H(t''') + \cdots \right\}$$
$$= \kappa H(t) U(t).$$

It should be noted that, in the integrals, $t > t' > t'' > \cdots > 0$ so that the factors H, taken at different times, form a time-ordered product with the later factor always standing in front of the earlier one.

Problem 12. Repeated measurement

The time-independent hamiltonian H of a system has eigenvectors $|\nu\rangle$ with non-degenerate eigenvalues $\hbar\omega_\nu$:

$$H|\nu\rangle = \hbar\omega_\nu|\nu\rangle\,. \tag{12.1}$$

Let an observable A be defined by the equally non-degenerate eigenvalue problem

$$A|n\rangle = a_n|n\rangle \tag{12.2}$$

in the same Hilbert space. Let the system initially have been in the state $|\nu\rangle$ and then let a measurement of the observable A be performed on this system. What is the expectation value of A and what is the probability of finding the value a_m of A by this measurement? – If the measurement leads to the value a_m and is then repeated a time interval t later, what is the probability that the value a_m will again be found?

Solution. The expectation value of A in the initial state is $\langle\nu|A|\nu\rangle$. With the expansions

$$|\nu\rangle = \sum_n |n\rangle\langle n|\nu\rangle\,; \qquad \langle\nu| = \sum_{n'} \langle\nu|n'\rangle\langle n'|$$

it becomes

$$\langle\nu|A|\nu\rangle = \sum_{nn'} \langle\nu|n'\rangle\langle n'|A|n\rangle\langle n|\nu\rangle\,.$$

With

$$\langle n'|A|n\rangle = a_n\delta_{nn'}$$

this simplifies to

$$\langle\nu|A|\nu\rangle = \sum_n a_n|\langle\nu|n\rangle|^2\,. \tag{12.3}$$

The probability of finding the value a_m as the result of the first measurement therefore is

$$P_m = |\langle m|\nu\rangle|^2\,. \tag{12.4}$$

Afterwards, the system is no longer in the initial state $|\nu\rangle$, but in the state $|m\rangle$. Its further development follows from the Schrödinger equation

$$i\hbar\frac{\partial}{\partial t}|t\rangle = H|t\rangle\,, \tag{12.5}$$

where $|t\rangle$ denotes the state vector at the time t, with the initial condition

$$|0\rangle = |m\rangle\,.$$

Since H does not depend on time, this leads to the solution

$$|t\rangle = \mathrm{e}^{-\frac{it}{\hbar}H}|m\rangle \tag{12.6}$$

or, if $|m\rangle$ is expanded in eigenfunctions of H,

$$|m\rangle = \sum_\mu |\mu\rangle\langle\mu|m\rangle,$$

and using

$$e^{-\frac{it}{\hbar}H}|\mu\rangle = e^{-i\omega_\mu t}|\mu\rangle,$$

it leads on to

$$|t\rangle = \sum_\mu e^{-i\omega_\mu t}|\mu\rangle\langle\mu|m\rangle. \tag{12.7}$$

Using now the same consideration that led us to Eq. (12.4) we find that the probability again to measure a_m at the time t is

$$P'_m = |\langle m|t\rangle|^2 \tag{12.8}$$

with

$$\langle m|t\rangle = \sum_\mu e^{-i\omega_\mu t}|\langle m|\mu\rangle|^2. \tag{12.9}$$

Problem 13. Curvilinear coordinates

To transform the Schrödinger equation for a system of point masses to general curvilinear coordinates.

Solution. The essential problem is the transformation of the operator of kinetic energy,

$$T = -\frac{1}{2}\hbar^2 \sum_\mu \frac{1}{m_\mu}\frac{\partial^2}{\partial x_\mu^2}, \tag{13.1}$$

where m_μ has the same value for groups of three terms belonging to one particle. The expression (13.1) is the quantum theoretical translation of the classical expression

$$T_{\text{class}} = \tfrac{1}{2}\sum_\mu m_\mu \dot{x}_\mu^2. \tag{13.2}$$

Let us use, instead of the x_μ's, coordinates

$$\xi_\mu = \sqrt{m_\mu}\, x_\mu, \tag{13.3}$$

then

$$T_{\text{class}} = \tfrac{1}{2}\sum_\mu \dot{\xi}_\mu^2. \tag{13.4}$$

Replacing now in ξ space with the line element

$$ds^2 = \sum_\mu d\xi_\mu^2 \tag{13.5}$$

the ξ_μ's by general coordinates, q^k, with the line element transforming into

$$ds^2 = \sum_i \sum_k g_{ik} dq^i dq^k, \tag{13.6}$$

we find

$$T_{\text{class}} = \tfrac{1}{2} \sum_i \sum_k g_{ik} \dot{q}^i \dot{q}^k. \tag{13.7}$$

Passing on to quantum theory, Eq. (13.1) is equivalent to

$$T = -\frac{1}{2} \hbar^2 \sum_\mu \frac{\partial^2}{\partial \xi_\mu^2} = -\frac{1}{2} \hbar^2 \nabla^2. \tag{13.8}$$

In differential geometry it is shown that, replacing the ξ_μ's by the q^i's, the Laplacian transforms into

$$\nabla^2 = \frac{1}{\sqrt{g}} \sum_i \sum_k \frac{\partial}{\partial q^i} \left(\sqrt{g}\, g^{ik} \frac{\partial}{\partial q^k} \right) \tag{13.9}$$

where g is the determinant of the metrical tensor g_{ik}, the g^{ik} being its contravariant components to be obtained from

$$g^{ik} = \frac{G_{ik}}{g} \tag{13.10}$$

with G_{ik} the minor of g_{ik} in the determinant g. Eq. (13.9) gives the solution of our problem, the kinetic energy operator corresponding to the classical expression (13.7) becoming

$$T = -\frac{1}{2} \hbar^2 \frac{1}{\sqrt{g}} \sum_i \sum_k \frac{\partial}{\partial q^i} \left(\sqrt{g}\, g^{ik} \frac{\partial}{\partial q^k} \right). \tag{13.11}$$

The computation of the potential energy will, of course, be obvious.

NB. The method may even be applied in such cases where no substratum of rectangular point-mass coordinates exists. It then simply links the classical expression (13.7) to the operator (13.11). This connection, however, is by no means obvious; cf. the remarks on the symmetrical top, Problem 46.

Problem 14. Momentum space wave functions

The Fourier transform $f(\boldsymbol{k})$ of the wave function $\psi(\boldsymbol{r})$ describes the momentum distribution in a quantum state. An integral equation shall be derived for $f(\boldsymbol{k})$ in which the Fourier transform of the potential plays the role of kernel.

Solution. Between $\psi(\boldsymbol{r})$ and $f(\boldsymbol{k})$ there hold the two reciprocal relations[1]

$$\psi(\boldsymbol{r}) = (2\pi)^{-\frac{3}{2}} \int d^3k\, \mathrm{e}^{i\boldsymbol{k}\cdot\boldsymbol{r}} f(\boldsymbol{k}); \tag{14.1}$$

$$f(\boldsymbol{k}) = (2\pi)^{-\frac{3}{2}} \int d^3x\, \mathrm{e}^{-i\boldsymbol{k}\cdot\boldsymbol{r}} \psi(\boldsymbol{r}). \tag{14.2}$$

Further let us write the potential

$$V(\boldsymbol{r}) = \int d^3k\, \mathrm{e}^{i\boldsymbol{k}\cdot\boldsymbol{r}} W(\boldsymbol{k}); \tag{14.3}$$

then its Fourier transform is

$$W(\boldsymbol{k}) = (2\pi)^{-3} \int d^3x\, \mathrm{e}^{-i\boldsymbol{k}\cdot\boldsymbol{r}} V(\boldsymbol{r}). \tag{14.4}$$

The space wave function $\psi(\boldsymbol{r})$ is assumed to satisfy the Schrödinger equation

$$-\frac{\hbar^2}{2m} \nabla^2 \psi + V(\boldsymbol{r})\psi = E\psi. \tag{14.5}$$

Writing (14.1) for ψ and (14.3) for V, we obtain instead

$$\begin{aligned}(2\pi)^{-\frac{3}{2}} \Big\{ \frac{\hbar^2}{2m} \int & d^3k\, \mathrm{e}^{i\boldsymbol{k}\cdot\boldsymbol{r}} k^2 f(\boldsymbol{k}) \\ &+ \int d^3k \int d^3k'\, \mathrm{e}^{i(\boldsymbol{k}+\boldsymbol{k}')\cdot\boldsymbol{r}} W(\boldsymbol{k}) f(\boldsymbol{k}') \\ &- E \int d^3k\, \mathrm{e}^{i\boldsymbol{k}\cdot\boldsymbol{r}} f(\boldsymbol{k}) \Big\} = 0.\end{aligned}$$

In the double integral let us then use $\boldsymbol{k}'' = \boldsymbol{k}+\boldsymbol{k}'$ instead of $\boldsymbol{k}$ as integration variable (and then again write $\boldsymbol{k}$ instead of $\boldsymbol{k}''$). Then, as the integral will only vanish for *any* value of $\boldsymbol{r}$ if the integrand vanishes, we arrive at

$$\left(\frac{\hbar^2 k^2}{2m} - E\right) f(\boldsymbol{k}) = -\int d^3k'\, W(\boldsymbol{k}-\boldsymbol{k}') f(\boldsymbol{k}'). \tag{14.6}$$

This is the integral equation wanted, with $W(\boldsymbol{k}-\boldsymbol{k}')$ as its kernel.

The integral equation (14.6) can, of course, be established only if the Fourier transform (14.4) of the potential exists, i.e. if $V(\boldsymbol{r})$ at large values of r vanishes at least as $r^{-1-\varepsilon}$ with infinitesimal $\varepsilon > 0$.

---

[1] If (14.1) is used to define $f(\boldsymbol{k})$, application of $\int d^3x\, \mathrm{e}^{-i\boldsymbol{k}\cdot\boldsymbol{r}}$ to (14.1) and use of the definition of the three-dimensional δ-function,

$$\delta(\boldsymbol{k}-\boldsymbol{k}') = \frac{1}{(2\pi)^3} \int d^3x\, \mathrm{e}^{i(\boldsymbol{k}-\boldsymbol{k}')\cdot\boldsymbol{r}},$$

lead easily to the inversion (14.2). —The same relation is used in deriving Eq. (14.8) below.

It should be noted that from the normalization

$$\int d^3x\,|\psi(\boldsymbol{r})|^2 = 1 \tag{14.7}$$

there follows

$$\int d^3k\,|f(\boldsymbol{k})|^2 = 1\,. \tag{14.8}$$

This can be shown by setting (14.1) into (14.7) for ψ:

$$\int d^3x\,|\psi(\boldsymbol{r})|^2 = (2\pi)^{-3}\int d^3x\int d^3k\int d^3k'\,\mathrm{e}^{i(\boldsymbol{k}-\boldsymbol{k}')\cdot\boldsymbol{r}} f(\boldsymbol{k})\,f^*(\boldsymbol{k}')\,.$$

If here the integration over coordinate space is first performed[2] we almost immediately arrive at the expression (14.8).

Problem 15. Momentum space: Periodic and aperiodic wave functions

To deduce the probability interpretation of momentum space wave functions in the continuous spectrum by starting from periodic wave functions $\psi(\boldsymbol{r})$ in ordinary space and investigating the limiting process for infinitely large periodicity cube.

Solution. Let L be the period in each of the three space directions x, y, z. Then the Fourier series

$$\psi(\boldsymbol{r},t) = L^{-\frac{3}{2}}\sum_{\boldsymbol{k}} c_{\boldsymbol{k}}(t)\,\mathrm{e}^{i(\boldsymbol{k}\cdot\boldsymbol{r}-\omega t)}\,; \qquad \omega = \frac{\hbar}{2m}k^2 \tag{15.1}$$

includes terms only with components

$$k_i = \frac{2\pi}{L}n_i\,; \qquad n_i = 0,\ \pm 1,\ \pm 2, \ldots \tag{15.2}$$

of each vector $\boldsymbol{k}$. This means that in $\boldsymbol{k}$ space, for large L, a volume element d^3k includes

$$d^3k\left(\frac{L}{2\pi}\right)^3 \tag{15.3}$$

states of different $\boldsymbol{k}$'s.

The normalization of series (15.1) can still be chosen by suitable choice of the coefficients $c_{\boldsymbol{k}}$. The square integral over the periodicity cube is

$$\int_{(L^3)} d^3x\,|\psi|^2 = L^{-3}\sum_{\boldsymbol{k}}\sum_{\boldsymbol{k}'} c_{\boldsymbol{k}}^*\,c_{\boldsymbol{k}'}\,\mathrm{e}^{i(\omega-\omega')t}\int d^3x\,\mathrm{e}^{i(\boldsymbol{k}'-\boldsymbol{k})\cdot\boldsymbol{r}}$$

---

[2] Cf. however, the remark at the end of the following problem.

where the last integral vanishes if $\boldsymbol{k}' \neq \boldsymbol{k}$ and becomes $= L^3$ if $\boldsymbol{k}' = \boldsymbol{k}$:

$$\int_{(L^3)} d^3x\, |\psi|^2 = \sum_{\boldsymbol{k}} |c_{\boldsymbol{k}}|^2 . \tag{15.4}$$

Let P_L now be the probability of finding the particle in the periodicity cube L^3, then $|c_{\boldsymbol{k}}|^2\, P_L$ will be the probability of finding there the particle with momentum $\hbar\boldsymbol{k}$, and $|c_{\boldsymbol{k}}|^2$ the probability that, if the particle is found within L^3, its momentum will turn out to be $\hbar\boldsymbol{k}$.

We now go on to infinitely large L. Then we may replace the Fourier sum (15.1) by a Fourier integral over $\boldsymbol{k}$ space which according to (15.2) and (15.3) may be done by using the rule

$$\sum_{\boldsymbol{k}} \to \left(\frac{L}{2\pi}\right)^3 d^3k . \tag{15.5}$$

Eq. (15.1) then yields

$$\psi(\boldsymbol{r},t) = \frac{L^{\frac{3}{2}}}{(2\pi)^3} \int d^3k\, c(\boldsymbol{k},t)\, \mathrm{e}^{i(\boldsymbol{k}\cdot\boldsymbol{r} - \omega t)} . \tag{15.6}$$

This Fourier integral describes a wave function of finite values independent of L if, and only if, the quantity

$$\left(\frac{L}{2\pi}\right)^{\frac{3}{2}} c(\boldsymbol{k}) = f(\boldsymbol{k}) \tag{15.7}$$

has a finite limit for $L \to \infty$. The wave function

$$\psi(\boldsymbol{r},t) = \frac{1}{(2\pi)^{\frac{3}{2}}} \int d^3k\, f(\boldsymbol{k},t)\, \mathrm{e}^{i(\boldsymbol{k}\cdot\boldsymbol{r} - \omega t)} \tag{15.8}$$

then may be normalized according to

$$\int d^3x\, |\psi|^2 = \frac{1}{(2\pi)^3} \int d^3k \int d^3k'\, f^*(\boldsymbol{k}) f(\boldsymbol{k}')\, \mathrm{e}^{i(\omega - \omega')t} \int d^3x\, \mathrm{e}^{i(\boldsymbol{k}' - \boldsymbol{k})\cdot\boldsymbol{r}} \tag{15.9}$$

where the last integral runs over infinite space and can be evaluated:

$$\int d^3x\, \mathrm{e}^{i(\boldsymbol{k}' - \boldsymbol{k})\cdot\boldsymbol{r}} = (2\pi)^3\, \delta(\boldsymbol{k}' - \boldsymbol{k}) , \tag{15.10}$$

so that we find for (15.9):

$$\int d^3x\, |\psi|^2 = \int d^3k\, |f(\boldsymbol{k},t)|^2 \tag{15.11}$$

which, by the way, is simply the same as translating the sum (15.4) with (15.7), according to the rule (15.5). Hence, the probability that a particle found anywhere has its momentum $\boldsymbol{k}$ within the element d^3k becomes

$$dP_{\boldsymbol{k}} = d^3k\, |f(\boldsymbol{k},t)|^2 . \tag{15.12}$$

NB. Some caution is necessary if $f(\boldsymbol{k})$ is not a continuous function of $\boldsymbol{k}$, with respect to exchange of integrals. Let e.g. be

$$f(\boldsymbol{k})=(2\pi)^{\frac{3}{2}}\,\delta(\boldsymbol{k}_0-\boldsymbol{k})\,. \tag{15.13}$$

According to (15.8) we then have

$$\psi(\boldsymbol{r},t)=\mathrm{e}^{i(\boldsymbol{k}_0\cdot\boldsymbol{r}-\omega_0 t)}\,. \tag{15.14}$$

The integrals (15.9) with exchanged order of integration and (15.11) would then be integrals over the square of a delta function and would be entirely meaningless. But if integration over ordinary space is performed last, Eq. (15.9) gives

$$\int d^3x\,|\psi|^2=\int d^3x\,,$$

which is consistent with the wave function (15.14).

II. One-Body Problems without Spin

A. One-Dimensional Problems

One-dimensional problems, though in a sense oversimplifications, may be used with advantage in order to understand the essential features of quantum mechanics. They may be derived from the three-dimensional wave equation,

$$-\frac{\hbar^2}{2m}\nabla^2\psi + V(x,t)\psi = -\frac{\hbar}{i}\frac{\partial\psi}{\partial t}, \tag{A.1}$$

if the potential depends upon only one rectangular coordinate x, by factorization:

$$\psi = e^{i(k_2 y + k_3 z)}\varphi(x,t). \tag{A.2}$$

The differential equation thus resulting for $\varphi(x,t)$,

$$-\frac{\hbar^2}{2m}\frac{\partial^2\varphi}{\partial x^2} + \frac{\hbar^2}{2m}(k_2^2 + k_3^2)\varphi + V(x,t)\varphi = -\frac{\hbar}{i}\frac{\partial\varphi}{\partial t}$$

may be further simplified by putting

$$\varphi(x,t) = e^{-i\omega_0 t}u(x,t); \qquad \hbar\omega_0 = \frac{\hbar^2}{2m}(k_2^2 + k_3^2). \tag{A.3}$$

We thus arrive at the one-dimensional wave equation,

$$-\frac{\hbar^2}{2m}\frac{\partial^2 u}{\partial x^2} + V(x,t)u = -\frac{\hbar}{i}\frac{\partial u}{\partial t}. \tag{A.4}$$

The exponential factors in (A.2) and (A.3) describe plane-wave qualities in the directions perpendicular to the x-axis which, however, do not affect the behaviour of the wave function in x-direction.

Problem 16. Force-free case: Basic solutions

The one-dimensional wave equation shall be solved in the case $V=0$ and the physical significance of the solutions shall be discussed.

Solution[1]. The wave equation

$$-\frac{\hbar^2}{2m}\frac{\partial^2\psi}{\partial x^2}=-\frac{\hbar}{i}\frac{\partial\psi}{\partial t} \tag{16.1}$$

permits factorization,

$$\psi(x,t)=u(x)g(t)\,, \tag{16.2}$$

because by putting (16.2) in (16.1) one arrives at

$$-\frac{\hbar^2}{2m}\frac{u''}{u}=-\frac{\hbar}{i}\frac{\dot g}{g}=\hbar\omega\,, \tag{16.3}$$

where $\hbar\omega$ stands as abbreviation for the separation parameter. Splitting (16.3) into two separate equations we obtain

$$\dot g=-i\omega g\,,\quad \text{i.e.}\quad g(t)=\mathrm{e}^{-i\omega t} \tag{16.4}$$

and

$$u''+\frac{2m\omega}{\hbar}u=0\,. \tag{16.5}$$

With real ω, the wave function is periodic in, and $|\psi|^2$ independent of, time *(stationary state)*; with ω positive, the constant

$$\frac{2m\omega}{\hbar}=k^2 \tag{16.6}$$

becomes positive too, and the solutions of (16.5) are as well periodic in space.

It is an essential feature of quantum mechanics that time dependence is of the complex form (16.4); the real functions $\sin\omega t$ and $\cos\omega t$ are not solutions of the differential equation (16.4). This behaviour, so different from classical physics, is a consequence of the Schrödinger equation being of the *first* order in time.

The physical meaning of the parameter ω may be further interpreted by considering the operator on the left-hand side of (16.1) to be the hamiltonian, consisting only of the kinetic energy operator in our case. It follows that $E=\hbar\omega$ is the kinetic energy of a particle and must hence be positive real. Our solution therefore is an eigenstate of the hamiltonian.

---

[1] In the following we shall write ψ again for the one-dimensional part of the wave function satisfying Eq. (A.4), and u for its space part.

Since k^2 is a positive constant, the complete solution of (16.5) or

$$u'' + k^2 u = 0 \tag{16.7}$$

is

$$u(x) = A\,e^{ikx} + B\,e^{-ikx} \tag{16.8 a}$$

so that the one-dimensional wave function

$$\psi(x,t) = A\,e^{i(kx-\omega t)} + B\,e^{-i(kx+\omega t)} \tag{16.8 b}$$

consists of two waves running in opposite directions, both with phase velocity $v_{ph} = \omega/k$.

The physical significance of the space part (16.8 a) of the wave function becomes clear when we derive density

$$\rho = \psi^* \psi \tag{16.9}$$

and flux

$$s = \frac{\hbar}{2im}\left(\psi^* \frac{\partial \psi}{\partial x} - \psi \frac{\partial \psi^*}{\partial x}\right) \tag{16.10}$$

from (16.8 b). We find

$$\rho = |A|^2 + |B|^2 + (A B^* e^{2ikx} + A^* B e^{-2ikx});$$

$$s = \frac{\hbar k}{m}(|A|^2 - |B|^2).$$

The two waves, of amplitudes A and B, apparently correspond to two opposite currents whose intensity is given by their respective normalization constants and is proportional to k. The density shows interference of the two (coherent) waves causing a space periodicity.

As long as no special reason (like boundary conditions) is given to achieve coherence, it will be reasonable to take either of the two waves, putting $B=0$ and obtaining $s>0$, or $A=0$ giving $s<0$. The result then corresponds to the linear motion of particles in one or the other direction. Admitting both signs of k, we may therefore summarize the final result as follows:

$$\left.\begin{aligned} &\psi(x,t) = C\,e^{i(kx-\omega t)}; \\ &E = \hbar\omega; \qquad k^2 = \frac{2m\omega}{\hbar}; \\ &\rho = |C|^2; \qquad s = \frac{\hbar k}{m}|C|^2. \end{aligned}\right\} \tag{16.11}$$

Elimination of ω yields

$$E = \frac{\hbar^2 k^2}{2m} \tag{16.12}$$

so that

$$p = \hbar k \tag{16.13}$$

is the momentum, and

$$v = \frac{\hbar k}{m} \tag{16.14}$$

the classical velocity of the particle. The latter is by no means identical with the phase velocity of the wave,

$$v_{\mathrm{ph}} = \frac{\omega}{k} = \frac{E}{p} = \frac{1}{2} v ;$$

it is, however, identical with the group velocity

$$v_{\mathrm{gr}} = \frac{d\omega}{dk} = \frac{dE}{dp} = v .$$

NB. The fundamental differential equation (16.1) may be interpreted as a *diffusion equation* with an imaginary constant of diffusion, D:

$$D \frac{\partial^2 \psi}{\partial x^2} = \frac{\partial \psi}{\partial t}; \qquad D = i \frac{\hbar}{2m} .$$

Whereas factorization plays rather an important role in quantum theory but not in diffusion problems, the typical source-type solutions of (real) diffusion theory,

$$\psi(x,t) = \frac{1}{\sqrt{t}} \int_{-\infty}^{+\infty} d\xi \, f(\xi) \, e^{i \frac{m}{2\hbar} \frac{(x-\xi)^2}{t}} ,$$

are of no importance in quantum theory.

Time reversal in (16.1) changes ψ in ψ^*.

Problem 17. Force-free case: Wave packet

A wave packet shall be constructed and its development as a function of time be investigated.

Solution. We start from the special solution of the wave equation found in (16.11):

$$\psi(k; x, t) = C(k) \, e^{i(kx - \omega t)} \tag{17.1}$$

with

$$\omega = \frac{\hbar}{2m} k^2 \tag{17.2}$$

and $C(k)$ an arbitrary amplitude constant. Here, k still is a free parameter so that the complete solution of the wave equation is obtained as any convergent integral of (17.1) over k:

$$\psi(x,t) = \int_{-\infty}^{+\infty} dk\, \psi(k; x,t). \tag{17.3}$$

Eq. (17.3) describes the most general form of an one-dimensional wave packet. The amplitude function $C(k)$ must, in order to make the integral converge, tend to zero at least as $1/k$ with $|k| \to \infty$. Every suitable choice of $C(k)$ yields a special solution.

Let us now construct the wave packet so that, for an initial time $t=0$, the probability of finding the particle it describes differs appreciably from zero only inside a small region around $x=0$, and that the particle moves with momentum $p_0 = \hbar k_0$. This can be accomplished if we choose

$$\psi(x,0) = A \exp\left[-\frac{x^2}{2a^2} + i k_0 x\right], \tag{17.4}$$

because then the density

$$\rho(x,0) = |\psi(x,0)|^2 = |A|^2 \exp\left[-\frac{x^2}{a^2}\right]$$

localizes the particle within $|x| \lesssim a$, and the flux (16.10) becomes

$$s(x,0) = \frac{\hbar}{2mi} 2 i k_0 |A|^2 \exp\left[-\frac{x^2}{a^2}\right] = \rho \frac{\hbar}{m} k_0$$

so that the particle velocity is $v_0 = \frac{\hbar}{m} k_0$ and $p_0 = m v_0 = \hbar k_0$ the momentum of the packet. Since the wave function represents one particle, the normalization condition

$$\int_{-\infty}^{+\infty} dx\, \rho = 1$$

holds, i. e.

$$|A|^2 = \frac{1}{a\sqrt{\pi}}. \tag{17.5}$$

The expression (17.4) may be decomposed into plane waves using (17.3) and (17.1):

$$\psi(x,0) = \int_{-\infty}^{+\infty} dk\, C(k)\, e^{ikx}. \tag{17.6}$$

This is a Fourier integral whose inversion is

$$C(k)=\frac{1}{2\pi}\int\limits_{-\infty}^{+\infty} dx\,\psi(x,0)\,e^{-ikx}$$

$$=\frac{A}{2\pi}\int\limits_{-\infty}^{+\infty} dx\exp\left[-\frac{x^2}{2a^2}+i(k_0-k)x\right].$$

This integral may be evaluated using the well-known formula

$$\int\limits_{-\infty}^{+\infty} dz\,e^{-z^2}=\sqrt{\pi}. \tag{17.7}$$

The result,

$$C(k)=\frac{Aa}{\sqrt{2\pi}}\exp\left[-\tfrac{1}{2}a^2(k-k_0)^2\right], \tag{17.8}$$

can easily be understood in terms of Heisenberg's uncertainty rule: In the initial state, the coordinate uncertainty of the particle is, according to Eq. (17.4), of the order of $\Delta x=a$; as (17.8) shows, to this wave function there contributes a spectrum of wave numbers k or of momenta $p=\hbar k$ around $k=k_0$ of a width $\Delta k=1/a$ or $\Delta p=\hbar/a$. Therefore, independently of the choice of a, there holds the relation

$$\Delta x\cdot\Delta p=\hbar, \tag{17.9}$$

which is in fact Heisenberg's principle of uncertainty.

Having determined $C(k)$ from the initial state at the time $t=0$, we are now prepared to evaluate the full integral (17.3) at *any* time, viz. the integral

$$\psi(x,t)=\frac{Aa}{\sqrt{2\pi}}\int\limits_{-\infty}^{+\infty} dk\exp\left[-\frac{1}{2}a^2(k-k_0)^2+ikx-i\frac{\hbar t}{2m}k^2\right].$$

The exponent is a quadratic form in k so that again reduction to the complete error integral (17.7) can be performed. The result is

$$\psi(x,t)=\frac{A}{\left(1+i\dfrac{\hbar t}{ma^2}\right)^{\frac{1}{2}}}\exp\left[-\frac{x^2-2ia^2k_0x+i\dfrac{\hbar t}{2m}k_0^2a^2}{2a^2\left(1+i\dfrac{\hbar t}{ma^2}\right)}\right]. \tag{17.10}$$

A good understanding of this rather cumbersome expression can be obtained by again discussing density ρ and flux s, but now at any time. The former becomes

$$\rho(x,t) = |\psi(x,t)|^2 = \frac{|A|^2}{\left[1+\left(\frac{\hbar t}{m a^2}\right)^2\right]^{\frac{1}{2}}} \exp\left[-\frac{\left(x-\frac{\hbar k_0}{m}t\right)^2}{a^2\left[1+\left(\frac{\hbar t}{m a^2}\right)^2\right]}\right]. \tag{17.11}$$

As a function of x, the density $\rho(x,t)$ still is a bell-shaped curve, whose maximum, however, has now shifted from $x=0$ to $x=\frac{\hbar k_0}{m}t$. The maximum of the 'wave group' represented by (17.10) therefore propagates with a velocity $v_0=\frac{\hbar k_0}{m}$ ('group velocity' = particle velocity). The denominator of the exponent in (17.11) shows that, at the same time, the wave packet has broadened from its initial width a at $t=0$ to

$$a' = a\left[1+\left(\frac{\hbar t}{m a^2}\right)^2\right]^{\frac{1}{2}} \simeq \frac{\hbar}{m a} t$$

at $t=t$. This effect can easily be explained from the spectral function (17.8): The wave number spectrum having the width $\Delta k = 1/a$, the velocities of the partial waves cover a region of width $\Delta v = \frac{\hbar}{m}\Delta k = \frac{\hbar}{m a}$ so that the packet broadens by $\Delta x = t\,\Delta v = \frac{\hbar}{m a} t$ as derived above.

The flux is obtained from (17.10) with the help of the relation

$$\frac{\partial \psi}{\partial x} = i k_0 \frac{1 + i\frac{x}{a^2 k_0}}{1 + i\frac{\hbar t}{m a^2}}\psi;$$

straightforward calculation yields by comparison with (17.11):

$$s(x,t) = \rho(x,t)\, v_0 \frac{1 + \frac{\hbar t x}{m a^4 k_0}}{1+\left(\frac{\hbar t}{m a^2}\right)^2}. \tag{17.12}$$

It follows that we by no means have $s = \rho v_0$ for all times, as we had for $t=0$. This again is a consequence of the finite width of the velocity spectrum: At the packet maximum, $x_0 = v_0 t$, Eq. (17.12) leads to the elementary relation $s=\rho v_0$; for $x \lessgtr x_0$ we find $s \lessgtr \rho v_0$, and this is reasonable because at a point $x<x_0$ $(x>x_0)$ there have arrived at the time t those parts of the packet whose velocity is smaller (bigger) than v_0.

It may finally be mentioned that the normalization condition $\int dx\,\rho=1$ holds for all times, thus reflecting the conservation of matter.

Problem 18. Standing wave

A particle is included between two impenetrable potential walls at $x=-a$ and $x=+a$. (The walls idealize a strong repulsion of the particle when it approaches these boundaries.) The eigenstates shall be determined and discussed.

Solution. We have

$$\psi(x,t)=u(x)\,e^{-i\frac{E}{\hbar}t} \tag{18.1}$$

for stationary states. The space function $u(x)$ satisfies the Schrödinger equation

$$u''+k^2u=0 \tag{18.2}$$

with

$$k^2=\frac{2mE}{\hbar^2}; \tag{18.3}$$

thence follows its most general form

$$u(x)=A\,e^{ikx}+B\,e^{-ikx}. \tag{18.4}$$

Finally, the potential walls are imposing boundary conditions

$$u(a)=0; \qquad u(-a)=0 \tag{18.5}$$

which, combined with the normalization condition

$$\int_{-a}^{+a} dx\,|u(x)|^2=1, \tag{18.6}$$

permit the complete determination of the eigenfunctions, always excepting the constant phase factor which remains free in quantum mechanics.

From (18.5) put into (18.4), we find two homogeneous linear equations for A and B,

$$A\,e^{ika}+B\,e^{-ika}=0,$$
$$A\,e^{-ika}+B\,e^{ika}=0,$$

permitting a non-zero solution only for vanishing determinant:

$$\begin{vmatrix} e^{ika} & e^{-ika} \\ e^{-ika} & e^{ika} \end{vmatrix} = 0 \quad \text{or} \quad \sin 2ka = 0. \tag{18.7}$$

This *eigenvalue condition* is satisfied only for the eigenvalues of k,

$$k_n = \frac{\pi}{2a} n, \qquad n = \pm 1, \pm 2, \pm 3, \dots \tag{18.8}$$

The value $k=0$, also satisfying (18.7), can be excluded since it contradicts the normalization condition (18.6). The eigenvalues of the energy follow from (18.3) and (18.8):

$$E_n = \frac{\hbar^2 k_n^2}{2m} = \frac{\hbar^2 \pi^2}{8ma^2} n^2. \tag{18.9}$$

From (18.8) we have

$$e^{ik_n a} = e^{i\frac{\pi}{2}n} = i^n$$

so that

$$B = (-1)^{n+1} A.$$

If n is an odd integer, $B=A$ and the normalized eigenfunctions are

$$u_n^+(x) = a^{-\frac{1}{2}} \cos k_n x = a^{-\frac{1}{2}} \cos \frac{\pi n x}{2a}, \qquad n = \pm 1, \pm 3, \dots \tag{18.10a}$$

If n is an even integer, $B=-A$ and we have

$$u_n^-(x) = a^{-\frac{1}{2}} \sin k_n x = a^{-\frac{1}{2}} \sin \frac{\pi n x}{2a}, \qquad n = \pm 2, \pm 4, \dots \tag{18.10b}$$

Since u_n, apart from an irrelevant sign in (18.10b), does not depend upon the sign of n, we can ignore negative values of n so that e.g. the four lowest states become

$$\left.\begin{aligned}
n&=1, & E_1 &= \frac{\hbar^2\pi^2}{8ma^2}, & u_1^+ &= a^{-\frac{1}{2}} \cos\frac{\pi x}{2a},\\
n&=2, & E_2 &= 4E_1, & u_2^- &= a^{-\frac{1}{2}} \sin\frac{\pi x}{a},\\
n&=3, & E_3 &= 9E_1, & u_3^+ &= a^{-\frac{1}{2}} \cos\frac{3\pi x}{2a},\\
n&=4, & E_4 &= 16E_1, & u_4^- &= a^{-\frac{1}{2}} \sin\frac{2\pi x}{a}.
\end{aligned}\right\} \tag{18.11}$$

It should be noted that the eigenfunctions are alternate symmetric (n odd) and antisymmetric (n even) with respect to reflection at the origin. This property is called the *parity* of the state; in the case of symmetry we speak of positive, in the opposite case of negative parity. The plus and minus signs (u_n^+, u_n^-) refer to this property.

The first four eigenfunctions have been drawn in Fig. 1.

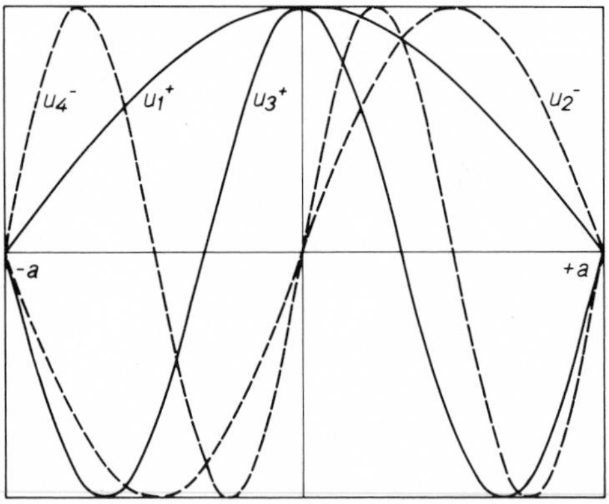

Fig. 1. The first four eigenfunctions of the one-dimensional potential well

Since the space part of the eigenfunctions is real, there can exist no resulting flux in any state. This is a consequence of $|A|=|B|$ in Eq. (18.4), cf. the discussion in Problem 16: The A wave and B wave in (18.4) contribute opposite currents and momenta. The eigenfunctions of the hamiltonian belonging to sharp energy values therefore are not eigenfunctions of the momentum operator

$$p = \frac{\hbar}{i}\frac{\partial}{\partial x}.$$

Indeed, differentiation of the functions (18.10a, b) does not reproduce them but exchanges cosine with sine solutions. The expectation value, however, of the momentum can be computed according to

$$\langle n|p|n\rangle = \frac{\hbar}{i}\int_{-a}^{+a} dx\, u_n(x)\frac{\partial}{\partial x}u_n(x).$$

Since the integrand is an odd function of x, this integral vanishes for all states: $\langle n|p|n\rangle=0$, in accordance with the vanishing flux.

NB. The mathematical problem is very much the same as the classical one of the vibrating string, the only difference being that here eigenvalues of the energy and there eigenvalues of the frequency follow the quadratic law (18.9). The classical vibration energy has no analogue, however, in the quantum mechanical problem since it derives from the possible excitation of the string vibration to arbitrary amplitudes, whereas the amplitudes of our wave functions are fixed by the normalization condition (18.6), i. e. by the fact that the particle number is one.

Problem 19. Opaque division wall

In the preceding problem an opaque wall of infinitely small width but infinitely large height shall be introduced at $x=0$ and its effect upon the eigenstates investigated.

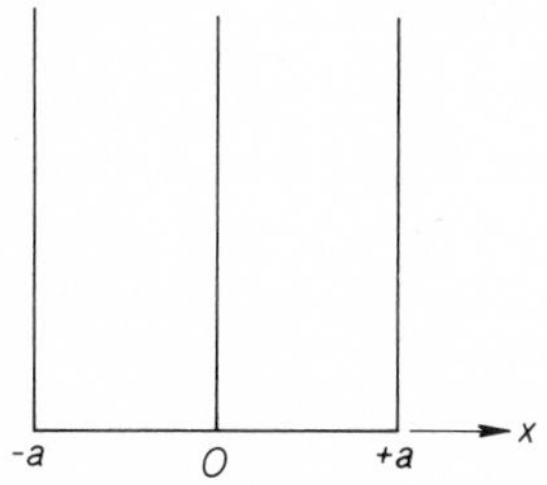

Fig. 2. Potential well with opaque wall

Solution. The opaque wall dividing the whole region into two equal parts may be obtained by idealization of a barrier of finite width 2ε (between $x=-\varepsilon$ and $x=+\varepsilon$) and height V_0. We introduce the abbreviations

$$\frac{2mE}{\hbar^2}=k^2\,; \qquad \frac{2m}{\hbar^2}(V_0-E)=\varkappa^2\,, \tag{19.1}$$

then we have a total of four boundary conditions at this barrier, viz. continuity of $u(x)$ and $u'(x)$ at $x=\pm\varepsilon$, besides the two boundary conditions $u(\pm a)=0$. Satisfying the latter two and writing the solution in real form, we have

$$u=\begin{cases} A_1 \sin k(x+a) & -a\leq x\leq -\varepsilon\,, \\ B\mathrm{e}^{-\varkappa x}+C\mathrm{e}^{\varkappa x} & -\varepsilon\leq x\leq +\varepsilon\,, \\ A_2 \sin k(x-a) & +\varepsilon\leq x\leq +a\,. \end{cases} \tag{19.2}$$

Then the continuity conditions at $x=\pm\varepsilon$ give

$$u(-\varepsilon)=A_1 \sin k(a-\varepsilon)=B\mathrm{e}^{\varkappa\varepsilon}+C\mathrm{e}^{-\varkappa\varepsilon}, \tag{19.3a}$$

$$u'(-\varepsilon)=kA_1 \cos k(a-\varepsilon)=\varkappa(-B\mathrm{e}^{\varkappa\varepsilon}+C\mathrm{e}^{-\varkappa\varepsilon}), \tag{19.3b}$$

$$u(+\varepsilon)=A_2 \sin k(\varepsilon-a)=B\mathrm{e}^{-\varkappa\varepsilon}+C\mathrm{e}^{\varkappa\varepsilon}, \tag{19.3c}$$

$$u'(+\varepsilon)=kA_2 \cos k(\varepsilon-a)=\varkappa(-B\mathrm{e}^{-\varkappa\varepsilon}+C\mathrm{e}^{\varkappa\varepsilon}). \tag{19.3d}$$

Eliminating A_1 from (19.3a, b) and A_2 from (19.3c, d), there remain two relations

$$\left.\begin{aligned} k\cot k(a-\varepsilon)&=\varkappa\frac{-B\mathrm{e}^{2\varkappa\varepsilon}+C}{B\mathrm{e}^{2\varkappa\varepsilon}+C};\\ k\cot k(a-\varepsilon)&=\varkappa\frac{B-C\mathrm{e}^{2\varkappa\varepsilon}}{B+C\mathrm{e}^{2\varkappa\varepsilon}}.\end{aligned}\right\} \tag{19.4}$$

The identity of the two left-hand sides leads to equality of the right-hand sides. These become equal if, and only if, $B=\pm C$. If $B=+C$, it follows from (19.3a, c) that $A_1=-A_2$ and we obtain a solution of even parity; if $B=-C$, we find $A_1=A_2$ and have odd parity. Hence again, as in the preceding problem, the eigenstates divide into two classes of different parities.

We now go to the limit $\varepsilon\to 0$, $\varkappa\to\infty$ in such a way that $\varkappa\varepsilon=0$ but

$$\varkappa^2\varepsilon=\Omega \tag{19.5}$$

remains finite. We shall then call Ω the *opacity* of the wall since the larger Ω, the more opaque it becomes. For even parity ($B=C$) Eq. (19.4) then leads by power expansion on the right-hand side to

$$k\cot ka=-\Omega. \tag{19.6$^+$}$$

For odd parity ($B=-C$) we find in the same manner

$$k\cot ka\to-\infty. \tag{19.6$^-$}$$

The second case is simplest. The eigenfunction has a zero at the wall, and the solution becomes

$$\left.\begin{aligned} &u_n^-(x)=\begin{cases} A\sin k_n^-(x+a), & -a\le x<0,\\ A\sin k_n^-(x-a), & 0<x\le +a,\end{cases}\\ &k_n^-a=n\pi;\quad n=1,2,3,\ldots;\\ &u_n^-(-x)=-u_n^-(x);\quad |A|^2=\frac{1}{a}.\end{aligned}\right\} \tag{19.7$^-$}$$

The eigenvalue condition (19.6⁺), on the other hand, permits only numerical determination of the eigenvalues k_n^+, except in the two limiting cases of an entirely opaque wall ($\Omega\to\infty$) where it leads to the same eigenvalues as (19.6⁻), viz. $k_n^+ = n\pi$, and of the entirely permeable wall ($\Omega=0$) with $k_n^+ = (n-\frac{1}{2})\pi$. For finite opacity, the value k_n^+ therefore lies between k_n^- and k_{n-1}^-, i.e. as in the preceding problem, here again levels of even and odd parities alternate. The eigenfunctions then may be written

$$\left.\begin{aligned}
&u_n^+(x) = \begin{cases} -A\sin k_n^+(x+a), & -a\le x<0,\\ +A\sin k_n^+(x-a), & 0<x\le +a,\end{cases}\\
&(n-\tfrac{1}{2})\pi < k_n^+ a < n\pi; \qquad n=1,2,3,\ldots;\\
&u_n^+(-x) = u_n^+(x); \qquad |A|^2 = \frac{2k_n^+}{2k_n^+ a - \sin 2k_n^+ a}.
\end{aligned}\right\} \qquad (19.7^+)$$

There is a finite amplitude at $x=0$ and a break in the curve.

Ωa	$n=1$	$n=2$	$n=3$	$n=4$	$n=5$
0	1.571	4.712	7.854	10.996	14.137
$\frac{1}{4}$	1.715	4.765	7.886	11.018	14.155
$\frac{1}{2}$	1.835	4.816	7.917	11.040	14.171
1	2.023	4.915	7.979	11.085	14.208
2	2.282	5.091	8.097	11.173	14.276
4	2.568	5.361	8.305	11.335	14.408
10	2.866	5.763	8.711	11.704	14.734
∞	3.142	6.283	9.425	12.566	15.708

In the accompanying table, for a few values of the dimensionless parameter Ωa, numerical values of $k_n^+ a$ have been computed for the lowest states according to Eq. (19.6⁺). The levels thus obtained have been drawn in Fig. 3 where full lines refer to even and broken lines to odd parity. On the left-hand side of Fig. 3, $\Omega=0$, we have full permeability of the wall, i.e. the situation of Problem 18. The more opaque the wall becomes towards the right-hand margin of the figure, the more the even-parity levels are raised, whereas the odd-parity levels remain unaffected. This behaviour is reflected by the even-parity eigenfunctions, as an example of which the one for $n=1$ has been drawn in Fig. 4. (The functions are not normalized.) For the impermeable wall ($\Omega\to\infty$), the amplitude at $x=0$ falls to zero and both halves become independent of one another. Except for the sign in the left half, this result is identical with the first odd eigenfunction shown at the bottom of Fig. 4, so that in the limit $\Omega\to\infty$ the energy levels become degenerate (right margin of Fig. 3).

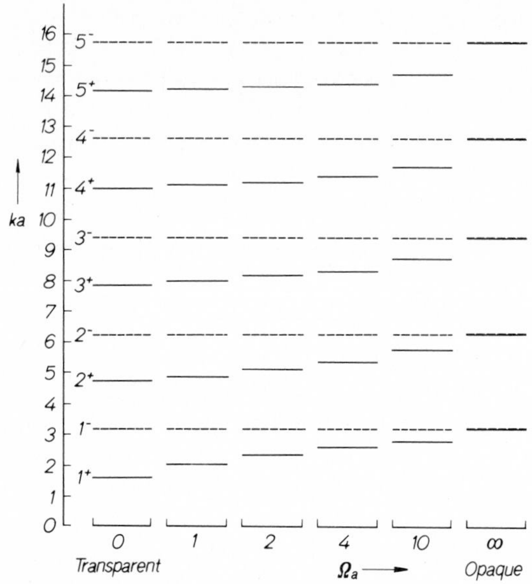

Fig. 3. Level positions for different values of wall opacity. Full lines even, broken lines odd parity

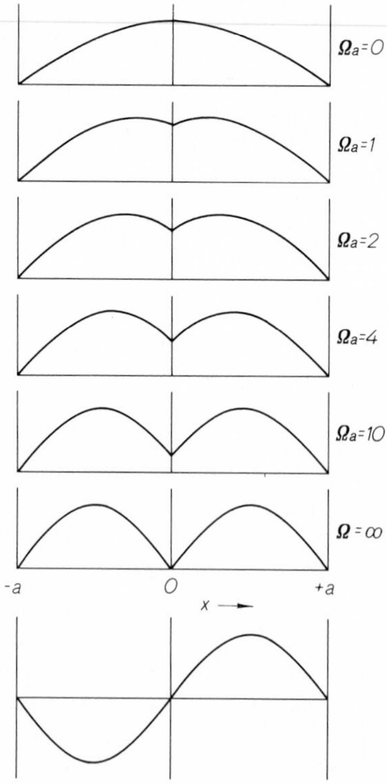

Fig. 4. Lowest eigenfunction for different wall opacities. Above 1^+, below 1^-, as limiting cases

Problem 20. Opaque wall described by Dirac δ function

The potential wall of the preceding problem may equally well be described by a Dirac δ function according to

$$V(x) = \frac{\hbar^2}{m} \Omega \delta(x) . \tag{20.1}$$

This shall be shown, and the boundary condition at the wall discussed.

Solution. The Schrödinger equation

$$u'' + [k^2 - 2\Omega\delta(x)]u = 0 \tag{20.2}$$

shall be solved. First of all, we state that the differential equation (20.2) as well as the boundary conditions $u(\pm a)=0$ are invariant with respect to the parity transformation $x \to -x$. The problem defined by both differential equation and boundary conditions can therefore be solved only by functions which are eigenfunctions of the parity operator (exchanging x with $-x$) as well. This can be seen as follows: any function $u(x)$ may be decomposed into an even part $f(x)=f(-x)$ and an odd part $g(x)=-g(-x)$ so that

$$u(x)=Af(x)+Bg(x); \quad u(-x)=Af(x)-Bg(x).$$

If $u(x)$ is a solution of our problem, so is $u(-x)$, and as there is no degeneracy involved, the solution is unambiguous save for a constant factor, say α. Hence,

$$u(-x)=\alpha u(x)$$

or

$$Af(x)-Bg(x)=\alpha[Af(x)+Bg(x)].$$

Since f and g are linearly independent this holds if, and only if, $A=\alpha A$ and $B=-\alpha B$, i.e. either if $\alpha=1$, $B=0$ so that $u(x)$ is of even parity, or if $A=0$, $\alpha=-1$ so that $u(x)$ is of odd parity.

We now consider the neighbourhood of the potential wall. Integration of Eq. (20.2) across the wall, u being continuous, yields

$$u'(+0)-u'(-0)=2\Omega u(0), \tag{20.3}$$

i.e. the logarithmic derivative

$$L(x) = \frac{u'(x)}{u(x)} \tag{20.4}$$

is discontinuous with a jump of 2Ω at the wall:

$$L(+0)-L(-0)=2\Omega. \tag{20.5}$$

Thus the rather cumbersome limiting process used in Problem 19 may be replaced by an admittedly rather artificial but very simple boundary condition at the wall. It should be noted that this boundary condition is derived from the behaviour of the differential equation in the wall neighbourhood only, so that it still holds for a superimposed non-singular potential $V(x)$ and any boundary conditions elsewhere.

We now turn to a brief discussion of the eigenfunctions. For a solution of *odd* parity, $u_n^-(x)$ has a zero at the wall: $u_n^-(0)=0$. Then according to Eq. (20.3) the derivative will be continuous without any jump. Hence the odd solutions turn out to be unaffected by the wall, however opaque. This is in perfect agreement with the results derived in Problem 19, Eq. (19.7⁻).

A solution of *even* parity, on the other hand, necessarily must be of the form

$$u_n^+(x)=\begin{cases}-A\sin k_n^+(x+a) & \text{in} \quad -a\leq x<0,\\ +A\sin k_n^+(x-a) & \text{in} \quad 0<x\leq +a\end{cases} \tag{20.6}$$

as given in Eq. (19.7⁺). From (20.6) we gather

$$L(+0)=-k_n^+\cot k_n^+ a\,; \qquad L(-0)=+k_n^+\cot k_n^+ a$$

and according to condition (20.5)

$$k_n^+\cot k_n^+ a=-\Omega\,,$$

in agreement with Eq. (19.6⁺). The further discussion thus becomes identical with the one in Problem 19.

Problem 21. Scattering at a Dirac δ function wall

A current of particles of energy E, arriving from the left, strikes a potential wall at $x=0$ given by

$$V(x)=\frac{\hbar^2}{m}\Omega\,\delta(x)\,. \tag{21.1}$$

It shall be shown that the effect of the wall consists of the production of a "scattered" wave, starting from the wall in both backward and forward directions.

Solution. Except for the infinitesimal neighbourhood of $x=0$, the complete solution of the Schrödinger equation can be written

$$u(x)=A\,e^{ikx}+B\,e^{-ikx}\,; \qquad k^2=\frac{2mE}{\hbar^2}\,; \qquad k>0 \tag{21.2}$$

where A and B are different constants for $x<0$ and $x>0$ which can be adjusted to make (21.2) satisfy the boundary conditions. These lead to the more special form

$$u(x)=\begin{cases} e^{ikx}+B e^{-ikx} & \text{for } x<0, \\ (1+F)e^{ikx} & \text{for } x>0 \end{cases} \tag{21.3}$$

if the amplitude of the incident particle current is normalized to 1. Then B is the backward, and F the forward scattering amplitude.

According to (20.3), the behaviour of $u(x)$ at $x=0$ is determined by

$$u(+0)=u(-0) \quad \text{and} \quad u'(+0)-u'(-0)=2\Omega u(0). \tag{21.4}$$

These relations yield $B=F$ and

$$ik(1+F)-ik(1-B)=2\Omega(1+B)$$

so that finally

$$B=F=\frac{\Omega}{ik-\Omega}. \tag{21.5}$$

In the solution (21.3) we can distinguish three different currents: the incident one has been normalized to unity, the reflected intensity is $|B|^2$ and the transmitted intensity $|1+F|^2$. From (21.5) there follows

$$|B|^2=\frac{\Omega^2}{\Omega^2+k^2}; \qquad |1+F|^2=\frac{k^2}{\Omega^2+k^2} \tag{21.6}$$

leading to the conservation law (equation of continuity)

$$1=|B|^2+|1+F|^2; \tag{21.7}$$

i.e. the sum of reflected and transmitted intensities is equal to the incident intensity of the current.

If the wall is very opaque ($\Omega\to\infty$), Eq. (21.5) leads to $B\simeq-1$ and $1+F\simeq 0$ so that we approach total reflection. If, on the contrary, the wall is almost completely transparent ($\Omega\to 0$), we find

$$B=F\simeq\frac{\Omega}{ik},$$

and the scattered intensity becomes inversely proportional to the particle energy. This does not, of course, hold for small energies ($k\lesssim\Omega$) but only within the high energy limit. It is a special case of Born's first approximation valid at high energies, and it should be noted that the condition for its validity is *not* (as is sometimes stated) that the kinetic energy remain large compared to the potential energy $V(x)$ *throughout*, since in the present example the latter becomes even infinitely large at $x=0$.

It is a special feature of the potential (21.1) that forward and backward scattering amplitudes are equal for all energies.

Problem 22. Scattering at a symmetric potential barrier

A current of particles of energy E strikes a potential barrier $V(x)$ restricted to the region $-a \leqq x \leqq +a$. The potential is supposed to be an even function of x:

$$V(x) = V(-x). \tag{22.1}$$

The amplitudes of backward and forward scattered waves shall be determined in terms of the logarithmic derivatives of the wave function at $x = \pm a$.

Solution. The symmetry condition (22.1) has the important consequence that for every energy E there exists a solution of the Schrödinger equation of even parity,

$$u_+(x) = u_+(-x); \qquad u'_+(x) = -u'_+(-x), \tag{22.2a}$$

and a solution of odd parity,

$$u_-(x) = -u_-(-x); \qquad u'_-(x) = u'_-(-x). \tag{22.2b}$$

Since both are, of course, linearly independent of one another, the general solution will be any linear combination of them. The two partial solutions u_+ and u_- may be determined inside the interval $-a \leq x \leq +a$, in the worst case by numerical procedures starting from $x=0$ with

$$u_+(0) = 1; \qquad u'_+(0) = 0$$

and

$$u_-(0) = 0; \qquad u'_-(0) = 1$$

in an arbitrary normalization of these two basic solutions. Thus we can compute their logarithmic derivatives at $x=a$ which we shall write for convenience in a dimensionless form

$$a u'_+(a)/u_+(a) = L_+; \qquad a u'_-(a)/u_-(a) = L_- \tag{22.3}$$

independent of their respective normalizations. The logarithmic derivatives $a u'_\pm(-a)/u_\pm(-a)$ at $x=-a$ then follow from (22.2) to be $-L_+$ and $-L_-$.

The solution, corresponding to a wave of unit amplitude incident from the left, can be written

$$u(x) = \begin{cases} e^{ikx} + B e^{-ikx}, & -\infty < x \leq -a, \\ C_1 u_+(x) + C_2 u_-(x), & -a \leq x \leq +a, \\ (1+F) e^{ikx}, & +a \leq x < +\infty. \end{cases} \tag{22.4}$$

Continuity of $u(x)$ and $u'(x)$ at $x=\pm a$ yields four conditions, viz.

$$e^{-ika}+Be^{ika} = C_1 u_+(a) - C_2 u_-(a), \tag{22.5a}$$

$$ik(e^{-ika}-Be^{ika}) = -C_1 u'_+(a) + C_2 u'_-(a), \tag{22.5b}$$

$$(1+F)e^{ika} = C_1 u_+(a) + C_2 u_-(a), \tag{22.5c}$$

$$ik(1+F)e^{ika} = C_1 u'_+(a) + C_2 u'_-(a). \tag{22.5d}$$

The sum of (22.5a) and (22.5c) gives $2C_1 u_+(a)$ and the difference of (22.5b) and (22.5d) $2C_1 u'_+(a)$. The ratio of both is

$$L_+ = ika\frac{-e^{-ika}+(1+F+B)e^{ika}}{e^{-ika}+(1+F+B)e^{ika}}. \tag{22.6a}$$

By the same procedure, with opposite signs, we find

$$L_- = ika\frac{e^{-ika}+(1+F-B)e^{ika}}{-e^{-ika}+(1+F-B)e^{ika}}. \tag{22.6b}$$

Solving (22.6a, b) with respect to $1+F\pm B$ and using the abbreviation

$$ka=q \tag{22.7}$$

we finally arrive at the amplitude relations

$$B = -\frac{1}{2}e^{-2iq}\left[\frac{L_+ + iq}{L_+ - iq} + \frac{L_- + iq}{L_- - iq}\right] \tag{22.8a}$$

and

$$1+F = -\frac{1}{2}e^{-2iq}\left[\frac{L_+ + iq}{L_+ - iq} - \frac{L_- + iq}{L_- - iq}\right]. \tag{22.8b}$$

According to the equation of continuity, we expect the sum of reflected and transmitted intensities to be equal to the incident one. Indeed, from (22.8a, b) there follow the formulae

$$|B|^2 = \frac{(L_+ L_- + q^2)^2}{(L_+ L_- + q^2)^2 + q^2(L_+ - L_-)^2} \tag{22.9a}$$

and

$$|1+F|^2 = \frac{q^2(L_+ - L_-)^2}{(L_+ L_- + q^2)^2 + q^2(L_+ - L_-)^2}. \tag{22.9b}$$

They evidently satisfy the expected relation

$$|B|^2 + |1+F|^2 = 1. \tag{22.10}$$

The problem of determining the scattering amplitudes in forward and backward directions has thus been reduced to finding the logarithmic derivatives (22.3) at $x=a$ of the even and the odd parity wave functions. This cannot, of course, be accomplished as long as no special form of the potential (22.1) has been introduced.

In contradistinction to Problem 21, we no longer have $B=F$. Forward scattering prevails if

$$q|L_+ - L_-| > |L_+ L_- + q^2|,$$

backward scattering in the opposite case.

Problem 23. Reflection at a rectangular barrier

The general formulae derived in Problem 22 shall be applied to a potential barrier with

$$\frac{2m}{\hbar^2} V = k_0^2 \quad \text{in } |x| \leq a, \tag{23.1}$$

and $V=0$ elsewhere. The transmittance of the barrier shall be determined.

Solution. Inside the barrier, the Schrödinger equation becomes

$$u'' + (k^2 - k_0^2) u = 0. \tag{23.2}$$

There are, therefore, solutions of different type for a kinetic energy below ($k<k_0$) and above ($k>k_0$) threshold. We begin with the first case and write

$$k_0^2 - k^2 = \varkappa^2; \qquad u'' - \varkappa^2 u = 0. \tag{23.3}$$

Then the even solution is

$$u_+(x) = \cosh \varkappa x; \qquad u_+(0)=1; \qquad u'_+(0)=0 \tag{23.4a}$$

and the odd solution

$$u_-(x) = \frac{1}{\varkappa} \sinh \varkappa x; \qquad u_-(0)=0; \qquad u'_-(0)=1. \tag{23.4b}$$

Hence,

$$L_+ = a u'_+(a)/u_+(a) = \varkappa a \tanh \varkappa a; \tag{23.5a}$$

$$L_- = a u'_-(a)/u_-(a) = \varkappa a \coth \varkappa a. \tag{23.5b}$$

The transmittance of the barrier then follows from (22.9b) by elementary reshaping:

$$T \equiv |1+F|^2 = \frac{1}{1 + \left(\dfrac{k_0^2}{2k\varkappa}\right)^2 \sinh^2 2\varkappa a} \tag{23.6}$$

whereas the reflectance, according to (22.10), is given by

$$R \equiv |B|^2 = 1 - T. \tag{23.7}$$

In classical mechanics, the whole flux arriving from the left side would be reflected at the barrier so that $|B|^2=1$ and $|1+F|^2=0$. This happens, according to Eq. (23.6) if, and only if, $\varkappa a \to \infty$, i.e. if there is a very great „potential mountain" above the energy level of the particles, the transparency of the barrier will become very small though still finite ("tunnel effect"). The transmittance of the barrier may then be written approximately

$$T = \frac{16 k^2 \varkappa^2}{k_0^4} e^{-4\varkappa a}, \tag{23.8}$$

its order of magnitude being mainly determined by the exponential factor.

The exponent

$$4\varkappa a = 2 \int_{-a}^{+a} dx \sqrt{\frac{2m}{\hbar^2}(V-E)}$$

will be generalized to this integral form for any potential $V(x)$, below (cf. Problem 116).

If, on the other hand, the kinetic energy exceeds the height of the barrier, the quantity $\varkappa$ defined by (23.3) becomes imaginary. With the abbreviation

$$K^2 = k^2 - k_0^2 = -\varkappa^2 \tag{23.9}$$

we may then write, instead of (23.6),

$$T = \frac{1}{1 + \left(\frac{k_0^2}{2kK}\right)^2 \sin^2 2Ka}. \tag{23.10}$$

Though in classical mechanics there should be $T=1$ and $R=0$ at these energies, the transparency following from (23.10) shows maxima of $T=1$ only at $2Ka = n\pi$ $(n=1,2,3,\ldots)$. Between these there are minima in the neighbourhood of $2Ka=(n+\frac{1}{2})\pi$ which lie the closer to $T=1$ the smaller the factor in front of the sine in (23.10), i.e. the higher the energy above threshold.

The general behaviour of T as a function of the energy (in units of the threshold height, say, U) is shown in Fig. 5 where T has been drawn as a function of E/U for the example $2k_0 a = 3\pi$. The wave function has been explained in Fig. 6 where $|u|^2$ has been drawn vs. x. On the right-hand side of the barrier we simply have $|u|^2 = |1+F|^2$, i.e. constant, whereas on its left there is interference of the incident with the reflected wave. Fig. 6 shows this feature for $k^2 = \varkappa^2 = \frac{1}{2} k_0^2$ and different widths of the barrier. The broader the latter, the smaller is the intensity transmitted and the more pronounced the interference phenomena become.

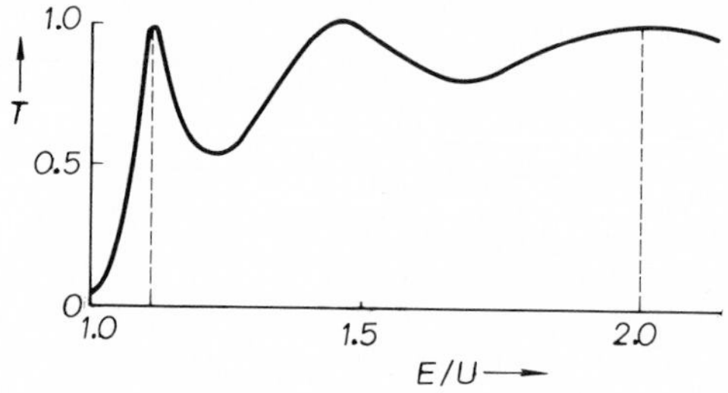

Fig. 5. Transmittance of potential barrier for $E > U$ in dependence upon energy

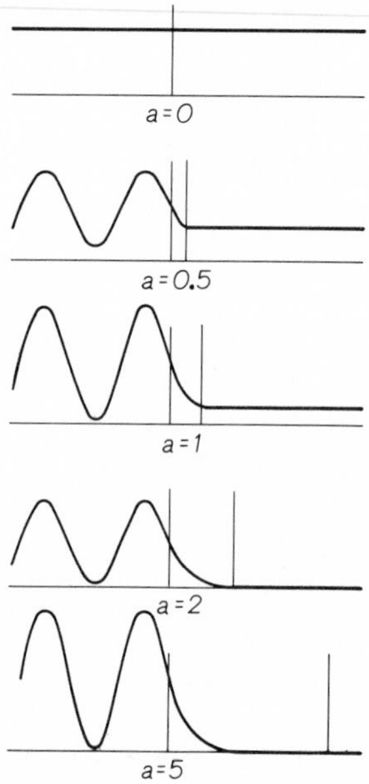

Fig. 6. Probability density $|u|^2$ in the current falling upon the barrier from the left, in the case $E < U$. The two vertical lines mark the width a of the barrier. The waves on the left are caused by interference between incident and reflected beams

Problem 24. Inversion of reflection

Let a potential threshold $V(x)>0$ in the interval $0<x<a$ form an obstacle to a wave incident from the left $(x<0)$. It shall be proved that the coefficient of reflection is the same when the wave falls on the obstacle from the right $(x>a)$, whatever the potential shape.

Solution. Let $u(x)$ and $v(x)$ be two real and independent solutions of the Schrödinger equation in the interval $0<x<a$ with the Wronskian

$$uv'-vu'=1. \tag{24.1}$$

The wave function in the case of the wave incident from the left is then of the form[2]

$$\psi=\left\{\begin{array}{ll} e^{ikx}+R\,e^{-ikx}, & x<0, \\ A\,u(x)+B\,v(x), & 0<x<a, \\ C\,e^{ik(x-a)}, & a<x \end{array}\right\} \tag{24.2}$$

and the conditions for continuity of ψ and ψ' at $x=0$ and $x=a$ are

$$\left.\begin{array}{r} 1+R=A\,u(0)+B\,v(0), \\ ik(1-R)=A\,u'(0)+B\,v'(0), \\ A\,u(a)+B\,v(a)=C, \\ A\,u'(a)+B\,v'(a)=ikC. \end{array}\right\} \tag{24.3}$$

From the last pair of equations, making use of (24.1), we find

$$\left.\begin{array}{l} A=C[v'(a)-ikv(a)], \\ B=-C[u'(a)-iku(a)]. \end{array}\right\} \tag{24.4}$$

Putting these expressions for A and B into the first pair of the Eqs. (24.3) we get

$$1+R=(p_{0a}-iq)C; \qquad 1-R=(p_{a0}-ir)C \tag{24.5}$$

with the abbreviations

$$\left.\begin{array}{r} p_{0a}=u(0)v'(a)-v(0)u'(a), \\ p_{a0}=u(a)v'(0)-v(a)u'(0), \\ q=k[u(0)v(a)-v(0)u(a)], \\ r=\dfrac{1}{k}[u'(0)v'(a)-v'(0)u'(a)]. \end{array}\right\} \tag{24.6}$$

From (24.5) we finally find

$$R=\frac{(p_{0a}-p_{a0})-i(q-r)}{(p_{0a}+p_{a0})-i(q+r)}, \tag{24.7}$$

---

[2] We write ψ for the space part of the wave function in this problem since u is used in another sense.

and the reflection coefficient becomes

$$|R|^2 = \frac{(p_{0a}-p_{a0})^2+(q-r)^2}{(p_{0a}+p_{a0})^2+(q+r)^2}. \tag{24.8}$$

Now take the opposite case with the wave incident from the right-hand side. The wave function (24.2) has then to be replaced by

$$\tilde{\psi} = \begin{cases} \tilde{C}\,\mathrm{e}^{-ikx}, & x<0, \\ \tilde{A}\,u(x)+\tilde{B}\,v(x), & 0<x<a, \\ \mathrm{e}^{-ik(x-a)}+\tilde{R}\,\mathrm{e}^{ik(x-a)}, & a<x. \end{cases} \tag{24.9}$$

The conditions of continuity run as follows:

$$\left.\begin{aligned} 1+\tilde{R} &= \tilde{A}\,u(a)+\tilde{B}\,v(a), \\ -ik(1-\tilde{R}) &= \tilde{A}\,u'(a)+\tilde{B}\,v'(a), \\ \tilde{A}\,u(0)+\tilde{B}\,v(0) &= \tilde{C}, \\ \tilde{A}\,u'(0)+\tilde{B}\,v'(0) &= -ik\tilde{C}. \end{aligned}\right\} \tag{24.10}$$

They have the same structure as Eqs. (24.3) from which they can be obtained by exchanging the two arguments $x=0$ and $x=a$ and replacing k by $-k$. This transformation, applied to (24.6) renders

$$p_{0a}\to p_{a0}; \quad p_{a0}\to p_{0a}; \quad q\to q; \quad r\to r, \tag{24.11}$$

i.e. the only difference in the final formulae (24.7) and (24.8) occurs by exchanging p_{0a} and p_{a0}. Therefore, since (24.8) is symmetrical in p_{0a} and p_{a0}, the reflection coefficient

$$|\tilde{R}|^2 = |R|^2 \tag{24.12}$$

is the same for waves incident from both sides, as was to be proved. This does not, however, hold for R, Eq. (24.7) which, written as a ratio $R=\alpha/\beta$ of two complex numbers α and β according to (24.7) is transformed into $\tilde{R}=-\alpha^*/\beta$.

Problem 25. Rectangular potential hole

For a rectangular potential hole

$$V(x) = \begin{cases} -U & |x|<a, \\ 0 & \text{elsewhere} \end{cases} \tag{25.1}$$

the bound-state solutions and their eigenvalues shall be determined.

Solution. Since for states of positive energy the general behaviour can be gathered without difficulty from the preceding problem, it suffices to discuss negative energies, i.e. bound states.

The potential is invariant with respect to inversion, $V(x)=V(-x)$, so that the solutions have either even or odd parity (cf. Problem 20). Putting

$$E=-\frac{\hbar^2\varkappa^2}{2m};\qquad U=\frac{\hbar^2 k_0^2}{2m};\qquad k^2=k_0^2-\varkappa^2 \tag{25.2}$$

these solutions are

even:

$$\left.\begin{aligned} &u_+(x)=\begin{cases}A_+\cos kx, & 0\le x\le a,\\ A_+\cos ka\,e^{\varkappa(a-x)}, & x>a,\end{cases}\\ &u_+(-x)=u_+(x),\\ &1/A_+^2=\frac{1}{k}\left[ka+\sin ka\cos ka\right]+\frac{1}{\varkappa}\cos^2 ka;\end{aligned}\right\} \tag{25.3e}$$

odd:

$$\left.\begin{aligned} &u_-(x)=\begin{cases}A_-\sin kx, & 0\le x\le a,\\ A_-\sin ka\,e^{\varkappa(a-x)}, & x>a,\end{cases}\\ &u_-(-x)=-u_-(x),\\ &1/A_-^2=\frac{1}{k}\left[ka-\sin ka\cos ka\right]+\frac{1}{\varkappa}\sin^2 ka.\end{aligned}\right\} \tag{25.3o}$$

Here the amplitudes inside and outside the potential hole have been adjusted to make $u(a)$ continuous. The normalization constants have been calculated from the condition

$$\int_{-\infty}^{+\infty} dx\,|u|^2=1.$$

Continuity of u' at $x=a$ adds another relation, viz.

even: $$-k\sin ka=-\varkappa\cos ka$$

or

$$\tan ka=\frac{\varkappa}{k}; \tag{25.4e}$$

odd: $$k\cos ka=-\varkappa\sin ka$$

or

$$\cot ka=-\frac{k}{\varkappa}. \tag{25.4o}$$

Using (25.4) and (25.2), we can recast the normalization expressions so that we obtain the same equation

$$1/A_{\pm}^2 = a + \frac{1}{\varkappa} \tag{25.5}$$

for both cases.

In order to find the eigenvalues from (25.4), we replace $\varkappa$ on the right-hand side according to (25.2) and introduce the abbreviation

$$C = k_0 a. \tag{25.6}$$

We thus obtain

even:
$$\tan ka = \frac{\sqrt{C^2 - (ka)^2}}{ka}; \tag{25.7e}$$

odd:
$$\tan ka = -\frac{ka}{\sqrt{C^2 - (ka)^2}}. \tag{25.7o}$$

For a given potential, C is a constant determining the size of the hole ($C^2 \propto U a^2$), and Eqs. (25.7e, o) can be used to find all values of ka and thus of the energy

$$E = -U\left[1 - \left(\frac{ka}{C}\right)^2\right] \tag{25.8}$$

compatible with the size of the hole.

Fig. 7 shows the function $\tan ka$ vs. ka as well as the expressions on the right-hand side of (25.7e) and (25.7o). The solution of these eigenvalue equations is obtained from the intersection of the curves of

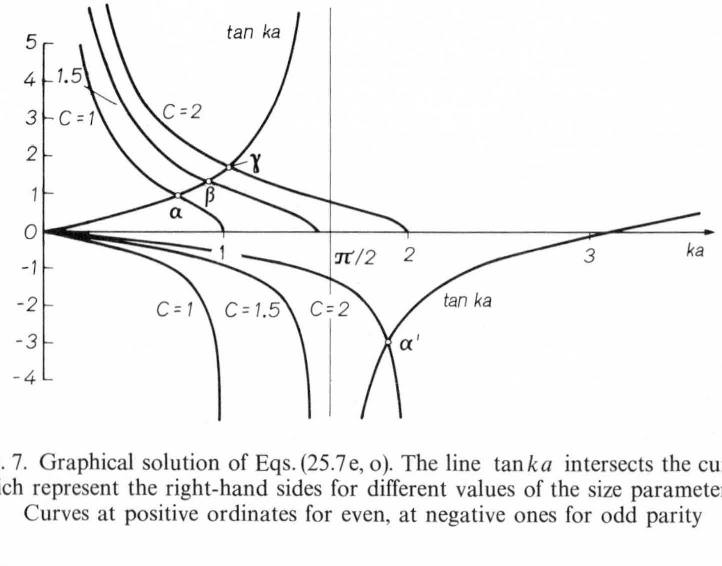

Fig. 7. Graphical solution of Eqs. (25.7e, o). The line $\tan ka$ intersects the curves which represent the right-hand sides for different values of the size parameter C. Curves at positive ordinates for even, at negative ones for odd parity

the last two with the tangent line. These curves, of course, still depend on the size parameter C. Starting with $C=1$, e. g., we have the intersection denoted by α in the even, and no intersection at all in the odd case. A hole of this size therefore contains no more than one eigenstate with even parity. In Fig. 8a the hole has been drawn with its eigen-

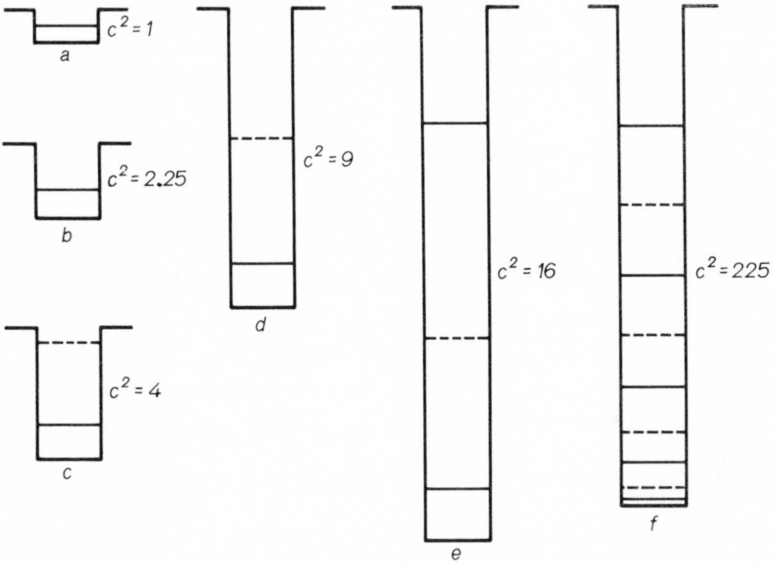

Fig. 8a—f. Energy levels in potential holes of different sizes determined by the size parameter C. Full lines even, broken lines odd parity

value. For a bigger hole, $C=1.5$, the intersection in Fig. 7 occurs at the point β; again we have only one even-parity state (Fig. 8b), and $E_\beta < E_\alpha$ because $(ka)_\beta > (ka)_\alpha$. If we further increase the size, taking e. g. $C=2$, the intersection γ leads to the lowest even-parity state ($E_\gamma < E_\beta$), but an odd-parity state is now added corresponding to the intersection at α' (cf. Fig. 8c). Further increase gives a growing "capacity" for eigenstates to the hole (Figs. 8d, e, f), the number of states growing linearly with C and forming an alternate series of even and odd states. The eigenfunctions follow the general rule that, the more zeros they have, the higher they lie in the energy scale. The four lowest states are shown in Fig. 9 for $C=5$.

In classical mechanics, the particle might oscillate with any energy between the two walls at $x=\pm a$ bounding the hole. Outside, it would have negative kinetic energy so that these regions would be inaccessible. In quantum mechanics, this condition becomes less rigorous. The probability of finding the particle inside the hole, P_i, is less than unity:

$$P_i = \int_{-a}^{+a} dx|u|^2 = 1 - \frac{k^2}{k_0^2(1+\varkappa a)}, \tag{25.9}$$

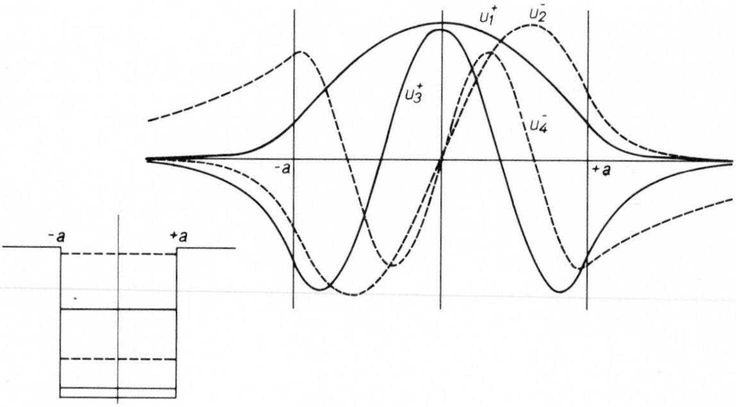

Fig. 9. Energy levels and eigenfunctions for $C=5$. Full lines even, broken lines odd parity

so that there remains a finite probability that it will stay outside. For any interval outside, this probability decreases exponentially as $e^{-2\varkappa(|x|-a)}$ with increasing distance $|x|-a$ from the hole.

Problem 26. Rectangular potential hole between two walls

The solutions of the Schrödinger equation shall be determined for the potential drawn in Fig. 10. Special consideration shall be given to the limit $l\to 0$ for the states of positive energy.

Solution. We begin with a brief discussion of the "bound" states, $E<0$. Using again the symbols k^2, k_0^2, and $\varkappa^2$ defined in (25.2) and normalizing so that

$$\int_{-l}^{+l} dx|u|^2 = 1,$$

we may write the wave functions as follows:

even parity:

$$u_+ = \begin{cases} A_+ \cos kx, & 0 \le x \le a, \\ A_+ \dfrac{\cos ka}{\sinh \varkappa(l-a)} \sinh \varkappa(l-x), & a < x \le l, \end{cases}$$

$$1/A_+^2 = \frac{1}{k}[ka + \sin ka \cos ka] + \frac{\cos^2 ka}{\varkappa}\left[\coth \varkappa(l-a) - \frac{\varkappa(l-a)}{\sinh^2 \varkappa(l-a)}\right] \tag{26.1e}$$

odd parity:

$$u_- = \begin{cases} A_- \sin kx, & 0 \le x \le a, \\ A_- \dfrac{\sin ka}{\sinh \varkappa(l-a)} \sinh \varkappa(l-x), & a < x \le l, \end{cases}$$

$$1/A_-^2 = \frac{1}{k}[ka - \sin ka \cos ka] + \frac{\sin^2 ka}{\varkappa}\left[\coth \varkappa(l-a) - \frac{\varkappa(l-a)}{\sinh^2 \varkappa(l-a)}\right]. \tag{26.1o}$$

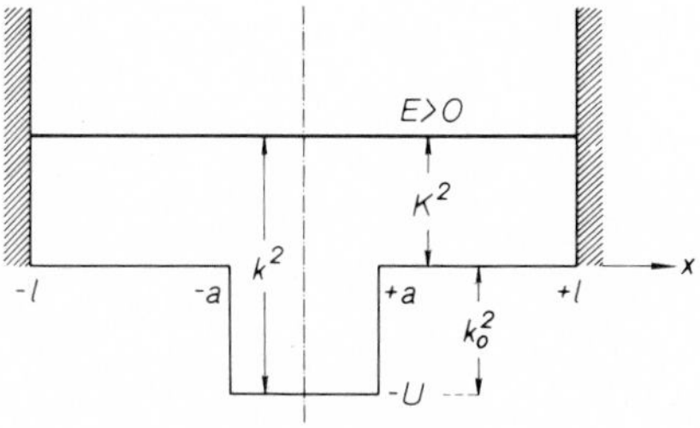

Fig. 10. Potential of Problem 26

Here again, as in Problem 25, $u(a)$ is made continuous, but continuity of $u'(a)$ imposes another condition in either case, viz. for

even parity: $$\tan ka = \frac{\varkappa}{k} \coth \varkappa(l-a), \tag{26.2e}$$

odd parity: $$\tan ka = -\frac{k}{\varkappa} \tanh \varkappa(l-a) \tag{26.2o}$$

from which the eigenvalues may be computed. We do not, however, follow this in detail. Suffice it to remark that if

$$\varkappa(l-a) \gg 1$$

the coth- and tanh-functions both rapidly tend towards unity. Eqs. (26.2e, o) then pass over into the eigenvalue conditions (25.4e, o) of the preceding problem, and the normalization relations for $1/A_{\pm}^2$ of (26.1e, o) into those of (25.3e, o). In the wave functions (26.1e, o) themselves, for $|x|<l$ so that $\varkappa(l-x) \gg 1$, one may write

$$\frac{\sinh \varkappa(l-x)}{\sinh \varkappa(l-a)} = e^{\varkappa(a-x)}$$

thus again falling back on the wave functions (25.3e, o).

It is of much more interest to study *positive energies* where for finite l there are discrete eigenvalues which with increasing l form an increasingly dense level system, in the limit $l \to \infty$ passing over into a continuum. Writing for $E>0$,

$$E = \frac{\hbar^2 K^2}{2m}; \quad K^2 = -\varkappa^2 = k^2 - k_0^2, \tag{26.3}$$

the wave functions then become for

even parity:

$$u_+ = \begin{cases} A_+ \cos kx, & 0 \le x \le a, \\ A_+ \dfrac{\cos ka}{\sin K(l-a)} \sin K(l-x), & a < x \le l, \end{cases}$$

$$1/A_+^2 = \frac{1}{k}[ka + \sin ka \cos ka] - \frac{\cos^2 ka}{K}\left[\cot K(l-a) - \frac{K(l-a)}{\sin^2 K(l-a)}\right]; \tag{26.4e}$$

odd parity:

$$u_- = \begin{cases} A_- \sin kx, & 0 \le x \le a, \\ A_- \dfrac{\sin ka}{\sin K(l-a)} \sin K(l-x), & a < x \le l, \end{cases}$$

$$1/A_-^2 = \frac{1}{k}[ka - \sin ka \cos ka] - \frac{\sin^2 ka}{K}\left[\cot K(l-a) - \frac{K(l-a)}{\sin^2 K(l-a)}\right]. \tag{26.4o}$$

Continuity of $u(a)$ is performed in these expressions, but continuity of $u'(a)$ again adds the eigenvalue conditions for

$$\textit{even parity:} \qquad \tan ka = \frac{K}{k} \cot K(l-a), \tag{26.5e}$$

$$\textit{odd parity:} \qquad \tan ka = -\frac{k}{K} \tan K(l-a). \tag{26.5o}$$

Making use of these relations, we may replace $\cot K(l-a)$ in the last bracket of the normalization expressions $1/A_\pm^2$ thus obtaining

$$1/A_+^2 = \frac{1}{k}[ka + \sin ka \cos ka] + (l-a)\left(\cos^2 ka + \frac{k^2}{K^2}\sin^2 ka\right) - \frac{k}{K^2}\sin ka \cos ka \tag{26.6e}$$

and

$$1/A_-^2 = \frac{1}{k}[ka - \sin ka \cos ka] + (l-a)\left(\sin^2 ka + \frac{k^2}{K^2}\cos^2 ka\right) + \frac{k}{K^2}\sin ka \cos ka. \tag{26.6o}$$

If $l \to \infty$, the second term will grow beyond all limits so that

$$1/A_+^2 = l\left(\cos^2 ka + \frac{k^2}{K^2}\sin^2 ka\right); \; 1/A_-^2 = l\left(\sin^2 ka + \frac{k^2}{K^2}\cos^2 ka\right). \tag{26.7}$$

The amplitudes outside the hole, however, can be determined from (26.4e, o):

$$\frac{\sin^2 K(l-a)}{A_+^2 \cos^2 ka} \to l; \qquad \frac{\sin^2 K(l-a)}{A_-^2 \sin^2 ka} \to l,$$

so that both wave functions, for $x > a$, become

$$u_\pm = \frac{1}{\sqrt{l}} \sin K(l-x). \tag{26.8}$$

Here, l still enters the phase of the wave functions from which it can be eliminated by again using the eigenvalue conditions (26.5e, o):

$$\textit{even:} \qquad Kl = \tan^{-1}\left(\frac{\frac{K}{k} + \tan ka \tan Ka}{\tan ka - \frac{K}{k}\tan Ka}\right); \tag{26.9e}$$

$$\textit{odd:} \qquad Kl = \tan^{-1}\left(\frac{\tan Ka - \frac{K}{k}\tan ka}{1 + \frac{K}{k}\tan ka \tan Ka}\right). \tag{26.9o}$$

The most striking feature, however, of the system of wave functions is their *level density* to be derived from (26.5e, o) for very large, but still finite values of l. The right-hand sides of these equations run through all real numbers between $-\infty$ and $+\infty$ at intervals of width π in the variable Kl. In each such interval there exists exactly one solution to either equation so that we obtain alternating even-parity and odd-parity levels at an average distance $\Delta K = \pi/(2l)$ in the K scale. From (26.3) the average level distance in the energy scale becomes

$$\Delta E = \frac{\hbar^2}{m} K \frac{\pi}{2l} = \frac{\pi}{l} \sqrt{\frac{\hbar^2}{2m} E}. \tag{26.10}$$

The average distance of two consecutive levels therefore only increases as $E^{\frac{1}{2}}$, and is inversely proportional to the length of the normalization interval. In the limit $l \to \infty$ therefore the discrete energy spectrum passes over into a continuous spectrum.

The amplitude structure of the continuum is shown for $k_0 a \equiv C = 2$ in Fig. 11. The dimensionless quantity $A^2 l$ is a measure of the amplitude square inside the hole when normalization outside stays the same

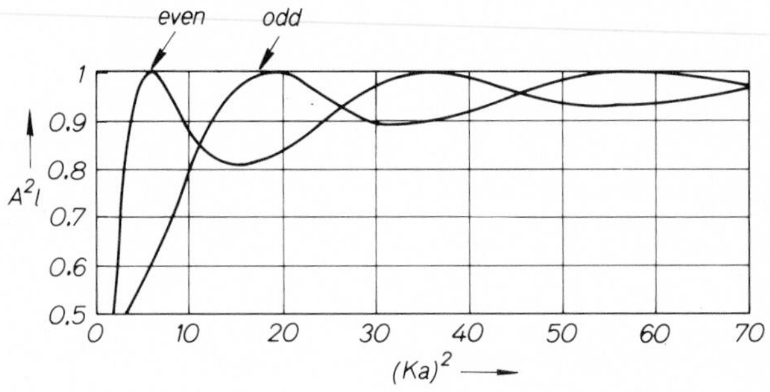

Fig. 11. Virtual states in the continuous spectrum

throughout and l is large. This quantity has been drawn according to (26.7) vs. $(Ka)^2$, i.e. vs. energy in a dimensionless scale. It is clear that $A^2 l$ reaches a maximum of 1 at an infinite sequence of energies between which there are amplitude minima which are the less pronounced the higher the energy. (Note that the lower half of the ordinate scale has been omitted in the Fig. 11.) The states at energies of maximum amplitude still preserve a remnant of the eigenvalue structure at positive energies, because they correspond to the greatest possible concentration of the wave function inside the hole. They are therefore frequently introduced as *virtual states* (in contrast to the "real" eigenstates at negative energies).

Problem 27. Virtual levels

A potential "cavity" between $x=0$ and $x=a$ is bounded on its right-hand side by an opaque wall (cf. Problem 20) as shown in Fig. 12 so that a small coupling exists between the wave function outside and inside this wall. It shall be shown that, for large wall opacity, there exist narrow bands of energy in which the coupling becomes rather strong. Numerical example: $\Omega a/\pi=50$.

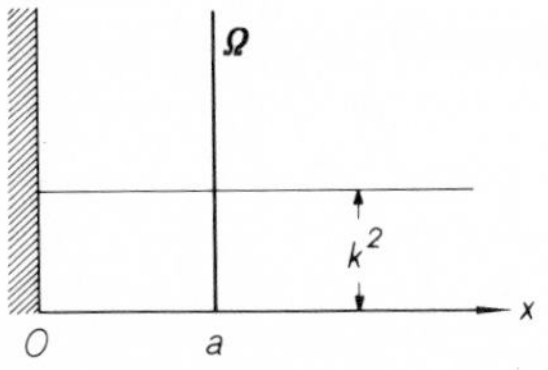

Fig. 12. Potential cavity bounded by a permeable wall

Solution. There is a continuous energy spectrum whose wave functions, normalized to unit amplitude outside the cavity, are

$$u(x)=\begin{cases} A\sin kx, & 0<x<a, \\ \sin(kx+\delta), & a<x<\infty. \end{cases} \tag{27.1}$$

At a wall of finite opacity Ω, the boundary conditions according to Problem 20 are

$$\left.\begin{aligned} u(a+0)&=u(a-0); \\ u'(a+0)&=u'(a-0)+2\Omega u(a) \end{aligned}\right\} \tag{27.2}$$

or, with (27.1),

$$\begin{aligned} \sin(ka+\delta)&=A\sin ka; \\ k\cos(ka+\delta)&=kA\cos ka+2\Omega A\sin ka. \end{aligned}$$

These two relations determine the phase angle δ from

$$\cot(ka+\delta)-\cot ka=2\frac{\Omega}{k} \tag{27.3}$$

and the wave amplitude A inside the cavity from

$$1/A^2=1+4\frac{\Omega}{k}\sin ka\cos ka+4\frac{\Omega^2}{k^2}\sin^2 ka \tag{27.4}$$

or

$$1/A^2=1+2\frac{\Omega}{k}\sin 2ka+2\frac{\Omega^2}{k^2}(1-\cos 2ka). \tag{27.4'}$$

For an impenetrable wall ($\Omega \to \infty$), the amplitude A vanishes, i.e. a vibration outside would not then be able to extend to the inside of the cavity; the latter would be wholly decoupled from the outside. A large but finite value of Ω, however, would allow for a small amplitude $A \sim k/\Omega$ which might even become quite large if the trigonometric functions in (27.4) are close to zero.

Let ka be a number large compared to π, then the trigonometric functions will vary over an entire period within an interval in which Ω/k remains nearly constant (viz., if $ka \gg 2\pi$). Maxima and minima of A^2 can then, in a reasonable approximation, be found by differentiating (27.4′) with respect to $2ka$, keeping Ω/k constant:

$$\left[\frac{\partial(1/A^2)}{\partial(2ka)}\right]_{\Omega/k} = 2\frac{\Omega}{k}\cos 2ka + 2\frac{\Omega^2}{k^2}\sin 2ka = 0$$

leads to

$$\tan 2ka = -\frac{k}{\Omega}. \tag{27.5}$$

Denoting the solutions of (27.5) by k_n and using the abbreviation

$$k_n/\Omega = \varepsilon_n \tag{27.6}$$

we obtain

$$2k_n a = n\pi - \tan^{-1}\varepsilon_n \tag{27.7}$$

with integer n, and therefore

$$\sin 2k_n a = (-1)^{n+1}\frac{\varepsilon_n}{\sqrt{1+\varepsilon_n^2}}; \qquad \cos 2k_n a = (-1)^n \frac{1}{\sqrt{1+\varepsilon_n^2}}.$$

Putting this into (27.4′), the extreme values of $1/A^2$ become for even $n = 2, 4, 6, \ldots$

$$1/A_n^2 = 1 - \frac{2}{\sqrt{1+\varepsilon_n^2}} + \frac{2}{\varepsilon_n^2}\left(1 - \frac{1}{\sqrt{1+\varepsilon_n^2}}\right) \tag{27.8a}$$

and for odd $n = 1, 3, 5, \ldots$

$$1/A_n^2 = 1 + \frac{2}{\sqrt{1+\varepsilon_n^2}} + \frac{2}{\varepsilon_n^2}\left(1 + \frac{1}{\sqrt{1+\varepsilon_n^2}}\right). \tag{27.8b}$$

Since $\varepsilon_n \ll 1$, these expressions may be expanded in powers of ε_n:

$$1/A_n^2 = \frac{1}{4}\varepsilon_n^2 = \frac{k_n^2}{4\Omega^2} \quad \text{for } n = 2, 4, 6, \ldots \tag{27.9a}$$

and

$$1/A_n^2 = \frac{4}{\varepsilon_n^2} + 2 - \frac{1}{4}\varepsilon_n^2 \simeq \frac{4\Omega^2}{k_n^2} \quad \text{for } n=1,3,5,\ldots, \tag{27.9b}$$

neglecting terms of the order ε_n^4 throughout.

For even n, i.e. for ka being close to an integer multiple of π, we therefore obtain a maximum amplitude

$$A_{n,\max} = \frac{2\Omega}{k_n} \gg 1, \tag{27.10}$$

and for odd n, i.e. if ka lies close to an half-integer multiple of π, a minimum amplitude

$$A_{n,\min} = \frac{k_n}{2\Omega} \ll 1. \tag{27.11}$$

Only around the energy values leading to (27.10) therefore there will exist appreciable coupling so that the outside wave function penetrates the barrier. In classical terms, this means the existence of *resonance* frequencies at which the cavity can be excited to eigenvibrations from outside.

These resonances occur, according to (27.7), if $k_n a$ deviates only a little from $n\pi$ with integer n, i.e. for such values of k that the wave function inside the cavity as given by (27.1) lies quite close to

$$u_n(x) = A\sin\frac{n\pi x}{a} \quad \text{with } u_n(a)=0,$$

the eigenfunction for an impenetrable wall ($\Omega\to\infty$). Hence, the resonance levels lie close to the energy eigenvalues of the cavity for $\Omega\to\infty$. They therefore are called *virtual levels* of the system.

For the numerical example $\Omega a = 50\pi = 157.08$, the amplitude A between $2ka = 19\pi$ and $2ka = 21\pi$ is shown in Fig. 13. In this case, $k/2\Omega \simeq 0.1$ and it can easily be seen that A is generally of this order, except in a very narrow region a little below $2ka = 20\pi$, where it has a typical resonance. Series expansion of $1/A^2$ around the resonance energy defined by

$$2k_0 a = 2n\pi - \tan^{-1}\frac{k}{\Omega}$$

yields approximately

$$\frac{1}{A^2} = \frac{k^2}{4\Omega^2}\left[1 + \frac{4\Omega^4}{k^4}(2ka - 2k_0 a)^2\right] \tag{27.12}$$

or a line-width of the order of

$$\Delta k \sim \frac{k^2}{4\Omega^2 a}. \tag{27.13}$$

It remains to discuss the phase angle from (27.3). A simple trigonometric reshaping yields

$$\tan\delta = -\frac{\Omega}{k}\,\frac{1-\cos 2ka}{1+\frac{\Omega}{k}\sin 2ka}. \tag{27.14}$$

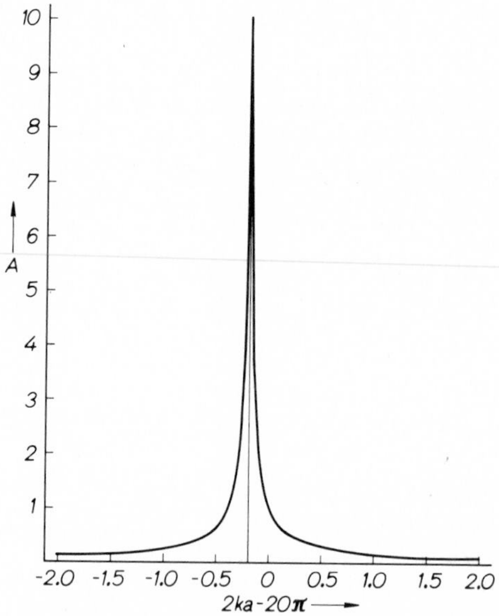

Fig. 13. Resonance level: Wave amplitude in dependence upon energy

If Ω/k is very large, we can in general neglect the 1 in the denominator so that (27.14) becomes independent of Ω:

$$\tan\delta \simeq -\frac{1-\cos 2ka}{\sin 2ka} = -\tan ka$$

which again corresponds to an impenetrable wall with the wave function

$$u(x) = \sin k(x-a)$$

for $x > a$. Only in the vicinity of

$$\sin 2ka = -\frac{k}{\Omega} \tag{27.15}$$

does this approximation no longer apply. There are two types of solution to (27.15), viz. where $2ka$ is either a little larger than $(2n+1)\pi$, or a little smaller than $2n\pi$. In the first case, the numerator $1-\cos 2ka$ is close to 2 and about constant. Then, $\tan\delta$ runs through a singularity or δ becomes an half-integer multiple of π, say, $\pi/2$. The phase curve has a steady slope and shows no very peculiar behaviour in this region. If, on the other hand, $2ka$ lies in the neighbourhood of $2n\pi$, the numerator has a zero quite close to, but not coincident with, the zero of the denominator defined by (27.15) so that $\tan\delta$, Eq. (27.14), passes infinity, i.e. phase angle $\pi/2$, and zero, i.e. phase angle π, within an extremely short interval of $2ka$. As we have seen above, resonances will occur in the regions close to $2ka = 2n\pi$ and it is quite usual for vibrating systems to exhibit a phase jump of about π when passing through resonance.

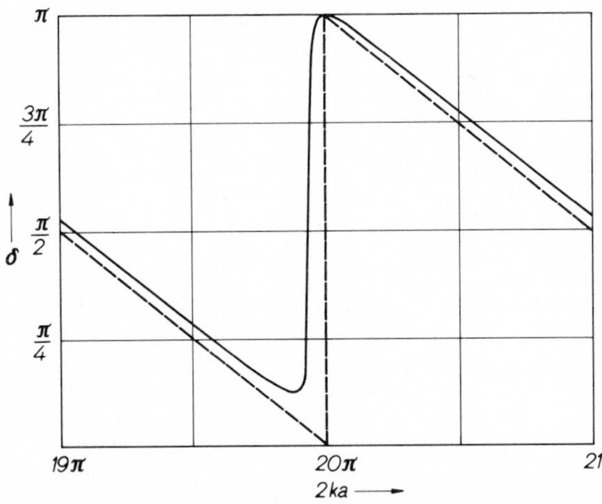

Fig. 14. Resonance level: Phase angle in dependence upon energy

As a numerical example, we have drawn in Fig. 14. the phase angle passing through the interval of $2ka$ shown in Fig. 13. If Ω were infinitely large, one would obtain the broken line where a phase jump π at $2ka = 20\pi$ has been arbitrarily introduced; this with increasing energy would suddenly invert the sign of the normalization constant in the wave function. It would, however, not have any physical significance because any integer which is a multiple of π may always be added to the phase. For the finite value of Ω used above, however, the full curve in Fig. 14 holds; this leads to a set of wave functions for varying energies, not departing very much in phase from those for the impenetrable wall but leading continuously through the resonance energy, so that the phase jump gains real significance.

Problem 28. Periodic potential

General conditions shall be derived for the wave functions and for the determination of the energy spectrum in a periodic potential.

Solution. If $V(x)$ is a periodic potential with period a, the Schrödinger equation is invariant with respect to all translations by integer multiples of a:

$$V(x+a)=V(x); \qquad x\to x+na, \qquad n=0,\,\pm 1,\,\pm 2,\ldots \tag{28.1}$$

Let us denote two linearly independent solutions of the Schrödinger equation by $u_1(x)$ and $u_2(x)$, then $u_1(x+a)$ and $u_2(x+a)$ must also be solutions. Since any solution may be represented as a linear combination of $u_1(x)$ and $u_2(x)$, this must also hold for $u_1(x+a)$ and $u_2(x+a)$, i.e.

$$\left.\begin{aligned} u_1(x+a) &= C_{11}u_1(x)+C_{12}u_2(x),\\ u_2(x+a) &= C_{21}u_1(x)+C_{22}u_2(x). \end{aligned}\right\} \tag{28.2}$$

It now can be proved *(Floquet's theorem)* that among these solutions there are two, say ψ_1 and ψ_2, with the property

$$\psi(x+a)=\lambda\psi(x) \tag{28.3}$$

with a constant factor λ. Then, of course,

$$\psi(x+na)=\lambda^n\psi(x); \qquad n=0,\,\pm 1,\,\pm 2,\ldots. \tag{28.3$'$}$$

The proof can be given by writing

$$\psi(x)=Au_1(x)+Bu_2(x) \tag{28.4}$$

and using (28.2):

$$\psi(x+a)=(AC_{11}+BC_{21})u_1(x)+(AC_{12}+BC_{22})u_2(x),$$

the latter expression becoming $\lambda\psi(x)$ if

$$\left.\begin{aligned} AC_{11}+BC_{21}&=\lambda A;\\ AC_{12}+BC_{22}&=\lambda B. \end{aligned}\right\} \tag{28.5}$$

This system of two homogeneous linear equations for A and B has a solution if, and only if, its determinant vanishes:

$$\begin{vmatrix} C_{11}-\lambda; & C_{21} \\ C_{12}; & C_{22}-\lambda \end{vmatrix} = 0. \tag{28.6}$$

That is a quadratic equation for λ with two solutions λ_1 und λ_2 leading to two functions ψ_1 and ψ_2.

From (28.3) it can be seen that the Wronskian determinant

$$D = \psi_1 \psi_2' - \psi_2 \psi_1'$$

satisfies the relation

$$D(x+a) = \lambda_1 \lambda_2 D(x).$$

Since, however, according to Green's theorem, D is a constant, independent of x, it follows that

$$\lambda_1 \lambda_2 = 1. \tag{28.7}$$

Even more can be said of the two parameters λ_1 and λ_2 when considering Eq. (28.3′). Let $|\lambda|>1$, then the amplitudes of ψ will grow beyond all limits for $x \to +\infty$ and decrease below all limits for $x \to -\infty$. The opposite will hold if $|\lambda|<1$. Such solutions cannot be normalized, not even in the wider sense of normalization of plane waves which we shall adopt here, so that physical solutions only exist if $|\lambda|=1$, say

$$\lambda_1 = e^{iKa}; \qquad \lambda_2 = e^{-iKa} \tag{28.8}$$

with real K. Since $e^{2\pi i n}=1$, we can reduce K from all positive and negative real numbers to an interval

$$-\frac{\pi}{a} \le K \le +\frac{\pi}{a} \tag{28.9}$$

which gives the complete set of all wave functions possible. Hence, for all limited solutions $\psi(x)$, we have

$$\psi(x+na) = e^{inKa}\psi(x) \tag{28.10}$$

which can only hold if

$$\psi(x) = e^{iKx} u_K(x) \tag{28.11}$$

with a periodic function

$$u_K(x) = u_K(x+a). \tag{28.12}$$

This result is called *Bloch's theorem*.

We now turn to discussing the energy spectrum. We build up ψ of any two solutions u_1 and u_2, as in (28.4), defined in the interval $0 \le x \le a$. Then in the next period, $a \le x \le 2a$, according to (28.10) we obtain

$$\psi(x) = e^{iKa}[A u_1(x-a) + B u_2(x-a)] \tag{28.13}$$

with the argument $x-a$ falling into the first interval. At the juncture $x=a$ the expressions (28.4) and (28.13) must coincide, and so must their derivatives, i. e., there will hold the two equations

$$\left.\begin{aligned} A u_1(a)+B u_2(a) &= e^{iKa}[A u_1(0)+B u_2(0)]; \\ A u_1'(a)+B u_2'(a) &= e^{iKa}[A u_1'(0)+B u_2'(0)]. \end{aligned}\right\} \tag{28.14}$$

This homogeneous system for A and B can be solved if, and only if, its determinant vanishes:

$$\begin{vmatrix} u_1(a)-e^{iKa}u_1(0); & u_2(a)-e^{iKa}u_2(0) \\ u_1'(a)-e^{iKa}u_1'(0); & u_2'(a)-e^{iKa}u_2'(0) \end{vmatrix} = 0.$$

Expanding the determinant, we finally arrive at the relation

$$\cos K a = \frac{[u_1(0)\,u_2'(a)+u_1(a)\,u_2'(0)]-[u_2(0)\,u_1'(a)+u_2(a)\,u_1'(0)]}{2(u_1 u_2' - u_2 u_1')} \tag{28.15}$$

where the denominator is the Wronskian, taken for any argument (omitted since the Wronskian is constant).

Eq. (28.15) gives the condition for the existence of eigenvalues. It can be satisfied only if the absolute value of its right-hand side is ≤ 1; in this case it enables K to be calculated. There are energy intervals which satisfy this condition alternating with intervals which do not. We therefore do not obtain sharp energy levels but an alternating sequence of allowed, and of forbidden *energy bands*. The band limits are obtained from (28.15) with $\cos Ka = \pm 1$.

NB. Since u_1 and u_2 can equally well be replaced by any other set of two linear combinations, v_1 and v_2, expression (28.15) may also be written in these functions. It must, however, lead to the same energy bands. This can readily be seen by inserting

$$u_1 = c_{11} v_1 + c_{12} v_2; \qquad u_2 = c_{21} v_1 + c_{22} v_2$$

into (28.15); a simple calculation shows that the same expression results in the v's as in the u's if the determinant of the c_{ik} does not vanish.

Problem 29. Dirac comb

Given a periodic potential formed by a sequence of Dirac functions, with a distance a between them:

$$V(x) = \frac{\hbar^2}{m}\,\Omega \sum_{n=-\infty}^{+\infty} \delta(x+na). \tag{29.1}$$

The energy bands for this potential shall be determined.

Solution. We start with the fundamental solutions

$$u_1(x) = e^{ikx}; \qquad u_2(x) = e^{-ikx}. \tag{29.2}$$

If, then,

$$u(x) = A\,e^{ikx} + B\,e^{-ikx} \quad \text{in } 0<x<a \tag{29.3}$$

is a limited solution in the sense of Problem 28, we have

$$u(x) = e^{iKa}\left[A\,e^{ik(x-a)} + B\,e^{-ik(x-a)}\right] \quad \text{in } a<x<2a. \tag{29.4}$$

Now, at $x=a$, there hold the boundary conditions

$$\left.\begin{aligned} u(a+0) &= u(a-0), \\ u'(a+0) &= u'(a-0) + 2\Omega u(a) \end{aligned}\right\} \tag{29.5}$$

so that

$$e^{iKa}(A+B) = A\,e^{ika} + B\,e^{-ika} \tag{29.6a}$$

and

$$ik\,e^{iKa}(A-B) = ik(A\,e^{ika} - B\,e^{-ika}) + 2\Omega(A\,e^{ika} + B\,e^{-ika}). \tag{29.6b}$$

Eqs. (29.6a, b) form a homogeneous linear system for A and B the determinant of which must vanish. Straightforward reshaping of the determinant leads to

$$\cos Ka = \cos ka + \frac{\Omega}{k} \sin ka. \tag{29.7}$$

The bands of energy eigenvalues are therefore defined by

$$\left|\cos ka + \frac{\Omega}{k} \sin ka\right| \leqq 1 \tag{29.8}$$

or

$$\left|\cos\left(ka - \tan^{-1}\frac{\Omega a}{ka}\right)\right| = \frac{1}{\sqrt{1+(\Omega a/ka)^2}}; \tag{29.9}$$

the energy then follows from

$$E = \frac{\hbar^2}{2ma^2}(ka)^2. \tag{29.10}$$

In the following figures, the results for the numerical example $\Omega a = 4$ are represented. The functions of ka standing on the left and on the right-hand sinde of Eq. (29.9) are shown in Fig. 15. Their intersections are set off by small circles, and the corresponding intervals in which the condition (29.9) holds are marked by heavy lines on the ka axis. The upper band limits appear at integer multiples of π where (29.7) simply leads to $\cos Ka = \cos ka$. In Fig. 16 the band limits determined in Fig. 15 are shown in the energy scale and the allowed bands

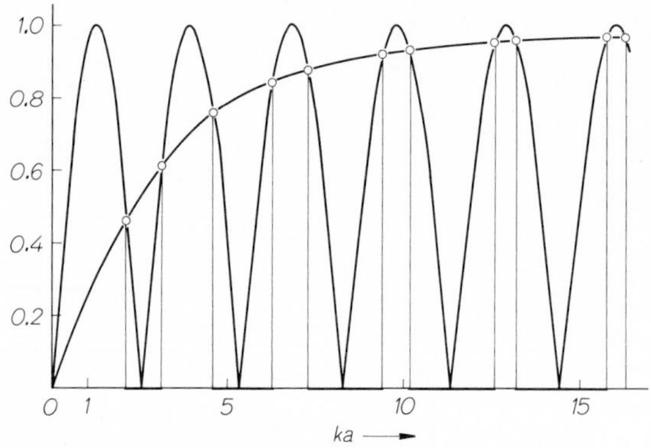

Fig. 15. Graphical determination of band structure

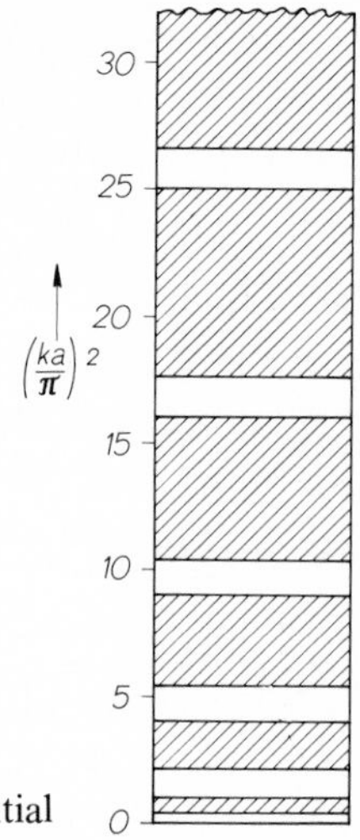

Fig. 16. Band structure of Dirac comb potential

are marked by hatching. The bands become broader with increasing energy so that the spectrum approaches (but never quite reaches) the continuum: even at highest energies there always remain forbidden zones above the upper band limits at $ka=n\pi$. In Figs. 17 and 18, for the first three bands, the energy (in dimensionless scale) has been drawn vs. Ka, with Ka in Fig. 17 monotonously growing from band to band,

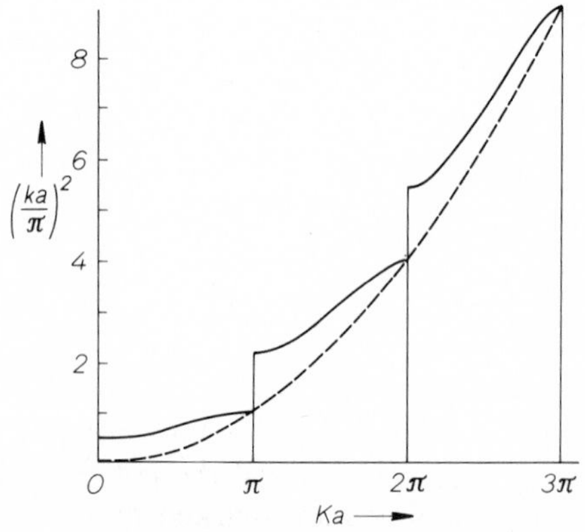

Fig. 17. Energy as a function of Ka for the first three bands (one-dimensional Brillouin zones). Broken line: Parabola of free particle energies

and in Fig. 18 reduced to the interval $-\pi \leq Ka \leq +\pi$. The broken line in Fig. 17 is the parabola $K=k$ which passes the upper band limits.

Since all figures have been drawn for the example $\Omega a=4$, they do not show the effect of wall opacity on the energy spectrum. For smaller values of Ωa, the right-hand side of (29.9) would shift closer to 1 so that the corresponding curve in Fig. 15 would intersect the cosine line much nearer its maxima. This means, of course, that the forbidden zones become narrower. Since the upper band limits of the allowed bands at $ka=n\pi$ are independent of Ωa, it is only the lower band limits which move downward in Fig. 16 with decreasing Ωa. If $\Omega a=0$, the forbidden zones vanish, but so does the Dirac comb potential and we arrive at the force-free case with its unbroken continuous spectrum.

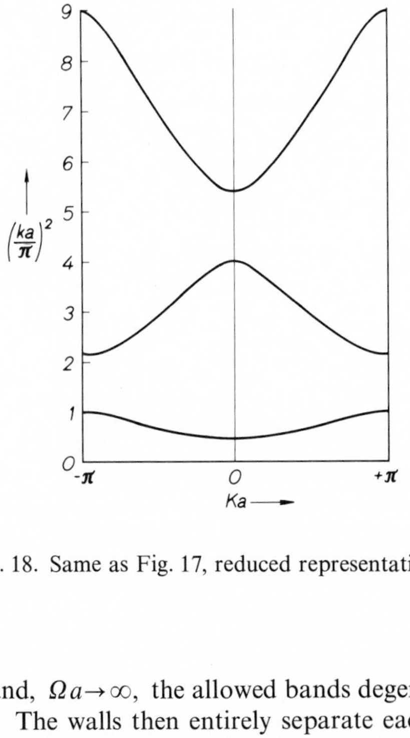

Fig. 18. Same as Fig. 17, reduced representation

If, on the other hand, $\Omega a \to \infty$, the allowed bands degenerate to discrete levels at $ka=n\pi$. The walls then entirely separate each potential hole between two walls from its neighbours thus reducing the spectrum to that of Problem 18 (with wall width here a instead of $2a$ in Problem 18).

Literature. Kronig, R. de L., Penney, W.: Proc. Roy. Soc., London, **130,** 499 (1931).

Problem 30. Harmonic oscillator

To find the eigenvalues and eigenfunctions for the oscillator potential

$$V(x) = \tfrac{1}{2} m \omega^2 x^2 . \tag{30.1}$$

Solution. Using the abbreviations

$$k^2 = \frac{2mE}{\hbar^2} \quad \text{and} \quad \lambda = \frac{m\omega}{\hbar} \tag{30.2}$$

the Schrödinger equation can be written

$$\frac{d^2 u}{dx^2} + (k^2 - \lambda^2 x^2) u = 0 . \tag{30.3}$$

The solutions of this differential equation behave asymptotically for $|x| \gg k/\lambda$ as $\exp(\pm \tfrac{1}{2} \lambda x^2)$. If therefore we split such a factor off the function u putting

$$u(x) = \mathrm{e}^{-\frac{1}{2}\lambda x^2} v(x) , \tag{30.4}$$

the remaining function $v(x)$ which satisfies the equation

$$v'' - 2\lambda x v' + (k^2 - \lambda) v = 0 \tag{30.5}$$

will either become proportional to $\mathrm{e}^{\lambda x^2}$ or a polynomial. Solving (30.5) by the series

$$v(x) = \sum_{j=0}^{\infty} a_j x^j \tag{30.6}$$

we find the recurrence relation

$$a_{j+2} = \frac{\lambda(2j+1) - k^2}{(j+2)(j+1)} a_j . \tag{30.7}$$

For $n \to \infty$ this leads to the asymptotic law $a_{j+2} = (2\lambda/j) a_j$ corresponding to the series expansion of $\mathrm{e}^{\lambda x^2}$. The solution (30.4) therefore cannot be normalized if the series (30.6) does not break off (polynomial method) which happens after the n-th power if $a_{n+2} = 0$, i.e. if

$$k^2 = \lambda(2n+1)$$

or, using (30.2), if

$$E = \hbar\omega(n + \tfrac{1}{2}); \qquad n = 0, 1, 2, \ldots . \tag{30.8}$$

The eigenfunctions $u_n(x)$ then can be determined from (30.7) and normalized by the condition

$$\int_{-\infty}^{+\infty} dx \, u_n(x)^2 = 1 . \tag{30.9}$$

The first few eigenfunctions run as follows:

$$\left.\begin{aligned} u_0 &= C_0 e^{-\frac{1}{2}\lambda x^2}, \\ u_1 &= C_1 x e^{-\frac{1}{2}\lambda x^2}, \\ u_2 &= C_2(1-2\lambda x^2) e^{-\frac{1}{2}\lambda x^2}, \\ u_3 &= C_3(x-\tfrac{2}{3}\lambda x^3) e^{-\frac{1}{2}\lambda x^2}, \\ u_4 &= C_4(1-4\lambda x^2+\tfrac{4}{3}\lambda^2 x^4) e^{-\frac{1}{2}\lambda x^2}, \\ u_5 &= C_5(x-\tfrac{4}{3}\lambda x^3+\tfrac{4}{15}\lambda^2 x^5) e^{-\frac{1}{2}\lambda x^2}, \end{aligned}\right\} \tag{30.10}$$

with normalization constants

$$\left.\begin{aligned} &C_n = \left(\frac{\lambda}{\pi}\right)^{\frac{1}{4}} c_n; \quad c_0 = 1; \quad c_1 = \sqrt{2\lambda}; \quad c_2 = \frac{1}{\sqrt{2}}; \\ &c_3 = \sqrt{3\lambda}; \quad c_4 = \sqrt{\frac{3}{8}}; \quad c_5 = \sqrt{\frac{15\lambda}{4}}. \end{aligned}\right\} \tag{30.11}$$

The eigenfunctions have been shown in Fig. 19.

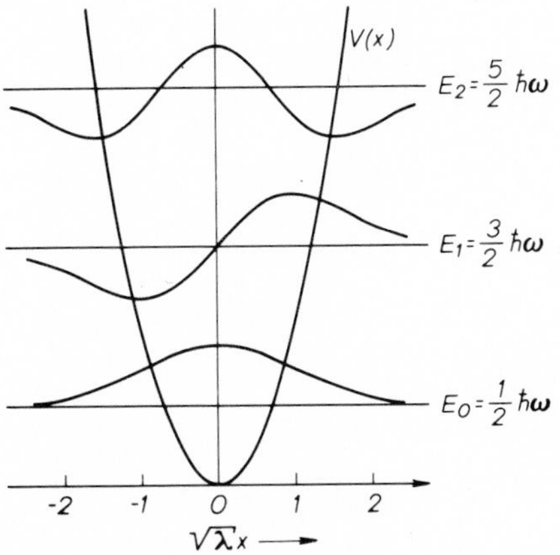

Fig. 19. Harmonic oscillator eigenfunctions, drawn over the potential and its energy levels

It will be seen that the eigenfunctions have either even or odd parity. This follows from the symmetry of the potential, $V(-x)=V(x)$, that makes $u_n(-x)$ a solution of the differential equation together with $u_n(x)$ to the same eigenvalue (30.8). As there is no degeneracy, the two solutions can differ only by a factor f fixed by the normalization to $|f|^2=1$. Since by twice reflecting the coordinate x at the origin we are brought back to the original solution, we find $f^2=1$ or $f=\pm 1$ which leaves us with either even or odd parity solutions.

This fact is borne out if we introduce instead of x the variable

$$y=\lambda x^2 \tag{30.12}$$

in Eq. (30.5); we then find $v(y)$ to satisfy the differential equation of the confluential hypergeometric functions,

$$y v'' + \left(\frac{1}{2} - y\right) v' + \left(\frac{k^2}{4\lambda} - \frac{1}{4}\right) v = 0$$

which with the abbreviation

$$a = \frac{1}{4} - \frac{k^2}{4\lambda} = \frac{1}{4} - \frac{E}{2\hbar\omega} \tag{30.13}$$

has the complete solution

$$v = A\,{}_1F_1(a, \tfrac{1}{2}; y) + B\,y^{\frac{1}{2}}\,{}_1F_1(a+\tfrac{1}{2}, \tfrac{3}{2}; y)\,. \tag{30.14}$$

Since the confluent series ${}_1F_1$ is an integral function, the A part of this solution is even, and the B part odd in the variable x. In the infinite point ($y=\infty$) both parts of v diverge as $e^y y^{a-\frac{1}{2}}$, i.e. u cannot be normalized if the series does not either vanish or break off. Since a confluent series breaks off if its first parameter is a negative integer, there are two possible cases. If

$$a = -n \quad \text{or} \quad E_{2n} = \hbar\omega(2n+\tfrac{1}{2}) \tag{30.15a}$$

the first series breaks off, and we find an eigenfunction

$$u_{2n}(x) = A\,{}_1F_1(-n, \tfrac{1}{2}; \lambda x^2)\,e^{-\frac{1}{2}\lambda x^2}\,. \tag{30.16a}$$

If, on the other hand,

$$a+\tfrac{1}{2} = -n \quad \text{or} \quad E_{2n+1} = \hbar\omega(2n+\tfrac{3}{2}) \tag{30.15b}$$

the second series breaks off, and the eigenfunction becomes

$$u_{2n+1}(x) = B x\,{}_1F_1(-n, \tfrac{3}{2}; \lambda x^2)\,e^{-\frac{1}{2}\lambda x^2}\,. \tag{30.16b}$$

This completely corroborates the eigenvalue formula (30.8) and the eigenfunctions (30.10).

The polynomials defined by Eqs. (30.16a, b) are known under the name of *Hermite polynomials* and defined either by

$$\left.\begin{aligned} H_{2n}(\xi) &= (-1)^n \frac{(2n)!}{n!}\,{}_1F_1\left(-n, \frac{1}{2}; \xi^2\right), \\ H_{2n+1}(\xi) &= (-1)^n \frac{(2n+1)!}{n!}\,2\xi\,{}_1F_1\left(-n, \frac{3}{2}; \xi^2\right) \end{aligned}\right\} \tag{30.17}$$

or by

$$H_n(\xi)=(-1)^n e^{\xi^2} \frac{d^n e^{-\xi^2}}{d\xi^n}. \tag{30.18}$$

The eigenfunctions in the normalization (30.9) are then given by the general formula

$$u_n(x)=\left\{\frac{1}{2^n n!}\sqrt{\frac{\lambda}{\pi}}\right\}^{\frac{1}{2}} H_n(\sqrt{\lambda}\,x)\,e^{-\frac{1}{2}\lambda x^2}. \tag{30.19}$$

The normalization factor of Eq. (30.19) can be determined in the following way. If we write

$$u_n=C_n e^{-\frac{1}{2}\xi^2} H_n(\xi); \quad \xi=\sqrt{\lambda}\,x,$$

we have from (30.9):

$$C_n^2 \int_{-\infty}^{+\infty} d\xi\, e^{-\xi^2}[H_n(\xi)]^2=\sqrt{\lambda}.$$

Replacing one of the two factors $H_n(\xi)$ by the representation (30.18) of this function,

$$C_n^2(-1)^n \int_{-\infty}^{+\infty} d\xi\, H_n(\xi) \frac{d^n e^{-\xi^2}}{d\xi^n}=\sqrt{\lambda}$$

and then performing n consecutive partial integrations we arrive at

$$C_n^2 \int_{-\infty}^{+\infty} d\xi\, e^{-\xi^2} \frac{d^n H_n(\xi)}{d\xi^n}=\sqrt{\lambda}.$$

As $H_n(\xi)$ is a polynomial in ξ of degree n, by n differentiations all terms vanish except the contribution of the highest power in H_n,

$$H_n(\xi)=(2\xi)^n+\cdots,$$

i. e.

$$\frac{d^n H_n(\xi)}{d\xi^n}=2^n n!$$

and

$$C_n^2\cdot 2^n n!\sqrt{\pi}=\sqrt{\lambda}$$

in agreement with Eq. (30.19).

Problem 31. Oscillator in Hilbert space

To construct the system of eigenvalues and eigenvectors of an harmonic oscillator in its Hilbert space using the operators

$$b = \frac{1}{\sqrt{2m\hbar\omega}}(p - i\omega m x); \qquad b^\dagger = \frac{1}{\sqrt{2m\hbar\omega}}(p + i\omega m x). \tag{31.1}$$

Solution. Setting the operator b and its hermitian conjugate $b^\dagger$ in the hamiltonian,

$$H = \frac{1}{2m}p^2 + \frac{m\omega^2}{2}x^2, \tag{31.2}$$

we get

$$H = \tfrac{1}{2}\hbar\omega(b b^\dagger + b^\dagger b). \tag{31.3}$$

Further it can easily be seen that, from the canonical commutation relation $px - xp = \hbar/i$, there follows

$$b b^\dagger - b^\dagger b = 1. \tag{31.4}$$

Eqs. (31.3) and (31.4) form the mathematical basis for the construction of the Hilbert space.

Let us suppose that there exists at least one ket vector $|\psi_\lambda\rangle$ to an eigenvalue λ of the operator $b^\dagger b$:

$$b^\dagger b|\psi_\lambda\rangle = \lambda|\psi_\lambda\rangle \tag{31.5}$$

which we shall suppose to be normalized:

$$\langle\psi_\lambda|\psi_\lambda\rangle = 1. \tag{31.6}$$

If then we multiply (31.5) with the bra vector $\langle\psi_\lambda|$,

$$\langle\psi_\lambda|b^\dagger b|\psi_\lambda\rangle = \lambda,$$

we may, on the left side, apply the definition of a hermitian conjugate operator,

$$\langle\chi|L^\dagger|\varphi\rangle = \langle L\chi|\varphi\rangle;$$

the result is

$$\langle b\psi_\lambda|b\psi_\lambda\rangle = \lambda. \tag{31.7}$$

The vector $|b\psi_\lambda\rangle$ therefore is normalized to λ which—as a norm—must then be real and positive:

$$\lambda \geqq 0, \tag{31.8}$$

with $\lambda = 0$ only if $b|\psi_\lambda\rangle = 0$. This holds for any eigenvalue since we have so far nothing else to distinguish the chosen λ from any other eigenvalue.

Another eigenvalue is found by multiplying Eq. (31.5) with b from the left and rearranging it by means of the associative law and the commutation rule (31.4):

$$b(b^\dagger b) = (b\,b^\dagger)b = (b^\dagger b + 1)b$$

so that

$$(b^\dagger b + 1)b|\psi_\lambda\rangle = \lambda\, b|\psi_\lambda\rangle$$

or

$$(b^\dagger b)|b\,\psi_\lambda\rangle = (\lambda - 1)|b\,\psi_\lambda\rangle.$$

Thus $|b\,\psi_\lambda\rangle$ turns out to be an eigenvector of the operator $b^\dagger b$ that belongs to the eigenvalue $\lambda - 1$. It is not yet normalized, as shown by Eq. (31.7), the normalized eigenvector being

$$|\psi_{\lambda-1}\rangle = \frac{1}{\sqrt{\lambda}}\, b|\psi_\lambda\rangle. \tag{31.9}$$

This principle of construction may be repeated, producing a descending sequence of eigenvalues:

$$|\psi_{\lambda-n}\rangle = \frac{1}{\sqrt{\lambda(\lambda-1)(\lambda-2)\dots(\lambda-n+1)}}\, b^n|\psi_\lambda\rangle. \tag{31.10}$$

These eigenvalues, for all $n > \lambda$, are negative, in contradiction to (31.8). Only if the eigenvalues are integers can this be avoided, the sequence (31.10) then breaking off at $|\psi_0\rangle$ because, according to (31.9), $b|\psi_0\rangle = 0$.

An ascending sequence of eigenvalues can be constructed by repeated application of the operator $b^\dagger$ according to

$$b^\dagger(b^\dagger b) = b^\dagger(b\,b^\dagger - 1) = (b^\dagger b)b^\dagger - b^\dagger$$

so that there follows from Eq. (31.5):

$$(b^\dagger b)|b^\dagger \psi_\lambda\rangle = (\lambda + 1)|b^\dagger \psi_\lambda\rangle.$$

Again, the vector $|b^\dagger \psi_\lambda\rangle$ is not yet normalized:

$$\langle b^\dagger \psi_\lambda | b^\dagger \psi_\lambda\rangle = \langle b\,b^\dagger \psi_\lambda | \psi_\lambda\rangle = \langle (1 + b^\dagger b)\psi_\lambda | \psi_\lambda\rangle = \lambda + 1.$$

We therefore find

$$|\psi_{\lambda+1}\rangle = \frac{1}{\sqrt{\lambda+1}}\, |b^\dagger \psi_\lambda\rangle \tag{31.11}$$

and the ascending sequence becomes

$$|\psi_{\lambda+n}\rangle = \frac{1}{\sqrt{(\lambda+1)(\lambda+2)\dots(\lambda+n)}}\, b^{\dagger n}|\psi_\lambda\rangle. \tag{31.12}$$

We now gather up our results. The eigenvalues of $b^\dagger b$ are the integers $n=0,1,2,\ldots$. It then follows from (31.4) that $b b^\dagger$ has the same system of eigenvectors, but with eigenvalues $n+1$. The hamiltonian, Eq. (31.3), therefore satisfies the relations

$$H|\psi_n\rangle = \tfrac{1}{2}\hbar\omega(2n+1)|\psi_n\rangle; \qquad n=0,1,2,\ldots. \tag{31.13}$$

These are the well-known oscillator eigenvalues, determined in Problem 30 within the framework of Schrödinger theory.

The Hilbert vectors $|\psi_n\rangle$ are not eigenvectors of the operators b and $b^\dagger$ whose matrices, however, can easily be constructed in the Hilbert coordinate system of the $|\psi_n\rangle$. We find from (31.9)

$$\langle\psi_{n-1}|b|\psi_n\rangle = \sqrt{n} \tag{31.14}$$

and from (31.11)

$$\langle\psi_{n+1}|b^\dagger|\psi_n\rangle = \sqrt{n+1}. \tag{31.15}$$

All other matrix elements vanish in consequence of the orthogonality of the $|\psi_n\rangle$. The latter can easily be checked:

$$\begin{aligned}\langle\psi_m|b^\dagger b\,\psi_n\rangle &= n\langle\psi_m|\psi_n\rangle = \langle b\,\psi_m|b\,\psi_n\rangle,\\ \langle b^\dagger b\,\psi_m|\psi_n\rangle &= m\langle\psi_m|\psi_n\rangle = \langle b\,\psi_m|b\,\psi_n\rangle.\end{aligned}$$

Hence, the difference,

$$(n-m)\langle\psi_m|\psi_n\rangle = 0$$

so that either $n=m$ or $\langle\psi_m|\psi_n\rangle=0$. The matrices (31.14) and (31.15) can be written as follows:

$$b=\begin{pmatrix} 0 & 0 & 0 & 0 & \ldots\\ \sqrt{1} & 0 & 0 & 0 & \ldots\\ 0 & \sqrt{2} & 0 & 0 & \ldots\\ 0 & 0 & \sqrt{3} & 0 & \ldots\\ \ldots & \ldots & \ldots & \ldots & \ldots \end{pmatrix}; \qquad b^\dagger=\begin{pmatrix} 0 & \sqrt{1} & 0 & 0 & \ldots\\ 0 & 0 & \sqrt{2} & 0 & \ldots\\ 0 & 0 & 0 & \sqrt{3} & \ldots\\ \ldots & \ldots & \ldots & \ldots & \ldots \end{pmatrix} \tag{31.16}$$

Literature. Becker, R., Leibfried, G.: Z. Physik **125**, 347 (1949).

Problem 32. Oscillator eigenfunctions constructed by Hilbert space operators

To find the oscillator eigenfunctions by translating the operators b and $b^\dagger$ of the preceding problem into the Schrödinger language.

Solution. In (31.1) we have introduced b and $b^\dagger$ as linear combinations of the operators p and x. In the Schrödinger language, x is a classical variable and

$$p = \frac{\hbar}{i}\frac{\partial}{\partial x} \tag{32.1}$$

a differential operator. It is convenient to choose an unit length,

$$l = \sqrt{\frac{\hbar}{m\omega}} \tag{32.2}$$

and to write all relations in the dimensionless variable

$$\xi = x/l \tag{32.3}$$

instead of in x. Then we get

$$b = -\frac{i}{\sqrt{2}}\left(\frac{d}{d\xi} + \xi\right); \qquad b^\dagger = \frac{i}{\sqrt{2}}\left(-\frac{d}{d\xi} + \xi\right). \tag{32.4}$$

The lowest eigenstate is, according to the preceding problem, defined by $b\psi_0 = 0$ or

$$\left(\frac{d}{d\xi} + \xi\right)\psi_0(\xi) = 0. \tag{32.5}$$

This is a differential equation, the complete solution of which

$$\psi_0 = C_0 \mathrm{e}^{-\frac{1}{2}\xi^2},$$

has still to be normalized by choosing the integration constant so that

$$\langle\psi_0|\psi_0\rangle = \int_{-\infty}^{+\infty} dx\,\psi_0^2 = l\int_{-\infty}^{+\infty} d\xi\,\psi_0(\xi)^2 = 1.$$

Thus we find

$$\psi_0(\xi) = \pi^{-\frac{1}{4}} l^{-\frac{1}{2}} \mathrm{e}^{-\frac{1}{2}\xi^2}. \tag{32.6}$$

The complete sequence of eigenfunctions then may be constructed by applying $b^\dagger$:

$$\psi_{n+1} = \frac{i}{\sqrt{2(n+1)}}\left(-\frac{d}{d\xi} + \xi\right)\psi_n. \tag{32.7}$$

They therefore follow by repeated differentiations from (32.6). If we write

$$\psi_n = C_n H_n(\xi)\mathrm{e}^{-\frac{1}{2}\xi^2} \tag{32.8}$$

then, according to (32.7), the function $H_n(\xi)$ will become a polynomial of degree n in the variable ξ. We find from (32.7) the recurrence relation

$$C_{n+1} H_{n+1} = -\frac{i}{\sqrt{2(n+1)}} C_n (H_n' - 2\xi H_n). \tag{32.9}$$

If we choose the normalization constants so that

$$C_{n+1} = \frac{i}{\sqrt{2(n+1)}} C_n, \tag{32.10}$$

then (32.9) becomes simply the recurrence relation of the Hermite polynomials,

$$H_n(\xi) = (-1)^n e^{\xi^2} \frac{d^n}{d\xi^n} e^{-\xi^2},$$

viz.

$$H_{n+1} = -(H_n' - 2\xi H_n).$$

Comparison with (30.19) shows that, apart from the factor i, Eq. (32.10) is satisfied by the normalization factor derived there:

$$C_n = \left\{ \frac{1}{2^n n!} \sqrt{\frac{\lambda}{\pi}} \right\}^{\frac{1}{2}} \quad \text{with } \lambda = 1/l^2.$$

The factor i is, of course, quite arbitrary since the norm leaves a phase factor undetermined in each eigenfunction.

Problem 33. Harmonic oscillator in matrix notation

To derive the matrices p_{mn} and x_{mn} of momentum p and coordinate x for the harmonic oscillator defined by the hamiltonian

$$H = \frac{1}{2M} p^2 + \frac{M\omega^2}{2} x^2 \tag{33.1}$$

in the Hilbert coordinate system diagonalizing H. What are the energy eigenvalues?

Solution. We have for a starting point the diagonality of the hamiltonian,

$$H_{mn} = \frac{1}{2M} \sum_k p_{mk} p_{kn} + \frac{M\omega^2}{2} \sum_k x_{mk} x_{kn} = E_n \delta_{mn}, \tag{33.2}$$

with E_n the energy eigenvalues. Furthermore, we have the relations (cf. Problem 8)

$$\frac{\partial H}{\partial x} = -\frac{i}{\hbar}(Hp - pH); \qquad \frac{\partial H}{\partial p} = +\frac{i}{\hbar}(Hx - xH) \tag{33.3}$$

or, with $\partial H/\partial x = M\omega^2 x$ and $\partial H/\partial p = p/M$, in matrix notation

$$M\omega^2 x_{mn} = -\frac{i}{\hbar}\sum_k (H_{mk} p_{kn} - p_{mk} H_{kn});$$

$$\frac{1}{M} p_{mn} = +\frac{i}{\hbar}\sum_k (H_{mk} x_{kn} - x_{mk} H_{kn})$$

which, by using (33.2), reduce to

$$\left.\begin{aligned} M\omega^2 x_{mn} &= -\frac{i}{\hbar}(E_m - E_n) p_{mn}; \\ \frac{1}{M} p_{mn} &= +\frac{i}{\hbar}(E_m - E_n) x_{mn}. \end{aligned}\right\} \tag{33.4}$$

These two homogeneous equations for x_{mn} and p_{mn} are compatible only if both matrix elements vanish, or if the determinant vanishes, i. e. if

$$(E_m - E_n)^2 = (\hbar\omega)^2. \tag{33.5}$$

Ordering the eigenvalues so that

$$\cdots < E_1 < E_2 < E_3 < \cdots,$$

the difference of two consecutive values therefore is $\hbar\omega$ and

$$E_n = \hbar\omega(n+\varepsilon); \qquad n = 0, \pm 1, \pm 2, \ldots \tag{33.6}$$

with ε a constant common to all eigenvalues. Then, only elements $p_{n,n\pm 1}$ and $x_{n,n\pm 1}$ differ from zero, and (33.4) establishes the relation

$$p_{n,n\pm 1} = \mp i M\omega x_{n,n\pm 1}. \tag{33.7}$$

We now try to obtain still more information on the hamiltonian. We start with the matrices for p^2 and x^2. According to (33.7) we have

$$(p^2)_{nk} = p_{n,n+1} p_{n+1,k} + p_{n,n-1} p_{n-1,k}$$

with only for $k = n, n+2, n-2$ not vanishing. The same holds for the matrix elements of x^2. Putting this into (33.2), we obtain the off-diagonal elements of the hamiltonian,

$$H_{n,n+2} = \frac{1}{2M} p_{n,n+1} p_{n+1,n+2} + \frac{M\omega^2}{2} x_{n,n+1} x_{n+1,n+2}$$

$$= x_{n,n+1} x_{n+1,n+2}\left[-\frac{1}{2M} M^2\omega^2 + \frac{M\omega^2}{2}\right] = 0,$$

and the same holds for the matrix elements $H_{n,n-2}$, in perfect agreement with the diagonality of H underlying Eq. (33.4). The remaining diagonal elements become

$$E_n = \frac{1}{2M}(p_{n,n+1}\,p_{n+1,n} + p_{n,n-1}\,p_{n-1,n}) + \frac{M\omega^2}{2}(x_{n,n+1}\,x_{n+1,n} + x_{n,n-1}\,x_{n-1,n}),$$

where according to (33.7) both lines are equal thus yielding

$$E_n = M\omega^2(x_{n,n+1}\,x_{n+1,n} + x_{n,n-1}\,x_{n-1,n}). \tag{33.8}$$

Eqs. (33.8) and (33.6) give a relation between the matrix elements of x. Another relation follows from the fact that x is a hermitian operator:

$$x_{n\pm 1,n} = x^*_{n,n\pm 1} \tag{33.9}$$

so that we get

$$\hbar\omega(n+\varepsilon) = M\omega^2(|x_{n,n+1}|^2 + |x_{n-1,n}|^2). \tag{33.10}$$

Using the abbreviation

$$\frac{M\omega}{\hbar}|x_{n-1,n}|^2 = f(n) \tag{33.11}$$

this equation may be written

$$n+\varepsilon = f(n+1) + f(n),$$

a functional equation to be solved by, and only by,

$$f(n) = \tfrac{1}{2}(n+\varepsilon-\tfrac{1}{2}). \tag{33.12}$$

This can easily be seen by supposing $f(n)$ to be a power series in n. It follows from (33.11) that

$$x_{n,n+1} = \sqrt{\frac{\hbar}{2M\omega}\left(n+\varepsilon+\frac{1}{2}\right)}. \tag{33.13}$$

We are now able to fix the constant ε. The right-hand side of Eq. (33.10) cannot be negative; the same therefore must hold for the energy value on its left-hand side. Suppose that $n=0$ is the smallest value of n; it then follows that

$$x_{-1,0} = \sqrt{\frac{\hbar}{2M}\left(\varepsilon-\frac{1}{2}\right)} \tag{33.14}$$

must be zero. Were it not so, Eq. (33.10) might be used as a recurrence to compute $x_{n-1,n}$ values down to $n=-\infty$, in contradiction to $n+\varepsilon>0$. The relation (33.14) determines

$$\varepsilon = \tfrac{1}{2} \tag{33.15}$$

and thus the energy eigenvalues (33.6),

$$E_n = \hbar\omega(n+\tfrac{1}{2}), \tag{33.16}$$

as well as, according to (33.9) and (33.13), the matrix elements,

$$x_{n,n+1} = \sqrt{\frac{\hbar}{2M\omega}(n+1)}\,; \qquad x_{n,n-1} = \sqrt{\frac{\hbar}{2M\omega}n}\,. \tag{33.17}$$

Problem 34. Momentum space wave functions of oscillator

To determine the oscillator wave functions in momentum space.

Solution. With

$$V(x) = \frac{m\omega^2}{2}x^2 \tag{34.1}$$

this is clearly a case where the Fourier transform of V does not exist, so that we cannot use the integral equation (14.6). Instead we fall back upon the definitions

$$u(x) = (2\pi)^{-\frac{1}{2}} \int\limits_{-\infty}^{+\infty} dk\, e^{ikx} f(k); \tag{34.2}$$

$$f(k) = (2\pi)^{-\frac{1}{2}} \int\limits_{-\infty}^{+\infty} dx\, e^{-ikx} u(x). \tag{34.3}$$

Putting (34.2) into the Schrödinger equation

$$\frac{d^2u}{dx^2} + \frac{2m}{\hbar^2}\left[E - \frac{m\omega^2}{2}x^2\right]u = 0 \tag{34.4}$$

we obtain by differentiations

$$\frac{d^2u}{dx^2} = -(2\pi)^{-\frac{1}{2}} \int\limits_{-\infty}^{+\infty} dk\, k^2 e^{ikx} f(k)$$

and by two consecutive integrations in part

$$x^2 u = -(2\pi)^{-\frac{1}{2}} \int\limits_{-\infty}^{+\infty} dk\, e^{ikx} f''(k).$$

Thus (34.4) becomes

$$(2\pi)^{-\frac{1}{2}} \int\limits_{-\infty}^{+\infty} dk\, e^{ikx} \left\{-k^2 f(k) + \frac{2mE}{\hbar^2} f(k) + \left(\frac{m\omega}{\hbar}\right)^2 f''(k)\right\} = 0. \tag{34.5}$$

Using the abbreviations

$$\lambda = \frac{m\omega}{\hbar}; \qquad \mu = \frac{E}{\hbar\omega} \tag{34.6}$$

already used in the oscillator treatment of Problem 30, we then gather from (34.4) and (34.5) an analogous behaviour of the oscillator problems in coordinate space and momentum space:

Coordinate space	Momentum space
$u'' + (2\mu\lambda - \lambda^2 x^2)u = 0$	$f'' + \left(\frac{2\mu}{\lambda} - \frac{k^2}{\lambda^2}\right) f = 0$
$\tilde{x} = \sqrt{\lambda}\, x$	$\tilde{k} = \frac{k}{\sqrt{\lambda}}$
$\frac{d^2 u}{d\tilde{x}^2} + (2\mu - \tilde{x}^2)u = 0$	$\frac{d^2 f}{d\tilde{k}^2} + (2\mu - \tilde{k}^2) f = 0$

Therefore, f is the same function of $\tilde{k}$ as u is of $\tilde{x}$, save of normalization. Since from

$$\int_{-\infty}^{+\infty} dx\, |u(x)|^2 = 1$$

there follows

$$\int_{-\infty}^{+\infty} dk\, |f(k)|^2 = 1,$$

as has been shown at the end of Problem 14, we find

$$u(x) = \lambda^{\frac{1}{4}}\, \tilde{u}(\tilde{x}); \qquad f(k) = \lambda^{-\frac{1}{4}}\, \tilde{f}(\tilde{k})$$

with $\tilde{f}$ the same function of $\tilde{k}$ as $\tilde{u}$ is of $\tilde{x}$, the latter having been determined in Problem 30.

Problem 35. Anharmonic oscillator

To approximate the energy levels of the anharmonic oscillator

$$V(x) = \frac{1}{2} m\omega^2 x^2 + \varepsilon_1 \left(\frac{x}{l}\right)^3 + \varepsilon_2 \left(\frac{x}{l}\right)^4; \qquad l = \sqrt{\hbar/m\omega} \tag{35.1}$$

by a perturbation method, in first and second order of the approximation.

Solution. The level of the harmonic oscillator with energy

$$E_n^0 = \hbar\omega(n + \tfrac{1}{2}) \tag{35.2}$$

with the normalized eigenfunction

$$u_n^0(x) = (2^n n!\, l\sqrt{\pi})^{-\frac{1}{2}} \mathrm{e}^{-\frac{1}{2}\xi^2} H_n(\xi); \qquad \xi = x/l \tag{35.3}$$

is, in first approximation, shifted by the diagonal matrix element of the perturbation energy,

$$\Delta^{(1)} E_n = E_n - E_n^0 = \int_{-\infty}^{+\infty} dx (\varepsilon_1 \xi^3 + \varepsilon_2 \xi^4)\, [u_n^0(x)]^2. \tag{35.4}$$

As $[u_n^0]^2$ is an even function of x, the potential term $\varepsilon_1 \xi^3$ does not contribute to $\Delta^{(1)} E_n$. The remaining shift

$$\Delta^{(1)} E_n = \varepsilon_2 \int_{-\infty}^{+\infty} d\xi\, \xi^4 [u_n^0(\xi)]^2$$

can be evaluated in a similar way as the normalization integral by leaving one of the two factors $H_n(\xi)$ in the integrand as polynomial and writing

$$H_n(\xi) = (-1)^n \mathrm{e}^{\xi^2} \frac{d^n}{d\xi^n} \mathrm{e}^{-\xi^2} \tag{35.5}$$

for the other. Then, by n consecutive partial integration steps, we arrive at

$$\Delta^{(1)} E_n = \frac{\varepsilon_2}{2^n n! \sqrt{\pi}} \int_{-\infty}^{+\infty} d\xi\, \mathrm{e}^{-\xi^2} \frac{d^n}{d\xi^n} (\xi^4 H_n(\xi)).$$

The polynomial of degree $n+4$ to be n times differentiated begins with the terms

$$\xi^4 H_n(\xi) = 2^n \left\{ \xi^{n+4} - \frac{1}{2}\binom{n}{2} \xi^{n+2} + \frac{3}{4}\binom{n}{4} \xi^n \ldots \right\} \tag{35.6}$$

so that

$$\frac{d^n}{d\xi^n}(\xi^4 H_n) = 2^n \left\{ \frac{(n+4)!}{4!} \xi^4 - \frac{1}{2}\binom{n}{2} \frac{(n+2)!}{2!} \xi^2 + \frac{3}{4}\binom{n}{4} n! \right\}.$$

With the well-known integrals

$$\int_{-\infty}^{+\infty} d\xi\, \mathrm{e}^{-\xi^2} = \sqrt{\pi}; \quad \int_{-\infty}^{+\infty} d\xi\, \xi^2 \mathrm{e}^{-\xi^2} = \tfrac{1}{2}\sqrt{\pi}; \quad \int_{-\infty}^{+\infty} d\xi\, \xi^4 \mathrm{e}^{-\xi^2} = \tfrac{3}{4}\sqrt{\pi} \tag{35.7}$$

we then finally arrive at the result

$$\Delta^{(1)} E_n = \tfrac{3}{4} \varepsilon_2 (2n^2 + 2n + 1). \tag{35.8}$$

The result of this first-order perturbation calculation may be qualitatively interpreted as follows. A *small* additive term $\varepsilon_1 \xi^3$ in the potential originates a little asymmetry of the parabola $V(x)$, though not yet changing its width at lower energies. The motion of the oscillating mass therefore remains limited within a domain of the same breadth as in the harmonic oscillator so that there is no reason why the energy should undergo an alteration. Addition of the term $\varepsilon_2 \xi^4$, however, raises (lowers) both sides of the parabola symmetrically for $\varepsilon_2 > 0$ ($\varepsilon_2 < 0$) even at smallest energies, thus narrowing (widening) the width of the parabola and of the domain of motion so that all energy levels will be raised (lowered) in agreement with the sign of (35.8).

In second approximation, another energy shift,

$$\Delta^{(2)} E_n = \sum_m \frac{|\langle n|\varepsilon_1 \xi^3 + \varepsilon_2 \xi^4 |m\rangle|^2}{E_n^0 - E_m^0}, \tag{35.9}$$

has to be added to (35.2) and (35.8). For the oscillator only the following off-diagonal matrix elements do not vanish and are evaluated in the same way as the diagonal contribution (35.8):

$$\left.\begin{aligned} &\langle n|\xi^3|n+1\rangle = 3\left(\frac{n+1}{2}\right)^{\frac{3}{2}}; \qquad \langle n|\xi^3|n-1\rangle = 3\left(\frac{n}{2}\right)^{\frac{3}{2}};\\ &\langle n|\xi^3|n+3\rangle = \{\tfrac{1}{8}(n+3)(n+2)(n+1)\}^{\frac{1}{2}};\\ &\langle n|\xi^3|n-3\rangle = \{\tfrac{1}{8}n(n-1)(n-2)\}^{\frac{1}{2}} \end{aligned}\right\} \tag{35.10}$$

and

$$\left.\begin{aligned} &\langle n|\xi^4|n+2\rangle = \sqrt{(n+2)(n+1)}\,(n+\tfrac{3}{2});\\ &\langle n|\xi^4|n-2\rangle = \sqrt{n(n-1)}\,(n-\tfrac{1}{2});\\ &\langle n|\xi^4|n+4\rangle = \tfrac{1}{4}\sqrt{(n+4)(n+3)(n+2)(n+1)};\\ &\langle n|\xi^4|n-4\rangle = \tfrac{1}{4}\sqrt{n(n-1)(n-2)(n-3)}. \end{aligned}\right\} \tag{35.11}$$

With these matrix elements, the general formula (35.9) leads to

$$\begin{aligned} \Delta^{(2)} E_n = &-\frac{\varepsilon_1^2}{\hbar\omega}\cdot\frac{15}{4}\left(n^2+n+\frac{11}{30}\right)\\ &-\frac{\varepsilon_2^2}{\hbar\omega}\cdot\frac{1}{8}(34n^3+51n^2+59n+21). \end{aligned} \tag{35.12}$$

If, in second approximation, we write

$$E_n = \hbar\omega\left\{\left(n+\frac{1}{2}\right) - \left(\frac{\varepsilon_1}{\hbar\omega}\right)^2 A_n^{(2)} + \frac{\varepsilon_1}{\hbar\omega} B_n^{(1)} - \left(\frac{\varepsilon_1}{\hbar\omega}\right)^2 B_n^{(2)}\right\} \tag{35.13}$$

with the abbreviations

$$\left.\begin{aligned} A_n^{(2)} &= \tfrac{15}{4}(n^2+n+\tfrac{11}{30});\\ B_n^{(1)} &= \tfrac{3}{4}(2n^2+2n+1);\\ B_n^{(2)} &= \tfrac{1}{8}(34n^3+51n^2+59n+21), \end{aligned}\right\} \tag{35.14}$$

then we obtain coefficients as shown in the table and can easily compute the energy levels. The latter have been shown in Fig. 20 for the perturbation $\varepsilon_1 \xi^3$ as a function of the dimensionless parameter $\varepsilon_1/\hbar\omega$.

n	$A_n^{(2)}$	$B_n^{(1)}$	$B_n^{(2)}$
0	$\frac{11}{8}$	$\frac{3}{4}$	$\frac{21}{8}$
1	$\frac{71}{8}$	$\frac{15}{4}$	$\frac{165}{8}$
2	$\frac{191}{8}$	$\frac{39}{4}$	$\frac{615}{8}$
3	$\frac{371}{8}$	$\frac{75}{4}$	$\frac{1575}{8}$

Since there is no first-order effect the curves are parabolae with the vertex at $\varepsilon_1 = 0$. The negative sign of the perturbation is caused by the broadening of the potential parabola for higher values of ε_1 where

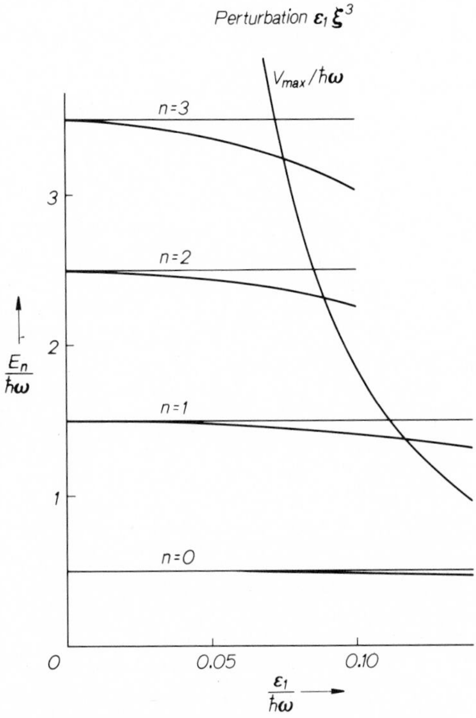

Fig. 20. Energy levels for the asymmetrical anharmonic oscillator. There are only contributions in second perturbation approximation. The line $V_{max}/\hbar\omega$ gives a rough limit for the suitability of the approximation

it can no longer be neglected. The broadening becomes increasingly important at higher energies where even the second-order theory becomes less and less reliable. At $\xi = -\hbar\omega/(3\varepsilon_1)$ a maximum of the potential curve of height $V_{\max} = (\hbar\omega)^3/(54\varepsilon_1^2)$ will occur so that the levels die out at higher energies, an effect which is not reproduced by our approximation.

Generally speaking, the curves show that the higher the energy of a level the smaller are the values of the perturbation parameter $\varepsilon_1/\hbar\omega$ up to which reliable results may be obtained by the approximation. This is corroborated in full by Fig. 21 where the linear first-order effect is shown by broken lines, the second-order correction by the full curves. The latter are certainly no longer reliable as soon as they begin to fall after having passed the energy maximum.

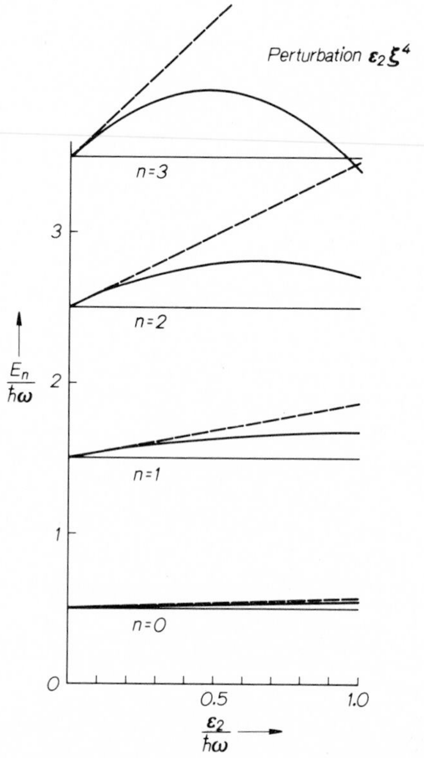

Fig. 21. Energy levels for the symmetrical anharmonic oscillator. The broken lines correspond to the first approximation only, full lines include the second

Problem 36. Approximate wave functions

A particle moves along the x axis between two rigid walls at $x = \pm a$. No other forces than those at the walls act upon it. The wave functions of the ground state and the first excited state shall be approximated by the simplest normalized polynomials giving accurate account of the zeros. Exact and approximate energy values shall be compared.

Solution. Let u_0 and u_1 be the exact wave functions and E_0 and E_1 the respective exact energy levels of the two states under consideration. Then we have (cf. Problem 18)

$$u_0 = \frac{1}{\sqrt{a}} \cos\frac{\pi x}{2a}; \quad E_0 = \frac{\hbar^2}{2ma^2} \cdot \frac{\pi^2}{4} \tag{36.1}$$

and

$$u_1 = \frac{1}{\sqrt{a}} \sin\frac{\pi x}{a}; \quad E_1 = \frac{\hbar^2}{2ma^2} \cdot \pi^2. \tag{36.2}$$

The simplest polynomial approximations with the same zeros are

$$\tilde{u}_0 = \sqrt{\frac{15}{16a}} \left[1 - \left(\frac{x}{a}\right)^2\right] \tag{36.3}$$

and

$$\tilde{u}_1 = \sqrt{\frac{15}{16a}} \left[\frac{x}{a} - \left(\frac{x}{a}\right)^3\right]. \tag{36.4}$$

All the wave functions (36.1) to (36.4) have been normalized according to

$$\int_{-a}^{+a} dx\, u^2 = 1.$$

The maximum of u_0 at $x=0$ is 1 (in units of $a^{-\frac{1}{2}}$), that of $\tilde{u}_0$ at the same place $(\frac{15}{16})^{\frac{1}{2}} = 0.970$ is only 3% smaller. The two maxima of u_1 lie at $x = \pm\frac{1}{2}a$ and amount to ± 1, whereas those of $\tilde{u}_1$ lying at $x = \pm\frac{1}{\sqrt{3}}a = 0.577\,a$ and amounting to $\pm(\frac{35}{36})^{\frac{1}{2}} = \pm 0.986$ are even better in height but shifted in position.

The energy can in all cases be determined from

$$E = \frac{\hbar^2}{2m} \int_{-a}^{+a} dx \left(\frac{du}{dx}\right)^2.$$

Elementary integrations yield the approximate values

$$\tilde{E}_0 = \frac{\hbar^2}{2ma^2} \cdot \frac{5}{2} \quad \text{and} \quad \tilde{E}_1 = \frac{\hbar^2}{2ma^2} \cdot \frac{21}{2}. \tag{36.5}$$

Comparing the numerical values of (36.1), (36.2) and (36.5) in units of $\hbar^2/(2ma^2)$, we obtain

$$E_0 = 2.4674, \qquad \tilde{E}_0 = 2.5000,$$
$$E_1 = 9.8696, \qquad \tilde{E}_1 = 10.5000.$$

The approximate energies lie throughout *above* the exact values; $\tilde{E}_0$ is very close to E_0 whereas $\tilde{E}_1$ compared to E_1 seems a little less good. However, even here, the error does not amount to more than 6 percent.

Problem 37. Potential step

To determine the coefficient of reflection for a wave normally incident upon a potential step defined by

$$V(x) = \frac{1}{2} V_0 \left(1 + \tanh \frac{x}{2a}\right). \tag{37.1}$$

Solution. The potential increases continuously from the value $V=0$ at $x=-\infty$ to the value $V=V_0$ at $x=+\infty$, the main rise occurring in the interval $-2a<x<+2a$:

$$V(-2a) = 0.119\, V_0; \qquad V(+2a) = 0.881\, V_0.$$

The wave function describing a wave coming from the left side must asymptotically become

$$\left.\begin{aligned} &u = e^{ikx} + R e^{-ikx} && \text{for } x \to -\infty; \\ &u = \begin{cases} C e^{-Kx} & \text{if } E < V_0 \\ C e^{ik'x} & \text{if } E > V_0 \end{cases} && \begin{aligned} &\text{for } x \to +\infty, \\ &\text{for } x \to +\infty. \end{aligned} \end{aligned}\right\} \tag{37.2}$$

The quantity $|R|^2$ then is the coefficient of reflection.

In order to solve the Schrödinger equation

$$\frac{d^2u}{dx^2} + \left[k^2 - \frac{2m}{\hbar^2} V(x)\right] u = 0 \tag{37.3}$$

with the potential (37.1) we use instead of x the variable

$$y = (1 + e^{x/a})^{-1} \tag{37.4}$$

and the abbreviations

$$\varkappa^2 = k^2 a^2 = \frac{2ma^2}{\hbar^2} E; \qquad \lambda^2 = \frac{2ma^2}{\hbar^2} V_0. \tag{37.5}$$

With

$$\frac{d}{dx} = -\frac{1}{a}y(1-y)\frac{d}{dy}; \qquad 1+\tanh\frac{x}{2a} = 2(1-y)$$

we then arrive at the differential equation

$$y(1-y)\frac{d^2u}{dy^2} + (1-2y)\frac{du}{dy} + \left[\frac{\varkappa^2}{y(1-y)} - \frac{\lambda^2}{y}\right]u = 0. \tag{37.6}$$

This equation has poles at $y=0,1,\infty$ and therefore leads to hypergeometric solutions. Putting

$$u(y) = y^\nu(1-y)^\mu f(y) \tag{37.7}$$

with

$$\nu^2 = \lambda^2 - \varkappa^2; \qquad \mu^2 = -\varkappa^2 \tag{37.8}$$

a simple calculation leads to the standard form of the Gaussian equation,

$$y(1-y)f'' + [(2\nu+1)-(2\mu+2\nu+2)y]f' - (\mu+\nu)(\mu+\nu+1)f = 0 \tag{37.9}$$

whose special solution

$$f(y) = C\,{}_2F_1(\mu+\nu, \mu+\nu+1, 2\nu+1; y) \tag{37.10}$$

with a constant C still to be fitted we now show just to satisfy the boundary conditions (37.2).

If we start with the limit $x \to +\infty$ or $y \simeq e^{-x/a} \to 0$, the solution (37.10) becomes $f(0) = C$ or

$$u(y) \to C y^\nu \simeq C e^{-\nu x/a}. \tag{37.11}$$

Now two cases have to be distinguished.

a) $\lambda > \varkappa$, $\nu > 0$ is real. The expression (37.11) vanishes exponentially as would be expected in the case $E < V_0$. We obtain

$$u \to C e^{-Kx} \quad \text{with } K^2 = \frac{\nu^2}{a^2} = \frac{2m}{\hbar^2}(V_0 - E). \tag{37.12a}$$

b) $\lambda < \varkappa$, $\nu = -ik'a$ is imaginary. Then

$$u \to C e^{ik'x} \quad \text{with } k'^2 = \frac{2m}{\hbar^2}(E - V_0). \tag{37.12b}$$

On the other hand, for $x \to -\infty$ or $y \to 1$, $1-y \simeq e^{+x/a} \to 0$ we apply the well-known transformation rules for passing over from the argument y to $1-y$ of the hypergeometric function,

$$
\begin{aligned}
&{}_2F_1(\mu+\nu,\mu+\nu+1,2\nu+1;y)\\
&=\frac{\Gamma(2\nu+1)\Gamma(-2\mu)}{\Gamma(\nu-\mu)\Gamma(\nu-\mu+1)}\,{}_2F_1(\mu+\nu,\mu+\nu+1,2\mu+1;1-y)\\
&\quad+(1-y)^{-2\mu}\frac{\Gamma(2\nu+1)\Gamma(2\mu)}{\Gamma(\mu+\nu)\Gamma(\mu+\nu+1)}\,{}_2F_1(\nu-\mu,\nu-\mu+1,-2\mu+1;1-y).
\end{aligned}
$$

With $1-y=\mathrm{e}^{x/a}$ this leads to

$$
u\to C\left\{\frac{\Gamma(2\nu+1)\Gamma(-2\mu)}{\Gamma(\nu-\mu)\Gamma(\nu-\mu+1)}\mathrm{e}^{\mu x/a}+\frac{\Gamma(2\nu+1)\Gamma(2\mu)}{\Gamma(\mu+\nu)\Gamma(\mu+\nu+1)}\mathrm{e}^{-\mu x/a}\right\}
$$

or, with

$$
\mu=+i\varkappa,\qquad \mu/a=ik \tag{37.13}
$$

we arrive at the form (37.2) of our solution with

$$
C=\frac{\Gamma(\nu-\mu)\Gamma(\nu-\mu+1)}{\Gamma(2\nu+1)\Gamma(-2\mu)} \tag{37.14}
$$

and

$$
R=\frac{\Gamma(+2\mu)\Gamma(\nu-\mu)\Gamma(\nu-\mu+1)}{\Gamma(-2\mu)\Gamma(\nu+\mu)\Gamma(\nu+\mu+1)}. \tag{37.15}
$$

Again, we distinguish the two cases $E \gtrless V_0$.

a) $E<V_0$; $\mu=i\varkappa$ is imaginary; $\nu>0$ is real. Numerator and denominator of (37.15) are complex conjugates so that

$$
|R|^2=1.
$$

This is the case of total reflection.

b) $E>V_0$; $\mu=i\varkappa$ and $\nu=-i\sigma$ are both imaginary. Only the factors $\Gamma(\pm 2\mu)$ are complex conjugates and do not contribute to $|R|^2$. Since

$$
\Gamma(z)=\frac{1}{z}\Gamma(z+1)
$$

we find

$$
|R|^2=\left|\frac{(\nu+\mu)\Gamma(\nu-\mu+1)^2}{(\nu-\mu)\Gamma(\nu+\mu+1)^2}\right|^2=\left(\frac{\varkappa-\sigma}{\varkappa+\sigma}\right)^2\left\{\left|\frac{\Gamma(1-i(\varkappa+\sigma))}{\Gamma(1+i(\varkappa-\sigma))}\right|^2\right\}^2.
$$

Here the elementary formula

$$
|\Gamma(1+i\eta)|^2=\frac{\pi\eta}{\sinh\pi\eta}
$$

may be applied so that we finally arrive at the reflection coefficient formula

$$|R|^2 = \left\{\frac{\sinh \pi(\varkappa - \sigma)}{\sinh \pi(\varkappa + \sigma)}\right\}^2 \tag{37.16}$$

or, using the wave numbers k on the left and k' on the right-hand side of the step,

$$|R|^2 = \left\{\frac{\sinh \pi(k - k')a}{\sinh \pi(k + k')a}\right\}^2. \tag{37.17}$$

NB 1. If a is very small, i. e. in the limit of a discontinuous potential jump at $x=0$, the last expression simply becomes

$$|R|^2 = \left(\frac{k-k'}{k+k'}\right)^2. \tag{37.18}$$

In this case, of course, it is no longer necessary to use hypergeometric functions but the two asymptotic conditions (37.2) will each exactly hold for one whole half-space:

$$\begin{aligned} u &= e^{ikx} + R e^{-ikx} \quad \text{for } x<0, \\ u &= C e^{ik'x} \qquad\qquad \text{for } x>0 \end{aligned} \tag{37.19}$$

in the only interesting case b). Continuity of both, u and u', at $x=0$ then yield

$$1+R=C; \qquad ik(1-R)=ik'C$$

with the solutions

$$C = \frac{2k}{k+k'}; \qquad R = \frac{k-k'}{k+k'}. \tag{37.20}$$

The continuity of flux at the point of jump requires

$$k(1-R^2) = k'C^2; \tag{37.21}$$

this condition is satisfied by the expressions (37.20).

NB 2. The point in introducing the new variable y, Eq. (37.4), is to render the coefficient functions in the differential equation rational and so to fall back upon a well-known class of differential equations. This purpose can also be attained by other transformations, e. g.

$$y = (1+e^{-x/a})^{-1}. \tag{37.4'}$$

Eq. (37.4) is chosen in such a way as to make the particular solution (37.10) suffice for the boundary conditions of our special problem to be satisfied, which with (37.4′) would not be possible. It is, however, instructive to perform the calculation using (37.4′).

Problem 38. Pöschl-Teller potential hole

To solve the Schrödinger equation for the potential

$$V(x) = \frac{1}{2} V_0 \left\{\frac{\varkappa(\varkappa-1)}{\sin^2 \alpha x} + \frac{\lambda(\lambda-1)}{\cos^2 \alpha x}\right\}; \qquad V_0 = \frac{\hbar^2 \alpha^2}{m} \tag{38.1}$$

with $\varkappa>1$ and $\lambda>1$, in the region bounded by the singularities of $V(x)$ at $x=0$ and at $x=\pi/2\alpha$.

Solution. The potential (38.1) is shown in Fig. 22 for $\lambda=2$ and several values of $\varkappa>1$. For $\varkappa=\lambda$, a potential hole symmetrical about $\alpha x=\frac{\pi}{4}$ is obtained; if $\varkappa>\lambda$ the hole becomes oblique with its minimum shifting to larger x values; if $1<\varkappa<\lambda$, the minimum shifts to the other side so that the singularity at $x=0$ becomes less and less pronounced

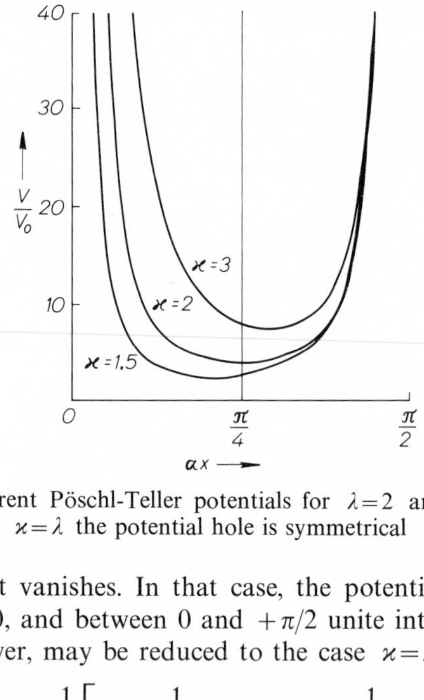

Fig. 22. Three different Pöschl-Teller potentials for $\lambda=2$ and $\varkappa \gtreqless \lambda$. Only for $\varkappa=\lambda$ the potential hole is symmetrical

until, at $\varkappa=1$, it vanishes. In that case, the potential holes between $\alpha x=-\pi/2$ and 0, and between 0 and $+\pi/2$ unite into one large hole. The latter, however, may be reduced to the case $\varkappa=\lambda$ by the identity

$$\frac{1}{\cos^2 \alpha x}=\frac{1}{4}\left[\frac{1}{\sin^2\left(\frac{\alpha x}{2}-\frac{\pi}{4}\right)}+\frac{1}{\cos^2\left(\frac{\alpha x}{2}-\frac{\pi}{4}\right)}\right],$$

replacing α by $\frac{1}{2}\alpha$, and V_0 by $\frac{1}{4}V_0$.

The potential (38.1) is, of course, periodical. This is of no particular importance for the solutions of the Schrödinger equation since the barriers put by the singularities between the holes are impenetrable. We therefore may pick out a single hole, say, the interval $0\leq x\leq\pi/2\alpha$, and solve the Schrödinger equation for this interval with boundary conditions

$$u=0 \quad \text{for } x=0 \text{ and } x=\frac{\pi}{2\alpha}. \tag{38.2}$$

Introducing the new variable

$$y = \sin^2 \alpha x \tag{38.3}$$

the Schrödinger equation is transformed into

$$y(1-y)u'' + \left(\frac{1}{2} - y\right)u' + \frac{1}{4}\left(\frac{k^2}{\alpha^2} - \frac{\varkappa(\varkappa-1)}{y} - \frac{\lambda(\lambda-1)}{1-y}\right)u = 0 \tag{38.4}$$

where

$$k^2 = \frac{2mE}{\hbar^2}.$$

This differential equation has three regular singularities at $y = 0, 1, \infty$ so that it can be reduced to the hypergeometric equation by putting

$$u = y^\mu (1-y)^\nu f(y) \tag{38.5}$$

with fitting values of μ and ν. Choosing

$$\mu = \frac{\varkappa}{2}, \qquad \nu = \frac{\lambda}{2} \tag{38.6}$$

we indeed arrive at

$$y(1-y) f'' + \left[\left(\varkappa + \frac{1}{2}\right) - y(\varkappa + \lambda + 1)\right] f' + \frac{1}{4}\left[\frac{k^2}{\alpha^2} - (\varkappa+\lambda)^2\right] f = 0. \tag{38.7}$$

The general solution of this equation is

$$f = C_1\, {}_2F_1(a,b,c;y) + C_2\, y^{1-c}\, {}_2F_1(a+1-c, b+1-c, 2-c; y) \tag{38.8}$$

with the parameters

$$a = \frac{1}{2}\left(\varkappa + \lambda + \frac{k}{a}\right), \qquad b = \frac{1}{2}\left(\varkappa + \lambda - \frac{k}{\alpha}\right), \qquad c = \varkappa + \frac{1}{2}. \tag{38.9}$$

We have yet to determine the integration constants C_1 and C_2 in (38.8) by taking account of the boundary conditions (38.2) which, in the variable y, run

$$u(0) = 0; \qquad u(1) = 0. \tag{38.2'}$$

At $y=0$, the two hypergeometric series are $=1$ so that

$$u \simeq y^\mu (C_1 + C_2 y^{\frac{1}{2}-\varkappa}) = C_1 y^{\frac{\varkappa}{2}} + C_2 y^{\frac{1-\varkappa}{2}}.$$

Since $\varkappa > 1$, the second term becomes singular so that there follows $C_2 = 0$. At $y=1$ we use the transform

$$
\begin{aligned}
&{}_2F_1(a,b,c;y)\\
&=\frac{\Gamma(c)\Gamma(c-a-b)}{\Gamma(c-a)\Gamma(c-b)}\,{}_2F_1(a,b,a+b-c+1;1-y)\\
&\quad+\frac{\Gamma(c)\Gamma(a+b-c)}{\Gamma(a)\Gamma(b)}(1-y)^{c-a-b}\,{}_2F_1(c-a,c-b,c-a-b+1;1-y).
\end{aligned}
$$

As $c-a-b=\frac{1}{2}-\lambda$, the second term contributes to u in the vicinity of $y=1$,

$$u=\cdots+C_1(1-y)^{\nu+c-a-b}\frac{\Gamma(c)\Gamma(a+b-c)}{\Gamma(a)\Gamma(b)}$$

with the exponent

$$\nu+c-a-b=\tfrac{1}{2}(1-\lambda)<0$$

leading to a singularity and thus preventing the satisfaction of the boundary condition if neither $a=-n$ nor $b=-n$ $(n=0,1,2,\ldots)$ so that the Γ-factor of the second term vanishes. If

$$a=-n \tag{38.10}$$

Eq. (38.9) leads to a positive

$$b=\varkappa+\lambda+n;$$

if, inversely, $b=-n$, we find $a=\varkappa+\lambda+n$, and since the wave function is invariant under exchange of the parameters a and b, both conditions lead to the same solution. Inserting a from Eq. (38.9) into (38.10), we then obtain the eigenvalues

$$k^2=\alpha^2(\varkappa+\lambda+2n)^2 \quad\text{or}\quad E_n=\tfrac{1}{2}V_0(\varkappa+\lambda+2n)^2 \tag{38.11}$$

and the wave functions

$$u_n(x)=C_1\sin^{\varkappa}\alpha x\cos^{\lambda}\alpha x\,{}_2F_1(-n,\varkappa+\lambda+n,\varkappa+\tfrac{1}{2};\sin^2\alpha x) \tag{38.12}$$

with the constant C_1 still to be fixed by normalization. In Fig. 23 the lowest eigenvalues have been shown for $\lambda=2$ as functions of $\varkappa$. Comparison with the potential curves of Fig. 22 shows that the narrower the hole the higher lie the energy levels, as is to be expected. If $2n\gg\varkappa+\lambda$, Eq. (38.11) tends towards the eigenvalue formula $k^2=4\alpha^2n^2$ $(n=1,2,\ldots)$ of a potential well. This too is reasonable, since the continuous curvature at the bottom of the potential will be of less and less influence with increasing height.

NB. Around its minimum, the potential curve may approximately be replaced by a parabola, i.e. by an oscillator potential. Confining ourselves to the case $\varkappa=\lambda$ with the minimum at $\alpha x=\pi/4$, and using the variable

$$z=x-\frac{\pi}{4\alpha}$$

we find

$$V = 4V_0\lambda(\lambda-1)\left[\tfrac{1}{2}+\alpha^2 z^2 + O(z^4)\right].$$

Putting the factor of z^2 equal to $\frac{1}{2}m\omega^2$, we arrive at

$$\omega = 2\sqrt{2\lambda(\lambda-1)}\,\frac{\hbar}{m}\,\alpha^2$$

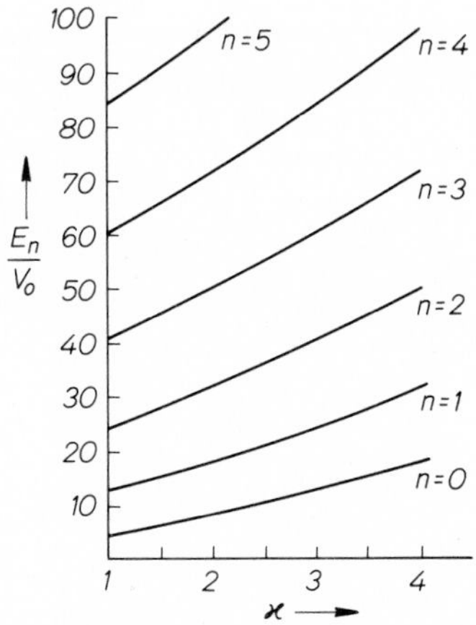

Fig. 23. Energy levels in Pöschl-Teller potentials for $\lambda=2$ in dependence on the asymmetry parameter $\varkappa$

and, neglecting the deviation of the potential from the parabolic shape,

$$E_n = \hbar\omega(n+\tfrac{1}{2}) + 2V_0\lambda(\lambda-1)$$

or

$$E_n/V_0 = 2\sqrt{2\lambda(\lambda-1)}\,(n+\tfrac{1}{2}) + 2\lambda(\lambda-1).$$

This approximation holds, of course, only if $|z|$ at the classical turning points (z_t) is much smaller than $1/\alpha$ or if

$$\alpha^2 z_t^2 = \frac{n+\frac{1}{2}}{\sqrt{2\lambda(\lambda-1)}} \ll 1.$$

This result will hold exactly in the limit $\alpha\to 0$, $\lambda\to\infty$, with $\lambda\alpha^2$ remaining finite.

Literature. Pöschl, G., Teller, E.: Z. Physik **83,** 143 (1933). — Rosen, N., Morse, P. M.: Phys. Rev. **42,** 210 (1932). — Lotmar, W.: Z. Physik **93,** 528 (1935).

Problem 39. Potential hole of modified Pöschl-Teller type

To solve the one-dimensional Schrödinger equation for the potential (Fig. 24)

$$V(x) = -\frac{\hbar^2}{2m}\alpha^2\frac{\lambda(\lambda-1)}{\cosh^2\alpha x} \tag{39.1}$$

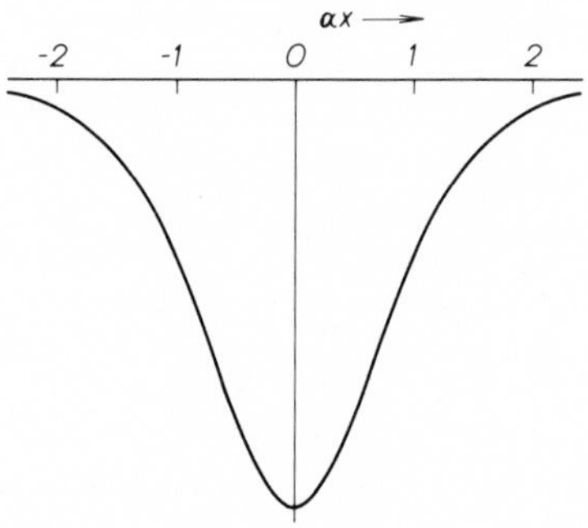

Fig. 24. Modified Pöschl-Teller potential hole

with $\lambda>1$. For positive energies the coefficients of reflection and transmission are to be found; for negative energies the eigenvalues shall be determined.

Solution. In the Schrödinger equation

$$u'' + \left\{k^2 + \frac{\alpha^2\lambda(\lambda-1)}{\cosh^2\alpha x}\right\}u = 0, \tag{39.2}$$

with $k^2 = 2mE/\hbar^2$, we introduce the new variable

$$y = \cosh^2\alpha x. \tag{39.3}$$

We then get

$$y(1-y)u'' + \left(\frac{1}{2} - y\right)u' - \left\{\frac{k^2}{4\alpha^2} + \frac{\lambda(\lambda-1)}{4y}\right\}u = 0.$$

Splitting off a fitting power of y by setting

$$u = y^{\lambda/2}v(y) \tag{39.4}$$

we arrive at the hypergeometric differential equation

$$y(1-y)v'' + \left\{\left(\lambda + \frac{1}{2}\right) - (\lambda+1)y\right\}v' - \frac{1}{4}\left(\lambda^2 + \frac{k^2}{\alpha^2}\right)v = 0. \tag{39.5}$$

With the abbreviations

$$a=\frac{1}{2}\left(\lambda+i\frac{k}{\alpha}\right); \quad b=\frac{1}{2}\left(\lambda-i\frac{k}{\alpha}\right) \tag{39.6}$$

the complete solution of (39.5) may be written[3]

$$v(x)=A\,{}_2F_1(a,b,\tfrac{1}{2};1-y)+B(1-y)^{\frac{1}{2}}\,{}_2F_1(a+\tfrac{1}{2},b+\tfrac{1}{2},\tfrac{3}{2};1-y), \tag{39.7}$$

so that at $x=0$ or $y=1$ the wave function approaches

$$u(0)=A+B(1-y)^{\frac{1}{2}}. \tag{39.8}$$

We shall pick out a fundamental system of two real standard solutions, u_e and u_o, even and odd in x,

$$u_e(-x)=u_e(x); \quad u_o(-x)=-u_o(x). \tag{39.9}$$

With $B=0$, $A=1$ we find the even standard solution,

$$u_e(x)=\cosh^{\lambda}\alpha x\,{}_2F_1(a,b,\tfrac{1}{2};-\sinh^2\alpha x) \tag{39.10a}$$

and with $A=0$, $B=i$ the odd standard solution,

$$u_o(x)=\cosh^{\lambda}\alpha x\,\sinh\alpha x\,{}_2F_1(a+\tfrac{1}{2},b+\tfrac{1}{2},\tfrac{3}{2};-\sinh^2\alpha x). \tag{39.10b}$$

These solutions we shall discuss in detail.

To answer the above questions, we first need the asymptotic behaviour of (39.10a, b) at negative infinitely large values of the argument

$$-\sinh^2\alpha x\to -2^{-2}e^{2\alpha|x|};$$

the well-known formulae yield

$$u_e(x)\to 2^{-\lambda}e^{\lambda\alpha|x|}\,\Gamma\left(\frac{1}{2}\right)\left\{\frac{\Gamma(b-a)}{\Gamma(b)\Gamma(\frac{1}{2}-a)}2^{2a}e^{-2a\alpha|x|} + \frac{\Gamma(a-b)}{\Gamma(a)\Gamma(\frac{1}{2}-b)}2^{2b}e^{-2b\alpha|x|}\right\} \tag{39.11a}$$

and

$$u_o(x)\to \pm 2^{-(\lambda+1)}e^{(\lambda+1)\alpha|x|}\,\Gamma\left(\frac{3}{2}\right)\left\{\frac{\Gamma(b-a)}{\Gamma(b+\frac{1}{2})\Gamma(1-a)}2^{2a+1}e^{-(2a+1)\alpha|x|} + \frac{\Gamma(a-b)}{\Gamma(a+\frac{1}{2})\Gamma(1-b)}2^{2b+1}e^{-(2b+1)\alpha|x|}\right\} \tag{39.11b}$$

---

[3] The solution is wanted for $0\le|x|<\infty$ or $1\le y<\infty$. The hypergeometric equation

$$y(1-y)v''+[c-(a+b+1)y]v'-abv=0$$

runs in the variable $z=1-y$,

$$z(1-z)v''+[c'-(a+b+1)z]v'-abv=0$$

with $c'=a+b+1-c$. Eq. (39.7) reduces to the usual form in the variable z. The domain wanted in z is $-\infty<z\le 0$.

where the $\pm$ signs correspond to $x \gtrless 1$. If the energy is *positive*, according to (39.6), a and b are complex conjugate so that

$$u_e(x) \to \Gamma\left(\frac{1}{2}\right)\left\{\frac{\Gamma(-ik/\alpha)\,e^{i\frac{k}{\alpha}\log 2}}{\Gamma\left(\frac{\lambda}{2}-i\frac{k}{2\alpha}\right)\Gamma\left(\frac{1-\lambda}{2}-i\frac{k}{2\alpha}\right)}\,e^{-ik|x|}\right.$$

$$\left.+\frac{\Gamma(ik/\alpha)\,e^{-i\frac{k}{\alpha}\log 2}}{\Gamma\left(\frac{\lambda}{2}+i\frac{k}{2\alpha}\right)\Gamma\left(\frac{1-\lambda}{2}+i\frac{k}{2\alpha}\right)}\,e^{ik|x|}\right\} \tag{39.12a}$$

and

$$u_o(x) \to \pm\Gamma\left(\frac{3}{2}\right)\left\{\frac{\Gamma(-ik/\alpha)\,e^{i\frac{k}{\alpha}\log 2}}{\Gamma\left(\frac{\lambda+1}{2}-i\frac{k}{2\alpha}\right)\Gamma\left(1-\frac{\lambda}{2}-i\frac{k}{2\alpha}\right)}\,e^{-ik|x|}\right.$$

$$\left.+\frac{\Gamma(ik/\alpha)\,e^{-i\frac{k}{\alpha}\log 2}}{\Gamma\left(\frac{\lambda+1}{2}+i\frac{k}{2\alpha}\right)\Gamma\left(1-\frac{\lambda}{2}+i\frac{k}{2\alpha}\right)}\,e^{ik|x|}\right\}. \tag{39.12b}$$

With the abbreviations

$$\varphi_e = \arg\frac{\Gamma(ik/\alpha)\,e^{-i\frac{k}{\alpha}\log 2}}{\Gamma\left(\frac{\lambda}{2}+i\frac{k}{2\alpha}\right)\Gamma\left(\frac{1-\lambda}{2}+i\frac{k}{2\alpha}\right)} \tag{39.13a}$$

and

$$\varphi_o = \arg\frac{\Gamma(ik/\alpha)\,e^{-i\frac{k}{\alpha}\log 2}}{\Gamma\left(\frac{\lambda-1}{2}+i\frac{k}{2\alpha}\right)\Gamma\left(1-\frac{\lambda}{2}+i\frac{k}{2\alpha}\right)} \tag{39.13b}$$

the two functions become

$$u_e \to C_e\cos(k|x|+\varphi_e); \qquad u_o \to \pm C_o\cos(k|x|+\varphi_o) \qquad (\text{at } x \gtrless 0), \tag{39.14}$$

with amplitude factors to be computed from (39.12a, b) but of no interest for our purpose.

We now compose a linear combination of the two fundamental solutions, say,

$$u = A u_e + B u_o$$

which for $x>0$ is

$$u=\frac{A}{2}C_e(e^{i\varphi_e}e^{ikx}+e^{-i\varphi_e}e^{-ikx})+\frac{B}{2}C_o(e^{i\varphi_o}e^{ikx}+e^{-i\varphi_o}e^{-ikx})$$

and for $x<0$,

$$u=\frac{A}{2}C_e(e^{i\varphi_e}e^{-ikx}+e^{-i\varphi_e}e^{ikx})-\frac{B}{2}C_o(e^{i\varphi_o}e^{-ikx}+e^{-i\varphi_o}e^{ikx}).$$

We want a solution of the asymptotic form

$$u=\begin{cases}e^{ikx}+Re^{-ikx} & \text{for } x<0,\\ Te^{ikx} & \text{for } x>0,\end{cases} \tag{39.15}$$

i.e.

$$\begin{aligned}
\frac{A}{2}C_e e^{i\varphi_e}+\frac{B}{2}C_o e^{i\varphi_o}&=T,\\
\frac{A}{2}C_e e^{-i\varphi_e}+\frac{B}{2}C_o e^{-i\varphi_o}&=0,\\
\frac{A}{2}C_e e^{i\varphi_e}-\frac{B}{2}C_o e^{i\varphi_o}&=R,\\
\frac{A}{2}C_e e^{-i\varphi_e}-\frac{B}{2}C_o e^{-i\varphi_o}&=1.
\end{aligned}$$

From the second and fourth of these equations, AC_e and BC_o can be determined, and then T and R computed from the first and third ones:

$$T=\tfrac{1}{2}(e^{2i\varphi_e}-e^{2i\varphi_o});\qquad R=\tfrac{1}{2}(e^{2i\varphi_e}+e^{2i\varphi_o}). \tag{39.16}$$

From these amplitudes there follow the intensity coefficients of transmission,

$$|T|^2=\sin^2(\varphi_e-\varphi_o), \tag{39.17}$$

and of reflection,

$$|R|^2=\cos^2(\varphi_e-\varphi_o). \tag{39.18}$$

They satisfy the conservation law

$$|T|^2+|R|^2=1$$

and depend upon only the phase angles and not upon any normalization factors in the eigenfunctions.

To compute the coefficients (39.17) and (39.18) let us go back to Eqs. (39.13) which, with the abbreviation

$$k/2\alpha=q,$$

yield

$$\varphi_e - \varphi_o = \arg\Gamma\left(iq + \frac{\lambda}{2} + \frac{1}{2}\right) + \arg\Gamma\left(iq + 1 - \frac{\lambda}{2}\right)$$
$$-\arg\Gamma\left(iq + \frac{\lambda}{2}\right) - \arg\Gamma\left(iq + \frac{1}{2} - \frac{\lambda}{2}\right). \tag{39.19}$$

From the general relation

$$\Gamma(z)\Gamma(1-z) = \frac{\pi}{\sin\pi z}$$

there follows

$$\arg\Gamma(z) - \arg\Gamma(1-z^*) = -\arg\sin\pi z\,. \tag{39.20}$$

Putting

$$iq + \frac{\lambda}{2} + \frac{1}{2} = z_1 \quad \text{and} \quad iq + \frac{\lambda}{2} = z_2$$

we have

$$iq + \frac{1}{2} - \frac{\lambda}{2} = 1 - z_1^* \quad \text{and} \quad iq + 1 - \frac{\lambda}{2} = 1 - z_2^*$$

so that pairs of arguments may be combined in (39.19) according to (39.20) thus leading to

$$\varphi_e - \varphi_o = -\arg\sin\pi\left(\frac{\lambda+1}{2} + iq\right) + \arg\sin\pi\left(\frac{\lambda}{2} + iq\right)$$
$$= \tan^{-1}\left(\tan\frac{\pi\lambda}{2}\tanh\pi q\right) + \tan^{-1}\left(\cot\frac{\pi\lambda}{2}\tanh\pi q\right)$$

or by an elementary reshaping,

$$\varphi_e - \varphi_o = \tan^{-1}\left\{\frac{\sinh\pi k/\alpha}{\sin\pi\lambda}\right\}. \tag{39.21}$$

If the depth parameter λ is an integer, the denominator vanishes so that $\varphi_e - \varphi_o = \dfrac{\pi}{2}$ and, according to (39.17) and (39.18), $|T|^2 = 1$, $|R|^2 = 0$. In these cases the entire incident energy will be transmitted for all energies of the particles and nothing will be reflected (which is obvious for $\lambda = 1$). On the other hand, in the limit $E = 0$ (i.e. $\pi k/\alpha = 0$) the numerator in (39.21) vanishes so that for non-integer values of λ we get $\varphi_e - \varphi_o = 0$ and total reflection: $|T|^2 = 0$, $|R|^2 = 1$. For all values of λ we may write

$$|R|^2 = \frac{1}{1+p^2}; \quad |T|^2 = \frac{p^2}{1+p^2} \tag{39.22}$$

with

$$p = \frac{\sinh \pi k/\alpha}{\sin \pi \lambda}$$

and find the reflection coefficients shown in Fig. 25 for two energies corresponding to $k/2\alpha = 0.1$ and 0.2. The same picture would be repeated in any interval $n < \lambda < n+1$ of integer n.

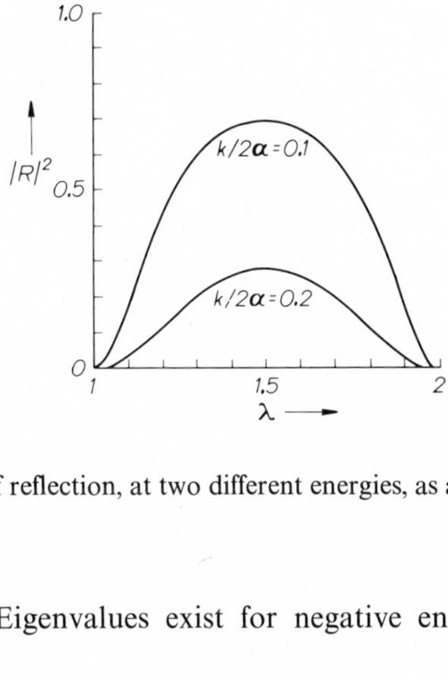

Fig. 25. Coefficient of reflection, at two different energies, as a function of hole size

Bound states. Eigenvalues exist for negative energies. Let us put $k = i\varkappa$ so that

$$E = -\frac{\hbar^2 \varkappa^2}{2m} \tag{39.23}$$

and the parameters (39.6) become real, viz.

$$a = \frac{1}{2}\left(\lambda - \frac{\varkappa}{\alpha}\right); \quad b = \frac{1}{2}\left(\lambda + \frac{\varkappa}{\alpha}\right). \tag{39.24}$$

Then we may again use the asymptotic formulae (39.11 a, b) in which, however, the first term now behaves as $e^{\varkappa|x|}$ and the second as $e^{-\varkappa|x|}$.

A normalizable solution is therefore possible for $\varkappa > 0$ if, and only if, the factor of the first term vanishes. Since the Γ functions are now all taken for real arguments where poles exist at negative integers $-n$ $(n = 0, 1, 2, \ldots)$ the eigenvalues follow from

$$\frac{1-\lambda}{2} + \frac{\varkappa}{2\alpha} = -n \quad \text{or} \quad \frac{\varkappa}{\alpha} = \lambda - 1 - 2n$$

for even, and

$$1 - \frac{\lambda}{2} + \frac{\varkappa}{2\alpha} = -n \quad \text{or} \quad \frac{\varkappa}{\alpha} = \lambda - 2 - 2n$$

for odd eigenstates. Hence the energy terms become, with a slight change in notation,

$$E_n = -\frac{\hbar^2 \varkappa^2}{2m} (\lambda - 1 - n)^2; \qquad n \leq \lambda - 1 \tag{39.25}$$

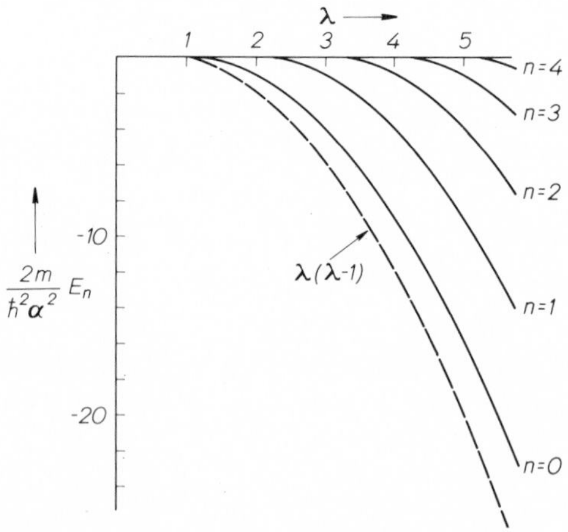

Fig. 26. Energy levels in modified Pöschl-Teller holes of different size. The broken line represents the depth of the potential hole

with even/odd states for even/odd $n = 0, 1, 2, \ldots$. These eigenvalues are shown in Fig. 26; it should be noted that for integer λ there is always one eigenvalue $(n = \lambda - 1)$ lying at zero energy.

Problem 40. Free fall of a body over earth's surface

A particle moves in the homogeneous gravitational field over the earth's surface, the latter being assumed to elastically reflect the particle (e.g. a steel ball dancing over an horizontal plate of glass). This basic problem of classical mechanics is to be treated according to quantum theory.

Solution. The dancing of the ball corresponds to a stationary state. With x the height over the earth's surface, and $V(x)=mgx$ the gravitation potential in its standard form, the Schrödinger equation

$$-\frac{\hbar^2}{2m}\frac{d^2u}{dx^2}+(mgx-E)u=0 \tag{40.1}$$

is to be solved for $x>0$. The assumption of elastic reflection at $x=0$ leads to the boundary condition

$$u(0)=0 \tag{40.2a}$$

to be supplemented by

$$u(\infty)\to 0\,. \tag{40.2b}$$

Using the abbreviations

$$\frac{2m^2g}{\hbar^2}=\frac{1}{l^3};\qquad \frac{2mE}{\hbar^2}=\frac{\lambda}{l^2} \tag{40.3}$$

and the variable

$$\xi=\frac{x}{l}-\lambda\,, \tag{40.4}$$

where l is a characteristic length and ξ dimensionless, Eq. (40.1) becomes

$$\frac{d^2u}{d\xi^2}-\xi u=0\,, \tag{40.5}$$

and the boundary conditions (40.2a, b)

$$u(-\lambda)=0;\qquad u(\infty)\to 0\,. \tag{40.6}$$

The interval of classically permitted motion lies between the two classical turning points at $\xi=-\lambda$ and $\xi=0$ (Fig. 27), i.e. entirely at negative values of ξ.

The differential equation (40.5) is solved by Bessel functions of the order 1/3. The solution satisfying the boundary condition $u(\infty)=0$ is the Airy function (Fig. 28)[4],

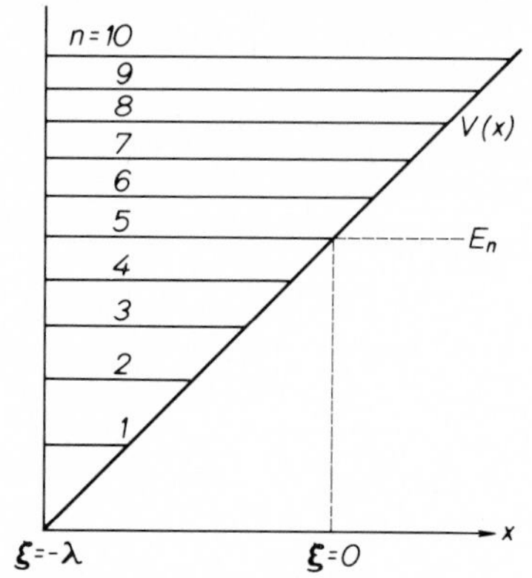

Fig. 27. Gravitation potential over Earth's surface with the ten lowest energy levels

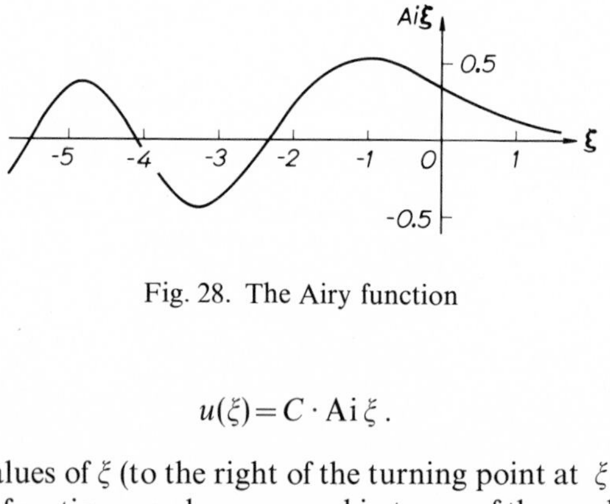

Fig. 28. The Airy function

$$u(\xi)=C\cdot \mathrm{Ai}\,\xi\,. \tag{40.7}$$

For *positive* values of ξ (to the right of the turning point at $\xi=0$ or $x=\lambda$ in Fig. 27) this function may be expressed in terms of the modified Hankel function

$$\mathrm{Ai}\,\xi=\frac{1}{\pi}\sqrt{\frac{\xi}{3}}\,K_{\frac{1}{3}}\left(\tfrac{2}{3}\xi^{\frac{2}{3}}\right)\quad \text{for } \xi>0\,. \tag{40.8}$$

[4] The Airy function is well tabulated today, and so are its zeros. Cf. e. g. Abramowitz, M., Stegun, I. E.: Handbook of Mathematical Functions. New York: Dover Publ. 1965. — The accompanying table of zeros has been taken from this source.

The asymptotic behaviour follows from the general formula

$$K_\nu(z) \to \sqrt{\frac{\pi}{2z}}\, e^{-z}; \qquad z \to \infty$$

so that

$$u(\xi) \to C \cdot \frac{1}{2}\sqrt{\frac{3}{\pi}}\, \xi^{-\frac{1}{4}} e^{-\frac{2}{3}\xi^{\frac{3}{2}}} \quad \text{for } \xi \to +\infty . \tag{40.9}$$

In the classical interval, i.e. for *negative* values of ξ, the Airy function is represented by Bessel functions,

$$\mathrm{Ai}(-\zeta) = \frac{1}{3}\sqrt{\zeta}\,\{J_{\frac{1}{3}}(\tfrac{2}{3}\zeta^{\frac{3}{2}}) + J_{-\frac{1}{3}}(\tfrac{2}{3}\zeta^{\frac{3}{2}})\}$$

or

$$u(\xi) = \frac{C}{3}\sqrt{|\xi|}\{J_{\frac{1}{3}}(\tfrac{2}{3}|\xi|^{\frac{3}{2}}) + J_{-\frac{1}{3}}(\tfrac{2}{3}|\xi|^{\frac{3}{2}})\} \quad \text{for } \xi < 0 . \tag{40.10}$$

According to the other boundary condition (40.6), this function must vanish at $\xi = -\lambda$. Since λ is the energy eigenvalue (in suitable units), the eigenvalues follow from either

$$\mathrm{Ai}(-\lambda) = 0 \tag{40.11 a}$$

or

$$J_{\frac{1}{3}}(\tfrac{2}{3}\lambda^{\frac{3}{2}}) + J_{-\frac{1}{3}}(\tfrac{2}{3}\lambda^{\frac{3}{2}}) = 0 . \tag{40.11 b}$$

Table of zeros

$\mathrm{Ai}(-\lambda_n)=0$	
n	λ_n
1	2.338
2	4.088
3	5.521
4	6.787
5	7.944
6	9.023
7	10.040
8	11.009
9	11.936
10	12.829

The first ten zeros are given in the accompanying table; the corresponding energy levels have been inserted into Fig. 27, above. For higher levels, $\lambda \gg 1$, we may use the asymptotic relations

$$J_{\frac{1}{3}}(z) \to \sqrt{\frac{2}{\pi z}} \cos\left(z - \frac{5\pi}{12}\right); \quad J_{-\frac{1}{3}}(z) \to \sqrt{\frac{2}{\pi z}} \cos\left(z - \frac{\pi}{12}\right)$$

leading to

$$u(\xi) \to \frac{C}{\sqrt{3\pi}} |\xi|^{-\frac{1}{4}} \cos\left(\frac{2}{3}|\xi|^{\frac{3}{2}} - \frac{\pi}{4}\right), \tag{40.12}$$

for large negative values of ξ, and to the eigenvalue condition

$$\frac{2}{3}\lambda_n^{\frac{3}{2}} = \left(2n - \frac{1}{2}\right)\frac{\pi}{2}, \tag{40.13}$$

supplementing the table for $n \gg 1$. The latter relation leads on to

$$E_n = \frac{\hbar^2}{2ml^2}\left\{\frac{3\pi}{4}\left(2n - \frac{1}{2}\right)\right\}^{\frac{2}{3}} \quad \text{for } n \gg 1. \tag{40.14}$$

NB 1. The asymptotic formulae (40.9) and (40.12) are in perfect agreement with those to be got by applying the WKB approximation, cf. Problem 115 where the analytic continuation of the wave function across a classical turning point is discussed without having recourse to the Airy function.

NB 2. The characteristic length in this problem, l, may be computed for any mass. For an electron we get $l = 0.088$ cm, and smaller for bigger masses ($l \propto m^{-\frac{2}{3}}$). The basic energy constant of Eq. (40.14),

$$\varepsilon = \frac{\hbar^2}{2ml^2},$$

is $\varepsilon = 1.03 \times 10^{-25}$ erg for electrons and smaller for bigger masses. These orders of magnitude give a fair impression of the vanishing quantum effects in macroscopic bodies. The wavelength of the wave function (40.12) then becomes extremely small so that only average values of u^2, taken over many wavelengths, retain any meaning:

$$\overline{u^2} = \frac{|C|^2}{6\pi}|\xi|^{-\frac{1}{2}} = \frac{|C|^2}{6\pi} \cdot \frac{1}{\sqrt{\lambda - \frac{x}{l}}}.$$

With $h = \lambda l$ the summit point of the classical motion,

$$\overline{u^2} \propto \frac{1}{\sqrt{h - x}},$$

and that exactly corresponds to the classical predictions, $\overline{u^2}\,dx$ being the probability of finding the particle (dancing, in the stationary state!) in the interval dx. As this probability is proportional to the time of stay, dt, in this interval, we have

$$\overline{u^2}\,dx \propto dt$$

or the velocity $v=dx/dt \propto \sqrt{h-x}$, in complete agreement with classical mechanics, giving

$$v=\sqrt{2g(h-x)}\,.$$

It is remarkable how great are the mathematical difficulties of this problem which, in classical mechanics, is one of the most basic and simple ones.

NB 3. The zero-point energy, $E_1=\varepsilon\lambda_1$, is of the order ε. This follows very simply from the uncertainty relation, $\Delta p\cdot\Delta x=\hbar$, with $\Delta p\sim\sqrt{2mE}$ and the (classical) interval of motion being $\Delta x=\dfrac{E}{mg}$. We then have

$$\Delta p\cdot\Delta x=\sqrt{2mE}\cdot\frac{E}{mg}=\hbar,$$

or

$$E=\sqrt[3]{\tfrac{1}{2}\hbar^2 mg^2}=\varepsilon,$$

in the order of magnitude expected.

Problem 41. Accelerating electrical field

Let a current of electrons of initial energy E be emitted from a metallic surface, in positive x direction, and an accelerating electrical field $\mathscr{E}$ be applied in the same direction. The classical formula

$$\tfrac{1}{2}mv^2=E+e\mathscr{E}x\,, \tag{41.1}$$

with v the velocity at the distance x from the emitting surface, shall be corroborated by quantum mechanics.

Solution. As $V(x)=-e\mathscr{E}x$, we have to solve the Schrödinger equation

$$-\frac{\hbar^2}{2m}\frac{d^2u}{dx^2}-e\mathscr{E}xu=Eu \tag{41.2}$$

for $x>0$ (Fig. 29), looking for such solutions which represent waves running in the positive x direction. In a similar way as in the preceding problem we now use the characteristic length l and the dimensionless energy parameter λ, defined by

$$\frac{2me\mathscr{E}}{\hbar^2}=\frac{1}{l^3}\,;\qquad \frac{2mE}{\hbar^2}=\frac{\lambda}{l^2}\,, \tag{41.3}$$

and the dimensionless variable

$$\xi=\frac{x}{l}+\lambda\,. \tag{41.4}$$

Then, we get the differential equation

$$\frac{d^2u}{d\xi^2} + \xi u = 0 \tag{41.5}$$

to be solved for $\xi > \lambda/l$. The solution, corresponding to an outgoing wave, is

$$u(\xi) = C\sqrt{\frac{\pi}{3}\,\xi}\,H^{(1)}_{\frac{1}{3}}\left(\frac{2}{3}\,\xi^{\frac{3}{2}}\right). \tag{41.6}$$

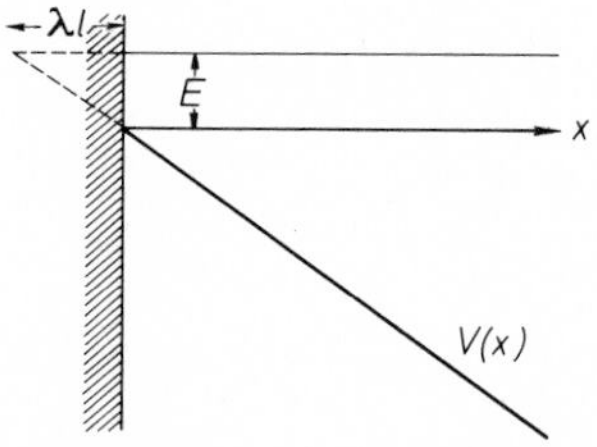

Fig. 29. Acceleration of electrons in homogeneous electrical field

Its asymptotic behaviour follows from

$$H^{(1)}_{\frac{1}{3}}(z) \to \sqrt{\frac{2}{\pi z}}\,\exp\left[i\left(z - \frac{5\pi}{12}\right)\right]$$

to be[5]

$$u(\xi) \to C\,e^{-\frac{5\pi i}{12}}\,\xi^{-\frac{1}{4}}\,e^{\frac{2i}{3}\xi^{\frac{3}{2}}}. \tag{41.7}$$

In order to corroborate (41.1) we now compare the particle flux,

$$s = \frac{\hbar}{2im}\left(u^*\frac{du}{dx} - u\,\frac{du^*}{dx}\right), \tag{41.8}$$

with the particle density,

$$\rho = u^* u, \tag{41.9}$$

[5] Note that (41.7) is identical with the WKB approximation. Here, as in the preceding problem, we have not used the Airy function, because no difficulties of analytic continuation occur, and furthermore, because the solution sought in this case is a linear combination of the Airy functions of the first and second kind,

$$u = C\sqrt{\pi}\,e^{-\frac{i\pi}{6}}\{\mathrm{Ai}(-\xi) - i\,\mathrm{Bi}(-\xi)\}.$$

the ratio of both giving the particle velocity v. From (41.7) we get

$$\frac{du}{dx}=\frac{1}{l}\left(-\frac{1}{4\xi}+i\sqrt{\xi}\right)u$$

so that, from (41.8), we find

$$s=\frac{\hbar}{2iml}\left\{\left(-\frac{1}{4\xi}+i\sqrt{\xi}\right)-\left(-\frac{1}{4\xi}-i\sqrt{\xi}\right)\right\}u^*u$$

or

$$v=\frac{s}{\rho}=\frac{\hbar}{ml}\sqrt{\xi}=\frac{\hbar}{m}\frac{\sqrt{x+\lambda l}}{l^{\frac{3}{2}}}$$

which, using (41.3), may be reshaped into

$$v=\sqrt{\frac{2e\mathscr{E}}{m}\left(x+\frac{E}{e\mathscr{E}}\right)}=\sqrt{\frac{2}{m}}\sqrt{e\mathscr{E}x+E},$$

in perfect agreement with (41.1).

In this proof we have used the asymptotic form of the wave function, i.e., we have supposed $\xi\gg 1$ or $x\gg l$. The characteristic length l may be determined from (41.3); for a reasonable field strength $\mathscr{E}=300$ volts/cm (or 1 electrostatic unit), and for electrons we find $l=3.69\times 10^{-6}$ cm. The condition $x\gg l$ is therefore generally very well satisfied.

B. Problems of Two or Three Degrees of Freedom without Spherical Symmetry

Problem 42. Circular oscillator

To study the motion of a point mass in the two-dimensional potential

$$V=\frac{m}{2}\omega^2(x^2+y^2). \tag{42.1}$$

Solution. The problem may be dealt with in rectangular or in polar coordinates with

$$r^2=x^2+y^2 \tag{42.2}$$

and the results of both ways compared.

a) *In rectangular coordinates*, the Schrödinger equation permits factorization of the wave function,

$$u(x,y)=f(x)\cdot g(y);$$

each of both factors then satisfies an one-dimensional oscillator equation (cf. Problem 30),

$$f'' + (k_1^2 - \lambda^2 x^2) f = 0; \qquad g'' + (k_2^2 - \lambda^2 y^2) g = 0$$

with

$$k_1^2 + k_2^2 = k^2; \qquad E = \frac{\hbar^2 k^2}{2m}; \qquad \lambda = \frac{m\omega}{\hbar}.$$

Hence, we have the eigenvalues

$$k_1^2 = 2\lambda(n_1 + \tfrac{1}{2}); \qquad k_2^2 = 2\lambda(n_2 + \tfrac{1}{2})$$

or

$$E = \hbar\omega(n_1 + n_2 + 1); \qquad n_1, n_2 = 0, 1, 2, \ldots \tag{42.3}$$

with wave functions

$$u_{n_1 n_2}(x, y) = f_{n_1}(x) f_{n_2}(y)$$

where the f_{n_i} denote one-dimensional oscillator functions as given in (30.10) with normalization constants (30.11). The energy levels are degenerate, because the sum

$$n_1 + n_2 = n \tag{42.4}$$

determining the energy may be decomposed into two integers in $n+1$ different ways; the general solution therefore is obtained by any linear combination

$$u_n(x, y) = \sum_{n_1=0}^{n} A_{n_1} f_{n_1}(x) f_{n-n_1}(y) \tag{42.5}$$

with

$$\sum_{n_1=0}^{n} |A_{n_1}|^2 = 1. \tag{42.6}$$

b) *In polar coordinates*, the potential depends upon r only, thus permitting factorization:

$$u(r, \varphi) = v(r) e^{iM\varphi}; \qquad M = 0, \pm 1, \pm 2, \ldots. \tag{42.7}$$

This leads to the differential equation

$$\left(v'' + \frac{1}{r} v' - \frac{M^2}{r^2} v\right) + (k^2 - \lambda^2 r^2) v = 0$$

where the first bracket is originated by the Laplacian. Putting

$$v = r^{|M|} e^{-\frac{\lambda}{2} r^2} F(r)$$

the differential equation

$$F'' + \left(\frac{2|M|+1}{r} - 2\lambda r\right) F' - [2\lambda(|M|+1) - k^2] F = 0$$

is obtained for $F(r)$ which by using, instead of r, the variable

$$t = \lambda r^2$$

is transformed into the Kummer equation

$$t \frac{d^2F}{dt^2} + [(|M|+1)-t]\frac{dF}{dt} - \frac{1}{2}\left[(|M|+1) - \frac{k^2}{4\lambda}\right]F = 0$$

whose solution regular at $t=0$ is the confluent series

$$F(t) = {}_1F_1(a, |M|+1; t)$$

where

$$a = \frac{1}{2}(|M|+1) - \frac{k^2}{4\lambda}. \tag{42.8}$$

For large values of t this function would diverge as e^t, thus preventing normalization. If, and only if

$$a = -n_r \quad \text{with } n_r = 0, 1, 2, \ldots, \tag{42.9}$$

the confluent series becomes a polynomial and the wave function can be normalized. We then have the solutions

$$u_{n_r,M}(r,\varphi) = C_{n_r,M}\, r^{|M|} e^{-\frac{\lambda}{2}r^2}\, {}_1F_1(-n_r, |M|+1; \lambda r^2) e^{iM\varphi} \tag{42.10}$$

and

$$E = \hbar\omega(|M|+1+2n_r). \tag{42.11}$$

As in Eq. (42.4), we may put

$$|M| + 2n_r = n$$

and again find degeneracy of the same multiplicities as above.

c) *Comparison.* Each solution (42.10) is an eigenfunction of the two-dimensional angular momentum operator

$$L = \frac{\hbar}{i}\frac{\partial}{\partial\varphi}$$

with the eigenvalue $\hbar M$:

$$L u_{n_r,M} = \hbar M u_{n_r,M}. \tag{42.12}$$

On the other hand, the product solutions in rectangular coordinates are not angular momentum eigenfunctions, except that of the ground state. One may, however, look for such values of the constants A_{n_1} in (42.5) so as to construct linear combinations of degenerate product functions leading to the type (42.10). A few examples, covering the three lowest energy states, are given below and may be checked by direct computation.

n=0 (ground state, non-degenerate); $n_r=0, M=0, n=0$. $E=\hbar\omega$.

$$u_{0,0}(r,\varphi)=\sqrt{\frac{\lambda}{\pi}}\,e^{-\frac{\lambda}{2}r^2}=u_0(x)\,u_0(y)\,.$$

n=1 (two degenerate states); $n_r=0$, $M=\pm 1$, $n=1$. $E=2\hbar\omega$.

$$u_{0,\pm 1}(r,\varphi)=\frac{\lambda}{\sqrt{\pi}}\,r\,e^{-\frac{\lambda}{2}r^2}\,e^{\pm i\varphi}=\frac{1}{\sqrt{2}}\{u_1(x)\,u_0(y)\pm i\,u_0(x)\,u_1(y)\}\,.$$

n=2 (three degenerate states); $n_r=1$, $M=0$ and $n_r=0$, $M=\pm 2$. $E=3\hbar\omega$.

$$u_{1,0}(r,\varphi)=\sqrt{\frac{\lambda}{\pi}}\,(1-\lambda r^2)\,e^{-\frac{\lambda}{2}r^2}$$
$$=\frac{1}{\sqrt{2}}\{u_2(x)\,u_0(y)+u_0(x)\,u_2(y)\}\,;$$
$$u_{0,\pm 2}(r,\varphi)=\sqrt{\frac{\lambda}{2\pi}}\,\lambda r^2\,e^{-\frac{\lambda}{2}r^2}\,e^{\pm 2i\varphi}$$
$$=\frac{1}{2}\{u_0(x)\,u_2(y)-u_2(x)\,u_0(y)\pm i\sqrt{2}\,u_1(x)\,u_1(y)\}\,.$$

NB. The problem should be compared with that of the spherical oscillator, Problems 65 and 66.

Problem 43. Stark effect of a two-dimensional rotator

A rigid rotator with the moment of inertia Θ and an electrical moment p is rotating in a plane. The perturbation of its energy terms by an homogeneous electric field $\mathscr{E}$ in x direction shall be investigated.

Solution. The unperturbed rotator satisfies the Schrödinger equation

$$-\frac{\hbar^2}{2\Theta}\frac{d^2u}{d\varphi^2}=E\,u \tag{43.1}$$

with eigenfunctions

$$|m\rangle=\frac{1}{\sqrt{2\pi}}\,e^{im\varphi}\,;\qquad m=0,\pm 1,\pm 2,\ldots \tag{43.2}$$

and eigenvalues

$$E_m=\frac{\hbar^2 m^2}{2\Theta}\,. \tag{43.3}$$

The perturbation potential is

$$V(\varphi) = -p\mathscr{E}\cos\varphi; \tag{43.4}$$

it has a minimum $(-p\mathscr{E})$ at $\varphi=0$ and a maximum $(+p\mathscr{E})$ at $\varphi=\pi$. Its first-order effect upon the energy shifts the eigenvalue (43.3) by

$$\Delta_1 E_m = \langle m|V|m\rangle = -\frac{p\mathscr{E}}{2\pi}\oint d\varphi\, e^{-im\varphi}\cos\varphi\, e^{im\varphi}.$$

Since this integral vanishes,

$$\Delta_1 E_m = 0, \tag{43.5}$$

there will only be a second-order Stark effect, proportional to $\mathscr{E}^2$,

$$\Delta_2 E_m = \sum_{m'}{}' \frac{|\langle m'|V|m\rangle|^2}{E_m - E_{m'}}. \tag{43.6}$$

The off-diagonal elements of V occurring in this sum are

$$\begin{aligned}\langle m'|V|m\rangle &= -\frac{p\mathscr{E}}{2\pi}\oint d\varphi\, e^{i(m-m')\varphi}\cos\varphi \\ &= -\tfrac{1}{2}p\mathscr{E}(\delta_{m',m+1}+\delta_{m',m-1}).\end{aligned} \tag{43.7}$$

The sum (43.6), therefore, consists of two terms $m'=m\pm 1$ only:

$$\Delta_2 E_m = \frac{1}{4}p^2\mathscr{E}^2\frac{2\Theta}{\hbar^2}\left\{\frac{1}{m^2-(m-1)^2}+\frac{1}{m^2-(m+1)^2}\right\}$$

or

$$\Delta_2 E_m = \mathscr{E}^2\cdot\frac{p^2\Theta}{\hbar^2}\cdot\frac{1}{4m^2-1}. \tag{43.8}$$

The eigenfunction in this approximation can be determined from the general formula

$$u_m = |m\rangle + \sum_{m'}{}' |m'\rangle \frac{\langle m'|V|m\rangle}{E_m - E_{m'}}; \tag{43.9}$$

again the sum reduces to only two terms:

$$u_m = \frac{1}{\sqrt{2\pi}}\left\{e^{im\varphi} + \frac{1}{2}p\mathscr{E}\left[\frac{1}{2m+1}e^{i(m+1)\varphi} - \frac{1}{2m-1}e^{i(m-1)\varphi}\right]\right\}. \tag{43.10}$$

In the same approximation, then

$$|u_m|^2 = 1 - \frac{2p\mathscr{E}}{4m^2-1}\cos\varphi \tag{43.11}$$

shows a variation along the circle. In the case $m=0$ we have

$$|u_m|^2 = 1 + 2p\mathscr{E}\cos\varphi\,,$$

i.e. the probability to find the dipole in field direction ($\varphi=0$) is a maximum, in the opposite position ($\varphi=\pi$) it is a minimum. This clearly corresponds to the classical static orientation of a dipole. On the other hand, if $m\neq 0$, the sign of the additional term in (43.11) changes and we get the opposite behaviour. Even this may be understood by the classical model: when the dipole rotates, it passes the angles around $\varphi=0$ with the potential minimum at a higher speed than those around the opposite direction. Therefore, if a statistical series of snapshots is obtained, the rotator will be found around $\varphi=0$ less frequently than around $\varphi=\pi$.

NB 1. The states $|m\rangle$ and $|-m\rangle$ are degenerate. It is, however, not necessary to start with linear combinations,

$$|\xi,|m|\rangle = \alpha_{\xi,|m|}|m\rangle + \beta_{\xi,|m|}|-m\rangle, \qquad \xi=1,2,$$

instead of with the functions $|m\rangle$ and $|-m\rangle$ themselves, because the matrix elements $\langle -m|V|m\rangle$ of the perturbation potential vanish so that there can be no coupling between such pairs of states.

NB 2. If a magnetic field $\mathscr{H}$ perpendicular to the plane of rotation is applied, it will cause another potential energy term

$$V_{\text{mag}} = -\mathscr{M}\cdot\mathscr{H}$$

where $\mathscr{M}$ is the magnetic moment connected with angular momentum,

$$\mathscr{M} = \frac{e}{2Mc}\boldsymbol{L}.$$

Hence, with M the mass and e the charge of the rotator,

$$V_{\text{mag}} = -\frac{e}{2Mc}\mathscr{H}\frac{\hbar}{i}\frac{\partial}{\partial\varphi} = i\mu\mathscr{H}\frac{\partial}{\partial\varphi} \quad \text{with } \mu = \frac{e\hbar}{2Mc}.$$

The Schrödinger equation (43.1) then has to be replaced by

$$-\frac{\hbar^2}{2\Theta}\frac{d^2u}{d\varphi^2} + i\mu\mathscr{H}\frac{du}{d\varphi} = Eu$$

with, again, the solutions (43.2), but the eigenvalues

$$E_m = \frac{\hbar^2}{2\Theta}m^2 - \mu\mathscr{H}m$$

no longer degenerate with respect to the sign of m (two-dimensional Zeeman effect).

Problem 44. Ionized hydrogen molecule

To approximate the dissociation energy of $H_2^+ \to H + H^+$ by an electron eigenfunction built up as a linear combination of

$$|a\rangle = \frac{\gamma^{\frac{3}{2}}}{\sqrt{\pi}} e^{-\gamma r_a} \quad \text{and} \quad |b\rangle = \frac{\gamma^{\frac{3}{2}}}{\sqrt{\pi}} e^{-\gamma r_b} \tag{44.1}$$

with r_a and r_b the distances of the electron from the two protons fixed a distance R apart (Fig. 30). The parameter γ may be used as a Ritz parameter.

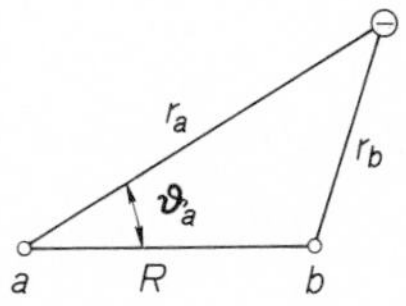

Fig. 30. Coordinates for H_2^+ molecule

The equilibrium distance R is to be determined and the zero-order energy of the relative vibration of the two protons may afterwards be taken into account.

Solution. We start with the two protons fixed at a distance R (Born-Oppenheimer approximation of fixed nuclei); then the Schrödinger equation for the electron motion, in atomic units, runs

$$H u \equiv -\frac{1}{2}\nabla^2 u - \left(\frac{1}{r_a} + \frac{1}{r_b}\right) u = E_e u \tag{44.2}$$

with E_e the energy of the electron. The energy of the molecule for fixed nuclei then is

$$E_{\text{mol}} = \frac{1}{R} + E_e . \tag{44.3}$$

We apply a variational method, making the latter expression a minimum:

$$E_{\text{mol}} = \langle u|H|u\rangle + \frac{1}{R} = \text{Minimum} \tag{44.4}$$

with the constraint

$$\langle u|u\rangle = 1 . \tag{44.5}$$

Using a linear combination

$$u = \alpha|a\rangle + \beta|b\rangle \tag{44.6}$$

where $|a\rangle$ is the same function of r_a as $|b\rangle$ is of r_b, we have with normalized functions,

$$\langle a|a\rangle = \langle b|b\rangle = 1; \qquad \langle a|b\rangle = \langle b|a\rangle = S \tag{44.7}$$

where S is called the *overlap integral*. More explicitly, we then have instead of (44.4) and (44.5),

$$E_{\text{mol}} = (\alpha^2+\beta^2)\langle a|H|a\rangle + 2\alpha\beta\langle b|H|a\rangle + \frac{1}{R} = \text{Min.} \tag{44.8}$$

and

$$(\alpha^2+\beta^2) + 2\alpha\beta S = 1. \tag{44.9}$$

It is evident from symmetry considerations that there are two solutions, a symmetrical one with

$$\alpha = \beta = \frac{1}{\sqrt{2(1+S)}}; \qquad E_e = \frac{\langle a|H|a\rangle + \langle b|H|a\rangle}{1+S} \tag{44.10}$$

and an antisymmetrical one with

$$\alpha = -\beta = \frac{1}{\sqrt{2(1-S)}}; \qquad E_e = \frac{\langle a|H|a\rangle - \langle b|H|a\rangle}{1-S}, \tag{44.11}$$

and that the symmetrical solution belongs to the lower energy level thus leading to the ground state of the molecule. We shall therefore restrict our further considerations to Eq. (44.10).

Using the special normalized functions $|a\rangle$ and $|b\rangle$ defined by Eq. (44.1), we find by direct evaluation

$$H|a\rangle = \left\{-\frac{1}{2}\gamma^2 + \frac{\gamma-1}{r_a} - \frac{1}{r_b}\right\}|a\rangle$$

so that

$$\langle a|H|a\rangle = \frac{\gamma^3}{\pi}\int d\tau\, e^{-2\gamma r_a}\left\{-\frac{1}{2}\gamma^2 + \frac{\gamma-1}{r_a} - \frac{1}{r_b}\right\}, \tag{44.12}$$

$$\langle b|H|a\rangle = \frac{\gamma^3}{\pi}\int d\tau\, e^{-\gamma(r_a+r_b)}\left\{-\frac{1}{2}\gamma^2 + \frac{\gamma-1}{r_a} - \frac{1}{r_b}\right\}. \tag{44.13}$$

Introducing the abbreviations

$$\mathscr{C} = \frac{\gamma^3}{\pi}\int d\tau\,\frac{e^{-2\gamma r_a}}{r_b}; \qquad \mathscr{E} = \frac{\gamma^3}{\pi}\int d\tau\,\frac{e^{-\gamma(r_a+r_b)}}{r_b} \tag{44.14}$$

and, corresponding to the definition (44.7),

$$S = \frac{\gamma^3}{\pi}\int d\tau\, e^{-\gamma(r_a+r_b)} \tag{44.15}$$

we get by elementary considerations

$$\langle a|H|a\rangle = -\tfrac{1}{2}\gamma^2 + \gamma(\gamma-1) - \mathscr{C} \tag{44.16}$$

and

$$\langle b|H|a\rangle = -\tfrac{1}{2}\gamma^2 S + (\gamma-2)\mathscr{E} \tag{44.17}$$

so that

$$E_{\text{mol}} = -\frac{1}{2}\gamma^2 + \frac{\gamma(\gamma-1) - \mathscr{C} + (\gamma-2)\mathscr{E}}{1+S} + \frac{1}{R}. \tag{44.18}$$

It remains to evaluate the integrals $\mathscr{C}$, $\mathscr{E}$, and S. The integral $\mathscr{C}$, the so-called *classical interaction* integral, describes the Coulomb repulsion by proton b of the electron charge centred according to the wave function $|a\rangle$ about the proton a. It can be computed by expanding (cf. Fig. 30)

$$\frac{1}{r_b} = \begin{cases} \dfrac{1}{R}\displaystyle\sum_{n=0}^{\infty}\left(\frac{r_a}{R}\right)^n P_n(\cos\vartheta_a) & \text{for } r_a < R\,, \\ \dfrac{1}{r_a}\displaystyle\sum_{n=0}^{\infty}\left(\frac{R}{r_a}\right)^n P_n(\cos\vartheta_a) & \text{for } r_a > R\,. \end{cases}$$

Only the term $n=0$ of this series then contributes to $\mathscr{C}$ which may be evaluated by elementary integrations:

$$\mathscr{C} = \frac{1}{R}\left[1-(1+\gamma R)e^{-2\gamma R}\right]. \tag{44.19}$$

The integral $\mathscr{E}$ is called the *exchange integral*. It has no classical equivalent and is a consequence of the symmetrization of the eigenfunction (exchange of the places at a and at b of the electron). It can be evaluated in prolate ellipsoidal coordinates over the two protons as foci, putting

$$r_a = \frac{R}{2}(\xi+\eta), \qquad r_b = \frac{R}{2}(\xi-\eta)$$

and using besides ξ and η the angle φ about the molecular axis as the third coordinate. Then the volume element is

$$d\tau = \left(\frac{R}{2}\right)^3 (\xi^2-\eta^2)\, d\xi\, d\eta\, d\varphi$$

and integration is to be performed within the bounds $1 \leq \xi \leq \infty$, $-1 \leq \eta \leq +1$, $0 \leq \varphi < 2\pi$. The same integration method may be applied to the overlap integral S. The results are

$$\mathscr{E} = \gamma(1+\gamma R)\,\mathrm{e}^{-\gamma R} \tag{44.20}$$

and

$$S = (1+\gamma R + \tfrac{1}{3}\gamma^2 R^2)\,\mathrm{e}^{-\gamma R}. \tag{44.21}$$

If we now write down the energy E_{mol} according to (44.18) we get a function of the two parameters γ and R. It is more convenient to use, instead, γ and

$$\rho = \gamma R \tag{44.22}$$

as Ritz parameters to minimize the energy. Using besides S, Eq. (44.21), instead of $\mathscr{C}$ and $\mathscr{E}$ the expressions

$$\mathscr{C}' = \frac{1}{\rho}\left[1-(1+\rho)\,\mathrm{e}^{-2\rho}\right] \tag{44.23}$$

and

$$\mathscr{E}' = (1+\rho)\,\mathrm{e}^{-\rho} \tag{44.24}$$

which all three depend upon only the parameter ρ, we arrive at

$$E_{\text{mol}} = -\frac{1}{2}\gamma^2 + \frac{\gamma(\gamma-1) - \gamma\mathscr{C}' + \gamma(\gamma-2)\mathscr{E}'}{1+S} + \frac{\gamma}{\rho}. \tag{44.25}$$

This is of the form

$$E_{\text{mol}} = f(\rho)\,\gamma^2 + g(\rho)\,\gamma$$

so that from

$$\frac{\partial E_{\text{mol}}}{\partial \gamma} = 2\gamma f(\rho) + g(\rho) = 0$$

we may determine the optimum

$$\gamma = -\frac{g(\rho)}{2f(\rho)}$$

yielding

$$E_{\text{mol}} = -\frac{g(\rho)^2}{4f(\rho)}.$$

We then have to determine the minimum of this expression as a function of the only variable ρ.

The numerical computations are not difficult but a little tedious. The main results have been assembled in the accompanying table.

Table of results

$\gamma R = \rho$	γ	$-E_{\text{mol}}$	R
1.0	0.832	0.300	1.200
1.5	1.113	0.500	1.345
2.0	1.216	0.573	1.645
2.3	1.235	0.5848	1.86
2.5	1.240	0.5865	2.02
2.8	1.228	0.5827	2.28
3.0	1.216	0.578	2.47
3.5	1.180	0.560	2.96

The minimum can be fixed even more accurately than shown in the table, if an energy parabola is put through the three points at $\rho = 2.3$, 2.5 and 2.8. We find by interpolation

$$E_{\text{mol}} = 0.0380(R - 2.08)^2 - 0.5866\,, \tag{44.26}$$

i.e. the energy minimum lies at the equilibrium distance

$$R_0 = 2.08 \text{ a.u.} = 1.10\,\text{Å}$$

and has a depth of

$$V_0 = -0.5866 \text{ a.u.}$$

In order to find the dissociation energy D we have to take account of the zero-point energy $\frac{1}{2}\hbar\omega$ of the molecular vibration in the nuclear oscillator potential (44.26) and to subtract from the binding energy of the molecule that of the hydrogen atom, equal to $-\frac{1}{2}$:

$$D = -\tfrac{1}{2} - (V_0 + \tfrac{1}{2}\hbar\omega) = 0.0866 - \tfrac{1}{2}\hbar\omega = 2.36\,\text{eV} - \tfrac{1}{2}\hbar\omega\,.$$

The zero-point energy may be roughly computed by considering only the harmonic approximation (44.26) to the potential with the obvious identification $\frac{1}{2}M\omega^2 = 0.0380$ (cf. Problem 30) and M the reduced mass of the vibration, i.e. one-half the proton mass. This leads, in atomic units, to $\hbar\omega = 0.00913$ and thus to a zero-point energy of

$$\tfrac{1}{2}\hbar\omega = 0.00457 = 0.124\,\text{eV}\,.$$

We thus arrive at a dissociation energy of $D = 2.24$ eV. This approximation is to be compared with the experimental value of 2.65 eV. It seems a rather good approximation considering that we have made a theory to approach not D but V_0; the experimental value of V_0 would be $V_0 = -0.6017$, so that our theoretical value of $V_0 = -0.5866$ is wrong by only 2.6 per cent.

Problem 45. Oblique incidence of a plane wave

A potential step may be idealized to

$$V = \begin{cases} 0 & \text{for } x < 0, \\ V_0 & \text{for } x > 0. \end{cases}$$

The laws of reflection and refraction shall be derived for an obliquely incident wave, and the intensities of reflected and refracted particle beam determined. Special consideration should be given to the case of total reflection.

Solution. We shall use throughout the abbreviations

$$\frac{2mE}{\hbar^2} = k^2\,; \qquad \frac{2m(E-V_0)}{\hbar^2} = K^2\,. \tag{45.1}$$

Let us first suppose $V_0 < E$, i.e. K to be real. (This includes the case of negative V_0.) In the half-space $x < 0$ the solution of the Schrödinger equation is, in standard normalization,

$$u = e^{i\boldsymbol{k}\boldsymbol{r}} + R\,e^{i\boldsymbol{k}'\boldsymbol{r}} \qquad (x < 0) \tag{45.2a}$$

with $\boldsymbol{k} = (k_x, k_y, k_z)$ the wave vector of the incident, R and $\boldsymbol{k}' = (k'_x, k'_y, k'_z)$ amplitude and wave vector of the reflected wave. In the half-space $x > 0$ we just have the transmitted wave

$$u = T\,e^{i\boldsymbol{K}\boldsymbol{r}} \qquad (x > 0) \tag{45.2b}$$

of amplitude T and wave vector $\boldsymbol{K} = (K_x, K_y, K_z)$. Here we have

$$\boldsymbol{k}^2 = \boldsymbol{k}'^2 = k^2; \qquad \boldsymbol{K}^2 = K^2\,. \tag{45.3}$$

At the surface $x = 0$ the two functions u and $\partial u/\partial x$ of the variables y and z must be the same if computed either from (45.2a) or (45.2b). It follows that their phases

$$k_y y + k_z z = k'_y y + k'_z z = K_y y + K_z z \tag{45.4}$$

as well as the amplitudes of u and $\partial u/\partial x$, viz.

$$1 + R = T; \qquad k_x + R k'_x = T K_x \tag{45.5}$$

must be equal.

We first investigate Eq. (45.4). The discussion is very much simplified if the z axis is chosen so that $k_z=0$, i. e. perpendicular to the plane of incidence. Then, according to (45.4), k'_z and K_z also vanish so that both vectors, $\boldsymbol{k}'$ and $\boldsymbol{K}$ lie in the y,z-plane. This is shown in Fig. 31, where the angles α, α', β are defined. They permit to write the y components of (45.4) in the form

$$k \sin\alpha = k \sin\alpha' = K \sin\beta .$$

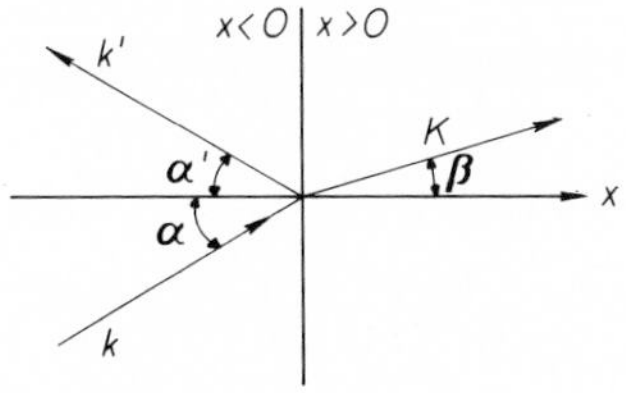

Fig. 31. Definitions of wave vectors and angles for oblique incidence

Hence, there follows the law of reflection, $\alpha=\alpha'$, and the Snellius refraction law,

$$\frac{\sin\alpha}{\sin\beta}=\frac{K}{k} \tag{45.6a}$$

allowing the definition of a refractive index,

$$n=\frac{K}{k}=\sqrt{1-\frac{V_0}{E}} \tag{45.6b}$$

in the same way as in optics.

From Eq. (45.5) we obtain the amplitudes T and R, viz.

$$\left.\begin{aligned} R&=\frac{K_x-k_x}{k'_x-K_x}=\frac{K\cos\beta-k\cos\alpha}{-k\cos\alpha-K\cos\beta}=-\frac{\sin(\alpha-\beta)}{\sin(\alpha+\beta)};\\ T&=\frac{k'_x-k_x}{k'_x-K_x}=\frac{-k\cos\alpha-k\cos\alpha}{-k\cos\alpha-K\cos\beta}=\frac{2\cos\alpha\sin\beta}{\sin(\alpha+\beta)}. \end{aligned}\right\} \tag{45.7}$$

These amplitudes gives access to the intensities which we define as current densities of the three waves,

$$s=\frac{\hbar}{2im}(u^*\nabla u-u\nabla u^*). \tag{45.8}$$

In the half-space of incidence ($x<0$) there follows from (45.2a)

$$\nabla u = i\boldsymbol{k}\,e^{i\boldsymbol{k}\boldsymbol{r}} + R\,i\boldsymbol{k}'\,e^{i\boldsymbol{k}'\boldsymbol{r}}$$

and thence by straightforward calculation the vector

$$\boldsymbol{s} = \frac{\hbar}{m}\{\boldsymbol{k} + R^2\boldsymbol{k}' + R(\boldsymbol{k}+\boldsymbol{k}')\cos(\boldsymbol{k}-\boldsymbol{k}')\boldsymbol{r}\}$$

with components

$$\left.\begin{aligned} s_x &= \frac{\hbar}{m}\,k\cos\alpha\,(1-R^2);\\ s_y &= \frac{\hbar}{m}\,k\sin\alpha\,\{1+R^2+2R\cos(2kx\cos\alpha)\}. \end{aligned}\right\} \tag{45.9}$$

On the other side of the potential discontinuity ($x>0$), we get from (45.2b),

$$\boldsymbol{s} = \frac{\hbar}{m}\,T^2\boldsymbol{K}$$

with components

$$s_x = \frac{\hbar}{m}\,K\cos\beta\cdot T^2; \qquad s_y = \frac{\hbar}{m}\,K\sin\beta\cdot T^2. \tag{45.10}$$

Continuity of the flux requires that s_x is the same on both sides of the discontinuity, or that

$$k\cos\alpha(1-R^2) = K\cos\beta\,T^2.$$

Using Eqs. (45.6a) and (45.7) it can easily be shown that this is indeed an identity. The component s_y, on the other hand, is not subject to any such condition except, of course, that it cannot depend upon y. It is interesting, however, that it depends upon x in consequence of the interference between incident and reflected waves. This periodic variation with the distance before the step nowhere leads to a change of sign since the curly bracket in (45.9) oscillates between $(1-R)^2$ and $(1+R)^2$, both positive.

If an average is taken over a domain of several wavelengths, the contribution of the interference terms will cancel out so that the three currents incident, reflected, and transmitted, become

$$\boldsymbol{s}_0 = \frac{\hbar}{m}\,\boldsymbol{k}; \qquad \boldsymbol{s}_r = \frac{\hbar}{m}\,R^2\boldsymbol{k}'; \qquad \boldsymbol{s}_t = \frac{\hbar}{m}\,T^2\boldsymbol{K}. \tag{45.11}$$

The absolute values of these vectors being called intensities, the ratio $s_r/s_0 = R^2$ becomes the reflection coefficient of optics, and $s_t/s_0 = nT^2$ the transmission coefficient.

Total reflection will occur for $V_0 > E$, i. e. if K, Eq. (45.1) is imaginary,

$$K = i\varkappa.$$

Since Eq. (45.4) still must hold, so that $K_y = k\sin\alpha$ and $K_z = 0$, it follows that

$$K_x^2 = K^2 - K_y^2 - K_z^2 = -(\varkappa^2 + k^2\sin^2\alpha)$$

is negative, $K_x = i\varkappa_x$ being imaginary. Eq. (45.5) then may be written

$$1 + R = T; \qquad k_x + R k_x' = T i\varkappa_x. \tag{45.5'}$$

The law of reflection is the same as before. There is, of course, no law of refraction now, and instead of (45.7) we get

$$R = \frac{i\varkappa_x - k\cos\alpha}{-k\cos\alpha - i\varkappa_x}; \qquad T = \frac{-2k\cos\alpha}{-k\cos\alpha - i\varkappa_x}. \tag{45.7'}$$

Thence $|R|^2 = 1$, so that the reflected intensity equals the incident one, in agreement with the behaviour of any waves in the case of total reflection.

NB. The considerations of this problem have the same structure as those of optics. The theory of a light wave, incident upon a refracting surface, is much more complicated, however, because of the vector character of light, whereas here we are simply dealing with scalar waves. Eq. (45.7) corresponds to Fresnel's formula for the electrical vector component perpendicular to the plane of incidence.

Problem 46. Symmetrical top

The rotations of a large class of molecules may be dealt with by treating the molecule as a rigid body, neglecting vibrations of the nuclei or electronic motion. Let the molecule have the shape of a symmetric top with the moments of inertia C about its figure axis, and A about all axes perpendicular to it through the centre of mass. The Schrödinger equation for free rotations shall then be derived from the classical hamiltonian, and the energy eigenvalues be determined.

Solution. If the position of the figure axis is described by polar angles ϑ and φ, and rotations of the top about this axis by an angle ψ, the classical expression of the kinetic energy is[6]

$$T_{\text{class}} = \tfrac{1}{2}\{A(\dot\varphi^2\sin^2\vartheta + \dot\vartheta^2) + C(\dot\varphi\cos\vartheta + \dot\psi)^2\}. \tag{46.1}$$

[6] For derivation of this expression, see textbooks of classical mechanics, e. g. the author's *Lehrbuch der theoretischen Physik*, vol. II, pp. 50, 75.

The three Eulerian angles ϑ, φ, ψ are generalized coordinates and, since T_{class} is a quadratic form in $\dot\vartheta, \dot\varphi, \dot\psi$, the method for its translation into quantum theory given in Problem 13 may be applied. The metrical tensor there defined has the covariant components

$$\left.\begin{array}{lll} g_{\vartheta\vartheta}=A, & g_{\vartheta\varphi}=0, & g_{\vartheta\psi}=0, \\ g_{\varphi\vartheta}=0, & g_{\varphi\varphi}=A\sin^2\vartheta+C\cos^2\vartheta, & g_{\varphi\psi}=C\cos\vartheta, \\ g_{\psi\vartheta}=0, & g_{\psi\varphi}=C\cos\vartheta, & g_{\psi\psi}=C \end{array}\right\} \tag{46.2}$$

with the determinant

$$g=A^2 C\sin^2\vartheta. \tag{46.3}$$

The contravariant components then follow from $g^{ik}=G_{ik}/g$:

$$\left.\begin{array}{lll} g^{\vartheta\vartheta}=\dfrac{1}{A}, & g^{\vartheta\varphi}=0, & g^{\vartheta\psi}=0, \\ g^{\varphi\vartheta}=0, & g^{\varphi\varphi}=\dfrac{1}{A\sin^2\vartheta}, & g^{\varphi\psi}=-\dfrac{\cos\vartheta}{A\sin^2\vartheta}, \\ g^{\psi\vartheta}=0, & g^{\psi\varphi}=-\dfrac{\cos\vartheta}{A\sin^2\vartheta}, & g^{\psi\psi}=\dfrac{1}{C}+\dfrac{\cos^2\vartheta}{A\sin^2\vartheta}. \end{array}\right\} \tag{46.4}$$

According to (13.11), the operator of kinetic energy may then be obtained by straightforward computation. We find

$$\begin{aligned} T=-\frac{\hbar^2}{2A}\Big\{&\frac{1}{\sin\vartheta}\frac{\partial}{\partial\vartheta}\left(\sin\vartheta\frac{\partial}{\partial\vartheta}\right)+\frac{1}{\sin^2\vartheta}\frac{\partial^2}{\partial\varphi^2}-\frac{2\cos\vartheta}{\sin^2\vartheta}\frac{\partial^2}{\partial\varphi\,\partial\psi} \\ &+\left(\frac{A}{C}+\frac{\cos^2\vartheta}{\sin^2\vartheta}\right)\frac{\partial^2}{\partial\psi^2}\Big\}. \end{aligned} \tag{46.5}$$

The Schrödinger equation for the force-free spinning top then becomes simply

$$Tu=Eu \tag{46.6}$$

since there is no external potential energy, so that T is identical with the hamiltonian.

Eq. (46.6) with the operator T, Eq. (46.5), may be solved by factorization:

$$u=f(\vartheta)\,e^{i(M\varphi+K\psi)}. \tag{46.7}$$

Since the angles φ and ψ both run through an interval 2π, the quantum numbers M and K must be integers to make u unique. They can very simply be interpreted: $\hbar M$ is the component of angular momentum in the direction of the space-fixed z axis (polar axis) and $\hbar K$ in the body-fixed direction of the figure axis.

With (46.7) for the wave function, we obtain for $f(\vartheta)$ the differential equation

$$-\frac{\hbar^2}{2A}\left\{\frac{1}{\sin\vartheta}\frac{d}{d\vartheta}\left(\sin\vartheta\frac{d}{d\vartheta}\right)-\frac{M^2}{\sin^2\vartheta}+\frac{2MK\cos\vartheta}{\sin^2\vartheta}\right.$$
$$\left.-\left(\frac{A}{C}+\frac{\cos^2\vartheta}{\sin^2\vartheta}\right)K^2\right\}f=Ef. \tag{46.8}$$

Using the abbreviation

$$\lambda=\frac{2A}{\hbar^2}E-K^2\left(\frac{A}{C}-1\right) \tag{46.9}$$

and introducing the variable

$$s=\frac{1}{2}(1+\cos\vartheta)=\cos^2\frac{\vartheta}{2};\qquad 0\le s\le 1, \tag{46.10}$$

the differential equation (46.8) is transformed into

$$s(1-s)\frac{d}{ds}\left[s(1-s)\frac{df}{ds}\right]+\left[\lambda s(1-s)-\tfrac{1}{4}(M+K)^2+MKs\right]f=0. \tag{46.11}$$

Instead of the quantum numbers M and K we then use the positive numbers

$$p=\tfrac{1}{2}|M+K|;\qquad q=\tfrac{1}{2}|M-K| \tag{46.12}$$

with

$$\tfrac{1}{4}(M+K)^2=p^2;\qquad MK=p^2-q^2.$$

Then, by writing

$$f(s)=s^p(1-s)^q F(s), \tag{46.13}$$

Eq. (46.11) is transformed into the differential equation of the hypergeometric series:

$$s(1-s)F''+[(2p+1)-(2p+2q+2)s]F'-[(p+q)(p+q+1)-\lambda]F=0 \tag{46.14}$$

solved by the function

$$F={}_2F_1(p+q+1+J,\,p+q-J,\,2p+1;\,s) \tag{46.15}$$

regular at $s=0$ ($\vartheta=\pi$) with

$$\lambda=J(J+1) \tag{46.16}$$

or, according to (46.9),

$$E=\frac{\hbar^2}{2A}J(J+1)+\frac{\hbar^2}{2}\left(\frac{1}{C}-\frac{1}{A}\right)K^2. \tag{46.17}$$

The second solution of (46.14) would be singular at $s=0$, viz. $F\propto s^{-2p}$ and $f\propto s^{-p}$ and can thus be excluded.

It should be emphasized that J so far is just another abbreviation and need not be an integer so that (46.1) as yet does not describe well-defined energy levels. We have, however, not yet used the second boundary condition. At $s=1$ ($\vartheta=0$) the hypergeometric series (46.15) makes $F\propto(1-s)^{-2q}$ and $f\propto(1-s)^{-q}$ singular if the series (46.15) does not break off. This is the case only if either the first or the second parameter in (46.15) is a negative integer, and since the first per definitionem is positive, it follows that

$$p+q-J=-n, \qquad n=1,2,3,\ldots \tag{46.18}$$

will make F a polynomial of degree n and the wave function finite at all physical angles. Now it follows from (46.12) that $p+q$ is a positive integer equal to either $|K|$ or $|M|$, according to which of the two is bigger, so that $J=n+p+q$ is an integer never smaller than the bigger of the two. This is reasonable if J is the total angular momentum of the top (which we have not yet shown), because a vector can never be smaller than any of its components. It is these results only which make Eq. (46.17) finally an eigenvalue formula.

That J really is the total angular momentum quantum number can most easily be seen from the classical expression

$$T_{\text{class}}=\frac{1}{2A}\left\{\boldsymbol{J}^2+\left(\frac{A}{C}-1\right)\boldsymbol{K}^2\right\}$$

which, of course, corresponds to (46.17).

NB 1. The eigenfunctions are, in standard notation, written

$$\mathscr{D}^J_{MK}=\frac{1}{2\pi}\,e^{i(M\varphi+K\psi)}\,d^J_{MK}(\vartheta) \tag{46.19}$$

with

$$d^J_{MK}=N^J_{MK}\,s^p(1-s)^q\,{}_2F_1(p+q+J+1,\,p+q-J,\,2p+1;\,s). \tag{46.20}$$

They are normalized so that

$$\int_0^\pi d\vartheta\sin\vartheta\int_0^{2\pi}d\varphi\int_0^{2\pi}d\psi\,|\mathscr{D}^J_{MK}|^2=1 \tag{46.21}$$

and

$$\int_0^\pi d\vartheta\sin\vartheta\,|d^J_{MK}|^2=1, \tag{46.22}$$

the latter leading to

$$N^J_{MK}=\frac{\sqrt{2J+1}}{2(2p)!}\sqrt{\frac{(J+p+q)!\,(J+p-q)!}{(J-p+q)!\,(J-p-q)!}}\,. \tag{46.23}$$

These functions are remarkable from a mathematical point of view as the matrices $\mathscr{D}^J$ with the elements $\mathscr{D}^J_{MK}$ form a $2J+1$ dimensional representation of the rotational group in 3-space. See also Problem 55.

NB 2. The method by which the operator T, Eq. (46.5), has been constructed has by no means been proved to be applicable to the rigid body, since in Problem 13 only the use of generalized coordinates for a system of point masses had been treated. What is even worse, there can be no rigid body in quantum mechanics because rigidity means a constraint to the internal motion of the constituents which leads to an infinitely large zero point energy. Only by a correct separation of these internal motions from the rotations of the system can the problem be solved. This has been done[7] in a paper by S. Flügge and A. Weiguny (Z. Physik **171,** 171, 1963) where the separation and the limiting process to infinitely small vibrations of the constituent particles about equilibrium positions have been performed.

The expression (46.5) and the quantum theory of the symmetrical top had, of course, already been found much earlier, their first treatment being due to F. Reiche, Z. Physik, **39,** 444 (1926).

C. The Angular Momentum

Problem 47. Infinitesimal rotation

To show that, under an infinitesimal rotation of the coordinate system, a scalar function $f(\boldsymbol{r})$ transforms in a way simply connected with angular momentum operators.

Solution. Let us describe the coordinate rotation by

$$\boldsymbol{r}' = R\,\boldsymbol{r}; \qquad R = 1 + A \tag{47.1a}$$

with R being a 3×3 matrix with

$$A = \begin{pmatrix} 0 & \varepsilon_z & -\varepsilon_y \\ -\varepsilon_z & 0 & \varepsilon_x \\ \varepsilon_y & -\varepsilon_x & 0 \end{pmatrix}. \tag{47.1b}$$

Then the three infinitesimal quantities ε_j are the angles of rotation with respect to the three axes x, y, z. We remark that the inverse transformation is obtained either by inversing the signs of all three ε_j or by using the transposed matrix $\tilde{A}$. Hence,

$$\tilde{A} = -A. \tag{47.1c}$$

[7] The problem was first attacked by Welker, H.: Z. Physik **101,** 95 (1936).

A function $f(\boldsymbol{r})$ will, under the transformation (47.1a–c), turn into a function $f'(\boldsymbol{r}')$ of the new coordinates. The latter function can be expressed in terms of the original coordinates $\boldsymbol{r}$ by a Taylor expansion,

$$f'(\boldsymbol{r}')=f'(\boldsymbol{r})+(\delta\boldsymbol{r}\cdot\nabla)f'(\boldsymbol{r}) \tag{47.2}$$

with $\delta\boldsymbol{r}=\boldsymbol{r}'-\boldsymbol{r}=A\boldsymbol{r}$, and higher-order derivatives neglected. Now, one finds from (47.1b) that

$$A\boldsymbol{r}=\begin{pmatrix}\varepsilon_z y-\varepsilon_y z\\ -\varepsilon_z x+\varepsilon_x z\\ \varepsilon_y x-\varepsilon_x y\end{pmatrix} \tag{47.3}$$

so that (47.2) becomes

$$f'(\boldsymbol{r}')=f'(\boldsymbol{r})+\left\{(\varepsilon_z y-\varepsilon_y z)\frac{\partial}{\partial x}+(-\varepsilon_z x+\varepsilon_x z)\frac{\partial}{\partial y}+(\varepsilon_y x-\varepsilon_x y)\frac{\partial}{\partial z}\right\}f'(\boldsymbol{r}).$$

Ordering the terms with respect to the ε_j's and introducing the angular momentum operators

$$L_x=\frac{\hbar}{i}\left(y\frac{\partial}{\partial z}-z\frac{\partial}{\partial y}\right)\quad \text{etc. cyclic} \tag{47.4}$$

we get, with $\boldsymbol{\varepsilon}=(\varepsilon_x,\varepsilon_y,\varepsilon_z)$,

$$f'(\boldsymbol{r}')=f'(\boldsymbol{r})-\frac{i}{\hbar}(\boldsymbol{\varepsilon}\cdot\mathbf{L})f'(\boldsymbol{r}). \tag{47.5}$$

Since the last term is infinitesimal, the infinitesimal difference between $f'(\boldsymbol{r})$ and $f'(\boldsymbol{r}')$ may there be neglected so that (47.5) may be rewritten in the form

$$f'(\boldsymbol{r})=\left\{1+\frac{i}{\hbar}(\boldsymbol{\varepsilon}\cdot\mathbf{L})\right\}f'(\boldsymbol{r}'). \tag{47.6}$$

So far, no use has been made of the function f being a scalar, i.e. of its invariance under rotation:

$$f'(\boldsymbol{r}')=f(\boldsymbol{r}) \tag{47.7}$$

which permits reducing Eq. (47.6) entirely to the original coordinates $\boldsymbol{r}$ in the form

$$f'(\boldsymbol{r})=\left\{1+\frac{i}{\hbar}(\boldsymbol{\varepsilon}\cdot\mathbf{L})\right\}f(\boldsymbol{r}). \tag{47.8}$$

Problem 48. Components in polar coordinates

To use the solution of the preceding problem in order to find the angular momentum components in spherical polar coordinates by infinitesimal rotations about the x, y, and z axes.

Solution. Writing ψ instead of f', Eq. (47.6) may be written in the form

$$\psi(\boldsymbol{r}') = \left\{1 - \frac{i}{\hbar}(\boldsymbol{\varepsilon}\cdot\boldsymbol{L})\right\}\psi(\boldsymbol{r}). \tag{48.1}$$

We first perform a rotation about the z axis. It corresponds to a transformation $r'=r$, $\vartheta'=\vartheta$, $\varphi'=\varphi-\varepsilon_z$. Thus it follows by Taylor expansion that

$$\psi(\boldsymbol{r}') = \psi(\boldsymbol{r}) + (\varphi'-\varphi)\frac{\partial\psi}{\partial\varphi} = \left\{1 - \varepsilon_z\frac{\partial}{\partial\varphi}\right\}\psi(\boldsymbol{r}).$$

On the other hand, according to (48.1),

$$\psi(\boldsymbol{r}') = \left\{1 - \frac{i}{\hbar}\varepsilon_z L_z\right\}\psi(\boldsymbol{r}),$$

so that there follows by comparison,

$$L_z = \frac{\hbar}{i}\frac{\partial}{\partial\varphi}. \tag{48.2}$$

Next, we rotate the coordinate system by ε_x about the x axis. Then again, $r'=r$, but both ϑ and φ will be changed into

$$\vartheta' = \vartheta + \delta\vartheta; \qquad \varphi' = \varphi + \delta\varphi.$$

In rectangular coordinates, according to (47.1a, b), this rotation is described by

$$x' = x; \qquad y' = y + \varepsilon_x z; \qquad z' = z - \varepsilon_x y.$$

Now,

$$z = r\cos\vartheta; \qquad y = r\sin\vartheta\sin\varphi$$

and

$$z' = r\cos\vartheta' = r(\cos\vartheta - \delta\vartheta\sin\vartheta),$$

whereas, on the other hand,

$$z' = z - \varepsilon_x y = r(\cos\vartheta - \varepsilon_x\sin\vartheta\sin\varphi)$$

so that by comparison we arrive at

$$\delta\vartheta = \varepsilon_x\sin\varphi. \tag{48.3}$$

In the same way from, on the one hand,

$$y' = r\sin\vartheta'\sin\varphi' = r(\sin\vartheta\sin\varphi + \delta\vartheta\cos\vartheta\sin\varphi + \delta\varphi\sin\vartheta\cos\varphi)$$

and, on the other hand,

$$y' = y + \varepsilon_x z = r(\sin\vartheta\sin\varphi + \varepsilon_x\cos\vartheta)$$

we obtain

$$\delta\vartheta \cos\vartheta \sin\varphi + \delta\varphi \sin\vartheta \cos\varphi = \varepsilon_x \cos\vartheta$$

and, using (48.3) for $\delta\vartheta$,

$$\delta\varphi = \varepsilon_x \cos\varphi \cot\vartheta. \tag{48.4}$$

We then have, with (48.3) and (48.4), the Taylor expansion

$$\begin{aligned}\psi(\boldsymbol{r}') &= \psi(\boldsymbol{r}) + \frac{\partial\psi}{\partial\vartheta}\delta\vartheta + \frac{\partial\psi}{\partial\varphi}\delta\varphi \\ &= \psi(\boldsymbol{r}) + \varepsilon_x\left(\sin\varphi\frac{\partial}{\partial\vartheta} + \cos\varphi\cot\vartheta\frac{\partial}{\partial\varphi}\right)\psi(\boldsymbol{r})\end{aligned}$$

and from (48.1),

$$\psi(\boldsymbol{r}') = \psi(\boldsymbol{r}) - \frac{i}{\hbar}\varepsilon_x L_x \psi(\boldsymbol{r})$$

so that by comparison it follows that

$$L_x = -\frac{\hbar}{i}\left(\sin\varphi\frac{\partial}{\partial\vartheta} + \cos\varphi\cot\vartheta\frac{\partial}{\partial\varphi}\right). \tag{48.5}$$

Lastly, we rotate the coordinate system by ε_y about the y axis. Again we have changes in ϑ and φ to be described by a $\delta\vartheta$ and a $\delta\varphi$ which, of course, have now another meaning determined from

$$\begin{aligned}x' &= r\sin\vartheta'\cos\varphi' = r(\sin\vartheta\cos\varphi + \delta\vartheta\cos\vartheta\cos\varphi - \delta\varphi\sin\vartheta\sin\varphi) \\ &= x - \varepsilon_y z = r(\sin\vartheta\cos\varphi - \varepsilon_y\cos\vartheta)\end{aligned} \tag{48.6a}$$

and

$$\begin{aligned}z' &= r\cos\vartheta' = r(\cos\vartheta - \delta\vartheta\sin\vartheta) \\ &= z + \varepsilon_y x = r(\cos\vartheta + \varepsilon_y\sin\vartheta\cos\varphi).\end{aligned} \tag{48.6b}$$

From (48.6b) we obtain

$$\delta\vartheta = -\varepsilon_y\cos\varphi \tag{48.7}$$

and then from (48.6a)

$$\delta\varphi = \varepsilon_y\sin\varphi\cot\vartheta. \tag{48.8}$$

Therefore,

$$\psi(\boldsymbol{r}') = \psi(\boldsymbol{r}) + \varepsilon_y\left\{-\cos\varphi\frac{\partial}{\partial\vartheta} + \sin\varphi\cot\vartheta\frac{\partial}{\partial\varphi}\right\}\psi(\boldsymbol{r})$$

which, according to (48.1), must agree with

$$\psi(\boldsymbol{r}') = \psi(\boldsymbol{r}) - \frac{i}{\hbar}\,\varepsilon_y L_y \psi(\boldsymbol{r})$$

so that

$$L_y = -\frac{\hbar}{i}\left(-\cos\varphi\,\frac{\partial}{\partial\vartheta} + \sin\varphi\cot\vartheta\,\frac{\partial}{\partial\varphi}\right). \tag{48.9}$$

It is usual to combine (48.5) and (48.9) to the more convenient expressions

$$L^+ = L_x + iL_y = -\frac{\hbar}{i}\,e^{i\varphi}\left(-i\frac{\partial}{\partial\vartheta} + \cot\vartheta\,\frac{\partial}{\partial\varphi}\right); \tag{48.10a}$$

$$L^- = L_x - iL_y = -\frac{\hbar}{i}\,e^{-i\varphi}\left(i\frac{\partial}{\partial\vartheta} + \cot\vartheta\,\frac{\partial}{\partial\varphi}\right). \tag{48.10b}$$

Eqs. (48.2) and either (48.5), (48.9) or (48.10a, b) form the solution to our problem.

Problem 49. Angular momentum and Laplacian

To express the operator L^2 in spherical polar coordinates and to compare it with the Laplacian and the kinetic energy operators.

Solution. Using the operators L^+ and L^- defined in (48.10a, b) we may write

$$L^2 = \tfrac{1}{2}(L^+L^- + L^-L^+) + L_z^2. \tag{49.1}$$

Now,

$$L^+L^- = -\hbar^2 e^{i\varphi}\left(-i\frac{\partial}{\partial\vartheta} + \cot\vartheta\,\frac{\partial}{\partial\varphi}\right)e^{-i\varphi}\left(i\frac{\partial}{\partial\vartheta} + \cot\vartheta\,\frac{\partial}{\partial\varphi}\right).$$

Using first the commutation relation

$$\frac{\partial}{\partial\varphi}\,e^{-i\varphi} = e^{-i\varphi}\left(-i + \frac{\partial}{\partial\varphi}\right)$$

in order to eliminate the exponentials, and then

$$\frac{\partial}{\partial\vartheta}\cot\vartheta = -\frac{1}{\sin^2\vartheta} + \cot\vartheta\,\frac{\partial}{\partial\vartheta},$$

we arrive at

$$L^+L^- = -\hbar^2\left(\frac{\partial^2}{\partial\vartheta^2} + \cot\vartheta\,\frac{\partial}{\partial\vartheta} + \cot^2\vartheta\,\frac{\partial^2}{\partial\varphi^2} + i\frac{\partial}{\partial\varphi}\right). \tag{49.2}$$

The operator $L^- L^+$ can in the same way be determined with the opposite sign of the last term, so that according to (49.1)

$$L^2 = -\hbar^2 \left\{ \frac{\partial^2}{\partial \vartheta^2} + \cot\vartheta \frac{\partial}{\partial \vartheta} + (1+\cot^2\vartheta) \frac{\partial^2}{\partial \varphi^2} \right\}$$

or

$$L^2 = -\hbar^2 \left\{ \frac{1}{\sin\vartheta} \frac{\partial}{\partial \vartheta} \left(\sin\vartheta \frac{\partial}{\partial \vartheta} \right) + \frac{1}{\sin^2\vartheta} \frac{\partial^2}{\partial \varphi^2} \right\}. \tag{49.3}$$

In the latter form, the bracket is identical with the angular part of the Laplacian so that we may write

$$\nabla^2 = \frac{\partial^2}{\partial r^2} + \frac{2}{r} \frac{\partial}{\partial r} - \frac{1}{\hbar^2 r^2} L^2 . \tag{49.4}$$

The kinetic energy of a particle, as we know from classical mechanics, may be written

$$\frac{p^2}{2m} = \frac{1}{2m} \left(p_r^2 + \frac{L^2}{r^2} \right).$$

Since it corresponds to the quantum mechanical operator $-\dfrac{\hbar^2}{2m}\nabla^2$, Eq. (49.4) leads to

$$p_r^2 = -\hbar^2 \nabla^2 - \frac{L^2}{r^2} = -\hbar^2 \left(\frac{\partial^2}{\partial r^2} + \frac{2}{r} \frac{\partial}{\partial r} \right), \tag{49.5}$$

i.e. to the radial part of the Laplacian. The operator corresponding to the classical radial momentum p_r then follows by factorization of (49.5):

$$p_r = \frac{\hbar}{i} \left(\frac{\partial}{\partial r} + \frac{1}{r} \right) \tag{49.6}$$

which indeed by iteration renders (49.5). The operator p_r can be shown to be hermitian (cf. Problem 59) and to obey canonical commutation with r.

Problem 50. Hilbert space transformations

To show that any quantum mechanical operator F transforms under infinitesimal rotation into

$$F' = U F U^\dagger \tag{50.1}$$

with U being the unitary operator

$$U = 1 - \frac{i}{\hbar} (\boldsymbol{\varepsilon} \cdot \mathbf{L}) . \tag{50.2}$$

The commutators of L_j with x_k, with p_k, with L_k, and with L^2 shall be deduced from this transformation relation.

Solution. From Eq. (47.8) we know that every wave scalar ψ or φ will transform into $\psi' = U\psi$ or $\varphi' = U\varphi$, U being defined by (50.2). Any measurable quantity, i.e. any matrix element, derived from an operator F must be invariant with respect to the choice of coordinates so that

$$\langle\psi'|F'|\varphi'\rangle = \langle\psi|F|\varphi\rangle. \tag{50.3}$$

Replacing ψ' and φ', we obtain

$$\langle\psi'|F'|\varphi'\rangle = \langle U\psi|F'|U\varphi\rangle = \langle\psi|U^{\dagger}F'U|\varphi\rangle, \tag{50.4}$$

so that by comparison with (50.3) there follows Eq. (50.1).

Introducing (50.2) into (50.1), we may write, in consequence of the infinitesimal character of the transformation,

$$F' = F - \boldsymbol{\varepsilon}\cdot[\mathbf{L}, F] \tag{50.5}$$

using the standard abbreviation

$$[a,b] = \frac{i}{\hbar}(ab - ba).$$

Hence, with F being rectangular coordinates x, y, z, we find on the one hand

$$x' = x - \boldsymbol{\varepsilon}\cdot[\mathbf{L}, x]; \quad y' = y - \boldsymbol{\varepsilon}\cdot[\mathbf{L}, y]; \quad z' = z - \boldsymbol{\varepsilon}\cdot[\mathbf{L}, z] \tag{50.6}$$

and on the other hand we know that, according to Eq. (47.3),

$$\begin{aligned} x' &= x + \varepsilon_z y - \varepsilon_y z, \\ y' &= -\varepsilon_z x + y + \varepsilon_x z, \\ z' &= \varepsilon_y x - \varepsilon_x y + z. \end{aligned} \tag{50.7}$$

Comparison of (50.6) and (50.7) yields

$$\begin{aligned} &[L_x, x] = 0; && [L_y, x] = +z; && [L_z, x] = -y; \\ &[L_x, y] = -z; && [L_y, y] = 0; && [L_z, y] = +x; \\ &[L_x, z] = +y; && [L_y, z] = -x; && [L_z, z] = 0. \end{aligned} \tag{50.8}$$

As for every vector the transformation formulae of its components have the same structure as (50.7), the same commutation relations with the components of angular momentum must necessarily hold, e.g.

$$[L_x, p_x] = 0; \quad [L_y, p_x] = p_z; \quad [L_z, p_x] = -p_y \quad \text{etc.} \tag{50.9}$$

Since angular momentum itself is a vector, too, the same must hold for its own components, i.e.

$$[L_x, L_x]=0; \quad [L_y, L_x]=L_z; \quad [L_z, L_x]=-L_y \quad \text{etc.} \tag{50.10}$$

A scalar operator, on the other hand, does not change under rotation, e.g. $(L^2)'=L^2$ so that it follows from (50.5) that it commutes with the angular momentum components:

$$[\boldsymbol{L}, L^2]=0 \tag{50.11}$$

and in the same way:

$$[\boldsymbol{L}, r^2]=0; \quad [\boldsymbol{L}, p^2]=0 \quad \text{etc.} \tag{50.12}$$

Problem 51. Commutators in Schrödinger representation

To verify directly in Schrödinger representation that for a system of N particles there will hold the commutation relations

$$[L_x, Y]=-Z; \quad [L_x, L_y]=-L_z \tag{51.1}$$

with $\boldsymbol{L}$ the angular momentum of the whole system, and X, Y, Z the coordinates of its centre of mass.

Solution. We have

$$\boldsymbol{L}=\sum_j \boldsymbol{L}_j; \quad X=\frac{1}{M}\sum_j m_j x_j; \quad j=1,2,\ldots,N \tag{51.2}$$

with m_j the particle masses and M the total mass of the system. We then obtain

$$[L_x, Y] = \frac{1}{M}\sum_j \sum_k m_k [L_{jx}, y_k];$$
$$[L_x, L_y]=\sum_j \sum_k [L_{jx}, L_{ky}].$$

The operators L_{jx} operating only on the coordinates of the j'th particle commute with all operators connected with the other particle coordinates so that the double sums reduce to simple ones:

$$[L_x, Y] = \frac{1}{M}\sum_k m_k [L_{kx}, y_k]; \tag{51.3}$$

$$[L_x, L_y]=\sum_k [L_{kx}, L_{ky}]. \tag{51.4}$$

We can now evaluate each term by directly performing the differentiations involved:

$$[L_{kx}, y_k]\psi = \left(y_k\frac{\partial}{\partial z_k} - z_k\frac{\partial}{\partial y_k}\right)(y_k\psi) - y_k\left(y_k\frac{\partial\psi}{\partial z_k} - z_k\frac{\partial\psi}{\partial y_k}\right) = -z_k\psi$$

and

$$\begin{aligned}[L_{kx}, L_{ky}]\psi &= \frac{\hbar}{i}\left\{\left(y_k\frac{\partial}{\partial z_k} - z_k\frac{\partial}{\partial y_k}\right)\left(z_k\frac{\partial\psi}{\partial x_k} - x_k\frac{\partial\psi}{\partial z_k}\right)\right.\\ &\quad\left. -\left(z_k\frac{\partial}{\partial x_k} - x_k\frac{\partial}{\partial z_k}\right)\left(y_k\frac{\partial\psi}{\partial z_k} - z_k\frac{\partial\psi}{\partial y_k}\right)\right\}\\ &= \frac{\hbar}{i}\left(y_k\frac{\partial\psi}{\partial x_k} - x_k\frac{\partial\psi}{\partial y_k}\right) = -L_{kz}\psi\,.\end{aligned}$$

Inserting these results in (51.3) and (51.4) yields

$$[L_x, Y] = -\frac{1}{M}\sum_k m_k z_k = -Z$$

and

$$[L_x, L_y] = -\sum_k L_{kz} = -L_z$$

thus proving the relations (51.1).

Problem 52. Particles of spin 1

To show by an infinitesimal coordinate rotation that a vector field ψ is suitable to represent particles of spin 1.

Solution. If, as in Problem 47, the rotation is described by

$$r' = (1 + A)r\,, \tag{52.1}$$

the vector field will undergo the same transformation as the vector r so that

$$\psi'(r') = (1 + A)\psi(r)\,. \tag{52.2}$$

On the other hand, we know from the solution of Problem 47 that by Taylor expansion of each component $\psi_i'(r')$ at r we obtain

$$\psi_i'(r') = \left\{1 - \frac{i}{\hbar}(\varepsilon\cdot L)\right\}\psi_i'(r)\,. \tag{52.3}$$

We shall solve the problem by comparison of these two expressions.

To this purpose, we use the description of A given in Eq. (47.1b) which we decompose into[8]

$$A=\varepsilon_x\begin{pmatrix}0&0&0\\0&0&1\\0&-1&0\end{pmatrix}+\varepsilon_y\begin{pmatrix}0&0&-1\\0&0&0\\1&0&0\end{pmatrix}+\varepsilon_z\begin{pmatrix}0&1&0\\-1&0&0\\0&0&0\end{pmatrix} \tag{52.4}$$

or, in a more appropriate form,

$$A=\frac{i}{\hbar}\sum_k \varepsilon_k S_k \tag{52.5}$$

with

$$S_x=\hbar\begin{pmatrix}0&0&0\\0&0&-i\\0&i&0\end{pmatrix};\quad S_y=\hbar\begin{pmatrix}0&0&i\\0&0&0\\-i&0&0\end{pmatrix};\quad S_z=\hbar\begin{pmatrix}0&-i&0\\i&0&0\\0&0&0\end{pmatrix}. \tag{52.6}$$

Then Eq. (52.2) becomes

$$\psi'(\boldsymbol{r}')=\left(1+\frac{i}{\hbar}\sum_k \varepsilon_k S_k\right)\psi(\boldsymbol{r}). \tag{52.7}$$

Combination of Eqs. (52.3) and (52.7) allows to eliminate $\psi'(\boldsymbol{r}')$:

$$\left\{1-\frac{i}{\hbar}\sum_k \varepsilon_k L_k\right\}\psi'(\boldsymbol{r})=\left\{1+\frac{i}{\hbar}\sum_k \varepsilon_k S_k\right\}\psi(\boldsymbol{r})$$

or

$$\psi'(\boldsymbol{r})=\left\{1+\frac{i}{\hbar}\sum_k \varepsilon_k J_k\right\}\psi(\boldsymbol{r}) \tag{52.8}$$

with

$$J_k=L_k+S_k. \tag{52.9}$$

It follows therefore that the three matrices S_k are the spin components, $\boldsymbol{J}$ being the total angular momentum and $\boldsymbol{L}$ its orbital part. It can easily be shown by direct calculation using the definitions in (52.6) that the S_k's obey the commutation relations of angular momenta,

$$[S_j,S_k]=-S_l \qquad (j,k,l=x,y,z \text{ or cyclic}).$$

Each of the three matrices (52.6) has eigenvalues $+1, 0, -1$ since it differs from the 2×2 Pauli matrix σ_y of eigenvalues $+1$ and -1 only by the insertion of one line and one column of zeros (save of a sign

---

[8] The three matrices in (52.4) are the generators of the rotational group SO(3) in 3-dimensional representation.

in S_y) which simply adds the eigenvalue 0 to those of σ_y. Hence, the spin component in any direction has eigenvalues $+\hbar, 0, -\hbar$. The absolute square, finally, becomes

$$S^2 = S_x^2 + S_y^2 + S_z^2 = \hbar^2 \left\{ \begin{pmatrix} 0 & 0 & 0 \\ 0 & 1 & 0 \\ 0 & 0 & 1 \end{pmatrix} + \begin{pmatrix} 1 & 0 & 0 \\ 0 & 0 & 0 \\ 0 & 0 & 1 \end{pmatrix} + \begin{pmatrix} 1 & 0 & 0 \\ 0 & 1 & 0 \\ 0 & 0 & 0 \end{pmatrix} \right\}$$

or $S^2 = 2\hbar^2$. According to the general formula $S^2 = \hbar^2 s(s+1)$ this yields the spin $s=1$ for the vector field.

Problem 53. Commutation with a tensor

To determine the commutation relations of a symmetrical tensor with the components of angular momentum by using an infinitesimal coordinate rotation.

Solution. It has been shown in Problem 50 that every quantum mechanical operator transforms under infinitesimal rotation according to

$$F' = F - \boldsymbol{\varepsilon} \cdot [\boldsymbol{L}, F]. \tag{53.1}$$

The transformation being defined by $\boldsymbol{r}' = R\,\boldsymbol{r}$, a tensor is defined by the transformation properties of its components,

$$T'_{ik} = \sum_\mu \sum_\nu R_{i\mu} R_{k\nu} T_{\mu\nu} = (R\,T\,\tilde{R})_{ik} \tag{53.2}$$

so that under infinitesimal rotation $R = \mathit{1} + A$, $\tilde{A} = -A$, we have

$$T' = (\mathit{1} + A)\,T(\mathit{1} - A) = T + (AT - TA). \tag{53.3}$$

Comparison with (53.1) leads to

$$\boldsymbol{\varepsilon} \cdot [\boldsymbol{L}, T] = TA - AT \tag{53.4}$$

which is the basic formula for the solution of the problem.

We restrict the calculation to a rotation about the x axis,

$$A = \varepsilon_x \begin{pmatrix} 0 & 0 & 0 \\ 0 & 0 & 1 \\ 0 & -1 & 0 \end{pmatrix}.$$

Then,

$$TA = \varepsilon_x \begin{pmatrix} 0 & -T_{13} & T_{12} \\ 0 & -T_{23} & T_{22} \\ 0 & -T_{33} & T_{32} \end{pmatrix}; \quad AT = \varepsilon_x \begin{pmatrix} 0 & 0 & 0 \\ T_{31} & T_{32} & T_{33} \\ -T_{21} & -T_{22} & -T_{23} \end{pmatrix},$$

so that for a symmetrical tensor the difference

$$TA - AT = \varepsilon_x \begin{pmatrix} 0; & -T_{13}; & T_{12} \\ -T_{13}; & -2T_{23}; & T_{22}-T_{33} \\ T_{12}; & T_{22}-T_{33}; & 2T_{23} \end{pmatrix} \tag{53.5}$$

again turns out to be symmetrical. This, according to (53.4), must equal $\varepsilon_x[L_1, T]$, hence

$$\begin{aligned} &[L_1, T_{11}] = 0; && [L_1, T_{12}] = -T_{13}; && [L_1, T_{13}] = T_{12}; \\ &[L_1, T_{22}] = -2T_{23}; && [L_1, T_{23}] = T_{22} - T_{33}; && [L_1, T_{33}] = 2T_{23}. \end{aligned} \tag{53.6}$$

The commutators of L_2 and L_3 may be found either by rotations about the y and z axes or, more easily, by cyclic permutation of symbols in (53.6), i. e.

$$\begin{aligned} &[L_2, T_{22}] = 0; && [L_2, T_{23}] = -T_{12}; && [L_2, T_{12}] = T_{23}; \\ &[L_2, T_{33}] = -2T_{13}; && [L_2, T_{13}] = T_{33} - T_{11}; && [L_2, T_{11}] = 2T_{13} \end{aligned} \tag{53.7}$$

and

$$\begin{aligned} &[L_3, T_{33}] = 0; && [L_3, T_{13}] = -T_{23}; && [L_3, T_{23}] = T_{13}; \\ &[L_3, T_{11}] = -2T_{12}; && [L_3, T_{12}] = T_{11} - T_{22}; && [L_3, T_{22}] = 2T_{12}. \end{aligned} \tag{53.8}$$

It should be noted that all three L_k's commute with the trace of the tensor:

$$[\boldsymbol{L}, \operatorname{tr} T] = 0. \tag{53.9}$$

This may be understood from the decomposition of a symmetrical tensor

$$T = \Theta + \tfrac{1}{3}(\operatorname{tr} T)\,\boldsymbol{1}$$

with $\operatorname{tr}\Theta = 0$. $\operatorname{tr} T$ therefore is a scalar (a tensor invariant) and it has been shown at the end of Problem 50 that a scalar always commutes with L_k.

Problem 54. Quadrupole tensor. Spherical harmonics

The quadrupole tensor Q defined by

$$Q_{ik} = Q_{ki} = 3x_i x_k - r^2 \delta_{ik} \tag{54.1}$$

is a symmetrical tensor with trace zero and therefore has five linearly independent components. These may be expressed (except for a factor r^2) as linear combinations of the five spherical harmonics of order 2.

Using the commutators found in Problem 53, the commutators of angular momentum operators with spherical harmonics shall be determined.

Solution. The spherical harmonics $Y_{2,m}$ with $m=0, \pm 1, \pm 2$ can be expressed in rectangular coordinates[9]:

$$r^2 Y_{2,0} = C(3z^2 - r^2); \qquad r^2 Y_{2,\pm 1} = \pm C\sqrt{6}(x \pm iy)z; \qquad (54.2)$$
$$r^2 Y_{2,\pm 2} = C\tfrac{1}{2}\sqrt{6}(x \pm iy)^2; \quad C = \sqrt{\frac{5}{16\pi}}; \quad \oint |Y_{2,m}|^2 d\Omega = 1.$$

These well-known definitions permit the spherical harmonics to be expressed in terms of the quadrupole tensor components:

$$r^2 Y_{2,0} = C Q_{33}; \quad r^2 Y_{2,\pm 1} = \pm C\sqrt{\tfrac{2}{3}}(Q_{13} \pm i Q_{23});$$
$$r^2 Y_{2,\pm 2} = C\sqrt{\tfrac{2}{3}}\{\tfrac{1}{2}(Q_{11} - Q_{22}) \pm i Q_{12}\}. \qquad (54.3)$$

Here we may introduce Eqs. (53.6)—(53.8) in order to find the commutators.

We start with (53.8) since the spherical harmonics refer to z as polar axis about which L_3 is the angular momentum component.

$$\begin{aligned}
r^2[L_z, Y_{2,0}] &= C[L_3, Q_{33}] = 0;\\
r^2[L_z, Y_{2,\pm 1}] &= C\sqrt{\tfrac{2}{3}}\{[L_3, Q_{13}] \pm i[L_3, Q_{23}]\}\\
&= C\sqrt{\tfrac{2}{3}}(-Q_{23} \pm i Q_{13}) = \pm i r^2 Y_{2,\pm 1};\\
r^2[L_z, Y_{2,\pm 2}] &= C\sqrt{\tfrac{2}{3}}\{\tfrac{1}{2}[L_3, Q_{11}] - \tfrac{1}{2}[L_3, Q_{22}] \pm i[L_3, Q_{12}]\}\\
&= C\sqrt{\tfrac{2}{3}}\{-Q_{12} - Q_{12} \pm i(Q_{11} - Q_{22})\} = \pm 2 i r^2 Y_{2,\pm 2}.
\end{aligned}$$

These three commutators may be gathered up in one relation

$$[L_z, Y_{2,m}] = i m Y_{2,m}. \qquad (54.4)$$

This result becomes rather obvious in the Schrödinger representation with $L_z = \frac{\hbar}{i}\frac{\partial}{\partial\varphi}$ and $Y_{2,m} \propto e^{im\varphi}$:

$$[L_z, e^{im\varphi}] = \frac{\partial}{\partial\varphi} e^{im\varphi} - e^{im\varphi}\frac{\partial}{\partial\varphi} = i m e^{im\varphi}.$$

We now use Eqs. (53.6) and (53.7) to construct the commutators with

$$L^+ = L_x + i L_y \quad \text{and} \quad L^- = L_x - i L_y.$$

---

[9] The choice of the front sign is arbitrary, but the different signs for $Y_{2,1}$ and $Y_{2,-1}$ simplify the final results.

We have,

$$\left.\begin{aligned}
&[L^+, Q_{11}] = 2iQ_{13}, && [L^-, Q_{11}] = -2iQ_{13},\\
&[L^+, Q_{12}] = -Q_{13} + iQ_{23}, && [L^-, Q_{12}] = -Q_{13} - iQ_{23},\\
&[L^+, Q_{13}] = Q_{12} + i(Q_{33} - Q_{11}), && [L^-, Q_{13}] = Q_{12} - i(Q_{33} - Q_{11}),\\
&[L^+, Q_{22}] = -2Q_{23}, && [L^-, Q_{22}] = -2Q_{23},\\
&[L^+, Q_{23}] = (Q_{22} - Q_{33}) - iQ_{12}, && [L^-, Q_{23}] = (Q_{22} - Q_{33}) + iQ_{12},\\
&[L^+, Q_{33}] = 2(Q_{23} - iQ_{13}), && [L^-, Q_{33}] = 2(Q_{23} + iQ_{13}).
\end{aligned}\right\} \quad (54.5)$$

Straightforward application of (54.3), then (54.5), and then (54.3) again, produces the commutators with spherical harmonics for which we shall give in more detail a few typical examples.

$$\begin{aligned}
r^2[L^+, Y_{2,1}] &= C\sqrt{\tfrac{2}{3}}\{[L^+, Q_{13}] + i[L^+, Q_{23}]\}\\
&= C\sqrt{\tfrac{2}{3}}\{Q_{12} + i(Q_{33} - Q_{11}) + i(Q_{22} - Q_{33}) + Q_{12}\}\\
&= -2iC\sqrt{\tfrac{2}{3}}\{\tfrac{1}{2}(Q_{11} - Q_{22}) + iQ_{12}\} = -2ir^2 Y_{2,2}.
\end{aligned}$$

Another typical example is

$$\begin{aligned}
r^2[L^-, Y_{2,1}] &= C\sqrt{\tfrac{2}{3}}\{[L^-, Q_{13}] + i[L^-, Q_{23}]\}\\
&= C\sqrt{\tfrac{2}{3}}\{Q_{12} - i(Q_{33} - Q_{11}) + i(Q_{22} - Q_{33}) - Q_{12}\}\\
&= C\sqrt{\tfrac{2}{3}}\, i(Q_{11} + Q_{22} - 2Q_{33}).
\end{aligned}$$

Since

$$\operatorname{tr} Q = Q_{11} + Q_{22} + Q_{33} = 0,$$

the bracket becomes $-3Q_{33}$ so that

$$r^2[L^-, Y_{2,1}] = -3i\sqrt{\tfrac{2}{3}}\, r^2 Y_{2,0} = -i\sqrt{6}\, r^2 Y_{2,0}.$$

Finally, at the ends of the m interval, the commutators vanish:

$$\begin{aligned}
r^2[L^+, Y_{2,2}] &= C\sqrt{\tfrac{2}{3}}\{\tfrac{1}{2}[L^+, Q_{11}] - \tfrac{1}{2}[L^+, Q_{22}] + i[L^+, Q_{12}]\}\\
&= C\sqrt{\tfrac{2}{3}}(iQ_{13} + Q_{23} - iQ_{13} - Q_{23}) = 0.
\end{aligned}$$

The results may be gathered up in two very simple formulae, viz.

$$[L^+, Y_{2,m}] = -i\sqrt{6 - m(m+1)}\, Y_{2,m+1} \tag{54.6}$$

and

$$[L^-, Y_{2,m}] = -i\sqrt{6 - (m-1)\,m}\, Y_{2,m-1}. \tag{54.7}$$

Eqs. (54.4), (54.6) and (54.7) form the complete set of commutators wanted.

Problem 55. Transformation of spherical harmonics

Let a *finite* coordinate rotation be defined by three Eulerian angles α, β, γ. The transformation properties of wave functions which are spherical harmonics shall be obtained.

Solution. We begin by defining the Eulerian angles according to the generally accepted way. The initial system x, y, z is first rotated by α about its z axis, with $0 \leq \alpha < 2\pi$. This results in a system x_1, y_1, z_1 which now is rotated through an angle β $(0 \leq \beta < \pi)$ about its y_1 axis. Thus another intermediary system x_2, y_2, z_2 is obtained which by rotation through another angle γ $(0 \leq \gamma < 2\pi)$ about its z_2 axis finally passes over into the system x', y', z'.

It can easily be seen from (48.1) by iteration that any function $\psi(\boldsymbol{r})$ under rotation about an axis A through a *finite* angle ε_A obeys the transformation law

$$\psi(\boldsymbol{r}') = D(\varepsilon_A)\psi(\boldsymbol{r}) \quad \text{with} \quad D(\varepsilon_A) = e^{-\frac{i}{\hbar}\varepsilon_A L_A}. \tag{55.1}$$

When the three Euler rotations are consecutively performed, three such transformations follow each other, thus giving the law

$$\psi(\boldsymbol{r}') = D(\alpha, \beta, \gamma)\psi(\boldsymbol{r}) \tag{55.2}$$

with

$$D(\alpha, \beta, \gamma) = e^{-\frac{i}{\hbar}\gamma L_{z_2}} \cdot e^{-\frac{i}{\hbar}\beta L_{y_1}} \cdot e^{-\frac{i}{\hbar}\alpha L_z}. \tag{55.3}$$

In (55.3) we next try to express the three rotations about the axes z, y_1, z_2 of three different coordinate systems by rotations about the original axes. In order to perform the rotation by β about y_1 we use the

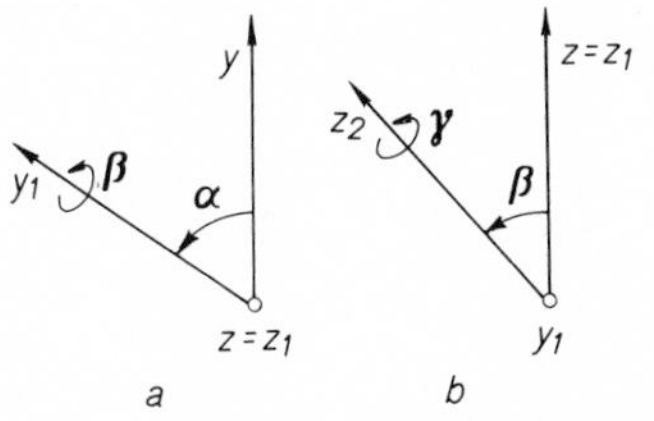

Fig. 32a and b. Rotations of coordinate systems

geometry of Fig. 32a with the axes z and z_1 coinciding and perpendicular to the paper plane, y and y_1 including the angle α. We replace the rotation by β about y_1 by first turning $x_1 y_1 z_1$ back by $-\alpha$ about z, then performing the rotation by β about the y instead of the y_1 axis, and finally turning the coordinate system back by $+\alpha$ about z, i.e.

$$e^{-\frac{i}{\hbar}\beta L_{y_1}} = e^{-\frac{i}{\hbar}\alpha L_z} \cdot e^{-\frac{i}{\hbar}\beta L_y} \cdot e^{+\frac{i}{\hbar}\alpha L_z}. \tag{55.4}$$

In a similar way we decompose the rotation through the angle γ about z_2 (Fig. 32b), first turning z_2 back by $-\beta$ about y_1, then turning by γ about z, and finally restituting the initial position of z_2 by turning through β about y_1:

$$e^{-\frac{i}{\hbar}\gamma L_{z_2}} = e^{-\frac{i}{\hbar}\beta L_{y_1}} \cdot e^{-\frac{i}{\hbar}\gamma L_z} \cdot e^{+\frac{i}{\hbar}\beta L_{y_1}}. \tag{55.5}$$

Combining Eqs. (55.4) and (55.5) with (55.3) we find

$$D(\alpha,\beta,\gamma) = e^{-\frac{i}{\hbar}\beta L_{y_1}} e^{-\frac{i}{\hbar}\gamma L_z} e^{\frac{i}{\hbar}\beta L_{y_1}} \cdot e^{-\frac{i}{\hbar}\beta L_{y_1}} \cdot e^{-\frac{i}{\hbar}\alpha L_z}$$

$$= e^{-\frac{i}{\hbar}\alpha L_z} e^{-\frac{i}{\hbar}\beta L_y} e^{\frac{i}{\hbar}\alpha L_z} \cdot e^{-\frac{i}{\hbar}(\gamma+\alpha) L_z}$$

or

$$D(\alpha,\beta,\gamma) = e^{-\frac{i}{\hbar}\alpha L_z} e^{-\frac{i}{\hbar}\beta L_y} e^{-\frac{i}{\hbar}\gamma L_z}. \tag{55.6}$$

This is a very general and purely geometrical formula which we will now apply especially to spherical harmonics. The function $Y_{L,M}$ referring to the polar axis z is an eigenfunction to the operator L_z with eigenvalue $\hbar M$. The operator L_y can, as may easily be seen from Problem 56, only affect M but not L so that in the new coordinate system ϑ', φ' we have, according to (55.2),

$$Y_{L,M}(\vartheta',\varphi') = D(\alpha,\beta,\gamma)\, Y_{L,M}(\vartheta,\varphi) = \sum_{M'=-L}^{+L} D^L_{M,M'}\, Y_{L,M'}(\vartheta,\varphi). \tag{55.7}$$

We proceed to calculating the coefficients

$$D^L_{M,M''} = \oint d\Omega\, Y^*_{L,M''}(\vartheta,\varphi)\, Y_{L,M}(\vartheta',\varphi')$$
$$= \oint d\Omega\, Y^*_{L,M''}(\vartheta,\varphi) D(\alpha,\beta,\gamma)\, Y_{L,M}(\vartheta,\varphi)$$

or briefly

$$D^L_{M,M''} = \langle LM'' | D(\alpha,\beta,\gamma) | LM \rangle \tag{55.8}$$

with $D(\alpha,\beta,\gamma)$ the operator (55.6). Now[10],

$$e^{-\frac{i}{\hbar}\gamma L_z} | LM \rangle = e^{-i\gamma M} | LM \rangle,$$

$$\langle LM'' | e^{-\frac{i}{\hbar}\alpha L_z} = e^{-i\alpha M''} \langle LM'' |,$$

---

[10] The sign may be doubtful in the bra contribution. We have

$$\langle LM'' | e^{-\frac{i}{\hbar}\alpha L_z} = \langle e^{\frac{i}{\hbar}\alpha L_z} Y_{L,M''} | = \langle e^{i\alpha M''} Y_{L,M''} | = e^{-i\alpha M''} \langle LM'' |$$

since $Y^*_{L,M''} \propto e^{-iM''\varphi}$ and $L_z Y^*_{L,M''} = -\hbar M'' Y^*_{L,M''}$.

hence

$$D^L_{M,M''} = e^{-i(\alpha M'' + \gamma M)} d^L_{M,M''}(\beta)$$

with (55.9)

$$d^L_{M,M''}(\beta) = \langle LM'' | e^{-\frac{i}{\hbar}\beta L_y} | LM \rangle$$

a rather involved function of the argument β. Therefore, the spherical harmonics transform according to the law

$$Y_{L,M}(\vartheta', \varphi') = \sum_{M'=-L}^{+L} e^{-i(\alpha M' + \gamma M)} d^L_{M,M'}(\beta)\, Y_{L,M'}(\vartheta, \varphi). \qquad (55.10)$$

Problem 56. Construction of Hilbert space for an angular momentum component

Let an atomic system be in a state with angular momentum quantum number l. Then the Hilbert eigenvectors of the component operator L_z shall be constructed. For the method cf. Problem 31.

Solution. We choose a Hilbert coordinate system in which L^2 is diagonal, i.e. (omitting all factors $\hbar$),

$$L^2 = \tfrac{1}{2}(L^+ L^- + L^- L^+) + L_z^2 = l(l+1) \qquad (56.1)$$

for all vectors to be discussed. Let further $|\psi_m\rangle$ be an eigenvector of L_z with the (still undetermined) eigenvalue m of this operator. Then we apply

$$L^+ = L_x + i L_y \qquad (56.2)$$

satisfying the commutation rule (cf. Problem 51)

$$L^+ L_z - L_z L^+ = -L^+ . \qquad (56.3)$$

From

$$L_z |\psi_m\rangle = m |\psi_m\rangle \qquad (56.4)$$

it then follows that

$$L^+ L_z |\psi_m\rangle = (L_z - 1) |L^+ \psi_m\rangle = m |L^+ \psi_m\rangle$$

or

$$L_z |L^+ \psi_m\rangle = (m+1) |L^+ \psi_m\rangle . \qquad (56.5)$$

$|L^+ \psi_m\rangle$ therefore is eigenvector for the eigenvalue $m+1$. It is not yet normalized, the scalar product

$$\langle L^+ \psi_m | L^+ \psi_m \rangle = \langle L^- L^+ \psi_m | \psi_m \rangle \quad (\text{with } L^- = L^{+\dagger})$$

not being $=1$. From (56.1), however, and from the commutation rule

$$L^+ L^- - L^- L^+ = 2 L_z \tag{56.6}$$

we find

$$L^- L^+ = L^2 - L_z^2 - L_z \tag{56.7}$$

so that

$$\langle L^- L^+ \psi_m | \psi_m \rangle = [l(l+1) - m(m+1)] \langle \psi_m | \psi_m \rangle .$$

The normalized Hilbert vector for the eigenvalue $m+1$ thus becomes

$$|\psi_{m+1}\rangle = \frac{1}{\sqrt{l(l+1) - m(m+1)}} |L^+ \psi_m\rangle . \tag{56.8}$$

This construction may be repeated up to $m=l$, the next step leading to $L^+ |\psi_l\rangle = 0$ so that the sequence breaks off and no $|\psi_{l+1}\rangle$ exists.

Besides this ascending sequence the descending sequence may be constructed by repeated application of the hermitian conjugate operator

$$L^- = L_x - i L_y \tag{56.9}$$

which satisfies the commutation rule

$$L^- L_z - L_z L^- = L^- . \tag{56.10}$$

Applying L^- to (56.4) yields

$$L_z |L^- \psi_m\rangle = (m-1) |L^- \psi_m\rangle . \tag{56.11}$$

Again, we still have to normalize the Hilbert vector:

$$\begin{aligned} \langle L^- \psi_m | L^- \psi_m \rangle &= \langle L^+ L^- \psi_m | \psi_m \rangle \\ &= \langle (L^2 - L_z^2 + L_z) \psi_m | \psi_m \rangle \\ &= [l(l+1) - m(m-1)] \langle \psi_m | \psi_m \rangle . \end{aligned}$$

Thus there follows

$$|\psi_{m-1}\rangle = \frac{1}{\sqrt{l(l+1) - m(m-1)}} |L^- \psi_m\rangle . \tag{56.12}$$

This sequence reaches down even to negative values of m until we arrive at $m = -l$. The next step then would lead on to

$$L^- |\psi_{-l}\rangle = 0 ,$$

so that no vector $|\psi_{-l-1}\rangle$ exists.

NB 1. From the Schrödinger theory we know that

$$|\psi_m\rangle = Y_{l,m}(\vartheta, \varphi) \tag{56.13}$$

is a normalized realization of the Hilbert vectors, save of an arbitrary phase factor in each. The relations (56.8) and (56.12), therefore, except of this phase factor (in the usual normalization, $(-1)^m$) are identical with the spherical harmonics recurrence relations:

$$\left.\begin{aligned} L^+ Y_{l,m} &= -\sqrt{l(l+1)-m(m+1)}\, Y_{l,m+1}, \\ L^- Y_{l,m} &= -\sqrt{l(l+1)-m(m-1)}\, Y_{l,m-1}. \end{aligned}\right\} \tag{56.14}$$

NB 2. In the preceding solution we have tacitly assumed l to be an integer. Since each application of the operators L^+ or L^- changes m by ± 1, from its maximum value $m=+l$ down to its minimum $m=-l$, the total difference between these two limits, $2l$, must be an integer. This is compatible with integer values of l and m as well as with *half-integer* values. The angular momentum commutation rules therefore allow for half-integer quantization. The latter does not yet occur in angular momenta derived from particle motion, but these do not exhaust the possibilities of theory. As soon as the spin of particles is included besides "orbital" momenta, half-integer values will occur.

Problem 57. Orthogonality of spherical harmonics

To prove that the eigenvectors of L_z for given l derived in the preceding problem form an orthogonal system. The matrix elements of the operators L^+ and L^- shall then be determined in this Hilbert coordinate system.

Solution. The vector $|\psi_m\rangle$ has been found to satisfy the eigenvalue relations

$$L^- L^+ |\psi_m\rangle = [l(l+1)-m(m+1)]|\psi_m\rangle; \tag{57.1a}$$

$$L^+ L^- |\psi_m\rangle = [l(l+1)-m(m-1)]|\psi_m\rangle. \tag{57.1b}$$

Now we form the two equal scalar products

$$\langle L^+ \psi_{m'}|L^+ \psi_m\rangle = \langle \psi_{m'}|L^- L^+ \psi_m\rangle \tag{57.2a}$$

and

$$\langle L^+ \psi_{m'}|L^+ \psi_m\rangle = \langle L^- L^+ \psi_{m'}|\psi_m\rangle. \tag{57.2b}$$

Applying (57.1a) to the right-hand sides of these relations we then find for these two expressions

$$[l(l+1)-m(m+1)]\langle \psi_{m'}|\psi_m\rangle$$

and

$$[l(l+1)-m'(m'+1)]\langle \psi_{m'}|\psi_m\rangle,$$

respectively. Their difference vanishes:

$$[m(m+1)-m'(m'+1)]\langle \psi_{m'}|\psi_m\rangle = 0. \tag{57.3}$$

The second factor then must be zero except when the first factor vanishes, i.e. orthogonality is proved except for the pairs of either $m'=m$ or $m'=-m-1$. To exclude the last possibility we use Eq. (57.1b) for the analogous procedure:

$$\begin{aligned}\langle L^-\psi_{m'}|L^-\psi_m\rangle &= \langle\psi_{m'}|L^+L^-\psi_m\rangle\\ &= [l(l+1)-m(m-1)]\langle\psi_{m'}|\psi_m\rangle\\ &= \langle L^+L^-\psi_{m'}|\psi_m\rangle\\ &= [l(l+1)-m'(m'-1)]\langle\psi_{m'}|\psi_m\rangle.\end{aligned}$$

Again taking the difference,

$$[m(m-1)-m'(m'-1)]\langle\psi_{m'}|\psi_m\rangle=0, \tag{57.4}$$

this time the first factor vanishes for $m'=m$ and $m'=-m+1$, but not for $m'=-m-1$. Both Eqs. (57.3) and (57.4) together therefore lead to the wanted orthogonality

$$\langle\psi_{m'}|\psi_m\rangle=\delta_{mm'} \tag{57.5}$$

with normalization as introduced in the preceding problem.

Now it is easy to construct the matrices from (56.8) and (56.12), using the arbitrary sign convention of (56.14):

$$\left.\begin{aligned}\langle\psi_{m+1}|L^+|\psi_m\rangle &= -\sqrt{l(l+1)-m(m+1)};\\ \langle\psi_{m-1}|L^-|\psi_m\rangle &= -\sqrt{l(l+1)-m(m-1)},\end{aligned}\right\} \tag{57.6}$$

all other matrix elements vanishing in consequence of (57.5).

D. Potentials of Spherical Symmetry

a) Bound States

In order to solve the Schrödinger equation for a potential depending only upon r (central force field) it is expedient to introduce spherical polar coordinates (Fig. 33) by the formulae

$$x=r\sin\vartheta\cos\varphi;\quad y=r\sin\vartheta\sin\varphi;\quad z=r\cos\vartheta. \tag{a.1}$$

The z axis then is called the polar axis, ϑ is the angle between $\boldsymbol{r}$ and the polar axis $(0\le\vartheta\le\pi)$ and φ measures a rotation about the polar axis $(0\le\varphi\le 2\pi)$. The Laplacian in these coordinates becomes

$$\nabla^2=\frac{\partial^2}{\partial r^2}+\frac{2}{r}\frac{\partial}{\partial r}+\frac{1}{r^2}\mathscr{A} \tag{a.2}$$

with $\mathscr{A}$ the angular operator

$$\mathscr{A} = \frac{1}{\sin\vartheta}\frac{\partial}{\partial\vartheta}\left(\sin\vartheta\frac{\partial}{\partial\vartheta}\right) + \frac{1}{\sin^2\vartheta}\frac{\partial^2}{\partial\varphi^2}. \tag{a.3}$$

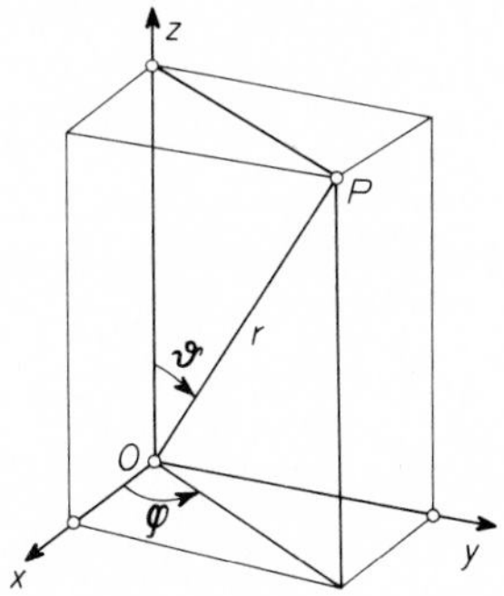

Fig. 33. Spherical polar coordinates

The Schrödinger equation

$$-\frac{\hbar^2}{2m}\left(\frac{\partial^2}{\partial r^2} + \frac{2}{r}\frac{\partial}{\partial r} + \frac{1}{r^2}\mathscr{A}\right)\psi + V(r)\psi = E\psi \tag{a.4}$$

then permits factorization

$$\psi(r,\vartheta,\varphi) = \frac{1}{r}\chi_l(r)\, Y_{l,m}(\vartheta,\varphi) \tag{a.5}$$

with the radial function $\chi_l(r)$ satisfying the *radial wave equation*

$$-\frac{\hbar^2}{2m}\left(\frac{d^2\chi_l}{dr^2} - \frac{l(l+1)}{r^2}\chi_l\right) + V(r)\chi_l = E\chi_l \tag{a.6}$$

and $Y_{l,m}$ satisfying the differential equation of the *spherical harmonics*,

$$\mathscr{A}\, Y_{l,m} + l(l+1)\, Y_{l,m} = 0. \tag{a.7}$$

Here, $l(l+1)$ comes in as separation parameter of radius and angles which leads to non-singular solutions only if l is an integer (see below), $l=0,1,2,\ldots$ For each integer l, there still exist $2l+1$ non-singular and unique solutions for $m=-l, -l+1, \ldots, l-1, l$.

In the following problems we shall always normalize the spherical harmonics according to

$$\oint d\Omega\,|Y_{l,m}|^2 = 1\,. \tag{a.8}$$

The special solution for $m=0$ shall be written in the form

$$Y_{l,0}(\vartheta) = \sqrt{\frac{2l+1}{4\pi}}\,P_l(\cos\vartheta)\,. \tag{a.9}$$

It depends upon ϑ only and is called a *zonal* spherical harmonic. The function P_l is normalized so that $P_l(1)=1$ and is a polynomial of l-th degree in $\cos\vartheta$ called Legendre polynomial.

The operators of the angular momentum components of a particle can be expressed in spherical polar coordinates as follows.

$$\begin{aligned} L^+ &= L_x + i L_y = -\frac{\hbar}{i}\,e^{i\varphi}\left(-i\frac{\partial}{\partial\vartheta} + \cot\vartheta\,\frac{\partial}{\partial\varphi}\right); \\ L^- &= L_x - i L_y = -\frac{\hbar}{i}\,e^{-i\varphi}\left(i\frac{\partial}{\partial\vartheta} + \cot\vartheta\,\frac{\partial}{\partial\varphi}\right); \\ L_z &= \frac{\hbar}{i}\,\frac{\partial}{\partial\varphi}\,. \end{aligned} \tag{a.10}$$

The operator of the absolute square of the angular momentum vector,

$$L^2 = L_x^2 + L_y^2 + L_z^2 = \tfrac{1}{2}(L^+L^- + L^-L^+) + L_z^2\,, \tag{a.11}$$

can explicitly be written

$$L^2 = -\hbar^2\left\{\frac{1}{\sin\vartheta}\,\frac{\partial}{\partial\vartheta}\left(\sin\vartheta\,\frac{\partial}{\partial\vartheta}\right) + \frac{1}{\sin^2\vartheta}\,\frac{\partial^2}{\partial\varphi^2}\right\} \tag{a.12}$$

or, using the abbreviation (a.3),

$$L^2 = -\hbar^2\mathscr{A}\,. \tag{a.13}$$

It then follows from Eq. (a.7) that $Y_{l,m}$ is an eigenfunction of the angular momentum operator L^2 with the eigenvalue $\hbar^2 l(l+1)$. Since $Y_{l,m}$ depends upon the angle φ as $e^{im\varphi}$, Eq. (a.10) makes $Y_{l,m}$ an eigenfunction too of the operator L_z with the eigenvalue $\hbar m$ where $|m| \le l$. These two properties of the wave functions in a potential field of central symmetry therefore reflect the classical conservation law of angular momentum in so far as it still holds in quantum mechanics.

Eqs. (a.2) and (a.10) can, of course, be derived by coordinate transformation according to (a.1). This is a straightforward but rather laborious way. A more elegant method, using the relation between angular momentum operators and infinitesimal coordinate rotations has been shown in Problems 47 to 49.

For mathematical details on spherical harmonics see Appendix, vol. II, p. 267.

Problem 58. Angular momentum expectation values

What can be said in the following two cases about the three angular momentum components L_x, L_y, L_z?

a) The wave function of a particle moving in a central field depends as $Y_{l,m}$ upon the angles.

b) For a given l, the wave functions $Y_{l,m}$ and $Y_{l,-m}$ show the same dependence upon ϑ. Both being degenerate solutions, linear combinations proportional to $\cos m\varphi$ and $\sin m\varphi$ may be constructed which yet remain solutions of the central force Schrödinger equation.

Solution. a) The spherical harmonic $Y_{l,m}$ is eigenfunction of L^2 for the eigenvalue $\hbar^2 l(l+1)$ and of L_z for the eigenvalue $\hbar m$. The latter component therefore has a sharp value in this state. The two components L_x and L_y then are bound to be undetermined in this state because of the non-commutability of angular momentum components (see Problems 50 and 51). This corresponds to the *classical picture* in the following way. For each individual motion, all three components of $\boldsymbol{L}$ have fixed values, viz.

$$L_x = L\sin\vartheta\cos\varphi; \qquad L_y = L\sin\vartheta\sin\varphi; \qquad L_z = L\cos\vartheta .$$

We are, however, using incomplete information which permits exact determination of L_z but none at all of the "phase" angle φ. Then, in any individual case, we can only determine phase averages as

$$\overline{L}_x = \frac{1}{2\pi}\int_0^{2\pi} d\varphi\, L_x \quad \text{etc.}$$

Since L_z does not depend on φ, it has a sharp value; the phase averages of L_x and L_y vanish. The difference of classical and quantum mechanics lies in the fact that in classical mechanics complete information is possible thus making phase averaging avoidable, whereas in quantum mechanics the non-commutability of operators necessarily makes information incomplete, leaving us with averages as best results.

Instead of phase averages, in a stationary state, quantum mechanics permits the calculation of the expectation values or state averages,

$$\langle L_x \rangle = \oint d\Omega\, Y^*_{l,m} L_x Y_{l,m} \quad \text{etc.}$$

The expectation value describes the average result obtained by a large number of independent measurements performed on systems all in the same state. If all these measurements give the same result (as in our case for L_z), then the value is sharp and the expectation value becomes an eigenvalue; if there occur different results, only their average can be derived from quantum theory.

Instead of calculating the averages of the components, it is expedient to use the combinations L^+ and L^-. Application of the operators (a.10) to $Y_{l,m}$ gives (see Appendix and Eq. (56.14))

$$\left.\begin{aligned} L^+ Y_{l,m} &= -\hbar\sqrt{(l+m+1)(l-m)}\, Y_{l,m+1}, \\ L^- Y_{l,m} &= -\hbar\sqrt{(l+m)(l-m+1)}\, Y_{l,m-1}. \end{aligned}\right\} \tag{58.1}$$

Therefore, in $\langle L^+\rangle$ and $\langle L^-\rangle$ there only occur integrals of the types

$$\oint d\Omega\, Y^*_{l,m}\, Y_{l,m\pm 1},$$

vanishing in consequence of orthogonality. The expectation values of L_x and L_y hence are zero, as were the phase averages of the classical picture.

b) The two real wave functions under consideration are

$$u_+ = \frac{1}{2}(Y_{l,m} + Y_{l,-m}) \propto \cos m\varphi,$$

and

$$u_- = \frac{1}{2i}(Y_{l,m} - Y_{l,-m}) \propto \sin m\varphi. \tag{58.2}$$

Both are eigenfunctions of L^2, but not of any component of $\mathbf{L}$, since

$$\begin{aligned} L^\pm u_+ &= -\frac{\hbar}{2}\{\sqrt{(l\pm m+1)(l\mp m)}\, Y_{l,m\pm 1} + \sqrt{(l\mp m+1)(l\pm m)}\, Y_{l,-m\pm 1}\}, \\ L^\pm u_- &= i\,\frac{\hbar}{2}\{\sqrt{(l\pm m+1)(l\mp m)}\, Y_{l,m\pm 1} - \sqrt{(l\mp m+1)(l\pm m)}\, Y_{l,-m\pm 1}\}, \end{aligned} \tag{58.3}$$

$$L_z u_+ = \hbar m i u_-; \qquad L_z u_- = -\hbar m i u_+. \tag{58.4}$$

Therefore, in all three cases, expectation values only can be derived. For L_x and L_y (or L^+ and L^-) we have essentially to repeat the procedure of a) and find expectation values zero in consequence of orthogonality. For L_z, Eqs. (58.4) also lead to zero expectation values because

$$\langle L_z\rangle = \oint d\Omega\, u^*_\pm\, L_z u_\pm = \pm\hbar m i \oint d\Omega\, u_\pm u_\mp = 0$$

since u_+ and u_- are orthogonal. This can easily be seen from the φ part of the integral

$$\oint d\Omega\, u_+ u_- \propto \int_0^{2\pi} d\varphi \cos m\varphi \sin m\varphi = 0.$$

There is, however, a striking difference between $\langle L_x\rangle$ and $\langle L_y\rangle$ on the one hand, and $\langle L_z\rangle$ on the other. In the classical picture, the wave functions u_+ and u_- still correspond to states with all phase angles φ

possible as before under a), the angle ϑ, however, now being restricted (not as before to one sharp, but) to two possible values $\vartheta_1 = \vartheta$ and $\vartheta_2 = \pi - \vartheta$ leading to $L_z = \pm L \cos\vartheta$ and

$$\overline{L}_z = L\cos\vartheta - L\cos\vartheta = 0 .$$

In quantum mechanics, the expectation value of L_z follows from a mixture of two states with $+\hbar m$ and $-\hbar m$ in equal parts. This can be seen from Eqs. (58.3) and (58.4): By application of L_x and L_y, quite different functions are formed instead of u_+ and u_-; by applying L_z we merely exchange u_+ and u_- so that by their linear combination we can again construct eigenfunctions of L_z, viz.

$$u_1 = u_+ + i u_- \quad \text{with } L_z u_1 = \hbar m u_1$$

and

$$u_2 = u_+ - i u_- \quad \text{with } L_z u_2 = -\hbar m u_2 .$$

NB. Real combinations of the type (58.2) play a role in quantum chemistry where a favoured direction to a neighbouring atom in the molecule often has precedence over considerations of angular momentum.

Problem 59. Construction of radial momentum operator

To construct the canonical conjugate operator to the radial coordinate r. What is its representation in the Schrödinger theory?

Solution. In classical mechanics, the conjugate momentum p_r of the radius r is defined as the projection of the vector $\boldsymbol{p}$ upon the direction of $\boldsymbol{r}$:

$$p_r = \frac{1}{r}(\boldsymbol{p}\,\boldsymbol{r}) . \tag{59.1}$$

In quantum mechanics, this definition becomes ambiguous since the components of $\boldsymbol{p}$ and $\boldsymbol{r}$ do not commute and any linear combination

$$p_r = \lambda\,\boldsymbol{p}\frac{\boldsymbol{r}}{r} + (1-\lambda)\frac{\boldsymbol{r}}{r}\boldsymbol{p} \tag{59.2}$$

corresponds to the classical expression (59.1).

Now, p_r should be a hermitian operator, i.e. $p_r = p_r^\dagger$, which leads to $\lambda = 1 - \lambda$ or $\lambda = \frac{1}{2}$,

$$p_r = \frac{1}{2}\left(\boldsymbol{p}\frac{\boldsymbol{r}}{r} + \frac{\boldsymbol{r}}{r}\boldsymbol{p}\right) . \tag{59.3}$$

In order to show that this symmetrical expression is indeed the canonical conjugate of r, we have to check whether the commutation relation

$$p_r r - r p_r = \frac{\hbar}{i} \tag{59.4}$$

will hold here if it holds for the rectangular components. This proof will be given by using the relation

$$\boldsymbol{p}\,\boldsymbol{r} - \boldsymbol{r}\,\boldsymbol{p} = 3\frac{\hbar}{i}$$

which follows from rectangular component decomposition of the scalar products. Hence,

$$\begin{aligned} p_r r - r p_r &= \frac{1}{2}\left\{(\boldsymbol{p}\,\boldsymbol{r}) + \frac{1}{r}(\boldsymbol{r}\,\boldsymbol{p})r - r(\boldsymbol{p}\,\boldsymbol{r})\frac{1}{r} - (\boldsymbol{r}\,\boldsymbol{p})\right\} \\ &= \frac{1}{2}\left\{3\frac{\hbar}{i} + \frac{1}{r}\left(\boldsymbol{p}\,\boldsymbol{r} - 3\frac{\hbar}{i}\right)r - r(\boldsymbol{p}\,\boldsymbol{r})\frac{1}{r}\right\} \\ &= \frac{1}{2r}\left\{(\boldsymbol{p}\,\boldsymbol{r}) - r^2(\boldsymbol{p}\,\boldsymbol{r})\frac{1}{r^2}\right\}r\,. \end{aligned}$$

Here, in the last term,

$$r^2(\boldsymbol{p}\,\boldsymbol{r}) = (x^2+y^2+z^2)(p_x x + p_y y + p_z z)$$

which may be commuted according to

$$x^2 p_x = -2\frac{\hbar}{i}x + p_x x^2 \quad \text{etc.}$$

thus arriving at

$$r^2(\boldsymbol{p}\,\boldsymbol{r}) = -2\frac{\hbar}{i}r^2 + (\boldsymbol{p}\,\boldsymbol{r})r^2\,.$$

We then obtain

$$p_r r - r p_r = \frac{1}{2r}\left\{(\boldsymbol{p}\,\boldsymbol{r}) + \left[2\frac{\hbar}{i}r^2 - (\boldsymbol{p}\,\boldsymbol{r})r^2\right]\frac{1}{r^2}\right\}r = \frac{\hbar}{i}$$

as postulated in Eq. (59.4).

In order to find the special representation of p_r for the Schrödinger theory, we rewrite Eq. (59.3) in rectangular components,

$$p_r = \frac{1}{2}\left\{p_x\frac{x}{r} + p_y\frac{y}{r} + p_z\frac{z}{r} + \frac{x}{r}p_x + \frac{y}{r}p_y + \frac{z}{r}p_z\right\}.$$

where

$$p_x = \frac{\hbar}{i}\frac{\partial}{\partial x} \quad \text{etc.}$$

and therefore

$$\left(p_x \frac{x}{r} + \frac{x}{r} p_x\right)\psi = \frac{\hbar}{i}\left\{\frac{\partial}{\partial x}\left(\frac{x}{r}\psi\right) + \frac{x}{r}\frac{\partial \psi}{\partial x}\right\}$$
$$= \frac{\hbar}{i}\left\{2\frac{x}{r}\frac{\partial \psi}{\partial x} + \left(\frac{1}{r} - \frac{x^2}{r^3}\right)\psi\right\}.$$

Addition of the analogous expressions in y and z, observing that

$$x\frac{\partial \psi}{\partial x} + y\frac{\partial \psi}{\partial y} + z\frac{\partial \psi}{\partial z} = (\boldsymbol{r}\nabla)\psi = r\frac{\partial \psi}{\partial r},$$

leads to the result (cf. Problem 49)

$$p_r = \frac{\hbar}{i}\left(\frac{\partial}{\partial r} + \frac{1}{r}\right). \tag{59.5}$$

The hermitian character of this operator can easily be checked by showing it to be self-adjoint, i.e. to satisfy the relation

$$\langle u | p_r v\rangle = \langle p_r u | v\rangle$$

for any pair of complex functions u and v for which these integrals exist:

$$\langle u | p_r v\rangle = \int d\tau\, u^* \frac{\hbar}{i}\left(\frac{\partial v}{\partial r} + \frac{1}{r} v\right);$$

$$\langle p_r u | v\rangle = \int d\tau \left\{\frac{\hbar}{i}\left(\frac{\partial u}{\partial r} + \frac{1}{r} u\right)\right\}^* v = \int d\tau \left\{-\frac{\hbar}{i}\left(\frac{\partial u^*}{\partial r} + \frac{1}{r} u^*\right)\right\} v.$$

These two integrals become equal if

$$\frac{\hbar}{i}\int d\tau \left\{u^* \frac{\partial v}{\partial r} + u^* \frac{1}{r} v + \frac{\partial u^*}{\partial r} v + u^* \frac{1}{r} v\right\} = 0$$

or if

$$\oint d\Omega \int_0^\infty dr\, r^2 \left\{\frac{\partial}{\partial r}(u^* v) + \frac{2}{r} u^* v\right\} = 0.$$

The inner integral may be reshaped into

$$\int_0^\infty dr \frac{\partial}{\partial r}(u^* v r^2) = [u^* v r^2]_0^\infty$$

which indeed vanishes if u^* and v remain finite at $r=0$ and vanish exponentially for $r \to \infty$. It should be noted, however, that the normalization integrals $\langle u|u\rangle$ and $\langle v|v\rangle$ will still exist if u and v become singular as $1/r$ at the origin. Normalization alone therefore does not always suffice to exclude solutions without physical significance, e.g. inside a spherical potential well for $l=0$.

Problem 60. Solutions neighbouring eigenfunctions

Given a potential well of radius a and of the dimensionless size parameter

$$W^2 = \frac{2m}{\hbar^2} V_0 a^2 = 9,$$

V_0 being the well depth. The wave functions shall be drawn for $l=0$ and $ka=2.20$; 2.28; 2.36, k being the wave number inside the well. These values of ka are chosen to lie rather close around the lowest eigenvalue.

Solution. Let

$$(\varkappa a)^2 = W^2 - (ka)^2 \tag{60.1}$$

be the energy parameter outside the well, then the wave function will be inside the well

$$u_i = \frac{1}{ka} \sin(kax) \tag{60.2a}$$

and outside

$$u_e = \frac{\cos ka}{\varkappa a} \sinh[\varkappa a(x-1)] + \frac{\sin ka}{ka} \cosh[\varkappa a(x-1)], \tag{60.2b}$$

with

$$x = \frac{r}{a} \tag{60.3}$$

the dimensionless radial coordinate in units of the well radius. The normalization has arbitrarily been chosen to satisfy the boundary conditions $u=0$, $du/dx=1$ at $x=0$. The coefficients of $u_e(x)$ are determined in such a way as to make u and u' continuous at the well surface $x=1$.

The eigenvalue is obtained by making the coefficients of the two hyperbolic functions in (60.2b) exactly opposite to one another,

$$\frac{\cos ka}{\varkappa a} = -\frac{\sin ka}{ka},$$

or, in combination with (60.1),

$$\frac{\tan ka}{ka} = -\frac{1}{\sqrt{W^2-(ka)^2}}, \tag{60.4}$$

cf. Problem 25. For $W^2=9$ the solution is

$$ka=2.2789. \tag{60.5}$$

The wave functions, according to (60.2a, b) are drawn in Fig. 34 where a sketch of the potential well with its respective energy levels

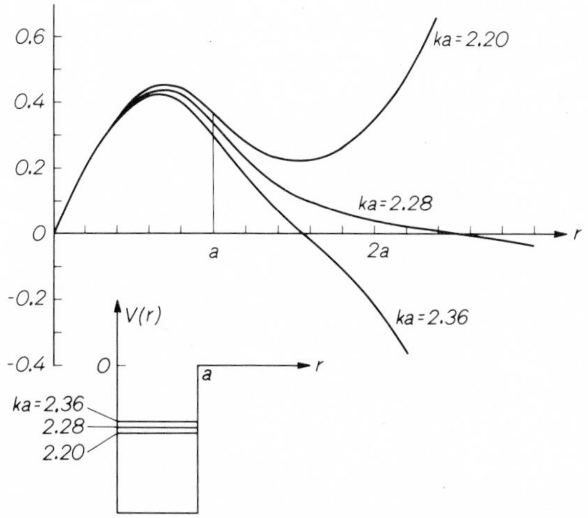

Fig. 34. Wave functions for energy levels in the immediate neighbourhood of an eigenstate

has been added. If the energy chosen is lower than the one of the eigenvalue ($ka=2.20$), the wavelength inside the well is too long and the decrease of wave function at the well surface is not steep enough to bring the amplitude asymptotically down to zero; the function then rises again at large distances. On the other hand, if the energy chosen is too large ($ka=2.36$) and hence the wavelength inside too short, the wave function at the surface falls off too steeply, thus arriving at zero at a finite distance, beyond which it again increases exponentially but with negative amplitude. The intermediate value ($ka=2.28$) lies so close to the eigenvalue (60.5) that the wave function almost tends towards zero. Since the energy is still just a little larger than it should be, there is a very slow rise again at large distance, with negative amplitude, which would vanish for the ka value (60.5).

Problem 61. Quadrupole moment

To determine the quadrupole moment of an angular momentum eigenstate for a potential of spherical symmetry.

Solution. An eigenfunction of angular momentum has the form

$$u = \frac{1}{r}\chi_l(r)\,Y_{l,m}(\vartheta,\varphi); \tag{61.1}$$

the quadrupole tensor has been defined in Problem 54 to be

$$Q_{ik} = 3x_i x_k - r^2\,\delta_{ik}. \tag{61.2}$$

We have to evaluate the quantum mechanical averages

$$\langle Q_{ik}\rangle = \int_0^\infty dr\,|\chi_l(r)|^2 \oint d\Omega\, Q_{ik}\,|Y_{l,m}|^2. \tag{61.3}$$

The non-diagonal elements of Q_{ik} depend upon the angle φ as $\cos\varphi\,\sin\varphi$, $\cos\varphi$, $\sin\varphi$, respectively, whereas $|Y_{l,m}|^2$ is independent of φ, so that the corresponding integrals vanish. In the diagonal elements,

$$\begin{aligned} Q_{xx} &= r^2(3\sin^2\vartheta\cos^2\varphi - 1),\\ Q_{yy} &= r^2(3\sin^2\vartheta\sin^2\varphi - 1),\\ Q_{zz} &= r^2(3\cos^2\vartheta - 1), \end{aligned}$$

integration over φ yields averages $\frac{1}{2}$ for $\cos^2\varphi$ and $\sin^2\varphi$. As

$$\tfrac{3}{2}\sin^2\vartheta - 1 = -\tfrac{1}{2}(3\cos^2\vartheta - 1),$$

the integrals $\langle Q_{xx}\rangle$ and $\langle Q_{yy}\rangle$ may be reduced to $\langle Q_{zz}\rangle$:

$$\langle Q_{xx}\rangle = \langle Q_{yy}\rangle = -\tfrac{1}{2}\langle Q_{zz}\rangle. \tag{61.4}$$

If we use the abbreviation

$$\langle r^2\rangle = \int_0^\infty dr\,r^2\,|\chi_l(r)|^2, \tag{61.5}$$

we then have

$$Q_{zz} = \langle r^2 \rangle \oint d\Omega (3\cos^2\vartheta - 1) |Y_{l,m}|^2 . \tag{61.6}$$

The integral (61.6) can easily be evaluated by using the general relation

$$\cos\vartheta \, Y_{l,m}(\vartheta,\varphi) = a_{l,m} Y_{l+1,m} + a_{l-1,m} Y_{l-1,m}$$

with (61.7)

$$a_{l,m} = \sqrt{\frac{(l+1+m)(l+1-m)}{(2l+1)(2l+3)}} .$$

Making use of orthogonality properties we therefore find from (61.6)

$$Q_{zz} = \langle r^2 \rangle \, (a_{l,m}^2 + a_{l-1,m}^2) .$$

Elementary evaluation of this expression finally leads to

$$Q_{zz} = \langle r^2 \rangle \cdot \frac{2l(l+1) - 6m^2}{(2l-1)(2l+3)} . \tag{61.8}$$

Only[11] S states ($l=0, m=0$) have no quadrupole moment, as befits their spherical symmetry. For P states, e. g., we find

$$Q_{zz} = \tfrac{4}{5}(1 - \tfrac{3}{2} m^2),$$

i. e. the $m=0$ state has a positive Q_{zz} corresponding to an oblong shape, twice as large as the negative Q_{zz} of the $m=\pm 1$ states which apparently are of oblate shape. This is quite reasonable, because the superposition of all three states leads to a configuration of spherical symmetry (closed-shell configuration). This closed-shell effect, indeed, holds for any value of l. If we sum up over a shell, we have

$$\sum_{m=-l}^{+l} Q_{zz} = \frac{2\langle r^2 \rangle}{(2l-1)(2l+3)} \sum_{m=-l}^{+l} [l(l+1) - 3m^2]$$

with the sum vanishing because

$$\sum_{m=-l}^{+l} m^2 = \tfrac{1}{3} l(l+1)(2l+1) .$$

Problem 62. Particle enclosed in a sphere

To determine the energy levels for a particle enclosed in a sphere of zero potential, with infinitely high potential walls defining its surface of radius R.

---

[11] The quadrupole moment vanishes also for $l=3$, $m=2$, but this is the only exception to the rule, at least up to $l=16$.

Solution. The Schrödinger equation may be factorized so that its solutions become

$$u(r,\vartheta,\varphi)=\frac{1}{r}\chi_l(r)\,Y_{l,m}(\vartheta,\varphi) \tag{62.1}$$

with radial equations

$$\chi_l''+\left[k^2-\frac{l(l+1)}{r^2}\right]\chi_l=0;\qquad k^2=\frac{2mE}{\hbar^2} \tag{62.2}$$

in $0\le r\le R$, and zero outside. This differential equation can, by using $z=kr$ as variable and splitting off a factor $z^{\frac{1}{2}}$,

$$\chi_l=z^{\frac{1}{2}}\,\varphi(z),$$

be reduced to

$$\varphi''+\frac{1}{z}\varphi'+\left\{1-\frac{(l+\frac{1}{2})^2}{z^2}\right\}\varphi=0,$$

i.e. to the Bessel equation whose solutions are $J_{\pm(l+\frac{1}{2})}(z)$. We therefore may write the complete solution of (62.2)

$$\chi_l(r)=\sqrt{\frac{\pi k r}{2}}\,\{C_1 J_{l+\frac{1}{2}}(kr)+C_2 J_{-(l+\frac{1}{2})}(kr)\}. \tag{62.3}$$

It has become usual to introduce the so-called spherical Bessel functions

$$\left.\begin{aligned} j_l(z)&=\sqrt{\frac{\pi z}{2}}\,J_{l+\frac{1}{2}}(z),\\ n_l(z)&=(-1)^{l+1}\sqrt{\frac{\pi z}{2}}\,J_{-(l+\frac{1}{2})}(z)\end{aligned}\right\} \tag{62.4}$$

with the asymptotic behaviour for large positive values of z:

$$j_l(z)\to\sin\left(z-\frac{l\pi}{2}\right);\qquad n_l(z)\to-\cos\left(z-\frac{l\pi}{2}\right) \tag{62.5}$$

and with the power series around the origin, approximated for $|z|\ll l+\frac{1}{2}$ each by its first term,

$$j_l=\frac{2^l\,l!}{(2l+1)!}\,z^{l+1};\qquad n_l(z)=-\frac{(2l)!}{2^l\,l!}\,z^{-l}. \tag{62.6}$$

The second term of the solution (62.3) leads, at $r=0$, to $\chi_l \propto r^{-l}$ and $u \propto r^{-l-1}$. The normalization integral,

$$\int d\tau\,|u|^2 = \int_0^\infty dr\,|\chi_l|^2 = 1\,, \tag{62.7}$$

therefore does not exist unless $C_2=0$ in (62.3). This does, however, not hold in the case $l=0$ where the singularity of u is not pronounced enough to let the integral diverge at $r=0$. Even in this case the singular solution must be excluded, because it would make the energy integral[12]

$$E = \frac{\hbar^2}{2m}\int d\tau(\nabla u)^2 \tag{62.8}$$

diverge at the origin (cf. also Problem 65). Hence, normalizable solutions are restricted to

$$\chi_l(r) = C j_l(kr)\,. \tag{62.9}$$

From this set of solutions we select the eigenfunctions by the condition

$$j_l(kR) = 0 \tag{62.10a}$$

or

$$J_{l+\frac{1}{2}}(kR) = 0\,. \tag{62.10b}$$

Since, for each given value of $l+\frac{1}{2}$, the Bessel function has an infinite number of zeros, we find an infinite number of values $k_{n_r,l}$ and of energy levels

$$E_{n_r,l} = \frac{\hbar^2}{2m}k_{n_r,l}^2 \tag{62.11}$$

for each l, with $n_r = 1, 2, 3, \ldots$ the radial quantum number counting the zeros.

For the lowest l values, the spherical Bessel functions are

$$\left.\begin{aligned} j_0(z) &= \sin z\,; \quad j_1(z) = \frac{\sin z}{z} - \cos z\,; \\ j_2(z) &= -3\frac{\cos z}{z} + \left(\frac{3}{z^2} - 1\right)\sin z\,, \end{aligned}\right\} \tag{62.12}$$

---

[12] For $l=0$ and small r, this solution would become $u \sim \frac{1}{r}$ and the kinetic energy integral $E \sim \int dr/r^2$.

and for higher values of l they may easily be constructed from the recurrence relation

$$j_l(z) = \frac{l}{z} j_{l-1}(z) - j'_{l-1}(z). \tag{62.13}$$

Their zeros may be determined from simple transcendental equations:

$$\left.\begin{aligned} j_0(z)=0 &\quad \text{if } \sin z = 0 \quad \text{or } z = n_r \pi; \\ j_1(z)=0 &\quad \text{if } \tan z = z; \\ j_2(z)=0 &\quad \text{if } \tan z = \frac{3z}{3-z^2} \quad \text{etc.} \end{aligned}\right\} \tag{62.14}$$

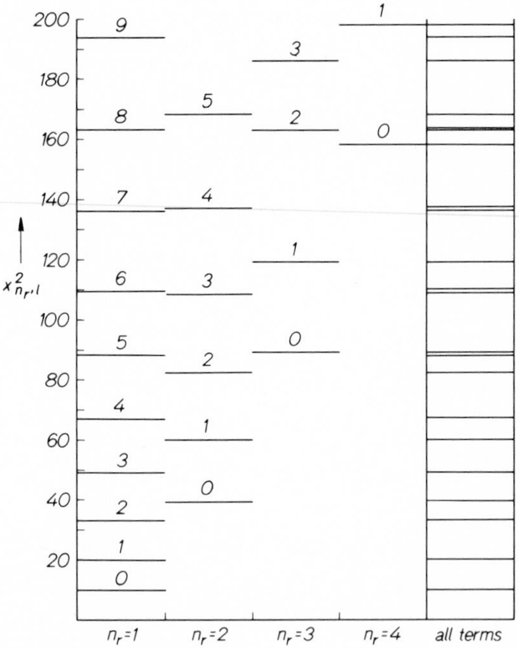

Fig. 35. Energy levels for a particle enclosed in a sphere

They all tend towards either $n\pi$ for even, or $(n+\frac{1}{2})\pi$ for odd l. In Fig. 35, the lowest energy values have been represented in multiples of $\hbar^2/2mR^2$, and the parameter

$$x_{n_r,l} = k_{n_r,l} R \tag{62.15}$$

has been tabulated completely up to 15.

Table of the values of $x_{n_r,l}$

l \ n_r	1	2	3	4
0	3.142	6.283	9.425	12.566
1	4.493	7.725	10.904	14.066
2	5.764	9.095	12.323	
3	6.988	10.417	13.698	
4	8.183	11.705		
5	9.356	12.967		
6	10.513	14.207		
7	11.657			
8	12.791			
9	13.916			

NB. This problem is closely related to the simpler, one-dimensional one dealt with in Problem 18. The solutions for $l=0$ exactly correspond to the antisymmetric wave functions of the one-dimensional problem.

Problem 63. Square well of finite depth

To determine the eigenvalues of bound states in the square well potential

$$V(r) = \begin{cases} -V_0 & \text{for } r<R, \\ 0 & \text{for } r>R \end{cases} \tag{63.1}$$

for $l=0,1,2$. Numerical results shall then be determined for the "well size" $2mV_0R^2/\hbar^2=100$.

Solution. With the abbreviations

$$\frac{2m|E|}{\hbar^2} = \varkappa^2; \quad \frac{2mV_0}{\hbar^2} = k_0^2; \quad \frac{2m(V_0-|E|)}{\hbar^2} = k^2 \tag{63.2}$$

the radial part of the Schrödinger equation becomes

$$\chi_l'' + \left(k^2 - \frac{l(l+1)}{r^2}\right)\chi_l=0 \tag{63.3a}$$

inside, and

$$\chi_l'' + \left(-\varkappa^2 - \frac{l(l+1)}{r^2}\right)\chi_l=0 \tag{63.3b}$$

outside the well, for bound states. Eq. (63.3a), with the obvious boundary condition $\chi_l(0)=0$, is solved by a spherical Bessel function,

$$\chi_l(r)=A j_l(kr) \qquad r<R; \tag{63.4a}$$

the solution of (63.3b) which decreases exponentially as $e^{-\varkappa r}$ for large values of r is a spherical Hankel function of imaginary argument,

$$\chi_l(r)=B h_l^{(1)}(i\varkappa r) \qquad r>R. \tag{63.4b}$$

Here A and B are constants still to be determined from continuity and normalization relations. Their ratio may be eliminated from the continuity relation of the logarithmic derivative at the well surface, $r=R$,

$$i\varkappa R\frac{h_l^{(1)\prime}(i\varkappa R)}{h_l^{(1)}(i\varkappa R)}=kR\frac{j_l'(kR)}{j_l(kR)} \tag{63.5}$$

where the primes denote differentiations to the respective arguments. Since (63.5) relates $\varkappa$ with k, it fixes, according to (63.2), the eigenvalues of the energy in any given well.

The functions only symbolized in Eq. (63.5) run as follows for the smallest values of l:

$$\left.\begin{aligned} j_0(z)&=\sin z, & h_0^{(1)}(z)&=-i\,e^{iz},\\ j_1(z)&=\frac{\sin z}{z}-\cos z, & h_1^{(1)}(z)&=\left(-\frac{i}{z}-1\right)e^{iz},\\ j_2(z)&=\left(\frac{3}{z^2}-1\right)\sin z-\frac{3}{z}\cos z, & h_2^{(1)}(z)&=\left(-\frac{3i}{z^2}-\frac{3}{z}+i\right)e^{iz}.\end{aligned}\right\} \tag{63.6}$$

For higher values of l, there may be used the recurrence formulae

$$\left.\begin{aligned} j_{l+1}(z)&=\frac{2l+1}{z}j_l(z)-j_{l-1}(z);\\ h_{l+1}^{(1)}(z)&=\frac{2l+1}{z}h_l^{(1)}(z)-h_{l-1}^{(1)}(z).\end{aligned}\right\} \tag{63.7}$$

With the further abbreviations

$$kR=x; \quad k_0R=x_0; \quad k/k_0=\xi; \quad \varkappa R=x_0\sqrt{1-\xi^2} \tag{63.8}$$

where x_0^2 is the "well size" as defined above, Eq. (63.5) after elementary though lengthy calculation leads to an eigenvalue relation which may be written

$$\tan(x_0\xi)=f_l(x_0,\xi) \tag{63.9}$$

with

$$f_0(x_0,\xi) = -\frac{\xi}{\sqrt{1-\xi^2}}\,; \tag{63.10a}$$

$$f_1(x_0,\xi) = \frac{x_0\xi}{1+\dfrac{\xi^2}{1-\xi^2}\left(1+x_0\sqrt{1-\xi^2}\right)}\,; \tag{63.10b}$$

$$f_2(x_0,\xi) = x_0\xi\cdot\frac{1+x_0\sqrt{1-\xi^2}+\frac{1}{3}x_0^2\xi^2(1-\xi^2)}{1+x_0\sqrt{1-\xi^2}\,[1-\frac{1}{3}x_0^2\xi^2(1-\xi^2)]}\,. \tag{63.10c}$$

The values of ξ, and thus of x, satisfying these equations are most easily found by a graphical procedure. In Fig. 36 the function $\tan x_0\xi$ has been drawn for $x_0=10$ in the physical interval $0<\xi<1$. Its intersections with the functions $f_l(\xi)$ are determined and by tabulation of the

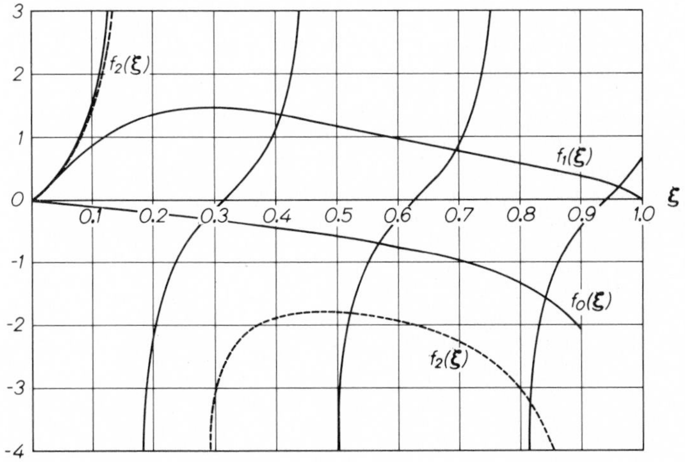

Fig. 36. Graphical determination of eigenvalues in the square well

functions in the immediate neighbourhood of the intersections corrected up to 4 digits. In the accompanying table, we have summarized the results for $x_0=10$ and compared them with those of a square well surrounded by an infinitely high wall as determined in the preceding problem. It should be noted that all level positions are lowered in the finite well so that it is to be expected, from the values of x given in the preceding problem for higher values of l, that bound states up to at least $l=5$ may occur in the example here discussed.

Well of finite depth, x values for			Infinitely deep well, x values for		
$l=0$	$l=1$	$l=2$	$l=0$	$l=1$	$l=2$
2.853	4.070	5.226	3.142	4.493	5.764
5.679	6.958	8.124	6.283	7.725	9.095
8.422	9.625		9.425	10.904	

NB. This problem is closely related to the simpler, one-dimensional one of Problem 25 whose antisymmetric solutions correspond to those for $l=0$ of the present problem.

Problem 64. Wood-Saxon potential

Let a potential hole of spherical symmetry be described by

$$V(r) = -\frac{V_0}{1+e^{\frac{r-R}{a}}} \tag{64.1}$$

with $a \ll R$. The bound states with $l=0$ in this potential shall be determined.

NB. This potential has been used to describe the interaction of a neutron with a heavy nucleus. The parameter R is the nuclear radius, and the parameter a determines the thickness of a surface layer in which the potential falls off from $V=0$ outside to $V=-V_0$ inside the nucleus (Fig. 37). For $a=0$ the simple potential well is obtained with a potential jump at the surface.

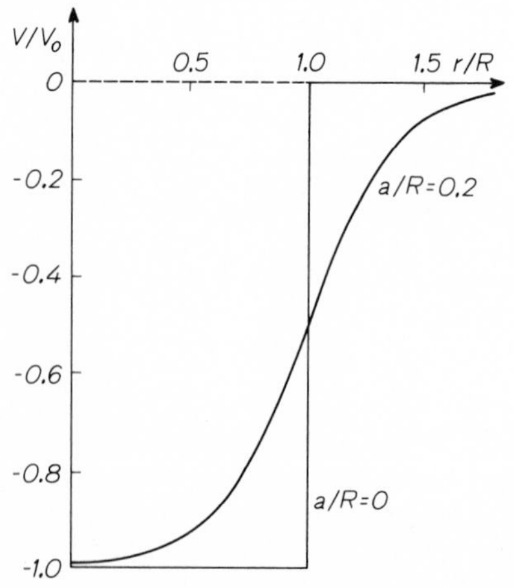

Fig. 37. Wood-Saxon potential for $a/R=0.2$. The rectangular hole corresponds to $a/R=0$

Solution. With the notations

$$u(r)=\frac{1}{r}\chi(r);\qquad \frac{2mE}{\hbar^2}a^2=-\beta^2;\qquad \frac{2mV_0}{\hbar^2}a^2=\gamma^2 \tag{64.2}$$

the Schrödinger equation

$$\frac{d^2u}{dr^2}+\frac{2}{r}\frac{du}{dr}+\frac{2m}{\hbar^2}(E-V)u=0$$

may be transformed to the independent variable

$$y=\frac{1}{1+e^{\frac{r-R}{a}}}. \tag{64.3}$$

It then runs

$$y(1-y)\frac{d^2\chi}{dy^2}+(1-2y)\frac{d\chi}{dy}+\frac{-\beta^2+\gamma^2 y}{y(1-y)}\chi=0.$$

This equation is to be solved with the boundary conditions

$$\begin{aligned}&\chi=0\quad\text{at}\quad y=0\quad(r=\infty)\\&\qquad\text{and}\ y\simeq 1-e^{-R/a}\simeq 1\quad(r=0).\end{aligned} \tag{64.4}$$

Writing

$$\chi(y)=y^\nu(1-y)^\mu f(y) \tag{64.5}$$

we find

$$\begin{aligned}y(1-y)f''&+[(2\nu+1)-y(2\nu+2\mu+2)]f'\\&+\left[\nu(\nu-1)\frac{1-y}{y}+\mu(\mu-1)\frac{y}{1-y}-2\mu\nu\right.\\&\left.+\nu\frac{1-2y}{y}-\mu\frac{1-2y}{1-y}+\frac{-\beta^2+\gamma^2 y}{y(1-y)}\right]f=0.\end{aligned} \tag{64.6}$$

If the parameters ν and μ are chosen to be

$$\nu=\beta;\qquad \mu^2=\beta^2-\gamma^2 \tag{64.7}$$

the factor of f in (64.6) does not depend on y and the equation becomes the hypergeometric differential equation:

$$y(1-y)f''+[(2\nu+1)-y(2\nu+2\mu+2)]f'-(\nu+\mu)(\nu+\mu+1)f=0. \tag{64.8}$$

This leads to the (unnormalized) solution

$$\chi=y^\nu(1-y)^\mu\,{}_2F_1(\mu+\nu,\mu+\nu+1,2\nu+1;y). \tag{64.9}$$

Another independent solution would be obtained by replacing v by $-v$ but would contradict the first boundary condition (64.4). At large values of $r(y\to 0)$, Eq. (64.9) yields

$$\chi \simeq y^{v}; \quad y \propto \exp\left(-\frac{r-R}{a}\right); \quad \chi \propto \mathrm{e}^{-\beta R/a} \exp\left(-\sqrt{\frac{2m|E|}{\hbar^2}}\, r\right).$$

This is the correct asymptotic behaviour.

The other boundary condition, at (or near) $y=1$, is not so easily satisfied. In order to describe the behaviour of (64.9) in the vicinity of $y=1$ we use the relation

$$\begin{aligned} &{}_2F_1(\mu+v, \mu+v+1, 2v+1; y) \\ &= \frac{\Gamma(2v+1)\Gamma(-2\mu)}{\Gamma(v-\mu+1)\Gamma(v-\mu)}\, {}_2F_1(v+\mu, v+\mu+1, 2\mu+1; 1-y) \\ &+ \frac{\Gamma(2v+1)\Gamma(2\mu)}{\Gamma(v+\mu+1)\Gamma(v+\mu)} (1-y)^{-2\mu}\, {}_2F_1(v-\mu, v-\mu+1, -2\mu+1; 1-y) \end{aligned}$$

which leads to

$$\chi \simeq \frac{\Gamma(2v+1)\Gamma(-2\mu)}{\Gamma(v-\mu+1)\Gamma(v-\mu)} (1-y)^{\mu} + \frac{\Gamma(2v+1)\Gamma(2\mu)}{\Gamma(v+\mu+1)\Gamma(v+\mu)} (1-y)^{-\mu}. \tag{64.10}$$

To discuss this expression we first note that $\gamma > \beta$ so that, according to (64.7), μ turns out to be imaginary:

$$\mu = i\lambda; \quad \lambda = \sqrt{\gamma^2 - \beta^2}. \tag{64.11}$$

We then write

$$\begin{aligned} \chi \simeq &\frac{\Gamma(2\beta+1)\Gamma(-2i\lambda)}{\Gamma(\beta+1-i\lambda)\Gamma(\beta-i\lambda)} \\ &\times \left\{(1-y)^{i\lambda} + \frac{\Gamma(2i\lambda)\Gamma(\beta+1-i\lambda)\Gamma(\beta-i\lambda)}{\Gamma(-2i\lambda)\Gamma(\beta+1+i\lambda)\Gamma(\beta+i\lambda)} (1-y)^{-i\lambda}\right\}. \end{aligned} \tag{64.12}$$

In the neighbourhood of $r=0$, the exponential in (64.3) is quite small so that approximately

$$1 - y = \mathrm{e}^{-R/a} \tag{64.13}$$

holds there. Let us further define

$$\varphi = \arg \Gamma(\beta + i\lambda); \quad \psi = \arg \Gamma(2i\lambda). \tag{64.14}$$

If now we omit the factor in front of the curly bracket in (64.12), since the function is unnormalized, and put (64.13) into (64.12) for $1-y$, the boundary condition leads to

$$e^{-i\frac{\lambda R}{a}}+\exp(2i\psi-4i\varphi-2i\tan^{-1}\lambda/\beta)\,e^{i\frac{\lambda R}{a}}=0.$$

This leads on to the eigenvalue condition

$$\exp\{i(\psi-2\varphi-\tan^{-1}\lambda/\beta)\}\cos\left(\frac{\lambda R}{a}+\psi-2\varphi-\tan^{-1}\lambda/\beta\right)=0$$

or

$$\frac{\lambda R}{a}+\psi-2\varphi-\tan^{-1}\frac{\lambda}{\beta}=(2n-1)\frac{\pi}{2};\qquad n=0,\pm1,\pm2,\dots \tag{64.15}$$

This is indeed a relation between β and λ, and therefore between E and V_0, for given values of the parameters R and a.

Formula (64.15) is not yet very well suited for practical purposes. We may, however, obtain the arguments φ and ψ of the Γ functions by using the expansion[13]

$$\Gamma(x+iy)=\xi e^{i\eta};$$

$$\eta=y\left\{-C+\sum_{n=1}^{\infty}\left(\frac{1}{n}-\frac{1}{y}\tan^{-1}\frac{y}{x+n-1}\right)\right\}$$

where $C=0{,}5772\dots$ is the Euler constant. We then find

$$\psi=-\frac{\pi}{2}+2\lambda\left\{-C+\sum_{n=1}^{\infty}\left(\frac{1}{n}-\frac{1}{2\lambda}\tan^{-1}\frac{2\lambda}{n}\right)\right\};$$

$$\varphi=-\tan^{-1}\frac{\lambda}{\beta}+\lambda\left\{-C+\sum_{n=1}^{\infty}\left(\frac{1}{n}-\frac{1}{\lambda}\tan^{-1}\frac{\lambda}{n+\beta}\right)\right\}.$$

Putting these expressions into (64.15), we get

$$\frac{\lambda R}{a}-\sum_{n=1}^{\infty}\left(\tan^{-1}\frac{2\lambda}{n}-2\tan^{-1}\frac{\lambda}{n+\beta}\right)+\tan^{-1}\frac{\lambda}{\beta}=n\pi. \tag{64.16}$$

To discuss this result it is expedient to introduce the parameters

$$\left.\begin{aligned} k&=\frac{\lambda}{a}=\left[\frac{2m}{\hbar^2}(V_0-|E|)\right]^{\frac{1}{2}};\qquad \varkappa=\frac{\beta}{a}=\left[\frac{2m}{\hbar^2}|E|\right]^{\frac{1}{2}};\\ k_0&=\frac{1}{a}\sqrt{\lambda^2+\beta^2}=\left[\frac{2m}{\hbar^2}V_0\right]^{\frac{1}{2}}.\end{aligned}\right\} \tag{64.17}$$

[13] This expansion may easily be derived from standard formulae to be found in the usual mathematical collections, such as Jahnke-Emde-Lösch or Magnus-Oberhettinger.

These parameters are independent of the thickness a of the surface layer so that they remain meaningful in the limit $a \to 0$ of the potential well. Eq. (64.16) may be rewritten in these parameters

$$\frac{kR}{\varkappa R} = -\tan\left\{kR - \sum_{n=1}^{\infty}\left(\tan^{-1}\frac{2ka}{n} - 2\tan^{-1}\frac{ka}{n+\varkappa a}\right)\right\}. \quad (64.18)$$

For $a \to 0$ the sum terms vanish and we fall back upon the eigenvalue condition of the potential well,

$$\frac{kR}{\varkappa R} = -\tan kR.$$

Series expansion of (64.18) into powers of the small parameter

$$\alpha = a/R \quad (64.19)$$

with the abbreviations $kR = x$, $\varkappa R = y$ finally leads to

$$\begin{aligned} -\frac{x}{y} = \tan\{&x - 2\alpha^2 x[\zeta(2)y - \alpha\zeta(3)(x^2+y^2) + \alpha^2\zeta(4)(y^3 - x^2 y) \\ &-\alpha^3\zeta(5)(y^4 - 2x^2y^2 - 3x^4) + \alpha^4\zeta(6)(y^5 - \tfrac{10}{3}x^2y^3 + x^4 y) \\ &-\alpha^5\zeta(7)(y^6 - 5x^2y^4 + 3x^4y^2 + 9x^6)\ldots]\}. \end{aligned}$$

Problem 65. Spherical oscillator

To solve the Schrödinger equation for the spherical oscillator potential

$$V(r) = \tfrac{1}{2}m\omega^2 r^2 \quad (65.1)$$

by factorizing the solution in spherical polar coordinates.

Solution. By the usual factorization

$$u(r,\vartheta,\varphi) = \frac{1}{r}\chi_l(r)\,Y_{l,m}(\vartheta,\varphi) \quad (65.2)$$

we get the radial equations

$$\frac{d^2\chi_l}{dr^2} + \left[\frac{2mE}{\hbar^2} - \frac{m^2\omega^2}{\hbar^2}r^2 - \frac{l(l+1)}{r^2}\right]\chi_l = 0. \quad (65.3)$$

For $l=0$ this again turns out to be identical with the linear oscillator equation (cf. Problem 30); for $l \neq 0$ the centrifugal force will drive the particle outwards so that we get different solutions.

Using the abbreviations

$$\frac{2mE}{\hbar^2}=k^2; \quad \frac{m\omega}{\hbar}=\lambda; \quad \frac{k^2}{2\lambda}=\frac{E}{\hbar\omega}=\mu, \tag{65.4}$$

Eq. (65.3) may be written in the standard form

$$\frac{d^2\chi_l}{dr^2}+\left[k^2-\lambda^2 r^2-\frac{l(l+1)}{r^2}\right]\chi_l=0. \tag{65.5}$$

The behaviour of the solution at $r=0$, determined by the centrifugal term, and its asymptotic behaviour, determined by the oscillator term, suggest to write

$$\chi_l=r^{l+1}\,\mathrm{e}^{-\frac{\lambda}{2}r^2}\,v(r), \tag{65.6}$$

and then to use, instead of r, the variable

$$t=\lambda r^2 \tag{65.7}$$

in analogy to the linear oscillator. This leads to

$$t\frac{d^2v}{dt^2}+\left[(l+\tfrac{3}{2})-t\right]\frac{dv}{dt}-\left[\tfrac{1}{2}(l+\tfrac{3}{2})-\tfrac{1}{2}\mu\right]v=0,$$

i.e. to a Kummer equation with the complete solution

$$\begin{aligned} v=\;&C_1\,{}_1F_1\!\left(\tfrac{1}{2}(l+\tfrac{3}{2}-\mu),\,l+\tfrac{3}{2};\,\lambda r^2\right)\\ &+C_2\,r^{-(2l+1)}\,{}_1F_1\!\left(\tfrac{1}{2}(-l+\tfrac{1}{2}-\mu),\,-l+\tfrac{1}{2};\,\lambda r^2\right). \end{aligned} \tag{65.8}$$

At $r=0$, the second part of the solution contradicts normalization[14] so that $C_2=0$. This clearly differs from the linear oscillator where no boundary condition exists at the origin.

A confluent series behaves asymptotically at large positive values of its argument as

$${}_1F_1(a,c;z)\to\frac{\Gamma(c)}{\Gamma(a)}\,\mathrm{e}^z z^{a-c};$$

with (65.6) and (65.8) this leads to an expression

$$\chi_l \propto r^{l+1}\,\mathrm{e}^{-\frac{1}{2}\lambda r^2}\cdot\mathrm{e}^{\lambda r^2}\,r^{-(l+\frac{3}{2}+\mu)}$$

exponentially divergent. The divergence cannot be avoided except by putting the parameter $a=-n_r$ with $n_r=0,1,2,\ldots$, thus transforming the series into a polynomial of degree n_r. Hence,

$$\tfrac{1}{2}(l+\tfrac{3}{2}-\mu)=-n_r, \tag{65.9a}$$

[14] For $l=0$ the second solution can be normalized, but not the energy integral. Cf. also Problem 62.

or, using (65.4), the energy levels become

$$E=\hbar\omega(2n_r+l+\tfrac{3}{2}); \qquad n_r=0,1,2,\dots \,. \tag{65.9b}$$

Here n_r may be called the radial quantum number. The energy levels start with a zero-point energy of $\frac{3}{2}\hbar\omega$ corresponding to the three degrees of freedom of the problem and are equidistant, as with the linear oscillator:

$$E_n=\hbar\omega(n+\tfrac{3}{2}) \tag{65.10}$$

with

$$n=2n_r+l. \tag{65.11}$$

Gathering the results from (65.6), (65.8) and (65.9a), the complete eigenfunctions may be written

$$u(r,\vartheta,\varphi)=C r^l e^{-\frac{\lambda}{2}r^2}\,{}_1F_1(-n_r,l+\tfrac{3}{2};\lambda r^2)\,Y_{l,m}(\vartheta,\varphi) \tag{65.12}$$

where C still has to be determined from the normalization condition.

The energy levels are *degenerate*, except for the ground state $n=0$, as, for even n, there are $\frac{1}{2}n+1$ partitions of n according to (65.11) and, for odd n, there are $\frac{1}{2}(n+1)$. Since, for each value of l, there are still $2l+1$ different values of m (ranging from $-l$ to $+l$), the degeneracy is again increased by this factor.

NB. This problem should be compared with that of the circular oscillator, Problem 42.

Problem 66. Degeneracy of the spherical oscillator

To show that the spherical oscillator eigenfunction with $l=2$, $m=0$, $n_r=1$ can be constructed by factorization in rectangular coordinates and linear combination of the degenerate solutions.

Solution. In rectangular coordinates the Schrödinger equation of the spherical oscillator may be written

$$\left[\frac{\partial^2 u}{\partial x^2}+(k_1^2-\lambda^2x^2)u\right]+\left[\frac{\partial^2 u}{\partial y^2}+(k_2^2-\lambda^2y^2)u\right]$$
$$+\left[\frac{\partial^2 u}{\partial z^2}+(k_3^2-\lambda^2z^2)u\right]=0 \tag{66.1}$$

with

$$k_1^2+k_2^2+k_3^2=k^2$$

and k^2 and λ the same abbreviations as in the preceding problem. Factorizing

$$u(x,y,z)=f(x)g(y)h(z) \tag{66.2}$$

we get three simple oscillator equations in x, y, and z, the first of which e.g. runs

$$\frac{d^2 f}{dx^2}+(k_1^2-\lambda^2 x^2)f=0. \tag{66.3}$$

It has eigenvalues $\hbar\omega(n_1+\frac{1}{2})$ with non-negative integer n_1. In Problem 30 it has been shown that the (unnormalized) eigenfunction of (66.3) for even $n_1=2n$ is

$$f_{2n}(x)=e^{-\frac{1}{2}\lambda x^2}\,{}_1F_1(-n,\tfrac{1}{2};\lambda x^2) \tag{66.4a}$$

and for odd $n_1=2n+1$

$$f_{2n+1}(x)=e^{-\frac{1}{2}\lambda x^2}\,x\,{}_1F_1(-n,\tfrac{3}{2};\lambda x^2). \tag{66.4b}$$

The same holds with quantum numbers n_2 for $g(y)$ and n_3 for $h(z)$ so that the total energy becomes

$$E_{n_1,n_2,n_3}=\hbar\omega(n_1+n_2+n_3+\tfrac{3}{2}). \tag{66.5}$$

On the other hand, in spherical polar coordinates we have found

$$E_{n_r,l,m}=\hbar\omega(2n_r+l+\tfrac{3}{2}) \tag{66.6}$$

so that the state with $l=2$, $m=0$, $n_r=1$ belongs to $2n_r+l=4$; its eigenfunction, according to (65.12), is

$$u=r^2 e^{-\frac{\lambda}{2}r^2}\,{}_1F_1(-1,\tfrac{7}{2};\lambda r^2)P_2(\cos\vartheta). \tag{66.7}$$

The problem now consists of expressing u, Eq. (66.7), by rectangular coordinates and replacing the result by a linear combination of such solutions $f(x)\,g(y)\,h(z)$ which belong to

$$n_1+n_2+n_3=4. \tag{66.8}$$

We get

$$\begin{aligned} u&=e^{-\frac{\lambda}{2}r^2}(1-\tfrac{2}{7}\lambda r^2)(\tfrac{3}{2}z^2-\tfrac{1}{2}r^2)\\ &=e^{-\frac{1}{2}\lambda r^2}\left\{\frac{\lambda}{7}x^4+\frac{\lambda}{7}y^4-\frac{2\lambda}{7}z^4+\frac{2\lambda}{7}x^2y^2\right.\\ &\quad\left.-\frac{\lambda}{7}y^2z^2-\frac{\lambda}{7}x^2z^2-\frac{1}{2}x^2-\frac{1}{2}y^2+z^2\right\}. \end{aligned} \tag{66.9}$$

Each product fgh has the same exponential factor which we omit when comparing the functions. Since $n_1 \leq 4$, according to (66.4a, b) the first parameter (n) of the confluent series cannot exceed 2. The same, of course, holds for the confluent series in $g(y)$ and $h(z)$. So there will occur only the polynomials

$$\left.\begin{array}{ll} {}_1F_1(0,\tfrac{1}{2};\lambda x^2) \;=1, & x\,{}_1F_1(0,\tfrac{3}{2};\lambda x^2)=x, \\ {}_1F_1(-1,\tfrac{1}{2};\lambda x^2)=1-2\lambda x^2, & x\,{}_1F_1(-1,\tfrac{3}{2};\lambda x^2)=x(1-\tfrac{2}{3}\lambda x^2), \\ {}_1F_1(-2,\tfrac{1}{2};\lambda x^2)=1-4\lambda x^2+\tfrac{4}{3}\lambda^2 x^4. & \end{array}\right\} \quad (66.10)$$

There are 15 different ways of decomposing 4 according to (66.8) into three integers, as is shown in the following table. Since in the polynomial (66.9) to be represented there occur only even powers of x, y, z, the nine rectangular product functions in which odd powers occur (i.e. those where at least one of the three numbers n_1, n_2, n_3 is odd) cannot contribute to the required linear combination. The last column gives the factors by which each rectangular function has to be multiplied in order that their sum may equal the polynomial (66.9). Since we have only six terms to add, but ten terms in (66.9) to explain, there remain four controls for corroborating the identity.

n_1	n_2	n_3	Rectangular coordinate eigenfunction (except for exponential factor)	To be multiplied by
0	0	4	$1-4\lambda z^2+\frac{4}{3}\lambda^2 z^4$	$-6/28\,\lambda$
0	1	3	$yz(1-\frac{2}{3}\lambda z^2)$	0
0	2	2	$(1-2\lambda y^2)(1-2\lambda z^2)$	$-1/28\,\lambda$
0	3	1	$yz(1-\frac{2}{3}\lambda y^2)$	0
0	4	0	$1-4\lambda y^2+\frac{4}{3}\lambda^2 y^4$	$+3/28\,\lambda$
1	0	3	$xz(1-\frac{2}{3}\lambda z^2)$	0
1	1	2	$xy(1-2\lambda z^2)$	0
1	2	1	$xz(1-2\lambda y^2)$	0
1	3	0	$xy(1-\frac{2}{3}\lambda y^2)$	0
2	0	2	$(1-2\lambda x^2)(1-2\lambda z^2)$	$-1/28\,\lambda$
2	1	1	$yz(1-2\lambda x^2)$	0
2	2	0	$(1-2\lambda x^2)(1-2\lambda y^2)$	$+2/28\,\lambda$
3	0	1	$xz(1-\frac{2}{3}\lambda x^2)$	0
3	1	0	$xy(1-\frac{2}{3}\lambda x^2)$	0
4	0	0	$1-4\lambda x^2+\frac{4}{3}\lambda^2 x^4$	$+3/28\,\lambda$

NB. Compare with the degeneracy of the circular oscillator, Problem 42.

Problem 67. Kepler problem

To solve the Schrödinger equation for the bound states of an electron in the field of a point nucleus of infinite mass and charge Ze. (For $Z=1$ this is the theory of the hydrogen atom).

Solution. Using the abbreviations

$$\frac{2mE}{\hbar^2} = -\gamma^2; \qquad \varkappa = \frac{Ze^2 m}{\gamma \hbar^2} = Z\frac{e^2}{\hbar c}\sqrt{-\frac{mc^2}{2E}} \tag{67.1}$$

and factorizing the solution in the usual way,

$$u = \frac{1}{r}\chi_l(r)\, Y_{l,m}(\vartheta, \varphi) \tag{67.2}$$

we find the radial differential equation

$$\chi_l'' + \left(-\gamma^2 + \frac{2\gamma\varkappa}{r} - \frac{l(l+1)}{r^2}\right)\chi_l = 0. \tag{67.3}$$

This equation has a regular singularity at $r=0$ and an irregular singularity at $r=\infty$. In the vicinity of $r=0$ it may be solved by a power series which leads to proportionality with r^{l+1} or r^{-l}. At $r=\infty$ it behaves as $e^{\pm\gamma r}$. Only behaviour as r^{l+1} at small, and as $e^{-\gamma r}$ at large values of r, simultaneously, leads to a result which may be normalized according to[15]

$$\int_0^\infty \chi_l^2(r)\,dr = 1. \tag{67.4}$$

We therefore put

$$\chi_l = r^{l+1} e^{-\gamma r} f(r). \tag{67.5}$$

Using, instead of r, the dimensionless variable

$$z = 2\gamma r \tag{67.6}$$

we thus arrive at the differential equation

$$z f'' + (2l+2-z) f' - (l+1-\varkappa) f = 0. \tag{67.7}$$

This is a Kummer equation of which we need only consider the solution

$$f = {}_1F_1(l+1-\varkappa, 2l+2; z) \tag{67.8}$$

[15] For $l=0$ the r^{-l} solution too could be normalized. The kinetic energy integral, however, would not exist so that it can be excluded. Cf. Problem 62.

regular at the origin, because by (67.5) we have already taken care of the boundary condition. For large values of z, the confluent series (67.8) diverges as e^{2z} thus destroying normalization, except for

$$l+1-\varkappa=-n_r \qquad (n_r=0,1,2,\dots) \tag{67.9}$$

when it becomes a polynomial. Eq. (67.9), because $\varkappa$ is connected through (67.1) with the energy, determines the eigenvalues. If we use the *principal quantum number*

$$n=n_r+l+1, \tag{67.10}$$

we find $\varkappa=n$, or

$$E_n=-\frac{Z^2}{2n^2}\cdot\frac{me^4}{\hbar^2}. \tag{67.11}$$

The levels are degenerate because there are several eigenfunctions

$$u_{n,l,m}=C r^l e^{-\gamma r}\,{}_1F_1(l+1-n,2l+2;2\gamma r)\,Y_{l,m}(\vartheta,\varphi) \tag{67.12}$$

corresponding to each level of given $n>1$, with n values $l=0,1,\dots,n-1$ for the angular momentum quantum number l and with $2l+1$ values

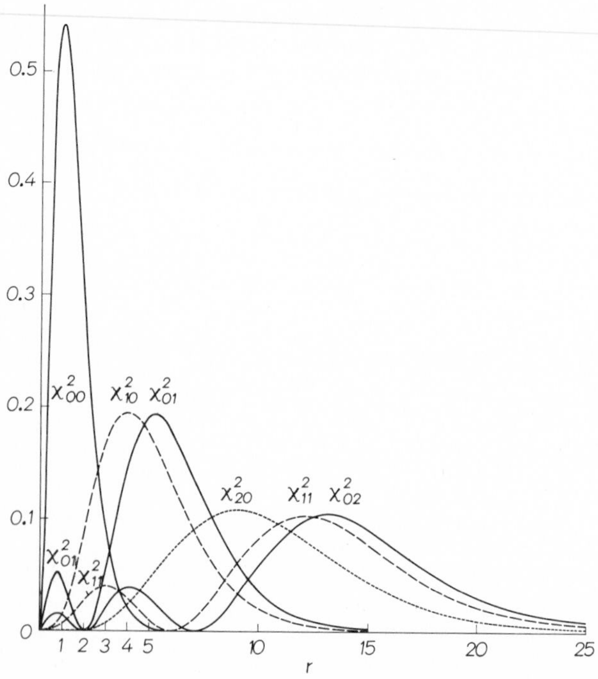

Fig. 38. Radial density in hydrogen atom represented by $|\chi_{l,n_r}|^2$

of the magnetic quantum number m from $m=-l$ to $m=+l$, so that there exist

$$\sum_{l=0}^{n-1}(2l+1)=n^2 \tag{67.13}$$

eigenfunctions for each value of n. Only the ground state, $n=1$, with $l=0$ and $m=0$, is not degenerate.

In the following table, the eigenfunctions for $n=1, 2$ and 3 are given in the normalization (67.4) of the radial part. We have used atomic units with $m=1$, $e=1$, $\hbar=1$ so that

$$E_n=-\frac{Z^2}{2n^2}; \quad \varkappa=\frac{Z}{\gamma}; \quad \gamma=\frac{Z}{n}. \tag{67.14}$$

A graphical representation of $|\chi_{l,n_r}|^2$ is given in Fig. 38. The spherical harmonics are added explicitly in the normalization

$$\oint d\Omega\,|Y_{l,m}|^2=1\,.$$

NB. Concerning the motion of both nucleus and electron about their centre of mass, cf. Problem 150.—Additional fine structure and relativistic effects are treated in Problems 202 and 203.—Atomic structures based on hydrogen-like wave functions, but with two or more electrons, are discussed in Problems 154ff.

Table of the hydrogen-like eigenfunctions

$$u_{n,l,m}=\frac{1}{r}\chi_{l,n_r}(r)\,Y_{l,m}(\vartheta,\varphi)$$

a) Radial parts for $n=1,2,3,4$.

n	l	n_r	Symbol	$\chi_{l,n_r}(r)$
1	0	0	$1s$	$2Z^{\frac{3}{2}}re^{-Zr}$
2	0	1	$2s$	$\frac{1}{\sqrt{2}}Z^{\frac{3}{2}}re^{-\frac{Z}{2}r}\left(1-\frac{1}{2}Zr\right)$
	1	0	$2p$	$\frac{1}{\sqrt{24}}Z^{\frac{5}{2}}r^2e^{-\frac{Z}{2}r}$
3	0	2	$3s$	$\frac{2}{3\sqrt{3}}Z^{\frac{3}{2}}re^{-\frac{Z}{3}r}\left(1-\frac{2}{3}Zr+\frac{2}{27}Z^2r^2\right)$
	1	1	$3p$	$\frac{4}{27}\sqrt{\frac{2}{3}}Z^{\frac{5}{2}}r^2e^{-\frac{Z}{3}r}\left(1-\frac{1}{6}Zr\right)$
	2	0	$3d$	$\frac{4}{81\sqrt{30}}Z^{\frac{7}{2}}r^3e^{-\frac{Z}{3}r}$

n	l	n_r	Symbol	$\chi_{l,n_r}(r)$
4	0	3	$4s$	$\frac{1}{4} Z^{\frac{3}{2}} r \mathrm{e}^{-\frac{Z}{4}r} \left(1 - \frac{3}{4} Z r + \frac{1}{8} Z^2 r^2 - \frac{1}{192} Z^3 r^3\right)$
	1	2	$4p$	$\frac{1}{16}\sqrt{\frac{5}{3}} Z^{\frac{5}{2}} r^2 \mathrm{e}^{-\frac{Z}{4}r} \left(1 - \frac{1}{4} Z r + \frac{1}{80} Z^2 r^2\right)$
	2	1	$4d$	$\frac{1}{64\sqrt{5}} Z^{\frac{7}{2}} r^3 \mathrm{e}^{-\frac{Z}{4}r} \left(1 - \frac{1}{12} Z r\right)$
	3	0	$4f$	$\frac{1}{768\sqrt{35}} Z^{\frac{9}{2}} r^4 \mathrm{e}^{-\frac{Z}{4}r}$

b) Angular parts for $l = 0, 1, 2, 3$.

l	m	$Y_{l,m}(\vartheta, \varphi)$
0	0	$\frac{1}{\sqrt{4\pi}}$
1	0	$\sqrt{\frac{3}{4\pi}} \cos\vartheta$
	± 1	$\pm\sqrt{\frac{3}{8\pi}} \sin\vartheta \, \mathrm{e}^{\pm i\varphi}$
2	0	$\sqrt{\frac{5}{4\pi}} \left(\frac{3}{2}\cos^2\vartheta - \frac{1}{2}\right)$
	± 1	$\pm\sqrt{\frac{15}{8\pi}} \sin\vartheta \cos\vartheta \, \mathrm{e}^{\pm i\varphi}$
	± 2	$\frac{1}{2}\sqrt{\frac{15}{8\pi}} \sin^2\vartheta \, \mathrm{e}^{\pm 2i\varphi}$
3	0	$\frac{1}{4}\sqrt{\frac{7}{\pi}} (5\cos^3\vartheta - 3\cos\vartheta)$
	± 1	$\pm\frac{1}{8}\sqrt{\frac{21}{\pi}} \sin\vartheta (5\cos^2\vartheta - 1) \mathrm{e}^{\pm i\varphi}$
	± 2	$\frac{1}{4}\sqrt{\frac{105}{2\pi}} \sin^2\vartheta \cos\vartheta \, \mathrm{e}^{\pm 2i\varphi}$
	± 3	$\pm\frac{1}{8}\sqrt{\frac{35}{\pi}} \sin^3\vartheta \, \mathrm{e}^{\pm 3i\varphi}$

Problem 68. Hulthén potential

To solve the Schrödinger equation for the so-called Hulthén potential,

$$V(r) = -V_0 \frac{e^{-r/a}}{1-e^{-r/a}} \tag{68.1}$$

in the case $l=0$, and to determine the energy eigenvalues.

Solution. We shall use the dimensionless variable

$$x = r/a \tag{68.2a}$$

and introduce the abbreviations

$$\alpha^2 = -\frac{2mE}{\hbar^2} a^2 > 0; \quad \beta^2 = \frac{2mV_0}{\hbar^2} a^2 > 0 \tag{68.2b}$$

with $\alpha > 0$ and $\beta > 0$. The radial differential equation for $\chi(r) = r\psi(r)$ may then be written

$$\frac{d^2\chi}{dx^2} + \left(-\alpha^2 + \beta^2 \frac{e^{-x}}{1-e^{-x}}\right)\chi = 0. \tag{68.3}$$

By the transformation

$$y = e^{-x} \tag{68.4}$$

the coefficient functions become rational,

$$y^2 \frac{d^2\chi}{dy^2} + y\frac{d\chi}{dy} + \left(-\alpha^2 + \beta^2 \frac{y}{1-y}\right)\chi = 0. \tag{68.5}$$

This equation is to be solved with the boundary conditions

$$\chi = 0 \quad \text{at } y = 0 \quad (r \to \infty); \tag{68.6a}$$

$$\chi = 0 \quad \text{at } y = 1 \quad (r = 0). \tag{68.6b}$$

The solution is found by putting

$$\chi = y^{\alpha}(1-y)\, w(y) \tag{68.7}$$

which transforms (68.5) into the hypergeometric equation

$$y(1-y)w'' + [(2\alpha+1)-(2\alpha+3)y]w' - (2\alpha+1-\beta^2)w = 0 \tag{68.8}$$

the complete solution of which runs as follows:

$$\begin{aligned} w(y) = {} & A\,{}_2F_1(\alpha+1+\gamma, \alpha+1-\gamma, 2\alpha+1; y) \\ & + B y^{-2\alpha}\,{}_2F_1(-\alpha+1+\gamma, -\alpha+1-\gamma, -2\alpha+1; y) \end{aligned} \tag{68.9}$$

with $\gamma = \sqrt{\alpha^2+\beta^2}$.

The first boundary condition, (68.6a), with $\alpha>0$ leads to $B=0$. The second boundary condition, (68.6b), requires closer investigation of the vicinity of $y=1$. This can easily be performed by using the identity

$$
\begin{aligned}
&{}_2F_1(\alpha+1+\gamma,\alpha+1-\gamma,2\alpha'+1;y)\\
&=\frac{\Gamma(2\alpha'+1)\,\Gamma(\varepsilon-1)}{\Gamma(\alpha-\gamma+\varepsilon)\,\Gamma(\alpha+\gamma+\varepsilon)}\,{}_2F_1(\alpha+1+\gamma,\alpha+1-\gamma,2\alpha'+1;1-y)\\
&\quad+(1-y)^{\varepsilon-1}\frac{\Gamma(2\alpha'+1)\,\Gamma(1-\varepsilon)}{\Gamma(\alpha+1+\gamma)\,\Gamma(\alpha+1-\gamma)}\,{}_2F_1(\alpha-\gamma+\varepsilon,\alpha+\gamma+\varepsilon,\varepsilon;1-y)
\end{aligned}
$$

with $\varepsilon=2(\alpha'-\alpha)$. Further,

$$
{}_2F_1(a,b,\varepsilon;1-y)=1+\frac{ab}{\varepsilon}(1-y)+\cdots.
$$

The first term, in the limit $\varepsilon\to 0$, becomes at $y=1$,

$$
\frac{\Gamma(2\alpha+1)}{\Gamma(\alpha+\gamma)}\lim_{\varepsilon\to 0}\frac{\Gamma(\varepsilon-1)}{\Gamma(\alpha-\gamma+\varepsilon)},
$$

and this has a finite value if, and only if

$$
\alpha-\gamma=-n;\qquad n=0,1,2,3,\ldots
$$

because

$$
\lim_{\varepsilon\to 0}\frac{\Gamma(\varepsilon-1)}{\Gamma(\varepsilon-n)}=(-1)^{n+1}\,n!\,.
$$

In the second term, however, a factor $\Gamma(\alpha+1-\gamma)=(-n)!\to\infty$ occurs in the denominator so that it vanishes, except for $\alpha-\gamma=0$ when the term becomes

$$
(1-y)^{-1}\frac{\Gamma(2\alpha+1)}{\Gamma(\alpha+1+\gamma)}\lim_{\varepsilon\to 0}\left\{1+\frac{\varepsilon(\alpha+\gamma+\varepsilon)}{\varepsilon}(1-y)+\cdots\right\}\propto(1-y)^{-1},
$$

i.e. it diverges for $y\to 1$. Thence we conclude that for $n=0$ the boundary condition (68.6b) is not satisfied so that the eigenvalues are determined by

$$
\alpha-\gamma=-n;\qquad n=1,2,3,\ldots \tag{68.10}
$$

with (unnormalized) wave functions

$$
\chi=\mathrm{e}^{-\alpha x}(1-\mathrm{e}^{-x})\,{}_2F_1(2\alpha+1+n,1-n,2\alpha+1;\mathrm{e}^{-x}). \tag{68.11}
$$

The hypergeometric series in this case degenerate into polynomials in the variable e^{-x}.

From (68.10) we obtain

$$\alpha = \frac{\beta^2 - n^2}{2n}. \tag{68.12}$$

Since $\alpha > 0$, we necessarily have

$$\beta^2 > n^2, \tag{68.13}$$

i.e. there exists a minimum size of potential hole before any energy eigenvalue at all can be obtained, viz. $\beta^2 = 1$. More exactly, Eq. (68.13) determines the number of eigenvalues in a potential hole of given size. Using Eq. (68.2b), the eigenvalue condition (68.12) can be written in the form

$$E_n = -V_0 \left(\frac{\beta^2 - n^2}{2n\beta} \right)^2. \tag{68.14}$$

The Hulthén potential at small values of r behaves like a Coulomb potential $V_C = -V_0 a/r$, whereas for large values of r it decreases exponentially so that its "capacity" for bound states is smaller than that of V_C. In Fig. 39 the two potentials are compared for the numerical

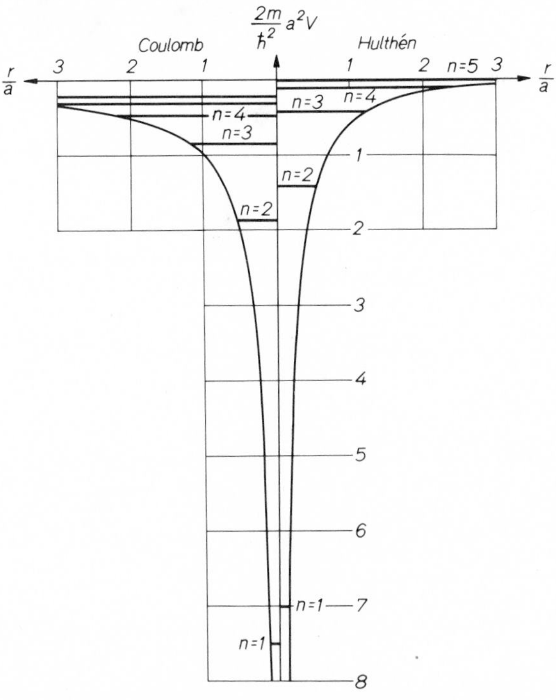

Fig. 39. Comparison of Coulomb potential and Hulthén potential with the same singularity. Note the shift of energy levels

example $\beta^2=30$, the Coulomb potential being drawn on the left, the Hulthén potential on the right-hand side. The energy levels are shown; they always lie lower in the Coulomb case than in the Hulthén case where there remains only space for a finite number, in our example for five levels.

Problem 69. Kratzer's molecular potential

To investigate the rotation-vibration spectrum of a diatomic molecule Kratzer has used the potential

$$V(r)=-2D\left(\frac{a}{r}-\frac{1}{2}\frac{a^2}{r^2}\right) \tag{69.1}$$

with a minimum $V(a)=-D$ (Fig. 40). Let one of the two atoms be much heavier than the other (as e.g. in HI) so that it may be considered at rest and be chosen as the coordinate centre. (Otherwise the equivalent one-body problem would have to be solved, cf. Problem 150.)

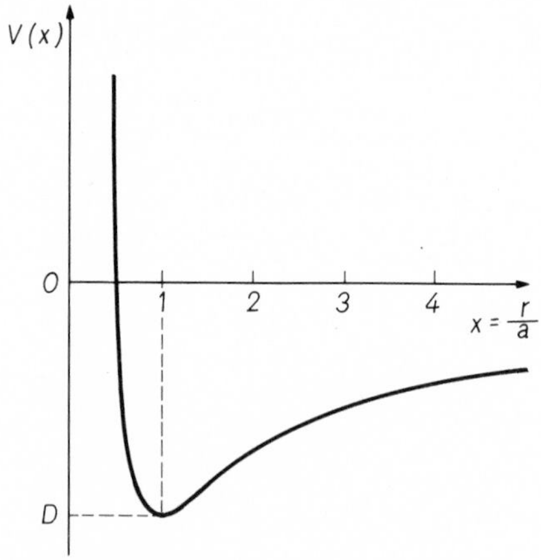

Fig. 40. Kratzer's molecular potential

Solution. The motion of the light atom of mass m satisfies the Schrödinger equation

$$\nabla^2 u+\frac{2m}{\hbar^2}\left[E+2D\left(\frac{a}{r}-\frac{1}{2}\frac{a^2}{r^2}\right)\right]u=0 \tag{69.2}$$

which may be factorized into

$$u = \frac{1}{r}\,\chi_l(r)\,Y_{l,m}(\vartheta,\varphi)\,. \tag{69.3}$$

Using the dimensionless abbreviations

$$x = \frac{r}{a}; \quad \beta^2 = -\frac{2ma^2}{\hbar^2}E; \quad \gamma^2 = \frac{2ma^2}{\hbar^2}D \tag{69.4}$$

with $\gamma>0$ and real $\beta>0$ for bound states, we get for the radial function χ_l the differential equation

$$\frac{d^2\chi_l}{dx^2} + \left[-\beta^2 + \frac{2\gamma^2}{x} - \frac{\gamma^2+l(l+1)}{x^2}\right]\chi_l = 0\,. \tag{69.5}$$

This differential equation has an irregular singularity at $x=\infty$ where its normalizable solutions in bound states behave as $e^{-\beta x}$. It further has a singularity at $x=0$ where $\chi_l \propto x^\lambda$ with an exponent λ following from the characteristic equation

$$\lambda(\lambda-1) = \gamma^2 + l(l+1)\,.$$

Of the two solutions of this quadratic equation we need the positive one:

$$\lambda = \tfrac{1}{2} + \sqrt{\gamma^2 + (l+\tfrac{1}{2})^2} \tag{69.6}$$

and may exclude the other (with negative sign of the radical). As $\lambda>1$, the wave function vanishes at $r=0$, corresponding to the strong repulsion between the two atoms (Fig. 40). Therefore, it is reasonable to set

$$\chi_l(r) = x^\lambda e^{-\beta x} f(x)\,. \tag{69.7}$$

Putting (69.7) into (69.5) leads to

$$x f'' + (2\lambda - 2\beta x) f' + (-2\lambda\beta + 2\gamma^2) f = 0 \tag{69.8}$$

which is of the general type of Kummer's equation of the confluent hypergeometric series into which it can be transformed by using the variable $z=2\beta x$ instead of x. This leads to

$$f = {}_1F_1\!\left(\lambda - \frac{\gamma^2}{\beta},\, 2\lambda;\, 2\beta x\right). \tag{69.9}$$

Let us now discuss, separately, the two cases of negative and positive energies.

a) Negative energies. For bound states, $\beta>0$, the solution becomes in arbitrary normalization

$$\chi_l = x^\lambda e^{-\beta x}\, {}_1F_1\left(\lambda - \frac{\gamma^2}{\beta}, 2\lambda; 2\beta x\right). \tag{69.10}$$

The confluent series, for large values of x, is proportional to $e^{2\beta x}$ so that χ_l diverges for $x\to\infty$ if the series ${}_1F_1$ does not break off. If it does ${}_1F_1$ is a polynomial and $\chi_l\to 0$ for $x\to\infty$ becomes normalizable. Hence,

$$\lambda - \frac{\gamma^2}{\beta} = -v; \qquad v=0,1,2,\dots. \tag{69.11}$$

This relation selects discrete eigenvalues β and leads thus to the energy levels

$$E = -\frac{\hbar^2}{2ma^2}\frac{\gamma^4}{(v+\lambda)^2}$$

or, using (69.6),

$$E = -\frac{\hbar^2}{2ma^2}\gamma^4\left\{v+\tfrac{1}{2}+\sqrt{(l+\tfrac{1}{2})^2+\gamma^2}\right\}^{-2}. \tag{69.12}$$

A brief discussion of this level formula may be added. Since the parameter $\gamma\gg 1$ for most molecules, we may expand (69.12) into powers of $1/\gamma$. This leads to

$$E=D\left[-1+\frac{2(v+\frac{1}{2})}{\gamma}+\frac{(l+\frac{1}{2})^2}{\gamma^2}-\frac{3(v+\frac{1}{2})^2}{\gamma^2}-\frac{3(v+\frac{1}{2})(l+\frac{1}{2})^2}{\gamma^3}\cdots\right]. \tag{69.13}$$

If $V(r)$, Eq. (69.1), is expanded about its minimum at $r=a$,

$$V(r)=D\frac{(r-a)^2}{a^2}-D, \tag{69.14}$$

we find the classical frequency for small harmonic vibrations,

$$\omega=\sqrt{\frac{2D}{ma^2}}. \tag{69.15}$$

Replacing D and γ in (69.13) by ω and the moment of inertia,

$$\Theta = ma^2 \tag{69.16}$$

we arrive at

$$E=-\tfrac{1}{2}\Theta\omega^2+\hbar\omega(v+\tfrac{1}{2})+\frac{\hbar^2}{2\Theta}(l+\tfrac{1}{2})^2-\frac{3\hbar^2}{2\Theta}(v+\tfrac{1}{2})^2$$
$$-\frac{3\hbar^3}{2\Theta^2\omega}(v+\tfrac{1}{2})(l+\tfrac{1}{2})^2\dots. \tag{69.17}$$

The first term is a constant not directly revealed by spectral line frequencies but, however, to be computed from them. The second describes harmonic vibrations with v the vibrational quantum number. The third term describes rotations with constant moment of inertia, except for another constant term because $(l+\frac{1}{2})^2 = l(l+1)+\frac{1}{4}$. The fourth term lowers the vibrational levels in consequence of the anharmonicity of the potential. The fifth term performs coupling between vibrations and rotations, again caused by the anharmonicity. Of course, this expansion will become worse the higher the quantum numbers, especially v, and cannot be used to the point of dissociation $(v \to \infty)$ where, however, the model breaks down in any case because of its rather unphysical behaviour at very large and very small r.

Finally, the energy of dissociation of the molecule follows from (69.17) with $v=0$, $l=0$ to be

$$E_{\text{diss}} = \tfrac{1}{2}\Theta\,\omega^2 - \tfrac{1}{2}\hbar\omega + \frac{\hbar^2}{4\Theta} + \frac{3\hbar^3}{16\Theta^2\omega}\cdots. \tag{69.18}$$

b) Positive energies. In this case β is no longer real but purely imaginary. Its definition shows that, using the wave number k, we may then write $\beta = -ika$ so that $\beta x = -ikr$. Instead of the solution (69.10) we then find, again in arbitrary normalization,

$$\chi_l = r^\lambda e^{ikr}\,{}_1F_1\left(\lambda - i\frac{\gamma^2}{ka},\, 2\lambda;\, -2ikr\right). \tag{69.19}$$

This wave function again vanishes at $r=0$. Its asymptotics may be determined from the well-known formula

$${}_1F_1(a,c;z) \to e^{-i\pi a}\frac{\Gamma(c)}{\Gamma(c-a)}z^{-a} + \frac{\Gamma(c)}{\Gamma(a)}e^z z^{a-c}$$

holding for the whole complex z plane cut along the positive imaginary axis. Applying this formula to (69.19), a somewhat lengthy computation leads to

$$\chi_l \to C_l \cos\left(kr + \frac{\gamma^2}{ka}\log 2kr - \frac{\lambda\pi}{2} - \eta_l\right) \tag{69.20}$$

with a phase angle η_l following from

$$e^{2i\eta_l} = \frac{\Gamma\left(\lambda + i\dfrac{\gamma^2}{ka}\right)}{\Gamma\left(\lambda - i\dfrac{\gamma^2}{ka}\right)} \tag{69.21}$$

and a (not very interesting) normalization factor C_l.

The radial wave functions (69.20) are periodic at large distances between the two atoms, with the exception of the logarithmic term. The latter is a consequence of the asymptotic Coulomb slope of the potential, cf. Problem 111. Interpreting the Kratzer potential as the interaction of two ions, each of electrical charge e, we get $2Da = e^2$ and $\gamma^2/ka = e^2/\hbar v$ (v velocity) which indeed makes the logarithmic term in (69.20) identical with the one in a Coulomb field.

NB. The potential of Kratzer played rather a large role in the early days of quantum mechanics because it permits of an exact solution, even for $l > 0$. It is, however, much less physical than the one of Morse which is still simple enough to give a survey of phenomena. In the following two problems we shall deal with the Morse potential. The results may be compared with those of the Kratzer potential, at least for bound states.

Literature. Kratzer, A.: Z. Physik **3**, 289 (1920) (classical treatment). — Fues, E.: Ann. Physik **80**, 367 (1926) (quantum mechanical treatment).

Problem 70. Morse potential

The vibrations of a two-atomic molecule are excellently described by the Morse potential

$$V(r) = D(\mathrm{e}^{-2\alpha x} - 2\,\mathrm{e}^{-\alpha x}); \qquad x = \frac{r - r_0}{r_0}. \tag{70.1}$$

The Schrödinger equation shall be solved for $l = 0$ bound states. The following numerical data for three typical molecules may be used for quantitative discussion. M is the reduced mass of the two atoms. Energies are given in cm^{-1} conforming to spectroscopic use.

Molecule	$\frac{\hbar^2}{2Mr_0^2}$ cm^{-1}	D cm^{-1}	α
H_2	60.8296	38292	1.440
HCl	10.5930	37244	2.380
I_2	0.0374	12550	4.954

$$E(\mathrm{eV}) = E(\mathrm{cm}^{-1}) \times 1.2398 \times 10^{-4}.$$

Solution. Let us begin with a brief remark on the Morse potential (see Fig. 41). It is attractive at large distances, comes to a minimum $-D$ at $x = 0$ or $r = r_0$, but produces a strong repulsion if the two nuclei approach even closer. Around $x = 0$ it may be expanded into a series,

$$V(r) = D(-1 + \alpha^2 x^2 \dots) = -D + \tfrac{1}{2} M \omega^2 (r - r_0)^2 + \cdots$$

$$\text{with } \omega^2 = \frac{2 D \alpha^2}{M r_0^2}. \tag{70.2}$$

For low-energy vibration terms we may therefore expect a spectrum not deviating very much from that of a harmonic oscillator,

$$E(v) = -D + \hbar\omega(v + \tfrac{1}{2}); \quad v = 0, 1, 2, \dots \tag{70.3}$$

with v the vibrational quantum number, the almost equidistant terms becoming increasingly denser with increasing energy, in consequence of the anharmonicity neglected in (70.2).

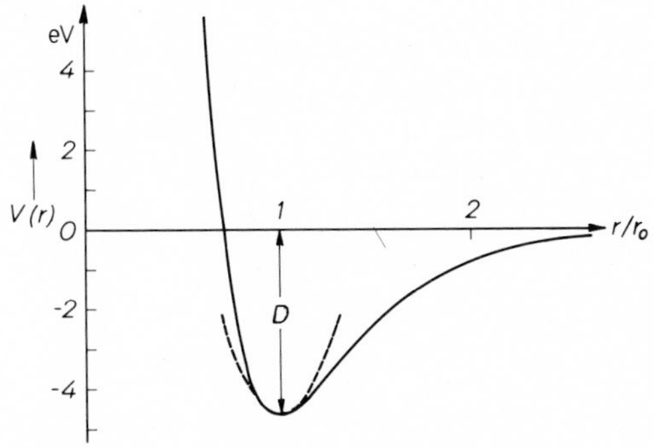

Fig. 41. Morse potential for HCl. Broken line: osculating parabola

We now write the Schrödinger radial equation for $l = 0$ in the variable x instead of in r and use the abbreviations

$$\beta^2 = -\frac{2 M E r_0^2}{\hbar^2} > 0; \quad \gamma^2 = \frac{2 M D r_0^2}{\hbar^2} \tag{70.4}$$

with $\beta > 0$, $\gamma > 0$ by definition. Then we get

$$\frac{d^2 \chi_0}{dx^2} + (-\beta^2 + 2\gamma^2 e^{-\alpha x} - \gamma^2 e^{-2\alpha x}) \chi_0 = 0. \tag{70.5}$$

In order to make the coefficients in the differential equation rational functions, some simple function of $e^{-\alpha x}$ should be used as variable. The equation becomes particularly simple if

$$y = \zeta e^{-\alpha x}; \quad \zeta = \frac{2\gamma}{\alpha} \tag{70.6}$$

is used; it then runs as follows:

$$y^2\chi_0''+y\chi_0'+\left(-\frac{\beta^2}{\alpha^2}+\frac{\gamma}{\alpha}y-\frac{1}{4}y^2\right)\chi_0=0.$$

It has singularities at $y=0$ and $y=\infty$; in the vicinity of $y=0$ we find χ_0 to be proportional to y^λ with $\lambda=\pm\beta/\alpha$, and at infinity we have $\chi_0\sim e^{\pm y/2}$. Therefore, if we put

$$\chi_0=y^{\beta/\alpha}e^{-\frac{1}{2}y}F(y)$$

and use the two parameters

$$c=2\frac{\beta}{\alpha}+1;\quad a=\frac{1}{2}c-\frac{\gamma}{\alpha}, \tag{70.7}$$

the differential equation for $F(y)$ will turn out to be Kummer's standard form of the confluent hypergeometric one so that the complete solution of (70.5) becomes

$$\chi_0=y^{\beta/\alpha}e^{-\frac{1}{2}y}\{A_1\,{}_1F_1(a,c;y)+A_2y^{1-c}\,{}_1F_1(a-c+1,2-c;y)\}. \tag{70.8}$$

To satisfy the boundary condition $\chi_0=0$ for $r\to\infty$ or $y=0$, the constant A_2 is bound to vanish since the exponent of y,

$$\frac{\beta}{\alpha}+1-c=-\frac{\beta}{\alpha},$$

is negative. The constant A_1 is to be fixed by the normalization.

The second boundary condition, $\chi_0=0$ at $r=0$, i.e. at $x=-1$ or at

$$y=y_0=\zeta e^\alpha \tag{70.9}$$

leads to a relation from which to determine the energy eigenvalues:

$${}_1F_1(a,c;y_0)=0. \tag{70.10}$$

This is an intricate transcendent equation the exact solution of which is difficult to take in at a glance, but the peculiar numerical values for all real molecules of the parameters α and γ permit a very simple and excellent approximation.

In the accompanying table we have collected some numerical values for the three typical molecules derived from the data given above. In all cases the values of $y_0\gg 1$. In order to use asymptotic series for the confluent function (70.10), $|y_0|$ must be rather large as compared to the parameters $|a|$ and $|c|$. Now, according to (70.7) both

	γ	α	ζ	y_0	y_0/ζ	G
H_2	25.09	1.440	34.9	147	4.22	5.6
HCl	59.30	2.380	49.8	539	10.80	270
I_2	579.51	4.954	234	33200	142	31240

depend upon β, i.e. upon the energy eigenvalue which we have not yet determined. We know, however, that

$$-D<E<0 \quad \text{or} \quad 0<\beta<\gamma, \tag{70.11}$$

because the energy must be negative in a bound state and cannot lie below the potential minimum. This limits the parameters a and c to the intervals

$$1<c<1+\zeta \quad \text{and} \quad \tfrac{1}{2}(1-\zeta)<a<\tfrac{1}{2}. \tag{70.12}$$

As the table shows, ζ throughout remains much smaller than y_0 even in the most unfavourable case of the hydrogen molecule. We are thus fully entitled to use the asymptotic expansion holding for real y_0:

$${}_1F_1(a,c;y_0)=e^{-i\pi a}\frac{\Gamma(c)}{\Gamma(c-a)}y_0^{-a}+\frac{\Gamma(c)}{\Gamma(a)}e^{y_0}y_0^{a-c}=0. \tag{70.13}$$

Putting $c=\zeta+2a$ and $y_0=\zeta e^{\alpha}$ in (70.13), this may be written,

$${}_1F_1(a,\zeta+2a;\zeta e^{\alpha})=\Gamma(\zeta+2a)e^{-a(\alpha+\log\zeta)}\left\{\frac{e^{-i\pi a}}{\Gamma(\zeta+a)}+\frac{1}{\Gamma(a)}e^{\zeta(e^{\alpha}-\alpha-\log\zeta)}\right\}=0$$

or

$$\Gamma(a)=-e^{i\pi a}\Gamma(a+\zeta)e^{\zeta(e^{\alpha}-\alpha-\log\zeta)}. \tag{70.14}$$

Now, $a+\zeta>\frac{1}{2}(1+\zeta)>\frac{1}{2}$. According to the table above, ζ even for hydrogen is so large that $a+\zeta>17.95$. We therefore can apply the Stirling formula to $\Gamma(\frac{1}{2}(1+\zeta))$ and find

$$|\Gamma(a)|>\sqrt{2\pi}\,e^{G}$$

with (70.15)

$$G=\tfrac{1}{2}\zeta\log\left[\tfrac{1}{2}(1+\zeta)\right]-\tfrac{1}{2}(1+\zeta)+\zeta(e^{\alpha}-\alpha-\log\zeta),$$

a quantity which can be computed independently of any further knowledge of the eigenvalues and has been given in the last column of the table. Even in the most unfavourable case of hydrogen, this gives $|\Gamma(a)|>675$. In all cases therefore $|\Gamma(a)|$ is very large and close to one of the singularities of the Γ function (since $a<\frac{1}{2}$) at negative integers,

$$a=-v;\qquad v=0,1,2,3,\ldots \tag{70.16}$$

This, however, is not an infinite series of v values since the condition $a>\frac{1}{2}(1-\zeta)$ sets an upper limit of

$$v<\tfrac{1}{2}(\zeta-1). \tag{70.17}$$

From (70.16) and (70.7) we now obtain,

$$-\beta^2=-\gamma^2+2\gamma\alpha(v+\tfrac{1}{2})-\alpha^2(v+\tfrac{1}{2})^2$$

and hence the energy in terms of the vibrational quantum number v,

$$E(v)=-D+\frac{\hbar^2}{2Mr_0^2}\{2\alpha\gamma(v+\tfrac{1}{2})-\alpha^2(v+\tfrac{1}{2})^2\}. \tag{70.18}$$

The first two terms of this formula are in complete agreement with the harmonic oscillator formula (70.3) since

$$\frac{\hbar^2}{2Mr_0^2}2\alpha\gamma=\hbar\omega.$$

The last term reflects the anharmonicity correction. If the formula is written

$$E(v)=-D+\hbar\omega\left\{(v+\tfrac{1}{2})-\frac{1}{\zeta}(v+\tfrac{1}{2})^2\right\}, \tag{70.19}$$

the condition (70.17) shows that the anharmonic can never exceed the harmonic term.

Literature. This potential was first used by Morse, P. M.: Phys. Rev. **34,** 57 (1929). — See also end of next problem.

Problem 71. Rotation correction of Morse formula

The rotation energy of a molecule is much smaller than that of vibration. In the preceding problem it has therefore been omitted. If it is to be included, this may either been done by perturbation or by replacing the centrifugal potential by an approximate one allowing the constants β and γ of the preceding problem to be included in it without meeting any serious new mathematical difficulties. This latter simple way shall be investigated.

Solution. If $l\neq 0$ the centrifugal potential

$$V'=\frac{\hbar^2 l(l+1)}{2Mr^2} \tag{71.1}$$

has to be added to the potential in the Schrödinger radial equation. Using the symbols of the last problem, we can write this

$$V'(r) = \frac{l(l+1)}{\gamma^2} D(r_0/r)^2 \tag{71.2}$$

which, for r not deviating in magnitude from r_0, is very small compared to D as long as l does not itself become unreasonably large. Even in the unfavourable case of hydrogen, where γ is only a little above 25, this holds as long as l remains small compared to 25. We may therefore take the centrifugal potential as a small correction in practically all cases.

Furthermore, the nuclear distance r is, of course, not a constant, but even for rather high vibrational levels it will not fluctuate very far from the equilibrium position at r_0, as a detailed study of the eigenfunctions (70.8) reveals. (The classical points of reversal lie at $x^2 = (2v+1)/\gamma\alpha^2 \ll 1$.) If, therefore, we expand (71.1) about $r = r_0$ in a series of powers of $x = (r - r_0)/r_0$, viz.

$$V' = \frac{l(l+1)}{\gamma^2} D \frac{1}{(1+x)^2} = \frac{l(l+1)}{\gamma^2} D(1 - 2x + 3x^2 - 4x^3 + \cdots) \tag{71.3}$$

the first few terms should be quite sufficient. Instead, we now replace $V'(r)$ by the potential

$$\tilde{V}'(r) = \frac{l(l+1)}{\gamma^2} D(C_0 + C_1 \mathrm{e}^{-\alpha x} + C_2 \mathrm{e}^{-2\alpha x}) \tag{71.4}$$

with

$$C_0 = 1 - \frac{3}{\alpha} + \frac{3}{\alpha^2}; \quad C_1 = \frac{4}{\alpha} - \frac{6}{\alpha^2}; \quad C_2 = -\frac{1}{\alpha} + \frac{3}{\alpha^2}. \tag{71.5}$$

The two expressions (71.3) and (71.4) deviate only in the order x^3, since expansion of (71.4) yields the series

$$\tilde{V}' = \frac{l(l+1)}{\gamma^2} D\{1 - 2x + 3x^2 - (3\alpha - \tfrac{2}{3}\alpha^2)x^3 + \cdots\}.$$

With the potential $\tilde{V}'$ instead of the true centrifugal potential V', the Schrödinger radial equation for $l \neq 0$ becomes

$$\frac{d^2\chi_l}{dx^2} + (-\beta_1^2 + 2\gamma_1^2 \mathrm{e}^{-\alpha x} - \gamma_2^2 \mathrm{e}^{-2\alpha x})\chi_l = 0 \tag{71.6}$$

with

$$\begin{aligned} \beta_1^2 &= \beta^2 + l(l+1)C_0; \\ \gamma_1^2 &= \gamma^2 - \tfrac{1}{2}l(l+1)C_1; \\ \gamma_2^2 &= \gamma^2 + l(l+1)C_2. \end{aligned} \tag{71.7}$$

The differential equation is almost the same as (70.5), except that β_1 replaces β, and γ_1 and γ_2 replace γ. The same procedure therefore applies to this equation. We introduce the new variable

$$y=\zeta_2\,\mathrm{e}^{-\alpha x};\qquad \zeta_2=\frac{2\gamma_2}{\alpha} \tag{71.8}$$

and the constants

$$\tilde{c}=\frac{2\beta_1}{\alpha}+1;\qquad \tilde{a}=\frac{1}{2}\tilde{c}-\frac{\gamma_1^2}{\alpha\gamma_2} \tag{71.9}$$

and find the solution

$$\chi_l=y^{\beta_1/\alpha}\,\mathrm{e}^{-\frac{1}{2}y}\,{}_1F_1(\tilde{a},\tilde{c};y) \tag{71.10}$$

leading to the equation of eigenvalue determination

$${}_1F_1(\tilde{a},\tilde{c};\tilde{y}_0)=0\quad\text{with }\tilde{y}_0=\zeta_2\,\mathrm{e}^{\alpha}. \tag{71.11}$$

All qualitative considerations of the approximate treatment of the last equation remain valid since the constants $\tilde{a}$ and $\tilde{c}$ differ from a and c only by slight corrections. Hence

$$\tilde{a}=-v;\qquad v=0,1,2,3,\ldots \tag{71.12}$$

and, according to (71.9),

$$-\beta_1^2=-\left[\frac{\gamma_1^2}{\gamma_2}-\alpha(v+\tfrac{1}{2})\right]^2. \tag{71.13}$$

In the energy formula then γ has to be replaced by

$$\frac{\gamma_1^2}{\gamma_2}\simeq\gamma\left(1-l(l+1)\frac{C_1+C_2}{2\gamma^2}\right),$$

and instead of β^2 we have to take β_1^2 on the left-hand side of (71.13). The result is, according to (71.7):

$$\begin{aligned}-\beta^2=&-\gamma^2+2\alpha\gamma(v+\tfrac{1}{2})-\alpha^2(v+\tfrac{1}{2})^2+l(l+1)(C_0+C_1+C_2)\\&-\frac{\alpha}{\gamma}(C_1+C_2)(v+\tfrac{1}{2})l(l+1)-\frac{(C_1+C_2)^2}{4\gamma^2}l^2(l+1)^2.\end{aligned}$$

Expressing the three C_i's in terms of α as given in (71.5) we finally arrive at the energy formula,

$$\begin{aligned}E=\frac{\hbar^2}{2Mr_0^2}\Big\{&-\gamma^2+2\alpha\gamma(v+\tfrac{1}{2})-\alpha^2(v+\tfrac{1}{2})^2\\&+l(l+1)-\frac{3(\alpha-1)}{\alpha\gamma}(v+\tfrac{1}{2})l(l+1)-\frac{9(\alpha-1)^2}{4\alpha^4\gamma^2}l^2(l+1)^2\Big\}.\end{aligned} \tag{71.14}$$

The terms in the first line of this formula are exactly the same as those derived in the preceding problem: potential depth $(-D)$, harmonic and anharmonic vibration terms. The second line comprises the three additional terms: the rotational energy at fixed distance r_0, a coupling term of rotation and vibration which is negative because at higher vibrational quantum numbers the average nuclear distance increases beyond r_0 in consequence of the anharmonicity, and a final term, negative as well, because larger rotational quantum numbers lead to an enlargement of the average distance even independently of vibrations.

NB. A generalization of the Morse potential and its practical application to the vibrations of diatomic molecules has been discussed by Flügge, S., Walger, P., Weiguny, A., in: J. Molec. Spectrosc. **23**, 243 (1967).

Problem 72. Yukawa potential hole

Given an attractive potential of Yukawa type,

$$V(r) = -V_0 \frac{e^{-r/a}}{r/a}, \tag{72.1}$$

acting between two particles (e.g. two nucleons). For $l=0$ an approximate wave function

$$u = \frac{1}{r}\chi(r) \quad \text{with} \quad \chi = C r e^{-\alpha r} \tag{72.2}$$

with a Ritz parameter α shall be used to find the optimum solution by variational procedure. What minimum size of the potential hole is required to make the existence of a bound state possible? (In the centre-of-mass system, the one-body Schrödinger equation will hold with the reduced mass m^*, cf. Problem 150.)

Solution. Let us start by determining C to satisfy the normalization condition

$$\int_0^\infty dr\, \chi^2 = 1.$$

That leads to

$$C = 2\alpha^{\frac{3}{2}}. \tag{72.3}$$

Then, the energy E can be calculated as the sum of the two integrals,

$$E_{\text{kin}} = \frac{\hbar^2}{2m^*}\int_0^\infty dr \left(\frac{d\chi}{dr}\right)^2; \qquad E_{\text{pot}} = -V_0 \int_0^\infty dr \frac{e^{-r/a}}{r/a}\chi^2. \tag{72.4}$$

Elementary integration, using Euler's integral, yields

$$E = \frac{\hbar^2}{2m^*}\alpha^2 - V_0 \frac{4a^3\alpha^3}{(1+2a\alpha)^3}. \tag{72.5}$$

We introduce the dimensionless abbreviations

$$2a\alpha = p; \qquad \frac{2m^* a^2}{\hbar^2} V_0 = K \tag{72.6}$$

the first of which gives the ratio of the average radii of potential hole and wave function, whereas the other is a measure for the hole size. Eq. (72.5) then may be written

$$\frac{E}{V_0} = \frac{1}{4K}p^2 - \frac{p^3}{2(1+p)^2}. \tag{72.7}$$

With p instead of α as variational parameter, the optimum solution follows from $dE/dp = 0$ which renders

$$K = \frac{(1+p)^3}{p(p+3)}. \tag{72.8}$$

Putting (72.8) into (72.7), the optimum energy value turns out to be

$$E = -V_0 \frac{p^3(p-1)}{4(p+1)^3}. \tag{72.9}$$

Since p is by definition a positive number, a bound state will only occur if $p > 1$ or, according to (72.8), if $K > 2$. This therefore is the minimum size of the potential hole necessary for binding, in the underlying approximation.

The following figures have been computed from Eqs. (72.8) and (72.9).

p	K	$-E/V_0$
1.00	2.000	0
1.05	2.025	0.0017
1.10	2.050	0.0036
1.15	2.08	0.0058
1.20	2.11	0.0081
1.30	2.18	0.0135
1.40	2.24	0.0198

p	K	$-E/V_0$
1.50	2.32	0.0270
1.60	2.39	0.0349
1.70	2.46	0.0438
1.80	2.54	0.0532
1.90	2.62	0.0633
2.00	2.70	0.0740
2.50	3.12	0.1364

NB. This approximation is rather good. Replacing the trial function (72.1) by the less favourable function

$$\chi' = C' r^2 \, e^{-\alpha r}; \qquad C'^2 = \tfrac{4}{3}\alpha^2$$

the same procedure as before leads to

$$K = \frac{2(p+1)^5}{3p^3(p+5)} \quad \text{and} \quad E = -V_0 \frac{p^5(p-3)}{8(p+1)^5}.$$

Bound states then obviously can exist only for $p > 3$, i.e. for $K > (\frac{4}{3})^4 = 3.16$.

Problem 73. Isotope shift in x-rays

The orbital of a K electron in a heavy atom has an r.m.s. radius about 100 times larger than the nuclear radius R. Its energy, therefore, should undergo a small shift due to the finite size of the nucleus. This energy perturbation shall be calculated for a nucleus of homogeneous charge distribution. The isotope shift in the K edge of thallium ($Z = 81$) x-rays shall then be determined for the two isotopes of atomic weights $A = 203$ and $A = 205$, using the nuclear radius formula

$$R = r_0 A^{\frac{1}{3}}; \qquad r_0 = 1.2 \times 10^{-13} \, \text{cm}. \tag{73.1}$$

Screening of the K electron is to be neglected.

Solution. The potential energy of an electron in the field of the nucleus is

$$V(r) = \begin{cases} -\dfrac{Ze^2}{r} & \text{for } r > R, \\[2ex] \dfrac{Ze^2}{R}\left(\dfrac{r^2}{2R^2} - \dfrac{3}{2}\right) & \text{for } r < R. \end{cases} \tag{73.2}$$

We begin with the unperturbed problem of the point nucleus ($R = 0$):

$$V_0 = -\frac{Ze^2}{r} \quad \text{throughout};$$

$$u_0 = \frac{\gamma^{\frac{3}{2}}}{\sqrt{\pi}} e^{-\gamma r}; \qquad \gamma = Z\frac{me^2}{\hbar^2}; \qquad E_0 = -\frac{1}{2} Z^2 \frac{me^4}{\hbar^2}, \tag{73.3}$$

with $u_0(r)$ the wave function and E_0 the energy of the ground state in the Kepler problem (see Problem 67). The perturbation potential then is the difference

$$V - V_0 = \begin{cases} 0 & \text{for } r > R, \\[2ex] \dfrac{Ze^2}{R}\left(\dfrac{r^2}{2R^2} - \dfrac{3}{2} + \dfrac{R}{r}\right) & \text{for } r < R. \end{cases} \tag{73.4}$$

Thus we arrive at the energy shift

$$\Delta E = E - E_0 = \int d\tau\, u_0^2 (V - V_0)$$

$$= 4\pi \int_0^R dr\, r^2 \frac{\gamma^3}{\pi} e^{-2\gamma r} \frac{Ze^2}{R} \left(\frac{r^2}{2R^2} - \frac{3}{2} + \frac{R}{r} \right). \qquad (73.5)$$

The argument of the exponential, $2\gamma r$, remains smaller than $2\gamma R \simeq 0.02$ within the domain of integration, so that we may put $e^{-2\gamma r} \simeq 1$ in (73.5). Using r/R as integration variable, elementary procedure then yields the simple result

$$\Delta E = \frac{2}{5} \gamma^3 Z e^2 R^2 = \frac{2}{5} Z^4 \frac{e^2}{a_0} \left(\frac{R}{a_0} \right)^2 = \frac{4}{5} Z^2 |E_0| \left(\frac{R}{a_0} \right)^2 \qquad (73.6)$$

with $a_0 = \hbar^2/(me^2)$ the Bohr radius.

For $Z=81$, Eq. (73.3) gives

$$E_0 = -Z^2 \cdot 13.60\,\mathrm{eV} = -88.3\,\mathrm{keV}. \qquad (73.7)$$

The total shift, ΔE for $A=203$, according to (73.1) and (73.6) turns out to be

$$\Delta E = |E_0| \cdot 0.935 \times 10^{-4} = 8.25\,\mathrm{eV}. \qquad (73.8)$$

The isotope shift, i.e. the difference between the ΔE values for $A=203$ and $A + \Delta A = 205$, is obtained by differentiation:

$$S = \frac{d\Delta E}{dR} \cdot \frac{dR}{dA} \Delta A = \frac{2}{3} \frac{\Delta A}{A} \Delta E.$$

This leads, for the present example, to

$$S = 0.00656\, \Delta E = 6.14 \times 10^{-7} |E_0|. \qquad (73.9)$$

Whereas, according to (73.8), the absolute value of the K shell energy is shifted by an appreciable amount of about $\Delta E/E \sim 10^{-4}$ towards longer wavelengths, the difference effect S/E for two neighbouring isotopes (which might be observed, if big enough, by the splitting of a K x-ray line) becomes less than 1 ppm., according to (73.9).

NB. The screening effect is, of course, much bigger than these shifts, but it is the same for both isotopes. The experimental position of the thallium K edge is at 6310 Rydbergs or 85.9 keV instead of its unscreened value, Eq. (73.7) above. The isotope splitting is almost independent of this correction.

Problem 74. Muonic atom ground state

The binding energy of a muon in the 1s state shall be calculated for the same nucleus ($Z=81$) as in the preceding problem. a) Why can the same perturbation procedure not be applied to the muon case? b) A variational method shall be used instead, working on the set of trial functions

$$u=C(1+\alpha r)\mathrm{e}^{-\alpha r} \tag{74.1}$$

with a Ritz parameter α.

Solution. a) The muon mass, μ, is 207 times larger than that of the electron, m. Instead of the atomic units of length and energy,

$$a_1 = \frac{\hbar^2}{me^2} = 5.29\times 10^{-9}\,\text{cm}; \qquad E_1 = \frac{me^4}{\hbar^2} = 27.2\,\text{eV}$$

we should obtain muonic units

$$\tilde{a}_1 = \frac{m}{\mu}a_1 = 2.56\times 10^{-11}\,\text{cm}; \qquad \tilde{E}_1 = \frac{\mu}{m}E_1 = 5.63\,\text{keV}. \tag{74.2}$$

In the unperturbed state (point nucleus) the average radius is a_1/Z for the electronic and $\tilde{a}_1/Z$ for the muonic orbital, i.e. in the respective units, $1/Z=1.236\times 10^{-2}$ in both cases. On the other hand, the nuclear radius of $R=7.05\times 10^{-13}\,\text{cm}$ becomes 1.33×10^{-4} in electronic, and 2.75×10^{-2} in muonic units. Whereas the electron orbital thus has a radius about 100 times larger than the nuclear radius, the muon orbital radius is running to about $\frac{1}{2}$ of it. The muon hence will stay mainly within, and the electron outside the nucleus, the muon wave function being determined by the oscillator potential inside, the electron wave function (as in the perturbation procedure of the last problem) by the Coulomb potential outside.

The geometrical situation is reflected by the energy values. The ratio $\Delta E/|E_0| = \frac{4}{5}Z^2(R/a_1)^2 = 0.935\times 10^{-4}$ of Eq. (73.8) for the electron would have to be replaced by $\frac{4}{5}Z^2(R/\tilde{a}_1)^2 = 3.98$ for the muon, so that the "perturbation" of the energy would no longer be small compared to its unperturbed value, but much larger.

b) Inside the nucleus, there holds the oscillator potential

$$V_0(r) = \frac{Ze^2}{2R^3}(r^2-3R^2) \tag{74.3}$$

whose ground state wave function (cf. Problem 65) is

$$u_0(r) = \left(\frac{\mu\omega}{\hbar\pi}\right)^{\frac{3}{2}} e^{-\frac{\mu\omega}{2\hbar}r^2} \tag{74.4a}$$

with

$$\omega^2 = \frac{Ze^2}{\mu R^3} \tag{74.4b}$$

and with the energy

$$E_0 = \frac{3}{2}\hbar\omega - \frac{3Ze^2}{2R}. \tag{74.4c}$$

The solution (74.4a) will be approximately correct for small values of r where it may be expanded into the series

$$u_0 = \left(\frac{\mu\omega}{\hbar\pi}\right)^{\frac{3}{2}} \left(1 - \frac{\mu\omega}{2\hbar} r^2 + \cdots\right). \tag{74.5}$$

On the other hand, for large values of r, the actual potential tends to zero instead of becoming infinitely large as (74.3), so that the correct wave function becomes dominated by a factor

$$e^{-\varkappa r} \quad \text{with } \varkappa^2 = \frac{2\mu|E|}{\hbar^2} \tag{74.6}$$

decreasing much more slowly than (74.4b). Expansion of the trial function (74.1) now yields

$$u = C\left(1 - \frac{\alpha^2}{2} r^2 + \cdots\right),$$

comparison with (74.5) making it probable that some value of α^2 in the neighbourhood of $\mu\omega/\hbar$ might be expected. On the other hand, instead of (74.6) we have an asymptotic behaviour as $e^{-\alpha r}$ in the trial function. This certainly lies closer to the actual wave function than (74.4a), but is a reasonable approximation only if $\alpha \simeq \varkappa$.

In muonic units we obtain $\omega = 1.98 \times 10^3$ and $E_0 = -1.46 \times 10^3$ from (74.4b, c), hence, according to (74.6), $\varkappa = 54$. If, in order to make the functions (74.4a) and (74.1) identical at small values of r, we put $\alpha^2 = \mu\omega/\hbar$, we find—in these units—$\alpha = \sqrt{\omega} = 44.5$. This is not so very far from $\varkappa = 54$. The trial function (74.1) therefore can be expected to be a fair approximation to reality.

It may be noted that the optimum value of α^2 derived below turns out to be $\alpha \simeq 64$ and $\varkappa \simeq 60$.

We are now prepared to start an approximate energy computation using muonic units ($\mu=1$, $\hbar=1$, $e=1$) throughout. Normalizing (74.1) we get

$$C^2 = \frac{\alpha^3}{7\pi}. \tag{74.7}$$

The expectation value of the kinetic energy becomes

$$E_{\text{kin}} = -\tfrac{1}{2}\int d\tau\, u \nabla^2 u = \tfrac{3}{14}\alpha^2. \tag{74.8}$$

Computation of the potential energy

$$E_{\text{pot}} = 4\pi \int_0^R dr\, r^2 u^2 \frac{Z}{2R}\left(\frac{r^2}{R^2} - 3\right) - 4\pi \int_R^\infty dr\, r^2 u^2 \frac{Z}{r}$$

turns out to be a little cumbersome. Using the dimensionless quantities

$$2\alpha r = x; \qquad 2\alpha R = \lambda, \tag{74.9}$$

however, the integration may be performed in quite an elementary way with the help of the generalized Euler integral,

$$\int_\lambda^\infty dx\, x^n e^{-x} = n!\, e^{-\lambda}\left(1 + \lambda + \frac{1}{2!}\lambda^2 + \cdots + \frac{1}{n!}\lambda^n\right).$$

The result may be written

$$E = \frac{3}{56R^2}\left[\lambda^2 + ZR\, f(\lambda)\right] \tag{74.10}$$

with

$$f(\lambda) = \left(\frac{216}{\lambda^2} - 28\right) - e^{-\lambda}\left(\frac{216}{\lambda^2} + \frac{216}{\lambda} + 80 + 14\lambda + \lambda^2\right). \tag{74.11}$$

The Ritz condition,

$$\frac{\partial E}{\partial \lambda} = 0,$$

leads on to the determination of λ from

$$\left.\begin{aligned} \frac{1}{ZR} &= \frac{e^{-\lambda}}{\lambda^4}\left[216 e^{\lambda} - \varphi(\lambda)\right] \\ \text{with}\quad \varphi(\lambda) &= 216 + 216\lambda + 108\lambda^2 + 33\lambda^3 + 6\lambda^4 + \tfrac{1}{2}\lambda^5. \end{aligned}\right\} \tag{74.12}$$

For $ZR=2.23$ a value of $\lambda \simeq 3.5$ is obtained from (74.12) with an energy minimum at about $E=-1808$ muonic units or $E=-10.18$ MeV.

Literature. The problem, and finer corrections, has been discussed in some detail by Flügge, S., Zickendraht, W.: Z. Physik **143,** 1 (1955).

Problem 75. Central-force model of deuteron

Let the neutron-proton interaction be idealized by the central-force potential

$$V(r) = -A\,e^{-r/a}. \tag{75.1}$$

The Schrödinger equation of the equivalent one-body problem (cf. Problem 150) shall be solved for the bound state with $l=0$ (deuteron). There exists only one bound state with a binding energy

$$E = -2.23\ \mathrm{MeV}$$

as determined by experiment. The relation between A and a leading to this value of E shall be found in the neighbourhood of $a=2$ fm (1 fm $=10^{-13}$ cm). To find this relation, three methods shall be applied:

a) exact solution of the Schrödinger equation,

b) approximation by use of a wave function

$$\tilde{u} = \frac{C}{r}\left(e^{-\gamma r} - e^{-2\gamma r}\right)$$

showing the correct asymptotic behaviour for large r by a suitable choice of γ,

c) abridged Ritz method with a one-parametric set of exponentials.

a) Exact solution. Writing for $l=0$

$$u = \frac{1}{r}\,\chi(r), \tag{75.2}$$

the Schrödinger equation of the equivalent one-body problem is

$$\frac{d^2\chi}{dr^2} + \frac{2m^*}{\hbar^2}\left(E + A\,e^{-r/a}\right)\chi = 0 \tag{75.3}$$

where $m^* = \frac{1}{2}m$ is the reduced mass (supposing equal masses m of both nucleons). Use of the new variable

$$y = e^{-r/2a} \tag{75.4}$$

leads to

$$\frac{d^2\chi}{dy^2}+\frac{1}{y}\frac{d\chi}{dy}+\left(c^2-\frac{q^2}{y^2}\right)\chi=0 \tag{75.5}$$

with the abbreviations

$$c^2=\frac{8m^*}{\hbar^2}Aa^2; \quad q^2=-\frac{8m^*}{\hbar^2}Ea^2>0. \tag{75.6}$$

Eq. (75.6) is Bessel's differential equation with the general solution

$$\chi=C_1 J_q(cy)+C_2 J_{-q}(cy). \tag{75.7}$$

According to (75.4), $y=0$ corresponds to $r\to\infty$ where χ must vanish. Therefore $C_2=0$, and the wave function becomes

$$u=\frac{C_1}{r}J_q(c\,\mathrm{e}^{-r/2a}). \tag{75.8}$$

On the other hand, $y=1$ corresponds to $r=0$ where u must be finite so that the Bessel function is bound to vanish:

$$J_q(c)=0. \tag{75.9}$$

Given now the numerical values of m^* and E, we obtain from (75.6)

$$q=0.458\,a; \qquad A=2.23\,c^2/q^2 \tag{75.10}$$

where a is taken in units of fm and A in MeV. For any given value of a, q may be determined from (75.10). Then, from (75.9), there follows c, and finally, from (75.10) again, the potential depth A. In (75.9) the smallest zero of J_q has to be taken since it will lead to the smallest value of A, i.e. to a potential hole holding one bound state only. Numerical values of A and a are listed in the accompanying table[16].

q	a	c	A
0.2	0.436	2.707	409
0.4	0.873	2.999	125
0.6	1.310	3.283	67.0
0.8	1.747	3.560	44.5
0.9	1.963	3.696	37.5
1.0	2.180	3.832	32.7
1.1	2.40	3.966	28.1
1.2	2.62	4.099	26.1
1.4	3.05	4.363	21.7

[16] Solutions of Eq. (75.9) may be taken from Jahnke-Emde-Lösch: Tables of functions.

b) Approximate solution. The wave function

$$\tilde{u} = \frac{C}{r}\left(e^{-\frac{qr}{2a}} - e^{-\frac{qr}{a}}\right) \tag{75.11}$$

has a finite value

$$u(0) = C\frac{q}{2a}$$

at $r=0$, and decreases, for large r, exponentially with the correct decay length:

$$\exp\left(-\frac{qr}{2a}\right) = \exp\left(-r\sqrt{\frac{2m^*|E|}{\hbar^2}}\right).$$

The normalization constant C follows from

$$4\pi \int_0^\infty dr\, r^2 \tilde{u}(r)^2 = 1;$$

one finds

$$C = \sqrt{\frac{3q}{2\pi a}}. \tag{75.12}$$

We now determine the expectation value of the energy with this wave function by using the Schrödinger variational principle (cf. Problem 2):

$$E = 4\pi \int_0^\infty dr\, r^2 \left\{\frac{\hbar^2}{2m^*}(du/dr)^2 - A\,e^{-r/a}u^2\right\}. \tag{75.13}$$

Were the exact solution (75.8) put into the integral (75.13), the exact eigenvalue E would result. Using instead the approximate function $\tilde{u}$, Eq. (75.11), we find an approximate eigenvalue $\tilde{E}$ which, according to the general rule, lies somewhat above the exact value.

The evaluation of (75.13) with (75.11) is elementary and yields

$$E = \frac{\hbar^2}{4m^* a^2}q^2 - 6A\left\{\frac{1}{1+\frac{1}{q}} - \frac{2}{\frac{3}{2}+\frac{1}{q}} + \frac{1}{2+\frac{1}{q}}\right\}. \tag{75.14}$$

Taking e.g. the numerical values $a=2.18$ fm and $A=32.7$ MeV, obtained above, we find $\tilde{E}=-2.18$ MeV. If, on the other hand, the correct value, $E=-2.23$ MeV, is to be found from (75.14), with $q=1$ and $a=2.18$, a value of $\tilde{A}=33.5$ MeV has to be chosen. Thus, either for the correct potential depth A the eigenvalue lies too high by about 2.5% or the correct eigenvalue requires a somewhat larger value $\tilde{A}$ of the potential depth.

c) Ritz approximation. We use the normalized set of trial wave functions

$$\bar{u} = \sqrt{\frac{\alpha^3}{8\pi a^3}}\, e^{-\frac{\alpha r}{2a}} \tag{75.15}$$

with the Ritz parameter α to be determined so that E, Eq. (75.13), becomes a minimum. Putting (75.15) into (75.13) and evaluating the elementary integral we arrive at

$$\bar{E} = \frac{\hbar^2}{4m^* a^2}\alpha - \frac{3A}{\alpha^2\left(1+\frac{1}{\alpha}\right)^4} = 0 \tag{75.16}$$

or

$$\frac{(\alpha+1)^4}{\alpha} = \frac{12m^* a^2 A}{\hbar^2}. \tag{75.17}$$

With $a = 2.18$ fm and $A = 32.7$ MeV, the right-hand side of (75.17) becomes equal to 22.3 thus yielding $\alpha = 1.34$ and with $\hbar^2/(8m^* a^2) = 2.21$ MeV, according to (75.16),

$$\bar{E} = 3.97\text{ MeV} - 6.15\text{ MeV} = -2.18\text{ MeV}.$$

Again, the approximate value is higher than the exact one.

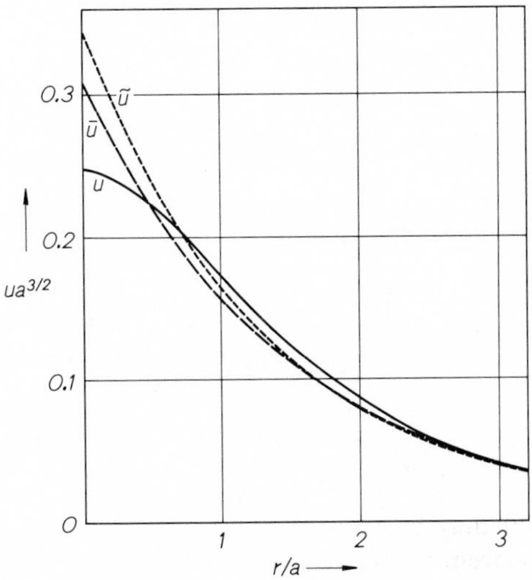

Fig. 42. Exact wave function and two approximations for a central-force deuteron model

The three wave functions u (exact), $\tilde{u}$, and $\bar{u}$ are shown in Fig. 42 in correct normalization (which, for u, can only be found by numerical computation). The two approximations are too large at small values of r which, however, have rather a small effect upon both, the normalization and the energy integral, since the volume element reduces their weight by a factor r^2. This deviation is compensated by values too small at larger r. The asymptotic behaviour shows no large differences between the three curves:

$$u a^{\frac{3}{2}} \simeq \bar{u} a^{\frac{3}{2}} \to 0.308\, \mathrm{e}^{-x}/x; \quad \tilde{u} a^{\frac{3}{2}} \to 0.346\, \mathrm{e}^{-x}/x$$

with $x = r/(2a)$.

Problem 76. Momentum space wave functions for central force potentials

To show that the splitting off of a spherical harmonic in the wave function in ordinary space permits the same factorization in momentum space.

Solution. In the Fourier transform

$$u(\boldsymbol{r}) = \frac{1}{(2\pi)^{\frac{3}{2}}} \int d^3k\, \mathrm{e}^{i\boldsymbol{k}\cdot\boldsymbol{r}} f(\boldsymbol{k}), \tag{76.1}$$

$$f(\boldsymbol{k}) = \frac{1}{(2\pi)^{\frac{3}{2}}} \int d^3x\, \mathrm{e}^{-i\boldsymbol{k}\cdot\boldsymbol{r}} u(\boldsymbol{r}) \tag{76.2}$$

it shall be supposed that u is factorized:

$$u(\boldsymbol{r}) = \frac{1}{r} \chi_l(r)\, Y_{l,m}(\vartheta, \varphi). \tag{76.3}$$

In order to find its Fourier transform according to (76.2), let us expand the exponential into spherical harmonics of the angle γ between the vectors $\boldsymbol{r}$ in the direction ϑ, φ and $\boldsymbol{k}$ in the direction Θ, Φ (cf. Problem 81):

$$\mathrm{e}^{-i\boldsymbol{k}\cdot\boldsymbol{r}} = \sum_{\lambda=0}^{\infty} \sqrt{4\pi(2\lambda+1)}\, i^{-\lambda} \frac{j_\lambda(kr)}{kr} Y^*_{\lambda,0}(\cos\gamma). \tag{76.4}$$

Here $Y_{\lambda,0}(\cos\gamma)$ may be expressed by the polar angles of $\boldsymbol{r}$ and $\boldsymbol{k}$ using the addition theorem of spherical harmonics,

$$Y_{\lambda,0}(\cos\gamma) = \sqrt{\frac{4\pi}{2\lambda+1}} \sum_{\mu=-\lambda}^{+\lambda} Y^*_{\lambda,\mu}(\Theta, \Phi)\, Y_{\lambda,\mu}(\vartheta, \varphi). \tag{76.5}$$

Setting (76.5) into (76.4), and the result and (76.3) into (76.2), we get

$$f(\boldsymbol{k}) = \frac{4\pi}{(2\pi)^{\frac{3}{2}}} \int_0^\infty dr\, r^2 \oint d\Omega_r \sum_{\lambda\mu} i^{-\lambda} \frac{j_\lambda(kr)}{kr}\, Y_{\lambda,\mu}(\Theta,\Phi)\, Y^*_{\lambda,\mu}(\vartheta,\varphi) \frac{\chi_l}{r}\, Y_{l,m}(\vartheta,\varphi).$$

The integration over all directions of the vector $\boldsymbol{r}$ can be performed:

$$\oint d\Omega_r\, Y^*_{\lambda,\mu}(\vartheta,\varphi)\, Y_{l,m}(\vartheta,\varphi) = \delta_{l\lambda}\,\delta_{m\mu}$$

so that only one term (l,m) remains of the double sum:

$$k f(\boldsymbol{k}) = \sqrt{\frac{2}{\pi}}\, i^{-l} \int_0^\infty dr\, j_l(kr)\,\chi_l(r)\cdot Y_{l,m}(\Theta,\Phi),$$

i.e. the momentum space function, $f(\boldsymbol{k})$, may be factorized in the form

$$f(\boldsymbol{k}) = \frac{1}{k}\, g_l(k)\, Y_{l,m}(\Theta,\Phi) \tag{76.6}$$

with

$$g_l(k) = \sqrt{\frac{2}{\pi}}\, i^{-l} \int_0^\infty dr\, j_l(kr)\,\chi_l(r). \tag{76.7}$$

The radial parts $g_l(k)$ and $i^{-l}\chi_l(r)$ therefore stand in the mutual relation of a Hankel integral transform, the inversion of (76.7) being

$$\chi_l(r) = \sqrt{\frac{2}{\pi}}\, i^{l} \int_0^\infty dk\, j_l(kr)\, g_l(k). \tag{76.8}$$

If we normalize the space function u in the usual way to permit its probability interpretation,

$$\int_0^\infty dr\, |\chi_l(r)|^2 = 1, \tag{76.9}$$

we find by putting (76.8) into (76.9):

$$\frac{2}{\pi} \int_0^\infty dr \int_0^\infty dk \int_0^\infty dk'\, j_l(kr)\, g_l(k)\, j_l(k'r)\, g_l^*(k') = 1.$$

Performing first the integration over r,

$$\int_0^\infty dr\, j_l(kr)\, j_l(k'r) = \frac{\pi}{2}\,\delta(k-k'), \tag{76.10}$$

we get

$$\int_0^\infty dk\,|g_l(k)|^2 = 1, \tag{76.11}$$

i.e. the same probability interpretation holds in momentum space: in the quantum state under consideration the particle will be found with the absolute value of momentum between k and $k+dk$ with a probability $|g_l(k)|^2\,dk$.

Problem 77. Momentum space integral equation for central force potentials

In Problem 14 a general integral equation has been established for the momentum space wave functions. It shall be shown that for a central force field the solutions can be factorized in the form

$$f(\boldsymbol{k}) = \frac{1}{k}\, g_l(k)\, Y_{l,m}(\Theta,\Phi). \tag{77.1}$$

The special integral equation for $g_l(k)$ shall then be derived for the hydrogen atom.

Solution. The integral equation (14.6), written in atomic units,

$$(\tfrac{1}{2}k^2 - E)\, f(\boldsymbol{k}) = -\int d^3k'\, W(\boldsymbol{k}-\boldsymbol{k}')\, f(\boldsymbol{k}') \tag{77.2}$$

with

$$W(\boldsymbol{k}) = \frac{1}{8\pi^3}\int d^3x\, e^{-i\boldsymbol{k}\cdot\boldsymbol{r}}\, V(\boldsymbol{r}) \tag{77.3}$$

shall be reduced to a radial equation for $g_l(k)$ if $V(\boldsymbol{r})$ depends on the absolute value r only of the vector $\boldsymbol{r}$. In that case, (77.3) may be integrated over the solid angle, the result depending only upon the absolute value of the vector $\boldsymbol{k}$:

$$W(k) = \frac{4\pi}{8\pi^3}\int_0^\infty dr\, r^2\, V(r)\,\frac{\sin kr}{kr}. \tag{77.4}$$

The kernel of the integral equation then becomes a function of

$$(\boldsymbol{k}-\boldsymbol{k}')^2 = k^2 + k'^2 - 2kk'\cos\gamma \tag{77.5}$$

where γ is the angle between the vectors $\boldsymbol{k}$ and $\boldsymbol{k}'$. The function W may then be expanded into a series of Legendre polynomials,

$$W(|\boldsymbol{k}-\boldsymbol{k}'|) = \sum_{n=0}^{\infty} a_n(k,k') P_n(\cos\gamma) \tag{77.6}$$

whose coefficients a_n depend on the absolute values k and k' only. This is the essential point leading to possible factorization.

Using (77.1) for $f(\boldsymbol{k})$, the integral equation (77.2) now becomes

$$(\tfrac{1}{2}k^2 - E)\frac{1}{k} g_l(k)\, Y_{l,m}(\Theta,\Phi)$$

$$= -\sum_n^{\infty} \int_0^{\infty} dk'\, k'^2\, a_n(k,k') \frac{1}{k'} g_l(k') \oint d\Omega'\, P_n(\cos\gamma)\, Y_{l,m}(\Theta',\Phi').$$

The angular integral can be evaluated using the addition theorem

$$P_n(\cos\gamma) = \frac{4\pi}{2n+1} \sum_{\mu=-n}^{n} Y^*_{n,\mu}(\Theta',\Phi')\, Y_{n,\mu}(\Theta,\Phi)$$

which reduces the sums to one term only with $n=l$, $\mu=m$:

$$\frac{1}{k} g_l(k)\, Y_{l,m}(\Theta,\Phi) = -\frac{4\pi}{2l+1} \int_0^{\infty} dk'\, k'\, a_l(k,k') g_l(k')\, Y_{l,m}(\Theta,\Phi).$$

This is an identity in the polar angles thus showing that the factorization (77.1) is correct, and leaving us with the radial integral equation

$$(\tfrac{1}{2}k^2 - E) g_l(k) = -\frac{4\pi k}{2l+1} \int_0^{\infty} dk'\, k'\, a_l(k,k') g_l(k'). \tag{77.7}$$

If the central force is the *Coulomb attraction* of the hydrogen atom,

$$V(r) = -\frac{1}{r}, \tag{77.8}$$

the integral (77.4) can be solved using the limiting relation[17]

$$\lim_{\varepsilon\to 0} \int_0^{\infty} dx\, e^{-\varepsilon x} \sin x = 1$$

[17] The Fourier transform (77.3) of the potential is essentially the Born amplitude the convergence of which is limited by Coulomb behaviour at large values of r. Cf. Problem 105.

so that we find

$$W(k) = -\frac{1}{2\pi^2 k^2}, \tag{77.9}$$

and, according to (77.5),

$$W(\boldsymbol{k}-\boldsymbol{k}') = -\frac{1}{2\pi^2 |\boldsymbol{k}-\boldsymbol{k}'|^2} = -\frac{1}{4\pi^2 k k'(z-\cos\gamma)}$$

with

$$z = \frac{k^2+k'^2}{2kk'}. \tag{77.10}$$

Putting $\cos\gamma = t$, Eq. (77.6) then is the well-known expansion of

$$\frac{1}{z-t} = \sum_{n=0}^{\infty} (2n+1) Q_n(z) P_n(t) \tag{77.11}$$

where the coefficients

$$Q_n(z) = \frac{1}{2}\int_{-1}^{+1} \frac{dt\, P_n(t)}{z-t} \tag{77.12}$$

are the Legendre functions of the second kind. We therefore find

$$a_n(k,k') = -\frac{1}{4\pi^2 k k'}(2n+1) Q_n\left(\frac{k^2+k'^2}{2kk'}\right) P_n(\cos\gamma), \tag{77.13}$$

so that finally we arrive at the radial integral equation

$$(\tfrac{1}{2}k^2 - E) g_l(k) = \frac{1}{\pi}\int_0^\infty dk'\, Q\left(\frac{k^2+k'^2}{2kk'}\right) g_l(k'). \tag{77.14}$$

Literature. The integral equation (77.14) was solved by Fock, V.: Z. Physik **98,** 145 (1935). This way means determination of the momentum space eigenfunctions without any recourse to coordinate space functions. The latter way, however, proves simpler for the Coulomb field and shall be given in the following problem.

Problem 78. Momentum space wave functions for hydrogen

To determine the momentum space wave functions for the lowest levels ($1s, 2s, 2p$) of the hydrogen atom.

Solution. Since in Problem 76 we have shown that, in a central field, a factorized coordinate wave function

$$u(\boldsymbol{r}) = \frac{1}{r}\chi_l(r)\, Y_{l,m}(\vartheta,\varphi) \tag{78.1}$$

leads to a factorized momentum space wave function

$$f(\boldsymbol{k}) = \frac{1}{k} g_l(k) Y_{l,m}(\Theta, \Phi), \tag{78.2}$$

we have only to determine the radial part $g_l(k)$ which follows from a Hankel transform of $\chi_l(r)$:

$$g_l(k) = \sqrt{\frac{2}{\pi}} i^{-l} \int_0^\infty dr j_l(kr) \chi_l(r). \tag{78.3}$$

For the three chosen states we have, in atomic units,

$1s$: $\chi_{10}(r) = 2r e^{-r}$,

$$g_{10} = \sqrt{\frac{2}{\pi}} 2 \int_0^\infty dr r \sin kr e^{-r};$$

$2s$: $\chi_{20}(r) = \frac{1}{\sqrt{2}} (r - \frac{1}{2} r^2) e^{-\frac{r}{2}}$,

$$g_{20} = \frac{1}{\sqrt{\pi}} \int_0^\infty dr (r - \tfrac{1}{2} r^2) \sin kr e^{-\frac{r}{2}};$$

$2p$: $\chi_{21}(r) = \frac{1}{\sqrt{24}} r^2 e^{-\frac{r}{2}}$,

$$g_{21} = -i \frac{1}{\sqrt{12\pi}} \int_0^\infty dr r^2 \left(\frac{\sin kr}{kr} - \cos kr \right) e^{-\frac{r}{2}}.$$

Evaluation of these integrals is elementary, though a little cumbersome. The results are

$$g_{10}(k) = \sqrt{\frac{2}{\pi}} \frac{4k}{(1+k^2)^2}; \tag{78.4a}$$

$$g_{20}(k) = \frac{32}{\sqrt{\pi}} \frac{k(1-4k^2)}{(1+4k^2)^3}; \tag{78.4b}$$

$$g_{21}(k) = -i \frac{128}{\sqrt{3\pi}} \frac{k^2}{(1+4k^2)^3}. \tag{78.4c}$$

In all cases there holds the normalization law

$$\int_0^\infty dk\,|g_{nl}(k)|^2 = 1 \tag{78.5}$$

which may be checked directly for each of the functions (78.4a–c), but which also follows from the general theory, cf. (76.11).

Problem 79. Stark effect of a three-dimensional rotator

To calculate the Stark effect in second approximation of perturbation for a three-dimensional free rotator with the electric dipole moment p.

Solution. The unperturbed Schrödinger equation is solved by spherical harmonics:

$$\frac{\hbar^2}{2\Theta} L^2\, Y_{l,m} = E_l\, Y_{l,m} \tag{79.1}$$

with eigenvalues

$$E_l = \frac{\hbar^2\, l(l+1)}{2\Theta}. \tag{79.2}$$

The perturbation energy for an electrical field $\mathscr{E}$ applied in arbitrary direction is

$$V = -p(\mathscr{E}_x \sin\vartheta\cos\varphi + \mathscr{E}_y \sin\vartheta\sin\varphi + \mathscr{E}_z\cos\vartheta). \tag{79.3}$$

To find the matrix elements of V,

$$\langle l',m'|V|l,m\rangle = \oint d\Omega\, Y^*_{l',m'}\, V(\vartheta,\varphi)\, Y_{l,m}, \tag{79.4}$$

we use the relations[18]

$$\begin{aligned} \sin\vartheta\, e^{i\varphi}\, Y_{l,m} &= a_{l,m}\, Y_{l+1,m+1} - a_{l-1,-m-1}\, Y_{l-1,m+1},\\ \sin\vartheta\, e^{-i\varphi}\, Y_{l,m} &= -a_{l,-m}\, Y_{l+1,m-1} + a_{l-1,m-1}\, Y_{l-1,m-1},\\ \cos\vartheta\, Y_{l,m} &= b_{l,m}\, Y_{l+1,m} + b_{l-1,m}\, Y_{l-1,m} \end{aligned} \tag{79.5}$$

with the abbreviations

$$a_{l,m} = \sqrt{\frac{(l+m+1)(l+m+2)}{(2l+1)(2l+3)}}\,; \qquad b_{l,m} = \sqrt{\frac{(l+m+1)(l-m+1)}{(2l+1)(2l+3)}}\,. \tag{79.6}$$

---

[18] Cf. e.g. Bethe, H.A., Salpeter, E.E., p. 432, in: *Encyclopedia of Physics*, vol. 35. Berlin-Göttingen-Heidelberg: Springer 1957.

Then, using the orthonormalization rules of spherical harmonics, we find for any given pair of quantum numbers l, m that there are only six matrix elements not vanishing, viz.

$$\begin{aligned}
\langle l+1,m+1|V|l,m\rangle &= -\tfrac{1}{2}(\mathscr{E}_x - i\mathscr{E}_y)p\,a_{l,m},\\
\langle l-1,m+1|V|l,m\rangle &= +\tfrac{1}{2}(\mathscr{E}_x - i\mathscr{E}_y)p\,a_{l-1,-m-1},\\
\langle l+1,m-1|V|l,m\rangle &= +\tfrac{1}{2}(\mathscr{E}_x + i\mathscr{E}_y)p\,a_{l,-m},\\
\langle l-1,m-1|V|l,m\rangle &= -\tfrac{1}{2}(\mathscr{E}_x + i\mathscr{E}_y)p\,a_{l-1,m-1},\\
\langle l+1,m|V|l,m\rangle &= -\mathscr{E}_z p\, b_{l,m},\\
\langle l-1,m|V|l,m\rangle &= -\mathscr{E}_z p\, b_{l-1,m}.
\end{aligned} \tag{79.7}$$

Among these there are no diagonal elements $l=l'$. This means that the first-order perturbation vanishes so that we get no linear Stark effect:

$$\Delta_1 E_l = \langle l,m|V|l,m\rangle = 0. \tag{79.8}$$

This even holds for matrix elements $\langle l,m'|V|l,m\rangle$ so that no mixing of the degenerate functions of equal l but different m's can occur. Therefore we need not start from linear combinations of the spherical harmonics.

The quadratic Stark effect follows from the second-order perturbation,

$$\Delta_2 E_{l,m} = \sum_{l'm'} \frac{|\langle l',m'|V|l,m\rangle|^2}{E_l - E_{l'}}. \tag{79.9}$$

This is a sum over six terms, according to (79.7), with the denominators following from (79.2):

$$\begin{aligned}
\Delta_2 E_{l,m} = p^2 \frac{2\Theta}{\hbar^2}\Bigg\{\frac{1}{4}(\mathscr{E}_x^2+\mathscr{E}_y^2)\Bigg[&\frac{a_{l,m}^2 + a_{l,-m}^2}{l(l+1)-(l+1)(l+2)}\\
&+\frac{a_{l-1,-m-1}^2 + a_{l-1,m-1}^2}{l(l+1)-(l-1)l}\Bigg] + \mathscr{E}_z^2\Bigg[\frac{b_{l,m}^2}{l(l+1)-(l+1)(l+2)}\\
&+\frac{b_{l-1,m}^2}{l(l+1)-(l-1)l}\Bigg]\Bigg\}.
\end{aligned} \tag{79.10}$$

Straightforward computation of the sum, using (79.6), leads to the final result:

$$\Delta_2 E_{l,m} = \frac{p^2\Theta}{2\hbar^2}(2\mathscr{E}_z^2 - \mathscr{E}_x^2 - \mathscr{E}_y^2)\frac{l(l+1)-3m^2}{(2l+3)(2l-1)l(l+1)}. \tag{79.11}$$

This formula does not hold for $l=0$. This is obvious, as the eigenfunction $Y_{0,0}$ does not distinguish any direction, so that the perturbation should simply become proportional to $\mathscr{E}^2$. An analogous calculation to the one above shows that then, of the six matrix elements (79.7), only the three elements $\langle 1,m'|V|0,0\rangle$ with $m'=+1,0,-1$ differ from zero. The final result becomes

$$\Delta_2 E_{0,0} = -\frac{p^2\Theta}{3\hbar^2}\mathscr{E}^2. \tag{79.12}$$

Literature. Meyenn, K. von: Z. Physik **231,** 154 (1970) where even high field-strengths are discussed and the limits of validity of the second-order perturbation theory have been shown.

b) Problems of Elastic Scattering

Problem 80. Interference of incident and scattered waves

The asymptotic boundary condition

$$u \to e^{ikz} + f(\vartheta)\frac{e^{ikr}}{r} \quad \text{for } kr \to \infty \tag{80.1}$$

has to be imposed upon the wave function of a scattering problem (so-called Sommerfeld radiation condition). It shall be shown that this leads to a superposition of incident and scattered currents without noticeable interference at large distances from the scattering object. The relation between scattering amplitudes and cross section is to be derived.

Solution. The current density formula

$$\mathbf{s} = \frac{\hbar}{2im} N(u^*\nabla u - u\nabla u^*) \tag{80.2}$$

with a normalization constant N determining the absolute intensity of the current may be applied to a wave function of the form (80.1). The gradient, in spherical polar coordinates, has the components

$$\nabla_r = \frac{\partial}{\partial r}; \quad \nabla_\vartheta = \frac{1}{r}\frac{\partial}{\partial\vartheta}; \quad \nabla_\varphi = \frac{1}{r\sin\vartheta}\frac{\partial}{\partial\varphi}. \tag{80.3}$$

Since u does not depend upon φ we have $s_\varphi = 0$ (an obvious result). The two other components of $\mathbf{s}$ follow from (80.2) by a simple calculation:

$$s_r = \frac{\hbar k}{m} N\left(\cos\vartheta + \frac{|f|^2}{r^2}\right)$$
$$+ \frac{\hbar}{2m} N\left\{ f[kr(1+\cos\vartheta)+i]\frac{e^{ik(r-z)}}{r^2} + f^*[kr(1+\cos\vartheta)-i]\frac{e^{ik(z-r)}}{r^2}\right\};$$

$$
\begin{aligned}
s_\vartheta = & -\frac{\hbar k}{m} N \sin\vartheta \\
& + \frac{\hbar}{2mi} N \left\{ (f' - ikr\sin\vartheta\, f) \frac{e^{ik(r-z)}}{r^2} - (f^{*\prime} + ikr\sin\vartheta\, f^*) \frac{e^{ik(z-r)}}{r^2} \right\} \\
& + \frac{\hbar}{2mir^3} N (f' f^* - f f^{*\prime}).
\end{aligned}
$$

Here the prime denotes differentiation with respect to ϑ. Since the formulae will hold only asymptotically for $kr \to \infty$, we may neglect the last term, proportional to r^{-3}, as compared to the scattered intensity, proportional to r^{-2}. In the curly brackets of the interference terms the members with kr are the only ones to be taken along. Thus for further discussion there remain the simpler expressions,

$$
\begin{aligned}
s_r &= \frac{\hbar k}{m} N \left\{ \cos\vartheta + \frac{|f|^2}{r^2} + \frac{1+\cos\vartheta}{2r} \left(f\, e^{ik(r-z)} + f^* e^{ik(z-r)} \right) \right\}; \\
s_\vartheta &= \frac{\hbar k}{m} N \left\{ -\sin\vartheta - \frac{\sin\vartheta}{2r} \left(f\, e^{ik(r-z)} + f^* e^{ik(z-r)} \right) \right\}.
\end{aligned}
\tag{80.3}
$$

In these formulae the first terms are independent of r. They represent the decomposition into r and ϑ components of the plane wave contribution with the current density $s_0 = \hbar k N/m$ in z direction. Since $\hbar k/m = v$ is the particle velocity, N turns out to be the number of particles per unit volume in the incident current. The second term of s_r is the radial current which we can identify as the scattered intensity. In any observation, in order to obtain a finite intensity, we necessarily have to use some detector of a finite though small solid angle $\delta\Omega$ at a distance r through whose surface $r^2\delta\Omega$ a scattered current of

$$
\delta S = \int_{\delta\Omega} \left(\frac{\hbar k}{m} N \frac{|f|^2}{r^2} \right) r^2\, d\Omega = \frac{\hbar k}{m} N \int_{\delta\Omega} |f|^2\, d\Omega \tag{80.4}
$$

particles per second will pass and be counted. If this current is divided by the plane wave current density of $(\hbar k/m)\, N$ particles per second and cm^2, a quantity

$$
\delta\sigma = \int_{\delta\Omega} |f(\vartheta)|^2\, d\Omega \simeq |f(\vartheta)|^2\, \delta\Omega \tag{80.5}
$$

independent of the primary intensity, and thus characteristic of the scattering properties of the scattering interaction, will emerge. This ratio has the dimensions of an area and is called the *differential cross section* of scattering into the solid angle $\delta\Omega$. The total scattering cross section is derived herefrom,

$$\sigma = \oint |f(\vartheta)|^2 \, d\Omega . \tag{80.6}$$

It remains to look at the interference terms in (80.3) which, being proportional to $1/r$, at first glance seem to be of even more importance than the scattered intensity. These terms are composed of factors varying slowly and varying rapidly with angle. If any one of them is integrated over a small though finite solid angle $\delta\Omega$, the slowly varying factors (as $\sin\vartheta$) may be considered constant so that there remain only integrals of the types

$$\int_{\delta\Omega} d\Omega \cos k(r-z) \quad \text{or} \quad \int_{\delta\Omega} d\Omega \sin k(r-z) . \tag{80.7}$$

The argument

$$k(r-z) = kr(1-\cos\vartheta)$$

will, if kr is very large, cause the periodic functions in (80.7) to perform a great number of oscillations, even if integrated over a modest interval $\delta\Omega$. These integrals therefore will practically vanish (at least they do not contribute proportionally to $\delta\Omega$) so that the interference terms may safely be omitted in the limit of extremely large kr and the two currents of incident and scattered waves treated independently.

It should perhaps be mentioned in this connection, that kr is $2\pi r/\lambda$ with λ being the de Broglie wavelength. The latter will generally be of the order of atomic (or even nuclear) dimensions, say e.g. 10^{-8} cm, whereas r is a macroscopic distance in the experimental device of the order at least of 10 cm. In this case we then have $kr = 6 \cdot 10^9$, which is indeed extremely large.

Problem 81. Partial wave expansion of plane wave

To decompose a plane wave into partial waves of well-defined angular momentum.

Solution. The plane wave

$$u = e^{ikz} = e^{ikr\cos\vartheta} \tag{81.1}$$

is a solution of the Schrödinger equation in the force-free case,

$$\nabla^2 u + k^2 u = 0; \qquad k^2 = \frac{2mE}{\hbar^2} \tag{81.2}$$

whose complete solution, obtained by factorization in spherical polar coordinates, is

$$u = \frac{1}{kr} \sum_{l=0}^{\infty} \sum_{m=-l}^{+l} \{A_{l,m} j_l(kr) + B_{l,m} n_l(kr)\} P_l^m(\vartheta) e^{im\varphi} \tag{81.3}$$

with

$$j_l(kr) = \sqrt{\frac{\pi kr}{2}} J_{l+\frac{1}{2}}(kr); \quad n_l(kr) = (-1)^{l+1} \sqrt{\frac{\pi kr}{2}} J_{-(l+\frac{1}{2})}(kr). \tag{81.4}$$

Each sum term in (81.3) then represents a contribution of angular momentum well-defined by its quantum numbers l and m.

To represent the plane wave (81.1) by (81.3), two specializations may immediately be performed:

1. Only contributions from $m=0$ occur since (81.1) does not depend on the angle φ or, speaking physically, a particle beam parallel to the z axis has no angular momentum component along the z axis.

2. There can be no contribution of the functions n_l, because they are not regular at the coordinate centre.

The angular momentum expansion of the plane wave therefore becomes

$$e^{ikr\cos\vartheta} = \frac{1}{kr} \sum_{l=0}^{\infty} A_l j_l(kr) P_l(\cos\vartheta). \tag{81.5}$$

We still have to determine the coefficients A_l. That can be done in a way analogous to the determination of Fourier coefficients using the orthonormality relations

$$\int_{-1}^{+1} dt\, P_l(t) P_{l'}(t) = \frac{2}{2l+1} \delta_{ll'} \tag{81.6}$$

with $t = \cos\vartheta$. Eq. (81.5) then yields

$$\int_{-1}^{+1} dt\, e^{ikrt} P_l(t) = \frac{1}{kr} A_l j_l(kr) \cdot \frac{2}{2l+1}. \tag{81.7}$$

To determine A_l from this equation the integral on the left-hand side should be evaluated and compared with the spherical Bessel function on the right-hand side. Such a calculation, though quite elementary, would be rather cumbersome. Fortunately we are spared the necessity, because it suffices to compare the expressions on both sides of (81.7) for very large values of kr. Then,

$$j_l(kr) \to \sin\left(kr - \frac{l\pi}{2}\right), \tag{81.8}$$

and the integral can, by repeated partial integrations, be represented by expansion into negative powers of kr as follows:

$$\int_{-1}^{+1} dt\, e^{ikrt} P_l(t) = \frac{1}{ikr}\left[e^{ikrt} P_l(t)\right]_{-1}^{+1} - \frac{1}{ikr}\left\{\frac{1}{ikr}\left[e^{ikrt} P_l'(t)\right]_{-1}^{+1} - \frac{1}{ikr}\int_{-1}^{+1} dt\, e^{ikrt} P_l''(t)\right\}.$$

Asymptotically, only the first term signifies in which we use

$$P_l(\pm 1) = (\pm 1)^l .$$

Thus we get,

$$\int_{-1}^{+1} dt\, e^{ikrt} P_l(t) \to \frac{1}{ikr}\left(e^{ikr} - (-1)^l e^{-ikr}\right) = i^l \frac{2}{kr} \sin\left(kr - \frac{l\pi}{2}\right). \tag{81.9}$$

Putting (81.8) and (81.9) into (81.7), we find

$$A_l = (2l+1)\, i^l , \tag{81.10}$$

and the expansion (81.5) becomes finally

$$e^{ikz} = \frac{1}{kr} \sum_{l=0}^{\infty} (2l+1)\, i^l j_l(kr) P_l(\cos\vartheta). \tag{81.11}$$

Using normalized spherical harmonics,

$$Y_{l,0} = \sqrt{\frac{2l+1}{4\pi}}\, P_l \tag{81.12}$$

the series can equally well by written

$$e^{ikz} = \frac{1}{kr} \sum_{l=0}^{\infty} \sqrt{4\pi(2l+1)}\, i^l j_l(kr)\, Y_{l,0}(\vartheta). \tag{81.13}$$

Eqs. (81.11) and (81.13) may both be used to advantage.

NB 1. From Eqs. (81.7) and (81.10), there follows an integral representation of the spherical Bessel functions,

$$j_l(z) = i^{-l} \frac{z}{2} \int_{-1}^{+1} dt\, e^{izt} P_l(t) . \tag{81.14}$$

Such integral representations are often found by the method applied to our special problem.

NB 2. For $m=0$ there are two non-vanishing components, L_x and L_y, of the angular momentum which, of course, cannot be diagonal simultaneously with $L_z=0$. This reflects the classical picture of dividing the particle-beam into concentric cylindrical shells, of radii ρ, about the z axis in each of which L has a definite value, $L=mv\rho$, but all partitions of L into L_x and L_y components still occur. An ingenuous quantization, $L\simeq\hbar l$, would render the relation

$$\rho \simeq \frac{\hbar l}{mv} = l\frac{\lambda}{2\pi}$$

with λ the de Broglie wavelength and ρ a rough measure of the distance at which a particle of the quantum number l passes the coordinate origin.

Such a discussion of one expansion term singled out from the whole series is, of course, incorrect insofar as observables cannot but be attained by bilinearly combining u and u^* (or their derivatives), thus necessarily originating interference terms. It is in the limit of large l only that a single partial wave may be interpreted in the classical particle picture, because the very high spherical harmonics produce rapidly oscillating and thus quenching interference terms, even when averaged over narrow angular intervals.

NB 3. Let the plane wave propagate not in z direction, but in any other direction $\boldsymbol{k}$ defined by polar angles Θ, Φ. If ϑ' denotes the angle between the vectors $\boldsymbol{k}$ and $\boldsymbol{r}$, we have, according to (81.13),

$$e^{i\boldsymbol{k}\boldsymbol{r}} = \frac{1}{kr}\sum_{l=0}^{\infty}\sqrt{4\pi(2l+1)}\,i^l j_l(kr)\,Y_{l,0}(\vartheta').$$

We then make use of the addition theorem of spherical harmonics,

$$\sqrt{\frac{2l+1}{4\pi}}\,Y_{l,0}(\vartheta') = \sum_{m=-l}^{+l} Y^*_{l,m}(\Theta,\Phi)\,Y_{l,m}(\vartheta,\varphi) \tag{81.15}$$

yielding the generalized formula

$$e^{i\boldsymbol{k}\boldsymbol{r}} = \frac{4\pi}{kr}\sum_{l=0}^{\infty}\sum_{m=-l}^{+l} i^l\,j_l(kr)\,Y^*_{l,m}(\Theta,\Phi)\,Y_{l,m}(\vartheta,\varphi). \tag{81.16}$$

Problem 82. Partial wave expansion of scattering amplitude

Inside the sphere $r=R$ there may be given a potential $V(r)$; outside the sphere the potential may vanish. A beam of particles, described by a plane wave, is scattered at this potential field. The scattering amplitude shall be computed by expansion into a series of partial waves and shall then be expressed in terms of their logarithmic derivatives on the sphere $r=R$.

Solution. In the domain $r<R$ the wave function can be written

$$u=\frac{1}{kr}\sum_{l=0}^{\infty} i^l(2l+1)\,\chi_l(k,r)\,P_l(\cos\vartheta) \tag{82.1}$$

with

$$\chi_l''+\left[k^2-\frac{l(l+1)}{r^2}-\frac{2m}{\hbar^2}V(r)\right]\chi_l=0;\qquad \chi_l(0)=0. \tag{82.2}$$

The boundary conditions determine the functions except for an amplitude factor of which the logarithmic derivatives

$$L_l=\left(\frac{d\log\chi_l}{d\log r}\right)_{r=R} \tag{82.3}$$

are independent. In the following, the quantities L_l will be supposed to be known.

Outside the sphere $r=R$ we write

$$u=\frac{1}{kr}\sum_{l=0}^{\infty} i^l(2l+1)\left[j_l(kr)+\tfrac{1}{2}\alpha_l h_l^{(1)}(kr)\right]P_l(\cos\vartheta). \tag{82.4}$$

Were all $\alpha_l=0$, this would be the plane wave of Problem 81; the terms proportional to spherical Hankel functions of the first kind represent additional outgoing spherical waves, since

$$h_l^{(1)}(kr)=j_l(kr)+i n_l(kr)\rightarrow i^{-(l+1)}e^{ikr}. \tag{82.5a}$$

Remembering the asymptotic behaviour of j_l,

$$j_l(kr)\rightarrow\sin\left(kr-\frac{l\pi}{2}\right), \tag{82.5b}$$

we find that u, Eq. (82.4), at large distances becomes

$$u\rightarrow\frac{1}{2ikr}\sum_{l=0}^{\infty}(2l+1)\left[(1+\alpha_l)e^{ikr}-(-1)^l e^{-ikr}\right]P_l(\cos\vartheta). \tag{82.6}$$

We may conclude from the conservation of the number of particles in elastic scattering that the absolute squares of the amplitudes of ingoing and outgoing waves must be equal or that

$$|1+\alpha_l|^2=1, \tag{82.7}$$

in other words, that

$$\alpha_l=e^{2i\delta_l}-1. \tag{82.8}$$

Eq. (82.6) then can be written in the more compact form

$$u \to \frac{1}{kr} \sum_{l=0}^{\infty} (2l+1) i^l e^{i\delta_l} \sin\left(kr - \frac{l\pi}{2} + \delta_l\right) P_l(\cos\vartheta) \tag{82.9}$$

where δ_l apparently is the asymptotic phase shift of the solution of Eq. (82.2) with respect to the force free case of Eq. (82.5 b).

The scattering amplitude $f(\vartheta)$ is defined from the scattered wave, u_s,

$$u_s \equiv u - e^{ikz} \to f(\vartheta)\frac{e^{ikr}}{r}; \tag{82.10}$$

hence

$$f(\vartheta) = \frac{1}{2ik} \sum_{l=0}^{\infty} (2l+1)\alpha_l P_l(\cos\vartheta) \tag{82.11 a}$$

or, using (82.8),

$$f(\vartheta) = \frac{1}{2ik} \sum_{l=0}^{\infty} (2l+1)(e^{2i\delta_l} - 1) P_l(\cos\vartheta). \tag{82.11 b}$$

It remains to express the coefficients α_l not only by the phase shifts δ_l, but by the logarithmic derivatives L_l, Eq. (82.3). This can be done with the help of the continuity of both χ_l and $d\chi_l/dr$ when passing through the sphere $r=R$, viz.

$$\chi_l(k,R) = j_l(kR) + \tfrac{1}{2}\alpha_l h_l^{(1)}(kR);$$

$$\frac{d\chi_l(k,R)}{dr} = k\left[j_l'(kR) + \tfrac{1}{2}\alpha_l h_l^{(1)\prime}(kR)\right].$$

Here the prime denotes differentiation with respect to the argument kr, not to r. Dividing the second by the first of these relations, we get

$$L_l = x\frac{j_l'(x) + \frac{1}{2}\alpha_l h_l^{(1)\prime}(x)}{j_l(x) + \frac{1}{2}\alpha_l h_l^{(1)}(x)}, \qquad x = kR, \tag{82.12 a}$$

or inversely,

$$\alpha_l = -2\frac{L_l j_l(x) - x j_l'(x)}{L_l h_l^{(1)}(x) - x h_l^{(1)\prime}(x)}. \tag{82.12 b}$$

Again, this result satisfies the law of conservation of particle number, Eq. (82.7), as can easily be seen if $j_l(x)$ is replaced by

$$j_l(x) = \tfrac{1}{2}\left[h_l^{(1)}(x) + h_l^{(2)}(x)\right] \tag{82.13}$$

thus introducing the spherical Hankel function of the second kind which for real argument is the complex conjugate of $h_l^{(1)}(x)$ so that in

$$1+\alpha_l = -\frac{L_l h_l^{(2)}(x) - x h_l^{(2)\prime}(x)}{L_l h_l^{(1)}(x) - x h_l^{(1)\prime}(x)} \tag{82.14}$$

the numerator is the complex conjugate of the denominator so that $|1+\alpha_l|=1$, in agreement with (82.7).

Problem 83. Scattering at low energies

The spherical harmonics expansion of the scattering amplitude converges the better, the smaller the parameter $x=kR$. This shall be shown by computing the coefficients α_l for $x \ll 1$. α_0 and α_1 shall then be expanded in series of powers of x assuming that α_2 is already small enough to be neglected.

Solution. We start from (82.12b) of the preceding problem,

$$\alpha_l = -2\frac{L_l j_l(x) - x j_l'(x)}{L_l h_l^{(1)}(x) - x h_l^{(1)\prime}(x)}. \tag{83.1}$$

Here we use the definitions

$$h_l^{(1)}(x) = j_l(x) + i n_l(x); \quad h_l^{(2)}(x) = j_l(x) - i n_l(x) \tag{83.2a}$$

to replace the spherical Hankel functions by j_l and n_l whose power expansions run as follows:

$$\begin{aligned} j_l &= A_l x^{l+1}\left[1 - \frac{x^2}{2(2l+3)} + \frac{x^4}{8(2l+3)(2l+5)} \cdots\right]; \\ n_l &= -B_l x^{-l}\left[1 + \frac{x^2}{2(2l-1)} + \frac{x^4}{8(2l-1)(2l-3)} \cdots\right] \end{aligned} \tag{83.2b}$$

with

$$A_l = \frac{2^l l!}{(2l+1)!}; \quad B_l = \frac{(2l)!}{2^l l!}. \tag{83.2c}$$

If $x \ll 1$, the first term in each series is sufficient; we then have $j_l \ll |n_l|$ and $h_l^{(1)} \simeq i n_l$, and therefore

$$L_l j_l - x j_l' \simeq (L_l - l - 1) A_l x^{l+1}; \quad i(L_l h_l^{(1)} - x h_l^{(1)'}) \simeq -i(L_l + l) B_l x^{-l}$$

so that from (83.1), making use of (83.2c), we get

$$\alpha_l = -\frac{2i}{2l+1}\left(\frac{2^l l!}{(2l)!}\right)^2 \frac{L_l - (l+1)}{L_l + l} x^{2l+1}. \tag{83.3}$$

In this formula only the leading term has been computed, all higher powers in x having been neglected.

The scattering amplitude is

$$f(\vartheta) = \frac{1}{2ik} \sum_{l=0}^{\infty} (2l+1)\alpha_l P_l(\cos\vartheta); \tag{83.4}$$

this series will excellently converge if $x \ll 1$ since according to (83.3)

$$\frac{(2l+3)\alpha_{l+1}}{(2l+1)\alpha_l} = \frac{1}{(2l+1)^2} \cdot \frac{L_{l+1}-(l+3)}{L_{l+1}+(l+1)} \cdot \frac{L_l+l}{L_l-(l+1)} x^2 \ll 1 .$$

If the convergence of (83.4) allows to neglect the term with $l=2$ already, this, according to Eq. (83.3), is equivalent to neglecting in the scattering amplitude x^5 and all higher powers of x. In this case, it becomes possible to expand α_0 and α_1 to x^4 inclusively, using the two series (83.2b). The same result is however arrived at much faster by starting from the special functions

$$h_0^{(1)}(x) = -i e^{ix}; \quad h_1^{(1)}(x) = -\left(1 + \frac{i}{x}\right) e^{ix} \tag{83.5a}$$

and

$$h_l^{(2)}(x) = h_l^{(1)*}(x) \quad \text{for real } x . \tag{83.5b}$$

Using instead of (83.1) the equivalent Eq. (82.14) of the preceding problem,

$$1+\alpha_l = -\frac{L_l h_l^{(2)}(x) - x h_l^{(2)\prime}(x)}{L_l h_l^{(1)}(x) - x h_l^{(1)\prime}(x)}, \tag{83.6}$$

we arrive at the exact expressions

$$1+\alpha_0 = e^{-2ix} \frac{1+ix/L_0}{1-ix/L_0} \tag{83.7a}$$

and

$$1+\alpha_1 = e^{-2ix} \frac{1+ix-x^2/(L_1+1)}{1-ix-x^2/(L_1+1)} . \tag{83.7b}$$

Power expansion of these two expressions leads on, within the approximation wanted, to

$$\alpha_0 = -2ix\left(\frac{L_0-1}{L_0}\right) - 2x^2\left(\frac{L_0-1}{L_0}\right)^2 + \frac{4}{3} i x^3 \left[\left(\frac{L_0-1}{L_0}\right)^3 - \frac{1}{2L_0^3}\right]$$
$$+ \frac{2}{3} x^4 \left[\left(\frac{L_0-1}{L_0}\right)^4 - \frac{2}{L_0^3}\left(\frac{L_0-1}{L_0}\right)\right] + \cdots \tag{83.8a}$$

and

$$\alpha_1 = -\frac{2}{3} i x^3 \cdot \frac{L_1-2}{L_1+1} . \tag{83.8b}$$

(The expansion of α_1 has no term with x^4.) In order to obtain sufficient accuracy when the influence of $l=1$ is just noticeable we have to take account of at least three, if not four, terms in (83.8a) so that it may often be simpler to use the exact formula (83.7a) instead.

Problem 84. Scattering by a constant repulsive potential

Given the potential $V=V_0>0$ for $r<R$, and $V=0$ for $r>R$. The phase shift δ_0 and the partial scattering cross section σ_0 for $l=0$ shall be determined as functions of the energy of the incident particles.

Solution. Using the abbreviations

$$K_0^2=\frac{2mV_0}{\hbar^2};\quad k^2=\frac{2mE}{\hbar^2};\quad \varkappa=\sqrt{K_0^2-k^2} \tag{84.1}$$

the radial wave function for $l=0$, $\chi_0(r)$, can be written as follows: If the energy E is below threshold V_0,

$$\chi_0(r)=\begin{cases} A\sinh\varkappa r & \text{for } r<R \\ \sin(kr+\delta_0) & \text{for } r>R \end{cases} \quad \text{if } E<V_0 \tag{84.2a}$$

and, if the energy is above, with $\varkappa=i\varkappa'$ becoming purely imaginary,

$$\chi_0(r)=\begin{cases} A'\sin\varkappa' r & \text{for } r<R \\ \sin(kr+\delta_0) & \text{for } r>R \end{cases} \quad \text{if } E>V_0. \tag{84.2b}$$

The normalization outside the potential sphere is the same in all cases, the amplitude constants A and A' therefore being a measure of how much the interior of the sphere will be excited to vibrations. The boundary condition, $\chi_0(0)=0$, has been taken care of in all expressions. Continuity of χ_0 and χ_0' at $r=R$ imposes continuity on the logarithmic derivative so that its value at $r=R$,

$$L_0\equiv R\chi_0'(R)/\chi_0(R), \tag{84.3}$$

can be formed from either of the two expressions for $r<R$ or $r>R$. Thus we get

$$L_0\equiv kR\cot(kR+\delta_0)=\varkappa R\coth\varkappa R \quad \text{if } E<V_0 \tag{84.4a}$$

and

$$L_0\equiv kR\cot(kR+\delta_0)=\varkappa' R\cot\varkappa' R \quad \text{if } E>V_0. \tag{84.4b}$$

Only in the latter case is there a periodic solution in the interior whose amplitude is of some interest:

$$A'^2 = \frac{1}{\sin^2 \varkappa' R + \dfrac{\varkappa'^2}{k^2} \cos^2 \varkappa' R}. \tag{84.5}$$

We start discussion of these formulae at lowest energies. In the limit $k \to 0$, $\varkappa$ is to be replaced by K_0 and, in order to keep L_0, Eq. (84.4a), finite, the argument of the cotangent on the left must tend towards zero as k. Eq. (84.4a) then becomes

$$L_0 = \frac{kR}{kR + \delta_0} = K_0 R \coth K_0 R$$

yielding

$$\delta_0 = kR\left(\frac{\tanh K_0 R}{K_0 R} - 1\right). \tag{84.6}$$

If the potential height K_0^2 is very large ($K_0 \to \infty$, *hard sphere*), k^2 can be neglected compared to K_0^2 for all energies, so that the relation $\varkappa = K_0$ and Eq. (84.6) will hold for all energies. This then means a linear increase of δ_0 with the parameter kR as represented by the straight line (1) in Fig. 43. The curve (2) in Fig. 43 has been computed from Eqs. (84.4a, b)

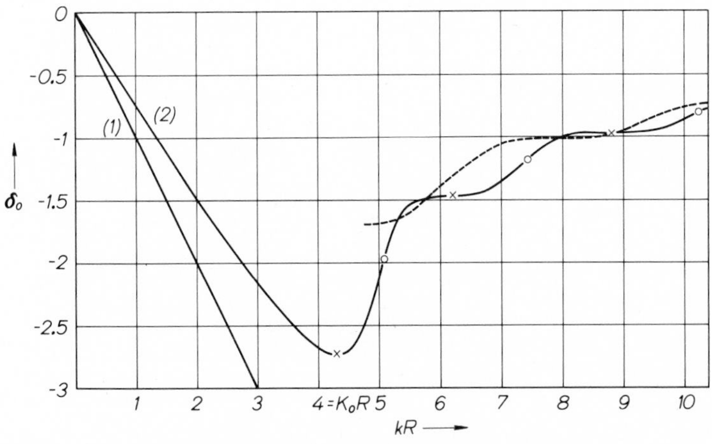

Fig. 43. Phase shift δ_0 as a function of kR. The straight line (1) holds for a hard sphere ($K_0 R = \infty$), curve (2) for $K_0 R = 4$. The broken line has been computed in first Born approximation. Resonances (amplitude maxima) marked by circles, minima of amplitude by crosses

for the numerical example $K_0R=4$. Even the start of this curve at $kR=0$ shows that for a finite potential height the inclination of the phase curve differs from that of the hard sphere curve. This is a characteristic trait of quantum mechanics where the wave penetrates into the potential sphere even at energies below threshold, the penetration depth depending upon the potential height. In classical mechanics, where no such penetration would be possible, the scattering phenomenon below threshold could not show any effect of the potential height.

If the energy E (or the parameter kR) increases to values above threshold, curve (2) reaches a minimum of δ_0, and then begins to rise again towards zero over a sequence of steps. At highest energies ($kR\to\infty$) the limit must, of course, be $\delta_0=0$ again since for $E\gg V_0$ the potential hill becomes quite an unimportant obstacle. The cause of the formation of steps may be elucidated by a computation of the amplitudes A', Eq. (84.5), of the interior. In Fig. 44a these amplitudes have been drawn for our example $K_0R=4$. There are values of kR, i.e. of wavelengths

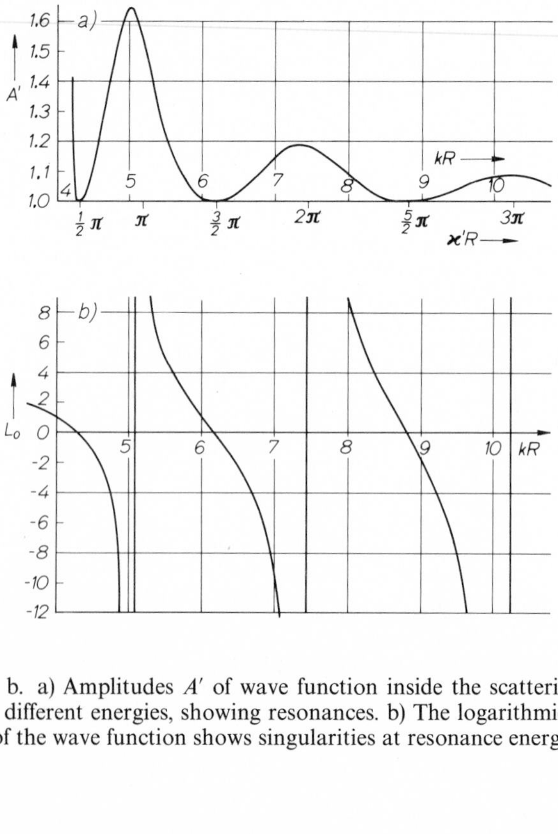

Fig. 44a and b. a) Amplitudes A' of wave function inside the scattering obstacle ($K_0R=4$) at different energies, showing resonances. b) The logarithmic derivative L_0 of the wave function shows singularities at resonance energies

of the steady particle beam exciting the interior of the potential sphere to vibrations, at which A' is maximum. Wavelength bands therefore exist in which the interior gets into *resonance* with the vibration impressed upon it from outside. Between any two resonances there lies a minimum amplitude $A'=1$ holding for $\cos\varkappa' R=0$ or at $\varkappa' R=(n+\frac{1}{2})\pi$.

Each of these minimum amplitudes coincides with a zero of L_0 (as drawn in Fig. 44b) so that the wave function, having horizontal tangent at $r=R$, has the same amplitude inside and outside of the scattering sphere. Between these energies there lie resonance energies in the neighbourhood of $\varkappa' R=n\pi$ or close to the singularities of L_0, with the wave function vanishing for $r=R$ but passing this point without break in the inclination of its tangent, thus in the interior with its longer wavelength rising to the highest possible amplitude.

Now we know it to be a general principle of resonance phenomena that on passing through a resonance there occurs a phase jump of π (in the ideal case). This accounts for the amplitude minima (marked by crosses on curve (2) of Fig. 43) between which a steep rise of phase occurs when passing resonances (marked by circles on curve (2) of Fig. 43).

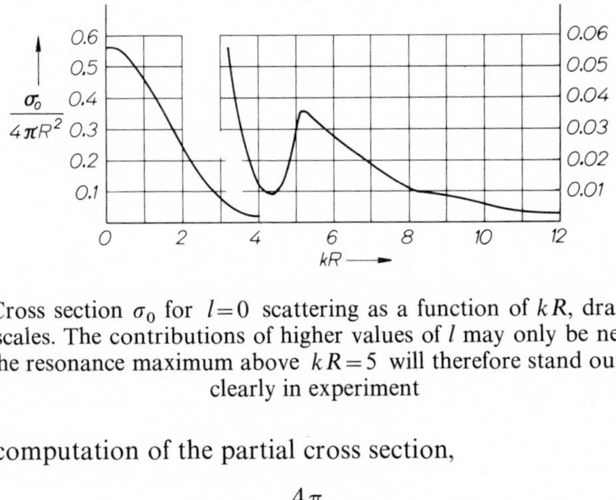

Fig. 45. Cross section σ_0 for $l=0$ scattering as a function of kR, drawn on two different scales. The contributions of higher values of l may only be neglected for $kR \ll 1$. The resonance maximum above $kR=5$ will therefore stand out much less clearly in experiment

The computation of the partial cross section,

$$\sigma_0 = \frac{4\pi}{k^2}\sin^2\delta_0 \tag{84.7}$$

as a function of kR may be performed from the δ_0 curve without difficulty. The result has been drawn in Fig. 45. In the low-energy limit, $kR\to 0$, δ_0 of Eq. (84.6) becomes infinitely small so that $\sin\delta_0$ in (84.7) may be replaced by its argument, and

$$\sigma_0(0)=4\pi a_0^2; \quad a_0=R\left(1-\frac{\tanh K_0 R}{K_0 R}\right). \tag{84.8}$$

The quantity a_0 defined by (84.8) is called the *scattering length* of the potential. In our numerical example it turns out to be $a_0 = 0.75R$. (Cf. also Problem 88 for the importance of the scattering length.) With increasing energy, the resonances show their influence on the cross section curve, but only the first one at about $kR = 5.1$ in our example leads to a pronounced effect. The higher resonances are almost invisible in the cross section curve and become even more so in the experimental evidence, where an increasing contribution to σ due to $l \neq 0$ has to be added. The information content of a cross section curve, therefore, is not very great.

NB. At very high energies the decrease towards zero of the phase shift δ_0 may be checked by applying the first Born approximation (cf. also Problem 105) for $l=0$,

$$\delta_0 = -\frac{2m}{\hbar^2} \cdot \frac{1}{k} \int_0^\infty dr \sin^2 kr \, V(r),$$

which yields in the special case of our constant repulsive potential,

$$\delta_0 = -\frac{K_0^2}{k} \int_0^R dr \sin^2 kr = -\frac{K_0^2}{4k^2}(2kR - \sin 2kR). \tag{84.9}$$

The phase shifts computed from this formula are shown as the broken line in Fig. 43. It is not to be expected that, for the rather large phase shifts in the energy region shown, the Born approximation should give quantitatively satisfactory results. It seems remarkable, however, that except for a shift in position of the resonances the correct picture is roughly obtained even at these energies, including the formation of steps in the curve as a consequence of the sine term in the Born formula (84.9).

Problem 85. Anomalous scattering

Let the potential in the preceding problem be altered by cutting out of its centre a potential hole so that

$$\frac{2m}{\hbar^2} V(r) = \begin{cases} -K_1^2 & \text{for } 0 \le r < r_1, \\ +K_0^2 & \text{for } r_1 \le r < R, \\ 0 & \text{for } R \le r. \end{cases} \tag{85.1}$$

The alteration of the phase shift δ_0 for $l=0$ compared to that determined in the preceding problem shall be calculated and discussed for the numerical example $K_0 R = 4$ (as before), $K_1 r_1 = 1.5$ and $r_1 = \frac{1}{2}R$.

Solution. The potential is sketched in Fig. 46 where the following abbreviations are explained:

$$\varkappa^2 = K_0^2 - k^2; \quad K^2 = K_1^2 + k^2 \tag{85.2}$$

for energies below threshold ($k^2 < K_0^2$, $\varkappa$ real). The radial wave function, $\chi_0(r)$, then becomes

$$\chi_0(r) = \begin{cases} B \sin K r & \text{for } \ 0 \le r < r_1, \\ A(\sinh \varkappa r + \gamma \cosh \varkappa r) & \text{for } \ r_1 \le r < R, \\ \sin(k r + \delta_0) & \text{for } \ R \le r. \end{cases} \tag{85.3}$$

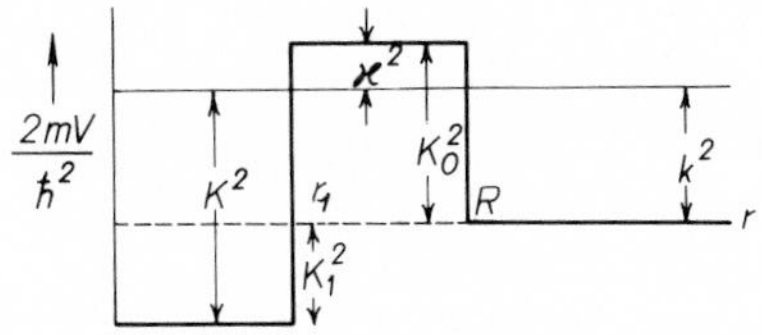

Fig. 46. Scattering potential of Problem 85

For energies above threshold, $\varkappa$ becomes imaginary, $\varkappa = i\varkappa'$ with real $\varkappa'$. With this change, (85.2) and (85.3) may still be used. The problem differs from the preceding one by the additional boundary condition at $r = r_1$. With $r_1 = 0$ it therefore becomes identical with the former problem that can formally be regained by $\gamma = 0$ from (85.3).

We now have two conditions of continuity of the logarithmic derivative of (85.3) at $r = r_1$ and $r = R$:

$$L_0(r_1) = K r_1 \cot K r_1 = \varkappa r_1 \frac{1 + \gamma \tanh \varkappa r_1}{\tanh \varkappa r_1 + \gamma}; \tag{85.4a}$$

$$L_0(R) = \varkappa R \frac{1 + \gamma \tanh \varkappa R}{\tanh \varkappa R + \gamma} = k R \cot(k R + \delta_0). \tag{85.4b}$$

From (85.4a) γ may be obtained and then δ_0 from (85.4b):

$$\gamma = \frac{\dfrac{\varkappa}{K} \tan K r_1 - \tanh \varkappa r_1}{1 - \dfrac{\varkappa}{K} \tan K r_1 \tanh \varkappa r_1}; \tag{85.5}$$

$$\delta_0 = -k R + \tan^{-1} \left\{ \frac{k}{\varkappa} \cdot \frac{\tanh \varkappa R + \gamma}{1 + \gamma \tanh \varkappa R} \right\}. \tag{85.6}$$

Eq. (85.5) shows that with $r_1 = 0$ we indeed find $\gamma = 0$, as surmised above. For the practical purpose of numerical computation, it is more comfortable to put γ from (85.5) into Eq. (85.6) which gives after some elementary reordering:

$$\delta_0 = -kR + \tan^{-1}\left\{\frac{kR}{\varkappa R}\cdot\frac{\tanh\varkappa(R-r_1)+\varkappa r_1\dfrac{\tan Kr_1}{Kr_1}}{1+\varkappa r_1\dfrac{\tan Kr_1}{Kr_1}\tanh\varkappa(R-r_1)}\right\}. \tag{85.7}$$

The phase shifts computed according to Eq. (85.7) for the numerical example $K_1 r_1 = 1.5$; $K_0(R-r_1)=2$; $K_0 R = 4$ have been drawn in Fig.

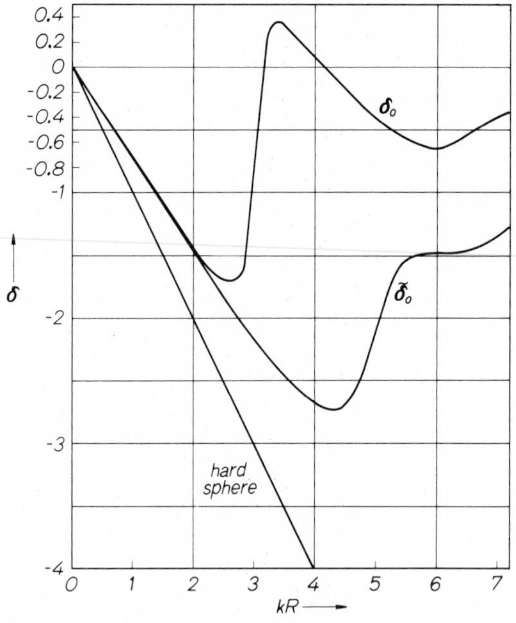

Fig. 47. Phase shifts δ_0 for $l=0$ for anomalous scattering originated by a potential hole inside the wall. They are compared with those without potential hole, $\tilde{\delta}_0$, as determined in the preceding problem. The straight line obtains for a hard sphere

47 where they have been compared with those obtained in the preceding problem (here denoted by $\tilde{\delta}_0$) and those of the hard sphere of radius R (the straight line). There are no significant differences between δ_0 and $\tilde{\delta}_0$ at energies below $kR \simeq 2$; we may conclude that this shows that the wave has not entered far enough into the interior of the scattering

region to be appreciably affected by the central hole. This changes grossly in the interval of about $3<kR<6$ where the behaviour of the two curves is totally different. The steep ascent of δ_0 about $kR=3$ indicates resonance at this energy.

Indeed it is along these lines that the deviation becomes understandable. Let R be infinitely large, then there would remain just a potential hole of the depth

$$\tilde{K}^2 = K_1^2 + K_0^2$$

and there would be eigenstates whose energies could be determined from

$$\tan K r_1 = -\frac{K}{\sqrt{\tilde{K}^2 - K^2}}.$$

In our numerical example this equation would be satisfied for $K r_1 = 2.125$ which, with our original radius R, would correspond to $kR = 3.14$, i.e. just to the place where the curve in Fig. 47 indicates the existence of a resonant level.

Problem 86. Scattering resonances

A spherical cavity of radius R is bounded by a thin wall of a given dimensionless opacity Ω defined by

$$V(r) = \frac{\hbar^2}{2m}\frac{\Omega}{R}\delta(r-R). \tag{86.1}$$

The scattering of the partial wave $l=0$ shall be investigated.

NB. This problem is closely related to the one-dimensional one dealt with in Problem 27. The definition of the opacity, however, differs by the factor $2R$ in the denominator of (86.1) from the one given in Problem 20 and used in Problem 27, because in that case there was no characteristic length that could have been used in the same sense as the radius of the cavity in the present problem. Note that the one-dimensional δ function has the dimension of a reciprocal length.

Solution. The Schrödinger equation for the radial wave function, $\chi_0(r)$, with the potential (86.1) is

$$\chi_0'' + \left[k^2 - \frac{\Omega}{R}\delta(r-R)\right]\chi_0 = 0. \tag{86.2}$$

The function χ_0 must be continuous at the potential singularity. Its derivative χ_0', however, undergoes a jump following from an integration of Eq. (86.2) across the wall,

$$\int_{R-\varepsilon}^{R+\varepsilon} dr\,\chi_0'' = \chi_0'(R+\varepsilon) - \chi_0'(R-\varepsilon) = \frac{\Omega}{R}\chi_0(R)$$

so that, in terms of the logarithmic derivative,

$$L_0(r)=r\,\frac{\chi_0'(r)}{\chi_0(r)}, \tag{86.3}$$

there emerges the boundary condition

$$L_0(R+\varepsilon)-L_0(R-\varepsilon)=\Omega. \tag{86.4}$$

The two parts of the wave function, inside and outside of the cavity,

$$\chi_0(r)=A\sin kr \qquad \text{for } r<R \tag{86.5a}$$

and

$$\chi_0(r)=\sin(kr+\delta_0) \quad \text{for } r>R, \tag{86.5b}$$

may then be connected by (86.4),

$$k[\cot(kR+\delta_0)-\cot kR]=\Omega, \tag{86.6}$$

or, in a more suitable shape for the determination of the phase shift,

$$\tan(x+\delta_0)=\frac{\tan x}{1+\Omega\,\dfrac{\tan x}{x}} \quad \text{with } x=kR. \tag{86.7}$$

The amplitude A, Eq. (86.5a), is easily determined from the continuity of $\chi_0(r)$,

$$A=\frac{\sin(x+\delta_0)}{\sin x}$$

or, using (86.7) to eliminate δ_0,

$$A^2=\frac{\tan^2 x+1}{\tan^2 x+\left(1+\Omega\,\dfrac{\tan x}{x}\right)^2}. \tag{86.8}$$

It can be seen from Eq. (86.7) that an impenetrable wall ($\Omega\to\infty$) leads to the limiting case of the hard sphere,

$$\tan(x+\delta_0)=0 \quad \text{or} \quad \delta_0=-x, \tag{86.9}$$

the amplitude A, according to Eq. (86.8), vanishing.

If Ω is very large but still finite, there are narrow energy bands around the zeros of the denominator of (86.7) where deviations occur from the hard sphere behaviour. These are the same narrow bands in which, as shown by Eq. (86.8), the amplitude A inside the cavity becomes large. We therefore obtain typical resonance phenomena with strong coupling between the vibrations on both sides of the wall restricted to narrow energy bands, and almost complete decoupling for all other energies. The more penetrable the wall, the greater the damping of this resonance phenomenon.

The resonances occur, if $\Omega \gg 1$, at about energies where $x = n\pi$. These are the energies at which $\chi_0(R) = 0$, i.e. of the eigenstates of the cavity in the case where it is surrounded by an impenetrable wall. Resonance levels of maximum coupling therefore lie at, or close to, the eigenvalues of the cavity.

Numerical discussion has been performed for $\Omega = 4$ and $\Omega = 10$. For the smaller of the two values, $\Omega = 4$, the phase shift is shown in Fig. 48 as a function of x. At small energies the curve does not differ

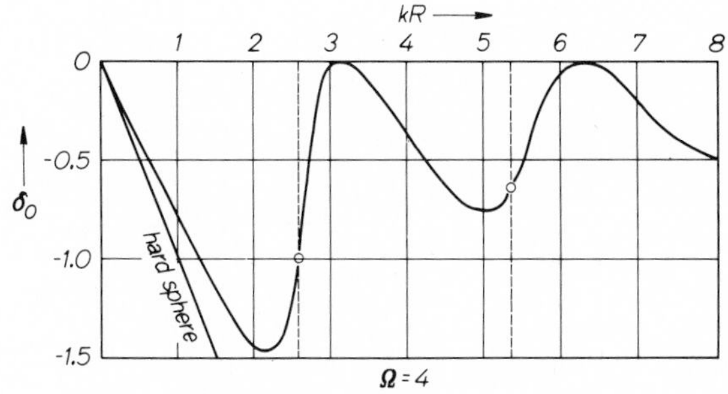

Fig. 48. Phase shifts δ_0 as a function of kR for a spherical cavity with $\Omega = 4$. The larger Ω, the more pronounced the resonances become (cf. Fig. 49a)

very much from the straight line of the hard sphere, its initial tangent inclination, however, being $-\Omega/(1+\Omega) = -0.8$ instead of -1 for the hard sphere. The first zero of the denominator in (86.7) occurs at $x_1 = 2.57$, the second at $x_2 = 5.35$; both have been marked in the figure by the vertical broken lines. These resonance energies still lie appreciably left of the values $x = n\pi$ where the phase falls back to zero. For a larger value of Ω they draw nearer to $x = n\pi$, e.g. for $\Omega = 10$, where the first resonance shifts to $x_1 = 2.86$ and the second to $x_2 = 5.76$ (see Fig. 49a). The rise of the phase angle to zero then becomes steeper and the resonance more pronounced. In the limit $\Omega \to \infty$ we finally get a sharp discontinuous jump of $\Delta\delta_0 = \pi$ at $x_n = n\pi$. This does not contradict the linear law (86.9) for the hard sphere since δ_0 is only defined modulo π.

The resonance phenomena are investigated in still more detail for the case $\Omega = 10$. In Fig. 49b the amplitude A, Eq. (86.8), is shown with the two first pronounced resonance maxima lying at the x values marked by the vertical broken lines in the phase shift diagram (Fig. 49a). From

Fig. 49c it can be seen that the cross section has a small but not very pronounced resonance peak, at least, at the first of these x values. On both sides of each resonance there are two places where $A=1$, viz. if

$$1+\Omega\frac{\tan x}{x}=\pm 1. \tag{86.10}$$

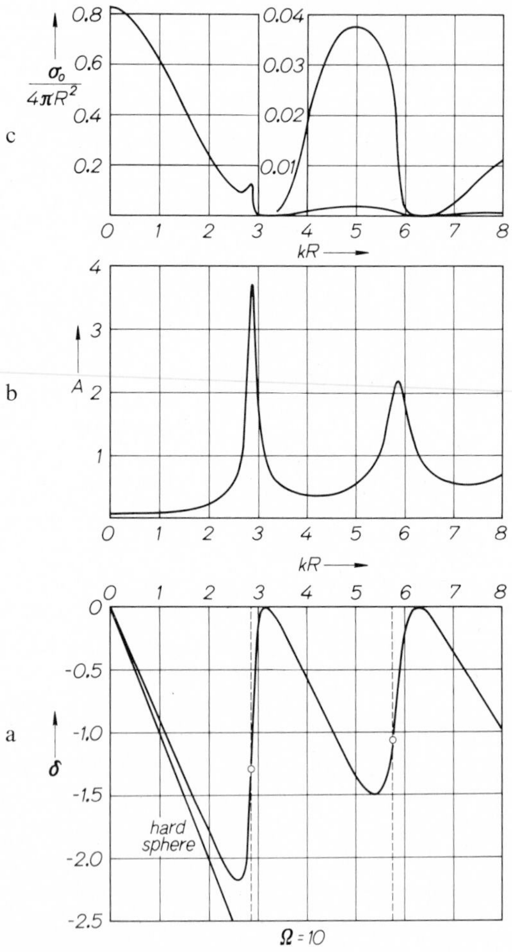

Fig. 49a—c. a) Phase shifts δ_0 as in Fig. 48, but for $\Omega=10$. b) Amplitudes A show pronounced resonances. c) Cross section for $l=0$ (drawn on two different scales). Even the pronounced resonance at $kR=2.86$ causes only a rather small peak in the cross section curve, even with higher angular momenta neglected

The upper sign just leads to $x = n\pi$ and thus to $\delta_0 = 0$ and $\sigma_0 = 0$. This is very understandable since the wave function inside the wall then exactly agrees with the wave function without wall and hence no scattered wave at all. The lower sign in (47.10) leads to the immediate neighbourhood of minimum phase angles, the phase curve becoming stationary at

$$\frac{\tan x}{x} = -\frac{2}{\Omega - 1},$$

as can easily be found from $d\delta_0/dx = 0$, i.e. at $x = 2.616$ and $x = 5.406$ for $\Omega = 10$, whereas (47.10) with the minus sign yields

$$\frac{\tan x}{x} = -\frac{2}{\Omega}$$

with the solutions $x = 2.654$ and $x = 5.454$ for $\Omega = 10$ lying very close to the minimum positions.

The information conveyed by the cross section curve, Fig. 49c, is rather scanty: The first resonance leads to a small peak only. Nothing can be seen of the second resonance except the two zeros, far left and right of it, from whose existence it may be deduced that somewhere between them there might be a resonance, though of unknown position, height and width. The results to be derived from an experimental cross section curve may give even less information in consequence of the background produced by the contribution of higher angular momenta, rising with increasing x, and obscuring the zeros of the σ_0 curve.

Problem 87. Contribution of higher angular momenta

To calculate the phase shifts and partial scattering cross sections, for the opaque cavity discussed in the preceding problem, up to $l = 2$.

Solution. The radial wave function for any l may be written

$$\chi_l(r) = A_l j_l(kr) \quad \text{for } r < R; \tag{87.1a}$$

$$\chi_l(r) = j_l(kr)\cos\delta_l - n_l(kr)\sin\delta_l \quad \text{for } r > R. \tag{87.1b}$$

Since the spherical cylinder functions for $kr \gg l$ approach asymptotically

$$j_l(kr) \to \sin\left(kr - \frac{l\pi}{2}\right); \quad n_l(kr) \to -\cos\left(kr - \frac{l\pi}{2}\right),$$

Eq. (87.1b) leads to the asymptotic behaviour of χ_l:

$$\chi_l \to \sin\left(kr - \frac{l\pi}{2} + \delta_l\right). \tag{87.2}$$

Therefore, δ_l is the phase shift of the l'th partial wave, yielding the partial cross section,

$$\sigma_l = 4\pi(2l+1)\sin^2\delta_l\,. \tag{87.3}$$

To determine the phase angles δ_l, we use the boundary conditions at $r=R$ where, again, χ_l has to be continuous, and the logarithmic derivative, $r\chi_l'(r)/\chi_l(r)$ jumps by Ω:

$$A_l j_l(x) = j_l(x)\cos\delta_l - n_l(x)\sin\delta_l\,; \tag{87.4}$$

$$x\left\{\frac{j_l'(x)-\tan\delta_l\, n_l'(x)}{j_l(x)-\tan\delta_l\, n_l(x)} - \frac{j_l'(x)}{j_l(x)}\right\} = \Omega\,. \tag{87.5}$$

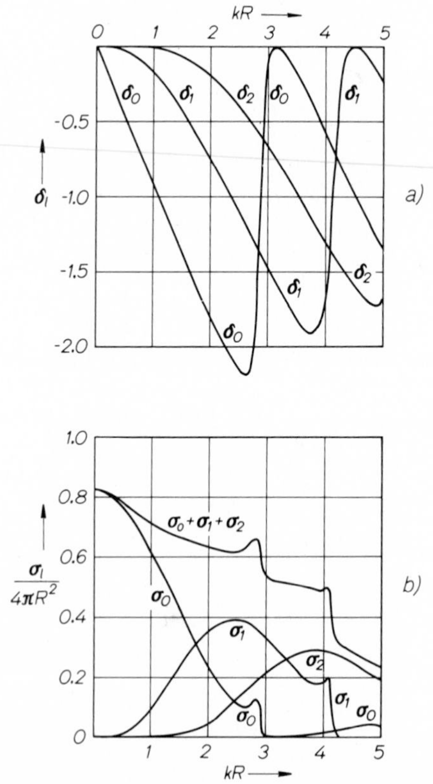

Fig. 50a and b. Phase shifts (a) and cross sections (b) for the spherical cavity, including higher angular momenta. The peak at $kR=2.86$ still may be detected

Here the primes denote differentiations with respect to the argument kr, and $x = kR$.

Eq. (87.5) enables the phase shifts δ_l to be computed in the form

$$\tan\delta_l = \frac{j_l(x)^2}{j_l(x)\,n_l(x) - \dfrac{x}{\Omega}}; \tag{87.6}$$

these may then be put in (87.4) to obtain the amplitudes A_l.

For $l=0$, the spherical cylinder functions reduce to $j_0(x) = \sin x$ and $n_0(x) = -\cos x$. Eqs. (87.4) and (87.6) then lead back to the preceding problem.

For the numerical example, $\Omega = 10$, the phase curves up to $x=5$ are shown in Fig. 50 a for $l=0,1,2$. They all depend upon x in a similar way, the main difference for small values of x being that, the higher l, the more slowly they deviate from zero. Thus, only head-on collisions with $l=0$ contribute at low energies. From about $x=1$ upwards, the contributions of $l=1$ become appreciable, and from about $x=2$ those of $l=2$. The partial cross sections (Fig. 48 b) show small resonance peaks for $l=0$ and $l=1$; in the sum curve $\sigma = \sigma_0 + \sigma_1 + \sigma_2$, however, they form no more than small humps over a continuous background. (The rather steep descent beyond $x=4$ is deceptive; it would be largely obliterated by the there rising contribution of $l=3$ omitted in the curve.)

Problem 88. Shape-independent approximation

It shall be shown that for small energies there holds an expansion in powers of k^2 beginning in the form

$$k\cot\delta_0 = -\frac{1}{a} + \frac{1}{2} r_0 k^2$$

with the scattering length a and the effective range r_0 the only parameters depending upon the potential.

The consequence of the existence of such an expansion is, of course, that for energies small enough to let these two first terms suffice (i.e. as long as no effect of $l=1$ is observed) no experimental cross section curve, $\sigma(E)$, gives any more information than the two constants. This means that the shape of the potential cannot be so determined. This is of special importance in low-energy nuclear physics and is one of the reasons, to put it in a nutshell, for our quantitative ignorance of nuclear forces.

Solution. At these low energies, we shall deal only with $l=0$. We write $\chi_k(r)$ for the radial wave function at the energy $\hbar^2 k^2/2m$, and $\chi_0(r)$ at zero energy. These two functions obey the differential equations

$$\chi_0'' - U(r)\,\chi_0 = 0 \tag{88.1a}$$

and

$$\chi_k'' + k^2\chi_k - U(r)\,\chi_k = 0 \tag{88.1b}$$

with

$$U(r) = \frac{2m}{\hbar^2}\,V(r),$$

and the boundary conditions

$$\chi_0(0) = 0; \qquad \chi_k(0) = 0. \tag{88.2}$$

Their behaviour at a radius $r > R_0$ where the potential has become zero will be

$$\chi_0 = C(r-a); \qquad \chi_k = \sin(kr + \delta_0) \tag{88.3}$$

where either sign of a is still possible.

Multiplication of (88.1 a) with χ_k, of (88.1 b) with χ_0 and subtraction yields

$$\chi_k\chi_0'' - \chi_0\chi_k'' = k^2\chi_0\chi_k\,.$$

Let this equation be integrated from 0 to r, using (88.2):

$$\chi_k(r)\,\chi_0'(r) - \chi_0(r)\,\chi_k'(r) = k^2\int_0^r dr\,\chi_0\chi_k$$

or

$$\frac{\chi_0'}{\chi_0} - \frac{\chi_k'}{\chi_k} = k^2\,\frac{\int_0^r dr\,\chi_0\chi_k}{\chi_0(r)\,\chi_k(r)}\,.$$

Choosing as the upper limit of the integral any radius $R > R_0$, we may use the expressions (88.3) on the left-hand side and get

$$\frac{1}{R-a} - k\cot(kR+\delta_0) = k^2\,\frac{\int_0^R dr\,\chi_0(r)\,\chi_k(r)}{\chi_0(R)\,\chi_k(R)}\,. \tag{88.4}$$

Here we may decompose the cotangent,

$$\cot(kR+\delta_0) = \frac{\cot\delta_0 - \tan kR}{1 + \cot\delta_0\tan kR}$$

which makes Eq. (88.4) a linear equation in $\cot\delta_0$ with the solution

$$k\cot\delta_0 = \frac{Q + k\tan kR}{1 - Q\dfrac{\tan kR}{k}} \tag{88.5}$$

where

$$Q = \frac{1}{R-a} - k^2 \frac{\int_0^R dr\,\chi_0(r)\,\chi_k(r)}{\chi_0(R)\,\chi_k(R)}. \tag{88.6}$$

In Eqs. (88.5) and (88.6) so far nothing has yet been neglected.

We now start expanding in powers of k^2, first replacing the tangent by its expansion,

$$k\cot\delta_0 = \frac{Q + k^2 R + \cdots}{1 - QR - \frac{1}{3}Qk^2R^3 \ldots},$$

and then replacing Q in (88.5) by the simpler expression

$$Q = \frac{1}{R-a} - k^2 \mathscr{K} \tag{88.7}$$

with

$$\mathscr{K} = \frac{\int_0^R dr\,\chi_0(r)^2}{\chi_0(R)^2}. \tag{88.8}$$

Thus we obtain in the numerator and denominator of (88.5) two series, each correct until the k^2 term. A final rearrangement then leads to the formula wanted

$$k\cot\delta_0 = -\frac{1}{a} + \frac{1}{2} r_0 k^2 \tag{88.9}$$

with the effective range

$$\frac{1}{2} r_0 = \frac{(R-a)^3 + a^3}{3a^2} - \mathscr{K}\left(\frac{R-a}{a}\right)^2. \tag{88.10}$$

There yet remain two questions to be answered.

1. As long as $R > R_0$ the choice of R is quite arbitrary. The effective range (88.10) then must not depend upon it.

2. If the explanation of r_0 as a range is to make sense, this constant must always be positive.

The first question can easily be answered by a differential shift from R to $R+dR$:

$$\frac{d}{dR}\left(\frac{1}{2}r_0\right)=\left(\frac{R-a}{a}\right)^2-\frac{d\mathscr{K}}{dR}\left(\frac{R-a}{a}\right)^2-2\mathscr{K}\frac{R-a}{a^2}. \tag{88.11}$$

From the definition (88.8) it follows that

$$\mathscr{K}(R+dR)=\frac{\mathscr{K}(R)\chi_0(R)^2+\chi_0(R)^2\,dR}{\chi_0(R)^2+2\chi_0(R)\chi_0'(R)dR}=\frac{\mathscr{K}(R)+dR}{1+\dfrac{2dR}{R-a}}$$

or

$$\frac{d\mathscr{K}}{dR}=1-\frac{2\mathscr{K}}{R-a}. \tag{88.12}$$

Putting (88.12) in (88.11) we at once see that

$$\frac{d}{dR}\left(\frac{1}{2}r_0\right)=0$$

whatever the value of R if only above R_0. Hence, r_0 is independent of R.

To answer the second question let $R>|a|$ be chosen; then the first term in (88.10) will be positive, the second, due to $\mathscr{K}>0$, negative. The range will turn out to be positive, therefore, if the first term outweighs the second, or if

$$\mathscr{K}<\frac{(R-a)^3+a^3}{3(R-a)^2}.$$

This is identical with the condition

$$\int_0^R dr\,\chi_0(r)^2<\int_0^R dr[C(r-a)]^2=\tfrac{1}{3}C^2\{(R-a)^3+a^3\}$$

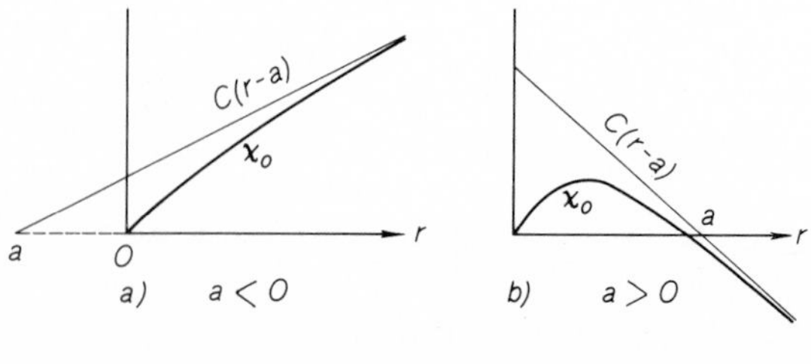

Fig. 51 a and b. Wave function $\chi_0(r)$ at zero energy and its asymptotic behaviour for different signs of the scattering length a

as can easily be seen comparing with (88.8). In the case $a<0$ (Fig. 51a) this follows from χ_0 everywhere being smaller than $C(r-a)$; it holds equally for $a>0$ (Fig. 51b) if the point $r=a$ is not too far away from the zero of $\chi_0(r)$.

NB. Because of

$$\int_0^R dr\,[C(r-a)]^2 = \tfrac{1}{3}C^2\{(R-a)^3+a^3\}$$

the range formula (88.10) may be reshaped into the form

$$\frac{1}{2}r_0 = \int_0^R dr\left\{\left(1-\frac{r}{a}\right)^2 - \frac{1}{a^2C^2}\chi_0(r)^2\right\}.$$

In the literature, in general, a normalization is used which makes $\chi_0 \to 1-\dfrac{r}{a}$ or $C=-1/a$. Furthermore, the difference of χ_0 and its asymptotic behaviour,

$$\varphi(r) = 1-\frac{r}{a}-\chi_0(r),$$

is introduced permitting us to write

$$\frac{1}{2}r_0 = \int_0^\infty dr\,\varphi(r)\left[2\left(1-\frac{r}{a}\right)-\varphi(r)\right].$$

Here we have written infinity as the upper limit but, of course, $\varphi(r)$ will be zero beyond $r=R_0$.

Problem 89. Rectangular hole: Low-energy scattering

For the rectangular hole sketched in Fig. 52 (potential depth $V_0=\hbar^2K_0^2/2m$, radius R) the low-energy expansion

$$k\cot\delta_0 = -\frac{1}{a}+\frac{1}{2}k^2 r_0 \tag{89.1}$$

shall be investigated

a) by deriving Eq. (89.1) from the boundary conditions at $r=R$.

b) by applying the general formula for r_0 derived in the preceding problem to this special potential.

Solution. For the present potential we have for $r<R$,

$$\chi_k = A_k \sin Kr;\qquad \chi_0 = A_0 \sin K_0 r \tag{89.2}$$

and for $r>R$,

$$\chi_k = \sin(kr+\delta_0);\qquad \chi_0 = C(r-a). \tag{89.3}$$

The logarithmic derivative at $r=R$, for zero energy, becomes

$$L_0 = R\,\chi_0'(R)/\chi_0(R) = K_0 R \cot K_0 R = \frac{R}{R-a}. \tag{89.4}$$

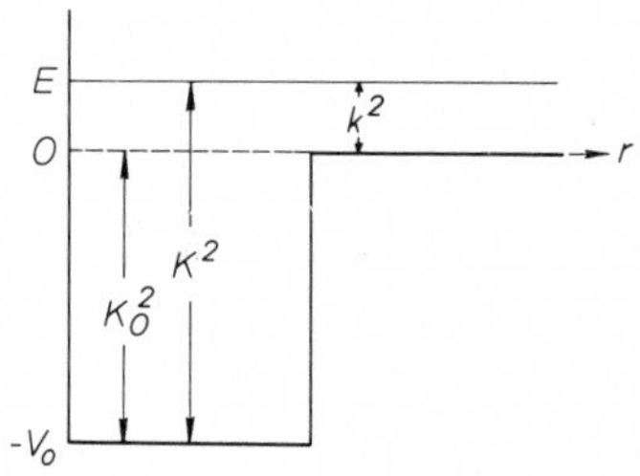

Fig. 52. Potential hole with energy notations, except for the common factor $2m/\hbar^2$

a) We form the same expression with χ_k:

$$L_k = K R \cot K R = k R \cot(kR + \delta_0). \tag{89.5}$$

Since

$$\cot(kR+\delta_0) = \frac{\cot\delta_0 - \tan kR}{1+\cot\delta_0 \tan kR},$$

this equation can be solved with respect to $\cot\delta_0$ and yields

$$\cot\delta_0 = \frac{L_k + kR\tan kR}{kR - L_k \tan kR}. \tag{89.6}$$

The expansion in powers of k may now best be performed in two steps. First we use $kR \ll 1$ and expand $\tan kR$. The result is

$$k\cot\delta_0 = \frac{L_k}{R(1-L_k)} + k^2 R\left[\frac{1}{1-L_k} + \frac{L_k^2}{3(1-L_k)^2}\right]. \tag{89.7}$$

It should be noted that Eqs. (89.6) and (89.7) still hold for any potential shape only if it vanishes for $r \geq R$. Only, when in a second step we now expand L_k, do we make any use of the special rectangular shape of the hole. Let us from now on use the abbreviations

$$X_0 = K_0 R; \quad X = KR; \quad x = kR; \tag{89.8}$$

then, from (89.5),

$$L_k = X \cot X \quad \text{with } X = \sqrt{X_0^2 + x^2} = X_0 + \frac{x^2}{2X_0} \cdots \tag{89.9}$$

and

$$L_k = L_0 \left\{ 1 + \frac{x^2}{2X_0} \left(\frac{1}{X_0} - \cot X_0 - \tan X_0 \right) \right\}$$

$$= L_0 \left\{ 1 + \frac{x^2}{2X_0^2} \left(1 - L_0 - \frac{X_0^2}{L_0} \right) \right\}.$$

This leads to

$$\frac{L_k}{R(1-L_k)} = \frac{L_0}{R(1-L_0)} \left\{ 1 + \frac{x^2}{2X_0^2} \left(1 - \frac{X_0^2}{L_0(1-L_0)} \right) \right\}$$

so that the first term in (89.7) may be decomposed into the scattering length contribution

$$-\frac{1}{a} = \frac{L_0}{R(1-L_0)} \tag{89.10}$$

and a contribution to the range correction term proportional to k^2. In the k^2 term in (89.7) we may replace L_k by L_0 in this approximation and thus arrive at

$$\frac{1}{2} r_0 = \frac{R}{2X_0^2} \cdot \frac{L_0}{1-L_0} \left(1 - \frac{X_0^2}{L_0(1-L_0)} \right) + R \left(\frac{1}{1-L_0} + \frac{L_0^2}{3(1-L_0)^2} \right).$$

In this formula let us express L_0 by the dimensionless quantity

$$\alpha = \frac{a}{R}, \tag{89.11}$$

then, with

$$L_0 = \frac{1}{1-\alpha}; \quad \frac{L_0}{1-L_0} = -\frac{1}{\alpha}; \quad \frac{1}{1-L_0} = -\frac{1-\alpha}{\alpha} \tag{89.12}$$

and after some simple recasting, we finally obtain

$$\frac{1}{2} r_0 = R \left\{ -\frac{1}{2X_0^2 \alpha} + \frac{1}{2} - \frac{1}{6\alpha^2} \right\}. \tag{89.13}$$

From (89.4) we yet have to construct the final expression of the scattering length,

$$a = R \left(1 - \frac{\tan X_0}{X_0} \right). \tag{89.14}$$

Eqs. (89.13) and (89.14) together with (89.1) then represent the complete solution of the problem.

b) The general method of the preceding problem allows for no other way of determining the scattering length than the use of Eq. (89.4), leading directly to (89.14). There is, however, no longer any necessity to introduce χ_k or L_k if we start from the general expression (88.10) which, with the abbreviation (89.11), becomes

$$\frac{1}{2} r_0 = R \left\{ \frac{(1-\alpha)^3 + \alpha^3}{3\alpha^2} - \mathscr{K} \frac{(1-\alpha)^2}{\alpha^2} \right\} \tag{89.15}$$

with

$$\mathscr{K} = \frac{\int_0^R dr\, \chi_0(r)^2}{\chi_0(R)^2}.$$

The constant $\mathscr{K}$ is to be calculated by elementary integration using $\chi_0(r)$, Eq. (89.2),

$$\mathscr{K} = \frac{R}{2X_0 \sin^2 X_0}(X_0 - \sin X_0 \cos X_0) = \frac{R}{2}\left(1 - \frac{\cot X_0}{X_0} + \cot^2 X_0\right)$$

or, since

$$\tan X_0 = X_0(1-\alpha),$$

this may be reshaped into

$$\mathscr{K} = \frac{1}{2} R \left(1 + \frac{\alpha}{X_0^2(1-\alpha)^2}\right). \tag{89.16}$$

Putting (89.16) into (89.15) we easily get again Eq. (89.13) for the effective range.

Problem 90. Low-energy scattering and bound state

To investigate the behaviour of $\cot\delta_0$ in the vicinity of $k=0$ in the complex k plane. It shall be discovered in particular whether a bound state ($k^2<0$) exists at some value where the $\cot\delta_0$ expansion still converges.

Solution. If the power expansion for small values of k,

$$k \cot\delta_0 = -\frac{1}{a} + \frac{1}{2} r_0 k^2 + \cdots, \tag{90.1}$$

exists, $\cot\delta_0$ is an analytical function in the complex k plane with a pole at $k=0$. Bound states are lying on the imaginary axis of this plane, let us say at

$$k = i\varkappa \tag{90.2}$$

with positive $\varkappa$. Putting this into (90.1) we have

$$\cot\delta_0 = i\left(\frac{1}{\varkappa a} + \frac{1}{2}\varkappa r_0\right) \tag{90.3}$$

so that $\cot\delta_0$ becomes purely imaginary on the imaginary axis. Now the phase angle δ_0 is connected with physical phenomena because of the asymptotic behaviour of the $l=0$ radial wave function as

$$\chi_0 \to C\sin(kr+\delta_0).$$

For negative energies, according to (90.3) and the analyticity, we then have

$$\chi_0 \to \frac{C}{2i}\left(e^{i\delta_0}e^{-\varkappa r} - e^{-i\delta_0}e^{\varkappa r}\right).$$

Such a wave function is the eigenfunction of a stationary state if, and only if, for $\varkappa > 0$

$$e^{-i\delta_0} = 0 \tag{90.4}$$

so that χ_0 may be normalized. The condition (90.4) may be written

$$\cos\delta_0 - i\sin\delta_0 = 0$$

or

$$\cot\delta_0 = i. \tag{90.5}$$

Putting (90.5) into (90.3), we arrive at the real relation

$$1 = \frac{1}{\varkappa a} + \frac{1}{2}\varkappa r_0, \tag{90.6}$$

a condition which connects the bound-state quantity $\varkappa$ with the scattering parameters a and r_0, and that may be used both ways.

1. *Determination of binding energy from scattering data:* Supposing the last term in (90.6) to be simply a correction, this equation yields the solution

$$\varkappa a = 1 + \frac{1}{2}\frac{r_0}{a} \tag{90.7}$$

so that the energy of the bound state becomes

$$E = -\frac{\hbar^2}{2m}\varkappa^2 = -\frac{\hbar^2}{2ma^2}\left(1+\frac{r_0}{a}\right). \tag{90.8}$$

Since here only $\varkappa^2$ occurs, it should be noted that, according to Eq. (90.7), a bound state can only exist for *positive* scattering length, within the limits of this approximation.

2. *Determination of scattering cross section from binding energy:* This can be performed less completely, as two parameters are needed whereas the binding energy provides only one. We write the scattering cross section

$$\sigma_0 = \frac{4\pi}{k^2}\sin^2\delta_0 = \frac{4\pi}{k^2}\cdot\frac{1}{1+\cot^2\delta_0} \tag{90.9}$$

and solve (90.6) with respect to a,

$$a = \frac{1}{\varkappa} + \frac{1}{2}r_0$$

in the same order of approximation as (90.7). Then we find from (90.1)

$$\cot^2\delta_0 = \left(-\frac{1}{ka}+\frac{1}{2}kr_0\right)^2 \simeq \frac{1}{k^2a^2} - \frac{r_0}{a} \simeq \frac{\varkappa^2}{k^2}(1-\varkappa r_0) - \varkappa r_0$$

and thus

$$1+\cot^2\delta_0 = \left(1+\frac{\varkappa^2}{k^2}\right)(1-\varkappa r_0)$$

so that finally there emerges the relation

$$\sigma_0 = \frac{4\pi}{k^2+\varkappa^2}(1+\varkappa r_0). \tag{90.10}$$

This cross-section formula holds for low energies and $\varkappa r_0 \ll 1$. It is the famous Bethe-Peierls formula of nuclear physics.

Literature. Bethe, H. A., Peierls, R.: Proc. Roy. Soc. Lond. (A) **148,** 146 (1935).

Problem 91. Deuteron potential with and without hard core

Using the bound-state condition $\cot\delta_0 = i$, the depth of an idealized rectangular potential hole of radius R shall be determined to give one bound state of the deuteron binding energy $E_D = -2.23\,\mathrm{MeV}$. The radius shall be taken to be $R=1.2\,\mathrm{fm}$ ($1\,\mathrm{fm}=10^{-13}\,\mathrm{cm}$).—What is the depth of the potential hole if a hard core of radius $a=0.2\,\mathrm{fm}$ is cut out of its centre?

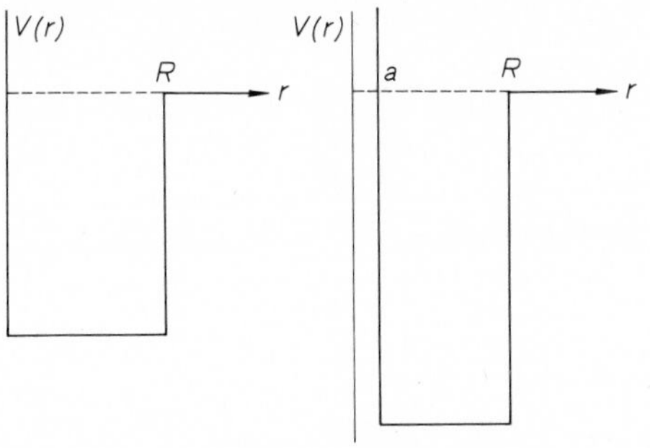

Fig. 53. Deuteron model potential, without and with hard core. Cutting out the core makes the well deeper if the binding energy remains the same

Solution. The $l=0$ wave function satisfying the boundary condition $\chi_0(a)=0$ is for *positive* energy

$$\chi_0 = C\sin K(r-a) \quad \text{for} \quad a<r\le R;$$
$$\chi_0 = \sin(kr+\delta_0) \quad \text{for} \quad R<r$$

and continuity of the logarithmic derivative at $r=R$ yields

$$X\cot\beta X = x\cot(x+\delta_0) \tag{91.1}$$

with the abbreviations

$$x=kR; \quad X=KR=\sqrt{X_0^2+x^2}; \quad X_0=K_0R; \quad \beta=\frac{R-a}{R}. \tag{91.2}$$

For a bound state of *negative* energy let us put

$$k=i\varkappa; \quad x=i\xi; \quad \xi=\varkappa R; \tag{91.3}$$

then

$$E_D=-\frac{\hbar^2}{2m^*R^2}\xi^2 \tag{91.4}$$

with m^* the reduced mass of the two nucleons ($m^*=\frac{1}{2}m$, cf. Problem 150). Instead of E_D, we may equally well use the characteristic length

$$l = \frac{\hbar}{\sqrt{2m^*|E_D|}} = 4.312\ \mathrm{fm}, \tag{91.5a}$$

thence

$$\xi = \frac{R}{l} = 0.2783 \tag{91.5b}$$

in our numerical example.

In Problem 90 it has been shown that a bound state is obtained for

$$e^{-i\delta_0} = 0 \quad \text{or} \quad \cot\delta_0 = i;$$

then, of course, we also have

$$e^{-i(x+\delta_0)} = 0 \quad \text{and} \quad \cot(x+\delta_0) = i$$

so that Eq. (91.1) yields

$$X \cot\beta X = -\xi. \tag{91.6}$$

Since here the right-hand side is negative, the argument of the cotangent is bound to lie[19] between $\pi/2$ and π, and, since ξ is small, not very far from $\pi/2$:

$$\beta X = \frac{\pi}{2} + \varepsilon; \qquad \cot\beta X = -\tan\varepsilon. \tag{91.7}$$

Wanted is the depth of the hole, i.e. the parameter X_0^2, for a given value of ξ, Eq. (91.5b). To get this, we first compute ε by putting (91.7) for X into (91.6) and expanding in powers of ε and ξ:

$$\varepsilon = \frac{2\beta}{\pi}\xi\left(1 - \frac{4\beta}{\pi^2}\xi \ldots\right). \tag{91.8}$$

Next, we put this result into the relation

$$X_0^2 = X^2 + \xi^2 = \left[\frac{\pi}{2\beta}\left(1 + \frac{2\varepsilon}{\pi}\right)\right]^2 + \xi^2;$$

this yields

$$X_0^2 = \frac{\pi^2}{4\beta^2} + \frac{2\xi}{\beta} + \left(1 - \frac{4}{\pi^2}\right)\xi^2 \ldots. \tag{91.9}$$

Let us now discuss this result. If there is no hard core, $\beta = 1$, and X_0 is just a little larger than $\pi/2$. Were $X_0 = \pi/2$, the eigenvalue would have binding energy zero ($\xi = 0$); since binding is stronger ($\xi > 0$), the

[19] This holds if, as in the deuteron case, the potential hole can have only one bound state. For two eigenvalues in the hole, Eq. (91.6) would have two solutions, the lower energy level belonging to the interval $\pi < \beta X < \frac{3}{2}\pi$.

hole must be deeper. The presence of a hard core shifts this limiting value to $\pi/2\beta > \pi/2$, because, if a hard core is cut out, the hole has to be deeper than before to provide the same binding energy.

With $\xi = 0.278$, according to Eq. (91.5 b), we find from (91.9)

$$X_0^2 = \frac{V_0}{|E_D|} \cdot \frac{R^2}{l^2} = \frac{2.467}{\beta^2} + \frac{0.557}{\beta} + 0.046 + \cdots$$

with V_0 the hole depth. Without hard core ($\beta = 1$) this gives $X_0^2 = 3.070$ and $V_0 = 88.4$ MeV. With core radius $a = 0.2$ fm ($\beta = 0.8333$) the values are shifted to $X_0^2 = 4.267$ and $V_0 = 122.9$ MeV. Fig. 53 is drawn according to these numerical results.

Problem 92. Low-energy cross section with and without hard core

For the two deuteron potentials of the last problem, there shall be computed the scattering length and the low-energy limit for the neutron-proton scattering cross section.

Solution. The scattering length a is connected with the limit of the logarithmic derivative at $r = R$ for $k \to 0$ by (89.10),

$$a = R \frac{L_0 - 1}{L_0}; \qquad L_0 = R \frac{\chi'(R)}{\chi(R)}. \tag{92.1}$$

On the other hand, according to (91.1), for the present potentials we have

$$L_0 = X_0 \cot \beta X_0 . \tag{92.2}$$

The low-energy cross section limit for $k \to 0$ is

$$\sigma(0) = 4\pi a^2 . \tag{92.3}$$

Since X_0 and β have been calculated in the preceding problem to match the correct binding energy of the deuteron, we may now simply compute a and $\sigma(0)$ from Eqs. (92.1) to (92.3). The numerical results are:

1. Potential hole without core: $\beta = 1$; $X_0^2 = 3.070$;

$$L_0 = -0.3214; \qquad a = 4.934 \text{ fm}; \qquad \sigma = 3.059 \text{ barn}.$$

2. Potential hole with core: $\beta = 0.8333$; $X_0^2 = 4.267$;

$$L_0 = -0.3135; \qquad a = 5.028 \text{ fm}; \qquad \sigma = 3.177 \text{ barn}.$$

These results are interesting as they show that, when gauged to the same deuteron binding energy and hole radius, there is a very small effect of only 3.8 per cent due to the existence of a hard core on the low-energy cross section of neutron-proton scattering. Thus it is almost impossible to use such measurements to decide whether a hard core really exists, or not.

It should be noted that the experimental cross section in fact amounts to about 21 barns in consequence of the spin dependence of nuclear forces and the much higher contribution of the ^1S state. The figures computed in our model are in good agreement, however, with the ^3S scattering length.

Problem 93. Low energy scattering by a modified Pöschl-Teller potential hole

Using the potential of Problem 39 in the spherical symmetrical form,

$$V(r) = -\frac{\hbar^2\alpha^2}{2m}\frac{\lambda(\lambda-1)}{\cosh^2\alpha r}. \tag{93.1}$$

the scattering cross section at low energies ($l=0$) shall be derived.

Solution. The one-dimensional Problem 39 provides the complete set of solutions of the three-dimensional case for $l=0$. The additional boundary condition at $r=0$, $\chi_0(0)=0$, however, leaves us with the odd solutions only, so that there remain Eq. (39.12 b) and the phase angles of (39.13 b), (39.14):

$$\chi_0(r) \to \sin(kr+\delta_0); \tag{93.2}$$

$$\delta_0 = \frac{\pi}{2} + \arg\frac{\Gamma(ik/\alpha)\,e^{-i\frac{k}{\alpha}\log 2}}{\Gamma\left(\frac{\lambda+1}{2}+i\frac{k}{2\alpha}\right)\Gamma\left(1-\frac{\lambda}{2}+i\frac{k}{2\alpha}\right)}. \tag{93.3}$$

There just remains the purely mathematical task of bringing (93.3) into a more suitable form for numerical evaluation.

With the abbreviations

$$\frac{k}{2\alpha} = q, \quad \frac{\lambda}{2} + iq = z \tag{93.4}$$

we have

$$\delta_0 - \frac{\pi}{2} = \arg\Gamma(2iq) - 2q\log 2 - \arg\Gamma(z+\tfrac{1}{2}) + \arg\Gamma(1-z). \tag{93.5}$$

From the general identities

$$\Gamma(2z) = \frac{1}{2\sqrt{\pi}} e^{2z\log 2}\,\Gamma(z)\,\Gamma(z+\tfrac{1}{2})$$

and

$$\Gamma(z)\,\Gamma(1-z) = \frac{\pi}{\sin \pi z}$$

there follow the phase relations

$$\arg\Gamma(2z) = 2\log 2\cdot \mathrm{Im}\,z + \arg\Gamma(z) + \arg\Gamma(z+\tfrac{1}{2})$$

and

$$\arg\Gamma(z) + \arg\Gamma(1-z) = -\arg\sin\pi z .$$

Eliminating from these two identities $\arg\Gamma(z)$, we have

$$-\arg\Gamma(z+\tfrac{1}{2}) + \arg\Gamma(1-z) = -\arg\Gamma(2z) + 2\log 2\cdot \mathrm{Im}\,z - \arg\sin\pi z .$$

Replacing z according to (93.4) we then obtain from (93.5),

$$\delta_0 - \frac{\pi}{2} = \{\arg\Gamma(2iq) - 2q\log 2\} + \left\{-\arg\Gamma(\lambda+2iq) + 2q\log 2 - \arg\sin\pi\left(\frac{\lambda}{2} + iq\right)\right\}$$

or

$$\delta_0 = \frac{\pi}{2} + \arg\Gamma(2iq) - \arg\Gamma(\lambda+2iq) - \tan^{-1}\left(\cot\frac{\pi\lambda}{2}\tanh\pi q\right). \qquad (93.6)$$

The remaining two arguments of Γ functions are represented by the two series

$$\arg\Gamma(\lambda+2iq) = 2q\left\{-C + \sum_{n=1}^{\infty}\left(\frac{1}{n} - \frac{1}{2q}\tan^{-1}\frac{2q}{\lambda+n-1}\right)\right\}$$

and

$$\arg\Gamma(2iq) = 2q\left\{-C + \sum_{n=1}^{\infty}\left(\frac{1}{n} - \frac{1}{2q}\tan^{-1}\frac{2q}{n-1}\right)\right\}$$

so that the difference becomes

$$\arg\Gamma(2iq) - \arg\Gamma(\lambda+2iq) = \sum_{n=1}^{\infty}\left\{\tan^{-1}\frac{2q}{\lambda+n-1} - \tan^{-1}\frac{2q}{n-1}\right\}$$

$$= \tan^{-1}\frac{2q}{\lambda} - \frac{\pi}{2} + \sum_{n=1}^{\infty}\left\{\tan^{-1}\frac{2q}{\lambda+n} - \tan^{-1}\frac{2q}{n}\right\}.$$

This leads to the final formula

$$\delta_0 = \tan^{-1}\frac{2q}{\lambda} - \tan^{-1}\left(\cot\frac{\pi\lambda}{2}\tanh\pi q\right) + \sum_{n=1}^{\infty}\left\{\tan^{-1}\frac{2q}{\lambda+n} - \tan^{-1}\frac{2q}{n}\right\}. \tag{93.7}$$

It is easily seen that for zero energy of the scattered particles (i.e. for $q=0$) the phase angle also tends to zero according to

$$\lim_{q\to 0}\frac{\delta_0}{2q} = \frac{1}{\lambda} - \frac{\pi}{2}\cot\frac{\pi\lambda}{2} + \sum_{n=1}^{\infty}\left(\frac{1}{\lambda+n} - \frac{1}{n}\right). \tag{93.8}$$

The resulting scattering cross section,

$$\sigma = \frac{4\pi}{k^2}\sin^2\delta_0 = \frac{4\pi}{\alpha^2}\left(\frac{\sin\delta_0}{2q}\right)^2 \tag{93.9}$$

then tends towards a finite limit at zero energy. The only exception $\sigma\to\infty$ occurs if $\cot\frac{\pi}{2}\lambda$ becomes infinitely large, i.e. for λ being an even integer. This is easily explained as a resonance phenomenon by the position of an energy level of the potential hole exactly at $E=0$ for these λ values.

The computation of δ_0 and σ from Eqs. (93.7) and (93.9) is purely technical and will not be dealt with in detail. It should, however, be mentioned that for integer λ (even or odd) the sum in (93.7) is finite and the second term on the right-hand side either zero (for odd λ) or $\pi/2$ (for even λ), so that for $\lambda=2n$,

$$\delta_0 = \frac{\pi}{2} - \sum_{n=1}^{2n-1}\tan^{-1}\frac{2q}{n}$$

and for $\lambda=2n+1$,

$$\delta_0 = -\sum_{n=1}^{2n}\tan^{-1}\frac{2q}{n}.$$

If λ is not an integer, the infinite sum in (93.7) may be expanded into a series of powers of $2q$. Denoting $\lambda-N=\mu$ where N is the next integer below λ $(0<\mu<1)$, we may write

$$\sum_{n=0}^{\infty}\tan^{-1}\frac{2q}{\lambda+n} = \sum_{n=N}^{\infty}\tan^{-1}\frac{2q}{n+\mu}$$

and

$$\sum_{n=N}^{\infty}\left\{\tan^{-1}\frac{2q}{n+\mu} - \tan^{-1}\frac{2q}{n}\right\} = \sum_{\nu=0}^{\infty}(-1)^{\nu}\frac{C_\nu}{2\nu+1}(2q)^{2\nu+1}$$

with

$$C_\nu = \sum_{n=N}^{\infty} \left\{ \frac{1}{(n+\mu)^{2\nu+1}} - \frac{1}{n^{2\nu+1}} \right\} = \sum_{n=N}^{\infty} \frac{1}{n^{2\nu+1}} \left[\left(1+\frac{\mu}{n}\right)^{-(2\nu+1)} - 1 \right]$$

$$= \sum_{n=N}^{\infty} \frac{1}{n^{2\nu+1}} \left\{ -(2\nu+1)\frac{\mu}{n} + \binom{-(2\nu+1)}{2} \frac{\mu^2}{n^2} + \cdots \right\}$$

where each term of the binomial expansion may be expressed by a ζ function of the arguments $2\nu+2$, $2\nu+3$, etc. This method leads to rapidly converging series expansions.

A few numerical results are shown in Fig. 54, where the cross section is drawn *versus* the energy of the scattered particles, both in suitable units. The hyperbola $1/(2q)^2$ corresponds to the limit $\sigma = 4\pi/k^2$ never

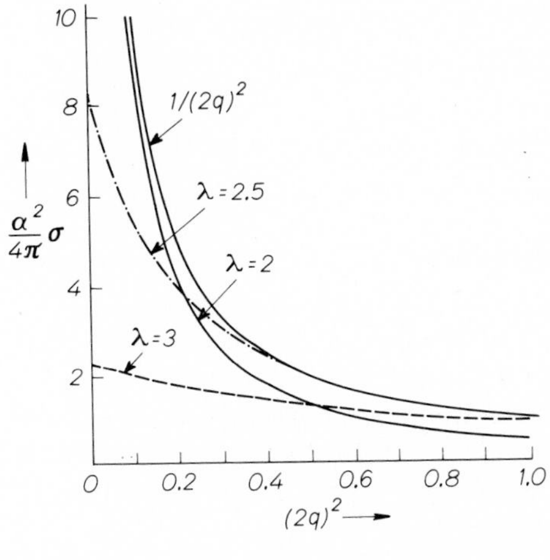

Fig. 54. Scattering cross sections for three modified Pöschl-Teller potential holes. The hyperbola defines the limit $4\pi/k^2$ never to be exceeded by σ_0

to be exceeded by $l=0$ scattering. For $\lambda=2$ we have resonance at zero-energy; the other two curves for $\lambda=2.5$ and $\lambda=3$ have finite cross-section limits. In the case of $\lambda=2.5$, we approach a virtual level $\left(\delta_0 = \frac{\pi}{2}\right)$ at about $(2q)^2=0.6$, so that the cross section comes very close to the limiting hyperbola around this energy. For $\lambda=4$ (not shown in the figure) the cross section drops to zero $(\delta_0=\pi)$ at $(2q)^2=1$ but recovers at larger energies.

Problem 94. Radial integral equation

To replace the radial differential equation

$$\chi_l'' + \left[k^2 - U(r) - \frac{l(l+1)}{r^2}\right]\chi_l = 0\,, \tag{94.1}$$

with $k^2 = 2mE/\hbar^2$ and $U(r) = 2mV(r)/\hbar^2$, by an integral equation. How can this integral equation be given the form of an equation with a symmetrical kernel? How can the asymptotic phase of χ_l be determined by an integral? How is the solution of the integral equation by successive steps connected with Born's approximation?

Solution. We may formally write Eq. (94.1) as an inhomogeneous differential equation

$$D\chi_l = F(r) \tag{94.2}$$

with

$$D = \frac{d^2}{dr^2} + k^2 - \frac{l(l+1)}{r^2}; \quad F(r) = U(r)\chi_l(r)\,, \tag{94.3}$$

where $F(r)$ may be considered the inhomogeneity. The homogeneous equation

$$D\chi_l = 0$$

has fundamental solutions

$$j_l(kr) \to \sin\left(kr - \frac{l\pi}{2}\right); \quad n_l(kr) \to -\cos\left(kr - \frac{l\pi}{2}\right). \tag{94.4}$$

The solution of the inhomogeneous equation (94.2) may be constructed as a superposition of that of the homogeneous equation and one special solution of the inhomogeneous one, the latter being determined by a Green's function, $G(r,r')$,

$$\chi_l(r) = A j_l(kr) + B n_l(kr) + \int_0^\infty dr'\, G(r,r') F(r')\,. \tag{94.5}$$

A Green's function is defined as simultaneous solution of

$$D\,G(r,r') = \delta(r-r') \quad \text{and} \quad D'\,G(r,r') = \delta(r-r');$$

such a solution is e.g.

$$G(r,r') = \begin{cases} \dfrac{1}{k} j_l(kr)\, n_l(kr') & \text{for } r<r'\,, \\[2ex] \dfrac{1}{k} n_l(kr)\, j_l(kr') & \text{for } r>r'\,. \end{cases} \tag{94.6}$$

Since we are interested only in wave functions $\chi_l(r)$ regular at the origin $r=0$, we may specialize (94.5) to $B=0$ and (in arbitrary normalization) $A=1$, thus obtaining

$$\chi_l(r)=j_l(kr)+\int_0^\infty dr'\, G(r,r')\,U(r')\,\chi_l(r'). \tag{94.7}$$

This is the integral equation wanted.

We now may turn to answer the special questions asked above.

a) *Symmetrical kernel.* The inhomogeneous integral equation (94.7) has an asymmetrical kernel $G(r,r')\,U(r')$. By the substitution of

$$\begin{aligned} y(r)&=\sqrt{U(r)}\,\chi_l(r); \qquad f(r)=\sqrt{U(r)}\,j_l(kr);\\ K(r,r')&=\sqrt{U(r)\,U(r')}\,G(r,r') \end{aligned} \tag{94.8}$$

it is, however, transformed into

$$y(r)=f(r)+\int_0^\infty dr'\, K(r,r')\,y(r') \tag{94.9}$$

with a symmetrical kernel

$$K(r,r')=K(r',r). \tag{94.10}$$

The advantage of this transformation will be seen in the following Problem 95 (p. 253).

b) *Asymptotic phase.* The solution $\chi_l(r)$ of the integral equation (94.7) for large values of r can be determined by using for the Green's function, Eq. (94.6), the expression for $r>r'$ in the whole integration domain since only a region $0\le r'<R$ with R of atomic dimensions contributes to the integral in (94.7) in consequence of the factor $U(r')$. Thus we find

$$\chi_l(r)\to\sin\left(kr-\frac{l\pi}{2}\right)-\frac{1}{k}\cos\left(kr-\frac{l\pi}{2}\right)\int_0^\infty dr'\, U(r')\,j_l(kr')\,\chi_l(r'). \tag{94.11}$$

This may be written

$$\chi_l(r)\to\frac{1}{\cos\delta_l}\sin\left(kr-\frac{l\pi}{2}+\delta_l\right) \tag{94.12}$$

with the asymptotic phase angle δ_l determined by the integral

$$\tan\delta_l=-\frac{1}{k}\int_0^\infty dr\, U(r)\,j_l(kr)\,\chi_l(r). \tag{94.13}$$

c) *Successive approximations* to the solution are obtained in the form of a Neumann series by the sequence

$$\chi_l^{(0)} = j_l(kr)\,, \tag{94.14a}$$

$$\chi_l^{(1)} = j_l(kr) + \int_0^\infty dr'\, G(r,r')\, U(r')\, j_l(kr')\,, \tag{94.14b}$$

$$\cdots\cdots\cdots\cdots\cdots\cdots\cdots\cdots\cdots\cdots$$

$$\chi_l^{(n)} = j_l(kr) + \int_0^\infty dr'\, G(r,r')\, U(r')\, \chi_l^{(n-1)}(r')\,. \tag{94.14c}$$

It can be shown that the recursion (94.14c) is, except for the normalization constants, identical with the Born approximation (cf. Problem 105),

$$u_n(\boldsymbol{r}) = e^{ikz} - \frac{1}{4\pi}\int d^3x'\, \frac{e^{ik|\boldsymbol{r}-\boldsymbol{r}'|}}{|\boldsymbol{r}-\boldsymbol{r}'|}\, U(r')\, u_{n-1}(\boldsymbol{r}')\,. \tag{94.15}$$

To show the equivalence of (94.14c) and (94.15) we have to expand the functions occurring in (94.15) in partial waves. We may write (cf. Problems 81 and 82)

$$u_n(\boldsymbol{r}) = \frac{1}{kr}\sum_{l=0}^{\infty} \sqrt{4\pi(2l+1)}\, i^l\, \tilde{\chi}_l^{(n)}(r)\, Y_{l,0}(\vartheta) \tag{94.16}$$

and

$$e^{ikz} = \frac{1}{kr}\sum_{l=0}^{\infty} \sqrt{4\pi(2l+1)}\, i^l\, j_l(kr)\, Y_{l,0}(\vartheta) \tag{94.17}$$

and use the expansion

$$\frac{e^{ik|\boldsymbol{r}-\boldsymbol{r}'|}}{|\boldsymbol{r}-\boldsymbol{r}'|} = \frac{1}{rr'}\sum_{l=0}^{\infty} \sqrt{4\pi(2+1)}\, \Gamma_l(r,r')\, Y_{l,0}(\gamma) \tag{94.18a}$$

where γ is the angle between the two vectors $\boldsymbol{r}$ and $\boldsymbol{r}'$, and

$$\Gamma_l(r,r') = \begin{cases} \dfrac{i}{k} j_l(kr)\, h_l^{(1)}(kr') & \text{for } r<r'\,, \\[2ex] \dfrac{i}{k} h_l^{(1)}(kr)\, j_l(kr') & \text{for } r>r'\,. \end{cases} \tag{94.18b}$$

We now project out of (94.15) the l-th partial wave by multiplying it with $Y_{l,0}(\vartheta)$ and integrating over all directions of the vector $\boldsymbol{r}$. In the second term on the right-hand side of (94.15) this is done by using the relation

$$\oint d\Omega\, Y_{l,0}(\vartheta)\, Y_{l,0}(\gamma) = \sqrt{\frac{4\pi}{2l+1}}\, Y_{l,0}(\vartheta')\,.$$

The result is

$$\tilde{\chi}_l^{(n)}(r) = j_l(kr) - \frac{k}{4\pi} i^{-l} \int_0^\infty d^3x' \, U(r') u_{n-1}(\mathbf{r}') \frac{1}{r'} \Gamma_l(r,r') \sqrt{\frac{4\pi}{2l+1}} \, Y_{l,0}(\vartheta') \, .$$

In the last integral we use the expansion analogous to (94.16) for $u_{n-1}(\mathbf{r}')$ and find

$$\tilde{\chi}_l^{(n)}(r) = j_l(kr) - \int_0^\infty dr' \, U(r') \, \Gamma_l(r,r') \tilde{\chi}_l^{(n-1)}(r') \, . \tag{94.19}$$

This recurrence equation is already very similar to (94.14c). To show its identity (except for normalization), we replace the spherical Hankel function in Γ_l according to its definition,

$$h_l^{(1)} = j_l + i n_l \, ,$$

which gives

$$\Gamma_l(r,r') = \frac{i}{k} j_l(kr) j_l(kr') - G(r,r') \, , \tag{94.20}$$

where G means the Green's function originally defined by Eq. (94.6). The first term of (94.20) then leads to an integral proportional to $j_l(kr)$ which may be joined with the first term on the right-hand side:

$$\tilde{\chi}_l^{(n)}(r) = A j_l(kr) + \int_0^\infty dr' \, U(r') \, G(r,r') \, \tilde{\chi}_l^{(n-1)}(r') \tag{94.21}$$

with the constant

$$A = 1 - \frac{i}{k} \int_0^\infty dr' \, U(r') j_l(kr') \, \tilde{\chi}_l^{(n-1)}(r') \, . \tag{94.22}$$

This is, indeed, identical with the wanted Eq. (94.14c), except for the factor A, leading to a somewhat changed (anyway arbitrary) normalization.

NB. If the exact integral equation (94.7) for $\chi_l(r)$ is compared with the exact integral equation

$$u(\mathbf{r}) = e^{ikz} - \frac{1}{4\pi} \int d^3x' \frac{e^{ik|\mathbf{r}-\mathbf{r}'|}}{|\mathbf{r}-\mathbf{r}'|} U(r') u(\mathbf{r}') \tag{94.15'}$$

the superscripts $n-1$, n, etc. have to be omitted. If we then put

$$\tilde{\chi}_l(r) = A \chi_l(r) \, ,$$

the integral equation analogous to (94.21) is transformed into (94.7), and (94.22) becomes

$$A = 1 - \frac{i}{k} A \int_0^\infty dr' U(r') j_l(kr') \, \chi_l(r') \, . \tag{94.22'}$$

Here Eq. (94.13) may now be used, simplifying (94.22′) into

$$A = 1 + iA\tan\delta_l$$

or

$$A = e^{i\delta_l}\cos\delta_l\,.$$

Taking further into account Eq. (94.12) for the asymptotic behaviour, the expansion (94.16) leads to

$$u(\mathbf{r}) \to \frac{1}{kr}\sum_{l=0}^{\infty}\sqrt{4\pi(2l+1)}\,i^l e^{i\delta_l}\sin\left(kr - \frac{l\pi}{2} + \delta_l\right) Y_{l,0}(\vartheta)$$

in perfect agreement with (82.9).

Problem 95. Variational principle of Schwinger

To construct a variational principle for the determination of the phase angle starting from the integral equation with symmetrical kernel established in the preceding problem.

Solution. Let us write the integral equation (94.9) in the form

$$y(r) - \int_0^\infty dr'\, K(r,r')\, y(r') = f(r) \tag{95.1}$$

with the inhomogeneity isolated on the right-hand side. Multiplying with $y(r)$ and integrating over r, this yields

$$I_1 = I_2 \tag{95.2}$$

with

$$I_1 = \int_0^\infty dr\, y(r)\left\{y(r) - \int_0^\infty dr'\, K(r,r')\, y(r')\right\} \tag{95.3}$$

and

$$I_2 = \int_0^\infty dr\, y(r)\, f(r)\,. \tag{95.4}$$

The two integrals are equal if $y(r)$ satisfies the integral equation (95.1). From the preceding problem we gather that then

$$I_2 = \int_0^\infty dr\, U(r)\, j_l(kr)\, \chi_l(r) = -k\tan\delta_l\,. \tag{95.5}$$

Let us now take a neighbouring function $y(r) + \delta y(r)$ instead of the correct solution $y(r)$. The values of the two integrals then shift by

$$\delta I_1 = 2\int_0^\infty dr\, \delta y(r)\left\{y(r) - \int_0^\infty dr'\, K(r,r')\, y(r')\right\} \tag{95.6}$$

and

$$\delta I_2 = \int_0^\infty dr\, \delta y(r)\, f(r)\,. \tag{95.7}$$

In Eq. (95.6) it is necessarily supposed that the kernel is symmetrical since in the variation of $y(r')$ originally $\delta y(r')$ occurs, and in (95.6) the integration variables are exchanged to fall back even here onto $\delta y(r)$ which would lead to $K(r', r)$.

Since $y(r)$, again, is meant to be the correct solution of the integral equation (95.1), the curly bracket in (95.6) may be replaced by $f(r)$ so that we get

$$\delta I_1 = 2\delta I_2 \,. \tag{95.8}$$

From (95.2) and (95.8) there follows

$$\frac{\delta I_1}{I_1} - 2\frac{\delta I_2}{I_2} = 0$$

or

$$\delta(I_1/I_2^2) = 0 \,. \tag{95.9}$$

This is the variational principle sought. Writing it in detail, we have to extremize the expression $S_l = I_1/I_2^2$, i.e.

$$S_l = \frac{\int\limits_0^\infty dr\, U(r)\,\chi_l(r)\left\{\chi_l(r) - \int\limits_0^\infty dr'\,\chi_l(r')\,U(r')\,G(r,r')\right\}}{\left[\int\limits_0^\infty dr\, U(r)\,\chi_l(r)\,j_l(kr)\right]^2} \tag{95.10}$$

and

$$\delta S_l = 0 \,. \tag{95.11}$$

The meaning of S_l is easily seen from (95.5) with $I_1 = I_2$, viz.

$$S_l = -\frac{1}{k}\cot\delta_l \,. \tag{95.12}$$

The extremal value of S_l therefore directly determines the correct phase angle δ_l.

It should be remarked that this variational principle is independent of the normalization of the wave function. This is easily seen by replacing in (95.10) the wave function χ_l by $C\chi_l$, which only leads to multiplying both numerator and denominator by C^2, thus leaving S_l unchanged.

NB. For very low energies ($k \to 0$) and an attractive potential, we are concerned only with the partial wave $l=0$ for scattering phenomena. As has been shown in Problem 88, we may then introduce the scattering length,

$$a_0 = -\frac{1}{k}\tan\delta_0 \,.$$

Eq. (95.12) shows that

$$\lim_{k\to 0} k^2\, S_0(k) = \frac{1}{a_0}$$

so that the variational principle may also be used with advantage to determine the scattering length.

Literature. Schwinger, J.: Phys. Rev. **72,** 742 A (1947). — Cf. also Blatt, J. M., Jackson, J. D.: Phys. Rev. **76,** 18 (1949).

Problem 96. Successive approximations to partial-wave phase shift

The radial Schrödinger equation for $l=0$ may be written formally as an inhomogeneous differential equation by treating the potential term as its inhomogeneity. The "solution" then is an integral equation for $\chi_0(r)$ which in its turn may be solved by successive approximations in the Born sense. The second approximation to $\tan\delta_0$ shall thus be determined in integral form.

Solution. We write the radial Schrödinger equation for $l=0$ (and omit in the following the subscript 0 in χ_0, δ_0 etc. referring to $l=0$) in the form

$$\chi'' + k^2 \chi = U\chi; \qquad U = \frac{2m}{\hbar^2} V(r). \tag{96.1}$$

Then, treating the right-hand side as inhomogeneity, the solution may be expressed in terms of a Green's function $G(r,r')$, i.e. of a solution of the equations

$$\frac{\partial^2 G}{\partial r^2} + k^2 G = \delta(r-r') = \frac{\partial^2 G}{\partial r'^2} + k^2 G\,. \tag{96.2}$$

We thus obtain

$$\chi(r) = A \sin kr + B \cos kr + \int_0^\infty dr'\, G(r,r')\, U(r')\, \chi(r')\,. \tag{96.3}$$

There are different ways of introducing Green's functions in (96.3); we will use

$$G(r,r') = \begin{cases} -\dfrac{1}{k} \sin k(r'-r) & \text{if } r' < r\,, \\ 0 & \text{if } r' > r\,, \end{cases} \tag{96.4}$$

because this allows the boundary condition $\chi(0)=0$ to be satisfied simply by putting $B=0$. In arbitrary normalization ($A=1$), which is irrelevant to phase shift determination, we then finally arrive at the Volterra integral equation

$$\chi(r) = \sin kr - \frac{1}{k} \int_0^r dr' \sin k(r'-r)\, U(r')\, \chi(r')\,. \tag{96.5}$$

Decomposing the factor $\sin k(r'-r)$, we may write instead

$$\chi(r) = C(r)\sin kr + S(r)\cos kr \tag{96.6a}$$

with

$$C(r) = 1 + \frac{1}{k}\int_0^r dr' \cos kr'\, U(r')\,\chi(r');$$
$$S(r) = -\frac{1}{k}\int_0^r dr' \sin kr'\, U(r')\,\chi(r'). \tag{96.6b}$$

The two amplitude functions $C(r)$ and $S(r)$ tend towards finite limits for $r\to\infty$ if the potential goes to zero as r^{-q} with $q>1$, when it follows from (96.6a) that $\chi(r)\sim\sin(kr+\delta)$. Therefore, the partial-wave phase shift δ can be determined from

$$\tan\delta = \frac{S(\infty)}{C(\infty)}. \tag{96.7}$$

Eqs. (96.6b) and (96.7) represent the rigorous solution to the problem.

A Born approximation by successive steps is defined by using $U(r)$ as a small quantity of the first order. Let us, the better to survey the approximation steps, write for a moment λU instead of U and expand χ into a power series with respect to the "ordering parameter" λ,

$$\chi = \chi_0 + \lambda\chi_1 + \lambda^2\chi_2 + \cdots \tag{96.8}$$

the subscripts denoting the orders of approximation. Eq. (96.5) in second-order approximation then will run as follows:

$$\chi_0(r) + \lambda\chi_1(r) + \lambda^2\chi_2(r)\cdots$$
$$= \sin kr - \frac{\lambda}{k}\int_0^r dr' \sin k(r'-r)\, U(r')\,\{\chi_0(r') + \lambda\chi_1(r')\cdots\}$$

or, decomposed into contributions of the orders λ^0, λ^1, λ^2:

$$\chi_0(r) = \sin kr \tag{96.9a}$$

$$\chi_1(r) = -\frac{1}{k}\int_0^r dr' \sin k(r'-r)\, U(r')\sin kr'; \tag{96.9b}$$

$$\chi_2(r) = \frac{1}{k^2}\int_0^r dr' \sin k(r'-r)\, U(r')\int_0^{r'} dr'' \sin k(r''-r')\, U(r'')\sin kr''. \tag{96.9c}$$

Here χ_0 is the unperturbed contribution of the plane wave. In first-order approximation we have $\chi = \chi_0 + \chi_1$ which, again decomposing $\sin k(r' - r)$ in the same way as in (96.6), leads to

$$\begin{aligned} C(r) &= 1 + \frac{1}{k} \int_0^r dr'\, U(r') \cos kr' \sin kr' ; \\ S(r) &= -\frac{1}{k} \int_0^r dr'\, U(r') \sin^2 kr' . \end{aligned} \tag{96.10}$$

In this order, $S(r)$ may be identified with the tangent expression (96.7) so that the first approximation to the phase angle becomes

$$\tan \delta = -\frac{1}{k} \int_0^\infty dr'\, U(r') \sin^2 kr' \tag{96.11}$$

which is indeed the first Born approximation (cf. Problem 106). In second-order approximation, χ_2 has to be added, and we find

$$\begin{aligned} C(r) = 1 + \frac{1}{k} \int_0^r dr'\, U(r') \cos kr' \Big\{ &\sin kr' \\ &- \frac{1}{k} \int_0^{r'} dr''\, U(r'') \sin k(r'' - r') \sin kr'' \Big\} ; \\ S(r) = -\frac{1}{k} \int_0^r dr'\, U(r') \sin kr' \Big\{ &\sin kr' \\ &- \frac{1}{k} \int_0^{r'} dr''\, U(r'') \sin k(r'' - r') \sin kr'' \Big\} . \end{aligned} \tag{96.12}$$

This shows that the approximation becomes an expansion in negative powers of k corresponding to the high-energy character of Born's method. This runs parallel to the increasing powers of U occurring in higher approximations.

If we now proceed to construct $\tan \delta$ from (96.12) according to (96.7), we may expand $1/C(\infty)$ in the same way. Since $S(\infty)$ is proportional to

$1/k$, in $C(\infty)$ we need the first approximation only in order to get results correct to $1/k^2$, inclusive. Thus,

$$\tan\delta = \left\{ -\frac{1}{k}\int_0^\infty dr\, U(r)\sin^2 kr \right.$$

$$\left. + \frac{1}{k^2}\int_0^\infty dr\, U(r)\sin kr \int_0^r dr'\, U(r')\sin k(r'-r)\sin kr' \right\}$$

$$\times \left\{ 1 - \frac{1}{k}\int_0^\infty dr'\, U(r')\cos kr' \sin kr' \right\}$$

$$= -\frac{1}{k}\int_0^\infty dr\, U(r)\sin^2 kr$$

$$+ \frac{1}{k^2}\int_0^\infty dr\, U(r)\sin^2 kr \int_0^\infty dr'\, U(r')\sin kr' \cos kr'$$

$$+ \frac{1}{k^2}\int_0^\infty dr\, U(r)\sin kr \int_0^r dr'\, U(r')\sin k(r'-r)\sin kr' .$$

A rather obvious reshaping of the $1/k^2$ terms finally yields either

$$\tan\delta = -\frac{1}{k}\int_0^\infty dr\, U(r)\sin^2 kr \left\{ 1 - \frac{1}{k}\int_r^\infty dr'\, U(r')\sin 2kr' \right\} \tag{96.12a}$$

or

$$\tan\delta = -\frac{1}{k}\int_0^\infty dr\, U(r)\sin kr \left\{ \sin kr - \frac{2}{k}\cos kr \int_0^r dr'\, U(r')\sin^2 kr' \right\}. \tag{96.12b}$$

This is the second-order approximation to the phase shift.

NB. The same method may be applied to partial waves with $l>0$, the only difference being that the trigonometric functions $\sin kr$ and $\cos kr$ have then to be replaced by the spherical Bessel functions $j_l(kr)$ and $-n_l(kr)$.

Problem 97. Calogero's equation

For the function

$$t(r) = \frac{S(r)}{C(r)} \tag{97.1}$$

with $S(r)$ and $C(r)$ defined in the preceding problem, a non-linear differential equation of the first order shall be constructed and a second-order approximation of its solution in the Born sense be derived.

Solution. It has been shown that in

$$\chi(r) = C(r) \sin kr + S(r) \cos kr \tag{97.2}$$

the amplitude functions and their derivatives are

$$\begin{aligned} C(r) &= 1 + \frac{1}{k}\int_0^r dr'\, U(r') \cos kr'\, \chi(r'); \qquad & C'(r) &= \frac{1}{k} U \cos kr\, \chi(r); \\ S(r) &= -\frac{1}{k}\int_0^r dr'\, U(r') \sin kr'\, \chi(r'); & S'(r) &= -\frac{1}{k} U \sin kr\, \chi(r). \end{aligned} \tag{97.3}$$

By differentiation of (97.1) we find

$$t' = \frac{C S' - S C'}{C^2}$$

with the numerator

$$C S' - S C' = -\frac{U}{k}\chi\{C \sin kr + S \cos kr\} = -\frac{U}{k}\chi^2(r).$$

Therefore,

$$\frac{dt}{dr} = -\frac{1}{k} U(r)\,\{\sin kr + t(r) \cos kr\}^2. \tag{97.4}$$

This is the differential equation sought for $t(r)$, the so-called Calogero equation.

Successive approximations in the Born sense are obtained if we again replace U by λU, expand $t(r)$ in powers of λ,

$$t = \lambda t_1 + \lambda^2 t_2 + \lambda^3 t_3 + \cdots, \tag{97.5}$$

and decompose (97.4) into equations with respect to equal powers of λ:

$$t_1' = -\frac{U}{k}\sin^2 kr\,; \tag{97.6a}$$

$$t_2' = -\frac{U}{k}t_1\sin 2kr\,; \tag{97.6b}$$

$$t_3' = -\frac{U}{k}\{t_2\sin 2kr + t_1\cos^2 kr\} \tag{97.6c}$$

etc. These equations can be solved successively by simple quadratures.

Eq. (97.6a) corresponds to the first Born approximation,

$$t_1(r) = -\frac{1}{k}\int_0^r dr'\, U(r')\sin^2 kr' \tag{97.7}$$

with the asymptotic phase-shift formula

$$\tan\delta = t_1(\infty) = -\frac{1}{k}\int_0^\infty dr\, U(r)\sin^2 kr\,. \tag{97.8}$$

The second Born approximation t_1+t_2 comes from (97.6a) and (97.6b):

$$t(r) = -\frac{1}{k}\int_0^r dr'\, U(r')\sin kr'\left\{\sin kr' - \frac{2}{k}\cos kr'\int_0^{r'} dr''\, U(r'')\sin^2 kr''\right\} \tag{97.9}$$

yielding a phase-shift formula

$$\tan\delta = -\frac{1}{k}\int_0^\infty dr\, U(r)\sin kr\left\{\sin kr - \frac{2}{k}\cos kr\int_0^r dr'\, U(r')\sin^2 kr'\right\} \tag{97.10}$$

in complete agreement with Eq. (96.12b) of the preceding problem.

Problem 98. Linearization of Calogero's equation

To solve approximately Calogero's differential equation (97.4) of the preceding problem under the assumption of $|t(\infty)| \ll 1$.

Solution. In Calogero's equation

$$\frac{dt}{dr} = -\frac{1}{k} U(r) \{\sin kr + t(r) \cos kr\}^2 \tag{98.1}$$

or

$$\frac{dt}{dr} = -\frac{U}{k} \{\sin^2 kr + t \sin 2kr + t^2 \cos^2 kr\}$$

the last term may be neglected for small $t(r)$ so that we arrive at the linear equation

$$\frac{d\tilde{t}}{dr} = -\frac{U}{k} \{\sin^2 kr + \tilde{t} \sin 2kr\} \tag{98.2}$$

for the approximate function $\tilde{t}(r)$. Comparing (98.2) with (97.6) of the preceding problem, it is immediately apparent that the second Born approximation

$$\frac{d}{dr}(t_1 + t_2) = -\frac{U}{k} \{\sin^2 kr + t_1 \sin 2kr\}$$

is worse than (98.2) which instead of t_1 has the same function t on the right as on the left-hand side, thus including the first term of (97.6c). The linearization therefore leads in a single step to a solution of a quality intermediate between the second and the third Born approximations.

The solution of (98.2) with the boundary condition

$$\tilde{t}(0) = 0 \tag{98.3}$$

can be found by standard methods to be

$$\tilde{t}(r) = -\frac{1}{k} \int_0^r dr' \, U(r') \sin^2 kr' \exp\left\{-\frac{1}{k} \int_{r'}^r dr'' \, U(r'') \sin 2kr''\right\}. \tag{98.4}$$

By expansion of the exponentials into powers of $1/k$ one falls back upon the second Born approximation. It should, however, be noted that (98.4) is far better because, even for very small values of k or very large $U(r)$, as in singular potentials, where the Born expressions would become meaningless, a well-defined solution still exists as long as we are dealing with repulsive potentials only.

Problem 99. Scattering length for a negative-power potential

To determine the scattering length for the repulsive potential

$$\frac{2m}{\hbar^2} V(r) \equiv U(r) = \frac{g^2}{r_0^2} \left(\frac{r_0}{r}\right)^n \tag{99.1}$$

a) by exact integration of the Schrödinger equation for $k=0$,
b) by solving the linearized Calogero equation.

Solution. a) The radial Schrödinger equation for $l=0$, in the limit of vanishing energy, runs

$$\chi_0'' - g^2 x^{-n} \chi_0 = 0; \qquad x = r/r_0 . \tag{99.2}$$

This equation is solved, with the boundary condition $\chi_0(0)=0$, by

$$\chi_0 = C\sqrt{x}\, K_{\frac{1}{2\lambda}}\left(\frac{g}{\lambda} x^{-\lambda}\right); \qquad \lambda = \frac{n-2}{2} \tag{99.3}$$

where $K_\nu(z)$ stands for the modified Hankel function

$$K_\nu(z) = \frac{\pi}{2\sin\nu\pi}\left[I_{-\nu}(z) - I_\nu(z)\right] = \frac{i\pi}{2} e^{\frac{i\nu\pi}{2}} H_\nu^{(1)}(iz) \tag{99.4}$$

and has the asymptotic form for $|z|\to\infty$,

$$K_\nu(z) \to \sqrt{\frac{\pi}{2z}}\, e^{-z} . \tag{99.5}$$

Since, for very small x, the argument of K in (99.3) becomes large, we then may apply Eq. (99.5) and obtain

$$\chi_0 \simeq C\sqrt{\frac{\pi\lambda}{2g}}\, x^{\frac{\lambda+1}{2}}\, e^{-\frac{g}{\lambda}x^{-\lambda}}$$

so that the boundary condition $\chi_0(0)=0$ is satisfied.

On the other hand, if x is large and z is small, we may use the power expansions of $I_{\pm\nu}(z)$ which, with Eq. (99.4), yield

$$K_\nu(z) = \frac{\pi}{2\sin\nu\pi}\left\{\frac{1}{\Gamma(1-\nu)}\left(\frac{z}{2}\right)^{-\nu}\left[1+\frac{z^2}{4(1-\nu)}+\cdots\right]\right.$$
$$\left. - \frac{1}{\Gamma(1+\nu)}\left(\frac{z}{2}\right)^{\nu}\left[1+\frac{z^2}{4(1+\nu)}+\cdots\right]\right\}.$$

To find the asymptotic behaviour of χ_0 at large distances r, we use only the first term of each of the two series, viz.

$$\chi_0(x) = C\sqrt{x}\,\frac{\pi}{2\sin\frac{\pi}{2\lambda}}\left\{\frac{1}{\Gamma\left(1-\frac{1}{2\lambda}\right)}\left(\frac{2\lambda}{g}\right)^{\frac{1}{2\lambda}}\sqrt{x}\right.$$
$$\left. - \frac{1}{\Gamma\left(1+\frac{1}{2\lambda}\right)}\left(\frac{g}{2\lambda}\right)^{\frac{1}{2\lambda}}\frac{1}{\sqrt{x}}\right\}. \tag{99.6}$$

This is a linear function of x and may be written

$$\chi_0(x) = A\left(1 - \frac{r_0 x}{a}\right) \tag{99.7}$$

where a is by definition the scattering length (cf. Problem 88). By comparing Eqs. (99.6) and (99.7) we arrive at the expression

$$a = r_0 \frac{\Gamma\left(1 - \frac{1}{2\lambda}\right)}{\Gamma\left(1 + \frac{1}{2\lambda}\right)} \left(\frac{g}{2\lambda}\right)^{\frac{1}{\lambda}}. \tag{99.8}$$

b) From Calogero's equation we find by linearization (cf. Eq. (98.4)),

$$\tan\delta_0 = -\frac{g^2}{k r_0^2}\int_0^\infty dr \left(\frac{r_0}{r}\right)^n \sin^2 kr \exp\left\{-\frac{g^2}{k r_0^2}\int_r^\infty dr' \left(\frac{r_0}{r'}\right)^n \sin 2kr'\right\}; \tag{99.9}$$

it determines the asymptotic behaviour of χ_0 which, in the limit $k \to 0$ is found from

$$\chi_0 \propto \sin(kr + \delta_0) = \sin kr \cos\delta_0 + \cos kr \sin\delta_0 \to \sin\delta_0 (1 + kr \cot\delta_0)$$

so that comparison with (99.7) leads to the definition of the scattering length by the phase angle:

$$a = -\lim_{k \to 0} \frac{\tan\delta_0}{k}. \tag{99.10}$$

Using this definition, and introducing $r/r_0 = x$ and $r'/r_0 = y$ as dimensionless integration variables, we obtain from (99.9) the approximation

$$\tilde{a} = g^2 r_0 \int_0^\infty dx\, x^{2-n} \exp\left(-2g^2 \int_x^\infty dy\, y^{1-n}\right)$$

of the scattering length. The exponent can easily be calculated to be $-2g^2 x^{2-n}/(n-2)$; by substituting

$$t = \frac{2g^2}{n-2} x^{2-n} = \frac{g^2}{\lambda} x^{-2\lambda}$$

the integral is reduced to Euler type so that we finally get

$$\tilde{a} = \tfrac{1}{2} r_0 (g^2/\lambda)^{\frac{1}{2\lambda}}\ \Gamma\left(1 - \frac{1}{2\lambda}\right). \tag{99.11}$$

The approximate result (99.11) may be compared with the exact formula (99.8). Both show the same dependence upon the dimensionless coupling constant g^2 so that their ratio

$$\frac{\tilde{a}}{a} = \frac{1}{2}\Gamma\left(1 + \frac{1}{2\lambda}\right)(4\lambda)^{\frac{1}{2\lambda}} \tag{99.12}$$

becomes a function of λ, and thus of the potential power exponent $n=2\ (\lambda+1)$, only. The quality of the approximation may be judged from the few values assembled in the accompanying table. There is no deviation in the case $n=3$, the error of the approximation increasing up to 50 per cent in the hard-core case ($n=\infty$).

n	$\tilde{a}/a$
3	1
4	0.886
5	0.810
6	0.762
8	0.702
10	0.665
12	0.642
∞	0.500

Problem 100. Second approximation to Calogero equation

To extend the linearization of Calogero's equation into a systematic approach by successive steps, and to compute the scattering length for the preceding problem in second approximation.

Solution. If, instead of the correct equation

$$\frac{dt}{dr} = -\frac{U}{k}\{\sin^2 kr + t\sin 2kr + t^2\cos^2 kr\}, \tag{100.1}$$

we use the set of consecutive equations

$$\frac{dt_n}{dr} = -\frac{U}{k}\{\sin^2 kr + t_n\sin 2kr + t_{n-1}^2\cos^2 kr\}; \quad t_0=0 \tag{100.2}$$

we shall, in first approximation, fall back upon the linear equation

$$\frac{dt_1}{dr} = -\frac{U}{k}\{\sin^2 kr + t_1\sin 2kr\} \tag{100.3}$$

which has been solved in Problem 98. Any further approximation then follows from the preceding one by the same procedure of integrating a linear equation by quadrature, only with different inhomogeneities, viz.

$$t_n(r) = -\frac{1}{k}\int_0^r dr'\, U(r')\left[\sin^2 kr' + t_{n-1}(r')^2\cos^2 kr'\right] \times \exp\left\{-\frac{1}{k}\int_{r'}^r dr''\, U(r'')\sin 2kr''\right\}. \tag{100.4}$$

In order to find the second approximation, we therefore only have to add a factor

$$[1+t_1(r')^2 \cot^2 k r'] \tag{100.5}$$

to the integrand of the linear solution. To calculate scattering lengths, however, this rule may even be more simplified. The "interpolating" scattering length, $a(r)$, defined by

$$a(r) = -\lim_{k\to 0} \frac{t(r)}{k}, \tag{100.6}$$

becomes in first approximation

$$a_1(r) = \int_0^r dr' \, U(r') \, r'^2 \exp\left\{-2 \int_{r'}^r dr'' \, U(r'') \, r''\right\}; \tag{100.7}$$

the second approximation of the scattering length, $a_2(\infty) \equiv a_2$, then is

$$a_2 = \int_0^\infty dr \, U(r) r^2 \left[1 + \frac{a_1(r)^2}{r^2}\right] \exp\left\{-2 \int_r^\infty dr' \, U(r') r'\right\}. \tag{100.8}$$

Eqs. (100.7) and (100.8) may now be used to determine the scattering length for the negative-power potential

$$U(r) = \frac{g^2}{r_0^2} \left(\frac{r_0}{r}\right)^n. \tag{100.9}$$

Using the abbreviations

$$x = r/r_0; \quad y = r'/r_0; \quad z = r''/r_0; \quad \lambda = \frac{n-2}{2}, \tag{100.10}$$

we have

$$a_1(r) = g^2 r_0 \int_0^x dy \, y^{2-n} \exp\left\{-2g^2 \int_y^x dz \, z^{1-n}\right\}$$

$$= g^2 r_0 \int_0^x dy \, y^{-2\lambda} \exp\left\{\frac{g^2}{\lambda}(x^{-2\lambda} - y^{-2\lambda})\right\}.$$

Substituting of the variables

$$s = \frac{g^2}{\lambda} x^{-2\lambda} \quad \text{and} \quad t = \frac{g^2}{\lambda} y^{-2\lambda} \tag{100.11}$$

will transform the expression into

$$a_1(r) = \tfrac{1}{2} r_0 (g^2/\lambda)^{\frac{1}{2\lambda}} e^s \int_s^\infty dt\, t^{-\frac{1}{2\lambda}} e^{-t}. \tag{100.12}$$

This may be put into Eq. (100.8) thus rendering

$$a_2 = \tfrac{1}{2} r_0 (g^2/\lambda)^{\frac{1}{2\lambda}} \int_0^\infty ds\, s^{-\frac{1}{2\lambda}} e^{-s} \left\{ 1 + \tfrac{1}{4} s^{\frac{1}{\lambda}} e^{2s} \left[\int_s^\infty dt\, t^{-\frac{1}{2\lambda}} e^{-t} \right]^2 \right\}. \tag{100.13}$$

The first term in the curly brackets leads back, of course, to the first approximation a_1, Eq. (99.11). The correction

$$\alpha = \frac{a_2 - a_1}{a_1}; \qquad a_2 = a_1(1+\alpha) \tag{100.14}$$

is then easily derived from (100.13),

$$\alpha = \frac{1}{4\Gamma\left(1 - \frac{1}{2\lambda}\right)} \int_0^\infty ds\, s^{\frac{1}{2\lambda}} e^s \left[\int_s^\infty dt\, t^{-\frac{1}{2\lambda}} e^{-t} \right]^2. \tag{100.15}$$

The integral converges and may be evaluated, in general, by numerical computation.

There are two cases permitting a rather simple evaluation, viz. $n=4$ and $n \to \infty$.

If $n=4$ or $\lambda=1$, we find

$$\alpha = \frac{1}{4\sqrt{\pi}} \int_0^\infty ds \sqrt{s}\, e^s \left[\int_s^\infty \frac{dt}{\sqrt{t}} e^{-t} \right]^2$$

or

$$\alpha = \frac{\sqrt{\pi}}{2} \int_0^\infty d\xi\, \xi^2 e^{\xi^2} (\operatorname{erfc} \xi)^2 .$$

Numerical integration with Simpson's rule and an estimate of the contribution beyond $\xi > 2.4$ using the asymptotic expression for the error function gives $\alpha = 0.0770$. Since, for $n=4$, we have found on p. 263 for the first approximation

$$a_1 = 0.886\, a,$$

we now get in second approximation,

$$a_2 = 0.886 \times 1.077\, a = 0.954\, a.$$

The error of 11.4% in the first is thus reduced to 4.6% in the second approximation.

The convergence of the method is worse in the case $n\to\infty$ or $1/\lambda=0$. Eq. (100.15) then yields

$$\alpha=\tfrac{1}{4}\int_0^\infty ds\,e^s\left[\int_s^\infty dt\,e^{-t}\right]^2=\tfrac{1}{4};$$

hence with

$$a_1=0.500\,a$$

we now arrive at

$$a_2=0.625\,a.$$

Problem 101. Square-well potential: Scattering length

To compare the scattering length for the square-well potential

$$U=\frac{2m}{\hbar^2}V(r)=\begin{cases}-K_0^2 & \text{for } 0\leqq r\leqq R,\\ 0 & \text{for } r>R\end{cases} \tag{101.1}$$

with its approximate value obtained from the linearized Calogero equation. The repulsive potential generated by a change of sign in (101.1) shall also be discussed.

Solution. With the abbreviations

$$kR=x;\quad K_0R=X_0;\quad KR=X=\sqrt{X_0^2+x^2} \tag{101.2}$$

we have, according to Problem 89 the exact relation

$$\tan\delta_0=\frac{x-L\tan x}{L+x\tan x} \tag{101.3}$$

where, for the potential (101.1), the logarithmic derivative (see (89.5)) is

$$L=X\cot X. \tag{101.4}$$

The scattering length is defined by

$$a=-\lim_{k\to 0}\frac{\tan\delta_0}{k} \tag{101.5}$$

or, in our special case (see also (89.14))

$$a=-R\lim_{\substack{x\to 0\\ X\to X_0}}\frac{\tan\delta_0}{x}=-R\left(\frac{\tan X_0}{X_0}-1\right). \tag{101.6}$$

On the other hand, the linearized Calogero equation, according to Problem 98 leads to

$$\tan\tilde{\delta}_0 = -\frac{1}{k}\int_0^\infty dr\, U(r)\sin^2 kr \exp\left[-\frac{1}{k}\int_r^\infty dr'\, U(r')\sin 2kr'\right]$$

and, with the definition (101.5), further to

$$\tilde{a} = \int_0^\infty dr\, r^2\, U(r)\exp\left[-2\int_r^\infty dr'\, r'\, U(r')\right]. \tag{101.7}$$

With the potential well (101.1), there follows

$$\tilde{a} = -K_0^2 \int_0^R dr\, r^2 \exp\left[K_0^2 (R^2 - r^2)\right]$$

or, with $t = X_0 r/R$,

$$\tilde{a} = -R\frac{e^{X_0^2}}{X_0}\int_0^{X_0} dt\, t^2\, e^{-t^2}. \tag{101.8}$$

Using the identity

$$\int_0^{X_0} dt\, t^2\, e^{-t^2} = \tfrac{1}{2}\int_0^{X_0} dt\, e^{-t^2} - \tfrac{1}{2} X_0\, e^{-X_0^2}$$

the integral in (101.8) may be reduced to the error function. Thus we arrive at the final result

$$\tilde{a} = -\frac{1}{2}R\left\{\frac{e^{X_0^2}}{X_0}\int_0^{X_0} dt\, e^{-t^2} - 1\right\}. \tag{101.9}$$

The results (101.6) of the exact, and (101.9) of the approximate calculations are compared in the accompanying table for a few values of the well size X_0. Below about $X_0 = 0.7$ the difference becomes very small. On the other hand, at $X_0 = \pi/2$, the exact values for the attractive

X_0	attraction a/R	attraction $\tilde{a}/R$	repulsion a/R	repulsion $\tilde{a}/R$
1.0	− 0.557	−0.515	0.238	0.231
1.2	− 1.142	−0.912	0.305	0.289
1.4	− 3.14	−1.632	0.368	0.337
1.6	+20.4	−3.02	0.424	0.375

potential become singular and change sign beyond this, a feature not reproduced by the approximation. It is therefore only safe to use the linearized approximation appreciably below resonance levels at zero energy.

For a repulsive potential

$$U = \begin{cases} +K_0^2 & \text{for } 0 \leqq r \leqq R, \\ 0 & \text{for } r > R \end{cases} \tag{101.10}$$

we find in a similar way the exact scattering length

$$a = -R\left(\frac{\tanh X_0}{X_0} - 1\right) \tag{101.11}$$

and its approximate value

$$\tilde{a} = -\frac{1}{2} R \left\{ \frac{e^{-X_0^2}}{X_0} \int_0^{X_0} dt\, e^{t^2} - 1 \right\}. \tag{101.12}$$

There is no resonance effect at repulsion and therefore the approximation is better, as is shown by the two last columns of the table above.

Problem 102. Scattering length for a Yukawa potential

To determine the scattering length for the attractive Yukawa potential

$$V(r) = -V_0 \frac{r_0}{r} e^{-r/r_0}; \quad V_0 = \frac{\hbar^2}{2m} \frac{g}{r_0^2} \tag{102.1}$$

a) by solving the linearized Calogero equation,
b) by numerical integration of the Schrödinger equation for $g=1$,
c) in the Born approximation.

Solution. Since both linearizing the Calogero equation and using the Born approximation are typical high-energy methods, it is not to be expected that they will give reliable results for the scattering length, i.e. at zero energy. As shown in Problem 98, the linearized Calogero equation will be better than the second Born approximation, and that, of course, better than the first. We may therefore expect to see in this example that such an order of approximations does indeed occur. It is amazing that the results obtained are at least sufficient to give a general impression of the behaviour of the scattering phenomena, even at zero energy.

a) Using Eq. (98.4) for $\tilde{t}(r)$ and the definition

$$a = -\lim_{k\to 0} \frac{\tan\delta_0}{k} \tag{102.2}$$

of the scattering length, and writing

$$U(x) = \frac{2m}{\hbar^2} V(r) = -\frac{g}{r_0^2}\cdot\frac{e^{-x}}{x} \quad \text{with } x = r/r_0 \tag{102.3}$$

we get the approximation

$$\tilde{a} = -g\, r_0 \int_0^\infty dx\, x\, e^{-x}\, e^{2g e^{-x}} . \tag{102.4}$$

Substituting

$$y = 2g\, e^{-x}$$

this can be written

$$\tilde{a} = -\tfrac{1}{2} r_0 \int_0^{2g} dy\, e^y (\log 2g - \log y)$$

or, with the exponential integral

$$\mathrm{Ei}\, y = \int_{-\infty}^{y} \frac{dt}{t} e^t$$

by partial integration

$$\tilde{a} = -\tfrac{1}{2} r_0 (\mathrm{Ei}\, 2g - \log 2g - C), \tag{102.5}$$

$C = 0.5772\ldots$ being the Euler constant. Numerical results for some values of the potential-strength parameter g are given in the accompanying table.

g	$-\tilde{a}/r_0$	g	$-\tilde{a}/r_0$
0.1	0.1052	0.9	1.5425
0.2	0.2220	1.0	1.8420
0.3	0.3518	1.1	2.1835
0.4	0.4967	1.2	2.5740
0.5	0.6589	1.3	3.0217
0.6	0.8413	1.4	3.5362
0.7	1.0468	1.5	4.1290
0.8	1.2790		

b) The scattering length can be determined from the solution of the radial Schrödinger equation at zero energy,

$$\chi_0'' = -g \frac{e^{-x}}{x} \chi_0 , \tag{102.6}$$

that is defined by the boundary conditions

$$\chi_0(0)=0; \quad \chi_0'(0)=1. \tag{102.7}$$

At large values of x, where the potential term on the right-hand side of (102.6) has become practically zero, the asymptotic form of the solution becomes a linear function,

$$\chi_0 \to A\left(1 - \frac{x}{\alpha}\right), \tag{102.8}$$

with

$$\alpha = a/r_0 \tag{102.9}$$

defining the scattering length a. Solving (102.8) with respect to α yields

$$-\alpha = \frac{\chi_0(x)}{\chi_0'(x)} - x . \tag{102.10}$$

In a numerical integration of (102.6), starting with the initial values (102.7), the combination (102.10) will slowly increase from zero at $x=0$ to a constant value at large x which determines the scattering length.

x	χ	χ'	$-\alpha$
0	0	1	0
1	0.638	0.457	0.348
2	1.040	0.298	1.484
3	1.312	0.255	2.15
4	1.559	0.241	2.47
5	1.798	0.237	2.60
5.5	1.916	0.236	2.63
6	2.033	0.235	2.65

The accompanying table shows a few figures for the example $g=1$ and gives the scattering length

$$a = -2.65\, r_0 \tag{102.11 a}$$

in this case, to be compared with the approximation (102.5) for $g=1$, viz.

$$\tilde{a} = -1.842\, r_0 . \tag{102.11 b}$$

c) The first Born approximation follows from (102.4) by using the linear term only in g:

$$a_1 = -g\, r_0 \int_0^\infty dx\, x\, e^{-x} = -g\, r_0 ,$$

i.e. for $g=1$ it leads to

$$a_1 = -r_0 . \tag{102.11c}$$

Including the quadratic term in g, by expansion of (102.4), we arrive at the second Born approximation,

$$a_2 = -g\, r_0 \int_0^\infty dx\, x\, e^{-x}(1+2g\, e^{-x}) = -g\, r_0 (1+\tfrac{1}{2} g) ,$$

correcting a_1 for $g=1$ to

$$a_2 = -1.500\, r_0 . \tag{102.11d}$$

The deviations from the exact value of the three approximations discussed, of course, become the larger the greater the potential strength. The ordering of the approximations is the one predicted at the beginning of this problem.

Literature to Problems 96—102. Calogero, F.: Nuovo Cimento **27**, 261 (1963).— Klar, H., Krüger, H.: Z. Physik **191**, 409 (1966); Krüger, H.: Z. Physik **204**, 114, **205**, 338 (1967).

Problem 103. Improvement of convergence in a spherical harmonics series

Using the recurrence relations between neighbouring Legendre polynomials, instead of the series

$$f(x) = \sum_{l=0}^{\infty} f_l\, P_l(x) \tag{103.1}$$

a better converging series shall be derived for $(1-x) f(x)$.

Solution. The recurrence relation

$$(l+1)\, P_{l+1}(x) + l\, P_{l-1}(x) = (2l+1)\, x\, P_l(x) \tag{103.2}$$

can be used to write

$$(1-x) f(x) = \sum_{l=0}^{\infty} f_l \left(P_l - \frac{l+1}{2l+1} P_{l+1} - \frac{l}{2l+1} P_{l-1} \right)$$

Renumbering the subscripts taking special care of the first terms leads to

$$(1-x)f(x)=f_0-\tfrac{1}{3}f_1+\sum_{l=1}^{\infty}\tilde{f}_l P_l(x) \tag{103.3}$$

with

$$\tilde{f}_l=f_l-\frac{l}{2l-1}f_{l-1}-\frac{l+1}{2l+3}f_{l+1}. \tag{103.4}$$

In order to get an idea of the improvement let

$$f_l=C\,l^{-\varkappa}$$

for large l. Then,

$$\begin{aligned}\tilde{f}_l&=C\left\{l^{-\varkappa}-\frac{l}{2l-1}(l-1)^{-\varkappa}-\frac{l+1}{2l+3}(l+1)^{-\varkappa}\right\}\\&=C\,l^{-\varkappa}\left\{1-\frac{l}{2l-1}\left(\frac{l}{l-1}\right)^{\varkappa}-\frac{l+1}{2l+3}\left(\frac{l}{l+1}\right)^{\varkappa}\right\}\end{aligned}$$

can be expanded for $l\gg 1$:

$$\tilde{f}_l=f_l\left\{1-\frac{1}{2}\left(1-\frac{1}{2l}\right)^{-1}\left(1-\frac{1}{l}\right)^{-\varkappa}-\frac{1}{2}\left(1+\frac{1}{l}\right)\left(1+\frac{3}{2l}\right)^{-1}\left(1+\frac{1}{l}\right)^{-\varkappa}\right\}.$$

The terms with l^0 and l^{-1} then cancel, and the leading term becomes

$$\tilde{f}_l=-f_l\frac{(\varkappa+1)^2}{2l^2}$$

which is indeed a considerable improvement.

NB. The formulae simplify even more if we write

$$f_l=(2l+1)\alpha_l$$

corresponding to the form such coefficients have in most cases. We then find,

$$(1-x)f(x)=\alpha_0-\alpha_1+\sum_{l=1}^{\infty}\{(2l+1)\alpha_l-l\alpha_{l-1}-(l+1)\alpha_{l+1}\}P_l(x).$$

Literature. Yennie, D. R., Ravenhall, D. G., Wilson, R. N.: Phys. Rev. **95,** 500 (1954).

Problem 104. Collision-parameter integral

If $kR\gg 1$, with R a measure for the extension of the scattering region, angular momenta of $l\gg 1$ will appreciably contribute to the scattering amplitude. The sum over l then shall be replaced by an integral over collision parameters b, a high-energy relation for $\delta(b)$ be derived from the WKB approach, and the total cross section determined.

Solution. Into the general formula

$$f(\vartheta)=\frac{1}{2ik}\sum_{l=0}^{\infty}(2l+1)(e^{2i\delta_l}-1)\,P_l(\cos\vartheta) \tag{104.1}$$

we introduce the collision parameter

$$b=(l+\tfrac{1}{2})/k \tag{104.2}$$

and use it as integration variable:

$$f(\vartheta)=-ik\int_0^{\infty}db\,b(e^{2i\delta(b)}-1)\,P_l(\cos\vartheta)\,. \tag{104.3}$$

The lower integration bound may be taken to be zero instead of $1/(2k)\ll R$. It is a little more difficult to replace P_l by a function of b. The definition of Legendre polynomials,

$$P_l(\cos\vartheta)={}_2F_1\left(l+1,-l,1;\ \sin^2\frac{\vartheta}{2}\right), \tag{104.4a}$$

by a hypergeometric series, viz.

$$P_l(\cos\vartheta)=1-\frac{l(l+1)}{1!^2}\sin^2\frac{\vartheta}{2}+\frac{(l-1)l(l+1)(l+2)}{2!^2}\sin^4\frac{\vartheta}{2}\cdots \tag{104.4b}$$

leads to a simple asymptotic expression for $l\gg1$. With the abbreviation

$$\lambda=l+\tfrac{1}{2} \tag{104.5}$$

we obtain

$$l(l+1)=\lambda^2-\tfrac{1}{4};\quad (l-1)(l+2)=\lambda^2-\tfrac{9}{4};\quad (l-n)\,(l+n+1)=\lambda^2-(n+\tfrac{1}{2})^2.$$

The coefficients of the series (104.4b) therefore differ from λ^{2n} only by terms smaller by the order of λ^{-2}. Neglecting this correction for large values of l, Eq. (104.4b) becomes

$$P_l(\cos\vartheta)=1-\frac{\lambda^2}{1!^2}\sin^2\frac{\vartheta}{2}+\frac{\lambda^4}{2!^2}\sin^4\frac{\vartheta}{2}\cdots.$$

This being the Bessel series, we arrive at

$$\lim_{l\to\infty}P_l(\cos\vartheta)=J_0\left((2l+1)\sin\frac{\vartheta}{2}\right). \tag{104.6}$$

Putting (104.6) into (104.3), taking account of (104.2), we find

$$f(\vartheta) = -ik \int_0^\infty db\, b(e^{2i\delta(b)} - 1)\, J_0\left(2kb \sin\frac{\vartheta}{2}\right). \tag{104.7}$$

Dealing with high energies, we may further use the WKB approximation to compute the phase function $\delta(b)$ using the quasipotential formalism of Problem 124. Eq. (104.7) then immediately yields

$$\delta(b) = -\frac{1}{2}k \int_b^\infty \frac{dt\, t}{\sqrt{t^2 - b^2}} \frac{Q(t)}{E},$$

where $Q(t)$ may be identified with $V(r)$ in the high-energy limit, i.e. if

$$|V(r)| \ll E$$

for all values of r. We then may write r again instead of t and have the simple result

$$\delta(b) = -\frac{k}{2E} \int_b^\infty \frac{dr\, r\, V(r)}{\sqrt{r^2 - b^2}}. \tag{104.8}$$

It should be noted that this expression is linear in the potential thus permitting addition of partial phase contributions for potentials composed as a sum of simple terms, as e.g. the intermolecular potentials composed of two or more negative-power contributions (Lennard-Jones potential etc.).

Finally, the total scattering cross section follows by applying the optical theorem

$$\sigma = \frac{4\pi}{k} \operatorname{Im} f(0)$$

to the integral (104.7). The result is

$$\sigma = 4\pi \int_0^\infty db\, b[1 - \cos 2\delta(b)]. \tag{104.9}$$

NB. Eq. (104.6) is a fit approximation for high-energy scattering phenomena and much better than the well-known asymptotic relation

$$P_l(\cos\vartheta) \to \sqrt{\frac{2}{\pi l \sin\vartheta}} \cos\left[\left(l + \frac{1}{2}\right)\vartheta - \frac{\pi}{4}\right] \tag{104.6'}$$

which has singularities at $\vartheta = 0$ and $\vartheta = \pi$. It is almost correct in the neighbourhood of $\vartheta = 0$ (forward scattering, i.e. the angle of greatest interest). On the other hand,

Eq. (104.6) becomes worse the greater ϑ, at $\vartheta=\pi$ leading to $J_0\,(2l+1)$ instead of $(-1)^l$. In Fig. 55, P_{10} is compared with its approximation by (104.6). As backscattering becomes less and less important at high energies ($kR \gg 1$), however, this error does not appreciably affect our purpose.

Literature. Molière, G.: Z. Naturforsch. **2a,** 133 (1947). — Blankenbecler, R., Goldberger, M. L.: Phys. Rev. **126,** 766 (1962). — Flügge, S., Krüger, H.: Z. Physik **216,** 213 (1968).

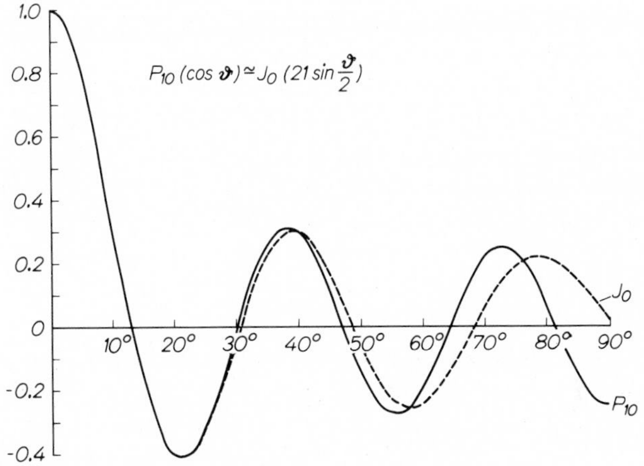

Fig. 55. The Legendre polynomial $P_{10}(\cos\vartheta)$ (full line) and its approximation by the Bessel function (104.6) (broken line)

Problem 105. Born scattering: Successive approximation steps

To solve the problem of potential scattering by treating the scattering potential as a perturbation. The case of central forces shall be dealt with in more detail.

Solution. We write the Schrödinger equation in the form

$$\nabla^2 u + k^2 u = g\,W(\boldsymbol{r})\,u \tag{105.1}$$

with

$$k^2 = \frac{2m}{\hbar^2}E\,; \qquad g\,W(\boldsymbol{r}) = \frac{2m}{\hbar^2}V(\boldsymbol{r})\,. \tag{105.2}$$

The left-hand side of (105.1) has a Green's function

$$G(\boldsymbol{r},\boldsymbol{r}') = -\frac{e^{ik|\boldsymbol{r}-\boldsymbol{r}'|}}{4\pi|\boldsymbol{r}-\boldsymbol{r}'|} \tag{105.3}$$

that permits the right-hand side of (105.1) to be formally considered as an inhomogeneity so that this equation can be solved according to the general method for inhomogeneous differential equations,

$$u(\boldsymbol{r}) = u_0(\boldsymbol{r}) + g\int d^3r'\, G(\boldsymbol{r},\boldsymbol{r}')\, W(\boldsymbol{r}')\, u(\boldsymbol{r}') \tag{105.4}$$

where $u_0(\boldsymbol{r})$ is a solution of the homogeneous equation. Our result is an integral equation for $u(\boldsymbol{r})$.

It can easily be seen that the integral term in (105.4) describes asymptotically, for large r, an outgoing spherical wave, because the potential function decreases rapidly enough with increasing r' to confine integration practically to a finite region, thus permitting us to consider the case of $r \gg r'$ with the Green's function (105.3) becoming

$$G(\boldsymbol{r},\boldsymbol{r}') \to -\frac{e^{ikr}}{4\pi r}\, e^{-ikr'\cos\Theta'} \tag{105.5}$$

where Θ' is the angle between the vectors $\boldsymbol{r}$ and $\boldsymbol{r}'$. The homogeneous solution, $u_0(\boldsymbol{r})$, in case of a scattering problem, must be the plane wave of the incident particle beam which we write in standard normalization

$$u_0(\boldsymbol{r}) = e^{i\boldsymbol{k}_0\cdot\boldsymbol{r}}; \quad |\boldsymbol{k}_0| = k. \tag{105.6}$$

The solution of (105.4) can be performed in successive steps:

$$u_n(\boldsymbol{r}) = u_0(\boldsymbol{r}) + g\int d^3r'\, G(\boldsymbol{r},\boldsymbol{r}')\, W(\boldsymbol{r}')\, u_{n-1}(\boldsymbol{r}'); \tag{105.7}$$

in other words, it may be written as a Neumann series,

$$u(\boldsymbol{r}) = u_0(\boldsymbol{r}) + g\int d^3r'\, G(\boldsymbol{r},\boldsymbol{r}')\, W(\boldsymbol{r}')\{\mu_0(\boldsymbol{r}')$$
$$+ g\int d^3r''\, G(\boldsymbol{r}',\boldsymbol{r}'')\, W(\boldsymbol{r}'')[u_0(\boldsymbol{r}'') + \cdots]\}.$$

This is a power expansion in the "coupling parameter" g. The n-th approximation is found by an n-fold volume integration which makes all terms but the first-order one rather cumbersome to evaluate.

The first-order problem is greatly simplified if $W(\boldsymbol{r})$ depends only on the absolute value of $\boldsymbol{r}$. We then find from (105.7) with (105.5) and (105.6) for the asymptotic behaviour of $u_1(\boldsymbol{r})$:

$$u_1(\boldsymbol{r}) = e^{i\boldsymbol{k}_0\cdot\boldsymbol{r}} - \frac{g\, e^{ikr}}{4\pi r}\int d^3r'\, e^{-ikr'\cos\Theta'}\, W(r')\, e^{i\boldsymbol{k}_0\cdot\boldsymbol{r}'}. \tag{105.8}$$

Let us define a vector $\boldsymbol{k}_1$ with $|\boldsymbol{k}_1|=k$ pointing in the direction of $\boldsymbol{r}$. Then $kr'\cos\Theta'=(\boldsymbol{k}_1\cdot\boldsymbol{r}')$, so that the integral in (105.8) becomes

$$\int_0^\infty dr'\, r'^2 \oint d\Omega'\, e^{i(\boldsymbol{k}_0-\boldsymbol{k}_1)\cdot\boldsymbol{r}'}\, W(r')\,. \tag{105.9}$$

The vector

$$\boldsymbol{K}=\boldsymbol{k}_1-\boldsymbol{k}_0 \tag{105.10}$$

is the momentum (in units of $\hbar$) transferred to a scattered particle in the direction $\boldsymbol{r}$. If we denote by ϑ the angle of deflection, i.e. the angle between $\boldsymbol{k}_0$ and $\boldsymbol{r}$, then (Fig. 56)

$$K=2k\sin\frac{\vartheta}{2}\,. \tag{105.11}$$

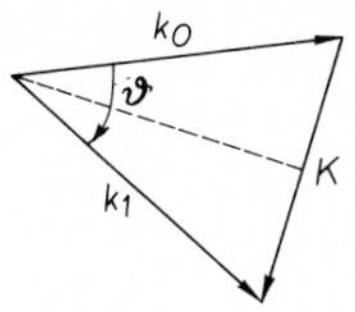

Fig. 56. Wave vectors and angle of deflection

The integration over all directions in (105.9) gives

$$\oint d\Omega'\, e^{-i\boldsymbol{K}\cdot\boldsymbol{r}'}=4\pi\frac{\sin Kr'}{Kr'} \tag{105.12}$$

so that finally Eq. (105.8) yields the first approximation

$$u_1(\boldsymbol{r})=e^{i\boldsymbol{k}_0\cdot\boldsymbol{r}}+f(\vartheta)\frac{e^{ikr}}{r} \tag{105.13}$$

with

$$f(\vartheta)=-g\int_0^\infty dr'\, r'^2\, W(r')\frac{\sin Kr'}{Kr'}\,. \tag{105.14}$$

The *scattering amplitude*, $f(\vartheta)$, is simply connected with the differential scattering cross section,

$$\frac{d\sigma}{d\Omega}=|f(\vartheta)|^2\,. \tag{105.15}$$

Eq. (105.14) for the scattering amplitude cannot be used if the integral diverges at either the lower or the upper integration bound. In the vicinity of $r'=0$ the potential must not therefore diverge stronger than $r'^{-3+\varepsilon}$, $\varepsilon>0$; for large r', on the other hand, it must decrease more steeply than the Coulomb potential, i.e. at least as $r'^{-1-\varepsilon}$.

Problem 106. Scattering by a Yukawa potential

The scattering amplitude for an attractive Yukawa potential,

$$V(r) = -V_0 \frac{e^{-\alpha r}}{\alpha r}; \qquad \alpha = 1/r_0 \tag{106.1}$$

is to be determined in first Born approximation. In order to judge the quality of the approximation, use shall be made of the phase angles of the partial wave expansion.

Solution. On the one hand, we have

$$f(\vartheta) = -\frac{2m}{\hbar^2}\int_0^\infty dr\, r^2\, V(r) \frac{\sin Kr}{Kr}; \quad K = 2k \sin\frac{\vartheta}{2}. \tag{106.2}$$

For the potential (106.1) this leads by elementary integration to

$$f(\vartheta) = P r_0 \frac{1}{1+K^2 r_0^2} \tag{106.3}$$

with the "hole size" constant

$$P = \frac{2m V_0 r_0^2}{\hbar^2}. \tag{106.4}$$

On the other hand, partial wave expansion leads to

$$f(\vartheta) = \frac{1}{2ik}\sum_{l=0}^{\infty} (2l+1)(e^{2i\delta_l}-1) P_l(\cos\vartheta). \tag{106.5}$$

To judge Eq. (106.3), the phase angles δ_l may be used only if we either expand (106.3) in a spherical harmonics series or, doing the same with (106.2), specialize the result afterwards to the Yukawa case. Both ways will, of course, lead to the same final result. They will be sketched in what follows.

1. Expansion of the scattering amplitude (106.3) may be written

$$\sum_{l=0}^{\infty}(2l+1)\frac{e^{2i\delta_l}-1}{2i}P_l(\cos\vartheta)=\frac{P\,x_0}{1+2x_0^2(1-\cos\vartheta)}$$

with the abbreviation

$$x_0=kr_0\,. \tag{106.6}$$

Taking advantage of the spherical harmonics' orthogonality, we find

$$\frac{e^{2i\delta_l}-1}{2i}=\frac{P}{4\,x_0}\int_{-1}^{+1}\frac{dt\,P_l(t)}{t_0-t};\quad t_0=1+\frac{1}{2x_0^2}\,. \tag{106.7}$$

Here, the integral

$$\frac{1}{2}\int_{-1}^{+1}\frac{dt\,P_l(t)}{t_0-t}=Q_l(t_0) \tag{106.8}$$

defines a spherical harmonic of the second kind $(t_0>1)$. The first two of these run as follows:

$$Q_0(t_0)=\frac{1}{2}\log\frac{t_0+1}{t_0-1};\quad Q_1(t_0)=\frac{1}{2}\,t_0\log\frac{t_0+1}{t_0-1}-1\,.$$

Introducing x_0 instead of t_0, we thus arrive at the final formulae

$$\left.\begin{aligned}&\delta_l\simeq\frac{1}{2i}(e^{2i\delta_l}-1)=P\,\varphi_l(x_0)\\ &\text{with}\\ &\varphi_0(x_0)=\frac{1}{4x_0}\log(1+4x_0^2);\\ &\varphi_1(x_0)=\frac{1}{4x_0}\left\{\left(1+\frac{1}{2x_0^2}\right)\log(1+4x_0^2)-2\right\}.\end{aligned}\right\} \tag{106.9}$$

2. Expanding the scattering amplitude (106.2), we make use of the identity

$$\frac{\sin Kr}{Kr}=\sum_{l=0}^{\infty}(2l+1)\,j_l(kr)^2\,P_l(\cos\vartheta); \tag{106.10}$$

then, by comparing (106.2) with (106.5), we get the general relation

$$\delta_l\simeq\frac{1}{2i}(e^{2i\delta_l}-1)=-\frac{2m}{\hbar^2}k\int_0^{\infty}dr\,r^2\,V(r)\,j_l(kr)^2\,. \tag{106.11}$$

If we here insert the Yukawa expression (106.1) for $V(r)$ and use the abbreviations $kr=z$ and $a=\alpha/k=1/(kr_0)=1/x_0$ we find

$$\frac{1}{2i}(e^{2i\delta_l}-1)=\frac{V_0}{Ea}\int_0^\infty dz\, z\, e^{-az}\, j_l(z)^2 .$$

The general formula

$$j_l(z)^2=\frac{2z^l}{l!}\int_0^{\pi/2} d\vartheta\, j_l(2z\sin\vartheta)\sin^{l+1}\vartheta\cos^{2l+1}\vartheta \tag{106.12}$$

then permits the integral (106.11) to be reshaped so that the Bessel function occurs linearly only in the integrand. The relation

$$\int_0^\infty dz\, z^{l+1} j_l(bz)\, e^{-az}=\frac{(2b)^l\, l!}{(a^2+b^2)^{l+1}} \tag{106.13}$$

finally leads to

$$\delta_l\simeq\frac{1}{2i}(e^{2i\delta_l}-1)=\frac{P}{2x_0}\int_0^{\pi/2} d\vartheta\frac{\sin^{2l+1}\vartheta\cos^{2l+1}\vartheta}{(\sin^2\vartheta+\beta^2)^{l+1}}. \tag{106.14}$$

Substituting the denominator bracket by

$$t=\sin^2\vartheta+\beta^2$$

the integral in the new variable becomes simply

$$\frac{1}{2}\int_{\beta^2}^{1+\beta^2} dt\frac{(t-\beta^2)^l(1+\beta^2-t)^l}{t^{l+1}} \tag{106.15}$$

and may be evaluated for each l in turn. The results will again be the ones summarized above in Eqs. (106.9).

In order now to judge the quality of the approximation, the functions $\varphi_0(x_0)$ and $\varphi_1(x_0)$ have been drawn in Fig. 57 versus a logarithmic x_0 scale. Both functions vanish at $x_0=0$; for low energies they start as

$$\begin{aligned}\varphi_0(x_0)&=x_0(1-2x_0^2+\cdots);\\ \varphi_1(x_0)&=\tfrac{2}{3}x_0(1-4x_0^2+\cdots)\end{aligned} \tag{106.16}$$

and, passing through a maximum, they tend to zero again at high energies. If, at the energy wanted, $P\varphi_0\ll 1$ and $kr_0\gg 1$, the Born approximation will be good. If $P\varphi_0$ does not satisfy this condition well enough, $P\varphi_1$ may still do so. Then we may write

$$f(\vartheta) = \frac{1}{2ik}(e^{2i\delta_0} - e^{2i\delta_{0B}}) + f_B(\vartheta) \tag{106.17}$$

where the subscript B refers to the Born approximations (106.9) and (106.3). Exact phase calculation has in this case to be performed for

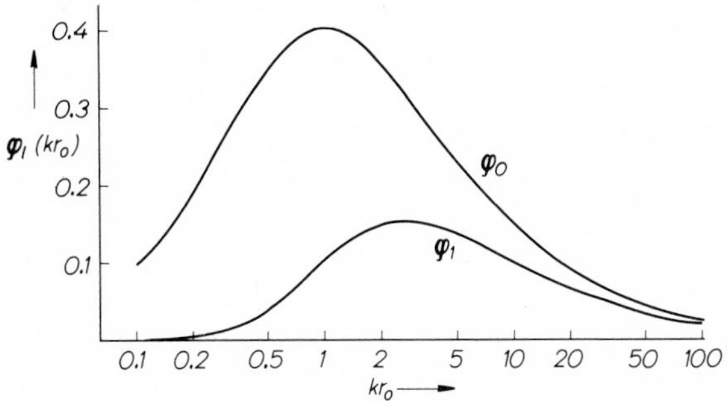

Fig. 57. Yukawa potential scattering phases in first approximation, $\delta_l = -P\varphi_l(kr_0)$ with P a potential size parameter. The abscissae describe the energy in logarithmic scale

$l=0$ only in order to find δ_0; the rest of $f(\vartheta)$ may be summed up in $f_B(\vartheta)$ according to (106.3).

NB. The scattering amplitude (106.3) as well as the phase expressions (106.9) for

$$\frac{1}{2i}(e^{2i\delta_l} - 1) = e^{i\delta_l}\sin\delta_l$$

turn out to be real in first Born approximation. This can only be satisfied by real phase angles $|\delta_l| \ll 1$, because power expansion leads to

$$e^{i\delta_l}\sin\delta_l = \delta_l + i\delta_l^2 + \cdots.$$

Since in the total scattering cross section

$$\sigma = \frac{4\pi}{k^2}\sum_{l=0}^{\infty}(2l+1)\sin^2\delta_l$$

only $\sin^2\delta_l = \delta_l^2 - \frac{1}{3}\delta_l^4 + \cdots$ occurs, here even the condition $\delta_l^2 \ll 1$ will suffice.

Problem 107. Scattering by an exponential potential

The scattering amplitude in the potential field

$$V(r) = -V_0 \, e^{-r/r_0} \tag{107.1}$$

shall be determined in first Born approximation and the phases δ_0 and δ_1 calculated by projection on the states with $l=0$ and $l=1$. The results shall be checked for δ_0 by computing this in second Born approximation according to the procedure of Problem 96.

Solution. The scattering amplitude in first Born approximation,

$$f(\vartheta) = -\frac{2m}{\hbar^2} \int_0^\infty dr\, r^2\, V(r) \frac{\sin Kr}{Kr}; \qquad K = 2k \sin\frac{\vartheta}{2}, \tag{107.2}$$

can with the dimensionless abbreviations

$$P = \frac{2m V_0 r_0^2}{\hbar^2}; \qquad x_0 = k r_0 \tag{107.3}$$

and with the integration variable $x = r/r_0$, be reshaped into

$$f(\vartheta) = \frac{P}{K} \int_0^\infty dx\, x\, e^{-x} \sin qx; \qquad q = 2x_0 \sin\frac{\vartheta}{2}. \tag{107.4}$$

Evaluation of this integral is elementary and yields

$$f(\vartheta) = \frac{P}{K} \frac{2q}{(1+q^2)^2}$$

or

$$f(\vartheta) = \frac{2 P r_0}{(a - b\cos\vartheta)^2}; \qquad a = 1 + 2x_0^2; \qquad b = 2x_0^2. \tag{107.5}$$

This amplitude permits first the total scattering cross section to be calculated:

$$\sigma = \oint d\Omega \, |f(\vartheta)|^2 = 4 P^2 r_0^2 \cdot 2\pi \int_{-1}^{+1} \frac{d\cos\vartheta}{(a - b\cos\vartheta)^4},$$

which, by an elementary integration, turns out to be

$$\sigma = \frac{32\pi}{3} P^2 r_0^2 \frac{1 + 2x_0^2}{(1 + 4x_0^2)^2}. \tag{107.6}$$

This approximation, of course, shows neither dependence on the sign of P (attraction or repulsion) nor resonance effects. So far, it is a typical high-energy formula and can, with $x_0^2 \gg 1$, be simplified to

$$\sigma = \frac{4\pi}{3} P^2 r_0^2 \cdot \frac{1}{x_0^2} = \frac{4\pi}{3} \frac{P^2}{k^2},$$

decreasing as $1/E$ with increasing energy E.

It further permits by using the expansion

$$f(\vartheta) = \frac{1}{2ik} \sum_{l=0}^{\infty} (2l+1)(e^{2i\delta_l} - 1) P_l(\cos\vartheta) \tag{107.7}$$

to find the phase angles δ_l in Born approximation by projection:

$$\frac{1}{2i}(e^{2i\delta_l} - 1) = \frac{k}{2} \int_{-1}^{+1} d\cos\vartheta\, f(\vartheta) P_l(\cos\vartheta). \tag{107.8}$$

With (107.5) for $f(\vartheta)$, these expressions become very simple integrals whose evaluation leads to

$$\frac{1}{2i}(e^{2i\delta_0} - 1) = P \frac{2x_0}{1+4x_0^2} \tag{107.9}$$

and

$$\frac{1}{2i}(e^{2i\delta_1} - 1) = P \left\{ \frac{1+2x_0^2}{x_0(1+4x_0^2)} - \frac{1}{4x_0^3} \log(1+4x_0^2) \right\}. \tag{107.10}$$

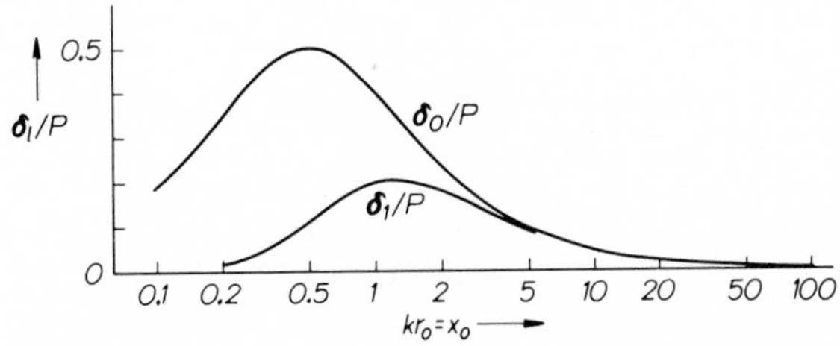

Fig. 58. Born scattering phases δ_0 and δ_1 for an exponential potential, in units of potential size parameter P, drawn over a logarithmic energy scale

These expressions may as well be written in the form $e^{i\delta_l} \sin\delta_l$ and will be equal to δ_l if small. They have been drawn in Fig. 58, except for the factor P. It is clearly apparent that for small values of x_0 the phase $\delta_1 \ll \delta_0$, i.e. P scattering, may be neglected compared with the S wave. This is a general feature at low energies preserved even by so typical a high-energy approximation as Born's. For $x_0 \gg 1$, both δ_0 and δ_1 are of the same order of magnitude so that forward scattering begins to prevail.

To get some insight into the bounds within which the Born approximation holds, we now investigate the phase angle δ_0 to second Born approximation, inclusively, by applying Eq. (96.12a),

$$\tan \delta_0 = -\frac{1}{k} \int_0^\infty dr\, U(r) \sin^2 kr \left\{ 1 - \frac{1}{k} \int_r^\infty dr'\, U(r') \sin 2kr' \right\} \tag{107.11}$$

where $U(r) = \frac{2m}{\hbar^2} V(r)$. With the potential (107.1), the abbreviations (107.3) and the dimensionless variables $x = r/r_0$, $y = r'/r_0$ we get

$$\tan \delta_0 = \frac{P}{x_0} \int_0^\infty dx\, e^{-x} \sin^2(x_0 x) \left\{ 1 + \frac{P}{x_0} \int_x^\infty dy\, e^{-y} \sin 2x_0 y \right\}.$$

The inner integral can easily be evaluated:

$$\int_x^\infty dy\, e^{-y} \sin 2x_0 y = \frac{e^{-x}}{1+4x_0^2} \{\sin 2x_0 x + 2x_0 \cos 2x_0 x\}.$$

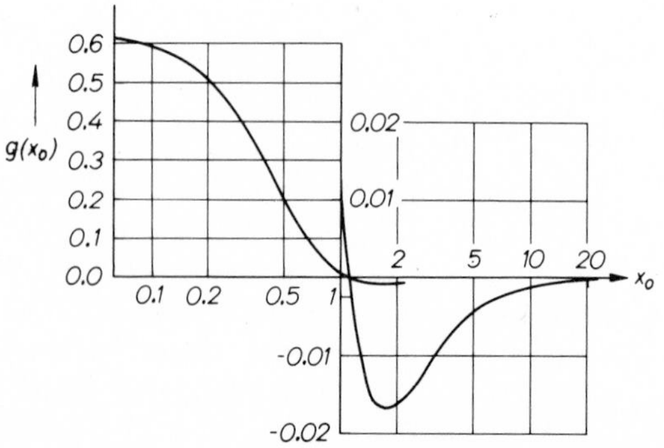

Fig. 59. The function $|Pg|$ must be small, if the first Born approximation shall be good

Complex decomposition of the trigonometric functions then allows the integrand to be broken up into a sum of (complex) exponentials. The evaluation of the integrals is a little cumbersome but quite elementary and leads to the result

$$\tan\delta_0 = P\cdot\frac{2x_0}{1+4x_0^2}\left\{1+P\frac{5-4x_0^2}{8(1+x_0^2)(1+4x_0^2)}\right\}. \tag{107.12}$$

The first term is identical with (107.9). As long as the second term in the curly bracket remains small compared to unity, the Born method will converge fairly well. Its validity, therefore, is bounded by the condition

$$|P\cdot g(x_0)| = \left|P\cdot\frac{5-4x_0^2}{8(1+x_0^2)(1+4x_0^2)}\right| \ll 1.$$

The function $g(x_0)$ has been drawn in Fig. 59 (on two different ordinate scales in the left and right half of the diagram).

Example. Let a potential be given with the size parameter $P=3$, and an approximation within an error of 2% in the amplitude (4% in scattered intensity) be sufficient. Then the condition of validity becomes $|Pg(x_0)|<0.02$ or $|g(x_0)|<0.0067$. This value will be reached at $x_0=4$ which then determines the smallest energy at which the first Born approximation may be used. The phase angle becomes $\delta_0\simeq 21°$ at this energy as can be derived from Fig. 58.

Problem 108. Born scattering by a charge distribution of spherical symmetry

To express the differential scattering cross section for electrons scattered at a nucleus in first Born approximation by an integral over the electrical density $\rho(r)$ of the spherically symmetrical charge distribution. The result may be applied to a nucleus of constant charge density.

Solution. The Poisson equation of classical electrostatics, connecting $V(r)$ with $\rho(r)$, may be written

$$\nabla^2 V \equiv \frac{1}{r}\frac{d^2}{dr^2}(rV) = 4\pi e\rho(r) \tag{108.1}$$

if the scattered particle is an electron of charge $-e$. The charge density is positive and normalized so that

$$4\pi\int_0^\infty dr\, r^2\rho(r) = Ze. \tag{108.2}$$

The scattering amplitude $f(\vartheta)$, in first Born approximation, is

$$f(\vartheta) = -\frac{2m}{\hbar^2}\int_0^\infty dr\, r^2 V(r)\frac{\sin Kr}{Kr}; \qquad K = 2k\sin\frac{\vartheta}{2}. \tag{108.3}$$

It should be noted that, outside the nucleus, $V(r) = -Ze^2/r$ and the integral does not converge at its upper limit, but becomes indefinite. This difficulty may easily be removed, as was first shown by Wentzel[20], if a convergence-generating factor $e^{-\alpha r}$ is added to the integrand and α put $=0$ in the final result. Screening by the atomic electrons may give physical justification to this mathematically somewhat dubious procedure.

Performing two consecutive integrations in part we find

$$\int_0^\infty dr(rV)\,e^{iKr} = \frac{1}{iK}\left[e^{iKr}\left(rV - \frac{1}{iK}(rV)'\right)\right]_0^\infty - \frac{1}{K^2}\int_0^\infty dr(rV)''\,e^{iKr}.$$

At the upper limit, the integrated term will vanish due to Wentzel's trick. At the lower limit, in the vicinity of $r=0$, the function $V(r)$ must be of the form $V = V_0 - V_1 r^2 + \cdots$ so that $rV \to 0$ and $(rV)' \to V_0$. Hence, the integrated term becomes $-V_0/K^2$. The same obtains for the complex conjugate to be subtracted in the sine integral (108.3) so that no contribution comes from the integrated term. In the remaining integral $(rV)''$ may be replaced according to (108.1) thus yielding

$$f(\vartheta) = \frac{2m}{\hbar^2}\cdot\frac{4\pi e}{K^3}\int_0^\infty dr\, r\rho(r)\sin Kr. \tag{108.4}$$

This integral will no longer cause any convergence difficulties.

If the nucleus can be treated as a point charge, only the vicinity of $r=0$ will contribute so that (108.2) may be applied directly and $f(\vartheta)$ becomes

$$f_0(\vartheta) = \frac{2mZe^2}{\hbar^2K^2} = \frac{mZe^2}{2\hbar^2k^2\sin^2\dfrac{\vartheta}{2}}. \tag{108.5}$$

With $\hbar k = mv$, $\frac{1}{2}mv^2 = E$ this may be written

$$f_0(\vartheta) = \frac{Ze^2}{4E\sin^2\dfrac{\vartheta}{2}}$$

---

[20] Wentzel, G.: Z. Physik **40**, 590 (1927).

and leads to the well-known Rutherford scattering formula

$$\frac{d\sigma_0}{d\Omega} = |f_0(\vartheta)|^2 = \left(\frac{Ze^2}{4E}\right)^2 \frac{1}{\sin^4 \frac{\vartheta}{2}}. \tag{108.6}$$

Eq. (108.4) may be written

$$f(\vartheta) = f_0(\vartheta) F(K) \tag{108.7}$$

with a "*form factor*"

$$F(K) = \frac{4\pi}{Ze} \int_0^\infty dr\, r^2 \rho(r) \frac{\sin Kr}{Kr} \tag{108.8}$$

determining the deviation of the scattering amplitude $f(\vartheta)$ from its Rutherford value:

$$\frac{d\sigma}{d\sigma_0} = F(K)^2. \tag{108.9}$$

An obvious example would be a nucleus of constant charge density within its radius R. The normalization (108.2) then becomes

$$\frac{4\pi}{3} \rho R^3 = Ze$$

and the form factor

$$F = \frac{3}{R^3} \int_0^R dr\, r^2 \frac{\sin Kr}{Kr} = \frac{3}{(KR)^3} \{\sin KR - KR \cos KR\} \tag{108.10}$$

or

$$F = \frac{3 j_1\left(2kR \sin \frac{\vartheta}{2}\right)}{4k^2 R^2 \sin^2 \frac{\vartheta}{2}}. \tag{108.11}$$

In the angle interval $0 \leq \vartheta \leq \pi$ the argument of the spherical Bessel function will vary from 0 to $2kR$ which, since the Born approximation should be applied only if $kR \gg 1$, is a rather large interval in which the Bessel function will have several zeros. Instead of the Rutherford scattering intensity, falling off monotonously with increasing angle, we shall therefore find a sequence of diffraction maxima, in a manner well known from similar patterns in classical optics. The number of maxima (if all are resolved) allows for a rough determination of the nuclear radius.

NB. If the potential energy of a neutron in the field of another nucleon is roughly determined by a Yukawa attraction,

$$V(r) = -\frac{g^2}{r} e^{-\varkappa r}, \tag{108.12}$$

a differential equation similar to that of Poisson may be set up for the interaction of a neutron with a nucleus of particle density ρ:

$$\frac{d^2}{dr^2}(rV) - \varkappa^2 (rV) = 4\pi g^2 \rho(r) r. \tag{108.13}$$

Here the normalization

$$4\pi \int_0^\infty dr\, r^2 \rho(r) = A \tag{108.14}$$

will hold with A the atomic number of the nucleus. The partial integration then leads to

$$\int_0^\infty dr (rV) \sin Kr = -\frac{1}{K^2} \int_0^\infty dr (rV)'' \sin Kr,$$

even without Wentzel's trick, or, if use is made of (108.13),

$$\int_0^\infty dr (rV) \sin Kr = -\frac{1}{K^2} \int_0^\infty dr \{4\pi g^2 \rho r + \varkappa^2 (rV)\} \sin Kr$$

leading to

$$\int_0^\infty dr (rV) \sin Kr = -\frac{4\pi g^2}{K^2 + \varkappa^2} \int_0^\infty dr\, r\rho(r) \sin Kr.$$

Thus we find for a point nucleus,

$$f_0(\vartheta) = \frac{2mg^2 A}{\hbar^2 (K^2 + \varkappa^2)} \tag{108.15}$$

and for an extended nucleus,

$$f(\vartheta) = f_0(\vartheta) F(K) \tag{108.16}$$

with the form factor

$$F = \frac{4\pi}{A} \int_0^\infty dr\, r^2 \rho(r) \frac{\sin Kr}{Kr} \tag{108.17}$$

essentially the same as above.

Problem 109. Hard sphere: High energy scattering

To show that the scattering cross section of a hard sphere of radius R tends towards $2\pi R^2$ for very high energies.

Solution. The Born approximation, usually to be applied to high-energy scattering, breaks down for all singular potentials where the Born integral diverges. We are therefore forced to apply the partial wave expansion method in spite of the fact that it converges worse and worse at high energies.

For any value of l, the radial wave function outside the hard sphere,

$$\chi_l(r) = j_l(kr)\cos\delta_l - n_l(kr)\sin\delta_l \to \sin\left(kr - \frac{l\pi}{2} + \delta_l\right) \tag{109.1}$$

has to satisfy the boundary condition

$$\chi_l(R) = 0 \tag{109.2}$$

so that

$$\tan\delta_l = \frac{j_l(x)}{n_l(x)}; \qquad x = kR. \tag{109.3}$$

The cross section therefore becomes

$$\sigma = \frac{4\pi}{k^2}\sum_{l=0}^{\infty}(2l+1)\sin^2\delta_l = \frac{4\pi}{k^2}\sum_{l=0}^{\infty}(2l+1)\frac{j_l^2(x)}{j_l^2(x)+n_l^2(x)}. \tag{109.4}$$

If the energy is high enough to make $x \gg 1$, the infinite sum (109.4) can be decomposed into two parts. As long as $l < x$, the sum terms correspond to particles hitting the sphere ($\hbar l < mvR$) and the functions j_l and n_l may be well represented by the asymptotic expressions,

$$j_l(x) = \sin\left(x - \frac{l\pi}{2}\right); \qquad n_l(x) = -\cos\left(x - \frac{l\pi}{2}\right). \tag{109.5}$$

If, on the other hand, $l > x$, the particles would in the classical picture pass by the sphere without hitting it ($\hbar l > mvR$). j_l and n_l might then be represented by the first term of their power expansions. Since these are $j_l \propto x^{l+1}$ and $n_l \propto x^{-l}$, the ratio in (109.4) becomes very small, and these sum terms may be neglected. In this approximation we may therefore write

$$\sigma \simeq \frac{4\pi}{k^2}\sum_{l=0}^{[x]}(2l+1)\sin^2\left(x - \frac{l\pi}{2}\right). \tag{109.6}$$

Of course this approximation does not hold for the terms around $l \simeq x$ where the transition from the one to the other of the two approximations of j_l and n_l is performed. The bigger x, however, and the more terms contributing to the sum (109.6), the smaller must be the effect of these few terms upon the sum and thus the smaller the error of the approximation.

Eq. (109.6) may be recast using the relation

$$\sin^2\left(x - \frac{l\pi}{2}\right) = \sin^2 x + \sin^2 \frac{l\pi}{2} \cos 2x;$$

it then runs

$$\sigma = \frac{4\pi}{k^2}\left\{\sin^2 x \sum_{l=0}^{[x]} (2l+1) + \cos 2x \sum_{l=1,3,5,\dots}^{[x]} (2l+1)\right\}.$$

The summations can be performed in an elementary way,

$$\sum_{l=0}^{[x]}(2l+1) = (x+1)^2; \qquad \sum_{l=1,3,5,\dots}^{[x]} (2l+1) = \tfrac{1}{2}(x+1)(x+2)$$

which finally leads to

$$\sigma = \frac{4\pi}{k^2}\left\{-(x+1)\sin^2 x + \tfrac{1}{2}(x+1)(x+2)\right\}. \tag{109.7}$$

Since this approximation will hold only for $x \gg 1$, it is sufficient to take the largest term in the bracket. That yields finally,

$$\sigma = \frac{4\pi}{k^2} \cdot \frac{1}{2} x^2 = 2\pi R^2, \tag{109.8}$$

i.e. the formula to be proved.

NB. This is a well-known result in classical optics and has an obvious explanation: all the particles hitting the geometrical cross section πR^2 form the scattered wave. Since they are then lacking in the primary beam behind the obstacle (shadow effect), the same cross section again obtains for this deficiency.—Note that the condition $x \gg 1$ of validity of this simple picture is identical with the wavelength being large compared to the radius of the obstacle.

Problem 110. Rutherford scattering formula

To solve the problem of scattering in the Coulomb field of two point charges by using parabolic coordinates.

Solution. Let us write the Schrödinger equation

$$\nabla^2 u + \left(k^2 - \frac{2k\varkappa}{r}\right) u = 0 \tag{110.1}$$

with the standard abbreviations

$$k = \frac{mv}{\hbar}; \qquad \varkappa = \frac{e_1 e_2}{\hbar v}; \qquad 2k\varkappa = \frac{2m}{\hbar^2} e_1 e_2 \tag{110.2}$$

where e_1 and e_2 are the two point charges, m is the mass of the scattered particles and v their velocity at infinity. The problem is of rotational symmetry about the z axis so that the solution depends on coordinates in a meridional plane only, as r and ϑ, or, using parabolic coordinates, on

$$\xi = r - z = 2r\sin^2\frac{\vartheta}{2}; \qquad \eta = r + z = 2r\cos^2\frac{\vartheta}{2}. \tag{110.3}$$

That to use these coordinates will be reasonable can easily be seen from the well-known result. Except for phase factors, the asymptotic solution must, according to Rutherford's law, become

$$u \to e^{ikz} + \frac{C}{\sin^2\frac{\vartheta}{2}}\,\frac{e^{ikr}}{r} = e^{ikz}\left\{1 + \frac{2C}{\xi}\,e^{ik\xi}\right\}$$

so that there should be, at least, some hope of reducing the problem by factorizing the wave function into

$$u = e^{ikz}\,v(\xi) \tag{110.4}$$

to a function of only one variable ξ.

Putting (110.4) into the Schrödinger equation (110.1) without any special assumption about the factor v we arrive at

$$\nabla^2 v + 2ik\frac{\partial v}{\partial z} - \frac{2k\varkappa}{r}\,v = 0.$$

From the definitions (110.3) one gets (omitting derivatives with respect to η and φ),

$$\nabla^2 v = \frac{4}{\xi+\eta}\,\frac{d}{d\xi}\left(\xi\frac{dv}{d\xi}\right); \qquad \frac{\partial v}{\partial z} = -\frac{2\xi}{\xi+\eta}\,\frac{dv}{d\xi}; \qquad r = \tfrac{1}{2}\,(\xi+\eta),$$

thus transforming the differential equation into

$$\xi\frac{d^2 v}{d\xi^2} + (1 - ik\xi)\frac{dv}{d\xi} - k\varkappa v = 0. \tag{110.5}$$

This is a Kummer equation whose solutions are the confluent hypergeometric functions, the solution regular at the origin $\xi=0$ being in arbitrary normalization

$$v = C\,{}_1F_1(-i\varkappa, 1; ik\xi). \tag{110.6}$$

The scattering problem will now be solved by

$$u = C\,e^{ikz}\,{}_1F_1(-i\varkappa, 1; ik\xi) \tag{110.7}$$

with a suitable choice of C. To prove this, we need the asymptotics of (110.7) for large values of the positive imaginary argument $ik\xi$. This implies a little mathematical difficulty. The confluent series has a branch cut from 0 to ∞, generally put along the positive imaginary axis, so that the well-known asymptotic expansion

$$
{}_1F_1(a,c;z)\to e^{-i\pi a}\frac{\Gamma(c)}{\Gamma(c-a)}z^{-a}+\frac{\Gamma(c)}{\Gamma(a)}e^z z^{a-c}
$$

$$
\text{if } |z|\gg a;\quad |z|\gg c;\quad c\neq -n,\quad n=0,1,2,\ldots \tag{110.8}
$$

cannot be used there without caution. On the other hand, we may apply (110.8) without any caution to the complex conjugate, u^*, where it renders

$$
{}_1F_1(i\varkappa,1;-ik\xi)\to\frac{e^{\pi\varkappa}}{\Gamma(1-i\varkappa)}(-ik\xi)^{-i\varkappa}+\frac{1}{\Gamma(i\varkappa)}e^{-ik\xi}(-ik\xi)^{i\varkappa-1}.
$$

Using such identities as

$$
(-ik\xi)^{-i\varkappa}=(e^{-\frac{i\pi}{2}}k\xi)^{-i\varkappa}=e^{-\frac{\pi\varkappa}{2}}e^{-i\varkappa\log k\xi}
$$

and

$$
\Gamma(1+i\varkappa)=i\varkappa\Gamma(i\varkappa),
$$

we then arrive at

$$
{}_1F_1(i\varkappa,1;-ik\xi)\to\frac{e^{\frac{\pi\varkappa}{2}}}{\Gamma(1-i\varkappa)}\left\{e^{-i\varkappa\log k\xi}-\varkappa\frac{\Gamma(1-i\varkappa)}{\Gamma(1+i\varkappa)}e^{-ik\xi}\frac{e^{i\varkappa\log k\xi}}{k\xi}\right\}. \tag{110.9}
$$

Taking again the complex conjugate of this formula, putting it into (110.7), making use of $z+\xi=r$, and finally choosing

$$
C=e^{-\frac{\pi\varkappa}{2}}\Gamma(1+i\varkappa) \tag{110.10}
$$

we find the asymptotic behaviour of the wave function (110.7):

$$
u\to e^{i\left[kz+\varkappa\log\left(2kr\sin^2\frac{\vartheta}{2}\right)\right]}-\frac{\varkappa e^{2i\eta_0}}{2k\sin^2\frac{\vartheta}{2}}\cdot\frac{e^{i\left[kr-\varkappa\log\left(2kr\sin^2\frac{\vartheta}{2}\right)\right]}}{r} \tag{110.11}
$$

with

$$
\eta_0=\arg\Gamma(1+i\varkappa). \tag{110.12}
$$

This, indeed, is the solution to the scattering problem in standard normalization, the only difference from other similar problems being the logarithmic phases characteristic for the Coulomb potential. From the amplitude of the outgoing spherical wave we get the differential scattering cross section

$$\frac{d\sigma}{d\Omega} = |f(\vartheta)|^2 = \left(\frac{\varkappa}{2k \sin^2 \frac{\vartheta}{2}} \right)^2 = \left(\frac{e_1 e_2}{4E} \right)^2 \cdot \frac{1}{\sin^4 \frac{\vartheta}{2}} . \tag{110.13}$$

This is the Rutherford law.

NB 1. At $r=0$, we gather from (110.7) that $u=C$ or, according to (110.10), that

$$|u(0)|^2 = e^{-\pi\varkappa} |\Gamma(1+i\varkappa)|^2 .$$

According to an elementary property of the Γ function we may write instead,

$$|u(0)|^2 = e^{-\pi\varkappa} \cdot \frac{\pi\varkappa}{\sinh \pi\varkappa} = \frac{2\pi\varkappa}{e^{2\pi\varkappa}-1} . \tag{110.14}$$

For positive $\varkappa$ this is always smaller than 1, and for $2\pi\varkappa \gg 1$ it becomes exponentially very small. Since

$$2\pi\varkappa = \frac{2\pi e_1 e_2}{\hbar v}$$

this means that, for strong repulsion and low energy, the wave will only penetrate the potential barrier at the origin with a rapidly decreasing amplitude (Gamow factor).

NB 2. The asymptotic solution (110.11) has been derived under the assumption of large $k\xi$, say, $k\xi \geq c$ with $c \gg 1$, not of large $kr \gg c$. It is valid, therefore, not outside the sphere $r=c/k$ but outside the paraboloid

$$2kr \sin \frac{\vartheta}{2} = c \quad \text{or} \quad r - z = c/k . \tag{110.15}$$

It may thus be wrong, even at very large r, if ϑ is very small. This restriction, however, is not very important. The wavelength will in most cases not be larger than the atomic dimensions, say, $1/k = 10^{-7}$ cm, and the distance of the counter detecting the scattered particles, say, at least $r = 10$ cm. This leads to $kr = 10^8$. Let c be about 10^3, then (110.15) leads to $\vartheta = 10^{-5}$ for any deviations, i.e. to an angle much too small to permit separating the scattered from the primary beam.

Problem 111. Partial wave expansion for the Coulomb field

To expand the Rutherford solution of the preceding problem into partial waves.

Solution. The Coulomb potential depending upon r only, the wave function may as well be constructed by factorization in spherical polar coordinates in the form

$$u = \frac{1}{r} \sum_{l=0}^{\infty} C_l \, \chi_l(r) \, P_l(\cos \vartheta) \tag{111.1}$$

with χ_l being the solution of the differential equation

$$\chi_l'' + \left[k^2 - \frac{2k\varkappa}{r} - \frac{l(l+1)}{r^2}\right]\chi_l = 0 \tag{111.2}$$

satisfying the boundary condition

$$\chi_l(0) = 0 \tag{111.3}$$

in a suitable normalization. If we put

$$\chi_l = (2kr)^{l+1}\, e^{ikr}\, F(\rho); \quad \rho = -2ikr \tag{111.4}$$

we get, instead of (111.2), the Kummer equation

$$\rho F'' + (2l+2-\rho)\, F' - (l+1+i\varkappa)\, F = 0 \tag{111.5}$$

whose regular solution is

$$F(\rho) = a_l\; {}_1F_1(l+1+i\varkappa\,, 2l+2;\, -2ikr)\,, \tag{111.6}$$

a_l being a constant still suitably to be chosen. The asymptotic behaviour of (111.6) may be directly determined from Eq. (110.8) of the preceding problem; using the abbreviation (generalising Eq. (110.12) for η_0)

$$\eta_l = \arg\Gamma(l+1+i\varkappa) \tag{111.7}$$

we find

$$F(\rho) \to a_l (2kr)^{-l-1} \frac{e^{\frac{\pi\varkappa}{2}}(2l+1)!}{\Gamma(l+1-i\varkappa)} \left\{ e^{-\frac{i\pi}{2}(l+1)}\, e^{-i\varkappa\log 2kr} + e^{\frac{i\pi}{2}(l+1)}\, e^{i\varkappa\log 2kr - 2i\eta_l - 2ikr} \right\}.$$

Setting this into (111.4) and choosing

$$a_l = e^{-\frac{\pi\varkappa}{2}} \frac{|\Gamma(l+1+i\varkappa)|}{2(2l+1)!} \tag{111.8}$$

we find

$$\chi_l(r) = a_l (2kr)^{l+1}\, e^{ikr}\; {}_1F_1(l+1+i\varkappa, 2l+2;\, -2ikr) \tag{111.9}$$

to have the asymptotic form

$$\chi_l \to \sin\left(kr - \varkappa\log 2kr + \eta_l - \frac{l\pi}{2}\right). \tag{111.10}$$

Our main problem now will become the expansion of the wave function u, determined in the preceding problem, into the series (111.1) of partial waves with χ_l as given by Eqs. (111.8–10). This wave function, rewritten in polar coordinates, runs as follows:

$$u(r,\vartheta)=e^{-\frac{\pi\varkappa}{2}}\,\Gamma(1+i\varkappa)\,e^{ikr\cos\vartheta}\,{}_1F_1(-i\varkappa,1;ikr(1-\cos\vartheta))\,. \quad (111.11)$$

By inverting (111.1) we obtain

$$C_l\frac{\chi_l(r)}{r}=\frac{2l+1}{2}\int_{-1}^{+1}d\cos\vartheta\,u(r,\vartheta)\,P_l(\cos\vartheta)\,. \quad (111.12)$$

Introducing into (111.11) the variable

$$x=-ik\xi=-ikr(1-\cos\vartheta)\,, \quad (111.13)$$

writing

$$u=e^{-\frac{\pi\varkappa}{2}}\,\Gamma(1+i\varkappa)\,e^{ikr}\,e^{x}\,{}_1F_1(-i\varkappa,1;\,-x)$$

which, using the identity

$$e^{x}\,{}_1F_1(a,c;\,-x)={}_1F_1(c-a,c;x)\,,$$

may be transformed into

$$u=e^{-\frac{\pi\varkappa}{2}}\,\Gamma(1+i\varkappa)\,e^{ikr}\,{}_1F_1(1+i\varkappa,1;x)$$

and putting this into (111.12), we get

$$C_l\frac{\chi_l(r)}{r}=\frac{2l+1}{2ik}\,e^{-\frac{\pi\varkappa}{2}}\,\Gamma(1+i\varkappa)\frac{e^{ikr}}{r}\,J_l \quad (111.14)$$

with

$$J_l=\int_{-2ikr}^{0}dx\,{}_1F_1(1+i\varkappa,1;x)\,P_l\left(1+\frac{x}{ikr}\right). \quad (111.15)$$

The problem is thus essentially reduced to evaluating the integrals J_l, especially under the assumption $kr\gg 1$. This can be performed by repeated partial integrations, making use of the relation[21]

$${}_1F_1(a,1;x)=\frac{d}{dx}\left[{}_1F_1(a,1;x)-{}_1F_1(a-1,1;x)\right]. \quad (111.16)$$

---

[21] This follows for $c=1$ from the general formula

$$\frac{d}{dx}\left[{}_1F_1(a,c;x)-{}_1F_1(a-1,c;x)\right]={}_1F_1(a,c;x)+\frac{1-c}{c}\,{}_1F_1(a,c+1;x)$$

which may easily be derived from the relations listed in the Appendix, p. 273f. of vol. II.

In each integration step, P_l then undergoes one differentiation with respect to x:

$$\frac{d^n P_l}{dx^n} = (ikr)^{-n} P_l^{(n)}$$

where $P_l^{(n)}$ denotes the n-th derivative of P_l with respect to its argument. Since it is a polynomial of l-th degree, it will be reduced to zero after the l-th step, thus putting an end to the integration procedure. The details of this calculation are given in the appendix to this problem. The result is, for $kr \gg 1$,

$$J_l \to -\sum_{n=0}^{l} \left(\frac{i}{kr}\right)^n P_l^{(n)}(-1) \left\{ \frac{1}{\Gamma(1+i\varkappa)} e^{-2ikr} (-2ikr)^{i\varkappa} + (-1)^{n+1} \frac{e^{-i\pi(i\varkappa - n)}}{\Gamma(1+n-i\varkappa)} (-2ikr)^{n-i\varkappa} \right\}. \tag{111.17}$$

In the curly bracket, only for $n=0$ both terms have the same order of magnitude, whereas for all $n \geqq 1$ only the second term contributes. With some elementary reshaping, using Eq. (111.22) of the appendix, we then find

$$J_l \to (-1)^{l+1} \frac{e^{\frac{\pi\varkappa}{2}}}{\Gamma(1+i\varkappa)} \left\{ e^{-2ikr + i\varkappa \log 2kr} - \left[\sum_{n=0}^{l} (-1)^n \frac{(l+n)!}{n!(l-n)!} \frac{\Gamma(1+i\varkappa)}{\Gamma(1+n-i\varkappa)} \right] e^{-i\varkappa \log 2kr} \right\}.$$

It can be shown that the coefficient in the square bracket is $e^{2i\left(\eta_l - \frac{l\pi}{2}\right)}$ so that we finally have

$$J_l \to (-1)^l \frac{e^{\frac{\pi\varkappa}{2}}}{\Gamma(1+i\varkappa)} 2i\, e^{i\left(\eta_l - \frac{l\pi}{2}\right)} e^{-ikr} \sin\left(kr - \varkappa \log 2kr + \eta_l - \frac{l\pi}{2}\right). \tag{111.18}$$

Setting this into Eq. (111.14) and using the asymptotic expression (111.10) for χ_l we obtain

$$C_l = \frac{1}{k} (2l+1)\, i^l\, e^{i\eta_l} \tag{111.19}$$

so that the expansion wanted becomes for large kr

$$u \to \frac{1}{kr} \sum_{l=0}^{\infty} (2l+1)\, i^l\, e^{i\eta_l} \sin\left(kr - \varkappa \log 2kr + \eta_l - \frac{l\pi}{2}\right) P_l(\cos\vartheta). \tag{111.20}$$

It should be noted that, in a sense, the wave function (111.11) and its expansion stand in strong analogy to the plane wave and its expansion. Omitting the Coulomb field effect by putting $\varkappa=0$, Eq. (111.7) gives $\eta_l=0$, thus transforming χ_l, Eq. (111.9), into a spherical Bessel function, the wave function u into the plane wave and the series (111.20) into its partial wave expansion.

Appendix. Let us write F_a instead of ${}_1F_1(a,1;x)$. Then, with $a=i\varkappa$, partial integration leads to

$$\int dx\, P_l F_{a+1} = \int dx\, P_l \frac{d}{dx}(F_{a+1}-F_a) = P_l(F_{a+1}-F_a) + \frac{i}{kr}\int dx\, P_l'(F_{a+1}-F_a)$$

and its being l times repeated to

$$\int dx\, P_l F_{a+1} = \sum_{n=0}^{l}\left(\frac{i}{kr}\right)^n P_l^{(n)}\left\{F_{a+1}-\binom{n+1}{1}F_a+\binom{n+1}{2}F_{a-1}\mp\cdots+(-1)^{n+1}F_{a-n}\right\}.$$

At the upper integration bound, $x=0$, all the $F_\nu=1$ and cancel in the curly bracket. Therefore there will only remain a contribution of the lower bound, $x=-2ikr$, where the spherical harmonics have the argument -1:

$$\left.\begin{aligned} J_l &= \sum_{n=0}^{l}\left(\frac{i}{kr}\right)^n P_l^{(n)}(-1)\sum_{\nu=-1}^{n}(-1)^\nu\binom{n+1}{\nu+1}F_{a-\nu} \\ \text{with} \qquad F_{a-\nu} &= {}_1F_1(i\varkappa-\nu,1;-2ikr)\,. \end{aligned}\right\} \qquad (111.21)$$

For $kr \gg 1$, there holds the asymptotic expression

$$F_{a-\nu} \to \frac{e^{-i\pi(i\varkappa-\nu)}}{\Gamma(1+\nu-i\varkappa)}(-2ikr)^{\nu-i\varkappa} + \frac{e^{-2ikr}}{\Gamma(-\nu+i\varkappa)}(-2ikr)^{-\nu-1+i\varkappa}$$

with its first term being proportional to $(kr)^\nu$, its second to $(kr)^{-\nu-1}$. In the sum $-1\leq\nu\leq n$ in (111.21), therefore, only $\nu=n$ remains of the first, and only $\nu=-1$ of the second term thus leading to Eq. (111.17), above.

The derivatives of the Legendre polynomials follow easily from

$$P_l(\cos\vartheta) = {}_2F_1(l+1,-l,1;t) = \sum_{m=0}^{l}(-1)^m\frac{(l+m)!}{(l-m)!\,m!^2}t^m \text{ with } t=\frac{1}{2}(1-\cos\vartheta)\,.$$

Differentiating this repeatedly with respect to t and then putting $t=0$ or $\cos\vartheta=1$, we find

$$\frac{d^n P_l(1)}{dt^n} = n!(-1)^n\frac{(l+n)!}{(l-n)!\,n!^2}$$

and therefore

$$P_l^{(n)}(1) = \frac{d^n P_l(1)}{(d\cos\vartheta)^n} = (-1)^{l+n}\frac{(l+n)!}{2^n(l-n)!\,n!},$$

thence

$$P_l^{(n)}(-1) = \frac{(l+n)!}{2^n(l-n)!\,n!}\,. \qquad (111.22)$$

Problem 112. Anomalous scattering

Let a short-range force be added to the Coulomb field of the preceding problems so that $kR \ll 1$ if R is the range of this force (e.g. nuclear forces or finite extension of a charge cloud). The scattering amplitude shall be expressed in terms of additional phase shifts.

Solution. If $kR \ll 1$, only the phase for $l=0$ will undergo an alteration in consequence of the additional short-range force so that, instead of Eq. (111.10) in the preceding problem, we now get the asymptotic expression

$$\tilde{\chi}_0 \to A \sin(kr - \varkappa \log 2kr + \eta_0 + \delta_0) \tag{112.1}$$

with an additional phase shift δ_0, to be compared with the original

$$\chi_0 \to \sin(kr - \varkappa \log 2kr + \eta_0) \tag{112.2}$$

for the point-charge field. Their difference,

$$\tilde{\chi}_0 - \chi_0 \to \frac{1}{2i}\{e^{i(kr - \varkappa \log 2kr + \eta_0)}(A\,e^{i\delta_0} - 1) - e^{-i(kr - \varkappa \log 2kr + \eta_0)}(A\,e^{-i\delta_0} - 1)\},$$

should then contribute only an *outgoing* spherical wave, hence

$$A = e^{i\delta_0} \tag{112.3}$$

and

$$\tilde{\chi}_0 - \chi_0 \to \frac{1}{2i}(e^{2i\delta_0} - 1)\,e^{i(kr - \varkappa \log 2kr + \eta_0)}\,. \tag{112.4}$$

Eq. (111.20) then has to be supplemented by a term

$$\frac{1}{kr}\,e^{i\eta_0}(\tilde{\chi}_0 - \chi_0)$$

so that, if (111.20) is replaced by (110.11), we get the scattered wave

$$u_s \to \frac{e^{i(kr - \varkappa \log 2kr + 2\eta_0)}}{2kr}\left\{-\frac{\varkappa}{\sin^2\frac{\vartheta}{2}}\,e^{-i\varkappa \log \sin^2\frac{\vartheta}{2}} - i(e^{2i\delta_0} - 1)\right\} \tag{112.5}$$

and the scattering amplitude

$$f(\vartheta) = -\frac{\varkappa}{2k\sin^2\frac{\vartheta}{2}}\,e^{-i\varkappa \log \sin^2\frac{\vartheta}{2}} + \frac{1}{2ik}(e^{2i\delta_0} - 1)\,. \tag{112.6}$$

It should be remarked (1) that the additional term, not depending upon the angle of deflection, will have its strongest influence at large scattering angles where the Rutherford term becomes small, and (2) that in cross section measurements $|f(\vartheta)|^2$ will show an interference term of Rutherford and anomalous scattering.

Problem 113. Sommerfeld-Watson transform

Let the scattering amplitude

$$f(\vartheta)=\sum_{l=0}^{\infty}(2l+1)\,f_l\,P_l(\cos\vartheta) \tag{113.1}$$

be transformed into an integral in the plane of the complex variable l by assuming each sum term in (113.1) to be the residue at an integer l. Under the assumption of $f_l \equiv f(l)$ to be a meromorphic function of l, the integral shall then be decomposed into pole contributions and a background integral, taken along a path parallel to the imaginary l axis through the point $l=-\frac{1}{2}$.

Solution. It is reasonable to replace l by the complex variable

$$\lambda = l+\tfrac{1}{2}\,. \tag{113.2}$$

Eq. (113.1) may then be replaced by

$$f(\vartheta)=\frac{1}{\pi i}\int_C d\lambda\,\lambda\,f(\lambda-\tfrac{1}{2})\,\frac{P_{\lambda-\frac{1}{2}}(-\cos\vartheta)}{\cos\pi\lambda} \tag{113.3}$$

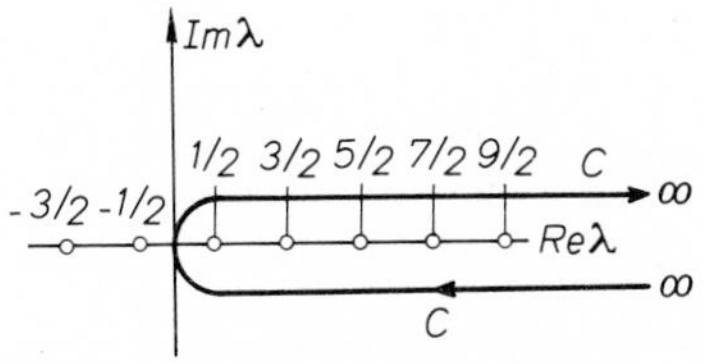

Fig. 60. Integration path which may either be replaced by loop integrals about the poles on the positive real axis or by a background integral along the imaginary λ axis plus Regge pole loops in the right half plane

where the integration path C is shown in Fig. 60. This can easily be proved. The function

$$\frac{1}{\cos\pi\lambda}=-\frac{1}{\sin\pi l}$$

has poles at all integer real points $l=0, \pm 1, \pm 2, \ldots$ or at $\lambda=\pm\frac{1}{2}, \pm\frac{3}{2}, \pm\frac{5}{2}, \ldots$; their residues are $(-1)^{l+1}$. If the other factors of the integrand in (113.3) are regular on the positive real axis, the integral over C may be decomposed into a sum of clockwise $(-2\pi i)$ loop integrals about the poles:

$$f(\vartheta) = \frac{1}{\pi i} \sum_{l=0}^{\infty} (-2\pi i)(-1)^{l+1}(l+\tfrac{1}{2})\, f(l)\, P_l(-\cos\vartheta). \qquad (113.4)$$

The parity property of the Legendre polynomials,

$$(-1)^l\, P_l(-\cos\vartheta) = P_l(\cos\vartheta) \qquad (l \text{ integer}) \qquad (113.5)$$

then reduces (113.4) to (113.1).

Let us now deform the integration path until it consists of the imaginary λ axis and an infinitely large semicircle in the right half of the λ plane. The latter does not contribute to the integral if the integrand falls off at least as $|\lambda|^{-1-\varepsilon}$ on the semicircle, as we shall suppose. Then there remains only the background integral along the imaginary λ axis, and the loop integrals around all the poles in the right half-plane the path sweeps over under deformation. Let these poles lie at the points $\lambda=\alpha_n$, and let the residues of $f(l)$ at these poles be β_n, then we get the pole contribution to (113.3):

$$f_P(\vartheta) = 2 \sum_n \beta_n \frac{\alpha_n P_{\alpha_n - \frac{1}{2}}(-\cos\vartheta)}{\cos\pi\alpha_n}. \qquad (113.6)$$

There remains the background integral, with $\lambda = iy$ and real y,

$$f_B(\vartheta) = -\frac{1}{\pi i} \int_{-\infty}^{+\infty} dy\, y\, f(iy - \tfrac{1}{2}) \frac{P_{iy-\frac{1}{2}}(-\cos\vartheta)}{\cosh\pi y}. \qquad (113.7)$$

The scattering amplitude $f(\vartheta)$ then consists of the sum of pole contributions and background contribution,

$$f(\vartheta) = f_P(\vartheta) + f_B(\vartheta). \qquad (113.8)$$

NB. A pole of $f(l)$ in the l plane is called a *Regge pole*. Its path through the l plane, if the energy varies, is called a *Regge trajectory*. The importance of Regge poles is that they offer an alternative to describing the interaction between two colliding particles by a potential. If the positions (α_n) and residues (β_n) of the poles are known, it may be possible to describe scattering by these parameters. This method seems to be of special advantage in elementary particle physics (where, of course, energies are high and the theory has to be rewritten in relativistic form), because the main features revealed by experiment there are rather sharp resonance states (intermediary particles or particle resonances). In nuclear physics the unrelativistic theory may be used to describe compound states.

Problem 114. Regge pole

Let a pole of $f(l)$ lie close to a physical value of l, say $l=L$ with integer L, for a given energy. What physical situation does the pole describe if the energy varies a little?

Solution. We shall consider only the contribution of one term in the sum (113.6) to the scattering amplitude which, omitting the subscript n, becomes

$$f_P(\vartheta)=2\beta\alpha\,\frac{P_{\alpha-\frac{1}{2}}(-\cos\vartheta)}{\cos\pi\alpha}. \tag{114.1}$$

Here,

$$\alpha=L+\tfrac{1}{2}+\xi+i\eta \tag{114.2}$$

with ξ and η real numbers, both supposed to be small ($|\xi|\ll 1, |\eta|\ll 1$). Then,

$$\cos\pi\alpha=(-1)^L\sin\pi(\xi+i\eta). \tag{114.3}$$

The Legendre function of complex index $\nu=\alpha-\frac{1}{2}$ can be expressed by the usual Legendre polynomials of integer indices:

$$P_\nu(t)=\frac{\sin\pi\nu}{\pi}\sum_{n=0}^{\infty}(2n+1)\frac{P_n(-t)}{(\nu-n)(\nu+n+1)} \tag{114.4}$$

so that, in our special case, we get

$$P_{\alpha-\frac{1}{2}}(-\cos\vartheta)=\frac{(-1)^L\sin\pi(\xi+i\eta)}{\pi}\sum_{n=0}^{\infty}\frac{(2n+1)P_n(\cos\vartheta)}{(L-n+\xi+i\eta)(L+n+1+\xi+i\eta)}. \tag{114.5}$$

This leads to

$$f_P(\vartheta)=\frac{2\beta}{\pi}\left(L+\frac{1}{2}+\xi+i\eta\right)\sum_{n=0}^{\infty}\frac{(2n+1)P_n(\cos\vartheta)}{(L-n+\xi+i\eta)(L+n+1+\xi+i\eta)}. \tag{114.6}$$

So far, nothing has been neglected for this pole contribution. Making use now of the smallness of ξ and η, the first factor in the denominator will become very small for $n=L$, thus making this term much bigger than the rest. We then may confine our attention, in reasonable approximation, to this main term and arrive at

$$f_P(\vartheta)\simeq\frac{\beta}{\pi}(2L+1)\frac{P_L(\cos\vartheta)}{\xi+i\eta}. \tag{114.7}$$

Let us now study the dependence of this expression upon the energy E of the scattered particle. The pole does not correspond to a state of integer momentum, but it is of course situated at α for a real energy E. On the other hand, a physical state of integer angular momentum L (or, $\lambda = L+\frac{1}{2}$) is supposed to lie close to the pole; its energy then can have only a very small imaginary part, its real part E_0 differing very little from E, say,

$$E_L = E_0 - \tfrac{1}{2} i\Gamma . \tag{114.8}$$

According to Eq. (114.2), this means for ξ and η as functions of the energy that

$$\xi(E_L) + i\eta(E_L) = 0 . \tag{114.9}$$

Let E be any real energy of a physical state, not far from E_L, then we have

$$\xi(E) + i\eta(E) = \left(\frac{d\alpha}{dE}\right)_{E_L} (E - E_L). \tag{114.10}$$

Putting this expression into (114.7) we get

$$f_P(\vartheta) = \frac{\beta}{\pi} \frac{2L+1}{(d\alpha/dE)_{E_L}} P_L(\cos\vartheta) \frac{1}{E - E_0 + \frac{1}{2} i\Gamma} . \tag{114.11}$$

This is the scattering amplitude for a typical resonance line of Breit-Wigner type with a differential cross section

$$|f_P(\vartheta)|^2 \propto \frac{1}{(E-E_0)^2 + \frac{1}{4}\Gamma^2} ; \tag{114.12}$$

E_0 being explained as the resonance energy and Γ as the line width defining the mean lifetime $\tau = \hbar/\Gamma$ of the intermediate resonance state.

Let now ξ_0 and η_0 be the values of ξ and η for the point of energy $E = E_0$ on the pole trajectory. Then we may write for $\lambda = L + \frac{1}{2}$,

$$L + \frac{1}{2} = \alpha(E_0) + \left(\frac{d\alpha}{dE}\right)_{E_0} (E_L - E_0)$$

or, using (114.8),

$$\xi_0 + i\eta_0 = \frac{1}{2} i\Gamma \left(\frac{d\alpha}{dE}\right)_{E_0} . \tag{114.13}$$

It can be shown (see below) that $(d\alpha/dE)_{E_0}$ is practically real so that $\xi_0 = 0$ and

$$\eta_0 = \frac{1}{2}\Gamma \left(\frac{d\alpha}{dE}\right)_{E_0} \tag{114.14}$$

is a measure for the width of the resonance line.

NB. A proof of the reality of $d\alpha/dE$ in the neighbourhood of a resonance can easily be obtained for potential scattering, cf. De Alfaro and Regge: *Potential Scattering*, Amsterdam 1965, p. 104. If R is a measure of the radius of the scattering domain, and v the velocity of the particles scattered, one gets $d\alpha/dE \simeq R/(\hbar v)$.

E. The Wentzel-Kramers-Brillouin (WKB) Approximation

Problem 115. Eikonal expansion

To solve the Schrödinger equation by successive steps following the optical method of solving the wave equation

$$\nabla^2 u + k^2 n(\boldsymbol{r})^2 u = 0 \tag{115.1}$$

with slowly varying $n(\boldsymbol{r})$ by introducing the eikonal function $S(\boldsymbol{r})$, putting

$$u = e^{\frac{2\pi i}{\lambda} S(\boldsymbol{r})} \tag{115.2}$$

where $\lambda = 2\pi/k$ is the vacuum wavelength, and expanding $S(\boldsymbol{r})$ into a series of powers of λ.

Solution. Let λ be the deBroglie wavelength at all points where $V(\boldsymbol{r}) = 0$:

$$\lambda = \frac{2\pi\hbar}{p}; \quad p = \sqrt{2mE} = \hbar k. \tag{115.3}$$

Then we may introduce the index of refraction,

$$n(\boldsymbol{r}) = \sqrt{1 - \frac{V(\boldsymbol{r})}{E}}. \tag{115.4}$$

Slow variation of n means, that it varies appreciably only over a distance $l \gg \lambda$, i.e.

$$\left|\frac{1}{n}\nabla n\right| \sim \frac{1}{l} \ll \frac{1}{\lambda}. \tag{115.5}$$

Putting (115.2) into (115.1) we arrive at the Riccati equation

$$\frac{\lambda}{2\pi i}\nabla^2 S + (\nabla S)^2 - n^2 = 0. \tag{115.6}$$

Were n a constant, then S would be a linear function of the coordinates and $\nabla^2 S = 0$. If n varies slowly, only a small effect of nonlinearity is to be expected in the exponent of (115.2) over a distance λ. Since, in any direction x, we have

$$S(x+\lambda)=S(x)+\lambda S'(x)+\tfrac{1}{2}\lambda^2 S''(x)+\cdots,$$

the non-linear term $\frac{1}{2}\lambda^2 S''$ can contribute very little to the exponent, i.e.

$$\left|\frac{2\pi}{\lambda}\cdot\frac{1}{2}\lambda^2 S''\right|\ll 1 \quad \text{or} \quad |\nabla^2 S|\ll\frac{1}{\lambda}.$$

The first term of the Riccati equation (115.6) therefore is small compared to the other two so that, in a first approximation, we get the eikonal equation instead:

$$(\nabla S)^2=n^2. \tag{115.7}$$

Let us now expand

$$S=S_0+\frac{\lambda}{2\pi i l}S_1+\left(\frac{\lambda}{2\pi i l}\right)^2 S_2+\cdots \tag{115.8}$$

with respect to the dimensionless parameter

$$\frac{\lambda}{2\pi l}\ll 1. \tag{115.9}$$

Eq. (115.6) then yields the following set of equations:

$$\left.\begin{aligned}(\nabla S_0)^2-n^2&=0,\\ l\nabla^2 S_0+2\nabla S_0\cdot\nabla S_1&=0,\\ l\,\nabla^2 S_1+(\nabla S_1)^2+2\nabla S_0\cdot\nabla S_2&=0,\\ l\nabla^2 S_2+2\nabla S_1\cdot\nabla S_2+2\nabla S_0\cdot\nabla S_3&=0,\quad\text{etc.}\end{aligned}\right\} \tag{115.10}$$

In these equations, the derivatives only occur so that we may introduce the dimensionless vectors

$$\boldsymbol{y}_n=\nabla S_n; \tag{115.11}$$

then

$$l\nabla^2 S_n=l(\nabla\cdot\boldsymbol{y}_n)$$

becomes the dimensionless divergence of $\boldsymbol{y}_n$, and the set of equations will run

$$\left.\begin{aligned}\boldsymbol{y}_0^2&=n^2,\\ \boldsymbol{y}_1\cdot\boldsymbol{y}_0&=-\tfrac{1}{2}l(\nabla\cdot\boldsymbol{y}_0),\\ \boldsymbol{y}_2\cdot\boldsymbol{y}_0&=-\tfrac{1}{2}[l(\nabla\cdot\boldsymbol{y}_1)+\boldsymbol{y}_1^2],\\ \boldsymbol{y}_3\cdot\boldsymbol{y}_0&=-\tfrac{1}{2}[l(\nabla\cdot\boldsymbol{y}_2)+2\boldsymbol{y}_1\cdot\boldsymbol{y}_2],\quad\text{etc.}\end{aligned}\right\} \tag{115.12}$$

These equations permit the successive determination of the y_n from whom, according to (115.11), the S_n may be determined by quadratures. We do not, however, go into details about boundary conditions here.

Problem 116. Radial WKB solutions

To apply the method developed in the preceding problem to the determination of the radial wave functions for a potential of spherical symmetry.

Solution. If the eikonal depends upon one variable r only, the wave equations

$$\chi_l'' + Q(r)^2 \chi_l = 0 \tag{116.1}$$

with

$$Q(r)^2 = k^2 \left(1 - \frac{V(r)}{E} - \frac{l(l+1)}{k^2 r^2}\right); \quad k^2 = \frac{2mE}{\hbar^2} \tag{116.2}$$

may be reduced by putting

$$\chi_l = \mathrm{e}^{\frac{2\pi i}{\lambda} S(r)}; \quad \lambda = \frac{2\pi}{k} \tag{116.3}$$

to Riccati equations

$$\frac{\lambda}{2\pi i} \frac{d^2 S}{dr^2} + \left(\frac{dS}{dr}\right)^2 - \frac{Q^2}{k^2} = 0. \tag{116.4}$$

The expansion

$$S = S_0 + \frac{\varepsilon}{i} S_1 + \left(\frac{\varepsilon}{i}\right)^2 S_2 + \cdots \tag{116.5}$$

with the dimensionless parameter

$$\varepsilon = \frac{\lambda}{2\pi R} \ll 1, \tag{116.6}$$

where R is a radius characteristic for the extension of the scattering region, then leads to the equations

$$\left.\begin{aligned} S_0'^2 &= Q^2/k^2, \\ S_1' S_0' &= -\tfrac{1}{2} R S_0'', \\ S_2' S_0' &= -\tfrac{1}{2}[R S_1'' + S_1'^2], \\ S_3' S_0' &= -\tfrac{1}{2}[R S_2'' + 2 S_1' S_2'], \quad \text{etc.} \end{aligned}\right\} \tag{116.7}$$

Introducing dimensionless functions

$$y_n(r) = S'_n(r) \tag{116.8}$$

we obtain

$$y_0 = \pm Q/k,$$
$$y_1 = -\frac{1}{2} R \frac{y'_0}{y_0},$$
$$y_2 = -\frac{R y'_1 + y_1^2}{2 y_0},$$
$$y_3 = -\frac{R y'_2 + 2 y_1 y_2}{2 y_0}$$

etc. From these equations we may consecutively express the functions $y_n(r)$ by $y_0(r)$ and its derivatives:

$$\begin{aligned} y_1 &= -\frac{R}{2}\frac{y'_0}{y_0}; \quad y_2 = \frac{R^2}{4}\left(\frac{y''_0}{y_0^2} - \frac{3}{2}\frac{{y'_0}^2}{y_0^3}\right); \\ y_3 &= -\frac{R^3}{8}\left(\frac{y'''_0}{y_0^3} - 6\frac{y'_0 y''_0}{y_0^4} + 6\frac{{y'_0}^3}{y_0^5}\right) \end{aligned} \tag{116.9}$$

etc. If

$$y_0 = \pm\sqrt{1 - \frac{V(r)}{E} - \frac{l(l+1)}{k^2 r^2}} \tag{116.10}$$

is real, all the functions $y_n(r)$ also become real, and since we have expanded S into powers of the purely imaginary parameter ε/i, the contributions of the S_n become alternately real and imaginary, leading to alternate corrections of phase and amplitude in the radial wave function $\chi_l(r)$:

$$\chi_l(r) = \exp \int^r \frac{dr}{R}\left\{\frac{i}{\varepsilon} y_0 + y_1 + \frac{\varepsilon}{i} y_2 + \left(\frac{\varepsilon}{i}\right)^2 y_3 + \cdots\right\}. \tag{116.11}$$

If we invert the sign of y_0, the odd functions y_{2n+1} remain unchanged and thus also the amplitude corrections, whereas the even functions y_{2n} invert their signs too, leading to the complex conjugate solution. Hence, the method yields a complete fundamental system of solutions.

Problem 117. WKB boundary condition of Langer

At the classical turning point r_t in a repulsive potential field (distance of closest approach) the WKB approximation has a singularity preventing formulation of a boundary condition. This difficulty can be circumvented by replacing the differential equation which has the WKB functions as exact solutions by another differential equation which (*a*) agrees with the Schrödinger equation in the vicinity of the classical turning point, and (*b*) agrees with the WKB differential equation elsewhere. This procedure is best performed using

$$x = \int_{r_t}^{r} dr\, Q(r) \tag{117.1}$$

instead of r as independent variable.

Solution. Since

$$Q^2 = k^2\left(1 - \frac{V(r)}{E} - \frac{l(l+1)}{k^2 r^2}\right) \equiv k^2 n(r)^2 \tag{117.2}$$

vanishes at the classical turning point, the WKB functions

$$u = Q^{-\frac{1}{2}} \exp\left\{\pm i \int_{r_t}^{r} dr\, Q(r)\right\} \tag{117.3}$$

have a singular amplitude at $r = r_t$. We want a solution remaining finite at r_t so that we may continue the oscillating solution of the domain $r > r_t$ by an exponentially decreasing solution in $r < r_t$.

The variable x defined by Eq. (117.1) can easily be introduced by the relation

$$\frac{d}{dr} = Q\,\frac{d}{dx}$$

which transforms the radial Schrödinger equation into

$$\ddot{\chi} + \frac{\dot{Q}}{Q}\dot{\chi} + \chi = 0 \tag{117.4}$$

where the dot denotes differentiation with respect to x. On the other hand, the WKB functions (117.3),

$$u = Q^{-\frac{1}{2}} e^{\pm ix}, \tag{117.5}$$

form a fundamental system of solutions to the differential equation

$$u'' + \left[Q^2 - \frac{3}{4}\frac{Q'^2}{Q^2} + \frac{1}{2}\frac{Q''}{Q}\right] u = 0, \tag{117.6}$$

the prime denoting differentiation with respect to r, or in the variable x,

$$\ddot{u} + \frac{\dot{Q}}{Q}\dot{u} + \left[1 - \frac{1}{4}\frac{\dot{Q}^2}{Q^2} + \frac{1}{2}\frac{\ddot{Q}}{Q}\right] u = 0. \tag{117.7}$$

In the vicinity of $r=r_t$ the function Q^2 becomes linear in the difference $r-r_t$ so that the integral (117.1) will become proportional to $(r-r_t)^{\frac{3}{2}}$:

$$Q \propto (r-r_t)^{\frac{1}{2}}; \quad x \propto (r-r_t)^{\frac{3}{2}};$$

hence,

$$Q \propto x^{\frac{1}{3}}. \tag{117.8}$$

In this infinitesimal region, therefore, we find

$$\frac{\dot{Q}}{Q} = \frac{1}{3x} \quad \text{and} \quad \frac{\ddot{Q}}{Q} = -\frac{2}{9x^2}.$$

The two differential equations (117.4) and (117.7) then can be replaced for $r \simeq r_t$ by

$$\ddot{\chi} + \frac{1}{3x}\dot{\chi} + \chi = 0 \tag{117.4'}$$

and

$$\ddot{u} + \frac{1}{3x}\dot{u} + \left(1 - \frac{5}{36x^2}\right)u = 0. \tag{177.7'}$$

Changing the notations into

$$\chi = x^{\frac{1}{3}}\varphi(x); \quad u = x^{\frac{1}{3}}f(x) \tag{117.9}$$

these two equations render

$$\ddot{\varphi} + \frac{1}{x}\dot{\varphi} + \left(1 - \frac{1}{9x^2}\right)\varphi = 0; \tag{117.4''}$$

$$\ddot{f} + \frac{1}{x}\dot{f} + \left(1 - \frac{1}{4x^2}\right)f = 0. \tag{117.7''}$$

Both these equations are of Bessel type with solutions

$$\varphi(x) = J_{\pm\frac{1}{3}}(x); \quad f(x) = J_{\pm\frac{1}{2}}(x).$$

Taking account of $J_\nu(x) \propto x^\nu$ for small x, this gives the fundamental solutions for χ behaving as x^0 and $x^{\frac{2}{3}}$, both finite or zero at $x=0$, and for u behaving as $x^{-\frac{1}{6}}$ and $x^{\frac{5}{6}}$, one singular and one zero at $x=0$. This is the singularity of the WKB solution at the classical turning point preventing the formulation of the boundary condition. It can be removed by replacing the term $1/4x^2$ in (117.7″) by $1/9x^2$ as in (117.4″), i. e. by adding $5/36x^2$ in (117.7″), (117.7′) and (117.7). This means an alteration of the WKB equation (117.7) only in the neighbourhood of $x=0$ and leaves it unchanged elsewhere. Let $v(x)$ be a function satisfying this differential equation, replacing the WKB equation (117.7), then

$$\ddot{v}+\frac{\dot{Q}}{Q}\dot{v}+\left[1+\frac{5}{36x^2}-\frac{1}{4}\frac{\dot{Q}^2}{Q^2}+\frac{1}{2}\frac{\ddot{Q}}{Q}\right]v=0 \tag{117.10}$$

is the differential equation wanted.

Writing the solution, in analogy to (117.3) or (117.5), in the form

$$v(x)=Q^{-\frac{1}{2}}F(x), \tag{117.11}$$

we arrive at

$$\ddot{F}+\left[1+\frac{5}{36x^2}\right]F=0 \tag{117.12}$$

leading to the complete solution

$$v(x)=\sqrt{\frac{x}{Q}}\{C_1 J_{\frac{1}{3}}(x)+C_2 J_{-\frac{1}{3}}(x)\}. \tag{117.13}$$

In consequence of (115.8), this solution has no singularity at $x=0$ and differs from the WKB equation only by replacing the factors $e^{\pm ix}$ by $\sqrt{x}\,J_{\pm\frac{1}{3}}(x)$ which, for $x \gg \frac{1}{3}$, again asymptotically becomes

$$\sqrt{x}\,J_{\pm\frac{1}{3}}(x)\rightarrow\sqrt{\frac{2}{\pi}}\cos\left(x-\frac{\pi}{4}\mp\frac{\pi}{6}\right). \tag{117.14}$$

We are now prepared to formulate the boundary value problem. Knowing the complete solution (117.13) for $r>r_t$ where $Q>0$, $x>0$, we need its continuation into the domain $r<r_t$ where

$$Q=i|Q|; \quad x=e^{\frac{3i\pi}{2}}\xi,\ \xi>0. \tag{117.15}$$

For small values of the complex variable x there hold the relations

$$J_{\frac{1}{3}}(x)=\frac{1}{\Gamma(\frac{4}{3})}\left(\frac{x}{2}\right)^{\frac{1}{3}}; \quad J_{-\frac{1}{3}}(x)=\frac{1}{\Gamma(\frac{2}{3})}\left(\frac{x}{2}\right)^{-\frac{1}{3}}$$

leading for the imaginary values of x, Eq. (117.15), to

$$J_{\frac{1}{3}}(x)=\frac{i}{\Gamma(\frac{4}{3})}\left(\frac{\xi}{2}\right)^{\frac{1}{3}} \quad J_{-\frac{1}{3}}(x)=-\frac{i}{\Gamma(\frac{2}{3})}\left(\frac{\xi}{2}\right)^{-\frac{1}{3}}$$

The so-called modified Bessel functions $I_\nu(\xi)$ defined by

$$J_{\frac{1}{3}}(x)=i I_{\frac{1}{3}}(\xi); \quad J_{-\frac{1}{3}}(x)=-i I_{-\frac{1}{3}}(\xi)$$

are real for real $\xi > 0$. So is the modified Hankel function

$$K_\nu(\xi) = \frac{\pi}{2\sin\nu\pi}\left[I_{-\nu}(\xi) - I_\nu(\xi)\right] = \frac{i\pi}{2} e^{\frac{i\nu\pi}{2}} H_\nu^{(1)}(i\xi),$$

its asymptotic form for large real $\xi > 0$ being

$$K_\nu(\xi) \to \sqrt{\frac{\pi}{2\xi}}\, e^{-\xi}. \tag{117.16}$$

Using this general mathematical apparatus, the solution (117.13) with the argument (117.15) may be written

$$v(x) = \left\{e^{i\pi}\frac{\xi}{|Q|}\right\}^{\frac{1}{2}} \{C_1 i I_{\frac{1}{3}}(\xi) - C_2 i I_{-\frac{1}{3}}(\xi)\} = \sqrt{\frac{\xi}{|Q|}}\{C_2 I_{-\frac{1}{3}}(\xi) - C_1 I_{\frac{1}{3}}(\xi)\}.$$

Since the solution wanted shall decrease when penetrating the potential barrier, i.e. with increasing ξ, as the combination (117.16) does, we have to put

$$C_1 = C_2, \quad \text{say,} \quad = C \tag{117.17}$$

and arrive at

$$v(x) = \sqrt{\frac{\xi}{|Q|}}\, C \frac{2\sin\frac{\pi}{3}}{\pi} K_{\frac{1}{3}}(\xi).$$

On the other hand, the phase angle at large values of $x > 0$ is thus determined. If we go into (117.14) and (117.13) with (117.17) we find

$$v(x) \to \sqrt{\frac{x}{Q}}\, C \sqrt{\frac{2}{\pi x}} \left\{\cos\left(x - \frac{\pi}{4} - \frac{\pi}{6}\right) + \cos\left(x - \frac{\pi}{4} + \frac{\pi}{6}\right)\right\}$$

or

$$v(x) \to C\sqrt{\frac{6}{\pi Q}} \sin\left(x + \frac{\pi}{4}\right). \tag{117.18}$$

This agrees asymptotically with the WKB solution (117.3) but definitely fixes the phase angle.

NB. The discussion of the solution in the vicinity of the classical turning point can be simplified by using the Airy functions, cf. Problem 40.

Literature to Problems 113—115. Wentzel, G.: Z. Physik **38,** 518 (1926). — Kramers, H. A.: Z. Physik **39,** 828 (1926). — Brillouin, L.: C. R. Acad. Sci. Paris **183,** 24 (1926). — Langer, R. E.: Phys. Rev. **51,** 669 (1937).

Problem 118. Oscillator according to WKB approach

The harmonic oscillator levels shall be found by applying the boundary conditions of Langer to the WKB wave functions.

Solution. In the oscillator potential

$$V(r)=\tfrac{1}{2}m\omega^2 x^2 \tag{118.1}$$

the turning points of classical motion at an energy

$$E=\tfrac{1}{2}mv^2\,, \tag{118.2}$$

with v the classical velocity at $x=0$, lie at

$$x=\pm x_0\,;\qquad x_0=\frac{v}{\omega}\,. \tag{118.3}$$

It follows that

$$Q^2=k^2\left(1-\frac{V(x)}{E}\right)=k^2\left(1-\frac{x^2}{x_0^2}\right). \tag{118.4}$$

The WKB wave functions, satisfying the Langer boundary condition at turning point $x=-x_0$, become

$$\tilde{u}(x)=Q^{-\frac{1}{2}}\cos\left\{\int\limits_{-x_0}^{x}dx\,Q(x)-\frac{\pi}{4}\right\}. \tag{118.5}$$

Evaluation of this integral for the function $Q(x)$ defined by (118.4) making use of

$$kx_0=\frac{mv}{\hbar}\cdot\frac{v}{\omega}=\frac{2E}{\hbar\omega}$$

yields

$$\tilde{u}(x)=Q^{-\frac{1}{2}}\cos\left\{\frac{E}{\hbar\omega}\left[\sin^{-1}\frac{x}{x_0}+\frac{x}{x_0}\sqrt{1-\frac{x^2}{x_0^2}}+\frac{\pi}{2}\right]-\frac{\pi}{4}\right\}. \tag{118.6}$$

This solution has still to be adapted to satisfy the other Langer boundary condition at $x=+x_0$ as well. That means either an odd or an even function. With the abbreviations

$$y(x)=\frac{E}{\hbar\omega}\left[\sin^{-1}\frac{x}{x_0}+\frac{x}{x_0}\sqrt{1-\frac{x^2}{x_0^2}}\right];\qquad \varphi=\frac{E}{\hbar\omega}\cdot\frac{\pi}{2}-\frac{\pi}{4} \tag{118.7}$$

where $y(x)$ is an odd function, $y(-x)=-y(x)$, we obtain

$$\tilde{u}(x)=Q^{-\frac{1}{2}}\cos\{y(x)+\varphi\}=Q^{-\frac{1}{2}}\{\cos y\cos\varphi-\sin y\sin\varphi\}$$

so that $\tilde{u}$ becomes even if

$$\varphi=2n\frac{\pi}{2};\qquad \tilde{u}_{2n}(x)=Q^{-\frac{1}{2}}(-1)^n\cos y \tag{118.8a}$$

and it becomes odd if

$$\varphi=(2n+1)\frac{\pi}{2}; \qquad \tilde{u}_{2n+1}(x)=Q^{-\frac{1}{2}}(-1)^{n+1}\sin y\,. \tag{118.8b}$$

Both conditions for φ may be summed up in one formula

$$\varphi=n\frac{\pi}{2}$$

or, according to (118.7),

$$E=\hbar\omega(n+\tfrac{1}{2})\,. \tag{118.9}$$

Negative values of n can be excluded, because they would not satisfy the condition $E>V_{\min}=0$ necessary for applying the WKB approach.

To get an estimate of the quality of the WKB approach to the wave functions it is interesting to compare the position of their zeros for the WKB functions $\tilde{u}_n$ and the exact solutions u_n given in Problem 30. They are compiled in the accompanying table.

n	x/x_0 values of the zeros of $\tilde{u}_n$	u_n
0	—	—
1	0	0
2	± 0.320	± 0.316
3	0	0
	± 0.466	± 0.462
4	± 0.1754	± 0.1750
	± 0.553	± 0.550

Problem 119. WKB eigenvalues in a homogeneous field

In the gravitational field

$$V(z)=mgz$$

above the surface of the earth at $z=0$ the eigenvalues of stationary motion shall be determined in WKB approximation. Cf. the exact solution in Problem 40.

Solution. There are two classical turning points at $z_1=0$ and $z_2=E/mg$. The Langer condition at the left turning point $z_1=0$ leads to

$$\tilde{u}(z)=Q^{-\frac{1}{2}}\cos\left\{\int_0^z dz\,Q(z)-\frac{\pi}{4}\right\} \tag{119.1}$$

with

$$Q(z)=k\sqrt{1-\frac{mg}{E}z}\,. \tag{119.2}$$

Integration is elementary and yields

$$\tilde{u}(z)=Q^{-\frac{1}{2}}\cos\left\{\frac{2}{3}k z_2\left[1-\left(\frac{z}{z_2}\right)^{\frac{3}{2}}\right]-\frac{\pi}{4}\right\}. \tag{119.3}$$

On the other hand, a Langer condition has to be satisfied at $z=z_2$. Using the variable $y=-z$, this again becomes a *left* turning point so that[22]

$$\tilde{u}(z)=\pm Q^{-\frac{1}{2}}\cos\left\{\int_{-z_0}^{-z} dy\,Q(-y)-\frac{\pi}{4}\right\}=\pm Q^{-\frac{1}{2}}\cos\left\{-\int_{z_2}^{z} dz\,Q(z)-\frac{\pi}{4}\right\}$$

or

$$\tilde{u}(z)=\pm Q^{-\frac{1}{2}}\cos\left\{\int_{z_2}^{z} dz\,Q(z)+\frac{\pi}{4}\right\}. \tag{119.4a}$$

Since the relation (119.1) may be generalized to

$$\tilde{u}(z)=Q^{-\frac{1}{2}}\cos\left\{\int_{z_1}^{z} dz\,Q(z)-\frac{\pi}{4}\right\} \tag{119.4b}$$

we find that the arguments of the cosines in (119.4a) and (119.4b) can only differ by an integer multiple of π:

$$\int_{z_1}^{z} dz\,Q(z)-\frac{\pi}{4}=\int_{z_2}^{z} dz\,Q(z)+\frac{\pi}{4}+n\pi$$

or

$$\int_{z_1}^{z_2} dz\,Q(z)=(n+\tfrac{1}{2})\pi\,. \tag{119.5}$$

This is the general WKB eigenvalue formula which we now apply to our gravitational field:

$$\tfrac{2}{3}k z_2=(n+\tfrac{1}{2})\pi\,. \tag{119.6}$$

---

[22] The double sign in this formula is left open in the derivation of the Langer formula because it is independent of normalization.

Replacing k by the energy and putting $z_2 = E/mg$, this leads to the final result

$$E_n = \tfrac{1}{2}[3\pi(n+\tfrac{1}{2})]^{\frac{2}{3}} \sqrt[3]{\hbar^2 mg^2}\,, \tag{119.7}$$

a formula which agrees with the asymptotic expression (40.14) of the exact solution.

NB. The result (119.5) can easily be applied to the harmonic oscillator of Problem 118 where

$$\int_{-x_0}^{+x_0} dx\, Q(x) = \frac{E}{\hbar\omega}\pi$$

leads directly to the level formula

$$E = \hbar\omega(n+\tfrac{1}{2})\,.$$

Problem 120. Kepler problem in WKB approach

To determine the eigenvalues in the attractive Coulomb field

$$V(r) = -\frac{e^2}{r}$$

by applying the WKB method to the radial wave equations.

Solution. The WKB eigenvalue condition

$$\int_{r_1}^{r_2} Q(r)\,dr = (n_r+\tfrac{1}{2})\pi; \qquad n_r = 0, 1, 2, \dots \tag{120.1}$$

with

$$Q^2 = \frac{2m}{\hbar^2}[E - V_{\text{eff}}(r)] \tag{120.2}$$

and $r_1 < r_2$ the two classical turning points may be applied to negative as well as to positive energies E. The symbol V_{eff} means inclusion of the centrifugal term, with $l(l+1)$ replaced by $(l+\frac{1}{2})^2$ (cf. Problem 121). In our case, therefore, we have

$$Q^2 = \frac{2m}{\hbar^2}\left\{-|E| + \frac{e^2}{r} - \frac{\hbar^2(l+\frac{1}{2})^2}{2mr^2}\right\} \tag{120.3}$$

or, using the abbreviations

$$k^2 = \frac{2m|E|}{\hbar^2}; \qquad a = \frac{\hbar^2}{me^2}; \qquad \lambda = l+\tfrac{1}{2} \tag{120.4}$$

with which the turning points become

$$r_1 = \frac{1}{k^2 a}\{1-\sqrt{1-(\lambda k a)^2}\}\,; \qquad r_2 = \frac{1}{k^2 a}\{1+\sqrt{1-(\lambda k a)^2}\} \qquad (120.5)$$

we have

$$Q^2 = \frac{k^2}{r^2}(r-r_1)(r_2-r)\,. \qquad (120.6)$$

The eigenvalue condition (120.1) therefore becomes

$$k\int_{r_1}^{r_2}\frac{dr}{r}\sqrt{(r-r_1)(r_2-r)}=(n_r+\tfrac{1}{2})\pi\,. \qquad (120.7)$$

The integral (120.7) may be evaluated in three steps. First substituting a new variable x by putting

$$r=\tfrac{1}{2}(r_2-r_1)x+\tfrac{1}{2}(r_2+r_1),$$

the quadratic form under the root is diagonalized. This may be followed by a second substitution of

$$y=\sqrt{\frac{1-x}{1+x}}$$

which leads to a rational integral:

$$k\frac{(r_2-r_1)^2}{r_1}\int_{-\infty}^{+\infty}dy\frac{y^2}{(y^2+1)^2(y^2+p^2)^2}=(n_r+\tfrac{1}{2})\pi$$

with

$$p^2=r_2/r_1\,.$$

Deformation of the integration path into an infinite half-circle in the upper half of the complex y plane leaves us with the poles at $y=i$ and $y=ip$, the latter being a second-order pole. Taking the residues, we finally arrive at

$$\frac{\pi}{2}k r_1(\sqrt{r_2/r_1}-1)^2=(n_r+\tfrac{1}{2})\pi\,. \qquad (120.8)$$

With $\lambda k a=t$ this may be written

$$\frac{\lambda}{2t}[1-\sqrt{1-t^2}]\left(\sqrt{\frac{1+\sqrt{1-t^2}}{1-\sqrt{1-t^2}}}-1\right)^2=n_r+\tfrac{1}{2}$$

or more simply,

$$\lambda\frac{1-t}{t}=n_r+\tfrac{1}{2}\,. \qquad (120.9)$$

It follows that

$$t=\frac{\lambda}{n_r+\frac{1}{2}+\lambda} \quad \text{or } ka=\frac{1}{n_r+\frac{1}{2}+\lambda}\,;$$

hence, according to Eq. (120.4):

$$E = -\frac{me^4}{2\hbar^2} \cdot \frac{1}{(n_r + \frac{1}{2} + \lambda)^2} \,. \tag{120.10}$$

Since $\lambda = l + \frac{1}{2}$, this renders the well-known exact eigenvalue relation with the principal quantum number

$$n = n_r + l + 1 \,. \tag{120.11}$$

NB. In order to get this result it was essential to use $(l+\frac{1}{2})^2$ in the centrifugal term instead of $l(l+1)$. This means that even for an S state this term does not vanish. Had we omitted it above and used

$$Q^2 = \frac{2m}{\hbar^2}\left\{-|E| + \frac{e^2}{r}\right\}$$

we should have arrived at half-integer quantum numbers n. This is easily seen. With $\lambda = 0$, the turning points follow from (120.5) to lie at $r_1 = 0$, $r_2 = 2/(k^2 a)$, and the condition (120.8) may be written

$$\frac{\pi}{2} k(\sqrt{r_2} - \sqrt{r_1})^2 = \frac{\pi}{ka} = (n_r + \tfrac{1}{2})\pi \,.$$

Problem 121. WKB phases in the force-free case

To show that, by replacing $l(l+1)$ in the radial wave functions by $(l+\frac{1}{2})^2$, the asymptotic behaviour of the wave functions of the forcefree case becomes correct.

Solution. According to Eq. (117.18) we find for $V=0$ in arbitrary normalization,

$$\chi_l^{\mathrm{WKB}} = \left(1 - \frac{\lambda^2}{k^2 r^2}\right)^{-\frac{1}{4}} \sin\left\{k \int_{r_t}^{r} dr \sqrt{1 - \frac{\lambda^2}{k^2 r^2}} + \frac{\pi}{4}\right\} \tag{121.1}$$

where $\lambda^2 = l(l+1)$ and the turning point is at $r_t = \lambda/k$. The amplitude factor of this function tends towards 1 at large r and may thus be omitted in a discussion of the asymptotic behaviour. The integral may easily be evaluated in the variable $x = r_t/r$:

$$\int_{r_t}^{r} dr \sqrt{1 - \frac{r_t^2}{r^2}} = r_t \left[\sqrt{\frac{r^2}{r_t^2} - 1} + \sin^{-1}\frac{r_t}{r} - \frac{\pi}{2}\right].$$

Expanding this expression into powers of $r_t/r \ll 1$ we get

$$r_t \left[\frac{r}{r_t}\left(1 - \frac{1}{2}\frac{r_t^2}{r^2}\right) + \frac{r_t}{r} + \cdots - \frac{\pi}{2}\right] \to r - \frac{\pi}{2} r_t \,.$$

Hence, the WKB function, at large values of r, becomes

$$\chi_l^{\mathrm{WKB}} \to \sin\left(kr - \frac{\pi}{2}kr_t + \frac{\pi}{4}\right) = \sin\left(kr - \frac{\pi}{2}\left(\lambda - \frac{1}{2}\right)\right). \quad (121.2)$$

On the other hand, the exact solution, in the same normalization, is

$$\chi_l = j_l(kr) \to \sin\left(kr - \frac{l\pi}{2}\right). \quad (121.3)$$

The expressions (121.2) and (121.3) become identical for large r if

$$\lambda - \tfrac{1}{2} = l \quad \text{or} \quad \lambda = l + \tfrac{1}{2} \quad (121.4)$$

thus replacing $l(l+1)$ by $(l+\frac{1}{2})^2$ in the centrifugal term as was to be proved.

Problem 122. Calculation of WKB phases

To prove that, when replacing $l(l+1)$ by

$$\lambda^2 = (l+\tfrac{1}{2})^2, \quad (122.1)$$

the asymptotic phase of the radial WKB wave function determined by the Langer condition (Problem 117) can equally well be found according to the formula

$$\delta_l = k \lim_{r\to\infty} \left\{ \int_{r_t}^{r} dr \sqrt{1 - \frac{V}{E} - \frac{\lambda^2}{k^2 r^2}} - \int_{\lambda/k}^{r} dr \sqrt{1 - \frac{\lambda^2}{k^2 r^2}} \right\}, \quad (122.2)$$

i.e. as difference of the WKB integrals with and without scattering potential, each integral taken from its own turning point.

Solution. The Langer expression for the phase δ_l follows from Eq. (117.18) giving asymptotically for large values of r,

$$\sin\left\{ k \int_{r_t}^{r} dr \sqrt{1 - \frac{V}{E} - \frac{\lambda^2}{k^2 r^2}} + \frac{\pi}{4} \right\} \to \sin\left(kr - \frac{l\pi}{2} + \delta_l\right),$$

i.e.

$$\delta_l = k \lim_{r\to\infty} \left\{ \int_{r_t}^{r} dr \sqrt{1 - \frac{V}{E} - \frac{\lambda^2}{k^2 r^2}} - r \right\} + \left(l + \frac{1}{2}\right)\frac{\pi}{2}. \quad (122.3)$$

The identity of the expressions (122.2) and (122.3) can easily be shown, since (see Problem 121)

$$\int_{\lambda/k}^{r} dr \sqrt{1-\frac{\lambda^2}{k^2 r^2}} = \frac{\lambda}{k}\left[\sqrt{\frac{r^2}{(\lambda/k)^2}-1}+\sin^{-1}\frac{\lambda/k}{r}-\frac{\pi}{2}\right]$$

tends towards

$$\frac{\lambda}{k}\left[\frac{r}{\lambda/k}-\frac{\pi}{2}\right]=r-\frac{\lambda}{k}\cdot\frac{\pi}{2}=r-\frac{1}{k}\left(l+\frac{1}{2}\right)\frac{\pi}{2}$$

for large r. Putting this into (122.2) for the second integral, it becomes identical with (122.3) as had to be proved.

Problem 123. Coulomb phases by WKB method

To determine the asymptotics of partial waves in a Coulomb field in WKB approximation and to compare it with the exact solutions.

Solution. Using the abbreviations

$$\frac{2mE}{\hbar^2}=k^2; \qquad \frac{Z_1 Z_2 e^2}{\hbar v}=\varkappa; \qquad l+\tfrac{1}{2}=\lambda, \tag{123.1}$$

we have the asymptotic behaviour (cf. Problem 111)

$$\chi_l \propto \sin\left(kr-\frac{l\pi}{2}-\varkappa\log 2kr+\eta_l\right) \tag{123.2}$$

with

$$\eta_l=\arg\Gamma(l+1+i\varkappa). \tag{123.3}$$

On the other hand, the WKB method yields

$$\chi_l \propto \sin\left(kr-\frac{l\pi}{2}+\delta_l\right) \tag{123.4}$$

with

$$\delta_l=\lim_{r\to\infty}\left\{\int_{r_t}^{r} dr\sqrt{k^2-\frac{2\varkappa k}{r}-\frac{\lambda^2}{r^2}}-\int_{\lambda/k}^{r} dr\sqrt{k^2-\frac{\lambda^2}{r^2}}\right\}. \tag{123.5}$$

We therefore have to show that (123.5) is an approximation to the exact expression,

$$\delta_l=-\varkappa\log 2kr+\eta_l \tag{123.6}$$

with η_l given by (123.3).

With the variable $x = kr$ and the abbreviation

$$X = x^2 - 2\varkappa x - \lambda^2$$

we may write

$$\delta_l = \lim_{x\to\infty}\left\{\int_{x_t}^{x}\frac{dx}{x}\sqrt{X} - \int_{\lambda}^{x}\frac{dx}{x}\sqrt{x^2-\lambda^2}\right\}$$

with

$$x_t = \varkappa + \sqrt{\varkappa^2+\lambda^2}.$$

The integration formula

$$\int\frac{dx}{x}\sqrt{X} = \sqrt{X} - \varkappa\log(x-\varkappa+\sqrt{X}) + \lambda\sin^{-1}\frac{\varkappa x+\lambda^2}{x\sqrt{\varkappa^2+\lambda^2}}$$

leads, with the correct integration bounds, to

$$\delta_l = \lim_{x\to\infty}\left\{\sqrt{X} - \sqrt{x^2-\lambda^2} - \varkappa\log\frac{x-\varkappa+\sqrt{X}}{\sqrt{\varkappa^2+\lambda^2}} + \lambda\left[\sin^{-1}\frac{\varkappa x+\lambda^2}{x\sqrt{\varkappa^2+\lambda^2}} - \sin^{-1}\frac{\lambda}{x}\right]\right\}.$$

Expanding into powers of $1/x$, the limit may be evaluated. The result is

$$\delta_l = -\varkappa + \varkappa\log\sqrt{\varkappa^2+\lambda^2} + \lambda\sin^{-1}\frac{\varkappa}{\sqrt{\varkappa^2+\lambda^2}} - \varkappa\log 2x$$

or, with (123.6), there holds the WKB approximation

$$\left.\begin{aligned}\eta_l &= \varkappa(\log\varkappa - 1) + \varkappa f\left(\frac{\lambda}{\varkappa}\right),\\ f(z) &= \log\sqrt{1+z^2} + z\sin^{-1}\frac{1}{\sqrt{1+z^2}}.\end{aligned}\right\} \tag{123.7}$$

We now have to compare the WKB result (123.7) with the correct expression (123.3).

In the accompanying table we give a few numerical values for the example $\varkappa = 2$. Computation of the WKB values according to (123.7) is straightforward; the computation of the η_l, Eq. (123.3), follows the relations

$$\eta_l = \tan^{-1}\frac{\varkappa}{l} + \eta_{l-1}; \tag{123.8a}$$

$$\eta_0 = \frac{\pi}{4} + \varkappa(\log\varkappa - 1) - \frac{1}{12\varkappa} - \frac{1}{360\varkappa^3}\cdots \tag{123.8b}$$

with (123.8 b) a fairly convergent series for $\varkappa \gtrsim 2$. The results are in good agreement. This may be seen more explicitly for $l=0$ when

l	η_l correct	η_l WKB
0	0.130	0.110
1	1.237	1.222
2	2.022	2.012
3	2.610	2.610
4	3.074	3.076

Eq. (123.7), with $z=\lambda/\varkappa=\frac{1}{4}\ll 1$, permits the expansion

$$\eta_0^{\text{WKB}} = \frac{\pi}{4} + \varkappa(\log\varkappa - 1) - \frac{1}{8\varkappa} + \frac{1}{192\varkappa^3} \cdots$$

differing from (123.8 b) only by

$$\eta_0^{\text{WKB}} - \eta_0 \simeq -\frac{1}{24\varkappa}.$$

If, on the other hand, $\lambda \gg \varkappa$, and $\lambda \gg 1$, Eq. (123.7) yields

$$(\eta_{l+1} - \eta_l)^{\text{WKB}} = \frac{\varkappa}{\lambda} - \frac{\varkappa}{2\lambda^2} + \frac{\varkappa}{3\lambda^3} - \frac{\varkappa^3}{3\lambda^3} \cdots$$

whereas it follows from (123.8 a) that

$$(\eta_{l+1} - \eta_l)^{\text{exact}} = \frac{\varkappa}{\lambda} - \frac{\varkappa}{2\lambda^2} + \frac{\varkappa}{4\lambda^3} - \frac{\varkappa^3}{3\lambda^3} \cdots$$

so that the difference of the two expressions tends to zero as $\dfrac{\varkappa}{12\lambda^3}$. This means that continuing our table beyond $l=4$ will not lead to any appreciable deviations of the WKB from the exact phase values.

Problem 124. Quasipotential

Instead of the variable r, it is often useful to introduce a new variable

$$t = r\sqrt{1 - \frac{V}{E}} \tag{124.1}$$

(Sabatier transform). Then the WKB phase can be written as a simple integral in this variable with a quasipotential

$$Q(t) = 2E \log \frac{r(t)}{t} \tag{124.2}$$

essentially replacing the potential $V(r)$.

Solution. With the abbreviation

$$(l+\tfrac{1}{2})/k = b \tag{124.3}$$

("collision parameter") the WKB phase may be written according to (122.2),

$$\delta_l = k \lim_{r\to\infty} \left\{ \int_{r_t}^{r} dr \sqrt{1 - \frac{V}{E} - \frac{b^2}{r^2}} - \int_{b}^{r} dr \sqrt{1 - \frac{b^2}{r^2}} \right\} \tag{124.4}$$

where r_t is the largest zero of the radical in the first integral. It is a little troublesome that the limit for $r \to \infty$ exists for the difference only of the two integrals, each single integral tending towards infinity and both having different lower bounds.

This difficulty, however, may be remedied by using the transformation (124.1) which alters the first integral in (124.4) into

$$\int_{r_t}^{r} dr \sqrt{1 - \frac{V}{E} - \frac{b^2}{r^2}} = \int_{b}^{t} dt \frac{d \log r(t)}{dt} \sqrt{t^2 - b^2}. \tag{124.5}$$

This, of course, can only be done if the transformation $t = t(r)$ may be uniquely inverted into $r = r(t)$, i. e. if t is a monotonous function of r. If we put (124.5) into (124.4) and replace the letter r in the second integral of (124.4) by the letter t, we arrive at

$$\delta_l = k \int_{b}^{\infty} dt \sqrt{t^2 - b^2} \left\{ \frac{d \log r(t)}{dt} - \frac{1}{t} \right\}. \tag{124.6}$$

This integral may even be more simplified by a partial integration, making use of the identity

$$\frac{d \log r}{dt} - \frac{1}{t} = \frac{d}{dt} \log \frac{r(t)}{t};$$

the result is

$$\delta_l = -\frac{1}{2} k \int_{b}^{\infty} \frac{dt\, t}{\sqrt{t^2 - b^2}} \frac{Q(t)}{E} \tag{124.7}$$

with $Q(t)$ the function defined by Eq. (124.2).

The function $Q(t)$ differs little from $V(r)$ as long as $|V(r)| \ll E$. This may be seen if we write instead of (124.2),

$$Q(t) = -E \log\left(1 - \frac{V(r)}{E}\right) \tag{124.8}$$

which, in first approximation, yields $Q(t) = V(r)$. In particular, the zeros of $Q(t)$ and their immediate neighbourhood will entirely coincide with $V(r)$ in the same regions.

Literature. Sabatier, P. C.: Nuovo Cimento **37,** 1180 (1965) introduced the transform (124.1). The quasipotential method was developed by Vollmer, G., Krüger, H.: Physics Letters **28 A,** no. 2 (1968). Details are given by Vollmer, G., in: Z. Physik **226,** 423 (1969).

F. The Magnetic Field

Problem 125. Introduction of a magnetic field

To construct the hamiltonian in presence of a magnetic field, and to show that the occurrence of the vector potential in the thus extended Schrödinger equation does not contradict gauge invariance.

Solution. A magnetic field, described by a vector potential $\boldsymbol{A}$, is shown[23] in classical mechanics to cause the momentum $\boldsymbol{p}$ of any particle of electric charge e, to be replaced by

$$\boldsymbol{P} = \boldsymbol{p} - \frac{e}{c}\boldsymbol{A} \tag{125.1}$$

so that the unrelativistic hamiltonian becomes

$$H = \frac{1}{2m}\left(\boldsymbol{p} - \frac{e}{c}\boldsymbol{A}\right)^2 + e\Phi(\boldsymbol{r}) + V(\boldsymbol{r}) \tag{125.2}$$

where Φ is the scalar electromagnetic potential, and $V(\boldsymbol{r})$ any other potential due to non-electromagnetic forces (e. g. nuclear forces). This

[23] This is shown by using the hamiltonian (125.2) to construct the canonical equations and, by eliminating momenta, deriving the equation of motion,

$$m\ddot{\boldsymbol{r}} = e\left(\mathscr{E} + \frac{1}{c}\boldsymbol{v} \times \mathscr{B}\right),$$

i. e. the correct expression for the Lorentz force acting on the particle of charge e.

classical expression may be translated into quantum mechanics replacing $\boldsymbol{p}$ by the operator $\frac{\hbar}{i}\nabla$. We thus arrive at the generalized Schrödinger equation

$$\left.\begin{aligned} -\frac{\hbar}{i}\frac{\partial\psi}{\partial t} &= H\psi \\ \text{with}\qquad & \\ H = -\frac{\hbar^2}{2m}\nabla^2 + \frac{e\hbar i}{2mc}(\nabla\cdot\boldsymbol{A}+\boldsymbol{A}\cdot\nabla) &+ \frac{e^2}{2mc^2}A^2 + e\Phi + V. \end{aligned}\right\} \tag{125.3}$$

As

$$\nabla\cdot\boldsymbol{A}\psi = \psi\,\mathrm{div}\,\boldsymbol{A} + \boldsymbol{A}\cdot\nabla\psi,$$

this may be simplified to

$$H = -\frac{\hbar^2}{2m}\nabla^2 + \frac{e\hbar}{mc}i(\boldsymbol{A}\cdot\nabla + \tfrac{1}{2}\mathrm{div}\,\boldsymbol{A}) + \frac{e^2}{2mc^2}A^2 + e\Phi + V. \tag{125.4}$$

In the classical theory of the Maxwell field it is further shown that the vector potential $\boldsymbol{A}$ and the scalar potential Φ may be subjected to the simultaneous gauge transformation

$$\boldsymbol{A}\to\boldsymbol{A}' = \boldsymbol{A}+\nabla\chi;\qquad \Phi\to\Phi' = \Phi - \frac{1}{c}\dot{\chi}, \tag{125.5}$$

where χ is an arbitrary function of space and time, without any change in the field strengths

$$\mathscr{B} = \mathrm{curl}\,\boldsymbol{A};\qquad \mathscr{E} = -\mathrm{grad}\,\Phi - \frac{1}{c}\dot{\boldsymbol{A}}. \tag{125.6}$$

This gauge invariance must also hold in quantum theory, if the field strengths only, and not the potentials, are to determine physical phenomena.

If we now simply put (125.5) into the hamiltonian (125.4) we would, of course, get a number of additional terms destroying gauge invariance of the Schrödinger equation. To dispose of these terms becomes possible only by letting the wave function take part in the gauge transformation. Since $\psi^*\psi$, however, has a physical meaning and should no more be altered by the transformation than the field strengths, there only remains to try

$$\psi\to\psi' = e^{i\alpha}\psi \tag{125.7a}$$

where α may still be any appropriate function of $\boldsymbol{r}$ and t. Then,

$$\begin{aligned} \nabla\psi' &= e^{i\alpha}\{\nabla\psi + i\psi\nabla\alpha\}; \\ \nabla^2\psi' &= e^{i\alpha}\{\nabla^2\psi + 2i\nabla\psi\cdot\nabla\alpha + i\psi\nabla^2\alpha - (\nabla\alpha)^2\psi\}; \\ \dot{\psi}' &= e^{i\alpha}\{\dot{\psi} + i\dot{\alpha}\psi\}. \end{aligned}$$

Putting these expressions into (125.4) and replacing A and Φ simultaneously by A' and Φ' according to (125.5), we find

$$
\begin{aligned}
e^{-i\alpha} H'\psi' = & -\frac{\hbar^2}{2m}\{\nabla^2\psi + 2i\nabla\alpha\cdot\nabla\psi + i\psi\nabla^2\alpha - (\nabla\alpha)^2\psi\} \\
& + \frac{e\hbar}{mc} i\{A\cdot\nabla\psi + i\psi A\cdot\nabla\alpha + \nabla\chi\cdot\nabla\psi + i\psi\nabla\chi\cdot\nabla\alpha + \tfrac{1}{2}\psi(\nabla\cdot A) + \tfrac{1}{2}\psi\nabla^2\chi\} \\
& + \frac{e^2}{2mc^2}\{A^2 + 2A\cdot\nabla\chi + (\nabla\chi)^2\}\psi + \left(e\Phi - \frac{e}{c}\dot{\chi}\right)\psi + V\psi .
\end{aligned}
$$

We rearrange the terms and get instead

$$
\begin{aligned}
e^{-i\alpha} H'\psi' = H\psi & + \frac{e\hbar}{mc} i\nabla\left(\chi - \frac{\hbar c}{e}\alpha\right)\cdot\nabla\psi \\
& + \frac{e\hbar}{2mc} i\left\{\nabla^2\left(\chi - \frac{\hbar c}{e}\alpha\right) + \frac{e^2}{mc^2} A\cdot\nabla\left(\chi - \frac{\hbar c}{e}\alpha\right)\right\}\psi \\
& + \frac{e^2}{2mc^2}\left\{(\nabla\chi)^2 - 2\frac{\hbar c}{e}\nabla\chi\cdot\nabla\alpha + \frac{\hbar^2 c^2}{e^2}(\nabla\alpha)^2\right\}\psi \\
& - \frac{e}{c}\dot{\chi}\psi + V\psi .
\end{aligned}
$$

This expression apparently contracts in the desired manner by setting

$$\alpha = \frac{e}{\hbar c}\chi , \tag{125.7b}$$

after which there remains

$$e^{-i\alpha} H'\psi' = H\psi - \frac{e}{c}\dot{\chi}\psi .$$

Since the other side of the Schrödinger equation (125.3), by the same transformation, renders

$$-\frac{\hbar}{i} e^{-i\alpha}\dot{\psi}' = -\frac{\hbar}{i}(\dot{\psi} + i\dot{\alpha}\psi) = -\frac{\hbar}{i}\dot{\psi} - \frac{e}{c}\dot{\chi}\psi ,$$

where the last term will cancel on both sides, Eq. (125.3) again emerges in the primed quantities. This proves the gauge invariance of the theory if ψ is transformed according to (125.7a, b).

Problem 126. Current in presence of a magnetic field

To derive the current density for the Schrödinger equation with vector potential, and to prove its gauge invariance.

Solution. The probability conversation law has been derived in Problem 1 by constructing an equation of continuity from the Schrödinger equation and its complex conjugate. We shall proceed in the same way here, starting from

$$-\frac{\hbar^2}{2m}\nabla^2\psi+\frac{e\hbar}{mc}i(\boldsymbol{A}\cdot\nabla+\tfrac{1}{2}\operatorname{div}\boldsymbol{A})\psi$$
$$+\left(\frac{e^2}{2mc^2}A^2+e\Phi+V\right)\psi=-\frac{\hbar}{i}\frac{\partial\psi}{\partial t} \qquad (126.1\,a)$$

and

$$-\frac{\hbar^2}{2m}\nabla^2\psi^*-\frac{e\hbar}{mc}i\left(\boldsymbol{A}\cdot\nabla+\frac{1}{2}\operatorname{div}\boldsymbol{A}\right)\psi^*$$
$$+\left(\frac{e^2}{2mc^2}A^2+e\Phi+V\right)\psi^*=+\frac{\hbar}{i}\frac{\partial\psi^*}{\partial t}. \qquad (126.1\,b)$$

Multiplying by ψ^* and ψ, respectively, and subtracting (126.1b) from (126.1a) we get

$$-\frac{\hbar^2}{2m}(\psi^*\nabla^2\psi-\psi\nabla^2\psi^*)+\frac{e\hbar}{mc}i\{\boldsymbol{A}\cdot\nabla(\psi^*\psi)+\psi^*\psi\operatorname{din}\boldsymbol{A}\}=-\frac{\hbar}{i}\frac{\partial}{\partial t}(\psi^*\psi).$$

Since

$$\psi^*\nabla^2\psi-\psi\nabla^2\psi^*=\operatorname{div}(\psi^*\nabla\psi-\psi\nabla\psi^*)$$

and

$$\boldsymbol{A}\cdot\nabla(\psi^*\psi)+\psi^*\psi\operatorname{div}\boldsymbol{A}=\operatorname{div}(\boldsymbol{A}\psi^*\psi),$$

the left-hand side may be written as the divergence of a vector field. Keeping the definition of probability density,

$$\rho=\psi^*\psi \qquad (126.2)$$

we thus arrive at an equation of continuity,

$$\operatorname{div}\boldsymbol{s}+\frac{\partial\rho}{\partial t}=0 \qquad (126.3)$$

with the probability current density

$$\boldsymbol{s}=\frac{\hbar}{2mi}\left(\psi^*\nabla\psi-\psi\nabla\psi^*-2i\frac{e}{\hbar c}\boldsymbol{A}\psi^*\psi\right). \qquad (126.4)$$

This is a generalization of Eq. (1.6) derived for $\boldsymbol{A}=0$.

Applying the gauge transformation derived in the preceding problem,

$$\boldsymbol{A}'=\boldsymbol{A}+\nabla\chi;\quad \psi'=e^{i\alpha}\psi;\quad \alpha=\frac{e}{\hbar c}\chi, \qquad (126.5)$$

we find

$$(\psi^* \nabla \psi - \psi \nabla \psi^*)' = \psi^* (\nabla \psi + i \psi \nabla \alpha) - \psi (\nabla \psi^* - i \psi^* \nabla \alpha)$$
$$= (\psi^* \nabla \psi - \psi \nabla \psi^*) + 2 i \psi^* \psi \nabla \alpha$$

and

$$-2i \frac{e}{\hbar c} (A \psi^* \psi)' = -2i \frac{e}{\hbar c} (A \psi^* \psi + \psi^* \psi \nabla \chi)$$

$$= -2i \frac{e}{\hbar c} A \psi^* \psi - 2 i \psi^* \psi \nabla \alpha$$

thus showing the gauge invariance of the current density (126.4). Obviously, this invariance will hold for ρ, Eq. (126.2) as well and therefore for the law of conservation.

Problem 127. Normal Zeeman effect

To determine the stationary states of an electron in a central field to which an homogeneous magnetic field $\mathscr{H}$ is applied. The electron spin shall not be taken account of.

Solution. Neglecting effects of a relativistic order, the differential equation for the electron charge $-e$ becomes

$$-\frac{\hbar^2}{2m} \nabla^2 u - \frac{e\hbar}{mc} i (A \cdot \nabla) u + V u = E u. \tag{127.1}$$

Here A has been taken in a gauge making $\operatorname{div} A = 0$. This can, for an homogeneous field $\mathscr{H}$ in z direction be done by setting

$$A_x = -\tfrac{1}{2} \mathscr{H} y; \quad A_y = \tfrac{1}{2} \mathscr{H} x; \quad A_z = 0. \tag{127.2}$$

Then,

$$A \cdot \nabla u = \frac{1}{2} \mathscr{H} \left(x \frac{\partial u}{\partial y} - y \frac{\partial u}{\partial x} \right) = \frac{1}{2} \mathscr{H} \frac{\partial u}{\partial \varphi} \tag{127.3}$$

and (127.1) can be written

$$-\frac{\hbar^2}{2m} \nabla^2 u - \frac{e\hbar}{2mc} \mathscr{H} i \frac{\partial u}{\partial \varphi} + V u = E u. \tag{127.4}$$

The magnetic term may as well be written in the form

$$\frac{e}{2mc} \mathscr{H} \cdot \mathbf{L} \tag{127.5}$$

with $\boldsymbol{L}$ the angular momentum operator whose z component is

$$L_z = \frac{\hbar}{i}\frac{\partial}{\partial\varphi}.$$

According to the Maxwell theory of the electromagnetic field, a charge $-e$ moving with angular momentum $\boldsymbol{L}$ originates a magnetic dipole field of magnetic moment

$$\boldsymbol{M} = \frac{-e}{2mc}\boldsymbol{L} \tag{127.6}$$

so that (127.5) may be written in the form $-\boldsymbol{M}\cdot\mathscr{H}$. This is in fact the well-known expression for the potential energy of a dipole $\boldsymbol{M}$ in a field $\mathscr{H}$.

The differential equation (127.4) may be solved by

$$u = f_l(r)\, Y_{l,\mu}(\vartheta,\varphi); \tag{127.7}$$

the magnetic term then yields

$$\frac{e\hbar}{2mc}\mathscr{H}\,\mu\cdot u$$

so that we may write, instead of (127.4),

$$-\frac{\hbar^2}{2m}\nabla^2 u + Vu = \left(E - \frac{e\hbar}{2mc}\mathscr{H}\,\mu\right)u. \tag{127.8}$$

This is the same differential equation as the one for $\mathscr{H}=0$ whose eigenvalues we will denote by $E^0_{n,l}$. The eigenvalues of the system under action of the magnetic field then shift to

$$E_{n,l,\mu} = E^0_{n,l} + \frac{e\hbar}{2mc}\mathscr{H}\,\mu \tag{127.9}$$

or, as $\hbar\mu = L_z$, according to (127.6),

$$E_{n,l,\mu} = E^0_{n,l} - \boldsymbol{M}\cdot\mathscr{H}, \tag{127.10}$$

as was to be expected even by classical considerations.

The characteristic magnetic moment $e\hbar/(2mc)$ is called the *Bohr magneton*, the quantum number μ the *magnetic quantum number*. We have the same eigenstates now as for $\mathscr{H}=0$, but the orientation degeneracy in the energy values is destroyed by the magnetic field.

NB. The energy formula (127.9) might as well be obtained by treating the magnetic energy operator (127.5),

$$W_{\text{mag}} = \frac{e}{2mc}\mathscr{H} L_z, \tag{127.11}$$

as perturbation. The first-order energy shift then is the diagonal element of W_{mag} formed with the unperturbed eigenfunctions,

$$\Delta E_{n,l,\mu} = \langle n,l,\mu | W_{\text{mag}} | n,l,\mu \rangle$$
$$= \frac{e}{2mc}\mathscr{H} \int d^3x\, u^*_{n,l,\mu} \frac{\hbar}{i}\frac{\partial}{\partial\varphi} u_{n,l,\mu} = \frac{e}{2mc}\mathscr{H}\hbar\mu. \tag{127.12}$$

This approximation is in perfect agreement with the exact result (127.9) because the zero-order eigenfunctions in the integral (127.12) are identical with the exact ones.—See also Problem 216 for selection rules.

Problem 128. Paramagnetic and diamagnetic susceptibilities without spin

For one bound electron without spin in a central field, formulae shall be derived for the paramagnetic and diamagnetic susceptibilities.

Solution. If a magnetic field $\mathscr{H} \| z$ is applied to the atomic electron under consideration, having the hamiltonian H_0, we get

$$H = H_0 - \frac{e\hbar}{mc} i\, \boldsymbol{A}\cdot\nabla + \frac{e^2}{2mc^2}\boldsymbol{A}^2 \tag{128.1}$$

if the vector potential is normalized to vanishing divergence. This can be achieved with

$$A_x = -\tfrac{1}{2}\mathscr{H} y; \quad A_y = +\tfrac{1}{2}\mathscr{H} x; \quad A_z = 0 \tag{128.2}$$

so that

$$H = H_0 + \frac{e\hbar}{mc} i \frac{\mathscr{H}}{2}\left(y\frac{\partial}{\partial x} - x\frac{\partial}{\partial y}\right) + \frac{e^2}{2mc^2}\cdot\frac{\mathscr{H}^2}{4}(x^2+y^2)$$

or

$$H = H_0 + \frac{e\mathscr{H}}{2mc} L_z + \frac{e^2\mathscr{H}^2}{8mc^2} r^2 \sin^2\vartheta . \tag{128.3}$$

The magnetic energy perturbation, up to second order, in a state $|n\rangle$ (where n stands for all quantum numbers of the electron state) becomes

$$\Delta E_n = \left\langle n \left| \frac{e\mathscr{H}}{2mc} L_z + \frac{e^2\mathscr{H}^2}{8mc^2} r^2 \sin^2\vartheta \right| n \right\rangle + \sum_{n'}{}' \frac{\left|\left\langle n' \left| \frac{e\mathscr{H}}{2mc} L_z \right| n \right\rangle\right|^2}{E_n - E_{n'}} . \tag{128.4}$$

Let all the atoms of a substance be in their ground state (which is a reasonable assumption since the excitation energy, in general, amounts to several eV's and is thus much larger than the thermal energy available). Let this ground state have an angular momentum $\hbar l$, then there may yet occur any orientation with component $\hbar m_l$ in field direction. The first term in (128.4) then, for one atom, contributes the magnetic energy

$$E^{(1)} = \frac{e\mathscr{H}}{2mc}\hbar m_l \,. \tag{128.5}$$

The number of atoms with angular momentum component m_l is, in thermal equilibrium, proportional to the Boltzmann factor $\exp(-E^{(1)}/kT)$ so that the thermal average of the magnetic energy term becomes, per atom,

$$\overline{E^{(1)}} = \frac{e\hbar}{2mc}\mathscr{H}\frac{\sum_{m_l=-l}^{+l} m_l \mathrm{e}^{-\alpha m_l}}{\sum_{m_l=-l}^{+l} \mathrm{e}^{-\alpha m_l}} \quad \text{with } \alpha = \frac{e\hbar}{2mc}\frac{\mathscr{H}}{kT}. \tag{128.6}$$

For all reasonable field strengths $\mathscr{H}$, the dimensionless quantity α is small ($\alpha \ll 1$) so that the exponentials may be expanded,

$$\frac{\sum m_l \mathrm{e}^{-\alpha m_l}}{\sum \mathrm{e}^{-\alpha m_l}} = \frac{\sum(m_l - \alpha m_l^2 + \frac{1}{2}\alpha^2 m_l^3 \ldots)}{\sum(1 - \alpha m_l + \frac{1}{2}\alpha^2 m_l^2 \ldots)}.$$

All the sums are symmetrical about $m_l = 0$ so that all odd-power sums vanish and the expression reduces to

$$\frac{-\alpha \sum m_l^2 + O(\alpha^3)}{\sum 1 + O(\alpha^2)} = -\alpha \frac{\frac{1}{6}l(l+1)(2l+1) + \cdots}{(2l+1) + \cdots}$$

where the neglected terms are of the order α^2 smaller than the leading ones. Thus we arrive at the magnetic energy per c.c., with N atoms per c.c.,

$$\overline{E^{(1)}}N = -N\frac{e\hbar}{2mc}\mathscr{H}\left(\frac{e\hbar}{2mc}\frac{\mathscr{H}}{kT}\right)\frac{1}{6}l(l+1)\,. \tag{128.7}$$

Note that the energy becomes proportional to the square of the field strength although the original $E^{(1)}$ was linear.

The second term in (128.4),

$$E^{(2)} = \frac{e^2\mathscr{H}^2}{8mc^2}\langle n|r^2 \sin^2 \vartheta|n\rangle \tag{128.8}$$

turns out to be much smaller than $E^{(1)}$ (cf. below) so that it really plays no large role if the first term does not vanish. The same holds for the third term in (128.4). For all $l>0$, therefore, the first term suffices, whereas for $l=0$ the term $E^{(2)}$ determines the magnetic behaviour of the substance. Since

$$\frac{1}{4\pi}\oint d\Omega \sin^2\vartheta = \tfrac{2}{3},$$

we may write

$$E^{(2)} = \frac{e^2\mathscr{H}^2}{12mc^2}\langle 0|r^2|0\rangle \tag{128.9}$$

where $|0\rangle$ denotes a groundstate S wave function and $\langle 0|r^2|0\rangle$ the expectation value of r^2 in this state. No thermal averaging is needed. The first and third terms of (128.4) vanish for an S state because of $L_z|0\rangle=0$; so we need not consider them.

According to Maxwellian electrodynamics a change of $\mathscr{H}$ by $\delta\mathscr{H}$ changes the magnetic energy density by

$$\delta E = -\mathscr{M}\,\delta\mathscr{H} \tag{128.10}$$

where $\mathscr{M}$ is the magnetization, i.e. the magnetic moment per c.c. induced by the field. It is supposed to be proportional to $\mathscr{H}$,

$$\mathscr{M} = \chi\mathscr{H}, \tag{128.11}$$

the factor χ being called the magnetic susceptibility. Combining (128.10) with (128.11) we get

$$\delta E = -\chi\mathscr{H}\,\delta\mathscr{H}$$

so that the magnetic energy density becomes

$$E = -\tfrac{1}{2}\chi\mathscr{H}^2. \tag{128.12}$$

If χ is positive we speak of paramagnetism, if negative, of diamagnetism.

The macroscopic expression (128.12) of the magnetic energy density must be identified either with (128.7) if $l\neq 0$ or with N times Eq. (128.9) if $l=0$. In the first case we thus get paramagnetic behaviour with the positive susceptibility.

$$\chi_{\text{para}} = \frac{1}{3}N\left(\frac{e\hbar}{2mc}\right)^2\frac{l(l+1)}{kT}, \tag{128.13}$$

in the second case diamagnetic behaviour with the negative susceptibility

$$\chi_{\text{dia}} = -N\frac{e^2}{6mc^2}\langle 0|r^2|0\rangle. \tag{128.14}$$

Eq. (128.13) is identical with the classical Langevin expression for the orientation statistics of a constant dipole moment

$$\boldsymbol{\mu} = \frac{e}{2mc}\boldsymbol{L} \tag{128.15}$$

as was to be expected. The diamagnetic susceptibility is much smaller than the paramagnetic one (so that we were allowed to neglect it in paramagnetic substances) as is seen from the ratio

$$\frac{\chi_{\text{para}}}{\chi_{\text{dia}}} = \frac{\hbar^2 l(l+1)}{2m\langle r^2\rangle} \Big/ kT, \tag{128.16}$$

i.e. the ratio of rotational electron energy and thermal energy, the first[24] attaining several eV's, the second being of the order of 0.03 eV at room temperature.

---

[24] With $\langle r^2\rangle \sim (\hbar^2/me^2)^2$ of the order of the square of the Bohr radius, and $l=1$, the numerator of (128.16) becomes $me^4/\hbar^2 = 27.2$ eV.

Practical Quantum Mechanics

Volume II

III. Particles with Spin

A. One-Body Problems

Problem 129. Construction of Pauli matrices

A particle of spin $\frac{1}{2}$ has three basic properties:

1. It bears an intrinsic vector property that does not depend upon space coordinates.

2. This vector is an angular momentum (= spin) to be added to the orbital momentum of the particle.

3. If one of the components of the spin is measured, the result can be only one of its two eigenvalues, $+\frac{1}{2}\hbar$ or $-\frac{1}{2}\hbar$.

These properties can be described by using a two-component wave function and correspondingly 2×2 matrices for the spin operators. These matrices shall be constructed.

Solution. Let $\boldsymbol{S}=\frac{\hbar}{2}\boldsymbol{\sigma}$ be the spin vector operator; then according to property 2 the three components have to obey the commutation relations of angular momentum operators,

$$S_x S_y - S_y S_x = \hbar i S_z \quad \text{(etc. cyclic)} \tag{129.1 a}$$

or, written in the dimensionless operators σ_i,

$$\sigma_x \sigma_y - \sigma_y \sigma_x = 2i\sigma_z \quad \text{(etc. cyclic)}. \tag{129.1 b}$$

Since, according to property 3, each σ_i has eigenvalues $+1$ and -1 only, it ought to be possible to represent these operators by 2×2 matrices in a two-dimensional Hilbert space. They cannot be all diagonal in the same Hilbert coordinate system since they do not commute. Let us choose the Hilbert coordinate system in such a way that

$$\sigma_z = \begin{pmatrix} 1 & 0 \\ 0 & -1 \end{pmatrix} \tag{129.2}$$

is diagonal; then the unit vectors of the coordinate directions are

$$\alpha=\begin{pmatrix}1\\0\end{pmatrix} \quad \text{and} \quad \beta=\begin{pmatrix}0\\1\end{pmatrix} \tag{129.3}$$

so that

$$\sigma_z\alpha=\alpha; \qquad \sigma_z\beta=-\beta. \tag{129.4}$$

If a particle is in the state with Hilbert vector $\alpha(\beta)$ its spin points in the positive (negative) z direction.

We now write σ_x and σ_y in the general form

$$\sigma_x=\begin{pmatrix}a_{11} & a_{12}\\ a_{21} & a_{22}\end{pmatrix}; \quad \sigma_y=\begin{pmatrix}b_{11} & b_{12}\\ b_{21} & b_{22}\end{pmatrix}. \tag{129.5}$$

To determine their matrix elements we first use the two commutators (129.1 b) linear in σ_x and σ_y, viz.

$$\sigma_x\sigma_z-\sigma_z\sigma_x=-2i\sigma_y, \qquad \text{or:} \begin{pmatrix}0 & -2a_{12}\\ 2a_{21} & 0\end{pmatrix}=-2i\begin{pmatrix}b_{11} & b_{12}\\ b_{21} & b_{22}\end{pmatrix}$$

and

$$\sigma_y\sigma_z-\sigma_z\sigma_y=+2i\sigma_x, \qquad \text{or:} \begin{pmatrix}0 & -2b_{12}\\ 2b_{21} & 0\end{pmatrix}=+2i\begin{pmatrix}a_{11} & a_{12}\\ a_{21} & a_{22}\end{pmatrix}.$$

This yields,

$$a_{11}=a_{22}=b_{11}=b_{22}=0; \qquad b_{12}=-ia_{12}; \qquad b_{21}=+ia_{21} \tag{129.6}$$

so that there yet remain only two matrix elements a_{12} and a_{21} still to be fixed. The third commutation relation,

$$\sigma_x\sigma_y-\sigma_y\sigma_x=2i\sigma_z, \qquad \text{or:} \begin{pmatrix}2ia_{12}a_{21} & 0\\ 0 & -2ia_{12}a_{21}\end{pmatrix}=2i\begin{pmatrix}1 & 0\\ 0 & -1\end{pmatrix},$$

leads to one relation between them, viz.

$$a_{12}a_{21}=1. \tag{129.7}$$

Eqs. (129.6) and (129.7) yet leave one complex number, say a_{12}, undetermined. We arbitrarily fix that parameter:

$$a_{12}=1; \tag{129.8}$$

then we obtain the representation by the three Pauli matrices,

$$\sigma_x=\begin{pmatrix}0 & 1\\ 1 & 0\end{pmatrix}; \quad \sigma_y=\begin{pmatrix}0 & -i\\ i & 0\end{pmatrix}; \quad \sigma_z=\begin{pmatrix}1 & 0\\ 0 & -1\end{pmatrix}. \tag{129.9}$$

In terms of the eigenvectors of σ_z, Eq. (129.3), we may replace (129.9) by the equivalent relations

$$\begin{aligned} &\sigma_x \alpha = \beta; \quad \sigma_y \alpha = i\beta; \quad \sigma_z \alpha = \alpha; \\ &\sigma_x \beta = \alpha; \quad \sigma_y \beta = -i\alpha; \quad \sigma_z \beta = -\beta. \end{aligned} \tag{129.10}$$

Problem 130. Eigenstates of Pauli matrices

To determine the eigenvectors of the operators σ_x and σ_y and to show that $|a_{12}|^2 = 1$ is a necessary condition. What are the properties of the two "shift operators"

$$\sigma_+ = \sigma_x + i\sigma_y; \; \sigma_- = \sigma_x - i\sigma_y \tag{130.1}$$

and of the absolute square of the spin vector operator

$$\boldsymbol{\sigma}^2 = \sigma_x^2 + \sigma_y^2 + \sigma_z^2 \, ? \tag{130.2}$$

Solution. Writing a instead of a_{12}, the last problem gave the results

$$\sigma_x = \begin{pmatrix} 0 & a \\ 1/a & 0 \end{pmatrix}; \quad \sigma_y = \begin{pmatrix} 0 & -ia \\ i/a & 0 \end{pmatrix} \tag{130.3}$$

and hence,

$$\sigma_+ = \begin{pmatrix} 0 & 2a \\ 0 & 0 \end{pmatrix}; \quad \sigma_- = \begin{pmatrix} 0 & 0 \\ 2/a & 0 \end{pmatrix}. \tag{130.4}$$

Let

$$\psi = \begin{pmatrix} u \\ v \end{pmatrix} = u \begin{pmatrix} 1 \\ 0 \end{pmatrix} + v \begin{pmatrix} 0 \\ 1 \end{pmatrix} = u\alpha + v\beta \tag{130.5}$$

be a two-component wave function, then

$$\begin{aligned} &\sigma_x \begin{pmatrix} u \\ v \end{pmatrix} = \begin{pmatrix} av \\ u/a \end{pmatrix}; \quad \sigma_y \begin{pmatrix} u \\ v \end{pmatrix} = \begin{pmatrix} -iav \\ iu/a \end{pmatrix}; \\ &\sigma_+ \begin{pmatrix} u \\ v \end{pmatrix} = \begin{pmatrix} 2v \\ 0 \end{pmatrix}; \quad \sigma_- \begin{pmatrix} u \\ v \end{pmatrix} = \begin{pmatrix} 0 \\ 2u \end{pmatrix}. \end{aligned} \tag{130.6}$$

The eigenvectors of σ_x satisfy the equation $\sigma_x \psi = \lambda \psi$ with λ an eigenvalue, i.e. in components

$$av = \lambda u \quad \text{and} \quad u/a = \lambda v.$$

These relations are compatible with one another only for $\lambda = \pm 1$. Thus we find the eigensolutions

$$\begin{aligned} &\text{for } \lambda_1 = +1: \ \psi_1 = u\begin{pmatrix} 1 \\ 1/a \end{pmatrix} = u\left(\alpha + \frac{1}{a}\beta\right); \\ &\text{for } \lambda_2 = -1: \ \psi_2 = u\begin{pmatrix} 1 \\ -1/a \end{pmatrix} = u\left(\alpha - \frac{1}{a}\beta\right). \end{aligned} \tag{130.7}$$

The probabilities that the spin may point upward (i.e. in the $+z$ direction) and downward are proportional to the absolute squares of the factors in front of the Hilbert vectors α and β, viz. $1:1/|a|^2$. Since there is no reason why the one should be greater than the other, it follows that

$$|a|^2 = 1. \tag{130.8}$$

The same consideration is possible for σ_y.

Let us, from now on, definitely choose $a=1$. Then we have eigenvalues $\lambda_1 = +1$ and $\lambda_2 = -1$ for each of the three σ_i's and the following eigenvectors.

$$\sigma_x = \begin{pmatrix} 0 & 1 \\ 1 & 0 \end{pmatrix}; \quad \psi_1 = 2^{-\frac{1}{2}}\begin{pmatrix} 1 \\ 1 \end{pmatrix}; \quad \psi_2 = 2^{-\frac{1}{2}}\begin{pmatrix} 1 \\ -1 \end{pmatrix}; \tag{130.9a}$$

$$\sigma_y = \begin{pmatrix} 0 & -i \\ i & 0 \end{pmatrix}; \quad \psi_1 = 2^{-\frac{1}{2}}\begin{pmatrix} 1 \\ i \end{pmatrix}; \quad \psi_2 = 2^{-\frac{1}{2}}\begin{pmatrix} 1 \\ -i \end{pmatrix}; \tag{130.9b}$$

$$\sigma_z = \begin{pmatrix} 1 & 0 \\ 0 & -1 \end{pmatrix}; \quad \psi_1 = \begin{pmatrix} 1 \\ 0 \end{pmatrix}; \quad \psi_2 = \begin{pmatrix} 0 \\ 1 \end{pmatrix}. \tag{130.9c}$$

The three Pauli matrices are hermitian operators, $\sigma_i^\dagger = \sigma_i$, with real eigenvalues. The two operators

$$\sigma_+ = \begin{pmatrix} 0 & 2 \\ 0 & 0 \end{pmatrix} \quad \text{and} \quad \sigma_- = \begin{pmatrix} 0 & 0 \\ 2 & 0 \end{pmatrix} \tag{130.10}$$

are not hermitian, $\sigma_+^\dagger = \sigma_-$ and $\sigma_-^\dagger = \sigma_+$. They have no eigenvalues or eigenvectors, because they cannot even be diagonalized. This may be seen as follows.

The most general unitary matrix in two dimensions may be written (apart from an irrelevant phase factor)

$$U = \begin{pmatrix} \cos\vartheta; & \sin\vartheta\, e^{i\xi} \\ -\sin\vartheta\, e^{i\eta}; & \cos\vartheta\, e^{i(\xi+\eta)} \end{pmatrix}$$

with real parameters ϑ, ξ, η. The unitary transformation of σ_+ then becomes

$$U^\dagger \sigma_+ U = 2\,e^{i\eta}\begin{pmatrix} -\sin\vartheta\cos\vartheta; & \cos^2\vartheta\, e^{i\xi} \\ -\sin^2\vartheta\, e^{-i\xi}; & \sin\vartheta\cos\vartheta \end{pmatrix},$$

and this cannot be made diagonal for any choice of the real parameters since $\sin\vartheta$ and $\cos\vartheta$ never vanish at the same argument ϑ.

If the operators σ_+ and σ_- are applied to the Hilbert vectors α and β we find according to (130.6):

$$\begin{aligned} &\sigma_+\alpha=0; && \sigma_-\alpha=2\beta; && \sigma_+\sigma_-\alpha=4\alpha; && \sigma_-\sigma_+\alpha=0; \\ &\sigma_+\beta=2\alpha; && \sigma_-\beta=0; && \sigma_+\sigma_-\beta=0; && \sigma_-\sigma_+\beta=4\beta. \end{aligned} \tag{130.11}$$

When commuted with σ_z, these operators are essentially reproduced,

$$\sigma_+\sigma_z - \sigma_z\sigma_+ = -2\sigma_+; \qquad \sigma_-\sigma_z - \sigma_z\sigma_- = +2\sigma_-. \tag{130.12}$$

The operators σ_+ and σ_-, in angular momentum normalization,

$$S_+ = \frac{\hbar}{2}\sigma_+; \qquad S_- = \frac{\hbar}{2}\sigma_- \tag{130.13}$$

have the same property as the analogous operators L_+ and L_- (Problem 56) to shift the z component by one (in units $\hbar$):

$$\begin{aligned} &S_+\alpha=0; && S_-\alpha=\hbar\beta; \\ &S_+\beta=\hbar\alpha; && S_-\beta=0. \end{aligned}$$

State β with spin component $-\frac{1}{2}\hbar$ is altered by S_+ into state α with $+\frac{1}{2}\hbar$, and inversely by S_-. Both, $S_+\alpha$ and $S_-\beta$ must vanish because, according to this shift rule, they would lead to states with $+\frac{3}{2}\hbar$ and $-\frac{3}{2}\hbar$ not existing in the Hilbert space used.

We finally investigate the absolute square of the spin operator,

$$\boldsymbol{\sigma}^2 = \sigma_x^2 + \sigma_y^2 + \sigma_z^2 = \tfrac{1}{2}(\sigma_+\sigma_- + \sigma_-\sigma_+) + \sigma_z^2. \tag{130.14}$$

It is easily seen that all three σ_i^2 are the unit matrix so that

$$\boldsymbol{\sigma}^2 = 3\begin{pmatrix} 1 & 0 \\ 0 & 1 \end{pmatrix}$$

is diagonal with its only value 3 for whatever Hilbert vector we choose, or—in less sophisticated wording—it is $=3$. This can also be seen from the second form in (130.14) using the relations for $\sigma_+\sigma_-$ and $\sigma_-\sigma_+$ given in (130.11).

Eq. (130.14) leads to $S^2 = \frac{3}{4}\hbar^2$ or, with S being the quantum number describing the spin, to $\hbar^2 S(S+1)$ with $S=\frac{1}{2}$. This is the meaning of a state to have "spin $\frac{1}{2}$".

Problem 131. Spin algebra

To show that the three Pauli matrices together with unity form the complete basis of an algebra.

Solution. If 1, $\sigma_x, \sigma_y, \sigma_z$ form a complete basis, no number outside the algebra can be generated either by adding or by multiplying any pair of numbers of the form

$$N = a_0 + a_1 \sigma_x + a_2 \sigma_y + a_3 \sigma_z \tag{131.1}$$

with complex number coefficients a_i. Obviously the rule obtains for addition, but it still has to be proved for a product of two numbers. For this purpose we shall construct a multiplication table of the Pauli matrices.

Let i, k, l be an arbitrary cyclic permutation of the three subscripts x, y, z, then the σ_i's satisfy the commutation relations

$$\sigma_i \sigma_k - \sigma_k \sigma_i = 2i\sigma_l \tag{131.2}$$

and the normalization relations

$$\sigma_i^2 = 1. \tag{131.3}$$

Besides, the Pauli matrices obey anticommutation rules,

$$\sigma_i \sigma_k + \sigma_k \sigma_i = 0 \quad (i \neq k) \tag{131.4}$$

as may easily be verified. Then, by addition and subtraction of (131.2) and (131.4), we find the products

$$\sigma_i \sigma_k = i\sigma_l; \qquad \sigma_k \sigma_i = -i\sigma_l. \tag{131.5}$$

Hence, any product of two basis elements leads back again, except for a complex number factor, to another basis element.

Multiplication table

first factor	second factor 1	σ_x	σ_y	σ_z
1	1	σ_x	σ_y	σ_z
σ_x	σ_x	1	$i\sigma_z$	$-i\sigma_y$
σ_y	σ_y	$-i\sigma_z$	1	$i\sigma_x$
σ_z	σ_z	$i\sigma_y$	$-i\sigma_x$	1

It should be noted that the product of all three Pauli matrices therefore simply becomes

$$\sigma_x \sigma_y \sigma_z = i. \tag{131.6}$$

Further, another argument may now be added for σ_+ and σ_- having no eigenvalues. From the multiplication table one gets

$$\sigma_\pm^2 = (\sigma_x \pm i\sigma_y)^2 = \sigma_x^2 - \sigma_y^2 \pm i(\sigma_x\sigma_y + \sigma_y\sigma_x) = 0.$$

This is an interesting result showing that in the Pauli algebra the square of a non-vanishing number may be zero. It is, by the way, corroborated by the fact that neither for σ_+ nor σ_- do any reciprocal numbers exist.

A number N for which the relation

$$N = N^2 \tag{131.7}$$

holds is called an idempotent number. Such numbers belong to our algebra, viz.

$$\tfrac{1}{2}(1+\sigma_i) \quad \text{and} \quad \tfrac{1}{2}(1-\sigma_i) \qquad (i = x, y, z). \tag{131.8}$$

In matrix representation we have e.g.

$$P_+ = \tfrac{1}{2}(1+\sigma_z) = \begin{pmatrix} 1 & 0 \\ 0 & 0 \end{pmatrix}; \qquad P_- = \tfrac{1}{2}(1-\sigma_z) = \begin{pmatrix} 0 & 0 \\ 0 & 1 \end{pmatrix};$$

applied to the basic Hilbert vectors α and β they give

$$P_+\alpha = \alpha; \qquad P_-\alpha = 0;$$
$$P_+\beta = 0; \qquad P_-\beta = \beta.$$

Thus they suppress either β or α in a state vector of mixed spin orientations:

$$P_+(u\alpha + v\beta) = u\alpha; \qquad P_-(u\alpha + v\beta) = v\beta$$

and leave us with the projection of the state vector upon one of the basic directions in Hilbert space. They are therefore called projection operators.

NB. The Pauli algebra is essentially the same as the algebra of quaternions which uses $i\sigma_k$ instead of σ_k as basis elements.

Problem 132. Spinor transformation properties

How can it be proved that the spin of a one-particle state,

$$\boldsymbol{s} = \int d^3x\, \psi^\dagger \boldsymbol{\sigma} \psi, \tag{132.1}$$

is a vector? It should be noted that the σ_i's shall not transform with the coordinates under space rotation, but the transformation properties of **s** shall rest entirely upon the wave function.

Solution. In consequence of the group property of space rotations, it suffices to investigate an infinitesimal rotation,

$$x'_i = x_i + \sum_k{}' \varepsilon_{ik} x_k; \qquad \varepsilon_{ki} = -\varepsilon_{ik} \tag{132.2}$$

with infinitesimal angles of rotation about the three axes,

$$\varepsilon_{12} = \alpha_3; \qquad \varepsilon_{23} = \alpha_1; \qquad \varepsilon_{31} = \alpha_2. \tag{132.3}$$

If **s** is a vector, it has to obey the same transformation rule,

$$s'_i = s_i + \sum_k{}' \varepsilon_{ik} s_k. \tag{132.4}$$

This has to be achieved by transforming only the wave function,

$$\psi' = (1+\xi)\psi; \qquad \psi'^{\dagger} = \psi^{\dagger}(1+\xi^{\dagger}) \tag{132.5}$$

with infinitesimal ξ. This transformation shall now be determined.

We begin by stating that $\psi^{\dagger}\psi$ is a scalar so that

$$\psi'^{\dagger}\psi' = \psi^{\dagger}(1+\xi^{\dagger})(1+\xi)\psi = \psi^{\dagger}\psi$$

or

$$\xi^{\dagger} = -\xi. \tag{132.6}$$

Putting (132.4) and (132.5) in (132.1), we get

$$s'_i = \int d^3x\, \psi^{\dagger}(1-\xi)\sigma_i(1+\xi)\psi = s_i + \int d^3x\, \psi^{\dagger}(\sigma_i\xi - \xi\sigma_i)\psi.$$

Comparison of this expression with (132.4) leads to the determining equations for the operator ξ:

$$\sigma_i\xi - \xi\sigma_i = \sum_k{}' \varepsilon_{ik}\sigma_k \qquad (i,k=1,2,3). \tag{132.7}$$

It can easily be shown that these equations are solved unambiguously by

$$\xi = \frac{i}{2}(\varepsilon_{12}\sigma_3 + \varepsilon_{23}\sigma_1 + \varepsilon_{31}\sigma_2) \tag{132.8}$$

because, using the commutation rules, we find for the left-hand side of Eq. (132.7), e. g. with $i=1$:

$$\frac{i}{2}[\sigma_1,\ \varepsilon_{12}\sigma_3 + \varepsilon_{23}\sigma_1 + \varepsilon_{31}\sigma_2] = \frac{i}{2}\{-\varepsilon_{12}\cdot 2i\sigma_2 + \varepsilon_{31}\cdot 2i\sigma_3\}$$

$$= \varepsilon_{12}\sigma_2 + \varepsilon_{13}\sigma_3$$

in agreement with the right-hand side. Analogous results are found for $i=2$ or 3.

The operator (132.8) can be written in a simpler form using the angles of rotation in the notation (132.3),

$$\xi = \frac{i}{2} \sum_k \alpha_k \sigma_k . \tag{132.9}$$

Applied to any two-component wave function,

$$\psi = \begin{pmatrix} u \\ v \end{pmatrix} = u\alpha + v\beta \tag{132.10}$$

this leads to the transformed wave function

$$\psi' = \begin{pmatrix} u' \\ v' \end{pmatrix} = u'\alpha + v'\beta \tag{132.10'}$$

with

$$\begin{aligned} u' &= \left(1 + \frac{i}{2}\alpha_3\right) u + \frac{i}{2}(\alpha_1 - i\alpha_2) v ; \\ v' &= \frac{i}{2}(\alpha_1 + i\alpha_2) u + \left(1 - \frac{i}{2}\alpha_3\right) v . \end{aligned} \tag{132.11}$$

A two-component function with these transformation properties is called a *spinor*.

Problem 133. Spin electron in a central field

To determine the wave functions of a spinning electron in a central spin-independent force field. The wave functions must be eigenfunctions of the two operators

$$\boldsymbol{J}^2 = (\boldsymbol{L} + \boldsymbol{S})^2 \quad \text{and} \quad J_z = L_z + S_z \tag{133.1}$$

with $\boldsymbol{L}$ the orbital momentum and $\boldsymbol{S}$ the spin of the electron.

Solution. We begin with the z component of the angular momentum. Since

$$L_z = \frac{\hbar}{i} \frac{\partial}{\partial \varphi} ; \quad S_z = \frac{\hbar}{2} \sigma_z ; \quad \sigma_z = \begin{pmatrix} 1 & 0 \\ 0 & -1 \end{pmatrix}$$

we can write

$$J_z = \hbar \begin{pmatrix} -i \dfrac{\partial}{\partial \varphi} + \dfrac{1}{2} ; & 0 \\ 0 ; & -i \dfrac{\partial}{\partial \varphi} - \dfrac{1}{2} \end{pmatrix} . \tag{133.2}$$

The eigenfunctions of this operator have the form

$$\psi=\begin{pmatrix} C_1\,e^{i(m_j-\frac{1}{2})\varphi} \\ C_2\,e^{i(m_j+\frac{1}{2})\varphi} \end{pmatrix} \tag{133.3}$$

with C_1 and C_2 still arbitrary functions of the variables r and ϑ. This can easily be seen by letting J_z operate on ψ; the result is

$$J_z\psi=\hbar m_j\psi \tag{133.4}$$

so that $\hbar m_j$ is an eigenvalue of the z component of the total angular momentum $\boldsymbol{J}$. A better understanding of Eq. (133.3) is achieved by using the notation

$$\psi=C_1\,e^{i(m_j-\frac{1}{2})\varphi}\alpha+C_2\,e^{i(m_j+\frac{1}{2})\varphi}\beta \tag{133.3$'$}$$

because it is then seen that the first term describes the dependence of ψ on the coordinates if the spin points upward, and the second term if it points downward. This coordinate dependence determines the component $\hbar m_l$ of the orbital momentum $\boldsymbol{L}$ which makes $m_l=0, \pm 1, \pm 2, \ldots$ an integer. With spin upward we have $m_j=m_l+\frac{1}{2}$ and with spin downward $m_j=m_l-\frac{1}{2}$ so that m_j becomes a half-integer. These are the addition rules well-known for the vector model. The characteristic feature of the wave function (133.3) or (133.3′) is that m_j is a "good quantum number" but that m_l is not, because the state vector ψ is a mixture of two parts with different values of m_l.

We now investigate the operator $\boldsymbol{J}^2$. We have

$$\boldsymbol{J}^2=\boldsymbol{L}^2+\boldsymbol{S}^2+2(\boldsymbol{L}\cdot\boldsymbol{S})=\boldsymbol{L}^2+\tfrac{3}{4}\hbar^2+\hbar\begin{pmatrix} L_z & L_+ \\ L_- & -L_z \end{pmatrix}$$

if the Pauli matrices are used for $\boldsymbol{S}=\dfrac{\hbar}{2}\boldsymbol{\sigma}$. We therefore have to solve the eigenvalue problem

$$\boldsymbol{J}^2\psi=\begin{pmatrix} \boldsymbol{L}^2+\frac{3}{4}\hbar^2+\hbar L_z; & \hbar L_+ \\ \hbar L_-; & \boldsymbol{L}^2+\frac{3}{4}\hbar^2-\hbar L_z \end{pmatrix}\psi=\hbar^2 j(j+1)\psi \tag{133.5}$$

if, in analogy to $\boldsymbol{L}^2$, we arbitrarily call the eigenvalue $\hbar^2 j(j+1)$. In order to make ψ simultaneously an eigenfunction of J_z and $\boldsymbol{J}^2$ it must have the form (133.3) where now the dependence of C_1 and C_2 upon the variable ϑ has to be determined. This can be done by putting

$$\psi=\begin{pmatrix} f(r)\,Y_{l,m_j-\frac{1}{2}}(\vartheta,\varphi) \\ g(r)\,Y_{l,m_j+\frac{1}{2}}(\vartheta,\varphi) \end{pmatrix}. \tag{133.6}$$

The spherical harmonics in (133.6) depend upon φ just in the way of Eq. (133.3). When $\boldsymbol{J}^2\psi$ is formed from (133.5) and (133.6), and we apply the general formulae (cf. Problem 56)

$$L_+ Y_{l,m} = -\hbar\sqrt{(l+m+1)(l-m)}\, Y_{l,m+1}; \tag{133.7a}$$
$$L_- Y_{l,m} = -\hbar\sqrt{(l+m)(l-m+1)}\, Y_{l,m-1}; \tag{133.7b}$$
$$L_z Y_{l,m} = \hbar m\, Y_{l,m}; \tag{133.7c}$$
$$L^2 Y_{l,m} = \hbar^2 l(l+1)\, Y_{l,m}, \tag{133.7d}$$

we obtain

$$J^2\psi = \hbar^2 \begin{pmatrix} f(r)\left[l(l+1)+\frac{3}{4}+(m_j-\frac{1}{2})\right] Y_{l,m_j-\frac{1}{2}} \\ -g(r)\sqrt{(l+m_j+\frac{1}{2})(l-m_j+\frac{1}{2})}\, Y_{l,m_j-\frac{1}{2}} \\ \\ -f(r)\sqrt{(l+m_j+\frac{1}{2})(l-m_j-\frac{1}{2})}\, Y_{l,m_j+\frac{1}{2}} \\ +g(r)\left[l(l+1)+\frac{3}{4}-(m_j+\frac{1}{2})\right] Y_{l,m_j+\frac{1}{2}} \end{pmatrix}$$

so that the eigenvalue problem (133.5) leads to two linear algebraic equations for $f(r)$ and $g(r)$, viz.

$$\begin{aligned} f\left[l(l+1)+\tfrac{3}{4}+(m_j-\tfrac{1}{2})-j(j+1)\right]-g\sqrt{(l+\tfrac{1}{2})^2-m_j^2}&=0; \\ -f\sqrt{(l+\tfrac{1}{2})^2-m_j^2}+g\left[l(l+1)+\tfrac{3}{4}-(m_j+\tfrac{1}{2})-j(j+1)\right]&=0. \end{aligned} \tag{133.8}$$

The possibility of thus eliminating the spherical harmonics shows that the function (133.6) suffices to solve the problem. Thus l is still a "good" quantum number even in spin theory.

The equations (133.8) are compatible with one another only if $f(r)$ and $g(r)$ differ only by a constant factor. We write

$$f(r) = A\,F(r); \qquad g(r) = B\,F(r). \tag{133.9}$$

Then (133.8) permits to determine the ratio B/A. As the linear equations (133.8) are homogeneous, their determinant must vanish,

$$\left[j(j+1)-(l+\tfrac{1}{2})^2\right]^2-m_j^2-\left[(l+\tfrac{1}{2})^2-m_j^2\right]=0. \tag{133.10}$$

Obviously, this condition is independent of m_j. That is one of the simplest consequences of a very general theorem of Wigner and Eckart. There are two different values of $j(j+1)$ satisfying (133.10), viz.

Solution I. $j=l+\frac{1}{2}$; $\quad B=-A\sqrt{\dfrac{l+\frac{1}{2}-m_j}{l+\frac{1}{2}+m_j}}$;

$$\psi_{\mathrm{I}} = \frac{F_l(r)}{\sqrt{2l+1}} \begin{pmatrix} \sqrt{l+\frac{1}{2}+m_j}\, Y_{l,m_j-\frac{1}{2}} \\ -\sqrt{l+\frac{1}{2}-m_j}\, Y_{l,m_j+\frac{1}{2}} \end{pmatrix}. \tag{133.11}$$

Solution II. $j=l-\frac{1}{2}$; $\quad B=A\sqrt{\dfrac{l+\frac{1}{2}+m_j}{l+\frac{1}{2}-m_j}}$;

$$\psi_{\mathrm{II}} = \frac{F_l(r)}{\sqrt{2l+1}} \begin{pmatrix} \sqrt{l+\frac{1}{2}-m_j}\, Y_{l,m_j-\frac{1}{2}} \\ \sqrt{l+\frac{1}{2}+m_j}\, Y_{l,m_j+\frac{1}{2}} \end{pmatrix}. \tag{133.12}$$

Both solutions are normalized. Since in all component functions a spherical harmonic of the same order l is split off, and since the potential is supposed not to depend upon spin, the function F_l is to be determined from the radial Schrödinger equation,

$$-\frac{\hbar^2}{2m}\left(\chi_l'' - \frac{l(l+1)}{r^2}\chi_l\right) + V(r)\chi_l = E_l\chi_l \tag{133.13}$$

with

$$\chi_l = r F_l(r). \tag{133.14}$$

NB. The formulae (133.11) and (133.12) may be applied to states with $l=0$ if attention is paid to the fact that the spherical harmonics then vanish if $m_j \mp \frac{1}{2}$ does not equal zero. Thus we get

$$\psi_{\mathrm{I}} = \frac{F_0(r)}{\sqrt{4\pi}}\begin{pmatrix}1\\0\end{pmatrix} \quad \text{for } l=0, m_j=+\tfrac{1}{2}$$

and

$$\psi_{\mathrm{I}} = -\frac{F_0(r)}{\sqrt{4\pi}}\begin{pmatrix}0\\1\end{pmatrix} \quad \text{for } l=0, m_j=-\tfrac{1}{2}.$$

The other function, ψ_{II}, vanishes identically in both cases so that no solutions with negative j are originated. The same results are found by applying the operators $\boldsymbol{J}^2$ and J_z to a two-component function that depends on the radius only.

Problem 134. Quadrupole moment of a spin state

To determine the quadrupole moment of a one-electron state in a potential field of spherical symmetry, taking the electron spin into account.

Solution. The eigenfunctions (cf. Eqs. (133.11, 12)) have the absolute squares

$$|\psi_{\mathrm{I}}|^2 = \frac{|F_l(r)|^2}{2l+1}\{(l+\tfrac{1}{2}+m_j)|Y_{l,m_j-\frac{1}{2}}|^2 + (l+\tfrac{1}{2}-m_j)|Y_{l,m_j+\frac{1}{2}}|^2\} \tag{134.1a}$$

if $j=l+\frac{1}{2}$, and

$$|\psi_{\mathrm{II}}|^2 = \frac{|F_l(r)|^2}{2l+1}\{(l+\tfrac{1}{2}-m_j)|Y_{l,m_j-\frac{1}{2}}|^2 + (l+\tfrac{1}{2}+m_j)|Y_{l,m_j+\frac{1}{2}}|^2\} \tag{134.1b}$$

if $j=l-\frac{1}{2}$. It should be noted that $|m_j| \leqq j$, and that for $l=0$ the function ψ_{II} vanishes identically.

Since (134.1a, b) do not depend upon the angle φ, the argument of Problem 61 still holds, so that the non-diagonal elements of the quadrupole tensor have vanishing averages and that

$$\langle Q_{xx}\rangle = \langle Q_{yy}\rangle = -\tfrac{1}{2}\langle Q_{zz}\rangle. \tag{134.2}$$

Thus again, we need only calculate $\langle Q_{zz}\rangle$ according to the formula

$$\langle Q_{zz}\rangle = \int d^3x\, |\psi|^2 r^2 (3\cos^2\vartheta - 1). \tag{134.3}$$

With the relation

$$\oint d\Omega (3\cos^2\vartheta - 1)|Y_{l,m}|^2 = \frac{2l(l+1)-6m^2}{(2l-1)(2l+3)} \tag{134.4}$$

proved in Problem 61, and the abbreviation

$$\int_0^\infty dr\, r^4 |F_l(r)|^2 = \langle r^2\rangle, \tag{134.5}$$

we get from (134.1a, b):

$$\langle Q_{zz}\rangle = \frac{\langle r^2\rangle}{2l+1}\left\{\left(l+\frac{1}{2}\pm m_j\right)\frac{2l(l+1)-6(m_j-\frac{1}{2})^2}{(2l-1)(2l+3)} + \left(l+\frac{1}{2}\mp m_j\right)\frac{2l(l+1)-6(m_j+\frac{1}{2})^2}{(2l-1)(2l+3)}\right\}$$

with the upper signs for $j=l+\frac{1}{2}$, the lower ones for $j=l-\frac{1}{2}$ and $l\geqq 1$. An elementary reordering in the curly bracket leads to

$$\langle Q_{zz}\rangle = \langle r^2\rangle \frac{2l(l+1)-6\left(m_j^2+\frac{1}{4}\right)\pm\frac{12}{2l+1}m_j^2}{(2l-1)(2l+3)}, \tag{134.6}$$

a formula to be compared with (61.8).

It is possible, by using j instead of l, to write instead of (134.6) one comprehensive formula,

$$\langle Q_{zz}\rangle = \langle r^2\rangle \cdot \frac{1}{2}\left(1-\frac{3m_j^2}{j(j+1)}\right) \tag{134.7}$$

wich obtains for both signs, $j=l\pm\frac{1}{2}$, equally. Numerical results for the lowest values of j are listed in the accompanying table.

States	j	$\langle Q_{zz}\rangle/\langle r^2\rangle$ for $m_j=\pm\frac{1}{2}$	$m_j=\pm\frac{3}{2}$	$m_j=\pm\frac{5}{2}$	$m_j=\pm\frac{7}{2}$
$S_{\frac{1}{2}}, P_{\frac{1}{2}}$	$\frac{1}{2}$	0	—	—	—
$P_{\frac{3}{2}}, D_{\frac{3}{2}}$	$\frac{3}{2}$	$+\frac{2}{5}$	$-\frac{2}{5}$	—	—
$D_{\frac{5}{2}}, F_{\frac{5}{2}}$	$\frac{5}{2}$	$+\frac{16}{35}$	$+\frac{4}{35}$	$-\frac{20}{35}$	—
$F_{\frac{7}{2}}, G_{\frac{7}{2}}$	$\frac{7}{2}$	$+\frac{10}{21}$	$+\frac{6}{21}$	$-\frac{2}{21}$	$-\frac{14}{21}$

It can be seen that for $j=\frac{1}{2}$ spherical symmetry obtains for S as well as for P states. Generally, it may be stated that with increasing values of $|m_j|$ the configuration passes from oblong to oblate figure of the electron distribution.

The sum of the moments in each horizontal line of the table vanishes, since it leads to a closed shell of spherical symmetry. This can quite generally be shown as

$$\sum_{m_j=\frac{1}{2}}^{j} 1 = j+\tfrac{1}{2} \quad \text{and} \quad \sum_{m_j=\frac{1}{2}}^{j} m_j^2 = \tfrac{1}{3}j(j+\tfrac{1}{2})(j+1)$$

which by combination lead to

$$\sum_{m_j=\frac{1}{2}}^{j} \langle Q_{zz}\rangle = \tfrac{1}{2}\langle r^2\rangle \left\{ \sum_{m_j=\frac{1}{2}}^{j} 1 - \frac{3}{j(j+1)} \sum_{m_j=\frac{1}{2}}^{j} m_j^2 \right\} = 0.$$

Problem 135. Expectation values of magnetic moments

For the spin electron in a central field there shall be derived the expectation values of all three components of the vectors $\boldsymbol{S}, \boldsymbol{L}, \boldsymbol{J}$ and of the magnetic moment.

Solution. If

$$u = \begin{pmatrix} u_1 \\ u_2 \end{pmatrix}$$

is the eigenspinor of a state, we have

$$u^\dagger S_x u = \frac{\hbar}{2}(u_1^* u_2 + u_2^* u_1); \qquad u^\dagger S_y u = \frac{\hbar}{2i}(u_1^* u_2 - u_2^* u_1);$$

$$u^\dagger S_z u = \frac{\hbar}{2}(u_1^* u_1 - u_2^* u_2). \tag{135.1}$$

The eigenspinors of $\boldsymbol{J}^2$ and J_z have been determined in Problem 133; they are

$$u_1 = \frac{F_l(r)}{\sqrt{2l+1}} A_{j,l} Y_{l,m_j-\frac{1}{2}}; \qquad u_2 = \frac{F_l(r)}{\sqrt{2l+1}} B_{j,l} Y_{l,m_j+\frac{1}{2}} \tag{135.2}$$

with

$$A_{l+\frac{1}{2},l} = \sqrt{l+\tfrac{1}{2}+m_j}; \qquad B_{l+\frac{1}{2},l} = \sqrt{l+\tfrac{1}{2}-m_j} \tag{135.3a}$$

and

$$A_{l-\frac{1}{2},l} = \sqrt{l+\tfrac{1}{2}-m_j}; \qquad B_{l-\frac{1}{2},l} = \sqrt{l+\tfrac{1}{2}+m_j}. \tag{135.3b}$$

In the expressions (135.1) underlying the expectation values of S_x and S_y this leads to products of different spherical harmonics so that $\langle S_x\rangle=0$ and $\langle S_y\rangle=0$. On the other hand, we find

$$\langle S_z\rangle = \frac{\hbar}{2}\int_0^\infty dr\, r^2 \frac{|F_l(r)|^2}{2l+1}(A_{j,l}^2 - B_{j,l}^2) \tag{135.4}$$

and for the normalization integral

$$\int_0^\infty dr\, r^2 \frac{|F_l(r)|^2}{2l+1}(A_{j,l}^2 + B_{j,l}^2) = \int_0^\infty dr\, r^2 |F_l(r)|^2 = 1. \tag{135.5}$$

Thus we get

$$\langle S_z\rangle = \frac{\hbar}{2}\,\frac{A_{j,l}^2 - B_{j,l}^2}{A_{j,l}^2 + B_{j,l}^2} \tag{135.6}$$

which, in a state with $j=l+\frac{1}{2}$, is

$$\langle S_z\rangle_+ = \hbar m_j \cdot \frac{1}{2l+1} = \hbar m_j \cdot \frac{1}{2j} \tag{135.7 a}$$

and in a state with $j=l-\frac{1}{2}$,

$$\langle S_z\rangle_- = -\hbar m_j \cdot \frac{1}{2l+1} = -\hbar m_j \cdot \frac{1}{2(j+1)}. \tag{135.7 b}$$

The expectation values of the three components of orbital momentum follow in a similar way from

$$u^\dagger \boldsymbol{L} u = u_1^* \boldsymbol{L} u_1 + u_2^* \boldsymbol{L} u_2 .$$

Since the operators $L_x \pm i L_y$ change the first subscript of the spherical harmonics from l into $l\pm1$, these components again have vanishing expectation values (cf. Problem 58). For the component L_z we have

$$\langle L_z\rangle = \hbar \frac{\int d^3x\,\{u_1^*(m_j-\frac{1}{2})u_1 + u_2^*(m_j+\frac{1}{2})u_2\}}{\int d^3x\,\{u_1^* u_1 + u_2^* u_2\}}$$

or

$$\langle L_z\rangle = \hbar \cdot \frac{A_{j,l}^2(m_j-\frac{1}{2}) + B_{j,l}^2(m_j+\frac{1}{2})}{A_{j,l}^2 + B_{j,l}^2}.$$

Using (135.6), this can be written much more simply

$$\langle L_z\rangle = \hbar m_j - \langle S_z\rangle, \tag{135.8}$$

a formula which we might well have started with because of u being eigenspinor of $J_z = L_z + S_z$ with the eigenvalue $\hbar m_j$.

The expectation values of J_x and J_y of course vanish, as do these two components of $\boldsymbol{L}$ and $\boldsymbol{S}$.

The *magnetic moment* operator is

$$\boldsymbol{M} = -\frac{e}{2mc}(\boldsymbol{L}+2\boldsymbol{S}) \tag{135.9}$$

for an electron of charge $-e$. Its expectation values in x and y direction vanish again, but

$$\langle M_z\rangle = -\frac{e}{2mc}\{\langle L_z\rangle + 2\langle S_z\rangle\} \tag{135.10}$$

may, according to (135.8), be written

$$\langle M_z\rangle = -\frac{e\hbar}{2mc}(m_j + \langle S_z\rangle)$$

which, with (135.7a, b) leads, for the states $j=l+\frac{1}{2}$, to

$$\langle M_z\rangle_+ = -\frac{e\hbar}{2mc}m_j\left(1+\frac{1}{2j}\right) = -\frac{e\hbar}{2mc}m_j\frac{2j+1}{2j} \tag{135.11a}$$

and, for the states $j=-\frac{1}{2}$, to

$$\langle M_z\rangle_- = -\frac{e\hbar}{2mc}m_j\left(1-\frac{1}{2(j+1)}\right) = -\frac{e\hbar}{2mc}m_j\frac{2j+1}{2j+2}. \tag{135.11b}$$

NB. The last formulae show that a closed subshell (n, l) with either $j=l+\frac{1}{2}$ or $j=l-\frac{1}{2}$ has no resultant magnetic moment.

The factor of $-\frac{e\hbar}{2mc}m_j$ in (135.11a, b) is called the Landé g-factor of the state. It permits $\langle M_z\rangle$ to be written in the form

$$\langle M_z\rangle = -\frac{e\hbar}{2mc}m_j g(j) = -\frac{e}{2mc}\langle J_z\rangle g(j)$$

thus describing the deviation, originated by the spin, from the classical Maxwell relation between magnetic moment and mechanical moment, i.e. angular momentum.

Problem 136. Fine structure

The interaction of the intrinsic magnetic moment of an electron,

$$\boldsymbol{\mu} = -g\frac{e}{mc}\boldsymbol{S}, \tag{136.1}$$

and its orbital momentum $\boldsymbol{L}$ is described by a hamiltonian term,

$$H' = \frac{g}{2m^2c^2}\,\frac{1}{r}\,\frac{dV(r)}{dr}\,(\boldsymbol{S}\cdot\boldsymbol{L})\,. \tag{136.2}$$

The level splitting due to this interaction shall be determined.

NB. The so-called g factor of the electron is almost 1. Its exact value has been found to be $g=1.001145$. Since the complete theory of fine structure cannot be given in this unrelativistic treatment, this g factor should not here be taken too seriously. The same holds for the factor 2 in the denominator of (136.2), the so-called Thomas factor, not to be explained by unrelativistic considerations.

Solution. The electron wave function in a central force field is a simultaneous eigenfunction of the operators J_z and $\boldsymbol{J}^2$, the angular structure of which has been given in the preceding problem. The operator $(\boldsymbol{S}\cdot\boldsymbol{L})$ occurring in the hamiltonian (136.2) is then to be reduced to the quantum numbers j and l of the state $\psi=|j,l\rangle$ by

$$\boldsymbol{J}^2|j,l\rangle=\{\boldsymbol{L}^2+\boldsymbol{S}^2+2(\boldsymbol{L}\cdot\boldsymbol{S})\}|j,l\rangle$$

or

$$\hbar^2\{j(j+1)-[l(l+1)+\tfrac{3}{4}]\}|j,l\rangle=2(\boldsymbol{L}\cdot\boldsymbol{S})|j,l\rangle\,.$$

The term (136.2) of the hamiltonian therefore simply adds to $V(r)$ an energy perturbation

$$V'(r) = \frac{g\hbar^2}{4m^2c^2}\,\frac{1}{r}\,\frac{dV(r)}{dr}\,\{j(j+1)-l(l+1)-\tfrac{3}{4}\}\,; \tag{136.3}$$

since it depends upon the quantum numbers j and l, it will differ for different values $j=l\pm\frac{1}{2}$ with the same l.

First-order perturbation calculation gives a contribution

$$E'_{j,l}=\langle j,l|V'|j,l\rangle \tag{136.4}$$

to the energy of a level. In the notation of Eqs. (133.11) and (133.12), in the normalization

$$\int_0^\infty dr\,r^2|F_l(r)|^2=1 \tag{136.5}$$

this leads to

$$E'_{j,l} = \frac{g\hbar^2}{4m^2c^2}\,\{j(j+1)-l(l+1)-\tfrac{3}{4}\}\int_0^\infty dr\,r^2|F_l(r)|^2\,\frac{1}{r}\,\frac{dV}{dr}\,. \tag{136.6}$$

The energy splitting between two levels of the same l, but with different j values then becomes proportional to the difference

$$(l+\tfrac{1}{2})(l+\tfrac{3}{2})-(l-\tfrac{1}{2})(l+\tfrac{1}{2})=2l+1$$

so that

$$\Delta E = \frac{g\hbar^2}{4m^2c^2}(2l+1)\int_0^\infty dr\, r^2 |F_l(r)|^2 \frac{1}{r}\frac{dV}{dr}. \tag{136.7}$$

The level of the smaller j value is the lower one (normal doublet).

Some estimate of the integral (136.7) may be made by using the potential

$$V=-\frac{Ze^2}{r}; \quad \frac{1}{r}\frac{dV}{dr}=\frac{Ze^2}{r^3}. \tag{136.8}$$

Since this expression holds in the neighbourhood of the nucleus in all atoms, and since $F_l \propto r^l$ in this domain, the integrand in (136.7) becomes $\propto r^{2l-1}$ so that the integral converges for $l=1,2,3,\ldots$ but diverges logarithmically for S states ($l=0$). As there is no level splitting but only a shift in an S state, this result is not of primary importance for the evaluation of spectroscopic data. The difficulty does not occur in a rigorous relativistic treatment of the problem (cf. Problem 203).

Without evaluating in detail integrals of the type (136.7) it may safely be said that the result is of the order of Ze^2/a^3 with a a length of the order of atomic radii. Since term energies in atoms are of the order Ze^2/a, we then have roughly

$$\Delta E/E \propto \lambda^2/a^2$$

with $\lambda=\hbar/mc$ the Compton wavelength. This is a small quantity, hence the effect is a fine structure only and may be treated as a first-order perturbation as has been done above.

Problem 137. Plane wave of spin $\frac{1}{2}$ particles

To expand a plane wave of spin $\frac{1}{2}$ particles with either positive or negative helicity into a series of spherical harmonics. Let the wave run in z direction.

Solution. The two-component wave spinors

$$\psi_+=\begin{pmatrix}1\\0\end{pmatrix}e^{ikz}; \quad \psi_-=\begin{pmatrix}0\\1\end{pmatrix}e^{ikz} \tag{137.1}$$

describe plane waves in z direction. In the state ψ_+ the particles have spin in the direction of propagation; we then speak of *helicity* $h=+1$. In ψ_- we have the opposite spin direction, $h=-1$. If we decompose these spinors into eigenspinors of angular momentum, we have $m_l=0$ in both cases and $m_j=+\frac{1}{2}$ for ψ_+, $m_j=-\frac{1}{2}$ for ψ_-.

It has been shown in problem 133 that, for a given orbital momentum l, there exist two eigenspinors of J_z and $\boldsymbol{J}^2$, viz.

$$u^{\mathrm{I}}_{l,m_j}=\frac{F_l(r)}{\sqrt{2l+1}}\begin{pmatrix}\sqrt{l+\frac{1}{2}+m_j}\,Y_{l,m_j-\frac{1}{2}}\\ -\sqrt{l+\frac{1}{2}-m_j}\,Y_{l,m_j+\frac{1}{2}}\end{pmatrix}\quad \text{if } j=l+\tfrac{1}{2} \tag{137.2a}$$

and

$$u^{\mathrm{II}}_{l,m_j}=\frac{F_l(r)}{\sqrt{2l+1}}\begin{pmatrix}\sqrt{l+\frac{1}{2}-m_j}\,Y_{l,m_j-\frac{1}{2}}\\ \sqrt{l+\frac{1}{2}+m_j}\,Y_{l,m_j+\frac{1}{2}}\end{pmatrix}\quad \text{if } j=l-\tfrac{1}{2} \tag{137.2b}$$

with $F_l(r)$ satisfying the radial Schrödinger equation. In the present force-free case the latter runs

$$F_l''+\frac{2}{r}F_l'+\left(k^2-\frac{l(l+1)}{r^2}\right)F_l=0 \tag{137.3}$$

and has solutions regular at the origin, in arbitrary normalization,

$$F_l=\frac{1}{kr}j_l(kr)\,. \tag{137.4}$$

The plane wave then shall be composed of solutions (137.2a) and (137.2b) in the form

$$\psi=\sum_{l=0}^{\infty}\left(A_l u^{\mathrm{I}}_{l,m_j}+B_l u^{\mathrm{II}}_{l,m_j}\right). \tag{137.5}$$

Let us begin with the case of helicity $h=+1$ where $m_j=+\frac{1}{2}$. Eq. (137.5) then may be written

$$\psi_+=\frac{1}{kr}\sum_{l=0}^{\infty}\frac{1}{\sqrt{2l+1}}j_l(kr)\begin{pmatrix}(A_l\sqrt{l+1}+B_l\sqrt{l})\,Y_{l,0}\\ (-A_l\sqrt{l}+B_l\sqrt{l+1})\,Y_{l,1}\end{pmatrix}. \tag{137.6}$$

In order to make the second line of the spinor vanish according to Eq. (137.1) we have to put

$$B_l=\sqrt{\frac{l}{l+1}}\,A_l \tag{137.7}$$

so that

$$\psi_+=\frac{1}{kr}\begin{pmatrix}1\\0\end{pmatrix}\sum_{l=0}^{\infty}A_l\sqrt{\frac{2l+1}{l+1}}\,j_l(kr)\,Y_{l,0}\,. \tag{137.8}$$

Comparing with the expansion of the plane wave (cf. (81.13)),

$$e^{ikz} = \frac{1}{kr}\sum_{l=0}^{\infty}\sqrt{4\pi(2l+1)}\, i^l j_l(kr)\, Y_{l,0} \tag{137.9}$$

there follows

$$A_l = \sqrt{4\pi(l+1)}\, i^l \,, \tag{137.10}$$

so that

$$\psi_+ = \sqrt{4\pi}\sum_{l=0}^{\infty} i^l\left(\sqrt{l+1}\, u^{\mathrm{I}}_{l,\frac{1}{2}} + \sqrt{l}\, u^{\mathrm{II}}_{l,\frac{1}{2}}\right) \tag{137.11}$$

is the correct expansion.

In the opposite case, $h=-1$ and $m_j=-\frac{1}{2}$, Eq. (137.5) yields

$$\psi_- = \frac{1}{kr}\sum_{l=0}^{\infty}\frac{1}{\sqrt{2l+1}}\, j_l(kr)\begin{pmatrix}(A_l\sqrt{l}+B_l\sqrt{l+1})\, Y_{l,-1}\\ (-A_l\sqrt{l+1}+B_l\sqrt{l})\, Y_{l,0}\end{pmatrix}. \tag{137.12}$$

Now, according to (137.1), the first line of the spinor should vanish, i.e.

$$B_l = -\sqrt{\frac{l}{l+1}}\, A_l \,. \tag{137.13}$$

Comparison with (137.9) now renders

$$A_l = -\sqrt{4\pi(l+1)}\, i^l \tag{137.14}$$

so that we arrive at the expansion

$$\psi_- = -\sqrt{4\pi}\sum_{l=0}^{\infty} i^l\left(\sqrt{l+1}\, u^{\mathrm{I}}_{l,-\frac{1}{2}} - \sqrt{l}\, u^{\mathrm{II}}_{l,-\frac{1}{2}}\right). \tag{137.15}$$

Problem 138. Free electron spin resonance

A free electron is put inside a cavity in which there exist two magnetic fields, viz. a constant homogeneous field $\mathcal{H}_0$ in z direction, and a field $\mathcal{H}'$ rotating in the x, y plane:

$$\left.\begin{array}{lll}\mathcal{H}_x = 0; & \mathcal{H}_y = 0; & \mathcal{H}_z = \mathcal{H}_0;\\ \mathcal{H}_x' = \mathcal{H}'\cos\omega t; & \mathcal{H}_y' = \mathcal{H}'\sin\omega t; & \mathcal{H}_z' = 0.\end{array}\right\} \tag{138.1}$$

At the time $t=0$, the electron has its spin in $+z$ direction, and the field $\mathcal{H}'$ is switched on. The probability P of the electron having its spin inverted into the $-z$ direction shall be determined as a function of time.

Solution. The hamiltonian of the problem runs

$$H=\mu(\sigma_z\mathscr{H}_0+\sigma_x\mathscr{H}'_x+\sigma_y\mathscr{H}'_y)$$

where $-\mu\boldsymbol{\sigma}$ is the intrinsic magnetic moment of the electron with $\mu=e\hbar/(2mc)$ except for quantum electrodynamical corrections. We write

$$\sigma_x\mathscr{H}'_x+\sigma_y\mathscr{H}'_y=\tfrac{1}{2}\mathscr{H}'(\sigma_+\mathrm{e}^{-i\omega t}+\sigma_-\mathrm{e}^{i\omega t})$$

with $\sigma_\pm=\sigma_x\pm i\sigma_y$. Thus the Schrödinger equation becomes

$$-\frac{\hbar}{i}\frac{\partial\psi}{\partial t}=\mu\{\mathscr{H}_0\sigma_z+\tfrac{1}{2}\mathscr{H}'(\mathrm{e}^{-i\omega t}\sigma_++\mathrm{e}^{i\omega t}\sigma_-)\}\psi. \tag{138.2}$$

The solution can be expressed by the eigenfunctions of σ_z, i. e.

$$\psi(t)=u(t)\alpha+v(t)\beta. \tag{138.3}$$

Putting (138.3) into (138.2) and using the relations [cf. (129.10)]

$$\begin{aligned}&\sigma_z\alpha=\alpha; &&\sigma_+\alpha=0; &&\sigma_-\alpha=2\beta;\\ &\sigma_z\beta=-\beta; &&\sigma_+\beta=2\alpha; &&\sigma_-\beta=0\end{aligned}$$

we obtain

$$-\frac{\hbar}{i}(\dot{u}\alpha+\dot{v}\beta)=\mu\mathscr{H}_0(u\alpha-v\beta)+\mu\mathscr{H}'(\mathrm{e}^{-i\omega t}v\alpha+\mathrm{e}^{i\omega t}u\beta).$$

Introducing the abbreviations

$$\frac{\mu\mathscr{H}_0}{\hbar}=\omega_0;\qquad \frac{\mu\mathscr{H}'}{\hbar}=\omega', \tag{138.4}$$

and separating into α and β parts, we find

$$\begin{aligned}i\dot{u}&=\ \ \omega_0u+\omega'\mathrm{e}^{-i\omega t}v;\\ i\dot{v}&=-\omega_0v+\omega'\mathrm{e}^{i\omega t}u.\end{aligned} \tag{138.5}$$

This system is solved by

$$u=A\mathrm{e}^{-i(\Omega+\frac{1}{2}\omega)t};\qquad v=B\mathrm{e}^{-i(\Omega-\frac{1}{2}\omega)t}. \tag{138.6}$$

A straightforward calculation leads to two solutions $\Omega_1=+\Omega$ and $\Omega_2=-\Omega$ with

$$\Omega=\sqrt{(\omega_0-\tfrac{1}{2}\omega)^2+\omega'^2} \tag{138.7}$$

with amplitudes A_1, B_1, respectively A_2, B_2:

$$\psi(t)=(A_1\mathrm{e}^{-i\Omega t}+A_2\mathrm{e}^{i\Omega t})\mathrm{e}^{-\frac{i\omega}{2}t}\alpha+(B_1\mathrm{e}^{-i\Omega t}+B_2\mathrm{e}^{i\Omega t})\mathrm{e}^{\frac{i\omega}{2}t}\beta \tag{138.8}$$

and

$$B_{1,2}=A_{1,2}\frac{\pm\Omega-(\omega_0-\frac{1}{2}\omega)}{\omega'}. \tag{138.9}$$

In (138.8) we introduce the initial condition $\psi(0)=\alpha$ or

$$A_1+A_2=1; \quad B_1+B_2=0, \tag{138.10}$$

which with (138.9) finally leads to

$$\psi(t)=\left\{\cos\Omega t-\frac{\omega_0-\frac{1}{2}\omega}{\Omega}\,i\sin\Omega t\right\}e^{-\frac{i\omega}{2}t}\alpha-\frac{\omega'}{\Omega}\,i\sin\Omega t\,e^{\frac{i\omega}{2}t}\beta. \tag{138.11}$$

The probability of spin flip at time t is therefore

$$P=\left(\frac{\omega'}{\Omega}\right)^2\sin^2\Omega t, \tag{138.12}$$

its time average being independent of time,

$$\overline{P}=\frac{1}{2}\frac{\omega'^2}{\Omega^2}=\frac{1}{2}\frac{\omega'^2}{(\omega_0-\frac{1}{2}\omega)^2+\omega'^2}. \tag{138.13}$$

If the homogeneous field $\mathscr{H}_0$ and thus, according to (138.4), the Larmor frequency ω_0 is continually varied, the average flip probability becomes a maximum if

$$\omega_0=\frac{1}{2}\omega, \quad \text{i.e. } \mathscr{H}_0=\frac{\hbar\omega}{2\mu}. \tag{138.14}$$

We call this the resonance field, $\mathscr{H}_{\text{res}}$, and find

$$\overline{P}=\frac{1}{2}\frac{\mathscr{H}'^2}{(\mathscr{H}_0-\mathscr{H}_{\text{res}})^2+\mathscr{H}'^2}. \tag{138.15}$$

At resonance, $\overline{P}=\frac{1}{2}$, independent of the strength of the rotating field $\mathscr{H}'$, the width of the resonance region, however, being determined by $\mathscr{H}'$.

NB. The method may either be used to determine μ from the resonance field strength or, if μ is sufficiently well known, to determine the difference between the field applied and the field acting on the electron inside a molecule. In a similar way, proton resonance may be used to detect molecular structures.

B. Two- and Three-Body Problems

Problem 139. Spin functions for two particles

To determine the spin eigenfunctions for a system of two particles of spin $\frac{1}{2}$ (say, a neutron and a proton) which for a total spin vector operator

$$\mathbf{S}=\frac{\hbar}{2}(\boldsymbol{\sigma}_n+\boldsymbol{\sigma}_p) \tag{139.1}$$

simultaneously diagonalize its component S_z and its absolute square, $\mathbf{S}^2$.

Solution. Let α_n, β_n be the Hilbert basis vectors for the neutron, and α_p, β_p for the proton. Then the spin function χ of the two-particle system is bound to be of the form

$$\chi = A\,\alpha_n\alpha_p + B\,\alpha_n\beta_p + C\,\beta_n\alpha_p + D\,\beta_n\beta_p. \tag{139.2}$$

From the definition of the spin operators (Problem 129) it follows that

$$\frac{2}{\hbar}S_z\chi = (\sigma_{nz}+\sigma_{pz})\chi = A\,\alpha_n\alpha_p + B\,\alpha_n\beta_p - C\,\beta_n\alpha_p - D\,\beta_n\beta_p$$
$$+A\,\alpha_n\alpha_p - B\,\alpha_n\beta_p + C\,\beta_n\alpha_p - D\,\beta_n\beta_p.$$

Each of the four terms of χ, Eq. (139.2), therefore is an eigenfunction of S_z, viz.

$$\begin{aligned}
&\alpha_n\alpha_p \quad \text{for the eigenvalue } +2 \text{ of } \tfrac{2}{\hbar}S_z \text{ or } +\hbar \text{ of } S_z;\\
&\alpha_n\beta_p \quad \text{for the eigenvalue } \ \ 0 \text{ of } \tfrac{2}{\hbar}S_z \text{ or } \ \ 0 \text{ of } S_z;\\
&\beta_n\alpha_p \quad \text{for the eigenvalue } \ \ 0 \text{ of } \tfrac{2}{\hbar}S_z \text{ or } \ \ 0 \text{ of } S_z;\\
&\beta_n\beta_p \quad \text{for the eigenvalue } -2 \text{ of } \tfrac{2}{\hbar}S_z \text{ or } -\hbar \text{ of } S_z.
\end{aligned} \tag{139.3}$$

The spin components of $+1, 0, -1$ in units $\hbar$ are in agreement with the half-classical vector model. The two functions $\alpha_n\beta_p$ and $\beta_n\alpha_p$ are still degenerate so that any linear combination of them still belongs to the eigenvalue zero.

We now proceed to investigate the operator

$$\left(\frac{2}{\hbar}\right)^2 \boldsymbol{S}^2 = \boldsymbol{\sigma}_n^2 + \boldsymbol{\sigma}_p^2 + 2(\boldsymbol{\sigma}_n\cdot\boldsymbol{\sigma}_p) = 6 + 2(\sigma_{nx}\sigma_{px} + \sigma_{ny}\sigma_{py} + \sigma_{nz}\sigma_{pz}).$$

We find

$$\begin{aligned}
\sigma_{nx}\sigma_{px}\chi &= \ \ A\,\beta_n\beta_p + B\,\beta_n\alpha_p + C\,\alpha_n\beta_p + D\,\alpha_n\alpha_p;\\
\sigma_{ny}\sigma_{py}\chi &= -A\,\beta_n\beta_p + B\,\beta_n\alpha_p + C\,\alpha_n\beta_p - D\,\alpha_n\alpha_p;\\
\sigma_{nz}\sigma_{pz}\chi &= \ \ A\,\alpha_n\alpha_p - B\,\alpha_n\beta_p - C\,\beta_n\alpha_p + D\,\beta_n\beta_p
\end{aligned}$$

and thence,

$$\left(\frac{2}{\hbar}\right)^2 \boldsymbol{S}^2\,\alpha_n\alpha_p = 8\,\alpha_n\alpha_p; \tag{139.4a}$$

$$\left(\frac{2}{\hbar}\right)^2 \boldsymbol{S}^2 (B\,\alpha_n\beta_p + C\,\beta_n\alpha_p) = 4(B+C)(\alpha_n\beta_p + \beta_n\alpha_p); \tag{139.4b}$$

$$\left(\frac{2}{\hbar}\right)^2 \boldsymbol{S}^2\,\beta_n\beta_p = 8\,\beta_n\beta_p. \tag{139.4c}$$

The functions $\alpha_n\alpha_p$ and $\beta_n\beta_p$ therefore are eigenfunctions to the eigenvalue $2\hbar^2$ of the operator $\boldsymbol{S}^2$. In the usual notation,

$$\boldsymbol{S}^2\chi = \hbar^2 S(S+1)\chi, \tag{139.5}$$

they belong to the quantum number $S=1$ or, in the half-classical language of the vector model, to the total spin $S=1$ (in units $\hbar$) with its component S_z either $+1$ or -1.

From (139.4b) we construct two more eigenfunctions of $\boldsymbol{S}^2$ with $S_z=0$. Let λ be the still unknown eigenvalue of $\boldsymbol{S}^2/\hbar^2$, then we have

$$(B+C)(\alpha_n\beta_p+\beta_n\alpha_p) = \lambda(B\alpha_n\beta_p+C\beta_n\alpha_p).$$

This yields two linear equations for B and C,

$$(B+C)=\lambda B \quad \text{and} \quad (B+C)=\lambda C.$$

Their determinant must vanish,

$$\begin{vmatrix} 1-\lambda; & 1 \\ 1; & 1-\lambda \end{vmatrix} = 0 \quad \text{or} \quad 1-\lambda = \pm 1.$$

The two eigenfunctions of $\boldsymbol{S}^2$ with $S_z=0$ therefore become

a) for $\lambda=2$: $B=C$; $\quad \chi=\alpha_n\beta_p+\beta_n\alpha_p$; $\quad S=1$; (139.6a)

b) for $\lambda=0$: $B=-C$; $\quad \chi=\alpha_n\beta_p-\beta_n\alpha_p$; $\quad S=0$. (139.6b)

The results have been collected in the following table where the functions have been normalized according to the rules

$$\langle\alpha|\alpha\rangle=\langle\beta|\beta\rangle=1; \qquad \langle\alpha|\beta\rangle=0.$$

Triplet, $S=1$ (symmetrical spin function)	$S_z=+1$	$\alpha_n\alpha_p$
	0	$\frac{1}{\sqrt{2}}(\alpha_n\beta_p+\beta_n\alpha_p)$
	-1	$\beta_n\beta_p$
Singlet, $S=0$ (antisymmetrical spin function)	$S_z=\ 0$	$\frac{1}{\sqrt{2}}(\alpha_n\beta_p-\beta_n\alpha_p)$

NB. From $(\boldsymbol{\sigma}_n+\boldsymbol{\sigma}_p)^2=6+2(\boldsymbol{\sigma}_n\cdot\boldsymbol{\sigma}_p)$ it follows that the triplet and singlet spin functions, say χ_t and χ_s, given in the table, are eigenfunctions also of the operator $(\boldsymbol{\sigma}_n\cdot\boldsymbol{\sigma}_p)$ so that

$$(\boldsymbol{\sigma}_n\cdot\boldsymbol{\sigma}_p)\chi_t=\chi_t; \qquad (\boldsymbol{\sigma}_n\cdot\boldsymbol{\sigma}_p)\chi_s=-3\chi_s.$$

Of these relations use will be made in the following problem.

Problem 140. Spin-dependent central force between nucleons

In a reasonable approximation, the interaction energy between a neutron and a proton in an S state may be described by a central force, different for symmetrical and antisymmetrical spin states. Such an interaction shall be expressed in terms of a spin-dependent potential

a) using the spin exchange operator Σ_{np},

b) using the spin vectors $\boldsymbol{\sigma}_n$ and $\boldsymbol{\sigma}_p$ of the two particles.

Solution. A central force means that the interaction energy must depend only upon the distance r between the two particles. This energy shall be different for different spin-state symmetry, say, $V_t(r)$ in the triplet case of parallel spins and $V_s(r)$ in the singlet case of antiparallel spins.

a) Let $\chi(s_n, s_p)$ be a two-particle function. Then the spin exchange operator is defined by

$$\Sigma_{np}\chi(s_n, s_p) = \chi(s_p, s_n). \tag{140.1}$$

For the symmetrical triplet state, $\chi_t(s_n, s_p) = \chi_t(s_p, s_n)$, we therefore have

$$\Sigma_{np}\chi_t = \chi_t \tag{140.2a}$$

and for the antisymmetrical singlet state, $\chi_s(s_n, s_p) = -\chi_s(s_p, s_n)$,

$$\Sigma_{np}\chi_s = -\chi_s. \tag{140.2b}$$

Hence both kinds of functions are eigenfunctions of the exchange operator, with its respective eigenvalues $+1$ and -1. As the three triplet and one singlet functions form a complete orthogonal set, Eqs. (140.2a, b) explain the exchange operator completely and uniquely.

An interaction energy

$$V = V_1(r) + V_2(r)\Sigma_{np}$$

yields, according to (140.2a, b),

$$V\chi_t = (V_1 + V_2)\chi_t; \qquad V\chi_s = (V_1 - V_2)\chi_s$$

so that

$$V_t = V_1 + V_2 \quad \text{and} \quad V_s = V_1 - V_2$$

are the interactions in the triplet and singlet states, respectively. Thence,

$$V = \tfrac{1}{2}(V_t + V_s) + \tfrac{1}{2}(V_t - V_s)\Sigma_{np}. \tag{140.3}$$

b) At the end of the preceding problem we have shown that the spin functions χ_t and χ_s are eigenfunctions also of the operator $(\boldsymbol{\sigma}_n \cdot \boldsymbol{\sigma}_p)$, viz.

$$(\boldsymbol{\sigma}_n \cdot \boldsymbol{\sigma}_p)\chi_t = \chi_t; \qquad (\boldsymbol{\sigma}_n \cdot \boldsymbol{\sigma}_p)\chi_s = -3\chi_s. \tag{140.4}$$

It follows that Σ_{np} may be expressed linearly by $(\boldsymbol{\sigma}_n \cdot \boldsymbol{\sigma}_p)$. Indeed,

$$\Sigma_{np} = \tfrac{1}{2}(1 + (\boldsymbol{\sigma}_n \cdot \boldsymbol{\sigma}_p)) \tag{140.5}$$

leads to the wanted eigenvalues (140.2a, b). Again, since there exist no other spin functions of the two-nucleon system, both operators are completely described by the eigenvalue problems (140.2a, b) and (140.4) so that (140.5) holds in full generality.

Replacing Σ_{np} in (140.3) according to (140.5), we arrive at the result

$$V = \tfrac{1}{4}(3V_t + V_s) + \tfrac{1}{4}(V_t - V_s)(\boldsymbol{\sigma}_n \cdot \boldsymbol{\sigma}_p). \tag{140.6}$$

Problem 141. Powers of spin operators

To show that the operator $(\boldsymbol{\sigma}_1 \cdot \boldsymbol{\sigma}_2)^n$ for two particles 1 and 2 can be linearly expressed by $(\boldsymbol{\sigma}_1 \cdot \boldsymbol{\sigma}_2)$.

Solution. The operator $(\boldsymbol{\sigma}_1 \cdot \boldsymbol{\sigma}_2)$ is completely described by the two eigenvalue relations

$$(\boldsymbol{\sigma}_1 \cdot \boldsymbol{\sigma}_2)\chi_t = \chi_t; \quad (\boldsymbol{\sigma}_1 \cdot \boldsymbol{\sigma}_2)\chi_s = -3\chi_s \tag{141.1}$$

for the three triplet and one singlet spin functions since these form a complete orthogonal set. It therefore suffices to investigate the application of $(\boldsymbol{\sigma}_1 \cdot \boldsymbol{\sigma}_2)^n$ to these four spin functions. By iteration of (141.1) we get at once,

$$(\boldsymbol{\sigma}_1 \cdot \boldsymbol{\sigma}_2)^n \chi_t = \chi_t; \quad (\boldsymbol{\sigma}_1 \cdot \boldsymbol{\sigma}_2)^n \chi_s = (-3)^n \chi_s. \tag{141.2}$$

It follows that

$$(\boldsymbol{\sigma}_1 \cdot \boldsymbol{\sigma}_2)^n = A + B(\boldsymbol{\sigma}_1 \cdot \boldsymbol{\sigma}_2) \tag{141.3}$$

can be linearly expressed by $(\boldsymbol{\sigma}_1 \cdot \boldsymbol{\sigma}_2)$. Putting (141.3) in (141.2) we find, according to (141.1),

$$(A+B)\chi_t = \chi_t; \quad (A-3B)\chi_s = (-3)^n \chi_s.$$

It follows that

$$A + B = 1; \quad A - 3B = (-3)^n$$

or that

$$A = \tfrac{1}{4}[3 + (-3)^n]; \quad B = \tfrac{1}{4}[1 - (-3)^n]. \tag{141.4}$$

Thus, e. g., we get

$$(\boldsymbol{\sigma}_1 \cdot \boldsymbol{\sigma}_2)^2 = 3 - 2(\boldsymbol{\sigma}_1 \cdot \boldsymbol{\sigma}_2); \quad (\boldsymbol{\sigma}_1 \cdot \boldsymbol{\sigma}_2)^3 = -6 + 7(\boldsymbol{\sigma}_1 \cdot \boldsymbol{\sigma}_2).$$

NB. The representation (140.6) of a spin-dependent force in the preceding problem is unique, because replacing it by a power series in $(\boldsymbol{\sigma}_n \cdot \boldsymbol{\sigma}_p)$ would change nothing in the result. – The solution of the problem becomes even simpler if powers of the exchange operator Σ_{12} are considered.

Problem 142. Angular momentum eigenfunctions for two spin particles

To construct the triplet state eigenfunctions of the operators J_z and $\boldsymbol{J}^2$ for a system of two spin $\frac{1}{2}$ particles. Use $\hbar=1$ as unit of angular momentum.

Solution. Any eigenfunction for a triplet state can certainly be written in the form

$$\psi = \sum_{l=0}^{\infty} \{f_l(r)\, Y_{l,m-1}\,\chi_{1,1} + g_l(r)\, Y_{l,m}\,\chi_{1,0} + h_l(r)\, Y_{l,m+1}\,\chi_{1,-1}\}, \tag{142.1}$$

each of the three possible spin functions multiplying by a space factor formally written as a spherical harmonics expansion, with only the one restriction to generality that the second subscript of each Y is so chosen as to make

$$J_z \psi = m\psi, \tag{142.2}$$

thus making ψ an eigenfunction of the operator J_z.

Let us now apply

$$\boldsymbol{J}^2 = \{\boldsymbol{L} + \tfrac{1}{2}(\boldsymbol{\sigma}_1 + \boldsymbol{\sigma}_2)\}^2 = \boldsymbol{L}^2 + \boldsymbol{L}\cdot(\boldsymbol{\sigma}_1 + \boldsymbol{\sigma}_2) + \tfrac{1}{4}[6 + 2(\boldsymbol{\sigma}_1 \cdot \boldsymbol{\sigma}_2)]$$

to the function (142.1). It is then suitable to define the operators

$$\sigma_+ = \sigma_x + i\sigma_y \quad \text{and} \quad \sigma_- = \sigma_x - i\sigma_y \tag{142.3}$$

in analogy to (cf. Problem 56)

$$L_+ = L_x + iL_y \quad \text{and} \quad L_- = L_x - iL_y. \tag{142.4}$$

Then $\boldsymbol{J}^2$ may equally well be written

$$\boldsymbol{J}^2 = \boldsymbol{L}^2 + \tfrac{1}{2}(L_+\sigma_- + L_-\sigma_+) + L_z\sigma_z + \tfrac{3}{2} + \tfrac{1}{2}(\boldsymbol{\sigma}_1 \cdot \boldsymbol{\sigma}_2), \tag{142.5}$$

where $\sigma_\pm = \sigma_{1\pm} + \sigma_{2\pm}$ and $\sigma_z = \sigma_{1z} + \sigma_{2z}$. Application of these operators to the triplet spin functions yields

$$\sigma_+ \begin{pmatrix} \chi_{1,1} \\ \chi_{1,0} \\ \chi_{1,-1} \end{pmatrix} = 2\sqrt{2} \begin{pmatrix} 0 \\ \chi_{1,1} \\ \chi_{1,0} \end{pmatrix}; \qquad \sigma_- \begin{pmatrix} \chi_{1,1} \\ \chi_{1,0} \\ \chi_{1,-1} \end{pmatrix} = 2\sqrt{2} \begin{pmatrix} \chi_{1,0} \\ \chi_{1,-1} \\ 0 \end{pmatrix};$$

$$\sigma_z \begin{pmatrix} \chi_{1,1} \\ \chi_{1,0} \\ \chi_{1,-1} \end{pmatrix} = 2 \begin{pmatrix} \chi_{1,1} \\ 0 \\ -\chi_{1,-1} \end{pmatrix}. \tag{142.6}$$

Hence, we obtain by straightforward computation

$$\left.\begin{aligned} \boldsymbol{J}^2\chi_{1,1} &= (L^2+2+2L_z)\chi_{1,1}+\sqrt{2}\,L_+\chi_{1,0};\\ \boldsymbol{J}^2\chi_{1,0} &= \sqrt{2}\,L_-\chi_{1,1}+(L^2+2)\chi_{1,0}+\sqrt{2}\,L_+\chi_{1,-1};\\ \boldsymbol{J}^2\chi_{1,-1} &= \sqrt{2}\,L_-\chi_{1,0}+(L^2+2-2L_z)\chi_{1,-1}. \end{aligned}\right\} \tag{142.7}$$

Using (142.7) and the well-known relations [cf. (56.14)]

$$\left.\begin{aligned} L_+Y_{l,m} &= -\sqrt{(l+m+1)(l-m)}\,Y_{l,m+1};\\ L_-Y_{l,m} &= -\sqrt{(l+m)(l-m+1)}\,Y_{l,m-1};\\ L_zY_{l,m} &= m\,Y_{l,m} \end{aligned}\right\} \tag{142.8}$$

we get

$$\begin{aligned} \boldsymbol{J}^2\psi = \sum_{l=0}^{\infty}\{&[f_l(l(l+1)+2m)-g_l\sqrt{2(l+m)(l-m+1)}]\,Y_{l,m-1}\chi_{1,1}\\ &+[-f_l\sqrt{2(l+m)(l-m+1)}+g_l(l(l+1)+2)-h_l\sqrt{2(l+m+1)(l-m)}]\,Y_{l,m}\chi_{1,0}\\ &+[-g_l\sqrt{2(l+m+1)(l-m)}+h_l(l(l+1)-2m)]\,Y_{l,m+1}\chi_{1,-1}\}. \end{aligned} \tag{142.9}$$

In order to make ψ an eigenfunction of $\boldsymbol{J}^2$ this must be

$$=j(j+1)\psi,$$

which gives three independent linear equations for f_l, g_l, h_l showing that these three radial functions must be of the same form but with different amplitudes:

$$f_l=A_lF_l(r);\qquad g_l=B_lF_l(r);\qquad h_l=C_lF_l(r) \tag{142.10}$$

where the constant amplitude factors A_l, B_l, C_l may be determined from the following set of linear equations

$$\begin{aligned} [l(l+1)+2m-j(j+1)]A_l-\sqrt{2(l+m)(l-m+1)}\,B_l&=0;\\ -\sqrt{2(l+m)(l-m+1)}\,A_l+[l(l+1)+2-j(j+1)]B_l-\sqrt{2(l+m+1)(l-m)}\,C_l&=0;\\ -\sqrt{2(l+m+1)(l-m)}\,B_l+[l(l+1)-2m-j(j+1)]C_l&=0. \end{aligned} \tag{142.11}$$

The determinant of these equations must vanish; if it is expanded, one finds that it becomes independent of m and has the form

$$[l(l+1)-j(j+1)]\{[l(l+1)-j(j+1)]^2+2[l(l+1)-j(j+1)]-4l(l+1)\}=0.$$

This leads to the (positive) solutions

$$j=l+1;\qquad j=l;\qquad j=l-1 \tag{142.12}$$

for which the amplitudes A_l, B_l, C_l then may be determined according to (142.11), except for a common normalization factor. Choosing arbitrarily the normalization

$$A_l^2 + B_l^2 + C_l^2 = 1 \tag{142.13}$$

we arrive at the results compiled in the table.

j	A_l	B_l	C_l
$l+1$	$-\sqrt{\dfrac{(l+m+1)(l+m)}{2(l+1)(2l+1)}}$	$\sqrt{\dfrac{(l+m+1)(l-m+1)}{(l+1)(2l+1)}}$	$-\sqrt{\dfrac{(l-m+1)(l-m)}{2(l+1)(2l+1)}}$
l	$\sqrt{\dfrac{(l-m+1)(l+m)}{2l(l+1)}}$	$\dfrac{m}{\sqrt{l(l+1)}}$	$-\sqrt{\dfrac{(l+m+1)(l-m)}{2l(l+1)}}$
$l-1$	$\sqrt{\dfrac{(l-m+1)(l-m)}{2l(2l+1)}}$	$\sqrt{\dfrac{(l+m)(l-m)}{l(2l+1)}}$	$\sqrt{\dfrac{(l+m+1)(l+m)}{2l(2l+1)}}$

Problem 143. Tensor force operator

The so-called tensor force between two particles 1 and 2 of spin $\frac{1}{2}$ is defined by the interaction energy

$$V = W(r) T_{12}$$

with the operator

$$T_{12} = \frac{(\boldsymbol{\sigma}_1 \cdot \boldsymbol{r})(\boldsymbol{\sigma}_2 \cdot \boldsymbol{r})}{r^2} - \frac{1}{3}(\boldsymbol{\sigma}_1 \cdot \boldsymbol{\sigma}_2). \tag{143.1}$$

To apply this operator to the spin eigenfunctions of the two-particle system.

Solution. The operator T_{12} is invariant under spin exchange. It therefore keeps the symmetry of the spin functions. Since there exists only one antisymmetrical spin function, $\chi_{0,0}$, this then must be an eigenfunction of the operator T_{12}. The three symmetrical spin functions, however, may be mixed up by its application. Since T_{12} is invariant also under exchange of the particle coordinates, i. e. under parity transformation, it will conserve parity. This means that only spherical harmonics of even order will enter the expression $T_{12}\chi$. No higher angular momenta than $l=2$ are to be expected.

In order to get details let us first apply the one-particle operator $(\boldsymbol{\sigma} \cdot \boldsymbol{r})$ to the one-particle spin functions:

$$(\boldsymbol{\sigma} \cdot \boldsymbol{r})\begin{pmatrix}\alpha\\ \beta\end{pmatrix} = (\sigma_x x + \sigma_y y + \sigma_z z)\begin{pmatrix}\alpha\\ \beta\end{pmatrix} = \begin{pmatrix}(x+iy)\beta + z\alpha\\ (x-iy)\alpha - z\beta\end{pmatrix}. \tag{143.2}$$

It then follows directly that

$$(\boldsymbol{\sigma}_1\cdot\boldsymbol{r})(\boldsymbol{\sigma}_2\cdot\boldsymbol{r})\begin{pmatrix}\alpha_1\alpha_2\\ \alpha_1\beta_2\\ \beta_1\alpha_2\\ \beta_1\beta_2\end{pmatrix}=\begin{pmatrix}[(x+iy)\beta_1+z\alpha_1]\,[(x+iy)\beta_2+z\alpha_2]\\ [(x+iy)\beta_1+z\alpha_1]\,[(x-iy)\alpha_2-z\beta_2]\\ [(x-iy)\alpha_1-z\beta_1]\,[(x+iy)\beta_2+z\alpha_2]\\ [(x-iy)\alpha_1-z\beta_1]\,[(x-iy)\alpha_2-z\beta_2]\end{pmatrix}.$$

With

$$(x\pm iy)^2=r^2\sin^2\vartheta\,\mathrm{e}^{\pm 2i\varphi};\qquad x^2+y^2=r^2\sin^2\vartheta;$$
$$(x\pm iy)z=r^2\sin\vartheta\cos\vartheta\,\mathrm{e}^{\pm i\varphi};\qquad z^2=r^2\cos^2\vartheta,$$

this leads to

$$\frac{(\boldsymbol{\sigma}_1\cdot\boldsymbol{r})(\boldsymbol{\sigma}_2\cdot\boldsymbol{r})}{r^2}\begin{pmatrix}\alpha_1\alpha_2\\ \alpha_1\beta_2\\ \beta_1\alpha_2\\ \beta_1\beta_2\end{pmatrix}$$
$$=\begin{pmatrix}\cos^2\vartheta\,\alpha_1\alpha_2+\sin\vartheta\cos\vartheta\,\mathrm{e}^{i\varphi}(\alpha_1\beta_2+\beta_1\alpha_2)+\sin^2\vartheta\,\mathrm{e}^{2i\varphi}\beta_1\beta_2\\ \sin\vartheta\cos\vartheta\,\mathrm{e}^{-i\varphi}\alpha_1\alpha_2-\cos^2\vartheta\,\alpha_1\beta_2+\sin^2\vartheta\,\beta_1\alpha_2-\sin\vartheta\cos\vartheta\,\mathrm{e}^{i\varphi}\beta_1\beta_2\\ \sin\vartheta\cos\vartheta\,\mathrm{e}^{-i\varphi}\alpha_1\alpha_2+\sin^2\vartheta\,\alpha_1\beta_2-\cos^2\vartheta\,\beta_1\alpha_2-\sin\vartheta\cos\vartheta\,\mathrm{e}^{i\varphi}\beta_1\beta_2\\ \sin^2\vartheta\,\mathrm{e}^{-2i\varphi}\alpha_1\alpha_2-\sin\vartheta\cos\vartheta\,\mathrm{e}^{-i\varphi}(\alpha_1\beta_2+\beta_1\alpha_2)+\cos^2\vartheta\,\beta_1\beta_2\end{pmatrix}.$$

Using the notation χ_{S,m_S}, i. e. for the triplet

$$\alpha_1\alpha_2=\chi_{1,1};\qquad \frac{1}{\sqrt{2}}(\alpha_1\beta_2+\beta_1\alpha_2)=\chi_{1,0};\qquad \beta_1\beta_2=\chi_{1,-1}\tag{143.3}$$

and for the singlet

$$\chi_{0,0}=\frac{1}{\sqrt{2}}(\alpha_1\beta_2-\beta_1\alpha_2),\tag{143.4}$$

we then have for the symmetrical functions of the triplet

$$\frac{(\boldsymbol{\sigma}_1\cdot\boldsymbol{r})(\boldsymbol{\sigma}_2\cdot\boldsymbol{r})}{r^2}\begin{pmatrix}\chi_{1,1}\\ \chi_{1,0}\\ \chi_{1,-1}\end{pmatrix}\tag{143.5}$$
$$=\begin{pmatrix}\cos^2\vartheta\,\chi_{1,1}+\sqrt{2}\sin\vartheta\cos\vartheta\,\mathrm{e}^{i\varphi}\chi_{1,0}+\sin^2\vartheta\,\mathrm{e}^{2i\varphi}\chi_{1,-1}\\ \sqrt{2}\sin\vartheta\cos\vartheta\,\mathrm{e}^{-i\varphi}\chi_{1,1}+(\sin^2\vartheta-\cos^2\vartheta)\chi_{1,0}-\sqrt{2}\sin\vartheta\cos\vartheta\,\mathrm{e}^{i\varphi}\chi_{1,-1}\\ \sin^2\vartheta\,\mathrm{e}^{-2i\varphi}\chi_{1,1}-\sqrt{2}\sin\vartheta\cos\vartheta\,\mathrm{e}^{-i\varphi}\chi_{1,0}+\cos^2\vartheta\,\chi_{1,-1}\end{pmatrix}$$

and for the antisymmetrical singlet function,

$$\frac{(\boldsymbol{\sigma}_1\cdot\boldsymbol{r})(\boldsymbol{\sigma}_2\cdot\boldsymbol{r})}{r^2}\chi_{0,0}=-\chi_{0,0}.\tag{143.6}$$

The second term of T_{12} has already been discussed in Problem 140:

$$(\boldsymbol{\sigma}_1 \cdot \boldsymbol{\sigma}_2)\begin{pmatrix}\chi_{1,m_s}\\ \chi_{0,0}\end{pmatrix} = \begin{pmatrix}\chi_{1,m_s}\\ -3\,\chi_{0,0}\end{pmatrix}. \tag{143.7}$$

Combination of Eqs. (143.6) and (143.7) then gives at once

$$T_{12}\,\chi_{0,0} = 0. \tag{143.8}$$

The tensor operator therefore cannot contribute any dynamical term to a spin singlet state.

There remains further discussion of the triplet. Introducing normalized spherical harmonics according to the definitions of the table of Problem 67, we obtain from (143.5) and (143.7),

$$\begin{aligned} T_{12}\,\chi_{1,1} &= \frac{2}{3}\sqrt{\frac{4\pi}{5}}\,(Y_{2,0}\chi_{1,1} + \sqrt{3}\,Y_{2,1}\,\chi_{1,0} + \sqrt{6}\,Y_{2,2}\,\chi_{1,-1});\\ T_{12}\,\chi_{1,0} &= \frac{2}{3}\sqrt{\frac{4\pi}{5}}\,(-\sqrt{3}\,Y_{2,-1}\,\chi_{1,1} - 2\,Y_{2,0}\,\chi_{1,0} - \sqrt{3}\,Y_{2,1}\,\chi_{1,-1});\\ T_{12}\,\chi_{1,-1} &= \frac{2}{3}\sqrt{\frac{4\pi}{5}}\,(\sqrt{6}\,Y_{2,-2}\,\chi_{1,1} + \sqrt{3}\,Y_{2,-1}\,\chi_{1,0} + Y_{2,0}\,\chi_{1,-1}). \end{aligned} \tag{143.9}$$

These formulae not only show spin exchange symmetry and parity to be conserved, but also the z component of the total angular momentum. The orbital momentum, however, as well as its z component, are not good quantum numbers in a two-particle system with tensor interaction.

Problem 144. Deuteron with tensor interaction

The interaction between a proton and a neutron consists in part of a central force, and in part of a tensor force,

$$V = V_c(r) + V_t(r)\,T_{pn}. \tag{144.1}$$

The deuteron groundstate therefore is a mixture of S and D state. The eigenfunction shall be constructed, except for radial S and D factors for which a set of two coupled differential equations shall be derived, under the assumption of nuclear spin orientation in z direction.

Solution. With the nuclear spin $i=1$ (in units $\hbar$) and its component in z direction also 1, we have for the most general $S-D$ mixture,

$$\psi = f(r)\,Y_{0,0}\,\chi_{1,1} + g(r)\{Y_{2,0}\,\chi_{1,1} + \lambda\,Y_{2,1}\,\chi_{1,0} + \mu\,Y_{2,2}\,\chi_{1,-1}\} \tag{144.2}$$

with the constants λ and μ to be adjusted so that

$$\boldsymbol{I}^2\psi = 2\psi. \tag{144.3}$$

Here $\boldsymbol{I}$ is the operator of total angular momentum (nuclear spin). According to the preceding problem we have

$$\begin{aligned}\boldsymbol{I}^2\psi = &\{\chi_{1,1}(L^2+2+2L_z)+\chi_{1,0}\sqrt{2}L_+\}(fY_{0,0}+gY_{2,0})\\ &+\lambda\{\chi_{1,1}\sqrt{2}L_- +\chi_{1,0}(L^2+2)+\chi_{1,-1}\sqrt{2}L_+\}gY_{2,1}\\ &+\mu\{\chi_{1,0}\sqrt{2}L_- +\chi_{1,-1}(L^2+2-2L_z)\}gY_{2,2}\end{aligned}$$

and further

$$\begin{aligned}\boldsymbol{I}^2\psi = &\chi_{1,1}[2fY_{0,0}+(8-2\sqrt{3}\lambda)gY_{2,0}]\\ &+\chi_{1,0}(-2\sqrt{3}+8\lambda-2\sqrt{2}\mu)gY_{2,1}+\chi_{1,-1}(-2\sqrt{2}\lambda+4\mu)gY_{2,2}.\end{aligned}$$

The last relation satisfies (144.3) if

$$\lambda=\sqrt{3}, \quad \mu=\sqrt{6}, \tag{144.4}$$

so that the angular momentum eigenfunction becomes

$$\psi = f(r)\,Y_{0,0}\chi_{1,1}+g(r)\{Y_{2,0}\chi_{1,1}+\sqrt{3}\,Y_{2,1}\chi_{1,0}+\sqrt{6}\,Y_{2,2}\chi_{1,-1}\}. \tag{144.5}$$

The curly bracket in (144.5) is the same combination of spherical harmonics and spin functions as was obtained in the first line of Eq. (143.9) of the preceding problem, so that we may write in a more compact form:

$$\psi = \frac{1}{\sqrt{4\pi}}\{f(r)+\tfrac{3}{2}\sqrt{5}g(r)\,T_{pn}\}\,\chi_{1,1}. \tag{144.6}$$

Let us now normalize this function. From (144.5) there follows at once

$$\int_0^\infty dr\,r^2[f^2+10g^2]=1.$$

It will be suitable to put

$$f(r)=\psi_S(r)\cos\omega; \quad g(r)=\frac{1}{\sqrt{10}}\psi_D(r)\sin\omega \tag{144.7}$$

so that

$$\int_0^\infty dr\,r^2\psi_S^2=1; \quad \int_0^\infty dr\,r^2\psi_D^2=1 \tag{144.8}$$

and

$$\psi=\frac{1}{\sqrt{4\pi}}\left\{\psi_S(r)\cos\omega+\frac{3}{2\sqrt{2}}\psi_D(r)\sin\omega\,T_{pn}\right\}\chi_{1,1}. \tag{144.9}$$

Now, the Schrödinger equation for the relative motion (with $\hbar=1$, $m_p=m_n=1$, reduced mass $=\frac{1}{2}$, cf. Problem 150),

$$(-\nabla^2+V_c+V_t\,T_{pn}-E)\psi=0,$$

has to be satisfied by (144.9):

$$\left\{(-\nabla^2+V_c-E)\cos\omega\,\psi_S+\left[\frac{3}{2\sqrt{2}}(-\nabla^2+V_c-E)\sin\omega\,\psi_D+V_t\cos\omega\,\psi_S\right\}T_{pn}\right.$$

$$\left.+\frac{3}{2\sqrt{2}}V_t\sin\omega\,\psi_D\,T_{pn}^2\right\}\chi_{1,1}=0. \qquad (144.10)$$

Here, if applied to a triplet spin function, χ_t, the operator T_{pn}^2 may be linearly expressed by T_{pn}, viz.

$$T_{pn}^2\chi_t=(\tfrac{8}{9}-\tfrac{2}{3}T_{pn})\chi_t\,. \qquad (144.11)$$

This can easily be shown using the identity

$$S\begin{pmatrix}\alpha\\ \beta\end{pmatrix}=\frac{1}{r}(\boldsymbol{\sigma}\cdot\boldsymbol{r})\begin{pmatrix}\alpha\\ \beta\end{pmatrix}=\begin{pmatrix}\cos\vartheta; & \sin\vartheta\,e^{i\varphi}\\ \sin\vartheta\,e^{-i\varphi}; & -\cos\vartheta\end{pmatrix}\begin{pmatrix}\alpha\\ \beta\end{pmatrix}$$

for one-particle spin states. It then follows that the square of this operator, $S^2=1$, so that

$$S_{pn}^2=\left\{\frac{1}{r^2}(\boldsymbol{\sigma}_p\cdot\boldsymbol{r})(\boldsymbol{\sigma}_n\cdot\boldsymbol{r})\right\}^2=1\,.$$

Since we already know (p. 24) that $(\boldsymbol{\sigma}_p\cdot\boldsymbol{\sigma}_n)\chi_t=\chi_t$ we find

$$T_{pn}^2=(S_{pn}-\tfrac{1}{3})^2=1-\tfrac{2}{3}S_{pn}+\tfrac{1}{9}=\tfrac{10}{9}-\tfrac{2}{3}(T_{pn}+\tfrac{1}{3})$$

in perfect agreement with (144.11).

Eq. (144.10) may now be written

$$\left\{\cos\omega(-\nabla^2+V_c-E)\psi_S+\frac{2\sqrt{2}}{3}\sin\omega\,V_t\psi_D\right\}\chi_{1,1}$$

$$+\left\{\sin\omega\left[\frac{3}{2\sqrt{2}}(-\nabla^2+V_c-E)-\frac{1}{\sqrt{2}}V_t\right]\psi_D+\cos\omega\,V_t\psi_S\right\}T_{pn}\chi_{1,1}=0.$$

(144.12)

The operator T_{pn} in the second line, when applied to $\chi_{1,1}$, yields only terms with $l=2$ orthogonal to those with $l=0$ in the first line. We may therefore decompose (144.12) into two coupled radial equations, viz.

$$\cos\omega\left[\psi_S''+\frac{2}{r}\psi_S'+(E-V_c)\psi_S\right]-\sin\omega\cdot\frac{2\sqrt{2}}{3}V_t\psi_D=0 \quad (144.13)$$

and

$$\sin\omega\left[\psi_D''+\frac{2}{r}\psi_D'-\frac{6}{r^2}\psi_D+\left(E-V_c+\frac{2}{3}V_t\right)\psi_D\right]-\cos\omega\cdot\frac{2\sqrt{2}}{3}V_t\psi_S=0.$$

(144.14)

This is the set of differential equations required.

Problem 145. Electrical quadrupole and magnetic dipole moments of deuteron

Given the deuteron wave function determined in the preceding problem.

a) The electrical quadrupole moment of the deuteron shall be expressed in terms of the two integrals

$$A = \int_0^\infty dr\, r^4 \psi_S \psi_D; \quad B = \int_0^\infty dr\, r^4 \psi_D^2. \tag{145.1}$$

b) The expectation value of the magnetic dipole moment shall be determined.

Solution. a) The quadrupole tensor (cf. Problem 61) can be defined by

$$Q_{ik} = \tfrac{1}{4}(3\,x_i x_k - r^2\,\delta_{ik}).$$

In the original definition of this tensor, the factor $\frac{1}{4}$ on the right-hand side had not been used. It occurs in the present problem in consequence of $\frac{1}{2}\boldsymbol{r}$ being the proton coordinate about the centre of mass, and only the proton contributing to the quadrupole moment since the neutron carries no electric charge. The deuteron charge distribution in the state $M_I = I$ being rotationally symmetrical about the z axis, averaging of the tensor elements over the angle φ leads to the relations

$$\bar{Q}_{xy} = \bar{Q}_{yz} = \bar{Q}_{zx} = 0; \quad \bar{Q}_{xx} = \bar{Q}_{yy} = -\tfrac{1}{2}\bar{Q}_{zz}. \tag{145.2}$$

We therefore need only evaluate the expectation value of the operator Q_{zz}, viz.

$$\begin{aligned}\langle Q_{zz}\rangle &= \sum_{\text{spin}} \int d\tau\, Q_{zz} |\psi|^2 \\ &= \frac{1}{4} \sum_{\text{spin}} \int d\tau\, r^2 (3\cos^2\vartheta - 1)|\psi|^2 \\ &= \frac{1}{2}\sqrt{\frac{4\pi}{5}} \sum_{\text{spin}} \int d\tau\, r^2\, Y_{2,0} |\psi|^2 .\end{aligned} \tag{145.3}$$

With the deuteron wave function determined in the preceding problem, viz.

$$\begin{aligned}\psi = \cos\omega\, \psi_S(r)\, Y_{0,0}\, \chi_{1,1} + \sin\omega \frac{1}{\sqrt{10}} \psi_D(r) \{ Y_{2,0}\, \chi_{1,1} + \\ + \sqrt{3}\, Y_{2,1}\, \chi_{1,0} + \sqrt{6}\, Y_{2,2}\, \chi_{1,-1} \},\end{aligned} \tag{145.4}$$

this yields by spin summation,

$$\langle Q_{zz}\rangle = \frac{1}{2}\sqrt{\frac{4\pi}{5}}\int_0^\infty dr\,r^4 \oint d\Omega\, Y_{2,0}\cdot\left\{\left|\cos\omega\,\psi_S\, Y_{0,0}+\sin\omega\frac{\psi_D}{\sqrt{10}}Y_{2,0}\right|^2+\right.$$

$$\left.+\sin^2\omega\frac{\psi_D^2}{10}(3|Y_{2,1}|^2+6|Y_{2,2}|^2)\right\}.$$

The term with ψ_S^2 vanishes in consequence of the orthogonality of the spherical harmonics. With the product $\psi_S\psi_D$ the obvious integral

$$\oint d\Omega\, Y_{0,0}|Y_{2,0}|^2 = \frac{1}{\sqrt{4\pi}}$$

is coupled. It is a little more laborious to evaluate the three remaining integrals occurring with ψ_D^2, viz.

$$\left.\begin{aligned}\oint d\Omega\, Y_{2,0}|Y_{2,0}|^2 &= \frac{2}{7}\sqrt{\frac{5}{4\pi}},\\ \oint d\Omega\, Y_{2,0}|Y_{2,1}|^2 &= \frac{1}{7}\sqrt{\frac{5}{4\pi}},\\ \oint d\Omega\, Y_{2,0}|Y_{2,2}|^2 &= -\frac{2}{7}\sqrt{\frac{5}{4\pi}}.\end{aligned}\right\} \tag{145.5}$$

Assembling all these details, we finally arrive at the simple formula

$$\langle Q_{zz}\rangle = \frac{1}{5\sqrt{2}}A\cos\omega\sin\omega - \frac{1}{20}B\sin^2\omega. \tag{145.6}$$

If the admixture of D to S state is a small percentage, the parameter ω is small and the second, negative term in (145.6) represents a small correction only to the first, positive part. Therefore, $\langle Q_{zz}\rangle$ is positive so that the deuteron has an oblong shape in z direction. This is borne out by experiment.

b) The magnetic dipole operator consists of a spin part, $\mu_p\sigma_{pz}+\mu_n\sigma_{nz}$, and of an orbital momentum part to which only the proton contributes. The orbital momentum component L_z for the two-particle problem is given by

$$L_z = \frac{\hbar}{i}\left(\frac{\partial}{\partial\varphi_p}+\frac{\partial}{\partial\varphi_n}\right)$$

of which only the first part will contribute to the magnetic moment whereas both terms contribute equal parts to the orbital momentum about the center of mass. Therefore only $\frac{1}{2}L_z$ enters the orbital part of the magnetic moment,

$$\mu_{\text{orbit}} = \frac{e}{2mc} \cdot \frac{1}{2} L_z .$$

The expectation value of the z component of the magnetic moment thus becomes

$$\langle\mu\rangle = \sum_{\text{spin}} \int d\tau \, \psi^* (\mu_p \sigma_{pz} + \mu_n \sigma_{nz} + \mu_{\text{orbit}}) \, \psi , \tag{145.7}$$

and the expectation values of x and y components vanish.

Application of the operators σ_{pz} and σ_{nz} to the triplet spin functions yields the relations,

$$\begin{aligned} \sigma_{pz}\chi_{1,1} &= \chi_{1,1}; & \sigma_{nz}\chi_{1,1} &= \chi_{1,1}; \\ \sigma_{pz}\chi_{1,0} &= \chi_{0,0}; & \sigma_{nz}\chi_{1,0} &= -\chi_{0,0}; \\ \sigma_{pz}\chi_{1,-1} &= -\chi_{1,-1}; & \sigma_{nz}\chi_{1,-1} &= -\chi_{1,-1} . \end{aligned} \tag{145.8}$$

If, therefore, we write a triplet wave function briefly in the form

$$\psi = u\chi_{1,1} + v\chi_{1,0} + w\chi_{1,-1}$$

we get

$$\begin{aligned} \langle\mu\rangle = \sum_{\text{spin}} \int d\tau (u^*\chi_{1,1} + v^*\chi_{1,0} + w^*\chi_{1,-1}) \\ \times \{\mu_p(u\chi_{1,1} + v\chi_{0,0} - w\chi_{1,-1}) \\ + \mu_n(u\chi_{1,1} - v\chi_{0,0} - w\chi_{1,-1}) \\ + \mu_{\text{orbit}}(u\chi_{1,1} + v\chi_{1,0} + w\chi_{1,-1})\} \end{aligned}$$

and, by using the orthonormality of the spin functions,

$$\langle\mu\rangle = \int d\tau \{u^*(\mu_p + \mu_n + \mu_{\text{orbit}})u + v^* \mu_{\text{orbit}} v + w^*(-\mu_p - \mu_n + \mu_{\text{orbit}}) w\} .$$

Now we have (cf. (144.5))

$$u = f\,Y_{0,0} + g\,Y_{2,0}; \quad v = \sqrt{3}\,g\,Y_{2,1}; \quad w = \sqrt{6}\,g\,Y_{2,2}$$

and therefore

$$L_z u = 0; \quad L_z v = \hbar v; \quad L_z w = 2\hbar w$$

so that

$$\begin{aligned} \langle\mu\rangle &= \int d\tau \left\{ (\mu_p + \mu_n)(u^* u - w^* w) + \frac{e\hbar}{4mc}(v^* v + 2 w^* w) \right\} \\ &= (\mu_p + \mu_n) \int_0^\infty dr\, r^2 (f^2 + g^2 - 6g^2) + \frac{e\hbar}{4mc} \int_0^\infty dr\, r^2 (3g^2 + 2\cdot 6g^2) . \end{aligned}$$

Measuring $\langle\mu\rangle$ in units of nuclear magnetons, $e\hbar/(2mc)$, and replacing f and g by the normalized functions ψ_S and ψ_D (144.7), we finally arrive at

$$\langle\mu\rangle=(\mu_p+\mu_n)-\tfrac{3}{2}\sin^2\omega(\mu_p+\mu_n-\tfrac{1}{2})\,. \tag{145.9}$$

The admixture of D state therefore only causes a second-order correction in the magnetic moment.

Problem 146. Spin functions of three particles

To construct the eigenfunctions of S_z and $\boldsymbol{S}^2$ for a system of three particles of spin $\frac{1}{2}$.

Solution. The total spin vector operator of the system is now

$$\boldsymbol{S}=\frac{\hbar}{2}(\boldsymbol{\sigma}_1+\boldsymbol{\sigma}_2+\boldsymbol{\sigma}_3)\,. \tag{146.1}$$

Its z component apparently has the following eigenfunctions:

$$\begin{aligned}\chi(\tfrac{3}{2})&=\alpha_1\alpha_2\alpha_3;\\ \chi(\tfrac{1}{2})&=A\alpha_1\alpha_2\beta_3+B\alpha_1\beta_2\alpha_3+C\beta_1\alpha_2\alpha_3;\\ \chi(-\tfrac{1}{2})&=A'\beta_1\beta_2\alpha_3+B'\beta_1\alpha_2\beta_3+C'\alpha_1\beta_2\beta_3;\\ \chi(-\tfrac{3}{2})&=\beta_1\beta_2\beta_3\,.\end{aligned} \tag{146.2}$$

The argument of χ denotes the eigenvalue of S_z in units of $\hbar$. Each of the two functions $\chi(\frac{1}{2})$ and $\chi(-\frac{1}{2})$ consists of three still degenerate functions. This degeneracy will now be dissolved by investigating the operator

$$\boldsymbol{S}^2=\left(\frac{\hbar}{2}\right)^2(\boldsymbol{\sigma}_1+\boldsymbol{\sigma}_2+\boldsymbol{\sigma}_3)^2=\left(\frac{\hbar}{2}\right)^2\{9+2(\boldsymbol{\sigma}_1\cdot\boldsymbol{\sigma}_2)+2(\boldsymbol{\sigma}_2\cdot\boldsymbol{\sigma}_3)+2(\boldsymbol{\sigma}_3\cdot\boldsymbol{\sigma}_1)\}\,. \tag{146.3}$$

In Problem 140 it has been shown that

$$(\boldsymbol{\sigma}_1\cdot\boldsymbol{\sigma}_2)\alpha_1\alpha_2=\alpha_1\alpha_2;\qquad(\boldsymbol{\sigma}_1\cdot\boldsymbol{\sigma}_2)\alpha_1\beta_2=2\beta_1\alpha_2-\alpha_1\beta_2;$$
$$(\boldsymbol{\sigma}_1\cdot\boldsymbol{\sigma}_2)\beta_1\alpha_2=2\alpha_1\beta_2-\beta_1\alpha_2;\qquad(\boldsymbol{\sigma}_1\cdot\boldsymbol{\sigma}_2)\beta_1\beta_2=\beta_1\beta_2;$$

or, even, more simply, that the operator

$$\Sigma_{12}=\tfrac{1}{2}(1+\boldsymbol{\sigma}_1\cdot\boldsymbol{\sigma}_2) \tag{146.4}$$

merely exchanges the spin functions of the two particles 1 and 2,

$$\Sigma_{12}\,\chi(1,2)=\chi(2,1), \tag{146.5a}$$

in detail:

$$\Sigma_{12}\alpha_1\alpha_2=\alpha_1\alpha_2;\qquad\Sigma_{12}\alpha_1\beta_2=\beta_1\alpha_2;\quad\text{etc.} \tag{146.5b}$$

The operator (146.4) is therefore called the *spin exchange operator* of particles 1 and 2.

We now can express (146.3) in terms of such exchange operators,

$$\boldsymbol{S}^2 = \left(\frac{\hbar}{2}\right)^2 \{3+4(\Sigma_{12}+\Sigma_{23}+\Sigma_{31})\}. \tag{146.6}$$

Application of this operator to the first and last of the four spin functions (146.2) leads to

$$\boldsymbol{S}^2 \chi(\tfrac{3}{2}) = \left(\frac{\hbar}{2}\right)^2 \cdot 15\,\chi(\tfrac{3}{2}); \quad \boldsymbol{S}^2 \chi(-\tfrac{3}{2}) = \left(\frac{\hbar}{2}\right)^2 \cdot 15\,\chi(-\tfrac{3}{2}). \tag{146.7}$$

These two functions therefore are non-degenerate eigenfunctions of $\boldsymbol{S}^2$ already to the eigenvalue $S(S+1)=\frac{15}{4}$ or $S=\frac{3}{2}$ for two different eigenvalues of S_z. In the vector model they correspond to parallel orientation of all three spins in z or $-z$ direction.

It is not so simple to deal with the degenerate functions $\chi(\frac{1}{2})$ and $\chi(-\frac{1}{2})$. Here, application of (146.6) to $\chi(\frac{1}{2})$ yields

$$\begin{aligned}\boldsymbol{S}^2 \chi(\tfrac{1}{2}) = \left(\frac{\hbar}{2}\right)^2 &\{3\chi(\tfrac{1}{2})+4[A\alpha_1\alpha_2\beta_3+B\beta_1\alpha_2\alpha_3+C\alpha_1\beta_2\alpha_3] \\ &+4[A\alpha_1\beta_2\alpha_3+B\alpha_1\alpha_2\beta_3+C\beta_1\alpha_2\alpha_3] \\ &+4[A\beta_1\alpha_2\alpha_3+B\alpha_1\beta_2\alpha_3+C\alpha_1\alpha_2\beta_3]\} \\ = \left(\frac{\hbar}{2}\right)^2 &\{(7A+4B+4C)\alpha_1\alpha_2\beta_3+(4A+7B+4C)\alpha_1\beta_2\alpha_3 \\ &+(4A+4B+7C)\beta_1\alpha_2\alpha_3\}.\end{aligned}$$

This shall become

$$=\hbar^2 S(S+1)\{A\alpha_1\alpha_2\beta_3+B\alpha_1\beta_2\alpha_3+C\beta_1\alpha_2\alpha_3\}.$$

Thus we arrive at a linear system of three homogeneous equations,

$$\begin{aligned}7A+4B+4C&=4S(S+1)\,A; \\ 4A+7B+4C&=4S(S+1)\,B; \\ 4A+4B+7C&=4S(S+1)\,C\end{aligned} \tag{146.8}$$

the determinant of which must vanish. This gives a cubic equation for the eigenvalues $S(S+1)$ possible with the solutions

$$S(S+1)=\tfrac{15}{4},\tfrac{3}{4},\tfrac{3}{4} \quad \text{or} \quad S=\tfrac{3}{2},\tfrac{1}{2},\tfrac{1}{2}. \tag{146.9}$$

The same result would be obtained by applying $\boldsymbol{S}^2$ to $\chi(-\frac{1}{2})$ if the symbols α and β are exchanged throughout.

The first eigenvalue (146.9) leads unambiguously to the solution $A=B=C$ of the system (146.8). Using a more complete notation for the spin functions, $\chi(S,S_z)$, we find a *quartet* of four completely symmetrical functions, viz.

$$\begin{aligned}
\chi\left(\frac{3}{2},\frac{3}{2}\right)&=\alpha_1\alpha_2\alpha_3;\\
\chi\left(\frac{3}{2},\frac{1}{2}\right)&=\frac{1}{\sqrt{3}}(\alpha_1\alpha_2\beta_3+\alpha_1\beta_2\alpha_3+\beta_1\alpha_2\alpha_3);\\
\chi\left(\frac{3}{2},-\frac{1}{2}\right)&=\frac{1}{\sqrt{3}}(\beta_1\beta_2\alpha_3+\beta_1\alpha_2\beta_3+\alpha_1\beta_2\beta_3);\\
\chi\left(\frac{3}{2},-\frac{3}{2}\right)&=\beta_1\beta_2\beta_3.
\end{aligned}\tag{146.10}$$

Besides this solution which corresponds to the four orientations of spin $\frac{3}{2}$ in the vector model, the double solution $S=\frac{1}{2}$ put in (146.8) leads three times to the same relation,

$$A+B+C=0;$$

we thus may express $C=-(A+B)$, but cannot then obtain separate information on A and B:

$$\begin{aligned}
\chi(\tfrac{1}{2},\tfrac{1}{2})&=A\alpha_1\alpha_2\beta_3+B\alpha_1\beta_2\alpha_3-(A+B)\beta_1\alpha_2\alpha_3;\\
\chi(\tfrac{1}{2},-\tfrac{1}{2})&=A'\beta_1\beta_2\alpha_3+B'\beta_1\alpha_2\beta_3-(A'+B')\alpha_1\beta_2\beta_3.
\end{aligned}\tag{146.11}$$

Two doublets, each with $S=\frac{1}{2}$, are still mixed up in these formulae and are still degenerate.

It is usual to decompose and normalize the doublets by the two assumptions

$$A=B=\frac{1}{\sqrt{6}}\quad\text{and}\quad A=-B=\frac{1}{\sqrt{2}}.\tag{146.12}$$

With the first assumption the doublet becomes

$$\begin{aligned}
\chi_1\left(\frac{1}{2},\frac{1}{2}\right)&=\frac{1}{\sqrt{6}}\{\alpha_1(\alpha_2\beta_3+\beta_2\alpha_3)-2\beta_1\cdot\alpha_2\alpha_3\};\\
\chi_1\left(\frac{1}{2},-\frac{1}{2}\right)&=\frac{1}{\sqrt{6}}\{\beta_1(\beta_2\alpha_3+\alpha_2\beta_3)-2\alpha_1\cdot\beta_2\beta_3\};
\end{aligned}\tag{146.13}$$

this doublet is symmetrical with respect to exchanging particles 2 and 3 (symbolic notation: $1,\overline{23}$). The other doublet will become

$$\chi_2\left(\frac{1}{2},\frac{1}{2}\right)=\frac{1}{\sqrt{2}}\alpha_1(\alpha_2\beta_3-\beta_2\alpha_3);$$

$$\chi_2\left(\frac{1}{2},-\frac{1}{2}\right)=\frac{1}{\sqrt{2}}\beta_1(\beta_2\alpha_3-\alpha_2\beta_3); \tag{146.14}$$

it is antisymmetrical in 2 and 3 (in symbols: $1,\widetilde{23}$).

Of course, it is quite arbitrary to select just particles 2 and 3 as affecting simple symmetry properties. By another choice of A and B, e.g. $B=-\frac{1}{2}A$, a function of symmetry $\overline{12},3$ would have been obtained. Only further conditions imposed on the solution in special problems can lead to the dissolution of this remaining degeneracy.

Problem 147. Neutron scattering by molecular hydrogen

Let the particles 1 and 2 of the preceding problem be the two protons of an hydrogen molecule and 3 be a slow neutron with its de Broglie wavelength large as compared to the nuclear distance. The scattering cross section shall be determined for para and orthohydrogen, separately, with the central-force n-p interaction (cf. Problem 140)

$$V=\tfrac{1}{4}(3V_t+V_s)+\tfrac{1}{4}(V_t-V_s)(\boldsymbol{\sigma}_n\cdot\boldsymbol{\sigma}_p). \tag{147.1}$$

To connect scattering lengths with potentials, the somewhat crude assumption may be made that the scattering length is proportional to the potential well depth.

Solution. The motion of the neutron will be governed by its interaction with the two protons. If its wavelength is large, both protons are practically at the same position and we have only one relative coordinate vector $\boldsymbol{r}$. Let us denote the neutron by subscript n (instead of 3), then the neutron-molecule interaction may be written, according to (147.1),

$$V=\tfrac{1}{2}(3V_t+V_s)+\tfrac{1}{4}(V_t-V_s)(\boldsymbol{\sigma}_n,\boldsymbol{\sigma}_1+\boldsymbol{\sigma}_2) \tag{147.2}$$

with $V_t(r)$ and $V_s(r)$.

Orthohydrogen is now defined by a symmetrical, parahydrogen by an antisymmetrical spin function so that, according to the results of the preceding problem, there exist the following eight spin functions of our three-body problem:

$$\left.\begin{aligned}
\chi\left(\frac{3}{2},+\frac{3}{2}\right)&=\alpha_1\alpha_2\alpha_n\\
\chi\left(\frac{3}{2},+\frac{1}{2}\right)&=\frac{1}{\sqrt{3}}(\alpha_1\alpha_2\beta_n+\alpha_1\beta_2\alpha_n+\beta_1\alpha_2\alpha_n)\\
\chi\left(\frac{3}{2},-\frac{1}{2}\right)&=\frac{1}{\sqrt{3}}(\beta_1\beta_2\alpha_n+\beta_1\alpha_2\beta_n+\alpha_1\beta_2\beta_n)\\
\chi\left(\frac{3}{2},-\frac{3}{2}\right)&=\beta_1\beta_2\beta_n
\end{aligned}\right\}\begin{array}{l}\text{Quartet,}\\ \text{spin } \tfrac{3}{2},\\ o-H_2,\\ 12\ n\end{array}\tag{147.3}$$

$$\left.\begin{aligned}
\chi_o\left(\frac{1}{2},+\frac{1}{2}\right)&=\frac{1}{\sqrt{6}}((\alpha_1\beta_2+\beta_1\alpha_2)\alpha_n-2\alpha_1\alpha_2\beta_n)\\
\chi_o\left(\frac{1}{2},-\frac{1}{2}\right)&=\frac{1}{\sqrt{6}}((\alpha_1\beta_2+\beta_1\alpha_2)\beta_n-2\beta_1\beta_2\alpha_n)
\end{aligned}\right\}\begin{array}{l}\text{Doublet,}\\ \text{spin } \tfrac{1}{2},\\ o-H_2,\\ 12, n\end{array}\tag{147.4}$$

$$\left.\begin{aligned}
\chi_p\left(\frac{1}{2},+\frac{1}{2}\right)&=\frac{1}{\sqrt{2}}(\alpha_1\beta_2-\beta_1\alpha_2)\alpha_n\\
\chi_p\left(\frac{1}{2},-\frac{1}{2}\right)&=\frac{1}{\sqrt{2}}(\alpha_1\beta_2-\beta_1\alpha_2)\beta_n
\end{aligned}\right\}\begin{array}{l}\text{Doublet,}\\ \text{spin } \tfrac{1}{2},\\ p-H_2,\\ \widetilde{12}, n\end{array}\tag{147.5}$$

These eight functions are eigenfunctions of the operator

$$\boldsymbol{S}^2=(\boldsymbol{\sigma}_1+\boldsymbol{\sigma}_2+\boldsymbol{\sigma}_n)^2$$

with the eigenvalues 15 in the quartet, and 3 in the doublet states. Since

$$\boldsymbol{S}^2=\boldsymbol{\sigma}_1^2+\boldsymbol{\sigma}_2^2+\boldsymbol{\sigma}_3^2+2(\boldsymbol{\sigma}_1\cdot\boldsymbol{\sigma}_2)+2(\boldsymbol{\sigma}_n,\boldsymbol{\sigma}_1+\boldsymbol{\sigma}_2)\tag{147.6}$$

with the three first terms all equal to 1, but the fourth contributing according to

$$(\boldsymbol{\sigma}_1\cdot\boldsymbol{\sigma}_2)=\begin{cases}+1 & \text{for orthohydrogen}\\ -3 & \text{for parahydrogen,}\end{cases}\tag{147.7}$$

we arrive at

$$\begin{aligned}15&=9+2+2(\boldsymbol{\sigma}_n,\boldsymbol{\sigma}_1+\boldsymbol{\sigma}_2)\\ 3&=9+2+2(\boldsymbol{\sigma}_n,\boldsymbol{\sigma}_1+\boldsymbol{\sigma}_2)\end{aligned}\quad\text{for orthohydrogen}\begin{cases}\text{quartet}\\ \text{doublet}\end{cases}$$

and

$$3=9-6+2(\boldsymbol{\sigma}_n,\boldsymbol{\sigma}_1+\boldsymbol{\sigma}_2)\quad\text{for parahydrogen doublet}$$

or

$$(\boldsymbol{\sigma}_n,\boldsymbol{\sigma}_1+\boldsymbol{\sigma}_2)=\begin{cases}2 & \text{for orthohydrogen quartet}\\ -4 & \text{for orthohydrogen doublet}\\ 0 & \text{for parahydrogen doublet.}\end{cases}\tag{147.8}$$

This, according to (147.2), leads to the following three different interactions between the neutron and the hydrogen molecule:

$$V=\tfrac{1}{2}(3V_t+V_s)+\begin{Bmatrix}\tfrac{1}{2}(V_t-V_s)\\ -(V_t-V_s)\\ 0\end{Bmatrix}=\begin{cases}2V_t & \text{for quartet, } o-H_2\\ \tfrac{1}{2}V_t+\tfrac{3}{2}V_s & \text{for doublet, } o-H_2\\ \tfrac{3}{2}V_t+\tfrac{1}{2}V_s & \text{for doublet, } p-H_2.\end{cases} \quad (147.9)$$

Turning to the scattering question, we need only consider the limiting case of zero energy where the scattering length a is linearly connected with the potential depth constant as long as the potential hole is "small". It should, however, be noted that this is rather a crude assumption in the real neutron-proton interaction case. Making this approximation, we get the elastic scattering cross sections

$$\sigma_{\text{ortho}}=4\pi\left\{\tfrac{2}{3}(2a_t)^2+\tfrac{1}{3}\left(\tfrac{1}{2}a_t+\tfrac{3}{2}a_s\right)^2\right\};$$
$$\sigma_{\text{para}}=4\pi\left(\tfrac{3}{2}a_t+\tfrac{1}{2}a_s\right)^2$$

or

$$\sigma_{\text{ortho}}=\pi\{(3a_t+a_s)^2+2(a_t-a_s)^2\};$$
$$\sigma_{\text{para}}=\pi(3a_t+a_s)^2. \quad (147.10)$$

NB. The best values for the scattering lengths a_t and a_s of the two-nucleon problem are[1] $a_t=+5.39$ fm and $a_s=-23.7$ fm. Inserting these values into Eq. (147.10) yields $\sigma_{\text{ortho}}=55$ barn and $\sigma_{\text{para}}=1.77$ barn. The characteristic feature of this somewhat rough result is the amazingly small value of the parahydrogen cross section. This is fully borne out by experimental evidence with thermal neutrons. The para cross section would vanish entirely with $a_s=-3a_t$; its smallness shows that anyhow a_s must be large and have the opposite sign to a_t. The triplet scattering length must be positive in order to allow for a ^3S bound state, the deuteron. Hence, $a_s<0$ so that no ^1S bound state can exist. It should be noted that this sign can only be determined by interference experiments of the kind described, not by scattering of neutrons at isolated protons producing incoherent waves.

Our results apply to the limit of energy zero whereas, in experiment, the neutrons still have a few hundredths of an eV energy. Their wavelength therefore is not so very large in comparison with the molecular distance between the two protons. This causes inelastic transitions with parity change in the molecule between the rotational states $J=1$ of ortho, and $J=0$ of parahydrogen. They occur because $r_{n1}\neq r_{n2}$ so that, with the abbreviation $\tfrac{1}{4}(V_t-V_s)=U(r)$, we have for the spin dependent part of the interaction,

$$U(r_{n1})(\boldsymbol{\sigma}_n\cdot\boldsymbol{\sigma}_1)+U(r_{n2})(\boldsymbol{\sigma}_n\cdot\boldsymbol{\sigma}_2)=\tfrac{1}{2}(U(r_{n1})+U(r_{n2}))(\boldsymbol{\sigma}_n,\boldsymbol{\sigma}_1+\boldsymbol{\sigma}_2)$$
$$+\tfrac{1}{2}(U(r_{n1})-U(r_{n2}))(\boldsymbol{\sigma}_n,\boldsymbol{\sigma}_1-\boldsymbol{\sigma}_2).$$

In consequence of the last term, the functions (147.3) to (147.5) no longer remain eigenfunctions of the potential; it is this last term that induces ortho-para transitions.

[1] 1 fm $=10^{-13}$ cm; 1 barn $=10^{-24}$ cm^2.

IV. Many-Body Problems

A. Few Particles

Problem 148. Two repulsive particles on a circle

Two particles are fixed on a circle with a mutual repulsion given by

$$V(\varphi_1, \varphi_2) = V_0 \cos(\varphi_1 - \varphi_2) \tag{148.1}$$

to simulate e.g. the Coulomb repulsion between the two helium electrons in the ground state. The conservation of angular momentum shall be derived, and the relative motion of the particles discussed.

Solution. The Schrödinger equation

$$-\frac{\hbar^2}{2mr^2}\left(\frac{\partial^2 U}{\partial \varphi_1^2} + \frac{\partial^2 U}{\partial \varphi_2^2}\right) + V_0 \cos(\varphi_1 - \varphi_2)\, U = E \cdot U \tag{148.2}$$

permits factorization by introducing the variables

$$\alpha = \varphi_1 - \varphi_2; \qquad \beta = \tfrac{1}{2}(\varphi_1 + \varphi_2) \tag{148.3}$$

of relative and absolute motion. Then we have

$$\frac{\partial}{\partial \varphi_1} = \frac{\partial}{\partial \alpha} + \frac{1}{2}\frac{\partial}{\partial \beta}; \qquad \frac{\partial}{\partial \varphi_2} = -\frac{\partial}{\partial \alpha} + \frac{1}{2}\frac{\partial}{\partial \beta}.$$

Putting this into (148.2) we get

$$-\frac{\hbar^2}{mr^2}\left(\frac{\partial^2 U}{\partial \alpha^2} + \frac{1}{4}\frac{\partial^2 U}{\partial \beta^2}\right) + V_0 \cos\alpha \cdot U = E \cdot U.$$

Factorization now becomes possible into

$$U(\alpha, \beta) = u(\alpha)\, v(\beta) \tag{148.4}$$

and leads to the separate equations of motion

$$-\frac{\hbar^2}{mr^2}\frac{d^2u}{d\alpha^2} + V_0 \cos\alpha \cdot u = E_\alpha \cdot u \tag{148.5}$$

and

$$-\frac{\hbar^2}{4mr^2}\frac{d^2v}{d\beta^2} = E_\beta \cdot v \tag{148.6}$$

with

$$E_\alpha + E_\beta = E. \tag{148.7}$$

The *absolute motion* can be determined from (148.6). The total orbital momentum operator of the two-particle system is

$$L = \frac{\hbar}{i}\left(\frac{\partial}{\partial\varphi_1} + \frac{\partial}{\partial\varphi_2}\right) = \frac{\hbar}{i}\frac{\partial}{\partial\beta}$$

so that (148.6) may as well be written

$$\frac{1}{2\Theta} L^2 v = E_\beta \cdot v \tag{148.6'}$$

with $\Theta = 2mr^2$ the total moment of inertia of both particles. Eq. (148.6') therefore is the eigenvalue problem of the operator of rotational energy. Since (148.6) is solved by

$$v = e^{iM\beta}; \quad M = 0, \pm 1, \pm 2, \dots, \tag{148.8}$$

the eigenvalues of the rotational energy become

$$E_\beta = \frac{(\hbar M)^2}{2\Theta}. \tag{148.9}$$

It is much more difficult to discuss the *relative motion* determined by the differential equation (148.5) of the Mathieu type. To alleviate the discussion we transform (148.5) to the standard form by putting

$$\alpha = 2\varphi; \quad 4\frac{mr^2}{\hbar^2} E_\alpha = \lambda; \quad 4\frac{mr^2}{\hbar^2} V_0 = 2q \tag{148.10}$$

so that we get

$$\frac{d^2v}{d\varphi^2} + (\lambda - 2q\cos 2\varphi)v = 0. \tag{148.11}$$

We are looking for periodic solutions[1] with period 2π in the variable α, or π in the variable φ. The coefficient of v in (148.11) being an even

[1] This makes a fundamental difference to the solutions in a periodic potential of lattice theory which are not periodical, but multiply with a phase factor in each period of the lattice potential, cf. Problems 28, 29. We therefore get discrete eigenvalues λ here, but a band structure in the lattice problem.

function of φ, there are two symmetry types of solutions, even and odd. Their periodicity permits Fourier expansion so that we have

$$v_{\text{even}} = A_0 + A_2 \cos 2\varphi + A_4 \cos 4\varphi + \cdots;$$
$$v_{\text{odd}} = B_2 \sin 2\varphi + B_4 \sin 4\varphi + \cdots.$$

It is usual to denote the eigenvalues of λ by $a_0, a_2, a_4, \ldots$ for even and by $b_2, b_4, \ldots$ for odd solutions[2]. It is shown in the theory of Mathieu equations that the eigenvalues may be ordered in a sequence

$$a_0 < b_2 < a_2 < b_4 < a_4 \ldots \tag{148.12}$$

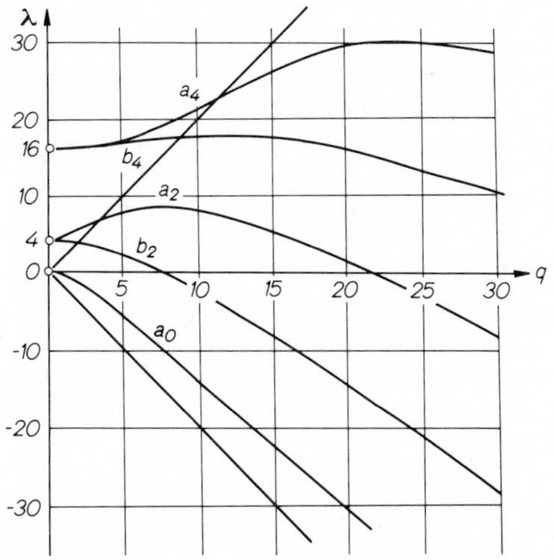

Fig. 61. Eigenvalues (λ) for different potential hole depths (q), both in dimensionless scale as defined in Eq. (148.10). The two straight lines mark the potential maximum and minimum

These eigenvalues are given, as functions of q, in Fig. 61; the relation (148.12) forbids intersections of the curves. At $q=0$ we have

$$a_0(0)=0; \quad b_2(0)=a_2(0)=2^2; \quad b_4(0)=a_4(0)=4^2 \quad \text{etc.}; \tag{148.13}$$

[2] We adopt, as far as reasonably possible, the mathematical notations of Abramowitz, M., Stegun, I.A., Handbook of Mathematical Functions, Chap. 20. New York: Dover Publ., 1965. – The curves of Fig. 61 are partially constructed with the help of this Handbook.

for very large values of q there hold the asymptotic laws

$$\left.\begin{aligned} a_{2r}(q) &\to -2q+2(2r+1)\sqrt{q}-\tfrac{1}{4}[(r+1)^2+r^2]; \\ b_{2r}(q) &\to -2q+2(2r-1)\sqrt{q}-\tfrac{1}{4}[(r-1)^2+r^2]. \end{aligned}\right\} \tag{148.14}$$

The relations (148.13) and (148.14) can largely be understood by elementary considerations.

If $q=0$, Eq. (148.11) simply becomes

$$\frac{d^2 v}{d\varphi^2}+\lambda v=0$$

with periodic solutions

$$v_{\text{even}}=A_{2r}\cos 2r\varphi; \qquad v_{\text{odd}}=B_{2r}\sin 2r\varphi$$

belonging to eigenvalues $\lambda=(2r)^2$. This exactly corresponds to Eq. (148.13).

If, on the other hand, q becomes very large, there will be a very deep potential hole around $\alpha=\pi$, almost fixing the two particles at opposite positions on the circle. It is then helpful to use, instead of φ, a variable

$$\eta=\tfrac{1}{2}(\varphi_1-\varphi_2-\pi)=\varphi-\frac{\pi}{2}. \tag{148.15}$$

From

$$\cos 2r\varphi=(-1)^r\cos 2r\eta; \qquad \sin 2r\varphi=(-1)^r\sin 2r\eta$$

it follows that the functions v_{even} and v_{odd} will be even and odd also with respect to the variable η. If the potential hole is very deep we may write in (148.11)

$$\cos 2\varphi=-\cos 2\eta\simeq -1+2\eta^2$$

and thus arrive, approximately, at the differential equation of the harmonic oscillator,

$$\frac{d^2 v}{d\eta^2}+[(\lambda+2q)-4q\eta^2]v=0. \tag{148.16}$$

Its well-known eigenvalues (cf. Problem 30) are

$$\lambda_n+2q=2\sqrt{q}(2n+1); \qquad n=0,1,2,\ldots \tag{148.17}$$

with the eigenfunctions for even/odd n being even/odd in η. Eq. (148.17) is almost identical with (148.14) if we identify solutions at small and at large values of q as follows:

	$n=0$	1	2	3	4	5	6 …	
even(a_{2r})	$r=0$		1		2		3 …	
odd(b_{2r})	$r=$	1		2		3		…

The additional constant in (148.14) may be corroborated if we expand $\cos 2\eta$ one term further on; then a perturbation calculation using oscillator functions in first approximation yields instead of (148.17):

$$\lambda_n + 2q = 2\sqrt{q}(2n+1) - \tfrac{4}{3}q\langle n|\eta^4|n\rangle, \tag{148.18}$$

where the additional term turns out to bring (148.18) to perfect agreement with (148.14).

NB 1. In Fig. 61 we have drawn two diagonals, the lower one marking the depths of the potential bottoms ($-2q$), the upper one the heights of potential summits ($+2q$). Eigenvalues must, of course, always lie above the bottom line. If they lie below the summit diagonal, they describe states of libration inside the potential hole. At $q=10$, e.g., there are 4 such libration states inside the hole, the fifth (a_4) eigenvalue leading to a vibration all around the circle, including coincidence of both particles. The first four states might be called anharmonic oscillator states; from the fifth state upwards they will correspond more and more to force-free motions of independent particles.

NB 2. The relative motion of the two particles occurs under the action of a potential energy which has the same form ($V_0 \cos\alpha$) as that of a pendulum. In classical mechanics this leads to no more complicated functions than elliptical integrals, whereas in quantum mechanics we need Mathieu functions. This again, as in Problem 40, shows how much more involved is the mathematical situation in quantum than in classical mechanics.

Problem 149. Three-atomic linear molecule

The carbon dioxide molecule has a linear O=C=O form in equilibrium. Let either equal or different oxygen isotopes be used, the C=O equilibrium distance be a and the force constant of the valence vibration f. The two valence vibration frequencies shall be determined in harmonic approximation using a one-dimensional model, thus neglecting bending vibrations.

Solution. Let x_1, x_2, x_3 be the positions along the x axis and m_1, m_2, m_3 the masses of the three atoms, then the Schrödinger equation for the linear harmonic model runs

$$-\frac{\hbar^2}{2}\sum_{i=1}^{3}\frac{1}{m_i}\frac{\partial^2\Psi}{\partial x_i^2} + \frac{1}{2}f\left[(x_2-x_1-a)^2 + (x_3-x_2-a)^2\right]\Psi = E\Psi. \tag{149.1}$$

In order to factorize the solution into centre-of-mass motion and internal motion we use the variables

$$\begin{aligned} u &= x_2 - x_1 - a,\\ v &= x_3 - x_2 - a,\\ X &= \frac{1}{M}(m_1x_1 + m_2x_2 + m_3x_3) \quad \text{with } M = m_1 + m_2 + m_3. \end{aligned} \tag{149.2}$$

It then follows that

$$\frac{\partial}{\partial x_1}=\frac{m_1}{M}\frac{\partial}{\partial X}-\frac{\partial}{\partial u};$$

$$\frac{\partial}{\partial x_2}=\frac{m_2}{M}\frac{\partial}{\partial X}+\frac{\partial}{\partial u}-\frac{\partial}{\partial v};$$

$$\frac{\partial}{\partial x_3}=\frac{m_3}{M}\frac{\partial}{\partial X}\qquad\quad+\frac{\partial}{\partial v}$$

and we find by iterating these operations

$$\frac{1}{m_i}\frac{\partial^2}{\partial x_i^2}=\frac{1}{M}\frac{\partial^2}{\partial X^2}+\left(\frac{1}{m_1}+\frac{1}{m_2}\right)\frac{\partial^2}{\partial u^2}+\left(\frac{1}{m_2}+\frac{1}{m_3}\right)\frac{\partial^2}{\partial v^2}-\frac{2}{m_2}\frac{\partial^2}{\partial u\,\partial v},$$

allowing for separation of the centre-of-mass motion. Changing the definition of E in (149.1) to mean the energy of only the internal motion, we arrive at a Schrödinger equation in two variables, u and v,

$$\left\{-\frac{\hbar^2}{2}\left[\left(\frac{1}{m_1}+\frac{1}{m_2}\right)\frac{\partial^2}{\partial u^2}+\left(\frac{1}{m_2}+\frac{1}{m_3}\right)\frac{\partial^2}{\partial v^2}-\frac{2}{m_2}\frac{\partial^2}{\partial u\,\partial v}\right]\right.$$
$$\left.+\frac{1}{2}f(u^2+v^2)-E\right\}\Psi=0. \tag{149.3}$$

The cross term, $\partial^2/\partial u\,\partial v$, in the kinetic energy makes factorization impossible in these variables. If, however, we introduce a "rotated" system,

$$\begin{aligned} u' &= \quad u\cos\alpha+v\sin\alpha,\\ v' &= -u\sin\alpha+v\cos\alpha \end{aligned} \tag{149.4}$$

it is possible by a suitable choice of α to make the term with $\partial^2/\partial u'\,\partial v'$ vanish, whereas the potential energy remains invariant under this transformation:

$$u^2+v^2=u'^2+v'^2.$$

The Schrödinger equation thus becomes

$$\left\{-\frac{\hbar^2}{2}\left[\left(\frac{1}{m_1}+\frac{1}{m_2}\right)\left(\cos^2\alpha\frac{\partial^2}{\partial u'^2}-2\cos\alpha\sin\alpha\frac{\partial^2}{\partial u'\,\partial v'}+\sin^2\alpha\frac{\partial^2}{\partial v'^2}\right)\right.\right.$$
$$+\left(\frac{1}{m_2}+\frac{1}{m_3}\right)\left(\sin^2\alpha\frac{\partial^2}{\partial u'^2}+2\sin\alpha\cos\alpha\frac{\partial^2}{\partial u'\,\partial v'}+\cos^2\alpha\frac{\partial^2}{\partial v'^2}\right)$$
$$\left.-\frac{2}{m_2}\left(\cos\alpha\sin\alpha\frac{\partial^2}{\partial u'^2}+(\cos^2\alpha-\sin^2\alpha)\frac{\partial^2}{\partial u'\,\partial v'}-\cos\alpha\sin\alpha\frac{\partial^2}{\partial v'^2}\right)\right]$$
$$\left.+\frac{1}{2}f(u'^2+v'^2)-E\right\}\Psi=0. \tag{149.5}$$

Here the factor of the mixed derivative will vanish if

$$\left(\frac{1}{m_3}-\frac{1}{m_1}\right)\sin 2\alpha = \frac{2}{m_2}\cos 2\alpha$$

or

$$\tan 2\alpha = \frac{2m_1 m_3}{m_2(m_1-m_3)}. \tag{149.6}$$

Using two mass constants A and B defined by

$$\left.\begin{aligned}\frac{1}{A} &= \left(\frac{1}{m_1}+\frac{1}{m_2}\right)\cos^2\alpha - \frac{2}{m_2}\cos\alpha\sin\alpha + \left(\frac{1}{m_2}+\frac{1}{m_3}\right)\sin^2\alpha \\ &\text{and} \\ \frac{1}{B} &= \left(\frac{1}{m_1}+\frac{1}{m_2}\right)\sin^2\alpha + \frac{2}{m_2}\cos\alpha\sin\alpha + \left(\frac{1}{m_2}+\frac{1}{m_3}\right)\cos^2\alpha\end{aligned}\right\} \tag{149.7}$$

the Schrödinger equation is now much simplified:

$$\left[-\frac{\hbar^2}{2A}\frac{\partial^2}{\partial u'^2}+\frac{1}{2}f\,u'^2\right]\Psi+\left[-\frac{\hbar^2}{2B}\frac{\partial^2}{\partial v'^2}+\frac{1}{2}f\,v'^2\right]\Psi=E\,\Psi. \tag{149.8}$$

Solution by factorization into

$$\Psi(u',v')=\psi(u')\,\varphi(v'); \qquad E=E_A+E_B \tag{149.9}$$

is now possible:

$$\left.\begin{aligned}-\frac{\hbar^2}{2A}\frac{d^2\psi}{du'^2}+\frac{1}{2}f\,u'^2\,\psi=E_A\,\psi;\\ -\frac{\hbar^2}{2B}\frac{d^2\varphi}{dv'^2}+\frac{1}{2}f\,v'^2\,\varphi=E_B\,\varphi.\end{aligned}\right\} \tag{149.10}$$

The eigenvalues of these two harmonic oscillators are well known (cf. Problem 30), viz.

$$\left.\begin{aligned}E_A=\hbar\omega_A(n_A+\tfrac{1}{2}); \quad \omega_A=\sqrt{f/A};\\ E_B=\hbar\omega_B(n_B+\tfrac{1}{2}); \quad \omega_B=\sqrt{f/B}.\end{aligned}\right\} \tag{149.11}$$

Eqs. (149.6) and (149.7) to determine A and B from the masses and (149.11) for the energies and frequencies form the solution of the problem.

We now proceed to a closer inspection of these formulae in the normal case $m_1=m_3$ (equal oxygen isotopes) for which, according to (149.6), $\alpha=\pi/4$. The mass constants A and B then follow from (149.7)

$$\frac{1}{A}=\frac{1}{m_1}; \quad \frac{1}{B}=\frac{1}{m_1}+\frac{2}{m_2}.$$

Thence,

$$\omega_A = \sqrt{f/m_1}\,; \qquad \omega_B = \omega_A \sqrt{1 + 2\frac{m_1}{m_2}}\,. \tag{149.12}$$

Since ω_A does not depend on m_2 it may be concluded that the carbon atom stays at rest in mode A vibrations and, since the centre of mass is supposed not to move, we must have a symmetrical vibration as indicated in Fig. 62a. On the other hand, it can be shown that mode B vibrations,

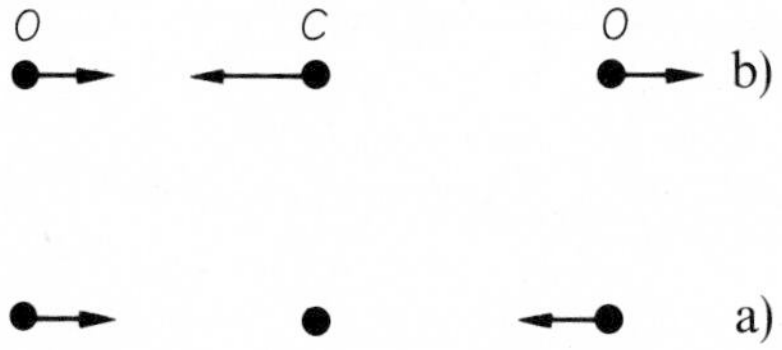

Fig. 62a and b. The two valence vibrational modes of CO_2. No bending vibrations are considered in this problem

for which the carbon atom takes part in the motion, will become antisymmetric as shown in Fig. 62b. To translate this classical description of normal vibrations into quantum mechanics, we construct the wave functions according to (149.9). In the ground state we have (in arbitrary normalization)

$$\Psi_0(u', v') = \exp\left(-\frac{A\omega_A}{2\hbar} u'^2\right) \cdot \exp\left(-\frac{B\omega_B}{2\hbar} v'^2\right), \tag{149.13}$$

both factors having a sole maximum at $u'=0$ and $v'=0$, respectively, where

$$u' = \frac{1}{\sqrt{2}}(x_3 - x_1 - 2a)\,; \qquad v' = \frac{1}{\sqrt{2}}(x_1 + x_3 - 2x_2)\,. \tag{149.14}$$

The zero-point vibration therefore occurs about the positions $x_1 + x_3 = 2x_2$ with the carbon atom halfway between the two oxygen atoms, and $x_3 - x_1 = 2a$, i.e. both oxygen atoms a distance $2a$ apart from each other. The most probable position of the ground state therefore is just the classical equilibrium position.

If the A mode is excited to, say, its first excited state, a factor u' is to be added to Ψ_0. Since a function

$$\psi(u') = u'\, e^{-\frac{1}{2}\lambda u'^2}\,; \qquad \lambda = \frac{A\omega_A}{\hbar}$$

has two maxima of opposite signs at $u' = \pm\lambda^{-\frac{1}{2}}$, we now have the most probable positions shifted to

$$x_3 - x_1 - 2a = \pm\sqrt{\frac{2}{\lambda}}; \qquad x_1 + x_3 - 2x_2 = 0. \tag{149.15}$$

The most probable position in which to find a vibrating particle has its classical equivalent in the two turning points where its time of stay is longest. The two maximum values of u' in (149.15) therefore mark something like the classical vibration amplitudes. The condition $v'=0$ shows that the carbon atom most probably still lies halfway between the two oxygen atoms, but these are now alternately at a smaller or at a larger distance apart from each other, just as indicated on Fig. 62a.

If, on the other hand, the B mode is excited we have inversely for the maximum

$$x_3 - x_1 - 2a = 0; \quad x_1 + x_3 - 2x_2 = \pm\sqrt{\frac{2}{\mu}}; \quad \mu = \frac{B\omega_B}{\hbar} \tag{149.16}$$

so that the most probable distance between the two oxygen atoms $(x_3 - x_1)$ remains $2a$ as in equilibrium, both being shifted to and fro with respect to the carbon atom as indicated on Fig. 62b.

Problem 150. Centre-of-mass motion

In classical mechanics the motion of the centre of mass in a many-body problem with only internal forces acting can be separated from the relative motion of the particles. It shall be shown that the same holds for quantum mechanics. Special attention shall be given to the case of only two particles.

Solution. We start with the hamiltonian of a system of N particles not subjected to external forces,

$$H = -\frac{\hbar^2}{2}\sum_{i=1}^{N}\frac{1}{m_i}\nabla_i^2 + \frac{1}{2}\sum_{i=1}^{N}\sum_{k=1}^{N}{}' V_{ik}(x_i - x_k, y_i - y_k, z_i - z_k) \tag{150.1}$$

and replace the $3N$ coordinates x_i, y_i, z_i by the position coordinates X, Y, Z of the centre of mass and the coordinates $\xi_\lambda, \eta_\lambda, \zeta_\lambda$ defining the position of particle λ $(\lambda = 1, 2, \ldots, N-1)$ relative to particle N:

$$\left.\begin{aligned} X &= \frac{1}{M}\sum_{i=1}^{N} m_i x_i; \qquad M = \sum_{i=1}^{N} m_i; \\ \xi_\lambda &= x_\lambda - x_N \qquad (\lambda = 1, 2, \ldots, N-1) \end{aligned}\right\} \tag{150.2}$$

and correspondingly for Y, Z, η_λ, ζ_λ. The use of these coordinates, of course, breaks up the natural symmetry of (150.1) by artificially distinguishing particle N from the rest.

We easily obtain from (150.2) the operators

$$\frac{\partial}{\partial x_\nu}=\frac{m_\nu}{M}\frac{\partial}{\partial X}+\frac{\partial}{\partial \xi_\nu} \qquad (\nu=1,2,\ldots,N-1);$$

$$\frac{\partial}{\partial x_N}=\frac{m_N}{M}\frac{\partial}{\partial X}-\sum_\lambda\frac{\partial}{\partial \xi_\lambda}$$

and

$$\sum_{i=1}^{N}\frac{1}{m_i}\frac{\partial^2}{\partial x_i^2}=\sum_{\lambda=1}^{N-1}\frac{1}{m_\lambda}\left(\frac{m_\lambda^2}{M^2}\frac{\partial^2}{\partial X^2}+2\frac{m_\lambda}{M}\frac{\partial^2}{\partial X\,\partial \xi_\lambda}+\frac{\partial^2}{\partial \xi_\lambda^2}\right)$$

$$+\frac{1}{m_N}\left(\frac{m_N^2}{M^2}\frac{\partial^2}{\partial X^2}-2\frac{m_N}{M}\sum_\lambda\frac{\partial^2}{\partial X\,\partial \xi_\lambda}+\sum_\mu\sum_\lambda\frac{\partial^2}{\partial \xi_\mu\,\partial \xi_\lambda}\right)$$

$$=\frac{1}{M}\frac{\partial^2}{\partial X^2}+\left\{\sum_\lambda\frac{1}{m_\lambda}\frac{\partial^2}{\partial \xi_\lambda^2}+\frac{1}{m_N}\sum_\mu\sum_\lambda\frac{\partial^2}{\partial \xi_\mu\,\partial \xi_\lambda}\right\},$$

where the sums over Greek subscripts run from 1 to $N-1$. The essential feature of this result is the cancelling of all mixed derivatives $\partial^2/\partial X\,\partial\xi_\lambda$ permitting separation of the hamiltonian,

$$H=H_0+H_r, \tag{150.3}$$

in a centre-of-mass part,

$$H_0=-\frac{\hbar^2}{2M}\left(\frac{\partial^2}{\partial X^2}+\frac{\partial^2}{\partial Y^2}+\frac{\partial^2}{\partial Z^2}\right), \tag{150.4}$$

and a part describing the relative motion of the particles,

$$H_r=-\frac{\hbar^2}{2}\left\{\sum_\lambda\frac{1}{m_\lambda}\nabla_\lambda^2+\frac{1}{m_N}\nabla_\lambda\cdot\nabla_\mu\right\}+V, \tag{150.5}$$

with the potential energy

$$V=\tfrac{1}{2}\sum_\lambda\sum_\mu{}' V_{\lambda\mu}(\xi_\lambda-\xi_\mu,\eta_\lambda-\eta_\mu,\zeta_\lambda-\zeta_\mu)+\sum_\lambda V_{\lambda N}(\xi_\lambda,\eta_\lambda,\zeta_\lambda) \tag{150.6}$$

independent of the centre-of-mass coordinates. The solution of the Schrödinger equation,

$$(H_0+H_r)\,U=E\cdot U \tag{150.7}$$

then permits factorization into

$$U=\varphi(X,Y,Z)\,u(\xi_\lambda,\eta_\lambda,\zeta_\lambda) \tag{150.8}$$

with

$$-\frac{\hbar^2}{2M}\nabla^2\varphi = E_0\varphi\,; \tag{150.9}$$

$$(H_r+V)\,u = E_r u\,; \tag{150.10}$$

$$E_0+E_r = E\,. \tag{150.11}$$

Eq. (150.9) is solved by the plane wave

$$\varphi = e^{i\boldsymbol{K}\cdot\boldsymbol{R}} \qquad E_0 = \frac{\hbar^2 K^2}{2M} \tag{150.12}$$

with $\boldsymbol{R}=(X,Y,Z)$. This is the centre-of-mass law as in classical mechanics: the total mass M of the system moves with constant momentum $\hbar\boldsymbol{K}$. The relative motion of the particles about the centre of mass, governed by (150.10), is quite independent thereof.

The third term in Eq. (150.5) prevents further factorization of $u(\xi_\lambda,\eta_\lambda,\zeta_\lambda)$. Only in the case of the two-body problem, $N=2$ with $\lambda=\mu=1$ only, the hamiltonian of the relative motion simplifies to

$$H_r = -\frac{\hbar^2}{2}\left\{\frac{1}{m_1}\nabla_1^2 + \frac{1}{m_2}\nabla_1^2\right\} + V_{12}(\xi_1,\eta_1,\zeta_1)\,. \tag{150.13}$$

Introducing, as in classical mechanics, the reduced mass m^* by putting

$$\frac{1}{m_1}+\frac{1}{m_2}=\frac{1}{m^*} \tag{150.14}$$

and omitting the subscripts of the relative coordinates (and of V_{12}), we arrive at

$$-\frac{\hbar^2}{2m}\nabla^2 u + V(\xi,\eta,\zeta)\,u = E_r u\,, \tag{150.15}$$

i.e. at the Schrödinger equation of an equivalent one-body problem.

NB. In Problem 67, the hydrogen atom has been treated as a one-body problem with its nucleus at rest. According to Eq. (150.15), we should more correctly introduce the reduced mass m^* of nucleus and electron, instead of m, the mass of the electron. No other change is required to take account of the participation of the nucleus in the relative motion about the centre of mass. Since the nuclear mass, say M, is very large as compared with m, Eq. (150.14) leads to

$$m^* = m\left(1-\frac{m}{M}\right),$$

approximately. Comparing e.g. the red spectral line $H_\alpha\,(n=3\rightarrow n=2)$ of the hydrogen atom with the frequency

$$\nu(H_\alpha) = \frac{5}{36}\,\frac{m_H^* e^4}{2\hbar^2 h}$$

and the corresponding line of the deuterium atom,

$$\nu(D_\alpha) = \frac{5}{36}\frac{m_D^* e^4}{2\hbar^2 h},$$

we find a line shift of

$$\nu(D_\alpha) - \nu(H_\alpha) = \frac{m_D^* - m_H^*}{m_H^*}\nu(H_\alpha) \simeq \frac{m}{2M_H}\nu(H_\alpha)$$

because $M_D \simeq 2M_H$. This difference is not very difficult to observe. It amounts to $4.12\,\mathrm{cm}^{-1}$ at a wavelength of 6563 Å. Heavy hydrogen was discovered in 1931 by Urey, Brickwedde and Murphy who observed this weak D_α satellite of the H_α line in natural hydrogen [Phys. Rev. **40**, 1 (1932)].

Problem 151. Virial theorem

To prove that the virial theorem

$$2E_{\mathrm{kin}} + E_{\mathrm{pot}} = 0$$

holds for any quantum mechanical system kept together by Coulomb forces only. The proof shall be performed by a scale transformation of the wave function of the system keeping normalization constant.

Solution. A system of N particles of masses m_i and electric charges e_i satisfies the Schrödinger equation

$$-\frac{\hbar^2}{2}\sum_{i=1}^{N}\frac{1}{m_i}\nabla_i^2\Psi + \frac{1}{2}\sum_{i=1}^{N}\sum_{k=1}^{N}\frac{e_i e_k}{r_{ik}}\Psi = E\Psi \tag{151.1}$$

with Ψ being normalized according to

$$\int d\tau_1 \int d\tau_2 \ldots \int d\tau_N\, \Psi^* \Psi = 1\,. \tag{151.2}$$

Kinetic and potential energy of the system in a state Ψ may be computed from the formulae

$$E_{\mathrm{kin}} = -\frac{\hbar^2}{2}\sum_{i=1}^{N}\frac{1}{m_i}\int d\tau_1 \int d\tau_2 \ldots \int d\tau_N\, \Psi^* \nabla_i^2 \Psi \tag{151.3a}$$

and

$$E_{\mathrm{pot}} = \frac{1}{2}\sum_{i=1}^{N}\sum_{k=1}^{N}{}' e_i e_k \int d\tau_1 \int d\tau_2 \ldots \int d\tau_N\, \Psi^* \frac{1}{r_{ik}}\Psi\,. \tag{151.3b}$$

A scale transformation,

$$\boldsymbol{r}_i' = \lambda \boldsymbol{r}_i \tag{151.4}$$

keeping (151.2) intact, means that the wave function

$$\Psi(\boldsymbol{r}_1, \boldsymbol{r}_2, \dots \boldsymbol{r}_N)$$

is replaced by

$$\Psi_\lambda = \lambda^{3N/2}\, \Psi(\lambda \boldsymbol{r}_1, \lambda \boldsymbol{r}_2, \dots \lambda \boldsymbol{r}_N)\,. \tag{151.5}$$

When introducing (151.5) into the energy expressions (151.3a, b) and passing over to the new variables (151.4), we find

$$\nabla_i^2 = \lambda^2\, \nabla_i'^2 \qquad \frac{1}{r_{ik}} = \lambda \frac{1}{r'_{ik}}$$

so that instead of the true energy of the system

$$E = E_{\text{kin}} + E_{\text{pot}}\,,$$

we obtain

$$E(\lambda) = \lambda^2\, E_{\text{kin}} + \lambda\, E_{\text{pot}}\,. \tag{151.6}$$

This function of λ apparently must be a minimum when we select from the set (151.5) of functions the correct solution of the Schrödinger equation, i.e. for $\lambda = 1$. Therefore,

$$\frac{\partial E(\lambda)}{\partial \lambda} = 2\lambda\, E_{\text{kin}} + E_{\text{pot}}$$

must vanish with $\lambda = 1$, i.e.

$$2\, E_{\text{kin}} + E_{\text{pot}} = 0\,. \tag{151.7}$$

This is the virial theorem which was to be proved.

NB. The theorem need not hold in approximate solutions. It is, therefore, remarkable that it can be proved for a Thomas-Fermi atom, cf. Problem 175.

Problem 152. Slater determinant

Let the wave function of a many-particle problem with N equal particles be factorized into a product of single-particle wave functions and antisymmetrized according to the Pauli principle. The expectation value of an operator describing the action of an external force field shall be reduced to single-particle integrals.

Solution. Let $u_i(\nu)$ be a single-particle wave function of the ν-th particle in state i, depending on its space coordinates and spin variable, all contracted into the symbol ν. The antisymmetrized product for a system of N equal particles may then be written as the Slater determinant,

$$\psi = C \begin{vmatrix} u_1(1)\, u_1(2) \dots u_1(N) \\ u_2(1)\, u_2(2) \dots u_2(N) \\ \dots\dots\dots\dots\dots\dots \\ u_N(1)\, u_N(2) \dots u_N(N) \end{vmatrix} \tag{152.1}$$

or, by expansion of the determinant,

$$\psi = C \sum_P (-1)^P P(u_1, u_2, \dots u_N), \tag{152.2}$$

where P means any permutation of the functions u_i with their arguments v in standard ordering $1, 2, \dots, N$. If P is an even (odd) permutation, the corresponding term in the sum over all permutations is positive (negative).

An operator Ω describing an external force will act on all particles in the same way, i.e.

$$\Omega = \sum_{v=1}^{N} \Omega_v . \tag{152.3}$$

Its expectation value then is

$$\langle \psi | \Omega | \psi \rangle = |C|^2 \sum_{P,P'} (-1)^{P+P'} \langle P'(u_1 \dots u_N) | \sum_v \Omega_v | P(u_1, \dots u_N) \rangle . \tag{152.4}$$

Let us now single out of (152.4) one term, Ω_v, acting only upon functions of the v-th particle coordinates and spin. Any other coordinate set, say μ, will then occur in some other function, say u_j, in both permutations P and P', because in any other combination the term would vanish by the orthogonality of the single-particle functions,

$$\langle u_j | u_k \rangle = \delta_{jk} . \tag{152.5}$$

This means identity of permutations P and P', the signature in (152.4) always being $+1$, and the term Ω_v thus contributing only one-particle integrals,

$$\langle P' | \Omega_v | P \rangle = \delta_{PP'} \sum_{i=1}^{N} \langle u_i(v) | \Omega_v | u_i(v) \rangle . \tag{152.6}$$

Now, in the wave function ψ a factor $u_i(v)$, with fixed i and v, is combined with a determinant of rank $N-1$. Therefore there still remain, among a total of $N!$ possible permutations, $(N-1)!$ permutations of the remaining $N-1$ functions except u_i over the remaining $N-1$ particles except v. Hence,

$$\langle \psi | \Omega_v | \psi \rangle = |C|^2 (N-1)! \sum_{i=1}^{N} \langle u_i(v) | \Omega_v | u_i(v) \rangle . \tag{152.7}$$

This result, of course, will hold for whatever term Ω_v we pick out of the sum (152.3), so that in $\langle \psi | \Omega | \psi \rangle$ we have a total of N equal expressions of the form (152.7). Hence,

$$\langle \psi | \Omega | \psi \rangle = |C|^2 N! \sum_{i=1}^{N} \langle u_i(v) | \Omega_v | u_i(v) \rangle . \tag{152.8}$$

It remains to determine the normalization constant C so that

$$\langle\psi|\psi\rangle = 1 \tag{152.9}$$

is the probability (viz. certainty) of finding N particles anywhere. This can be formally achieved by putting $\Omega_\nu = 1/N$ in Eq. (152.3) thus making $\Omega = 1$. Making use of the single-particle normalization, according to (152.5), Eq. (152.8) then yields

$$\langle\psi|\psi\rangle = |C|^2 N! \sum_{i=1}^{N} \frac{1}{N} = |C|^2 N!$$

or, using (152.9),

$$C = N!^{-\frac{1}{2}} \,. \tag{152.10}$$

Eq. (152.8) may then finally be written

$$\langle\psi|\Omega|\psi\rangle = \sum_{i=1}^{N} \langle u_i(\nu)|\Omega_\nu|u_i(\nu)\rangle \tag{152.11}$$

as the simple sum of the expectation values of the single-particle states.

NB. If we neglect symmetrization and replace (152.1) by the simple product

$$\mathring{\psi} = u_1(1)u_2(2)\ldots u_N(N) \tag{152.12}$$

we get

$$\langle\mathring{\psi}|\Omega|\mathring{\psi}\rangle = \sum_{\nu=1}^{N} \langle u_\nu|\Omega_\nu|u_\nu\rangle \tag{152.13}$$

and

$$\langle\mathring{\psi}|\mathring{\psi}\rangle = 1, \tag{152.14}$$

i.e. essentially the same results as we found in (152.11) and (152.9) for the antisymmetrized wave function. Neither for interaction between particles, contradicting the structure of (152.3), nor for the use of non-orthogonal single-particle functions, contradicting (152.5), do these relations hold.

Problem 153. Exchange in interaction terms with Slater determinant

For the factorized, antisymmetrized wave function of the preceding problem the expectation value of a particle-pair interaction,

$$\Omega = \tfrac{1}{2}\sum_{\mu\nu}{}' \Omega_{\mu\nu} \tag{153.1}$$

shall be determined.

Solution. Using the same notation and normalization as in the preceding problem, the expectation value of one term of (153.1) may be written,

$$\langle\psi|\Omega_{\mu\nu}|\psi\rangle=\frac{1}{N!}\sum_{P,P'}(-1)^{P+P'}\langle P'(u_1,\ldots,u_N)|\Omega_{\mu\nu}|P(u_1,\ldots,u_N)\rangle\,. \tag{153.2}$$

All functions u_n with arguments neither μ nor ν must be identical pairs in P and P' if the corresponding term in the sum is not to vanish. There being $N-2$ such pairs of functions of as many arguments, there still remain $(N-2)!$ permutations among them. If these all are identical in P and P', only one pair of functions, say $u_i u_j$, will remain for arguments μ and ν in each non-vanishing term of (153.2),

$$\langle\psi|\Omega_{\mu\nu}|\psi\rangle=\frac{(N-2)!}{N!}\sum_{ij}{}'\{\langle u_i(\mu)u_j(\nu)|\Omega_{\mu\nu}|u_i(\mu)u_j(\nu)\rangle$$
$$-\langle u_j(\mu)u_i(\nu)|\Omega_{\mu\nu}|u_i(\mu)u_j(\nu)\rangle\}\,. \tag{153.3}$$

In the first, classical term of the curly bracket, the permutations P and P' coincide completely, even with respect to μ and ν, in the second, exchange term one different permutation ($ij\to ji$) just changes the sign.

Let us now consider the sum (153.1) of such operators. Then,

$$\langle\psi|\Omega|\psi\rangle=\frac{1}{2N(N-1)}\sum_{\mu\nu}{}'\sum_{ij}{}'\{\langle u_i(\mu)u_j(\nu)|\Omega_{\mu\nu}|u_i(\mu)u_j(\nu)\rangle$$
$$-\langle u_j(\mu)u_i(\nu)|\Omega_{\mu\nu}|u_i(\mu)u_j(\nu)\rangle\}\,. \tag{153.4}$$

Here μ and ν are dummies so that the sum $\sum_{\mu\nu}'$ consists of $N(N-1)$ equal terms (μ, ν and ν, μ here being counted as different terms). Thence, the expectation value wanted, becomes

$$\langle\psi|\Omega|\psi\rangle=\tfrac{1}{2}\sum_{ij}{}'\{\langle u_i(1)u_j(2)|\Omega_{12}|u_i(1)u_j(2)\rangle$$
$$-\langle u_j(1)u_i(2)|\Omega_{12}|u_i(1)u_j(2)\rangle\}\,. \tag{153.5}$$

Use of the symbols 1 and 2 is, of course, quite arbitrary.

It should be emphasized again that here and in the preceding problem each single-particle wave function u_i comprises space as well as spin state.

Problem 154. Two electrons in the atomic ground state

The K shell of an atom is composed of two electrons in the $1s$ state. Its energy shall be approximated by using screened hydrogen wave functions in the field of a nucleus of charge Ze and infinitely large mass.

Solution. The hamiltonian of the problem is (in atomic units, $e=\hbar=m=1$):

$$H = -\frac{1}{2}(\nabla_1^2+\nabla_2^2)-Z\left(\frac{1}{r_1}+\frac{1}{r_2}\right)+\frac{1}{r_{12}}; \tag{154.1}$$

the approximate wave function is according to the table of Problem 67

$$U=u(r_1)\,u(r_2)=\frac{\alpha^3}{\pi}\,\mathrm{e}^{-\alpha(r_1+r_2)} \tag{154.2}$$

with

$$\alpha=Z-\sigma \tag{154.3}$$

and σ the screening constant. It is to be expected that $0<\sigma<1$, because the nuclear charge, in its effect on each electron, is only partly screened by the other electron. The two factors of (154.2) satisfy the wave equations

$$\left(-\frac{1}{2}\nabla_1^2-\frac{\alpha}{r_1}\right)u(r_1)=-\frac{1}{2}\alpha^2 u(r_1);$$

$$\left(-\frac{1}{2}\nabla_2^2-\frac{\alpha}{r_2}\right)u(r_2)=-\frac{1}{2}\alpha^2 u(r_2)$$

so that

$$HU=\left\{\left(-\frac{1}{2}\alpha^2-\frac{Z-\alpha}{r_1}\right)+\left(-\frac{1}{2}\alpha^2-\frac{Z-\alpha}{r_2}\right)+\frac{1}{r_{12}}\right\}U$$

and the energy becomes

$$E=\iint d\tau_1\,d\tau_2\left\{-\alpha^2-\frac{\sigma}{r_1}-\frac{\sigma}{r_2}+\frac{1}{r_{12}}\right\}U^2. \tag{154.4}$$

With U according to (154.2) and using the normalization of each factor u, this yields

$$E=-\alpha^2-2\sigma\frac{\alpha^3}{\pi}\int d\tau_1\,\frac{\mathrm{e}^{-2\alpha r_1}}{r_1}+\frac{\alpha^6}{\pi^2}\iint d\tau_1\,d\tau_2\,\frac{\mathrm{e}^{-2\alpha(r_1+r_2)}}{r_{12}}. \tag{154.5}$$

The first integral in (154.5) can be evaluated in an elementary way:

$$\int d\tau_1\,\frac{\mathrm{e}^{-2\alpha r_1}}{r_1}=4\pi\int_0^\infty dr_1\,r_1\,\mathrm{e}^{-2\alpha r_1}=\frac{\pi}{\alpha^2}. \tag{154.6}$$

To calculate the double integral

$$J=\iint d\tau_1\,d\tau_2\,\frac{\mathrm{e}^{-2\alpha(r_1+r_2)}}{r_{12}}$$

we expand $1/r_{12}$ into Legendre polynomials of the angle Θ between the vectors $\boldsymbol{r}_1$ and $\boldsymbol{r}_2$:

$$\frac{1}{r_{12}}=\begin{cases}\dfrac{1}{r_2}\sum\limits_{n=0}^{\infty}\left(\dfrac{r_1}{r_2}\right)^n P_n(\cos\Theta) & 0\le r_1\le r_2;\\[2ex] \dfrac{1}{r_1}\sum\limits_{n=0}^{\infty}\left(\dfrac{r_2}{r_1}\right)^n P_n(\cos\Theta) & r_2\le r_1<\infty.\end{cases}$$

Then only the first term of the expansion ($n=0$) contributes to J, and we obtain

$$J=\int d\tau_2\, e^{-2\alpha r_2}\cdot 4\pi\left\{\frac{1}{r_2}\int_0^{r_2} dr_1\, r_1^2\, e^{-2\alpha r_1}+\int_{r_2}^{\infty} dr_1\, r_1\, e^{-2\alpha r_1}\right\}.$$

With $d\tau_2=4\pi r_2^2\,dr_2$ and $0\le r_2<\infty$, all the integrations become elementary again and lead to

$$J=\frac{5\pi^2}{8\alpha^5}. \tag{154.7}$$

The energy expression (154.5) with the integrals (154.6) and (154.7) then becomes

$$E=-\alpha^2-2(Z-\alpha)\alpha+\tfrac{5}{8}\alpha. \tag{154.8}$$

Up to this point, we have not yet disposed of the value of α that we now choose optimally in the sense of variational calculus putting

$$\frac{dE}{d\alpha}=0. \tag{154.9}$$

That leads to

$$\alpha=Z-\tfrac{5}{16} \tag{154.10}$$

and

$$E=-\left(Z-\tfrac{5}{16}\right)^2. \tag{154.11}$$

It should be remarked that this value of α makes (154.2) the exact solution of the hamiltonian

$$H^0=-\frac{1}{2}(\nabla_1^2+\nabla_2^2)-\alpha\left(\frac{1}{r_1}+\frac{1}{r_2}\right) \tag{154.12}$$

which permits factorization. Comparing (154.12) with (154.2) we find

$$H'=H-H^0=-\sigma\left(\frac{1}{r_1}+\frac{1}{r_2}\right)+\frac{1}{r_{12}}.$$

Defining H' as perturbation, the energy shift of a first-order perturbation theory,

$$\Delta E=\iint d\tau_1\, d\tau_2\, U H' U,$$

would vanish if the screening constant σ is chosen according to (154.10).

We add a few numerical remarks. The theory describes atoms stripped of all electrons except the two in the K shell. There exist experimental values for

$$\begin{array}{lllll} Z=2 & 3 & 4 & 6 & 8 \\ \mathrm{He} & \mathrm{Li}^{+} & \mathrm{Be}^{++} & \mathrm{C}^{4+} & \mathrm{O}^{6+}. \end{array}$$

In all these cases, it is not E itself which has been observed, but the ionization energy I necessary to strip the ion of only one of the two K electrons. The remaining ion, keeping only its last electron, then has the energy

$$E' = -\tfrac{1}{2} Z^2$$

so that the ionization energy becomes

$$I = (Z - \tfrac{5}{16})^2 - \tfrac{1}{2} Z^2 . \qquad (154.13)$$

The accompanying table shows that the agreement of Eq. (154.13) with experiment improves continuously with increasing Z. This is reasonable, because the role of the interaction term $1/r_{12}$ becomes less important as the coupling of each electron to the centre becomes stronger with increasing Z.

Z	I in eV theor.	I in eV exper.
2	23,2	24,5
3	74,1	75,6
4	152,2	153,6
6	390	393
8	737	738

Problem 155. Excited states of the helium atom

In a neutral helium atom let one electron be in the $1s$ ground state and the other in an n, l excited state ($n \geqq 2$, $l \geqq 1$). The ionization energy for the (n, l) electron shall then be determined for both, ortho and parahelium using hydrogen-like wave functions with screening of one nuclear charge by the $1s$ electron. The method shall be applied numerically to the $2p$ state ($n=2$, $l=1$).

Solution. If the $1s$ electron is exposed to the full nuclear charge $2e$, but the (n, l) electron only to the screened charge e, we may describe the two one-electron states by solution of the differential equations

$$\left(-\frac{1}{2}\nabla^2 - \frac{2}{r}\right) u = E_1 u ; \quad \left(-\frac{1}{2}\nabla^2 - \frac{1}{r}\right) v_{nl} = E_n v_{nl} \qquad (155.1)$$

with

$$u \equiv |1\rangle = \sqrt{\frac{8}{\pi}}\, e^{-2r}; \qquad E_1 = -2;$$

$$v_{nl} \equiv |n\rangle = R_{nl}(r)\, Y_{l,m}(\vartheta, \varphi); \qquad E_n = -\frac{1}{2n^2} \tag{155.2}$$

and R_{nl} the normalized radial functions of the hydrogen atom (see Problem 67).

The Schrödinger equation of the two-electron problem

$$\left\{ -\frac{1}{2}\nabla_1^2 - \frac{1}{2}\nabla_2^2 - \frac{2}{r_1} - \frac{2}{r_2} + \frac{1}{r_{12}} \right\} \psi = E\psi \tag{155.3}$$

shall be approximately solved by the symmetrized product wave function

$$\psi = u(1)\, v_n(2) + \varepsilon\, v_n(1)\, u(2) = |1\,n\rangle + \varepsilon |n\,1\rangle \tag{155.4}$$

with $\varepsilon = +1$ for parahelium (spins antiparallel) and $\varepsilon = -1$ for orthohelium (spins parallel). The function ψ is normalized according to $\langle\psi|\psi\rangle = 2$.

In order to satisfy (155.3) as well as possible by (155.4) we apply $\langle 1\,n|$ to (155.3):

$$\langle 1\,n| -\frac{1}{2}\nabla_1^2 - \frac{1}{2}\nabla_2^2 - \frac{2}{r_1} - \frac{2}{r_2} + \frac{1}{r_{12}} |\psi\rangle = E\langle 1\,n|\psi\rangle\,. \tag{155.5}$$

Since

$$\langle 1|1\rangle = 1; \qquad \langle n|n\rangle = 1; \qquad \langle 1|n\rangle = 0$$

(if $l \neq 0$) the integrals will partially split up and partially vanish. For instance we have

$$\langle 1\,n| -\frac{1}{2}\nabla_1^2 - \frac{2}{r_1} |1\,n\rangle = E_1; \qquad \langle 1\,n| -\frac{1}{2}\nabla_1^2 - \frac{2}{r_1} |n\,1\rangle = 0;$$

$$\langle 1\,n| -\frac{1}{2}\nabla_2^2 - \frac{2}{r_2} |1\,n\rangle = E_n - \langle n|\frac{1}{r}|n\rangle \qquad \text{etc.}$$

Since for all one-electron states the virial theorem (cf. Problem 151) leads to $E_{\text{pot}} = 2E$, we obtain

$$\langle n|\frac{1}{r}|n\rangle = -E_{\text{pot}}^{(n)} = -2E_n = \frac{1}{n^2}\,.$$

Thus, finally, Eq. (155.5) leads to the energy expression

$$E = -2 - \frac{3}{2n^2} + \mathscr{C} + \varepsilon\mathscr{E} \tag{155.6}$$

with the abbreviations

$$\mathscr{C} = \langle 1\,n|\frac{1}{r_{12}}|1\,n\rangle \tag{155.7}$$

for the classical, and

$$\mathscr{E} = \langle 1\,n|\frac{1}{r_{12}}|n\,1\rangle \tag{155.8}$$

for the exchange integral of the electron-electron interaction. It remains to evaluate these two integrals.

In both cases we expand $1/r_{12}$ into spherical harmonics of the angle ϑ_{12} between the position vectors $\boldsymbol{r}_1$ and $\boldsymbol{r}_2$ of the electrons:

$$\frac{1}{r_{12}} = \begin{cases} \dfrac{1}{r_2} \sum\limits_{\lambda=0}^{\infty} \left(\dfrac{r_1}{r_2}\right)^{\lambda} P_\lambda(\cos\vartheta_{12}) & \text{if } r_1 < r_2; \\ \dfrac{1}{r_1} \sum\limits_{\lambda=0}^{\infty} \left(\dfrac{r_2}{r_1}\right)^{\lambda} P_\lambda(\cos\vartheta_{12}) & \text{if } r_1 > r_2. \end{cases} \tag{155.9}$$

When first performing the integrations over the polar angles, we have to calculate the integrals

$$\mathscr{C}_{\text{ang}} = \oint d\Omega_2 |Y_{l,m}(2)|^2 \oint d\Omega_1 \, P_\lambda(\cos\vartheta_{12}) \tag{155.10}$$

and

$$\mathscr{E}_{\text{ang}} = \oint d\Omega_2 \, Y^*_{l,m}(2) \oint d\Omega_1 \, Y_{l,m}(1)\, P_\lambda(\cos\vartheta_{12}). \tag{155.11}$$

In $\mathscr{C}_{\text{ang}}$, the inner integral becomes $4\pi\delta_{\lambda,0}$ so that of the series (155.9) there remains only a contribution from the term $\lambda=0$. The classical interaction therefore is

$$\mathscr{C} = 4\pi \int_0^\infty dr_2\, r_2^2 |u(r_2)|^2 \left\{ \frac{1}{r_2} \int_0^{r_2} dr_1\, r_1^2 |R_{nl}(r_1)|^2 + \int_{r_2}^\infty dr_1\, r_1 |R_{nl}(r_1)|^2 \right\}. \tag{155.12}$$

In order to calculate the inner integral of (155.11) we use the spherical harmonics addition theorem,

$$P_\lambda(\cos\vartheta_{12}) = \frac{4\pi}{2\lambda+1} \sum_{\mu=-\lambda}^{+\lambda} Y^*_{\lambda,\mu}(1)\, Y_{\lambda,\mu}(2); \tag{155.13}$$

then

$$\oint d\Omega_1 \, Y_{l,m}(1)\, P_\lambda(\cos\vartheta_{12}) = \frac{4\pi}{2l+1} Y_{l,m}(2)\, \delta_{l,\lambda}$$

and

$$\mathscr{E}_{\text{ang}} = \frac{4\pi}{2l+1} \delta_{l,\lambda}. \tag{155.14}$$

The exchange integral therefore receives a contribution only from the term $\lambda = l$ of the series (155.9) so that

$$\mathscr{E} = \frac{4\pi}{2l+1} \int_0^\infty dr_2\, r_2^2\, u(r_2)\, R_{nl}(r_2) \left\{ \frac{1}{r_2} \int_0^{r_2} dr_1\, r_1^2 \left(\frac{r_1}{r_2}\right)^l u(r_1)\, R_{nl}(r_1) + \int_{r_2}^\infty dr_1\, r_1 \left(\frac{r_2}{r_1}\right)^l u(r_1)\, R_{nl}(r_1) \right\}. \tag{155.15}$$

The higher the values of n and l of the excited electron, the more will our approximation improve because of decreasing overlap of the two one-electron wave functions. Except therefore for S states, the method will be worst for $n=2$ and $l=1$, but may be well trusted elsewhere if in that special case it produces reasonable results. We therefore now proceed to calculate the energy for this special excited state of the helium atom and compare the result with experimental evidence.

The normalized radial function R_{nl} then becomes

$$R_{2,1} = \frac{1}{\sqrt{24}}\, r e^{-\frac{r}{2}}, \tag{155.16}$$

and the radial integrals (155.12) and (155.15) can easily, though in a somewhat cumbersome way, be evaluated using (155.2) for u and (155.16) for $R_{2,1}$. The results are

$$\mathscr{C} = \tfrac{1}{4}\left(1 - \tfrac{13}{3125}\right) = 0.24896$$

and

$$\mathscr{E} = \tfrac{14}{3}\left(\tfrac{4}{5}\right)^3 \left(\tfrac{1}{5}\right)^4 = 0.00382.$$

This leads to

$$E = -2.12604 + \varepsilon \cdot 0.00382$$

in atomic units. The ionization energy is the difference of E and the energy $E^+ = -2$ of He^+ in the ground state (i.e. with one electron still in the $1s$ state and the other removed),

$$I = E^+ - E = 0.12604 - \varepsilon \cdot 0.00382$$

or

$$I = (3.429 - \varepsilon \cdot 0.104)\ \text{eV}.$$

This result may be compared with experiment as shown in the table. The results fit quite nicely, and even the splitting between para and ortho states is not as bad as might be expected from its being rather sensitive to the overlap and mutual polarization of the two one-electron functions. It should be noted that the para state with the space symmetric wave function lies *above* the antisymmetric ortho state, the

	ε	ionization energy in eV	
		theory	experiment
para	+1	3.325	3.368
ortho	−1	3.533	3.623
difference		0.208	0.255

situation thus being the opposite of the one in the H_2 molecule (Problem 163). This can easily be verified in our calculation where the integral $\mathscr{E}$, Eq. (155.8), gives the only contribution depending on the sign ε; and since it derives from the mutual repulsion of the two electrons, $\mathscr{E}$ is positive, thus raising the energy level for $\varepsilon=+1$.

Problem 156. Excited S states of the helium atom

The method of the preceding problem shall be extended to the configuration $1s$, ns using again the undisturbed wave function for the $1s$ state, but making no specializing assumptions on the ns wave function. It shall be shown that, if overlap and exchange integrals are small, an effective potential field can be constructed in which the ns electron moves.

Solution. The wave function will be written as a symmetrized product of one-electron states,

$$\psi=u(1)\,v_n(2)+\varepsilon\,v_n(1)\,u(2)=|1\,n\rangle+\varepsilon|n\,1\rangle \tag{156.1}$$

with $\varepsilon=\pm 1$ and the $1s$ state function in atomic units

$$u\equiv|1\rangle=\sqrt{\frac{8}{\pi}}\,e^{-2r};\quad \left(-\frac{1}{2}\nabla^2-\frac{2}{r}\right)u=-2u;\quad \langle 1|1\rangle=1\,. \tag{156.2}$$

Of the function for the ns state we know only that it does not depend on angles, and that it too is supposed to be normalized,

$$v_n(r)\equiv|n\rangle;\quad \langle n|n\rangle=1\,. \tag{156.3}$$

No further specialization of $|n\rangle$ will be attempted.

The wave function ψ is an approximate solution of the Schrödinger equation

$$(H-E)\psi=0 \tag{156.4}$$

with the atomic hamiltonian

$$H=-\frac{1}{2}\nabla_1^2-\frac{2}{r_1}-\frac{1}{2}\nabla_2^2-\frac{2}{r_2}+\frac{1}{r_{12}}. \tag{156.5}$$

These are the basic equations of our problem. We start by forming the Hilbert product of (156.4) with $\langle 1\,n|$,

$$\langle 1\,n|H-E|1\,n\rangle+\varepsilon\langle 1\,n|H-E|n\,1\rangle=0 \tag{156.6}$$

and determining the integrals occurring with H given by (156.5). So far the formulae do not yet appreciably differ from those of the preceding problem. One main difference, however, is seen immediately since the two functions $|1\rangle$ and $|n\rangle$ are no longer orthogonal because they both belong to $l=0$ but have different potential fields. Hence we have to introduce the overlap integral

$$S=\langle 1|n\rangle=\langle n|1\rangle\,. \tag{156.7}$$

Further, let us again use the abbreviations

$$\langle 1\,n|\frac{1}{r_{12}}|1\,n\rangle=\mathscr{C}\,;\qquad \langle 1\,n|\frac{1}{r_{12}}|n\,1\rangle=\mathscr{E}\,. \tag{156.8}$$

It then remains to evaluate the following integrals

$$\langle 1\,n|-\frac{1}{2}\nabla_1^2-\frac{2}{r_1}|1\,n\rangle=\langle 1|-\frac{1}{2}\nabla^2-\frac{2}{r}|1\rangle=-2\,;$$

$$\langle 1\,n|-\frac{1}{2}\nabla_2^2-\frac{2}{r_2}|1\,n\rangle=\langle n|-\frac{1}{2}\nabla^2-\frac{2}{r}|n\rangle=K_n;$$

$$\langle 1\,n|-\frac{1}{2}\nabla_1^2-\frac{2}{r_1}|n\,1\rangle=\langle 1\,n|-\frac{1}{2}\nabla_2^2-\frac{2}{r_2}|n\,1\rangle=-2S^2\,. \tag{156.9}$$

In the last line, the identity

$$\langle 1|\nabla^2|n\rangle=\langle n|\nabla^2|1\rangle$$

has been used. Eq. (156.6) then may be written

$$-2+K_n+\mathscr{C}-E+\varepsilon(-4S^2+\mathscr{E}-E\,S^2)=0$$

or

$$E=-2+\frac{K_n+\mathscr{C}+\varepsilon(\mathscr{E}-2S^2)}{1+\varepsilon S^2}. \tag{156.10}$$

Since $E^+ = -2$ is the energy of the He^+ ion ground state, the ionization energy $I = E^+ - E$ becomes

$$I = -\frac{K_n + \mathscr{C} + \varepsilon(\mathscr{E} - 2S^2)}{1 + \varepsilon S^2}. \tag{156.11}$$

Either E or I may then be determined by evaluating the integrals S, K_n, $\mathscr{C}$, $\mathscr{E}$ for a set of sufficiently pliable functions $|n\rangle$ defined by Ritz parameters and extremizing E or I by their suitable choice.

If the overlap integral S and the exchange integral $\mathscr{E}$ are very small, Eqs. (156.10) and (156.11) simplify to

$$E = -2 + K_n + \mathscr{C}; \qquad I = -(K_n + \mathscr{C}). \tag{156.12}$$

The same expressions would be achieved by neglecting symmetrization (i.e. with $\varepsilon = 0$), and it is in this sense only that symmetrization in many-body problems may occasionally be omitted.

Falling back upon the definitions (156.8) and (156.9) of the integrals K_n and $\mathscr{C}$, (156.12) may be written in more detail

$$E = -2 + \langle n|\Omega|n\rangle \tag{156.13}$$

with the operator

$$\Omega = -\frac{1}{2}\nabla^2 - \frac{2}{r} + \int d\tau' \frac{u(r')^2}{|\boldsymbol{r} - \boldsymbol{r}'|}. \tag{156.14}$$

The choice of such a normalized function $|n\rangle$ as makes E a minimum is performed by variation,

$$\delta(\langle n|\Omega|n\rangle + \lambda\langle n|n\rangle) = 0$$

with a Lagrange multiplicator λ. Since

$$\delta\langle n|\Omega|n\rangle = 2\langle \delta n|\Omega|n\rangle; \quad \delta\langle n|n\rangle = 2\langle \delta n|n\rangle$$

we arrive at

$$\langle \delta n|\Omega + \lambda|n\rangle = 0$$

or, $|\delta n\rangle$ being an arbitrary function, at the differential equation

$$(\Omega + \lambda)|n\rangle = 0.$$

Rewriting (156.13) in the form

$$\langle n|\Omega - E - 2|n\rangle = 0$$

we see that $\lambda = -E - 2$. Hence, $|n\rangle$ is to satisfy the one-electron Schrödinger equation

$$-\tfrac{1}{2}\nabla^2|n\rangle + V_{\mathrm{eff}}(r)|n\rangle = (E+2)|n\rangle \tag{156.15}$$

with the effective potential

$$V_{\text{eff}}(r) = -\frac{2}{r} + \int d\tau' \frac{u(r')^2}{|\boldsymbol{r}-\boldsymbol{r}'|}. \tag{156.16}$$

This is exactly the electrostatic potential to be derived from the Poisson equation $\nabla^2 V_{\text{eff}} = +4\pi\rho$ with the charge density ρ composed of the nuclear point charge $+2$ and the negative space charge $-u^2$ of the $1s$ electron. Evaluation of (156.16) with the wave function u of Eq. (156.2) renders

$$V_{\text{eff}} = -\frac{1}{r} - \left(\frac{1}{r} + 2\right) e^{-4r}. \tag{156.17}$$

If $|n\rangle$ is determined from (156.15), this may be considered a sufficient approximation to *all* integrals of the energy expression (156.10) as long as $S \ll 1$ and $\mathscr{E} \ll \mathscr{C}$ are no more than slight corrections.

Appendix. Numerical calculations along the general outline sketched before may become rather involved. As an example, we give numerical results for an abridged variational procedure making (156.12) extremal, which we have performed for the $2s$ state using the set of trial functions

$$|2\rangle = A(e^{-2r} - pre^{-\frac{1}{2}r}) \tag{156.18}$$

normalized according to $\langle 2|2\rangle = 1$ and defined by the Ritz parameter p. These functions have finite value at $r=0$, they have a zero as necessary for a $2s$ state, and show the correct asymptotic behaviour determined by the second term. The first term describes deviations from hydrogen-like behaviour at small distances where the nuclear charge becomes less and less completely screened; since the $1s$ distribution is described by e^{-2r}, this effect should show approximately the same dependence on distance.

The approximate function (156.18) leads to the following numerical results.

$$\pi A^2 = \left(\tfrac{1}{8} - \tfrac{768}{625}p + 96p^2\right)^{-1};$$

$$K_2 = 4\pi A^2\left(-\tfrac{1}{16} + \tfrac{384}{625}p - 11p^2\right);$$

$$\mathscr{C} = \pi A^2\left(\tfrac{5}{32} - \tfrac{512\cdot 6359}{125\cdot 28561}p + \tfrac{74688}{3125}p^2\right);$$

$$S = 8\sqrt{2\pi}\, A\left(\tfrac{1}{32} - \tfrac{96}{625}p\right);$$

$$\mathscr{E} = 128\pi A^2\left(\tfrac{5}{4096} - \tfrac{25436}{3570125}p + \tfrac{1056}{78125}p^2\right).$$

The energy expression (156.12) has then been minimized by suitable choice of p. This leads to a quadratic equation in p with one positive solution $p=0.1105$. Without symmetrization this yields an ionization energy $I = -(K_2 + \mathscr{C}) = 0.145$ atomic units or $I = 3.94$ eV. If, with the same value of p, the full energy expression (156.11) is evaluated we find

$$I = \frac{0{,}145 - 0{,}021\,\varepsilon}{1 + 0{,}0225\,\varepsilon} \text{ atomic units}$$

and arrive at the tabulated results. The theoretical approximation,

symmetry	ionization energy for $2s$ in eV	
	theory	experiment
para, $\varepsilon = +1$	3.30	3.97
ortho, $\varepsilon = -1$	4.62	4.76
difference	1.32	0.79

as always in variational procedures, leads to somewhat higher energy term values than the correct ones. The rather large ortho-para splitting is reproduced with an accuracy of about 35% even by this very simplified approximation.

Problem 157. Lithium ground state

To calculate the binding energy of a lithium atom ($Z = 3$) in its ground state. For the two $1s$ electrons the screened hydrogen-like functions of Problem 154 may be used. Exchange is to be neglected.

Solution. The hamiltonian of the problem is

$$H = \left\{ -\frac{1}{2}(\nabla_1^2 + \nabla_2^2) - 3\left(\frac{1}{r_1} + \frac{1}{r_2}\right) + \frac{1}{r_{12}} \right\} + \left\{ -\frac{1}{2}\nabla_3^2 - \frac{1}{r_3} \right\} + \left\{ \frac{1}{r_{13}} + \frac{1}{r_{23}} - \frac{2}{r_3} \right\} \tag{157.1}$$

where the first curly bracket corresponds to the two-electron problem of Li^+, the second bracket leads to the $2s$ function of the third electron in the field of the screened nucleus of rest charge 1, and in the third term the remaining interactions are assembled. The treatment of the third electron thus indicated would be quite correct, were the radius of the K shell very small compared to the extension of the $2s$ wave function; since it is not, the use of a hydrogen function with central charge 1 for the third electron is an approximation.

We write the eigenfunction in product form

$$U(1,2,3) = u(1)\,u(2)\,v(3). \tag{157.2}$$

Here $u(r)$ means the $1s$ function

$$u(r)=\frac{\alpha^{\frac{3}{2}}}{\sqrt{\pi}}\,e^{-\alpha r} \tag{157.3}$$

and

$$\alpha=Z-\tfrac{5}{16}=2.6875 \tag{157.4}$$

the effective nuclear charge originated by the mutual screening effect of the two $1s$ electrons in Li^+ (cf. Problem 154). From Problem 154 we further gather that the energy of the $(1s)^2$ state,

$$E^+=\iint d\tau_1\,d\tau_2\,u(1)\,u(2)\left\{-\frac{1}{2}(\nabla_1^2+\nabla_2^2)-3\left(\frac{1}{r_1}+\frac{1}{r_2}\right)+\frac{1}{r_{12}}\right\}u(1)\,u(2)\,, \tag{157.5}$$

becomes

$$E^+=-\alpha^2\,. \tag{157.6}$$

For the third electron we take the eigenfunction from the table of Problem 67 (hydrogen problem); in its lowest state, $2s$, it is

$$v(r)=\frac{1}{\sqrt{8\pi}}\left(1-\tfrac{1}{2}r\right)e^{-\frac{1}{2}r} \tag{157.7}$$

and satisfies the differential equation

$$\left(-\frac{1}{2}\nabla_3^2-\frac{1}{r_3}\right)v(r_3)=-\tfrac{1}{8}v(r_3)\,. \tag{157.8}$$

If we put the functions (157.3) and (157.7) into the energy expression,

$$E=\iiint d\tau_1\,d\tau_2\,d\tau_3\,u(1)\,u(2)\,v(3)\,H\,u(1)\,u(2)\,v(3) \tag{157.9}$$

with the hamiltonian (157.1), the first bracket of (157.1) will contribute E^+, Eq. (157.6), and the second bracket $-\frac{1}{8}$, according to (157.8), so that

$$E=E^+-\frac{1}{8}+\iiint d\tau_1\,d\tau_2\,d\tau_3\,u(1)^2\,u(2)^2\,v(3)^2\left(\frac{1}{r_{13}}+\frac{1}{r_{23}}-\frac{2}{r_3}\right). \tag{157.10}$$

The last integral has still to be evaluated. It may be simplified into

$$J=2\int d\tau_3\,v(3)^2\left\{\int d\tau_1\,\frac{u(1)^2}{r_{13}}-\frac{1}{r_3}\right\}. \tag{157.11}$$

Using [cf. Problem 44, Eq. (44.19)]

$$\int d\tau_1\,\frac{u(1)^2}{r_{13}}=\frac{1}{r_3}\left\{1-(1+\alpha r_3)\,e^{-2\alpha r_3}\right\}$$

we may combine

$$J = -2\int d\tau_3 \frac{v(3)^2}{r_3}(1+\alpha r_3)\,e^{-2\alpha r_3}.$$

With v according to (157.7) this integral is elementary though cumbersome and yields

$$J = -\frac{\frac{1}{2}+3\alpha+16\alpha^3}{(1+2\alpha)^5}. \tag{157.12}$$

The energy of the ground state of the lithium atom then becomes

$$E = -\alpha^2 - \frac{1}{8} - \frac{\frac{1}{2}+3\alpha+16\alpha^3}{(1+2\alpha)^5}. \tag{157.13}$$

Numerical computation with the value (157.4) for α gives the ionization energy

$$I = E^+ - E = \frac{1}{8} - \frac{\frac{1}{2}+3\alpha+16\alpha^3}{(1+2\alpha)^5} \tag{157.14}$$

as

$$I = 0.1553 = 4.23\ \text{eV}.$$

This is to be compared with the experimental value of 5.37 eV. The approximation, of course, is not very good. The reason for the difference is to be sought neither in the use of complete screening of the third electron by the K shell, nor in neglecting any small difference of the α values between atom and ion. Both these corrections are far too small to account for a discrepany of more than 1 eV. There remain two features of the wave function which may still account for it: its product form, and the neglect of symmetrization and hence of exchange binding.

Problem 158. Exchange correction to lithium ground state

To correct the energy of the lithium ground state, found in the preceding problem, by taking account of the correct symmetry of the eigenfunction.

Solution. The eigenfunction must again describe a state with two electrons in the $1s$ state $u(r)$ and one electron in the $2s$ state $v(r)$, as defined in the preceding problem. Symmetrization requires the inclusion of spins. A totally antisymmetrical eigenfunction is then obtained by the Slater determinant (see Problem 152)

$$\psi = \frac{1}{\sqrt{6}}\begin{vmatrix} u(1)\,\alpha(1); & u(2)\,\alpha(2); & u(3)\,\alpha(3) \\ u(1)\,\beta(1); & u(2)\,\beta(2); & u(3)\,\beta(3) \\ v(1)\,\alpha(1); & v(2)\,\alpha(2); & v(3)\,\alpha(3) \end{vmatrix} \tag{158.1}$$

in which the spin functions α and β describe opposite spin directions. The determinant (158.1) is an approximate solution of the Schrödinger equation

$$(H-E)\psi=0 \tag{158.2}$$

with H the hamiltonian of the preceding problem. From

$$\langle\psi|H-E|\psi\rangle=0,$$

where the scalar Hilbert product includes spin summation, we then are led on, by performing the sum over spins, to

$$\int\int\int d\tau_1\, d\tau_2\, d\tau_3 \{u(1)u(2)v(3)-v(1)u(2)u(3)\}(H-E)u(1)u(2)v(3)=0 \tag{158.3}$$

or, with the abbreviations

$$\tilde{E}=\int\int\int d\tau_1\, d\tau_2\, d\tau_3\, u(1)u(2)v(3)\, H\, u(1)u(2)v(3)\,; \tag{158.4}$$

$$\mathscr{E}=\int\int\int d\tau_1\, d\tau_2\, d\tau_3\, v(1)u(2)u(3)\, H\, u(1)u(2)v(3)\,; \tag{158.5}$$

$$S=\int d\tau_1\, v(1)u(1)\,, \tag{158.6}$$

to the corrected energy formula

$$E=\frac{\tilde{E}-\mathscr{E}}{1-S^2}. \tag{158.7}$$

Here, by $\tilde{E}$ we denote the uncorrected energy as determined in Eq. (157.13) of the preceding problem, viz.

$$\tilde{E}=-(\alpha^2+\tfrac{1}{8})+J; \quad J=-\frac{\frac{1}{2}+3\alpha+16\alpha^3}{(1+2\alpha)^5}. \tag{158.8}$$

$\mathscr{E}$ is the exchange energy, and S the overlap integral of the functions u and v which, as we know, are not orthogonal. The main problem still remaining will then be the evaluation of the exchange energy (158.5). If we write the hamiltonian in the form

$$H=\left(-\frac{1}{2}\nabla_1^2-\frac{\alpha}{r_1}\right)+\left(-\frac{1}{2}\nabla_2^2-\frac{\alpha}{r_2}\right)+\left(-\frac{1}{2}\nabla_3^2-\frac{1}{r_3}\right)$$
$$-\frac{3-\alpha}{r_1}-\frac{3-\alpha}{r_2}-\frac{2}{r_3}+\frac{1}{r_{12}}+\frac{1}{r_{13}}+\frac{1}{r_{23}} \tag{158.9}$$

then its application to $u(1)u(2)v(3)$ leads in the first line of (158.9) simply to multiplication by $-\frac{1}{2}\alpha^2$, $-\frac{1}{2}\alpha^2$, $-\frac{1}{8}$, respectively, so that we get

$$\mathscr{E} = -(\alpha^2+\tfrac{1}{8})\,S^2-(3-\alpha)\,S\int d\tau_1\frac{u(1)\,v(1)}{r_1}-(3-\alpha)\,S^2\int d\tau_2\frac{u(2)^2}{r_2}$$
$$-2S\int d\tau_3\frac{u(3)\,v(3)}{r_3}+S\iint d\tau_1\,d\tau_2\frac{u(1)\,v(1)\cdot u(2)^2}{r_{12}}$$
$$+\iint d\tau_1\,d\tau_3\frac{u(1)\,v(1)\cdot u(3)\,v(3)}{r_{13}}+S\iint d\tau_2\,d\tau_3\frac{u(2)^2\cdot u(3)\,v(3)}{r_{23}}.$$

Using the further abbreviations

$$U=\int d\tau\frac{u^2}{r};\qquad V=\int d\tau\frac{u\,v}{r};\tag{158.10}$$

$$X=\iint d\tau\,d\tau'\frac{u(r)\,v(r)\cdot u(r')\,v(r')}{|\boldsymbol{r}-\boldsymbol{r}'|};\tag{158.11}$$

$$Y=\iint d\tau\,d\tau'\frac{u(r)^2\cdot u(r')\,v(r')}{|\boldsymbol{r}-\boldsymbol{r}'|},\tag{158.12}$$

the exchange energy becomes

$$\mathscr{E}=-(\alpha^2+\tfrac{1}{8})\,S^2-(5-\alpha)\,S\,V-(3-\alpha)\,S^2\,U+2\,S\,Y+X\tag{158.13}$$

and the corrected energy (158.7),

$$E=\tilde{E}+\frac{[J+(3-\alpha)\,U]\,S^2+[(5-\alpha)\,V-2\,Y]\,S-X}{1-S^2}.\tag{158.14}$$

As a last step, we may now proceed to evaluate the integrals S, Eq. (158.6), U, V, X, Y, Eqs. (158.10–12), with the functions

$$u(r)=\frac{\alpha^{\frac{3}{2}}}{\sqrt{\pi}}\,e^{-\alpha r};\quad v(r)=\frac{1}{\sqrt{8\pi}}(1-\tfrac{1}{2}r)\,e^{-\frac{1}{2}r}.$$

A little difficulty arises only in the two-particle integrals X and Y where $1/|\boldsymbol{r}-\boldsymbol{r}'|$ can be expanded into Legendre polynomials for $\cos(\boldsymbol{r},\boldsymbol{r}')$. Only the term P_0 of this series contributes, since neither u nor v depend upon polar angles, hence the inner integral in X and Y becomes

$$\int d\tau'\frac{u(r')\,v(r')}{|\boldsymbol{r}-\boldsymbol{r}'|}=4\pi\left\{\frac{1}{r}\int_0^r dr'\,r'^2\,u(r')\,v(r')+\int_r^\infty dr'\,r'\,u(r')\,v(r')\right\}.$$

All other computations are elementary and yield the following results:

$$S=\frac{2\sqrt{2}\,\alpha^{\frac{3}{2}}(\alpha-1)}{(\alpha+\frac{1}{2})^4};\qquad U=\alpha;\qquad V=\frac{\alpha^2-\frac{1}{4}}{2(\alpha-1)}S;$$

$$X=\frac{\alpha^3(\frac{5}{2}\alpha^2-\frac{15}{4}\alpha+\frac{13}{8})}{(\alpha+\frac{1}{2})^7};$$

$$Y=\frac{66\alpha^6+26\alpha^5-25\alpha^4-16\alpha^3-\frac{23}{8}\alpha^2-\frac{1}{8}\alpha}{2(\alpha-1)(3\alpha+\frac{1}{2})^4}S.$$

Numerical values with $\alpha=2.6875$ are

$$\begin{aligned}
&S=0.203; \\
&J=-0.030; \\
&U=2.6875 \qquad [J+(3-\alpha)U]\,S^2=+0.0334; \\
&V=0.419 \qquad [(5-\alpha)\,V-2\,Y]\,S=+0.0735; \\
&X=0.0558; \\
&Y=0.303.
\end{aligned}$$

The two positive contributions to the numerator in (158.14) therefore exceed the negative term $-X$, so that the exchange correction yields a smaller binding energy of the lithium atom. The reason for this rather unhappy result lies in the choice of $v(r)$ which is far too small in the overlap zone, as the increasing effective charge to which the $2s$ electron is subjected when penetrating into the $1s$ core has been neglected. This causes a small error only in the uncorrected energy of the preceding problem, but will have a large effect upon the exchange correction. Since V and Y are linear, but X is quadratic in the overlap product uv, the third (negative) term in the numerator of (158.14) should be much larger whereas the second (positive) term would be only moderately increased by using a better approximation for $v(r)$, so that the entire expression may easily change its sign.

Problem 159. Dielectric susceptibility

Let the states of an atom be described by the Schrödinger equation $H|n\rangle=E_n\cdot|n\rangle$, its ground state being $|0\rangle$. The dielectric polarizability α of the atom (or the susceptibility χ of a substance consisting of N atoms per cm^3) shall be determined. What can, in general, be said on the polarizability of alkali atoms?

Solution. If an electrical field $\mathscr{E}$ is applied to the atom in z direction, it causes a perturbation energy

$$W = e\mathscr{E} \sum_\lambda z_\lambda \tag{159.1}$$

with $-e$ the electron charge and the subscript λ numbering the atomic electrons. The equation

$$(H+W)\,\psi = E\psi \tag{159.2}$$

is then approximately solved by

$$\psi = |0\rangle + \sum_n{}' \frac{\langle n|W|0\rangle}{E_0 - E_n} |n\rangle$$

or

$$\psi = |0\rangle + e\mathscr{E} \sum_n{}' \frac{\langle n|\sum z_\lambda|0\rangle}{E_0 - E_n} |n\rangle \,. \tag{159.3}$$

The component in field direction of the dipole moment of this state has the expectation value

$$p_z = -e\langle\psi|\sum_\lambda z_\lambda|\psi\rangle \,. \tag{159.4}$$

In first order this is composed of two terms:

$$p_z = -e\left\{\langle 0|\sum_\lambda z_\lambda|0\rangle + \sum_n{}'\left[\frac{\langle n|W|0\rangle}{E_0 - E_n}\langle 0|\sum_\lambda z_\lambda|n\rangle + \frac{\langle 0|W|n\rangle}{E_0 - E_n}\langle n|\sum_\lambda z_\lambda|0\rangle\right]\right\}.$$

The first term is the moment of the undisturbed state (if any). The second term describes the moment induced by the field. Denoting this latter moment by p_{ind} the polarizability α is defined by

$$p_{\text{ind}} = \alpha\mathscr{E} \,. \tag{159.5}$$

We thus find

$$\alpha = 2e^2 \sum_n{}' \frac{|\langle n|\sum z_\lambda|0\rangle|^2}{E_n - E_0} \,. \tag{159.6}$$

Since E_0 is supposed to be the ground state, the denominator will be positive. So therefore will be the polarizability, too.

The susceptibility χ connects the dielectric polarization $P = N p_{\text{ind}}$ with the field,

$$P = \chi\mathscr{E} \tag{159.7}$$

so that

$$\chi = 2Ne^2 \sum_n{}' \frac{|\langle n|\sum z_\lambda|0\rangle|^2}{E_n - E_0} \tag{159.8}$$

and $\chi > 0$.

For alkali atoms consisting of a core plus one single outer electron, the excitation of a core electron requires much energy, thus leading to a large denominator in (159.8). For a rough orientation it therefore suffices to study excitations of the outer electron only, moving in the field of an unaffected core. Its eigenfunctions may be written

$$|0\rangle = u(r); \quad |n\rangle = v_n(r)\, Y_{l,m}(\vartheta, \varphi)$$

because the ground state $|0\rangle$ is an s state not depending on polar angles. With

$$z = r\cos\vartheta = r\sqrt{\frac{4\pi}{3}}\, Y_{1,0}$$

the matrix element then becomes

$$\langle n|z|0\rangle = \int_0^\infty dr\, r^3 v_n(r)\, u(r) \cdot \oint d\Omega\, Y_{l,m}^* \cos\vartheta\,.$$

This vanishes, except for excitation of states with $l=1$ and $m=0$, when it becomes

$$\langle n|z|0\rangle = \sqrt{\frac{4\pi}{3}} \int_0^\infty dr\, r^3 v_n(r)\, u(r)\,.$$

Further computations imply detailed knowledge of the radial parts of the eigenfunctions.

If dimensionless units are not being used, it is easily seen that the polarizability α has the dimension of a volume so that it will, roughly speaking, be of the order of $(\hbar^2/me^2)^3$.

NB. One can as well calculate the second-order Stark effect with an energy shift

$$\Delta_2 E = \sum_n{}' \frac{|\langle 0|W|n\rangle|^2}{E_0 - E_n}$$

which must be $= -\frac{1}{2}\alpha\mathscr{E}^2$. The result for α is the same as above.

Problem 160. Diamagnetic susceptibility of neon

To compute the diamagnetic susceptibility of neon ($Z=10$) using hydrogen-like wave functions with different screening constants $\sigma_{n,l}$. The following screening constants may be used:

$$\sigma_{1,0} = 0.23 \qquad \sigma_{2,0} = 3.26 \qquad \sigma_{2,1} = 4.11\,.$$

Solution. The diamagnetic susceptibility per mole is given by the formula [cf. (128.14)]

$$\chi = -\frac{e^2}{6mc^2} N \sum \langle r^2 \rangle \tag{160.1}$$

where N is the Loschmidt number ($N = 6.02 \times 10^{23}$) and the sum is to be extended over all electrons of one atom (or molecule). The expectation values of r^2 in states with eigenfunctions

$$u_{n,l,m} = \frac{1}{r} \chi_{n,l}(r)\, Y_{l,m}(\vartheta, \varphi)$$

are given by the integrals

$$\langle r^2 \rangle = \int_0^\infty dr\, r^2 |\chi_{n,l}|^2 . \tag{160.2}$$

The radial parts of hydrogen-like wave functions may be taken from the table of Problem 67, with Z replaced by $Z - \sigma$; the resulting values of the integrals (160.2), in units of $(\hbar^2/me^2)^2$ then become[3]

for $(n, l) =$	$(1,0)$	$(2,0)$	$(2,1)$
$(Z-\sigma)^2 \langle r^2 \rangle =$	3	42	30

as may be checked by elementary integrations. The respective numbers of electrons in the three (n, l) states are 2, 2, 6. The order of magnitude of the susceptibilities will be determined by the factor

$$\chi_0 = \frac{e^2}{6mc^2} N \left(\frac{\hbar^2}{me^2} \right)^2 = 0.790 \times 10^{-6}\ \text{cm}^3/\text{mole}. \tag{160.3}$$

We thus obtain for the susceptibility of neon:

$$\chi_{\text{Ne}} = -\chi_0 \left\{ \frac{2 \cdot 3}{(10-\sigma_{1,0})^2} + \frac{2 \cdot 42}{(10-\sigma_{2,0})^2} + \frac{6 \cdot 30}{(10-\sigma_{2,1})^2} \right\}$$

$$= -5.61 \times 10^{-6}\ \text{cm}^3/\text{mole}. \tag{160.4}$$

This result may be compared with the experimental value of

$$\chi_{\text{Ne}} = -6.7 \times 10^{-6}\ \text{cm}^3/\text{mole}.$$

---

[3] These are special cases of the general relation

$$\langle r^2 \rangle = \frac{n^2}{2Z^2} \{5n^2 + 1 - 3l(l+1)\}$$

the deduction of which is cumbersome and not very interesting. For details see Bethe, H. A., Salpeter, E. E., *Encyclopedia of Physics*, vol. **35** (1957), p. 103.

It should be remarked that the contribution of the outermost subshell (n,l) is by far the biggest of the three terms in (160.4) contributing $\chi(1s) = -0.05$, $\chi(2s) = -1.46$ and $\chi(2p) = -4.10\,\mathrm{cm}^3/\mathrm{mole}$. Unfortunately the screening effect is not only very big for the outermost electrons but also rather uncertain experimentally.

Problem 161. Van der Waals attraction

Two hydrogen atoms in their *ground states* are at a distance R from each other. The nuclei may be supposed at rest. It shall be shown that the interaction between the two atoms vanishes in first order of a perturbation calculation, and that a second-order approximation leads to Van der Waals attraction. Of the interaction part of the hamiltonian, only the leading term proportional to the lowest negative power of R is to be used (large distance approximation).

Solution. Let us denote the position of electron 1 relative to nucleus a by $\boldsymbol{r}_1$ with components $x_1\ y_1\ z_1$ and the position of electron 2 relative to nucleus b by $\boldsymbol{r}_2$ with components $x_2\ y_2\ z_2$. The z direction shall

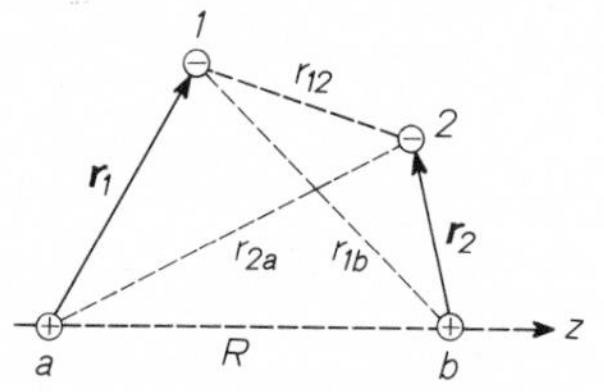

Fig. 63. Notations. The distances marked by broken lines enter the interaction (161.3)

coincide with the nuclear axis (Fig. 63). With the nuclei at rest, i.e. in Born-Oppenheimer approximation, we then have the hamiltonian

$$H = H^0 + H' \tag{161.1}$$

with

$$H^0 = -\frac{\hbar^2}{2m}(\nabla_1^2 + \nabla_2^2) - \frac{e^2}{r_1} - \frac{e^2}{r_2} \tag{161.2}$$

describing two independent atoms, and

$$H' = \frac{e^2}{R} + \frac{e^2}{r_{12}} - \frac{e^2}{r_{1b}} - \frac{e^2}{r_{2a}} \tag{161.3}$$

their mutual interaction. We shall use H' as perturbation energy. If H' is expanded in a series of negative powers of R (i.e. for $r_1 \ll R$ and $r_2 \ll R$), the large-distance main term comes from the interaction of the two dipoles a1 and b2 with dipole moments $\boldsymbol{p}_1 = -e\boldsymbol{r}_1$ and $\boldsymbol{p}_2 = -e\boldsymbol{r}_2$ respectively, viz.

$$H' = \frac{\boldsymbol{p}_1 \boldsymbol{p}_2}{R^3} - 3\frac{(\boldsymbol{p}_1 \boldsymbol{R})(\boldsymbol{p}_2 \boldsymbol{R})}{R^5}. \tag{161.4}$$

In coordinate formulation, (161.4) becomes

$$H' = \frac{e^2}{R^3}(x_1 x_2 + y_1 y_2 - 2 z_1 z_2). \tag{161.5}$$

This expression shall be used in the following calculations.

Let $u_0(r)$ be the wave function of the atomic ground state. The zero-order wave function of the entire system then is the product

$$U(1,2) = u_0(r_1)\, u_0(r_2) \tag{161.6}$$

where symmetrization has been omitted since exchange contributions tend exponentially towards zero at large distances R and may thus be neglected.

In zero-order approximation the sum of the two atom energies is the energy of the system. In first order we have to add

$$E' = \langle U|H'|U\rangle = 0.$$

It can easily be seen that this term needs must vanish. Taking, e.g., the first term of (161.5), we have

$$\frac{e^2}{R^3}\langle U|x_1 x_2|U\rangle = \frac{e^2}{R^3}\langle u_0|x|u_0\rangle^2 = \frac{e^2}{R^3}\{\int d\tau\, u_0^2(r)\, x\}^2,$$

and these integrals describing dipole moments of atomic states with spherical symmetry indeed vanish[4].

The second-order energy perturbation is

$$E'' = \sum_n{}' \frac{|\langle 0|H'|n\rangle|^2}{E_0 - E_n}, \tag{161.7}$$

---

[4] They vanish not only for S states but always, if both atoms are in the same state, see the following problem. Even for two excited states, $|u_0|^2$ depends upon polar angles as the square of a spherical harmonic. This can be decomposed into a sum of spherical harmonics of *even* orders only. In the integrand it is multiplied by a coordinate x, y, or z, i.e. with a spherical harmonic of first, hence of *odd* order. Orthogonality properties of spherical harmonics thus cause the product integral to vanish.

the sum being extended over all excited states, and 0 referring to the ground state. As $E_n > E_0$, all denominators of the sum are negative so that $E'' < 0$ and there arises an attraction. The matrix elements depend upon R only through the constant factor R^{-3} in Eq. (161.5) which stands in front of each integral. E'' therefore must be of the form

$$E'' = -\frac{C}{R^6}$$

with C a positive constant. This is, however, the well-known distance dependence of the Van der Waals attraction.

Literature. Schiff, L. I.: *Quantum Mechanics.* New York 1949. pp. 174—178.

Problem 162. Excitation degeneracy

Two hydrogen atoms at rest, a distance R apart, shall be in *different* quantum states, one in the ground state, the other in a P state. Dipole-dipole interaction exists between them, as deduced in the preceding problem. It shall be shown that now there is a non-vanishing first-order contribution to the energy of the system, even at large distances where overlap of eigenfunctions may again be neglected. This first-order energy perturbation shall be calculated.

Solution. Let $|l\,m\rangle$ be written for an atomic wave function with quantum numbers l and m. The ground state may then be described by $|00\rangle$, and the three possible P states by $|1\,m\rangle$ with $m = 1, 0, -1$. Zero-order wave functions of the system, in product form, are then

$$|00, 1\,m\rangle \quad \text{and} \quad |1\,m, 00\rangle, \tag{162.1}$$

the first pair of quantum numbers referring to the first, the second pair to the second atom.

The perturbation energy (161.5) may be used. Introducing the symbols

$$\xi = x + i\,y, \qquad \xi^\dagger = x - i\,y \tag{162.2}$$

it may be reshaped into

$$H' = \frac{e^2}{2R^3}\left(\xi_1 \xi_2^\dagger + \xi_1^\dagger \xi_2 - 4 z_1 z_2\right). \tag{162.3}$$

This operator, being linear in the coordinates of either electron, has matrix elements with functions of type (162.1) different from zero only if for each of the two electrons an S and a P state are combined[5]. These are the matrix elements

$$\langle 1\,m_1, 00|H'|00, 1\,m_2\rangle \quad \text{and} \quad \langle 00, 1\,m_2|H'|1\,m_1, 00\rangle\,.$$

The bilinear structure of H', Eq. (162.3), allows decomposition into atomic matrix elements:

$$\begin{aligned}\langle 1\,m_1, 00|H'|00, 1\,m_2\rangle = \frac{e^2}{2R^3}\{&\langle 1\,m_1|\xi|00\rangle\,\langle 00|\xi^\dagger|1\,m_2\rangle\\ &+\langle 1\,m_1|\xi^\dagger|00\rangle\,\langle 00|\xi|1\,m_2\rangle\\ &-4\langle 1\,m_1|z|00\rangle\,\langle 00|z|1\,m_2\rangle\}\,.\end{aligned} \tag{162.4}$$

Denoting the radial parts of the atomic wave functions by $f_l(r)$ and using the abbreviation

$$r_0 = \int_0^\infty dr\, r^3 f_0(r) f_1(r)\,, \tag{162.5}$$

the following results are obtained for the non-vanishing integrals occurring in (162.4):

$$\left.\begin{aligned}\langle 1\,1|\xi|00\rangle &= \sqrt{\tfrac{2}{3}}\,r_0\,; & \langle 00|\xi|1-1\rangle &= -\sqrt{\tfrac{2}{3}}\,r_0\,;\\ \langle 1-1|\xi^\dagger|00\rangle &= -\sqrt{\tfrac{2}{3}}\,r_0\,; & \langle 00|\xi^\dagger|1\,1\rangle &= \sqrt{\tfrac{2}{3}}\,r_0\,;\\ \langle 1\,0|z|00\rangle &= \sqrt{\tfrac{1}{3}}\,r_0\,; & \langle 00|z|1\,0\rangle &= \sqrt{\tfrac{1}{3}}\,r_0\,.\end{aligned}\right\} \tag{162.6}$$

All other combinations of quantum numbers lead to vanishing matrix elements. Eq. (162.4) then becomes

$$\begin{aligned}\langle 1\,m_1, 00|H'|00, 1\,m_2\rangle = \frac{e^2 r_0^2}{2R^3}\{&\tfrac{2}{3}\,\delta_{1,m_1}\,\delta_{1,m_2} + \tfrac{2}{3}\,\delta_{-1,m_1}\,\delta_{-1,m_2}\\ &-\tfrac{4}{3}\,\delta_{0,m_1}\,\delta_{0,m_2}\}\,.\end{aligned} \tag{162.7}$$

Thus, only with $m_1 = m_2$ the matrix element (162.7) does not vanish.

We are now prepared to write down the secular determinant of the six degenerate zero-order wave functions (162.1). If E' is the first-order perturbation and the six functions are used in the order

$$|00,11\rangle\,;\quad |11,00\rangle\,;\quad |00,10\rangle\,;\quad |10,00\rangle\,;\quad |00,1-1\rangle\,;\quad |1-1,00\rangle$$

---

[5] Coupling of any states with even l and odd $l' = l \pm 1$ would do, e.g. $P-D$ coupling. This however exceeds the limitations set by Eq. (162.1).

the secular equation becomes

$$\begin{vmatrix} -E' & \frac{2}{3}\varepsilon & 0 & 0 & 0 & 0 \\ \frac{2}{3}\varepsilon & -E' & 0 & 0 & 0 & 0 \\ 0 & 0 & -E' & -\frac{4}{3}\varepsilon & 0 & 0 \\ 0 & 0 & -\frac{4}{3}\varepsilon & -E' & 0 & 0 \\ 0 & 0 & 0 & 0 & -E' & \frac{2}{3}\varepsilon \\ 0 & 0 & 0 & 0 & \frac{2}{3}\varepsilon & -E' \end{vmatrix} = 0 \tag{162.8}$$

with

$$\varepsilon = \frac{e^2 r_0^2}{2R^3}. \tag{162.9}$$

The determinant can be decomposed into three 2×2 ones, thus extremely simplifying its evaluation. The results are therefore immediately shown in the accompanying table. Here Λ stands for the sum of the m values of both atoms, i.e. for the component of the total electron orbital momentum along the nuclear axis. The classification symbols used are those of molecular spectroscopy, the signs Σ and Π referring to $\Lambda=0$ and ± 1, respectively, and the subscripts g and u to wave functions even and odd. The two Π_g states having the same energy are still degenerate; so are the two Π_u states. The last column gives the interaction energy E' in multiples of ε.

State	Λ	Wave function (unnormalized)	E'
Π_g	1	$\lvert 00, 11\rangle + \lvert 11, 00\rangle$	$+\frac{2}{3}\varepsilon$
Π_u	1	$\lvert 00, 11\rangle - \lvert 11, 00\rangle$	$-\frac{2}{3}\varepsilon$
Σ_g	0	$\lvert 00, 10\rangle + \lvert 10, 00\rangle$	$-\frac{4}{3}\varepsilon$
Σ_u	0	$\lvert 00, 10\rangle - \lvert 10, 00\rangle$	$+\frac{4}{3}\varepsilon$
Π_g	-1	$\lvert 00, 1-1\rangle + \lvert 1-1, 00\rangle$	$+\frac{2}{3}\varepsilon$
Π_u	-1	$\lvert 00, 1-1\rangle - \lvert 1-1, 00\rangle$	$-\frac{2}{3}\varepsilon$

Introducing the hydrogen wave functions (see Problem 67)

$$f_0 = 2\,e^{-r}; \qquad f_1 = \frac{\sqrt{6}}{12}\, r\, e^{-r/2}$$

in atomic units, the integral r_0, Eq. (162.5), can be evaluated:

$$r_0 = \sqrt{6}\cdot\tfrac{128}{243} \tag{162.10}$$

and, from Eq. (162.9),

$$\varepsilon = \frac{16384}{19683} \frac{e^2 a_0^2}{R^3} \tag{162.11}$$

with a_0 the Bohr radius. In all states, therefore, the interaction energy E' becomes proportional to R^{-3}, decreasing more slowly than, and thus dominating, the Van der Waals energy ($\propto R^{-6}$) at large distances. The sign of the interaction still depends upon the state of the system: in the states Σ_u and Π_g there is repulsion between the two atoms and attraction only in the states Σ_g and Π_u.

Literature. Herzberg, G.: *Spectra of diatomic molecules*, p. 378. — Landau-Lifschitz: *Quantum Mechanics*, p. 302. — Margenau, H.: Rev. Modern Phys. **11**, 1 (1939). — King, G. W., Van Vleck, J. H.: Phys. Rev. **55**, 1165 (1939).

Problem 163. Neutral hydrogen molecule

To find the binding energy and equilibrium distance of the neutral hydrogen molecule by a method analogous to that of the H_2^+ treatment given in Problem 44.

Solution. This is a two-body problem in the Born-Oppenheimer approximation of fixed nuclei. Let the two nuclei (protons) be denoted by a and b, and the two electrons by 1 and 2, then we have the hamiltonian, in atomic units,

$$H = -\tfrac{1}{2}(\nabla_1^2 + \nabla_2^2) + \frac{1}{r_{12}} - \left(\frac{1}{r_{a1}} + \frac{1}{r_{b1}} + \frac{1}{r_{a2}} + \frac{1}{r_{b2}}\right) + \frac{1}{R} \tag{163.1}$$

with R the distance between the nuclei. At large distances R, the wave function should pass over into the product of the separate atoms, either becoming of the form $f(\mathbf{r}_{a1})\, f(\mathbf{r}_{b2})$ if electron 1 forms an atom with nucleus a, and 2 with b, or of the form $f(\mathbf{r}_{b1})\, f(\mathbf{r}_{a2})$ if the two electrons are exchanged. A reasonable approach at finite distances R will be a linear combination of two such products, and symmetry considerations lead to the choice of the symmetrical solution

$$U(1,2) = \alpha[f(\mathbf{r}_{a1})\, f(\mathbf{r}_{b2}) + f(\mathbf{r}_{b1})\, f(\mathbf{r}_{a2})] \tag{163.2}$$

for the ground state (with antiparallel electron spins, according to the Pauli principle). The antisymmetric combination, which is an eigenfunction as well, would lead to a larger energy with no attraction and no formation of a molecule at all.

If we put (163.2) into the Schrödinger equation,

$$H U = E \cdot U \tag{163.3}$$

with H the hamiltonian (163.1) we get

$$\begin{aligned} F(\boldsymbol{r}_{a1}) f(\boldsymbol{r}_{b2}) + f(\boldsymbol{r}_{a1}) F(\boldsymbol{r}_{b2}) + \left[\frac{1}{r_{12}} - \frac{1}{r_{b1}} - \frac{1}{r_{a2}} - E + \frac{1}{R}\right] f(\boldsymbol{r}_{a1}) f(\boldsymbol{r}_{b2}) \\ + F(\boldsymbol{r}_{b1}) f(\boldsymbol{r}_{a2}) + f(\boldsymbol{r}_{b1}) F(\boldsymbol{r}_{a2}) \\ + \left[\frac{1}{r_{12}} - \frac{1}{r_{a1}} - \frac{1}{r_{b2}} - E + \frac{1}{R}\right] f(\boldsymbol{r}_{b1}) f(\boldsymbol{r}_{a2}) = 0 \end{aligned} \tag{163.4}$$

where F stands as an abbreviation:

$$F(\boldsymbol{r}_{a1}) = \left(-\frac{1}{2}\nabla_1^2 - \frac{1}{r_{a1}}\right) f(\boldsymbol{r}_{a1}). \tag{163.5}$$

We now apply the operator

$$\int d\tau_1 \int d\tau_2\, f^*(\boldsymbol{r}_{a1}) f^*(\boldsymbol{r}_{b2}) \ldots$$

to Eq. (163.4). The function f is supposed to be normalized. Using the following abbreviations:

$$S = \int d\tau_1\, f^*(\boldsymbol{r}_{a1}) f(\boldsymbol{r}_{b2}) \tag{163.6}$$

for the overlap integral,

$$\mathscr{C} = \int d\tau_1 \frac{1}{r_{b1}} |f(\boldsymbol{r}_{a1})|^2 \tag{163.7}$$

and

$$\mathscr{C}' = \iint d\tau_1\, d\tau_2 \frac{1}{r_{12}} |f(\boldsymbol{r}_{a1})|^2 |f(\boldsymbol{r}_{b2})|^2 \tag{163.8}$$

for the classical interaction integrals,

$$\mathscr{E} = \int d\tau_1 \frac{1}{r_{a1}} f^*(\boldsymbol{r}_{a1}) f(\boldsymbol{r}_{b1}) \tag{163.9}$$

and

$$\mathscr{E}' = \iint d\tau_1\, d\tau_2 \frac{1}{r_{12}} f^*(\boldsymbol{r}_{a1}) f(\boldsymbol{r}_{b1}) f(\boldsymbol{r}_{a2}) f^*(\boldsymbol{r}_{b2}) \tag{163.10}$$

for the exchange integrals, and finally

$$A = \int d\tau_1 f^*(\boldsymbol{r}_{a1}) F(\boldsymbol{r}_{a1}) \tag{163.11}$$

and

$$A' = \int d\tau_1 f^*(\boldsymbol{r}_{a1}) F(\boldsymbol{r}_{b1}) \tag{163.12}$$

for the two remaining integrals, we find by this procedure

$$2(A+A'S)-2(\mathscr{C}+\mathscr{E}S)+(\mathscr{C}'+\mathscr{E}')=\left(E-\frac{1}{R}\right)(1+S^2) \quad (163.13)$$

or

$$E=2\frac{A+A'S}{1+S^2}-\frac{2(\mathscr{C}+\mathscr{E}S)-(\mathscr{C}'+\mathscr{E}')}{1+S^2}+\frac{1}{R}. \quad (163.14)$$

In a way analogous to that used in Problem 44 for H_2^+ we now put

$$f(\mathbf{r})=\sqrt{\frac{\gamma^3}{\pi}}\,\mathrm{e}^{-\gamma r}. \quad (163.15)$$

For $\gamma=1$ this would be the wave function of the atomic ground state; with γ a variational Ritz parameter we may still get a better approximation. Specializing to (163.15) we find, according to (163.5),

$$F(r_{a1})=\left(-\frac{1}{2}\gamma^2+\frac{\gamma-1}{r_{a1}}\right)f(r_{a1})$$

so that (163.11) and (163.12) yield

$$A=-\tfrac{1}{2}\gamma^2+\gamma(\gamma-1); \qquad A'=-\tfrac{1}{2}\gamma^2 S+(\gamma-1)\mathscr{E}. \quad (163.16)$$

It can further be shown that the overlap integral S depends only upon the combination

$$\rho=\gamma R \quad (163.17)$$

which we may use as a second Ritz parameter besides γ, and that the four remaining interaction integrals $\mathscr{C}$, $\mathscr{C}'$, $\mathscr{E}$, $\mathscr{E}'$ all become proportional to γ so that we may write

$$\mathscr{C}=\gamma\overline{\mathscr{C}}(\rho); \quad \mathscr{C}'=\gamma\overline{\mathscr{C}}'(\rho); \quad \mathscr{E}=\gamma\overline{\mathscr{E}}(\rho); \quad \mathscr{E}'=\gamma\overline{\mathscr{E}}'(\rho). \quad (163.18)$$

The energy expression (163.14) thus becomes

$$E=-a\gamma+b\gamma^2 \quad (163.19)$$

with

$$a(\rho)=\frac{2(1+\overline{\mathscr{C}})+4S\overline{\mathscr{E}}-(\overline{\mathscr{C}}'+\overline{\mathscr{E}}')}{1+S^2}-\frac{1}{\rho} \quad (163.20)$$

and

$$b(\rho)=\frac{1-S^2+2S\overline{\mathscr{E}}}{1+S^2} \quad (163.21)$$

depending only on ρ. We then get an energy minimum if

$$\frac{\partial E}{\partial \gamma} = -a + 2b\gamma = 0$$

or

$$\gamma = \frac{a}{2b}; \tag{163.22}$$

it amounts to

$$E = -\frac{a^2}{4b}. \tag{163.23}$$

It remains to evaluate the five integrals (163.6–10). Three of them have been computed in the H_2^+, Problem 44, viz.

$$\begin{aligned} S &= (1+\rho+\tfrac{1}{3}\rho^2)\mathrm{e}^{-\rho}, \\ \overline{\mathscr{C}} &= \frac{1}{\rho}[1-(1+\rho)\mathrm{e}^{-2\rho}], \\ \overline{\mathscr{E}} &= (1+\rho)\mathrm{e}^{-\rho}. \end{aligned}$$

The integral $\mathscr{C}'$ is a little more difficult but can still be evaluated in an elementary way if integration over the coordinates of the electron 2 is first performed using spherical polar coordinates with the origin in b and $b1$ the polar axis:

$$\frac{1}{\pi}\int d\tau_2 \frac{1}{r_{12}} \mathrm{e}^{-2r_{b2}} = \overline{\mathscr{C}}(r_{b1}).$$

The second integration, over electron 1, then leads to integrals already evaluated, only some having factors 2 in the exponential. The result is

$$\overline{\mathscr{C}}' = \frac{1}{\rho}[1-(1+\tfrac{11}{8}\rho+\tfrac{3}{4}\rho^2+\tfrac{1}{6}\rho^3)\mathrm{e}^{-2\rho}].$$

Real difficulties, however, are encountered in the last integral, $\mathscr{E}'$, which can no longer be reduced to elementary integrations. The result, first obtained by Sugiura, is

$$\overline{\mathscr{E}}' = [\tfrac{5}{8} - \tfrac{23}{20}\rho - \tfrac{3}{5}\rho^2 - \tfrac{1}{15}\rho^3]\mathrm{e}^{-2\rho} + \frac{6}{5}\frac{\varphi(\rho)}{\rho}$$

with

$$\varphi(\rho) = S(\rho)^2(\log\rho + C) - S(-\rho)^2 E_1(4\rho) + 2S(\rho)S(-\rho)E_1(2\rho),$$

where

$$E_1(z) = \int_z^\infty \frac{dt}{t}\mathrm{e}^{-t}$$

is the exponential integral.

It is easily seen that, for large r, this integral tends to zero as $e^{-2\rho}$ and that, at $\rho=0$, it has a limit $\overline{\mathscr{E}}'(0)=\frac{5}{8}$ in agreement with the theory of the helium ground state emerging for nuclear distance zero (cf. Problem 154).

Numerical computation yields the following table. The binding energy of the molecule has a maximum at about $R=1.46$ a.u. corresponding to an equilibrium distance of $R_0=0.77$ Å (instead of the experimental value 0.742 Å). The energy then is $E=-1.139$ which is to be compared with the binding energy of two separate hydrogen atoms in the ground state, $2E_0=-1$. If the zero-point energy of the molecular vibrations is denoted by $\frac{1}{2}\hbar\omega$, we therefore have a dissociation energy of

$$D=2E_0-(E+\tfrac{1}{2}\hbar\omega)=0.139-\tfrac{1}{2}\hbar\omega.$$

The zero-point energy can be determined by the same procedure as in the case of H_2^+ (Problem 44), but the energy parabola in the neighbourhood of the minimum is much less accurately determined by the table.

ρ	γ	$-E$	R
1.3	1.145	1.120	1.133
1.4	1.152	1.127	1.214
1.5	1.160	1.131	1.293
1.6	1.164	1.137	1.374
1.7	1.166	1.139	1.458
1.8	1.164	1.137	1.546
1.9	1.161	1.134	1.635
2.0	1.156	1.129	1.730

However, we find a value close to 0.010 a.u. or 0.27 eV, with an error of about $\pm 5\%$, in perfect agreement with experiment ($\hbar\omega=0.54$ eV). Thus, the dissociation energy turns out to be

$$D=0.138 \text{ a.u.}=3.75 \text{ eV}$$

whereas its experimental value is $D=4.45$ eV. The agreement is not so bad, as we have explained in the similar situation obtaining in the case of H_2^+.

NB. It should be mentioned that, for large distances R or ρ, the parameter γ tends towards 1 and the function f to the atomic ground state wave function. In the original Heitler-London method, this value was used throughout, so that there was no Ritz parameter γ. In that rougher picture the dissociation energy becomes only 2.90 eV and the equilibrium distance 0.88 Å. The increasing values of γ which, according to the table, obtain during an adiabatic approach of the two atoms describe a contraction of their electron wave functions in binding.

Literature. Heitler, W., London, F.: Z. Physik **44**, 455 (1927). — *Evaluation of the integral $\mathscr{E}'$:* Sugiura, Y., Z. Physik **45**, 484 (1927). — *Variation of γ:* Wang, S. C., Phys. Rev. **31**, 579 (1928); Rosen, N.: Phys. Rev. **38**, 2099 (1931). — *Better variational approximations:* James, H. M., Coolidge, A. S., J. chem. Phys. **1**, 825 (1933); **3**, 129 (1935).

Problem 164. Scattering of equal particles

A beam of particles of charge e collides with a target consisting of particles of the same kind at rest. How does the angular distribution by scattering compare with that expected of classical physics, if allowance is made for the correct symmetry of the wave function? This shall be discussed for unpolarized particle beams of spins 0, $\frac{1}{2}$, and 1.

Solution. The Rutherford amplitude has been derived in Problem 110 in the centre-of-mass system. It is

$$f(\vartheta) = -\frac{\varkappa^*}{2k^*} e^{2i\eta_0} \frac{e^{-i\kappa^* \log \sin^2 \frac{\vartheta}{2}}}{\sin^2 \frac{\vartheta}{2}}; \qquad \eta_0 = \Gamma(1+i\varkappa^*); \qquad (164.1)$$

here the abbrevations κ^* and k^* refer to centre-of-mass values,

$$\varkappa^* = \frac{e^2}{\hbar v}; \qquad k^* = \frac{m^* v}{\hbar}; \qquad E^* = \frac{\hbar^2 k^{*2}}{2m^*} \qquad (164.2)$$

and m^* is the reduced mass. For two equal particles, $m^* = \frac{1}{2}m$. The relative velocity of the two particles, v, is independent of the frame of reference. Therefore, if κ, k and E refer to the laboratory frame, we get

$$\varkappa^* = \varkappa; \qquad k^* = \tfrac{1}{2}k; \qquad E^* = \tfrac{1}{2}E \qquad (164.3)$$

so that

$$f(\vartheta) = -\frac{\varkappa}{k} e^{2i\eta_0} \frac{e^{-i\kappa \log \sin^2 \frac{\vartheta}{2}}}{\sin^2 \frac{\vartheta}{2}}; \qquad \eta_0 = \Gamma(1+i\varkappa). \qquad (164.4)$$

Further, the angle of deflection in the laboratory frame, Θ, for equal masses becomes

$$\Theta = \tfrac{1}{2}\vartheta \qquad (164.5)$$

with ϑ the angle in the centre-of-mass system. The solid angle element therefore transforms according to

$$d\omega = 2\pi \sin\vartheta\, d\vartheta = 2\pi \cdot 4\cos\Theta \sin\Theta\, d\Theta = 4\cos\Theta\, d\Omega, \qquad (164.6)$$

and the differential Rutherford cross section in the laboratory frame becomes

$$\frac{d\sigma_R}{d\Omega} = 4\cos\Theta \left(\frac{\varkappa}{k}\right)^2 \frac{1}{\sin^4\Theta} \quad \text{with} \quad \frac{\varkappa}{k} = \frac{e^2}{mv^2} = \frac{e^2}{2E}. \qquad (164.7)$$

Even in classical mechanics, this formula has to undergo an essential correction. Since it is impossible to distinguish a scattered particle from a recoil particle, if both are of equal kind, both have to be added in the cross section. As, according to (164.5), the two paths will be perpendicular to each other, the recoil particle emerges from the target under the angle $\frac{\pi}{2} - \Theta$ and we get, instead of (164.7),

$$\frac{d\sigma_{\text{class}}}{d\Omega} = 4\cos\Theta \left(\frac{\varkappa}{k}\right)^2 \left\{\frac{1}{\sin^4\Theta} + \frac{1}{\cos^4\Theta}\right\}. \qquad (164.8)$$

This is the classical expression with which we have to compare the quantum mechanical result now to be derived.

According to quantum theory, we have not to superimpose intensities (i.e. cross sections) but amplitudes. Let $u(\boldsymbol{r})$ be the unsymmetrized wave function in the centre-of-mass frame with $\boldsymbol{r}$ the relative coordinate vector. Its asymptotic behaviour, apart from logarithmic phases, is

$$u(\boldsymbol{r}) \to e^{ik^*z} + f(\vartheta)\frac{e^{ik^*r}}{r}.$$

The plane wave part can be written

$$e^{ik^*(z_1 - z_2)},$$

describing one particle of velocity $+\frac{1}{2}v$ and the other of velocity $-\frac{1}{2}v$ in z direction. If centre-of-mass motion is superimposed by a factor

$$e^{ik^*(z_1 + z_2)},$$

particle 2 is brought to rest (target particle), and there remains the motion of particle 1 according to $e^{2ik^*z_1} = e^{ikz_1}$ (colliding particle). This holds for the unsymmetrized wave function of the two-particle system. Its symmetrization means replacing $u(\boldsymbol{r})$ by

$$u(\boldsymbol{r}) + \varepsilon u(-\boldsymbol{r}) \quad \text{with } \varepsilon = \pm 1.$$

For the asymptotic spherical wave part this means replacing $f(\vartheta)$ by

$$f(\vartheta)+f(\pi-\vartheta)$$

leaving r unchanged, so that, with (164.4), instead of $f(\vartheta)$ we get

$$f(\vartheta)+\varepsilon f(\pi-\vartheta)=-\frac{e^2}{mv^2}e^{2i\eta_0}\left\{\frac{e^{-i\kappa\log\sin^2\frac{\vartheta}{2}}}{\sin^2\frac{\vartheta}{2}}+\varepsilon\frac{e^{-i\kappa\log\sin^2\frac{\vartheta}{2}}}{\cos^2\frac{\vartheta}{2}}\right\} \tag{164.9}$$

and, instead of the classical cross section formula (164.8),

$$\frac{d\sigma}{d\Omega}=4\cos\Theta\left(\frac{e^2}{mv^2}\right)^2\left|\frac{e^{-i\varkappa\log\sin^2\Theta}}{\sin^2\Theta}+\frac{e^{-i\varkappa\log\cos^2\Theta}}{\cos^2\Theta}\right|^2. \tag{164.10}$$

Evaluation of the last expression leads to

$$\frac{d\sigma}{d\Omega}=4\cos\Theta\left(\frac{e^2}{mv^2}\right)^2\left\{\frac{1}{\sin^4\Theta}+\frac{1}{\cos^4\Theta}+\varepsilon\frac{2\cos(\frac{e^2}{\hbar v}\log\tan^2\Theta)}{\sin^2\Theta\cos^2\Theta}\right\}, \tag{164.11}$$

with an interference term added to the classical expression. For comparing both, quantum theoretical and classical cross sections, the ratio

$$\frac{d\sigma}{d\sigma_{\text{class}}}=1+\varepsilon\frac{2\tan^2\Theta\cos(\frac{e^2}{\hbar v}\log\tan^2\Theta)}{1+\tan^4\Theta} \tag{164.12}$$

may be useful.

As a last step we now have to decide which part of an unpolarized beam will be symmetrical, and which antisymmetrical. If the scattering particles are *fermions* of spin $\frac{1}{2}$ (two protons or two electrons) whose total wave function must be antisymmetrical, we have weights $\frac{3}{4}$ for the spin-symmetrical, space-antisymmetrical triplet and $\frac{1}{4}$ for the opposite symmetry of the singlet state so that we get

$$d\sigma=\tfrac{3}{4}d\sigma_-+\tfrac{1}{4}d\sigma_+$$

with the subscript signs referring to the two signs of ε in (164.11). Therefore, unpolarized beam experiments in this case yield

$$\varepsilon_{\text{eff}}=-\tfrac{3}{4}+\tfrac{1}{4}=-\tfrac{1}{2}$$

and

$$\frac{d\sigma}{d\sigma_{\text{class}}}=1-\frac{\tan^2\Theta\cos(\frac{e^2}{\hbar v}\log\tan^2\Theta)}{1+\tan^4\Theta}. \tag{164.13}$$

This function is shown in Fig. 64 for proton-proton scattering at an energy $E=100$ keV. This is about the highest energy for which no very appreciable scattering anomaly occurs due to the short-range nuclear attraction between the protons (cf. the following problem). At this energy, we have $\varkappa=0.50$ for protons. The function (164.13) is invariant under the transformation $\Theta \to \frac{\pi}{2} - \Theta$ so that it need only be computed for values $0<\Theta\leqq 45°$. Passing on to much lower energies, $\varkappa$ becomes so large that $\cos(\varkappa \log \tan^2 \Theta)$ has several oscillations in this interval.

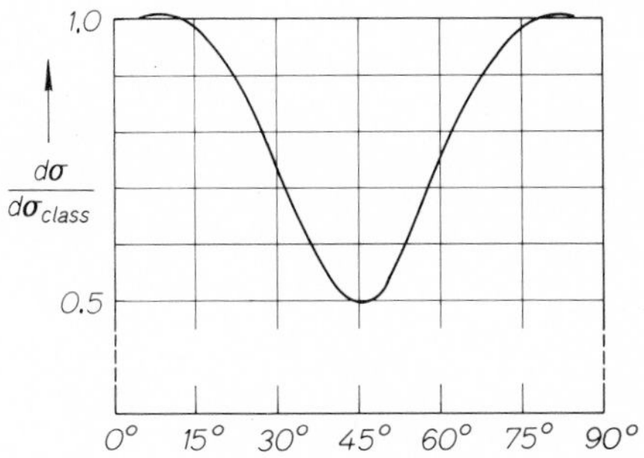

Fig. 64. Scattering of two equal fermions. The figure shows the ratio of quantum theoretical to classical scattering intensity as a function of scattering angle. The curve has an infinite number of oscillations in the vicinities of 0° and 90°, with decreasing amplitudes

Finally, for very large $\varkappa$, these oscillations will be so rapid that they can no longer be resolved experimentally so that the classical expression remains.

If the particles are *bosons* without spin (e.g. two α particles or two pions), only the space symmetrical state with $\varepsilon=+1$ occurs. Of course, for α particles e^2 has to be replaced by $4e^2$. If the particles are bosons of spin 1 (e.g. two deuterons), the total spins possible are 2 (weight $\frac{5}{9}$), 1 (weight $\frac{3}{9}$), and 0 (weight $\frac{1}{9}$) with space symmetry for total spins 2 and 0. Then we get

$$\varepsilon_{\text{eff}}=(\tfrac{5}{9}+\tfrac{1}{9})-\tfrac{3}{9}=+\tfrac{1}{3}$$

and

$$\frac{d\sigma}{d\sigma_{\text{class}}}=1+\frac{2}{3}\,\frac{\tan^2\Theta\cos(\frac{e^2}{\hbar v}\log\tan^2\Theta)}{1+\tan^4\Theta}. \qquad (164.14)$$

Literature. The classical symmetrized expression (164.8) was first used by Darwin, C. G.: Proc. Roy. Soc. London, **A 120,** 631 (1928). The quantum mechanical formula was derived by Mott, N. F., ibid: **126,** 259 (1930). This formula was corroborated experimentally with α particles, as described in the papers by Chadwick, J.: Proc. Roy. Soc. London, **A 128,** 114 (1930) and by Blackett, P. M. S., Champion, F. C., ibid.: **130,** 380 (1931), and with protons by Gerthsen, C.: Ann. Physik **9,** 769 (1931).

Problem 165. Anomalous proton-proton scattering

The short-range nuclear force between two protons gives rise to an attraction which causes a scattering anomaly above energies of about 100 keV. This anomaly shall be described by an additional phase shift δ_0 in the partial wave $l=0$.

Solution. In Problem 112 we have treated anomalous scattering for charged particles without symmetrization and derived Eq. (112.5), in the centre-of-mass frame,

$$f(\vartheta) = -\frac{\varkappa}{2k^* \sin^2\frac{\vartheta}{2}} e^{-i\varkappa \log \sin^2\frac{\vartheta}{2}} + \frac{1}{2ik^*}(e^{2i\delta_0}-1). \qquad (165.1)$$

Let us instead for what follows use the abbreviation

$$f(\vartheta) = f_R(\vartheta) + f_a \qquad (165.2)$$

where the first term describes the Rutherford scattering of the Coulomb field and the second,

$$f_a = \frac{1}{2ik^*}(e^{2i\delta_0}-1) = \frac{1}{k^*} e^{i\delta_0} \sin\delta_0, \qquad (165.3)$$

is the anomaly amplitude. Symmetrization, then, according to the preceding problem, with

$$k^* = \tfrac{1}{2}k; \quad \tfrac{1}{2}\vartheta = \Theta, \qquad (165.4)$$

leads to the relation

$$\frac{d\sigma}{d\Omega} = 4\cos\Theta\,\{\tfrac{3}{4}|f(\vartheta)-f(\pi-\vartheta)|^2 + \tfrac{1}{4}|f(\vartheta)+f(\pi-\vartheta)|^2\}, \qquad (165.5)$$

or

$$\frac{d\sigma}{d\Omega} = 4\cos\Theta\,\{\tfrac{3}{4}|f_R(\vartheta)-f_R(\pi-\vartheta)|^2 + \tfrac{1}{4}|f_R(\vartheta)+f_R(\pi-\vartheta)+2f_a|^2\}. \qquad (165.6)$$

In the first square of (165.6), the anomaly not depending upon ϑ cancels out. It is remarkable that even for higher energies, where higher angular momenta contribute to f_a so that

$$f_a = \frac{1}{2ik^*} \sum_{l=0}^{\infty} (2l+1)\, e^{2i(\eta_l - \eta_0)} (e^{2i\delta_l} - 1)\, P_l(\cos\vartheta),$$

in consequence of

$$P_l(\cos(\pi - \vartheta)) = (-1)^l P_l(\cos\vartheta),$$

all contributions with even l would cancel out in the triplet, and all with odd l in the singlet term. Thus, contributions would arise only from the terms

$${}^1S,\ {}^3P,\ {}^1D,\ {}^3F,\ {}^1G, \ldots$$

This is in complete agreement with the Pauli principle applied separately to each of these partial waves (forbidding ${}^3S$, ${}^1P, \ldots$).

If the absolute squares are evaluated, Eq. (165.6) yields

$$\frac{d\sigma}{d\Omega} = 4\cos\Theta \{ |f_R(\vartheta)|^2 + |f_R(\pi - \vartheta)|^2 - \mathrm{Re}[f_R(\vartheta) f_R^*(\pi - \vartheta)]$$
$$+ \mathrm{Re}[f_a f_R^*(\vartheta)] + \mathrm{Re}[f_a f_R^*(\pi - \vartheta)] + |f_a|^2 \}. \qquad (165.7)$$

In the first line of this equation there stand the terms contributing to Coulomb scattering as discussed in the preceding problem; in the second line there stand two interference terms of Coulomb and anomalous scattering, and the anomaly itself. With the explanations of the symbols f_R and f_a given above, these last three terms can easily be computed:

$$\mathrm{Re}[f_a f_R^*(\vartheta)] = -\frac{2\varkappa}{k^2} \frac{\sin\delta_0}{\sin^2\Theta} \cos(\delta_0 + \varkappa \log \sin^2\Theta);$$

$$\mathrm{Re}[f_a f_R^*(\pi - \vartheta)] = -\frac{2\varkappa}{k^2} \frac{\sin\delta_0}{\cos^2\Theta} \cos(\delta_0 + \varkappa \log \cos^2\Theta);$$

$$|f_a|^2 = \frac{4}{k^2} \sin^2\delta_0 .$$

It is usual to give the so-called scattering ratio,

$$R = \frac{d\sigma}{d\sigma_C}, \qquad (165.8)$$

where $d\sigma_C$ means the Coulomb term only (as derived in the preceding problem or from the first line only of (165.7)). This becomes

$$R = 1 + \frac{-\frac{2}{\varkappa}\sin\delta_0 \left[\frac{\cos(\delta_0 + \varkappa \log\sin^2\Theta)}{\sin^2\Theta} + \frac{\cos(\delta_0 + \varkappa \log\cos^2\Theta)}{\cos^2\Theta} \right] + \frac{4}{\varkappa^2}\sin^2\delta_0}{\frac{1}{\sin^4\Theta} + \frac{1}{\cos^4\Theta} - \frac{\cos(\varkappa \log\tan^2\Theta)}{\sin^2\Theta\cos^2\Theta}} \qquad (165.9)$$

This formula holds upwards to such energies where ^3P scattering begins to play a role, i.e. upwards to a few MeV proton energy.

Eq. (165.9) gives $R=1$ for $\Theta=0°$ and for $\Theta=90°$ where the singularity of Coulomb scattering outgrows the finite anomaly.

More significance attaches to the value of R at the angle $\Theta=45°$ where we get

$$R(45°)=1-\frac{2}{\varkappa}\sin\delta_0\cos(\delta_0-\log 2)+\frac{1}{\varkappa^2}\sin^2\delta_0. \qquad (165.10)$$

Let us first discuss this expression for a rather small proton energy, say, of 250 keV. Then $\varkappa=0.316$ is still rather large and δ_0 very small, so that the second term in (165.10) by far outweighs the third one. Observation at 45° then easily decides on the sign of the additional force: If it is an attraction, $\delta_0>0$ and $R(45°)<1$; if it is a repulsion, $\delta_0<0$ and $R(45°)>1$. Experiment shows that the nuclear force is attractive.

Let us now go on to higher energies, say, to 1 MeV ($\varkappa=0.158$). Fig. 65 shows the value of $R(45°)$ then to be expected for different positive and negative values of δ_0, according to (165.10). Since we

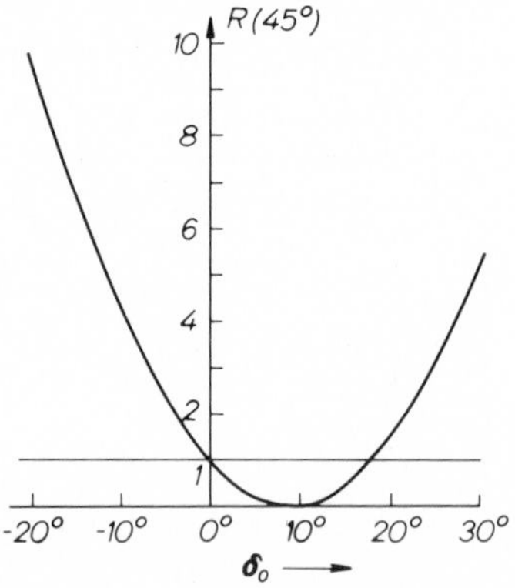

Fig. 65. Anomalous proton-proton scattering at 1 MeV. The ratio of actual to pure Coulomb scattering at $\Theta=45°$ is shown as a function of the phase angle δ_0. Positive (negative) values of δ_0 correspond to short-range attraction (repulsion)

have already decided in favour of $\delta_0>0$ (attraction), there is a unique determination of δ_0 if $R(45°)>1$. This really is observed at 1 MeV, viz. $R(45°)=4.6$. The phase angle then turns out to be $\delta_0=32°$. In this way δ_0 can be uniquely determined as a function of energy.

The ratios R at other angles of deflection may then be computed from (165.9) with the value of δ_0 so determined, and the whole curve of angular distribution for each energy be compared with experiment. Thus the theory can be corroborated in more detail. Perfect agreement has thus been obtained.

Literature. Blatt, J. M., Jackson, J. D.: Rev. Modern Phys. **22,** 77 (1950). — Flügge, S.: Ergebn. exakt. Naturwiss. **26,** 165 (1952).

Problem 166. Inelastic scattering

A beam of protons is passing a target consisting of alkali atoms. Excitation cross sections for the outermost atomic electron, originally in its ground state, shall be determined by treating the proton-atom interaction as a perturbation. Momentum transfer to the atomic core shall be neglected (infinitely heavy nucleus).

Solution. Let us use atomic units throughout ($\hbar=1$, $e=1$, $m=1$) and let $\boldsymbol{r}_1$ be the position vector of the proton, $\boldsymbol{r}_2$ of the electron. The hamiltonian may then be decomposed into three parts,

$$H = H_1 + H_2 + H_{12} \tag{166.1}$$

with

$$H_1 = -\frac{1}{2M}\nabla_1^2 \tag{166.2}$$

describing force-free motion of a proton of mass M,

$$H_2 = -\tfrac{1}{2}\nabla_2^2 + V(r_2) \tag{166.3}$$

describing the motion of the outermost electron in the field of the atomic core, and

$$H_{12} = -V(r_1) - \frac{1}{r_{12}} \tag{166.4}$$

describing the interaction of the proton with the atomic core and the electron to be excited. This last term of the hamiltonian shall be regarded as a perturbation. Such a formulation of the problem is reasonable as long as the energy of the proton is not big enough to excite any one of the core electrons, so that we may simply deal with only the outermost electron and a rigid core.

Let the eigenfunctions of the operator H_2 be u_ν (where the subscript ν comprises the three quantum numbers n, l, m, and $\nu=0$ denotes the ground state) with eigenvalues W_ν,

$$H_2 u_\nu(\boldsymbol{r}_2) = W_\nu(\boldsymbol{r}_2), \tag{166.5}$$

and let $\hbar \boldsymbol{k}$ be the momentum of an undeflected beam proton. Then the zero-order solution of the Schrödinger equation, neglecting H_{12}, is

$$U^0(\boldsymbol{r}_1, \boldsymbol{r}_2) = \mathrm{e}^{i\boldsymbol{k}\cdot\boldsymbol{r}_1} u_0(\boldsymbol{r}_2). \tag{166.6}$$

The first-order solution may be written as an expansion with respect to the complete orthogonal set $\{u_\nu\}$, viz.

$$U(\boldsymbol{r}_1, \boldsymbol{r}_2) = \mathrm{e}^{i\boldsymbol{k}\cdot\boldsymbol{r}_1} u_0(\boldsymbol{r}_2) + \sum_\mu{}' F_\mu(\boldsymbol{r}_1) u_\mu(\boldsymbol{r}_2) \tag{166.7}$$

where $\mu=0$ is excluded from, and the integral over the continuum functions included in, the sum.

Putting (166.7) in the Schrödinger equation we arrive at

$$\sum_\mu{}' \{\nabla_1^2 F_\mu + [k^2 - 2M(W_\mu - W_0)] F_\mu - 2MH_{12} F_\mu\} u_\mu(\boldsymbol{r}_2) = 2MH_{12} \mathrm{e}^{i\boldsymbol{k}\cdot\boldsymbol{r}_1} u_0(\boldsymbol{r}_2).$$

In the first order of perturbation we neglect H_{12} on the left-hand side. With the notation

$$k_\mu^2 = k^2 - 2M(W_\mu - W_0) \tag{166.8}$$

we then have

$$\sum_\mu{}' \{\nabla_1^2 F_\mu(\boldsymbol{r}_1) + k_\mu^2 F_\mu(\boldsymbol{r}_1)\} u_\mu(\boldsymbol{r}_2) = 2MH_{12} \mathrm{e}^{i\boldsymbol{k}\cdot\boldsymbol{r}_1} u_0(\boldsymbol{r}_2). \tag{166.9}$$

Multiplying (166.9) by $u_\nu^*(\boldsymbol{r}_2)$ and integrating over $\boldsymbol{r}_2$ we obtain a set of independent differential equations for the F_ν's:

$$\nabla_1^2 F_\nu + k_\nu^2 F_\nu = \Phi_\nu(\boldsymbol{r}_1) \tag{166.10}$$

with

$$\Phi_\nu(\boldsymbol{r}_1) = 2M \mathrm{e}^{i\boldsymbol{k}\cdot\boldsymbol{r}_1} \int d^3 r_2\, u_\nu^*(\boldsymbol{r}_2) H_{12} u_0(\boldsymbol{r}_2). \tag{166.11}$$

Eq. (166.10) is an inhomogeneous equation and can be solved by using a Green's function,

$$F_\nu(\boldsymbol{r}_1) = -\frac{1}{4\pi} \int d^3 r' \frac{\mathrm{e}^{ik_\nu|\boldsymbol{r}_1 - \boldsymbol{r}'|}}{|\boldsymbol{r}_1 - \boldsymbol{r}'|} \Phi_\nu(\boldsymbol{r}'). \tag{166.12}$$

In order to derive cross sections from the solution, we next have to study the asymptotic behaviour of (166.12) for $r_1 \to \infty$. The integral (166.11) in $\Phi_\nu(\boldsymbol{r}_1)$ decreases as $1/r_1^2$ with large r_1 since H_{12} is the interaction of a proton with the neutral atom, which according to (166.4), for $r_1 \gg r_2$, tends towards—$(\boldsymbol{r}_1 \cdot \boldsymbol{r}_2)/r_1^3$. The factor $\Phi_\nu(\boldsymbol{r}')$ in (166.12) thus practically limits the integration domain in (166.12) to atomic dimensions. For $r_1 \to \infty$ therefore $r_1 \gg r'$ may be supposed so that

$$F_\nu(\boldsymbol{r}_1) \to -\frac{\mathrm{e}^{ik_\nu r_1}}{4\pi r_1} \int d^3 r'\, \mathrm{e}^{-ik_\nu r' \cos(\boldsymbol{r}_1, \boldsymbol{r}')} \Phi_\nu(\boldsymbol{r}'). \tag{166.13}$$

This is an outgoing spherical wave,

$$F_\nu(\boldsymbol{r}_1) \to f(\vartheta_1) \frac{e^{ik_\nu r_1}}{r_1} \tag{166.14}$$

with a scattering amplitude

$$f(\vartheta_1) = -\frac{1}{4\pi} \int d^3r' \, e^{-ik_\nu r' \cos(\boldsymbol{r}_1, \boldsymbol{r}')} \Phi_\nu(\boldsymbol{r}') \tag{166.15}$$

still depending upon the direction in which the proton is finally deflected. It follows that the differential inelastic cross section for the proton, under excitation of the outermost alkali electron to the state ν, then is

$$d\sigma = \frac{k_\nu}{k} |f(\vartheta_1)|^2 \, d\Omega_1 \tag{166.16}$$

because the velocity of the outgoing protons, and thence their current, are lowered by the factor k_ν/k. This is an immediate consequence of interpreting (166.8) as energy conservation law.

Let us now go into some more details with respect to the angular distribution of the inelastically scattered protons. In the exponent of (166.15) we introduce a vector $\boldsymbol{k}_\nu$ in the direction of $\boldsymbol{r}_1$ so that

$$k_\nu r' \cos(\boldsymbol{r}_1, \boldsymbol{r}') = \boldsymbol{k}_\nu \cdot \boldsymbol{r}'$$

and, using (166.11) and (166.4), we get

$$f(\vartheta_1) = -\frac{M}{2\pi} \int d^3r' \, e^{i(\boldsymbol{k}-\boldsymbol{k}_\nu)\cdot \boldsymbol{r}'} \int d^3r_2 \, u_\nu^*(\boldsymbol{r}_2) \left\{ -V(r') - \frac{1}{|\boldsymbol{r}'-\boldsymbol{r}_2|} \right\} u_0(\boldsymbol{r}_2).$$

It is obvious that the term with $V(r')$ does not contribute to inelastic collisions ($\nu \neq 0$). This follows from orthogonality as well as from the physical fact that interaction of the proton with the core cannot excite an electron which forms no part of the core. Further introducing a vector $\boldsymbol{K}_\nu = \boldsymbol{k} - \boldsymbol{k}_\nu$ for the change of momentum we then arrive at

$$f(\vartheta_1) = \frac{M}{2\pi} \int d^3r_2 \, u_\nu^*(\boldsymbol{r}_2) u_0(\boldsymbol{r}_2) \int d^3r' \, e^{i\boldsymbol{K}_\nu \cdot \boldsymbol{r}'} \frac{1}{|\boldsymbol{r}'-\boldsymbol{r}_2|}. \tag{166.17}$$

The integrand of the inner integral in this formula consists of two factors, both of which may be expanded into spherical harmonics of the angles ϑ' between $\boldsymbol{K}_\nu$ and $\boldsymbol{r}'$, and ϑ_2' between $\boldsymbol{r}'$ and $\boldsymbol{r}_2$:

$$e^{i\boldsymbol{K}_\nu\cdot\boldsymbol{r}'}=\frac{1}{K_\nu r'}\sum_{l=0}^{\infty}\sqrt{4\pi(2l+1)}\,i^l j_l(K_\nu r')\,Y_{l,0}(\vartheta')$$

and

$$\frac{1}{|\boldsymbol{r}'-\boldsymbol{r}_2|}=\sum_{n=0}^{\infty}\sqrt{\frac{4\pi}{2n+1}}\,R_n\,Y_{n,0}(\vartheta_2')$$

with

$$\left.\begin{aligned} R_n&=\frac{1}{r_2}\left(\frac{r'}{r_2}\right)^n \quad \text{if } r'<r_2,\\ R_n&=\frac{1}{r'}\left(\frac{r_2}{r'}\right)^n \quad \text{if } r'>r_2.\end{aligned}\right\} \tag{166.18}$$

Referring all angles to the direction of $\boldsymbol{K}_\nu$ as polar axis, we may apply the addition theorem of spherical harmonics to $Y_{n,0}(\vartheta_2')$, viz.

$$Y_{n,0}(\vartheta_2')=\sqrt{\frac{4\pi}{2n+1}}\sum_{m=-n}^{+n}Y^*_{n,m}(\vartheta',\varphi')\,Y_{n,m}(\vartheta_2,\varphi_2)$$

with ϑ',φ' defining the direction of $\boldsymbol{r}'$ and ϑ_2,φ_2 that of $\boldsymbol{r}_2$ with respect to $\boldsymbol{K}_\nu$. Then, in the inner integral of (166.17), the angular integrations may be performed and we get

$$\int d^3r'\,e^{i\boldsymbol{K}_\nu\cdot\boldsymbol{r}'}\frac{1}{|\boldsymbol{r}'-\boldsymbol{r}_2|}=\sum_{l=0}^{\infty}\int_0^{\infty}dr'\,r'^2\frac{4\pi}{K_\nu r'}\sqrt{\frac{4\pi}{2l+1}}\,i^l j_l(K_\nu r')R_l\,Y_{l,0}(\vartheta_2).$$

Using the abbreviation

$$g_l(r_2)=\sqrt{\frac{4\pi}{2l+1}}\,i^l\int_0^{\infty}dr'\,r'^2\frac{j_l(K_\nu r')}{K_\nu r'}R_l \tag{166.19}$$

Eq. (166.17) becomes

$$f(\vartheta_1)=2M\int d^3r_2\,u^*_\nu(\boldsymbol{r}_2)u_0(\boldsymbol{r}_2)\sum_{l=0}^{\infty}g_l(r_2)\,Y_{l,0}(\vartheta_2). \tag{166.20}$$

The integral (166.19) can be explicitly evaluated. Using $y=K_\nu r'$ as integration variable and putting $x=K_\nu r_2$, we find with (166.18),

$$g_l(r_2)=\sqrt{\frac{4\pi}{2l+1}}\,i^l\frac{1}{K_\nu^2}\left\{x^{-l-1}\int_0^x dy\,y^{l+1}j_l(y)+x^l\int_x^{\infty}dy\,y^{-l}j_l(y)\right\}.$$

These integrals are well known from the theory of Bessel functions:

$$\int_0^x dy\, y^{l+1} j_l(y) = x^{l+1} j_{l+1}(x);$$

$$\int_x^\infty dy\, y^{-l} j_l(y) = x^{-l} j_{l-1}(x)$$

so that the curly bracket above becomes

$$j_{l+1}(x) + j_{l-1}(x) = (2l+1)\frac{j_l(x)}{x}.$$

Hence,

$$g_l(r_2) = \sqrt{4\pi(2l+1)}\, i^l \frac{j_l(K_\nu r_2)}{K_\nu^3 r_2}. \tag{166.21}$$

In the evaluation of the scattering amplitude (166.20), we can go one step farther. We know that the states u_ν may be factorized with a spherical harmonic. Since u_0, the ground state, does not depend on angles, there are, therefore, in (166.20) products of two spherical harmonics to be integrated. Orthogonality then selects only one term of the sum (166.20). With

$$u_\nu \equiv \frac{1}{r_2} \chi_{n,l}(r_2)\, Y_{l,m}(\vartheta_2, \varphi_2) \tag{166.22a}$$

where $\boldsymbol{K}_\nu$ has been used as axis of quantization, and

$$u_0 = \frac{1}{r_2} \chi_0(r_2) \tag{166.22b}$$

as the ground state, the scattering amplitude becomes

$$f(\vartheta_1) = \frac{2M}{K_\nu^2} \delta_{m0} \sqrt{4\pi(2l+1)}\, i^l \int_0^\infty dr_2\, \chi_{n,l}(r_2)\, \chi_0(r_2) \frac{j_l(K_\nu r_2)}{K_\nu r_2}. \tag{166.23}$$

The states of different m are degenerate. Only such a linear combination of them can be excited that its angular momentum has no component about the direction of the momentum transfer vector K_ν. No selection rules exist for l. It should further be remarked that the expression (166.23) still depends upon the angle of deflection, ϑ_1, of the proton since K_ν does,

$$K_\nu^2 = k^2 + k_\nu^2 - 2k k_\nu \cos\vartheta_1 \tag{166.24}$$

k_ν being derived from the energy law (166.8) and independent of ϑ_1.

NB. The same results can be obtained by applying the Golden Rule (Problem 183) to this process.

B. Very Many Particles: Quantum Statistics

Problem 167. Electron gas in a metal

In a rough approximation, the conduction electrons in a metal may be treated as freely moving with potential walls at the surface which prevent their leaving the metal. In a cube of silver (density $\rho = 10.5\ \mathrm{gm/cm^3}$, atomic weight 108, one conduction electron per silver ion) there shall be determined

a) the highest electron energy occurring in the ground state,
b) the average kinetic electron energy,
c) the pressure of the electron gas.

Temperature excitation may be neglected.

Solution. In a silver cube of volume L^3 the possible electron energies, according to Problem 18, are given by

$$E = \frac{\hbar^2 \pi^2}{2mL^2}(n_1^2 + n_2^2 + n_3^2) \tag{167.1}$$

with n_1, n_2, n_3 positive integers $(= 1, 2, 3, \ldots)$. In each state described by a triple of quantum numbers (n_1, n_2, n_3) there are two electrons of opposite spin directions, according to the Pauli principle. Since in our metal cube a great many electrons have to be distributed, we shall essentially have large values of the quantum numbers.

Let us consider a space with coordinates n_1, n_2, n_3. Each lattice point with integer coordinates in its first octant corresponds to a state of energy (167.1). If we denote the radius from the origin in this space by n, so that

$$n_1^2 + n_2^2 + n_3^2 = n^2, \tag{167.2}$$

a spherical shell between radii n and $n + dn$ in this octant will contain

$$\frac{1}{8} \cdot 4\pi n^2\, dn = \frac{\pi}{2} n^2\, dn$$

integer lattice points. Occupying each of them with two electrons of opposite spins, we have $\pi n^2\, dn$ electrons between n and $n + dn$. As, on the other hand, the energy (167.1) depends only upon n,

$$E = \frac{\hbar^2 \pi^2}{2mL^2} n^2; \qquad dE = \frac{\hbar^2 \pi^2}{mL^2} n\, dn \tag{167.3}$$

there will be, in the interval dn,

$$dN = \pi n^2 \, dn = \pi \sqrt{\frac{2 m L^2 E}{\hbar^2 \pi^2}} \cdot \frac{m L^2}{\hbar^2 \pi^2} \, dE$$

or

$$dN = \sqrt{2m} \, \frac{m L^3}{\pi^2 \hbar^3} \sqrt{E} \, dE \qquad (167.4)$$

electrons with energies between E and $E + dE$.

a) The highest electron energy in the ground state of this electron gas, ζ, follows from the total number of electrons, N, in the cube to be a given constant:

$$N = \sqrt{2m} \, \frac{m L^3}{\pi^2 \hbar^3} \int_0^{\zeta} \sqrt{E} \, dE . \qquad (167.5)$$

Let us denote the electron density by

$$\mathcal{N} = \frac{N}{L^3}, \qquad (167.6)$$

then we arrive at a formula independent of the volume of the metal considered:

$$\mathcal{N} = \frac{1}{3\pi^2} \left(\frac{2 m \zeta}{\hbar^2} \right)^{\frac{3}{2}} \qquad (167.7\text{a})$$

or

$$\zeta = \frac{\hbar^2}{2m} (3 \pi^2 \mathcal{N})^{\frac{2}{3}} . \qquad (167.7\text{b})$$

As $\mathcal{N} = \rho / M$ with M the mass of a silver atom ($M = 1.80 \times 10^{-22}$ gm) we have

$$\mathcal{N} = 5.85 \times 10^{22} \, \text{cm}^{-3}$$

so that Eq. (167.7b) leads to an upper energy limit of

$$\zeta = 8.80 \times 10^{-12} \, \text{erg} = 5.55 \, \text{eV} .$$

This is an energy so large compared to thermal energies ($kT = 0.026$ eV at 300 °K) that thermal excitation may indeed only slightly change the distribution of electrons over the energy states. The reason for this effect, called degeneracy of the electron gas (Fermi gas), is, of course, the very small mass of the electron in the denominator of (167.7b).

The energy limit ζ is, in general, called the *Fermi energy* of the electron gas.

b) The average energy of a conduction electron is determined by

$$\bar{E} = \int dN\, E \Big/ \int dN \tag{167.8}$$

or, according to (167.4),

$$\bar{E} = \int_0^{\zeta} dE \sqrt{E}\, E \Bigg/ \int_0^{\zeta} dE \sqrt{E} = \tfrac{3}{5}\zeta . \tag{167.9}$$

c) Pressure may always be defined, without using any thermodynamics, by the work of compression if a volume V is diminished by dV:

$$dW = p\, dV .$$

This work is used to increase the total energy content U of the gas by dU,

$$dW = dU .$$

This total energy (at $T = 0$) is

$$U = N\bar{E} = \tfrac{3}{5} N \zeta . \tag{167.10}$$

According to (167.7b), the Fermi energy ζ depends on $\mathcal{N} = N/V$ and therefore on V:

$$U \propto V^{-\frac{2}{3}}; \qquad \frac{dU}{U} = -\frac{2}{3}\frac{dV}{V} .$$

Thus we find for the pressure,

$$p = -\frac{dU}{dV} = \frac{2}{3}\frac{U}{V} = \frac{2}{5}\mathcal{N}\zeta . \tag{167.11}$$

With the numerical values of $\mathcal{N}$ and ζ determined above, this leads to a pressure of

$$p = 2.06 \times 10^{11}\ \text{dyn/cm}^2$$

or about 200,000 atmospheres. This immense pressure is counterbalanced by the strong Coulomb attraction between the conduction electrons and the lattice ions.

Problem 168. Paramagnetic susceptibility of a metal

To determine the paramagnetic susceptibility of a metal, for zero temperature, by treating the conduction electrons as a Fermi gas, neglecting the polarizability of the lattice ions.

Solution. According to the preceding problem the conduction electrons form a Fermi gas whose energy limit (the Fermi energy) is

$$\zeta = \frac{\hbar^2}{2m}(3\pi^2 \mathcal{N})^{\frac{2}{3}}, \tag{168.1}$$

with $\mathcal{N}$ the number of conduction electrons per unit volume. The energy difference, ΔE, between two consecutive electron levels follows from

$$4\pi p^2 \Delta p = 4\pi\sqrt{2mE}\, m\Delta E = \frac{(2\pi\hbar)^3}{V}. \tag{168.2}$$

In the vicinity of the Fermi energy this yields (writing $\Delta E_0 \propto \zeta^{-\frac{1}{2}} = \zeta/\zeta^{\frac{3}{2}}$ and putting (168.1) for $\zeta^{\frac{3}{2}}$),

$$\Delta E_0 = \frac{4}{3}\frac{\zeta}{\mathcal{N}V}. \tag{168.3}$$

All levels $E<\zeta$ are occupied by pairs of electrons of opposite spin directions, all levels $E>\zeta$ are unoccupied, at zero temperature.

If now a magnetic field is applied to the metal, energy can be gained by separating pairs of electrons and directing the spins of each pair so that both are parallel to the field strength $\mathcal{H}$. If ν pairs are separated, the energy gain apparently becomes

$$2\nu\cdot\mu\mathcal{H} \quad \text{with} \quad \mu = \frac{e\hbar}{2mc}. \tag{168.4}$$

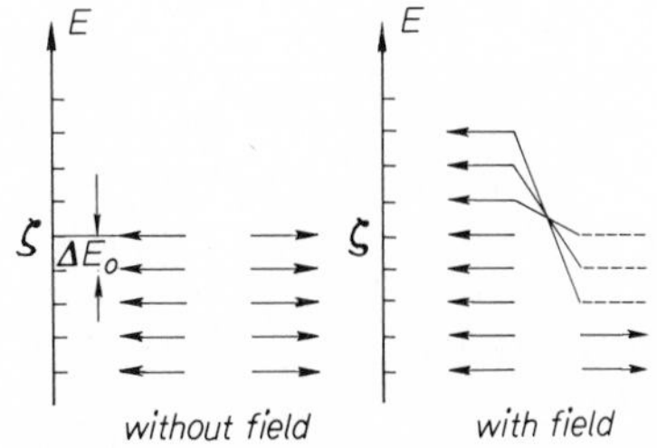

Fig. 66. Spin-flip near Fermi energy under action of a magnetic field

Such a separation is, of course, possible only by raising one electron of the pair to an unoccupied level beyond $E=\zeta$. This means, however, an expenditure of kinetic energy counteracting the gain (168.4). Fig. 66

shows that the amount of energy expended in separating the first pair (by transferring an electron from the uppermost occupied to the lowest unoccupied level) is ΔE_0, for the second pair it is $3\Delta E_0$, for the third pair $5\Delta E_0$, etc. In general, the separation of ν pairs requires an expenditure of

$$[1+3+5+\cdots+(2\nu-1)]\Delta E_0=\nu^2\Delta E_0. \tag{168.5}$$

Equilibrium will be reached if the total energy change W of the gas affected by the magnetic field,

$$W=-2\nu\cdot\mu\mathcal{H}+\nu^2\Delta E_0, \tag{168.6}$$

is a minimum:

$$\frac{dW}{d\nu}=-2\mu\mathcal{H}+2\nu\Delta E_0=0,$$

i.e. for

$$\nu=\frac{\mu\mathcal{H}}{\Delta E_0} \tag{168.7}$$

with

$$W_{\min}=-\frac{(\mu\mathcal{H})^2}{\Delta E_0}. \tag{168.8}$$

If more pairs were separated, the total energy of the gas would increase again. In equilibrium, the total magnetic moment of the metal becomes

$$\mathcal{M}=2\nu\cdot\mu=\frac{2\mu^2\mathcal{H}}{\Delta E_0}$$

and, by definition, its paramagnetic susceptibility per unit volume,

$$\chi=\frac{\mathcal{M}}{\mathcal{H}V}=\frac{2\mu^2}{V\Delta E_0}, \tag{168.9}$$

according to (168.3) and (168.1),

$$\chi=\frac{e^2}{4\pi mc^2}\left(\frac{3\mathcal{N}}{\pi}\right)^{\frac{1}{3}}. \tag{168.10}$$

To evaluate this expression numerically we may write

$$\mathcal{N}=\frac{z\rho}{m_H A},$$

expressing the electron density by the mass density ρ of the metal, the mass of one of its atoms $m_H A$ (with A the atomic weight) and its valence z. Then we find,

$$\chi_{\text{para}}=1.86\times10^{-6}\left(\frac{z\rho}{A}\right)^{\frac{1}{3}}. \tag{168.11}$$

To compare this result with experimental values, the diamagnetic susceptibility of the lattice ions has to be subtracted.

NB. In problem 160 we have calculated the diamagnetic susceptibility of neon which must be practically identical with that of Na^+. It turned out to be

$$\chi_{\text{dia}} = -5.61 \times 10^{-6}\ \text{cm}^3/\text{mole}$$

which, with a density of sodium metal of about $1\ \text{gm/cm}^3 = \frac{1}{23}\ \text{mole/cm}^3$ is the same as

$$\chi_{\text{dia}} = -0.25 \times 10^{-6}$$

in the dimensionless scale used in the present problem. From (168.11), on the other hand, we obtain the electron contribution

$$\chi_{\text{para}} = +0.66 \times 10^{-6}.$$

The expressions are of the same order of magnitude so that in some metals (e.g. in caesium) even a resultant diamagnetism is observed.

Literature. Frenkel, J.: Z. Physik **49**, 31 (1928).

Problem 169. Field emission, uncorrected for image force

To determine the electron current emitted from a metallic surface under the action of a high electric field strength $\mathscr{E}$. The temperature can be supposed to be low; image force and lattice structure shall be neglected.

Solution. Let $z=0$ be the metallic surface. The interior $(z<0)$ may have constant potential energy, $V=0$, whereas the exterior $(z>0)$ has potential V_0. Inside the metal the conduction electrons form the ground state of a Fermi gas, occupying all levels up to the Fermi energy ζ. Outside, there holds the potential

$$V(z) = V_0 - e\mathscr{E} z. \tag{169.1}$$

As shown in Fig. 67 a potential barrier is formed beyond the surface. Let E_z be the part of the electron energy corresponding to its velocity

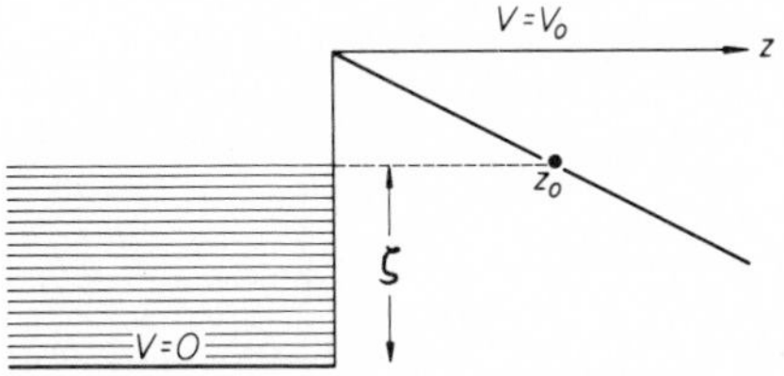

Fig. 67. Field emission of electrons from a metal surface. Left-hand side: densely lying electron levels inside the metal up to the Fermi energy ζ. Right-hand side: Potential outside the metal under action of an electrical field

component in z direction; then the potential barrier has a transmission coefficient T, to be calculated in WKB approximation,

$$T = \exp\left\{-2\frac{\sqrt{2m}}{\hbar}\int_0^{z_0} dz\sqrt{V(z)-E_z}\right\}, \tag{169.2}$$

rapidly decreasing with decreasing E_z. Here, $V(z)$ is the expression (169.1) and

$$E_z = \frac{m}{2}v_z^2; \qquad z_0 = \frac{V_0 - E_z}{e\mathscr{E}}. \tag{169.3}$$

The integration in (169.2), performed in an elementary way, yields

$$T = \exp\left\{-\frac{4}{3}\frac{\sqrt{2m}}{\hbar e\mathscr{E}}(V_0 - E_z)^{\frac{3}{2}}\right\}. \tag{169.4}$$

The electric current density (per cm^2) is

$$j = e\int dn\, v_z T \tag{169.5}$$

with dn the number of conduction electrons per cm^3 and per momentum space element $dp_x dp_y dp_z$. For the Fermi gas, there is

$$dn = 2\frac{dp_x dp_y dp_z}{h^3} \qquad (h = 2\pi\hbar)$$

inside the Fermi sphere, i.e. for

$$p_x^2 + p_y^2 + p_z^2 \le 2m\zeta, \tag{169.6}$$

and, outside it, $dn = 0$. With cylindrical coordinates ρ, φ, p_z in momentum space according to

$$p_x = \rho\cos\varphi, \qquad p_y = \rho\sin\varphi, \qquad \rho^2 + p_z^2 \le 2m\zeta,$$

the integral (169.5) then may be written

$$j = \frac{2e}{h^3}2\pi\int_0^{\sqrt{2m\zeta}} dp_z \int_0^{\sqrt{2m\zeta - p_z^2}} d\rho\, \rho(p_z/m)T,$$

the integration being extended over all electrons with $v_z > 0$. If we use the auxiliary variable

$$\varepsilon = \zeta - E_z \tag{169.7}$$

the integral simplifies to

$$j = \frac{4\pi e m}{h^3} \int_0^{\zeta} d\varepsilon\, \varepsilon\, T(\varepsilon) \tag{169.8}$$

with

$$T = \exp\left\{ -\frac{4}{3} \frac{\sqrt{2m}}{\hbar e \mathscr{E}} (V_0 - \zeta + \varepsilon)^{\frac{3}{2}} \right\}. \tag{169.9}$$

To evaluate the integral (169.8) we use the fact that, starting from $\varepsilon = 0$ (maximum energy $E_z = \zeta$), the transmission coefficient $T(\varepsilon)$ decreases rapidly with increasing ε. Therefore, mainly the electrons with small values of ε contribute to the integral (169.8) and we may expand

$$(V_0 - \zeta + \varepsilon)^{\frac{3}{2}} = (V_0 - \zeta)^{\frac{3}{2}} + \tfrac{3}{2}\varepsilon (V_0 - \zeta)^{\frac{1}{2}} + \cdots$$

With the abbreviation

$$2 \frac{\sqrt{2m}}{\hbar e \mathscr{E}} (V_0 - \zeta)^{\frac{3}{2}} = q \tag{169.10}$$

we then obtain

$$T = e^{-\frac{2}{3}q} \exp\left(-\frac{q\varepsilon}{V_0 - \zeta} \right)$$

and

$$j = \frac{4\pi e m}{h^3} e^{-\frac{2}{3}q} \int_0^{\zeta} d\varepsilon\, \varepsilon \exp\left(-\frac{q\varepsilon}{V_0 - \zeta} \right).$$

Again, the integrand falls off rapidly with increasing ε so that we may extend the integration to infinity, without noticeable error, and finally arrive at

$$j = \frac{4\pi e m}{h^3} \frac{(V_0 - \zeta)^2}{q^2} e^{-\frac{2}{3}q}. \tag{169.11}$$

Numerical values. According to Eqs. (169.10) and (169.11) the electric current falls rapidly off with decreasing field strength $\mathscr{E}$, and with increasing work function $V_0 - \zeta$. If we measure the field strength in volts/cm, the work function in eV and the current density in amp/cm^2, then we get the following numerical relations:

$$\left. \begin{aligned} q &= 1.047 \times 10^8 (V_0 - \zeta)^{\frac{3}{2}} / \mathscr{E}; \\ j &= 1.59 \times 10^{10} \frac{(V_0 - \zeta)^2}{q^2} e^{-\frac{2}{3}q}. \end{aligned} \right\} \tag{169.12}$$

For q of the order 1, a current density of the order 10^{10} amp/cm^2 might be expected, i.e. almost every electron hitting the surface would leave the metal. Of course, model and approximations are equally untenable under these extreme conditions. For larger values of q the current rapidly falls off. Thus it may not be unreasonable to ask at which field strength we may expect a current density of 1 amp/cm^2, with different values of the work function. One finds the following pairs of values:

$$\begin{array}{lll} \mathscr{E}=10^6 \text{ volt/cm} & \text{and} & V_0-\zeta=0.083 \text{ eV} \\ \quad 10^7 & & \quad 0.43 \\ \quad 10^8 & & \quad 2.19 \end{array}$$

As $V_0-\zeta$ is always of the order of several eV's in metals, one should not observe any appreciable field emission below 10^8 volts/cm. In fact, experiment shows a threshold field strength of only about 10^6 volts/cm. This wide discrepancy can certainly not be explained by temperature excitation of the Fermi gas, which would only help to lower the work function by an amount between $\frac{1}{10}$ and $\frac{1}{100}$ eV ($kT\sim\frac{1}{30}$ eV at normal temperatures). It can, however, be explained by taking account of the image force, as is shown in the next problem.

Problem 170. Field emission, corrected for image force

The potential threshold for field emission of electrons from a metal surface is essentially lowered by the image force. Its effect upon the electron current emitted shall be investigated.

Solution. The image force is originated by the distortion of the surface charge in the neighbourhood of any electron at a distance $z>0$ outside the metal. It can be calculated from classical electrostatics, neglecting effects of the metal structure if z is appreciably larger than the lattice constant, so that the metal may be treated as a continuum. It then turns out to be

$$V_{\text{image}} = -\frac{e^2}{4z}. \tag{170.1}$$

For smaller values of z the expression is rather bad, as is seen by its unphysical singularity at $z=0$. This error, however, does not affect the following considerations which depend entirely upon the height and breadth of the potential wall *above* electron energy.

This potential wall then becomes

$$V(z)=V_0-\frac{e^2}{4z}-e\mathscr{E}z \tag{170.2}$$

with the notations of the preceding problem (Fig. 68). We are interested in that part of it that extends between z_1 and z_2, the two solutions of the quadratic equation $V(z)=E_z$. This yields

$$z_{1,2}=\frac{V_0-E_z}{2e\mathscr{E}}\pm\sqrt{\frac{(V_0-E_z)^2}{4e^2\mathscr{E}^2}-\frac{e}{4\mathscr{E}}}. \qquad (170.3)$$

Both solutions are real if

$$e\mathscr{E}<\left(\frac{V_0-E_z}{e}\right)^2,$$

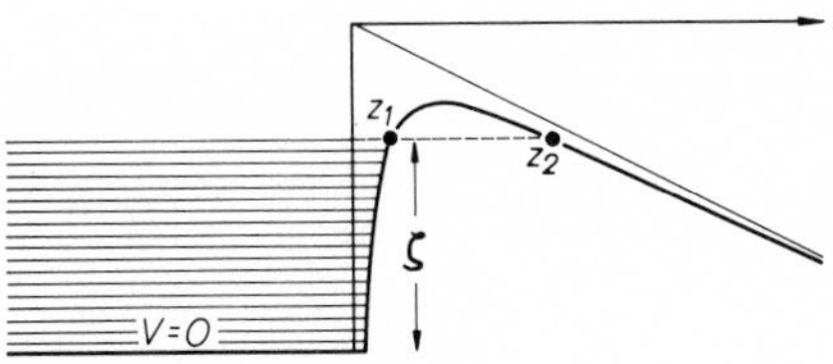

Fig. 68. Same as Fig. 67 but corrected for image force

a condition which is satisfied even for field strengths of about 10^9 volts/cm; for higher values the threshold would be submerged under the Fermi level of the electron sea. For the field strengths used in experiments, which are even below 10^7 volts/cm, we may safely assume

$$e\mathscr{E}\ll\left(\frac{V_0-E_z}{e}\right)^2 \qquad (170.4)$$

and accordingly expand the radical in (170.3). The results are

$$z_1=\frac{e^2}{4(V_0-E_z)} \quad \text{and} \quad z_2=z_0-z_1 \qquad (170.5)$$

with

$$z_0=\frac{V_0-E_z}{e\mathscr{E}} \qquad (170.6)$$

and $z_1 \ll z_2$. The summit of the barrier, according to Eq. (170.2), is now, due to the image force, shifted from $z=0$ to $z=\sqrt{e/4\mathscr{E}}$ and instead of V_0 its height is only

$$V_0-e\sqrt{e\mathscr{E}}.$$

According to (170.4) this does not involve much lowering of the threshold but rather a flattening of the summit, so that we may expect a much larger transmission coefficient. Only the neighbourhood of $E_z = \zeta$ will contribute appreciably to the emission current (as in the preceding problem); for this energy $z_1 = e^2/[4(V_0 - \zeta)]$ is at least of the order of the lattice constant, so that the singularity of the image force potential at $z=0$ will become a matter of indifference.

Again using the WKB approximation, we find the transmission coefficient

$$T = \exp\left\{-2\frac{\sqrt{2m}}{\hbar}\int_{z_1}^{z_2} dz\sqrt{V(z)-E_z}\right\}$$

or, writing $V(z)$ in the form

$$V(z) = \frac{e\mathscr{E}}{z}(z-z_1)(z_2-z),$$

we have

$$-\frac{\hbar}{2\sqrt{2me\mathscr{E}}}\log T = \int_{z_1}^{z_2} dz\sqrt{\frac{(z-z_1)(z_2-z)}{z}}. \tag{170.7}$$

This integral is of elliptic type and may be reduced to standard integrals as follows: use, instead of z, the variable $x=(z-z_1)/(z_2-z_1)$, thus transforming the integral into

$$(z_2-z_1)^{\frac{3}{2}}\int_0^1 dx\sqrt{\frac{x(1-x)}{x+c}} \quad \text{with } c = \frac{z_1}{z_2-z_1}.$$

Next put

$$\frac{z_2-z_1}{z_2} = k^2 \tag{170.8}$$

and

$$x = \frac{(1-k^2)\sin^2\varphi}{1-k^2\sin^2\varphi};$$

then the right-hand side of (170.7) becomes

$$\frac{2k^4}{\sqrt{1-k^2}}z_1^{\frac{3}{2}}\int_0^{\frac{\pi}{2}}\frac{d\varphi\sin^2\varphi\cos^2\varphi}{(1-k^2\sin^2\varphi)^{\frac{5}{2}}}. \tag{170.9}$$

The last integral may be reduced to the two complete elliptic integrals $E(k)$ and $K(k)$:

$$\int_0^{\frac{\pi}{2}} \frac{d\varphi \sin^2\varphi \cos^2\varphi}{(1-k^2\sin^2\varphi)^{\frac{5}{2}}} = \frac{1}{3k^4}\left[\frac{2-k^2}{1-k^2} E(k) - 2K(k)\right]. \qquad (170.10)$$

Eq. (170.10) may be proved as follows. Putting $1-k^2\sin^2\varphi = \Delta^2$, the two complete integrals in standard form are defined by

$$K(k) = \int_0^{\frac{\pi}{2}} \frac{d\varphi}{\Delta}; \qquad E(k) = \int_0^{\frac{\pi}{2}} d\varphi\, \Delta.$$

Now, there can be proved the following identity by simple differentiation,

$$3k^2 \frac{\sin^2\varphi\cos^2\varphi}{\Delta^5} = \frac{2-k^2}{k^2(1-k^2)}\Delta - \frac{2}{k^2\Delta} + \frac{d}{d\varphi}\left\{\sin\varphi\cos\varphi\left(\frac{1}{\Delta^3} - \frac{2-k^2}{(1-k^2)\Delta}\right)\right\}.$$

Integration of this identity yields directly

$$3k^2 \int_0^{\frac{\pi}{2}} d\varphi \frac{\sin^2\varphi\cos^2\varphi}{\Delta^5} = \frac{2-k^2}{k^2(1-k^2)} E(k) - \frac{2}{k^2} K(k),$$

in agreement with (170.10).

Eqs. (170.7) and (170.10) may then be unified into

$$-\frac{\hbar}{2\sqrt{2m e\mathscr{E}}} \log T = \frac{2}{3} z_2^{\frac{3}{2}} [(2-k^2)E(k) - 2(1-k^2)K(k)]. \qquad (170.11)$$

This formula may be brought into a much simpler shape. Remembering that $z_1 \ll z_2$ so that $k^2 \simeq 1$, we may replace k^2 by the parameter

$$k'^2 = 1 - k^2 = z_1/z_2 \ll 1 \qquad (170.12)$$

and expand (170.11) into an extremely well converging series in this parameter[6] according to

$$K(k) = \Lambda + \tfrac{1}{4}(\Lambda - 1)k'^2 + \cdots; \qquad E(k) = 1 + \tfrac{1}{2}(\Lambda - \tfrac{1}{2})k'^2 + \cdots$$

---

[6] Cf. Jahnke-Emde, 2nd edition (1933), p. 145.

with $\Lambda = \log(4/k')$. If these expansions are put into (170.11), the right-hand side becomes

$$\frac{2}{3} z_2^{\frac{3}{2}} \left[1 + k'^2 \left(\frac{3}{4} - \frac{3}{2} \log \frac{4}{k'}\right)\right] = \frac{2}{3} z_0^{\frac{3}{2}} \left[1 - \frac{3}{2} k'^2 \left(\frac{1}{2} + \log \frac{4}{k'}\right)\right].$$

Were $k' = 0$, we would fall back on the expression (170.9) for T without image force. Let us denote this by T_0; then the result is

$$T = T_0^{1-\lambda} \tag{170.13}$$

with

$$\lambda = \frac{3}{2} k'^2 \left(\frac{1}{2} + \log \frac{4}{k'}\right). \tag{170.14}$$

It remains to evaluate the current integral (169.8) with the new expression for T. Again, as in the preceding problem, it is essentially the vicinity of $E_z = \zeta$ that contributes to the current, so that we may expand λ at $E_z = \zeta$ or $\varepsilon = 0$ and confine ourselves to the linear term in ε. This is practically the same as putting

$$k'^2 \simeq (z_1/z_0)_{E_z=\zeta} = \frac{e^3 \mathscr{E}}{4(V_0 - \zeta)^2} \tag{170.15}$$

and then performing the integration as in Problem 169. Instead of (169.11), which we will denote by j_0, we then find

$$j = j_0 \, e^{\frac{2}{3}\lambda q} \tag{170.16}$$

with $\lambda \ll 1$; the next better approximation would add a factor

$$[1 - \tfrac{7}{3}\lambda + k'^2]^2$$

in the denominator. The essential thing, of course, is the exponential in (170.16).

Let us finally discuss a few numerical consequences. Besides the relations (169.12) we now get

$$k'^2 = 3.58 \times 10^{-8} \mathscr{E}/(V_0 - \zeta)^2$$

in the same units of volts/cm for $\mathscr{E}$ and eV's for $V_0 - \zeta$. With a reasonable value of the work function, $V_0 - \zeta = 3$ eV, and a field strength $\mathscr{E} = 10^7$ volts/cm, we then arrive at

$$q = 54.5, \quad k'^2 = 0.0397, \quad \lambda = 0.208, \quad e^{\frac{2}{3}\lambda q} = 1860,$$
$$j_0 = 0.9 \times 10^{-8} \, \text{amp/cm}^2,$$
$$j = 1.7 \times 10^{-5} \, \text{amp/cm}^2.$$

Problem 171. White dwarf

Let the temperature of a white dwarf be high enough to ionize its atoms practically completely, and low enough to neglect gas pressure and radiation pressure compared to zero-point pressure of the degenerate electron gas. (The latter assumption is rather bad.) The distribution of density through the star shall be calculated for a given total mass of the star, from the equilibrium of zero-point pressure and gravitational pressure.

Solution. In a spherical mass of gas the radial pressure gradient must be in equilibrium with the gravitational force density (barometric formula):

$$\frac{dp}{dr} = -\frac{GM_r}{r^2}\rho . \tag{171.1}$$

Here G is the gravitational constant, M_r the mass inside a sphere of radius r,

$$M_r = 4\pi \int_0^r dr' \, r'^2 \rho(r'), \tag{171.2}$$

and $\rho(r)$ is the mass density, i.e. the mass of all ions and free electrons inside $1\,\mathrm{cm}^3$ of star matter. For complete ionization with $\mathcal{N}$ electrons per c.c., there are $\mathcal{N}/Z$ ions (nuclei) so that

$$\rho = \frac{\mathcal{N}}{Z} m_H A ,$$

with $m_H A$ the mass of one neutral atom. If there are different elements, A and Z represent average values. It should be noted, however, that the ratio

$$\frac{A}{Z} = 2\alpha \tag{171.3}$$

is almost independent of the chemical composition, α varying from 1.0 to 1.3 from light to heavy elements with the one exception of hydrogen where $\alpha = \frac{1}{2}$. Therefore,

$$\rho = 2\alpha m_H \mathcal{N}, \tag{171.4}$$

depends essentially only upon the density of electrons.

The pressure of the electron gas is, according to Problem 167, its zero-point pressure,

$$p_e = \frac{2}{5} \mathcal{N} \frac{\hbar^2}{2m} (3\pi^2 \mathcal{N})^{\frac{2}{3}} . \tag{171.5}$$

The zero-point pressure of the ions, p_i, according to the proportionality with $\mathcal{N}^{\frac{5}{3}}/m$, would be much smaller:

$$p_i/p_e = (\mathcal{N}_i/\mathcal{N})^{\frac{5}{3}}\, m/m_i = Z^{-\frac{5}{3}}\,\frac{m}{m_H A}.$$

Even in hydrogen ($Z=1$, $A=1$), we would have $p_i/p_e = 1/1838$, the ratio being much smaller for all other elements. We shall, therefore, neglect p_i and identify p_e, Eq. (171.5), with the total pressure, p.

It is not quite so easy to dispense with temperature effects. Only if $\zeta \gg kT$ can we assume the gas to be extremely degenerate so that its pressure is mainly zero-point pressure. For the electrons, the Fermi energy is

$$\zeta = \frac{\hbar^2}{2m}\left(\frac{3\pi^2\rho}{2\alpha m_H}\right)^{\frac{2}{3}} = 1.64\,\mathrm{eV}\left(\frac{\rho}{\alpha}\right)^{\frac{2}{3}}.$$

This is to be compared with $kT \simeq 100$ eV at 10^6 degrees. Even with $\rho = 10^3$, both quantities would be of the same order of magnitude. The ion gas would not at all be degenerate, then, and contribute in the same order, too.—The radiation pressure is

$$p_R = 2.52 \times 10^{-15}\, T^4\ \mathrm{dyn/cm^2},$$

whereas from (171.5) and (171.4) there follows

$$p_e = 3.16 \times 10^{12} (\rho/\alpha)^{\frac{5}{3}}\ \mathrm{dyn/cm^2}.$$

With $T = 10^6$ degrees, therefore, the radiation pressure is of the order of 10^9 dyn/cm^2 which may, indeed, be neglected under the density conditions of a white dwarf.

From (171.5) and (171.4) we get the equation of state,

$$p = f\rho^{\frac{5}{3}}; \quad f = \frac{\hbar^2}{10m}\left(\frac{3\pi^2}{2}\right)^{\frac{2}{3}} (\alpha m_H)^{-\frac{5}{3}}$$
$$= 3.17 \times 10^{12}\, \alpha^{-\frac{5}{3}}\, \mathrm{gm}^{-\frac{2}{3}}\, \mathrm{cm}^4\, \mathrm{sec}^{-2}. \qquad (171.6)$$

Any connection between pressure and density of the form

$$p = f\rho^{1+\frac{1}{n}}$$

is called a *polytrope* of index n. The white dwarf, therefore, is built according to a polytrope of index $n = \frac{3}{2}$.

Putting (171.6) in the equilibrium condition (171.1) we get

$$\frac{5}{3} f \rho^{\frac{2}{3}} \frac{d\rho}{dr} = -\frac{G\rho}{r^2} M_r$$

or, by differentiation,

$$\frac{5f}{3G}\frac{d}{dr}\left(r^2 \rho^{-\frac{1}{3}} \frac{d\rho}{dr}\right) = -4\pi r^2 \rho. \qquad (171.7)$$

Using, instead of ρ, the dimensionless function

$$\varphi = (\rho/\rho_0)^{\frac{2}{3}} \tag{171.8}$$

with a constant ρ_0 and, instead of the radius r, the dimensionless variable

$$x = r/r_1 \tag{171.9}$$

choosing the unit of length, r_1, according to

$$r_1^2 = \frac{5f}{8\pi G} \rho_0^{-\frac{1}{3}}, \tag{171.10}$$

the differential equation (171.7) is reshaped into

$$\frac{d^2\varphi}{dx^2} + \frac{2}{x}\frac{d\varphi}{dx} + \varphi^{\frac{3}{2}} = 0, \tag{171.11}$$

independent of all physical constants. If we choose ρ_0 to be the density at the centre of the star, we have to solve Eq. (171.11) with the boundary conditions

$$\varphi(0) = 1; \quad \varphi'(0) = 0. \tag{171.12}$$

The solution of the non-linear equation (171.11) with boundary conditions (171.12) is uniquely to be obtained by numerical integration. This solution decreases monotonously, reaching $\varphi = 0$ at

$$x = X = 3.6537 \tag{171.13a}$$

where its derivative is

$$\left(\frac{d\varphi}{dx}\right)_{x=X} = -D = -0.206. \tag{171.13b}$$

According to (171.8) this zero of φ corresponds to the surface of the star, say, $r = R$. The total mass of the star therefore becomes

$$M = 4\pi \int_0^R dr\, r^2 \rho(r) = 4\pi \rho_0 r_1^3 \int_0^X dx\, x^2 \varphi^{\frac{3}{2}}.$$

This integral can be evaluated without detailed numerical knowledge of the function $\varphi(x)$ by replacing its integrand according to (171.11):

$$x^2 \varphi^{\frac{3}{2}} = -\frac{d}{dx}\left(x^2 \frac{d\varphi}{dx}\right).$$

The result is

$$M = 4\pi \rho_0 r_1^3 X^2 D = 34.5\, \rho_0 r_1^3. \tag{171.14}$$

If the mass of the star is known by observation, there exist two relations between ρ_0 and r_1, viz. Eqs. (171.10) and (171.14):

$$\left.\begin{aligned} &\rho_0^{\frac{1}{3}} r_1^2 = a \quad \text{with } a = \frac{5f}{8\pi G} = 9.46 \times 10^{18}\,\alpha^{-\frac{5}{3}}\,\text{gm}^{\frac{1}{3}}\,\text{cm}, \\ &\rho_0 r_1^3 = b \quad \text{with } b = 0.0290\,M. \end{aligned}\right\} \tag{171.15}$$

This leads to

$$r_1 = a b^{-\frac{1}{3}}; \qquad \rho_0 = b^2 a^{-3} \tag{171.16}$$

and for the radius R of the star to

$$R = r_1 X = 3.6537\, r_1 . \tag{171.17}$$

Finally, the average density is

$$\bar{\rho} = \frac{3D}{X}\rho_0 = 0.169\,\rho_0, \tag{171.18}$$

i.e. about $\frac{1}{6}$ of the central density.

Numerical example. The companion of Sirius (α Can maj), Sirius B, has a mass determined from the motions of Sirius about the centre-of-mass of this binary system. It is about the same as the mass of the sun, viz. $M = 1.94 \times 10^{33}$ gm. This leads to the following numerical values:

$$\begin{aligned} r_1 &= 2.47 \times 10^8\,\alpha^{-\frac{5}{3}}\,\text{cm}; \\ R &= 8.98 \times 10^8\,\alpha^{-\frac{5}{3}}\,\text{cm}; \\ \rho_0 &= 3.73 \times 10^6\,\alpha^5\,\text{gm cm}^{-3}; \\ \bar{\rho} &= 6.15 \times 10^5\,\alpha^5\,\text{gm cm}^{-3}. \end{aligned}$$

Since observations lead to a radius about $\frac{1}{20}$ of that of the sun, $R_0 = 6.95 \times 10^{10}$ cm, our model leads to $\alpha = 0.445$. This is not very far from $\alpha = 0.5$ for a hydrogen star, but on the wrong side. Our radius, however, is certainly too small. The neglect of a rather important part of the pressure accounts for that: the star will be inflated to a larger radius if temperature effects are accounted for.

Problem 172. Thomas-Fermi approximation

To calculate the electron density of an atom or a positive ion. To obtain a suitable approximation, it shall be supposed that regions within which the electrostatic potential varies but little contain enough electrons to justify treating them statistically.

Solution. There are two basic ideas underlying the model, one electrostatic and the other quantum statistical. Let us start with the electrostatic side of the problem. If $n(r)$ electrons are contained in a unit volume at a distance r from the atomic nucleus, the electrostatic potential originated by both nucleus and electrons satisfies the Poisson equation, the first fundamental equation of our problem:

$$\nabla^2 \Phi = 4\pi e \cdot n(r) \tag{172.1}$$

with $\rho(r) = -en(r)$ the charge density of the electron cloud. The solution of this equation is subject to the boundary conditions

$$\Phi = \frac{Ze}{r} \quad \text{for } r \to 0 \tag{172.2}$$

in the vicinity of the nuclear charge Ze, and

$$\Phi = \frac{ze}{r} \quad \text{for } r \geqq R \tag{172.3}$$

if R is the radius of the positive ion of charge ze. This radius has still to be determined.

There can be no singularity of charge density at $r = R$, so that not only the potential but the field strength as well must be continuous there. This, instead of (172.3), permits the boundary condition to be written in the form

$$\Phi(R) = \frac{ze}{R} \quad \text{and} \quad \left(\frac{d\Phi}{dr}\right)_R = -\frac{ze}{R^2}\,. \tag{172.4}$$

We now come to the second principle underlying the calculation, viz. the quantum statistical part of the problem. Considering any volume element inside the atom or ion, we find that the momentum p of an electron found there will be connected with its energy by the relation

$$E = \frac{p^2}{2m} - e\Phi(r).$$

In order to bind the electron, this energy must apparently be smaller inside the atom than the potential energy $-e\Phi(R)$ at its surface. Hence, at a distance r from the nucleus, no electron can have a momentum larger than $p_{\max}$ given by

$$\frac{p_{\max}^2}{2m} = e[\Phi(r) - \Phi(R)]. \tag{172.5}$$

Quantum statistics now couples p_{max} with the electron density $n(r)$ by the relation (cf. Problem 167)

$$n = 2 \cdot \frac{4\pi}{3} p_{max}^3 / (2\pi\hbar)^3 . \tag{172.6}$$

By comparing (172.5) and (172.6), we then arrive at the other fundamental equation

$$n(r) = \frac{1}{3\pi^2 \hbar^3} \{2me[\Phi(r) - \Phi(R)]\}^{\frac{3}{2}} . \tag{172.7}$$

From Eqs. (172.1) and (172.7), the two functions $n(r)$ and $\Phi(r)$ may, in principle, both be determined. Elimination of $n(r)$ and use of the central symmetry of the system lead to

$$\nabla^2 \Phi \equiv \frac{1}{r} \frac{d^2}{dr^2}(r\Phi) = \frac{4e}{3\pi\hbar^3} \{2me[\Phi(r) - \Phi(R)]\}^{\frac{3}{2}} .$$

Using instead of $\Phi(r)$ the dimensionless function

$$\varphi(r) = \frac{r}{Ze} [\Phi(r) - \Phi(R)] , \tag{172.8}$$

and instead of r the dimensionless variable

$$x = r/a \quad \text{with} \quad a = \left(\frac{9\pi^2}{128 Z}\right)^{\frac{1}{3}} \frac{\hbar^2}{me^2} = 0.88534 Z^{-\frac{1}{3}} \frac{\hbar^2}{me^2} \tag{172.9}$$

we obtain the universal differential equation

$$\frac{d^2 \varphi}{dx^2} = \frac{\varphi^{\frac{3}{2}}}{\sqrt{x}} , \tag{172.10}$$

the boundary conditions (172.2) and (172.4) passing over into

$$\varphi(0) = 1 \tag{172.11}$$

and with $X = R/a$,

$$\varphi(X) = 0; \quad X \varphi'(X) = -\frac{z}{Z} . \tag{172.12}$$

It should be noted that with these boundary conditions all $Z - z$ electrons are indeed enclosed within the sphere of radius R. This can easily be shown by first deriving from (172.7) and (172.9),

$$4\pi \int_0^R n(r) r^2 dr = Z \int_0^X dx \sqrt{x}\, \varphi(x)^{\frac{3}{2}} ,$$

and then using the differential equation (172.10) to replace $\varphi^{\frac{3}{2}}$ by φ'':

$$= Z\int_0^X dx\, x\,\varphi'' = Z[x\varphi' - \varphi]_0^X = Z\{\varphi(0) + X\,\varphi'(X)\}$$

or with the boundary conditions (172.11) and (172.12),

$$= Z\left(1 - \frac{z}{Z}\right) = Z - z$$

which is indeed the number of electrons.

The problem is thus reduced to the integration of (172.10) with the boundary conditions (172.11) and (172.12). To obtain a general survey of the diversity of solutions, the differential equation may be integrated with $\varphi(0)=1$ and different initial tangent inclinations $\varphi'(0)<0$. In Fig. 69 four such solutions have been drawn. The lines (1) and (2) lead to finite radii X_1, X_2. Since $\varphi'(X)<0$ for these solutions, according to Eq. (172.12), they belong to positive ions. For a neutral atom, (172.12) gives $\varphi'(X)=0$, which is impossible for a finite X, thus leading to line (3) of Fig. 69 with an infinitely large atomic radius. Solutions of type (4)

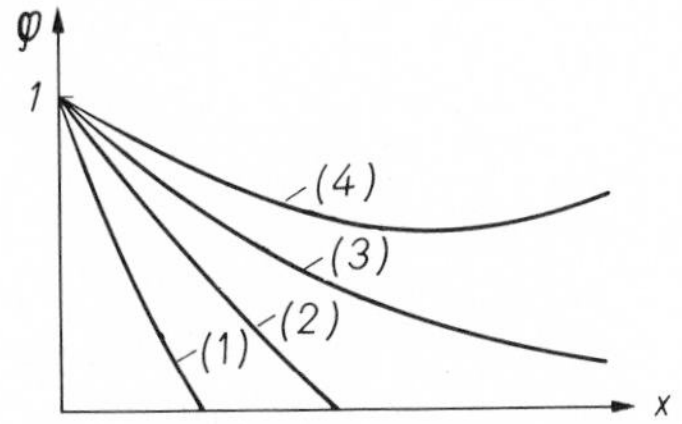

Fig. 69. Solutions of Thomas-Fermi equation (172.10) with different initial tangents

have no direct physical significance for free atoms or ions but may well serve for the description of atoms bound in a crystal lattice under changed boundary conditions.

Our main interest will be concentrated on line (3) of neutral atoms. We shall call this the *standard solution* $\varphi_0(x)$. It is numerically given in the accompanying table. It belongs to the initial tangent inclination $\varphi_0'(0) = -1.58807$. Its asymptotic behaviour is given by $\varphi_0(x) \to 144/x^3$ (which, by the way, exactly satisfies the differential equation but has a singularity at $x=0$); for practical purposes, however, this expression is rather useless since, even at so large a value as $x=100$, is still differs about 40% from $\varphi_0(x)$. On the other hand, φ_0 should show a much steeper decrease at large x, something certainly of an exponential type. The error of the Thomas-Fermi approximation, as of any statistical model, rapidly increases with small particle numbers, and since the latter decrease below every limit at large distances the method cannot be expected to hold there any longer, whatever good results it may give for the inner parts of the atom.

To obtain other solutions, not too far removed from the standard one, we write

$$\varphi(x) = \varphi_0(x) + k\eta_0(x) \tag{172.13}$$

and linearize in the small deviation $k\eta_0$. From (172.10) we then find

$$\frac{d^2\eta_0}{dx^2} = \frac{3}{2}\left(\frac{\varphi_0}{x}\right)^{\frac{1}{2}} \eta_0 . \tag{172.14}$$

In order to satisfy the boundary condition (172.11), we have $\eta_0(0)=0$. We further standardize η_0 by choosing $\eta_0'(0)=1$, satisfying the boundary conditions (172.12) by a suitable choice of the parameter k to be gathered from any one of the following relations:

$$k = -\frac{\varphi_0(X)}{\eta_0(X)}; \tag{172.15a}$$

$$k = -\frac{1}{\eta_0'(X)}\left(\frac{1}{ZX} + \varphi_0'(X)\right); \tag{172.15b}$$

$$k = \varphi'(0) - \varphi_0'(0). \tag{172.15c}$$

Eq. (172.15a) permits a simple relation to be found between k and the ionic radius X. The function $\eta_0(x)$ and its derivative $\eta_0'(x)$ are shown on the table.

x	$\varphi_0(x)$	$-\varphi_0'(x)$	$\eta_0(x)$	$\eta_0'(x)$
0.00	1.0000	1.5881	0.0000	1.0000
0.02	0.9720	1.3093	0.0200	1.0028
0.04	0.9470	1.1991	0.0401	1.0079
0.06	0.9238	1.1177	0.0604	1.0144
0.08	0.9022	1.0516	0.0807	1.0220
0.10	0.8817	0.9954	0.1012	1.0306
0.2	0.7931	0.7942	0.2069	1.0846
0.3	0.7206	0.6618	0.3186	1.1528
0.4	0.6595	0.5646	0.4378	1.2321
0.5	0.6070	0.4894	0.5654	1.3210
0.6	0.5612	0.4292	0.7023	1.4187
0.7	0.5208	0.3798	0.8494	1.5246
0.8	0.4849	0.3386	1.0075	1.6384
0.9	0.4529	0.3038	1.1773	1.7599
1.0	0.4240	0.2740	1.3597	1.8890

x	$\varphi_0(x)$	$-\varphi_0'(x)$	$\eta_0(x)$	$\eta_0'(x)$
1.2	0.3742	0.2259	1.7650	2.1696
1.4	0.3329	0.1890	2.2296	2.4805
1.6	0.2981	0.1601	2.7593	2.8222
1.8	0.2685	0.1370	3.3605	3.1954
2.0	0.2430	0.1182	4.0396	3.6012
2.2	0.2210	0.1028	4.8032	4.0406
2.4	0.2017	0.0900	5.6582	4.5149
2.6	0.1848	0.0793	6.6116	5.0253
2.8	0.1699	0.0702	7.6708	5.5730
3.0	0.1566	0.0625	8.8434	6.1594
3.2	0.1448	0.0558	10.137	6.7858
3.4	0.1343	0.0501	11.561	7.4538
3.6	0.1247	0.0451	13.122	8.1646
3.8	0.1162	0.0408	14.829	8.9198
4.0	0.1084	0.0369	16.693	9.7208
4.5	0.0919	0.0293	22.09	11.93
5.0	0.0788	0.0236	28.68	14.47
5.5	0.0682	0.0192	36.62	17.34
6.0	0.0594	0.0159	46.08	20.59
6.5	0.0522	0.0132	57.27	24.23
7.0	0.0461	0.0111	70.39	28.30
7.5	0.0410	0.0095	85.64	32.81
8.0	0.0366	0.0081	103.27	37.80
8.5	0.0328	0.0070	123.52	43.29
9.0	0.0296	0.0060	146.66	49.32
9.5	0.0268	0.0053	172.94	55.92
10.0	0.0243	0.0046	202.67	63.11

Problem 173. Amaldi correction for a neutral atom

In the Poisson equation underlying the Thomas-Fermi model, it would be more correct to introduce the charge density of all but one of the electrons on the right-hand side, because the equation serves to determine the potential field in which one of the electrons moves. This correction leads to an alteration of the Thomas-Fermi model to be investigated for a neutral atom.

Solution. Instead of (172.1) we now write

$$\nabla^2 \Phi = 4\pi e \cdot \frac{Z-1}{Z} n(r) \tag{173.1}$$

where Φ is the potential field originated by $Z-1$ electrons, acting on the Z'th one. This simple correction makes no difference as to which of the electrons is to be taken to be the probe, a neglect corresponding to a nicety to the statistical picture. The boundary condition for small r is determined by the nucleus and therefore remains unchanged:

$$\Phi(r) = \frac{Ze}{r} \quad \text{for } r \to 0, \tag{173.2}$$

whereas at the atomic surface we now have

$$\Phi(R) = \frac{e}{R}; \quad \left(\frac{d\Phi}{dr}\right)_R = -\frac{e}{R^2} \tag{173.3}$$

for a neutral atom, because there remains a surplus charge e of the nucleus not screened away by the other $Z-1$ electrons, thus still acting on the one considered.

The other fundamental equation originating from quantum statistics remains unaltered, so that we still have

$$n(r) = \frac{1}{3\pi^2 \hbar^3} \{2me[\Phi(r) - \Phi(R)]\}^{\frac{3}{2}}. \tag{173.4}$$

Eliminating $n(r)$ from (173.1) and (173.4) and using again

$$\varphi = \frac{r}{Ze}[\Phi(r) - \Phi(R)], \tag{173.5}$$

the same universal differential equation

$$\frac{d^2\varphi}{dx^2} = \frac{\varphi^{\frac{3}{2}}}{\sqrt{x}} \tag{173.6}$$

will be obtained if only the variable $x = r/\tilde{a}$ is now defined by

$$\tilde{a} = a\left(1 - \frac{1}{Z}\right)^{-\frac{2}{3}} \tag{173.7}$$

with a the characteristic length of Eq. (172.9).

Since $\tilde{a} > a$ this at first sight gives the impression that the atom has been made larger by the correction, in contrast to the physical meaning of the correction in the Poisson equation that lowers the electron-electron repulsive interaction, thus leading to a stronger bond and smaller atom. This discrepancy is, however, resolved by the alteration in the boundary condition. Eq. (173.3), expressed in terms of the function $\varphi(x)$, makes the boundary conditions of the atomic surface become

$$\varphi(X) = 0 \quad \text{and} \quad X\varphi'(X) = -\frac{1}{Z}. \tag{173.8}$$

These conditions can no longer be satisfied except by a finite radius, thus more than compensating for the stretching effect of $\tilde{a}$.

Problem 174. Energy of a Thomas-Fermi atom

The total energy content of a neutral Thomas-Fermi atom shall be calculated. A differential equation for the electron density $n(r)$, resp. for the electrostatic potential $\Phi(r)$ is to be derived by a variational procedure minimizing the energy.

Solution. There are three contributions to the total energy of the atom, viz. the kinetic energy of the electrons, the potential energy $E_{\text{pot}}^{(1)}$ of their interaction with the nucleus of charge Ze, and the potential energy $E_{\text{pot}}^{(2)}$ of their mutual interaction.

The kinetic energy follows from the basic considerations of Problem 167. If $n(r)$ is the density of electrons at some place at a distance r from the nucleus, then the average kinetic energy of an electron at this place is

$$\bar{E}=\tfrac{3}{5}\zeta=\varkappa n^{\frac{2}{3}} \quad \text{with } \varkappa=\frac{3\hbar^2}{10m}(3\pi^2)^{\frac{2}{3}}. \tag{174.1}$$

The total kinetic energy of the electrons then becomes

$$E_{\text{kin}}=\int d\tau\, n(r)\bar{E}(r)$$

or

$$E_{\text{kin}}=\varkappa\int d\tau\, n^{\frac{5}{3}}. \tag{174.2}$$

The two parts of potential energy follow from electrostatics,

$$E_{\text{pot}}^{(1)}=-Ze^2\int\frac{d\tau}{r}n(r) \tag{174.3}$$

and

$$E_{\text{pot}}^{(2)}=\frac{1}{2}e^2\iint d\tau\, d\tau'\frac{n(r)n(r')}{|\mathbf{r}-\mathbf{r}'|}. \tag{174.4}$$

The total energy, i.e. the sum of (174.2), (174.3), and (174.4),

$$E=\int d\tau\left\{\varkappa n^{\frac{5}{3}}-\frac{Ze^2}{r}n+\frac{1}{2}e^2 n\int d\tau'\frac{n(r')}{|\mathbf{r}-\mathbf{r}'|}\right\}\equiv\int d\tau\,\eta \tag{174.5}$$

must be minimized by a suitable choice of $n(r)$ with the constraint

$$\int d\tau\, n(r)=Z, \tag{174.6}$$

the latter integral being the total number of electrons. This is a variational problem to be solved by

$$\delta\int d\tau(\eta+\lambda n)=0 \tag{174.7}$$

with a multiplicator λ. Putting Eq. (174.5) into (174.7) we find

$$\int d\tau\, \delta n(r) \left\{ \frac{5}{3} \varkappa n^{\frac{5}{3}} - \frac{Ze^2}{r} + e^2 \int d\tau' \frac{n(r')}{|\mathbf{r}-\mathbf{r}'|} + \lambda \right\} = 0. \quad (174.8)$$

Here, in the last term, use has been made of the fact that variation of $n(r)$ as well as of $n(r')$ in the double integral twice leads to the same result. The extremal value of E will be obtained if the curly bracket in (174.8) vanishes.

Since for an atom n only depends upon r, not upon direction, the third term in the curly bracket can by expansion into spherical harmonics be written

$$\int d\tau' \frac{n(r')}{|\mathbf{r}-\mathbf{r}'|} = \frac{4\pi}{r} \int_0^r dr'\, r'^2\, n(r') + 4\pi \int_r^\infty dr'\, r'\, n(r').$$

The variational result from (174.8) therefore becomes

$$\frac{5}{3} \varkappa n^{\frac{2}{3}} - \frac{Ze^2}{r} + \frac{4\pi e^2}{r} \int_0^r dr'\, r'^2\, n(r') + 4\pi e^2 \int_r^\infty dr'\, r'\, n(r') + \lambda = 0.$$

By differentiation with respect to r, λ can be eliminated,

$$\frac{10}{9} \kappa n^{-\frac{1}{3}} \frac{dn}{dr} + \frac{Ze^2}{r^2} - \frac{4\pi e^2}{r^2} \int_0^r dr'\, r'^2\, n(r') = 0,$$

the contributions from differentiating the integrals with respect to their limits cancelling each other out. Multiplying by r^2 and again differentiating removes the integral and leaves us with the differential equation

$$\frac{10}{9} \varkappa \frac{d}{dr} \left(r^2 n^{-\frac{1}{3}} \frac{dn}{dr} \right) = 4\pi e^2 r^2 n. \quad (174.9)$$

It is advantageous to pass over from $n(r)$ to the electrostatic potential $\Phi(r)$, as introduced in Problem 172, according to

$$n = \frac{1}{3\pi^2} \left(\frac{2me\Phi}{\hbar^2} \right)^{\frac{3}{2}} \quad (174.10)$$

which satisfies the Poisson equation $\nabla^2 \Phi = 4\pi e n$. With the further abbreviation

$$C = \frac{8}{3\pi} \frac{me^2}{\hbar^3} \sqrt{2me} \quad (174.11)$$

we then find

$$\nabla^2 \Phi = C\Phi^{\frac{3}{2}}, \tag{174.12}$$

an equation which, with

$$\nabla^2 \Phi = \frac{1}{r}\frac{d^2}{dr^2}(r\Phi)$$

has already been derived in Problem 172.

The relations (174.10) and (174.11) enable the fractional powers of n to be eliminated in the energy expressions (174.2), (174.3), (174.4) by putting

$$\text{either } n = \frac{1}{4\pi e}\nabla^2\Phi \text{ or } n = \frac{C}{4\pi e}\Phi^{\frac{3}{2}}. \tag{174.13}$$

We thus obtain

$$E_{\text{kin}} = \frac{3}{20\pi}\int d\tau\, \Phi\nabla^2\Phi; \tag{174.14a}$$

$$E_{\text{pot}}^{(1)} = -\frac{Ze}{4\pi}\int d\tau \frac{\nabla^2\Phi}{r}; \tag{174.14b}$$

$$E_{\text{pot}}^{(2)} = \frac{1}{32\pi^2}\iint d\tau\, d\tau_1 \frac{\nabla^2\Phi\,\nabla_1^2\Phi}{|\boldsymbol{r}-\boldsymbol{r}_1|}. \tag{174.14c}$$

These integrals can be considerably simplified by making use of the central symmetry and by introducing instead of $\Phi(r)$ the function

$$\varphi(r) = \frac{r}{Ze}\Phi(r). \tag{174.15}$$

We then have,

$$\nabla^2\Phi = \frac{Ze}{r}\varphi''(r)$$

and from (174.14a),

$$E_{\text{kin}} = \frac{3}{5}\int_0^\infty dr\, r^2 \cdot \frac{Ze}{r}\varphi \cdot \frac{Ze}{r}\varphi'' = \frac{3}{5}Z^2e^2\left\{-(\varphi\varphi')_{r=0} - \int_0^\infty dr\, \varphi'^2\right\}. \tag{174.16a}$$

In the same way we find from (174.14 b),

$$E_{\text{pot}}^{(1)} = -\frac{Ze}{4\pi}\int \frac{d\tau}{r}\,\frac{Ze}{r}\,\varphi'' = Z^2 e^2 \varphi'(0)\,. \qquad (174.16\,\text{b})$$

In order to evaluate the integral (174.14c), we begin with the inner integral,

$$I(r) = \frac{1}{4\pi}\int d\tau_1 \frac{\nabla_1^2 \Phi(r_1)}{|\boldsymbol{r}-\boldsymbol{r}_1|} = \frac{1}{r}\int_0^r dr_1\, r_1^2 \nabla_1^2 \Phi + \int_r^\infty dr_1\, r_1 \nabla_1^2 \Phi$$

$$= \frac{Ze}{r}\int_0^r dr_1\, r_1\, \varphi''(r_1) + Ze\int_r^\infty dr_1\, \varphi''(r_1) = \frac{Ze}{r}\{(\varphi - r\varphi')_{r=0} - \varphi(r)\}\,.$$

Then, (174.14c) leads to

$$E_{\text{pot}}^{(2)} = \frac{1}{8\pi}\int d\tau\, I(r)\nabla^2\Phi = \frac{1}{2}\,Ze\int_0^\infty dr\, r^2\, I(r)\,\frac{\varphi''}{r}$$

$$= \frac{1}{2}\,Z^2 e^2 \left\{\int_0^\infty dr\, \varphi'^2 + [r\varphi'^2]_{r=0}\right\}. \qquad (174.16\,\text{c})$$

It is convenient to use instead of r the dimensionless variable

$$x = \frac{r}{a}; \qquad a = \left(\frac{9\pi^2}{128\,Z}\right)^{\frac{1}{3}} \frac{\hbar^2}{me^2} \qquad (174.17)$$

defined in Problem 172. Since, for small r or x, we have

$$\varphi(x) = 1 - \mu x + \cdots,$$

we may, in Eqs. (174.16a–c), put

$$-(\varphi\varphi')_{r=0} = \mu/a; \qquad \left(\frac{d\varphi}{dr}\right)_{r=0} = -\mu/a; \qquad \left[r\left(\frac{d\varphi}{dr}\right)^2\right]_{r=0} = 0$$

so that finally there remain the expressions

$$\left.\begin{aligned} E_{\text{kin}} &= \frac{3}{5}\,\frac{Z^2 e^2}{a}(\mu - J);\\ E_{\text{pot}}^{(1)} &= -\frac{Z^2 e^2}{a}\,\mu;\\ E_{\text{pot}}^{(2)} &= \frac{1}{2}\,\frac{Z^2 e^2}{a}\,J \end{aligned}\right\} \qquad (174.18)$$

with

$$J = \int_0^\infty dx \left(\frac{d\varphi}{dx}\right)^2 . \tag{174.19}$$

Here, the integral J and the derivative μ are independent of Z since they depend only upon properties of the universal function $\varphi(x)$. Their numerical values can be determined from $\varphi(x)$, Problem 172:

$$\mu = -\varphi'(0) = 1{,}588 \qquad J = 0{,}454 . \tag{174.20}$$

The total energy of the atom, i.e. the sum of the three expressions (174.18),

$$E = -\frac{Z^2 e^2}{a}\left(\frac{2}{5}\mu + \frac{1}{10}J\right) = -0{,}680\frac{Z^2 e^2}{a}, \tag{174.21}$$

then becomes proportional to $Z^{\frac{7}{3}}$ because $a \propto Z^{-\frac{1}{3}}$. Numerically,

$$E = -0{,}7687\, Z^{\frac{7}{3}}\,\mathrm{Ry} = -20{,}93\, Z^{\frac{7}{3}}\,\mathrm{eV}. \tag{174.22}$$

Problem 175. Virial theorem for the Thomas-Fermi atom

To prove the virial theorem for a neutral Thomas-Fermi atom, following the procedure of Problem 151. What relation follows between μ and J of Eq. (174.20) from the virial theorem, and what can be concluded on the ratios of the three parts of energy?

Solution. By scale transformation we replace $n(r)$ by the set of functions

$$n_\lambda(r) = \lambda^3 n(\lambda r)$$

all satisfying the normalization condition

$$\int d\tau\, n_\lambda(r) = Z .$$

The energy expressions (174.2), (174.3), and (174.4) then transform as

$$E_{\mathrm{kin}}(\lambda) = \lambda^2 E_{\mathrm{kin}}; \qquad E^{(1)}_{\mathrm{pot}}(\lambda) = \lambda E^{(1)}_{\mathrm{pot}}; \qquad E^{(2)}_{\mathrm{pot}}(\lambda) = \lambda E^{(2)}_{\mathrm{pot}},$$

thus yielding

$$E(\lambda) = \lambda^2 E_{\mathrm{kin}} + \lambda E_{\mathrm{pot}}$$

which leads through $\partial E(\lambda)/\partial\lambda = 0$ for $\lambda = 1$ as in Problem 151 to the virial theorem

$$2E_{\mathrm{kin}} + E_{\mathrm{pot}} = 0. \tag{175.1}$$

If the three energy expressions (174.18) are put into this relation, we arrive at

$$\mu = \tfrac{7}{2} J \tag{175.2}$$

which is corroborated by the numerical values (174.20). The energy expressions (174.18) then may all be written in terms of J only, with the result

$$E_{\text{kin}} = \tfrac{3}{2} U; \quad E_{\text{pot}}^{(1)} = -\tfrac{7}{2} U; \quad E_{\text{pot}}^{(2)} = +\tfrac{1}{2} U \tag{175.3}$$

and

$$U = \frac{Z^2 e^2}{a} J. \tag{175.4}$$

The total energy thus becomes the sum

$$E = -\tfrac{3}{2} U \tag{175.5}$$

which again leads to the numerical results given at the end of the preceding problem. A comparison of (175.5) with the kinetic energy (175.3) corroborates the virial theorem.

Problem 176. Tietz approximation of a Thomas-Fermi field

The function

$$\tilde{\varphi}(x) = \frac{1}{(1+\alpha x)^2} \tag{176.1}$$

with a suitable value of α, independent of Z, may be used as a fair approximation to the Thomas-Fermi function $\varphi_0(x)$ for a neutral atom. The constant α shall be determined in such a way as to permit exact normalization of $\tilde{\varphi}$, and a numerical comparison shall be made of $\tilde{\varphi}$ and φ_0.

Solution. In Problem 172 it has been shown that the electron density $n(r)$ and the atomic potential

$$V(r) = -\frac{Z}{r} \varphi_0(r) \tag{176.2}$$

(in atomic units) are coupled by the relation

$$n(r) = \frac{1}{3\pi^2} (-2V)^{\frac{3}{2}}. \tag{176.3}$$

The normalization condition,

$$4\pi \int_0^\infty dr\, r^2 n(r) = Z, \tag{176.4}$$

therefore may be written

$$\frac{4}{3\pi}(2Z)^{\frac{3}{2}}\int_0^\infty dr\, r^{\frac{1}{2}}\,\varphi_0^{\frac{3}{2}} = Z. \tag{176.5}$$

This equation is satisfied exactly by the Fermi function $\varphi_0(x)$ with

$$x = r/a; \qquad a = 0.88534\, Z^{-\frac{1}{3}}. \tag{176.6}$$

We now replace φ_0 by the approximate function $\tilde{\varphi}$, Eq. (176.1), but still keep this normalization. Introduction of the integration variable $y = \alpha x = \frac{\alpha}{a} r$ then leads to

$$\frac{8\sqrt{2}}{3\pi}\sqrt{Z}\left(\frac{a}{\alpha}\right)^{\frac{3}{2}}\int_0^\infty \frac{dy\sqrt{y}}{(1+y)^3} = 1. \tag{176.7}$$

The integral can be solved by the substitution of $u = y^2$; one easily verifies

$$\int \frac{dy\sqrt{y}}{(1+y)^3} = 2\int du \frac{u^2}{(u^2+1)^3} = \frac{1}{4}\left[\frac{u(u^2-1)}{(u^2+1)^2} + \tan^{-1} u\right];$$

in the limits $0 \le y < \infty$ therefore, the integral becomes $\pi/8$ and Eq. (176.7) yields

$$\alpha = \left(\frac{2Z}{9}\right)^{\frac{1}{3}} a = 0.60570\, Z^{\frac{1}{3}}\, a, \tag{176.8}$$

or with Eq. (176.6),

$$\alpha = 0.53625. \tag{176.9}$$

In the accompanying table, the functions φ_0 and $\tilde{\varphi}$ have been compared, using this value of α.

x	$\tilde{\varphi}$	φ_0	$\tilde{\varphi}-\varphi_0$
0	1	1	0
0.1	0.9008	0.8817	+0.0191
0.2	0.8156	0.7931	+0.0225
0.5	0.6219	0.6070	+0.0149
1.0	0.4237	0.4240	−0.0003
2.0	0.2328	0.2430	−0.0102
5.0	0.0738	0.0788	−0.0050
10.0	0.0247	0.0243	+0.0004

Literature. Tietz, T.: J. Chem. Physics **25,** 787 (1956); Z. Naturforsch. **23a,** 191 (1968).—In Tietz's original papers a factor 0.64309 has been used instead of our normalizing factor 0.60570 in Eq. (176.8). Tietz's approximation therefore does not satisfy normalization. His deviations $\tilde{\varphi}-\varphi_0$, however, are somewhat smaller in the most significant region $0<x<0.5$, but are much bigger for $x>1$.

Problem 177. Variational approximation of Thomas-Fermi field

To use a set of Tietz functions

$$\tilde{\varphi}(x) = \frac{1}{(1+\alpha x)^2} \tag{177.1}$$

as trial functions with the Ritz parameter α for the approximate solution of a variational problem equivalent to the Thomas-Fermi differential equation.

Solution. The differential equation

$$\varphi'' = x^{-\frac{1}{2}}\varphi^{\frac{3}{2}} \tag{177.2}$$

may be replaced by the variational problem, to make the integral

$$J = \int_0^\infty dx\left(\tfrac{1}{2}\varphi'^2 + \tfrac{2}{5}x^{-\frac{1}{2}}\varphi^{\frac{5}{2}}\right) \tag{177.3}$$

an extremum with fixed boundary conditions $\varphi(0)=1$ and $\varphi(\infty)=0$. Putting the trial function (177.1) satisfying the boundary conditions in the integral (177.3) we have

$$J = \int_0^\infty dx\left\{\frac{2\alpha^2}{(1+\alpha x)^6} + \frac{2}{5}x^{-\frac{1}{2}}\frac{1}{(1+\alpha x)^5}\right\}.$$

For the evaluation of the integral we set $\alpha x = t^2$ in the second term; then we may use the formula

$$\int \frac{dt}{(1+t^2)^5} = \frac{1}{8}\left\{\frac{t}{(1+t^2)^4} + \frac{7}{6}\frac{t}{(1+t^2)^3} + \frac{35}{24}\frac{t}{(1+t^2)^2} + \frac{35}{16}\frac{t}{1+t^2} + \frac{35}{16}\tan^{-1}t\right\}$$

which can easily be verified. We find

$$J = \tfrac{2}{5}\left(\alpha + \tfrac{35}{128}\alpha^{-\frac{1}{2}}\right). \tag{177.4}$$

The extremum condition $dJ/d\alpha = 0$ leads to

$$\alpha = \left(\tfrac{35}{256}\right)^{\frac{2}{3}} = 0.570. \tag{177.5}$$

This value of α deviates only slightly from $\alpha=0.536$ which is found to satisfy the normalization condition

$$\int_0^\infty dx\sqrt{x}\,\varphi^{\frac{3}{2}}=1 \tag{177.6}$$

of the preceding problem; with the present value of α it yields

$$\alpha^{-\frac{3}{2}}\cdot\frac{\pi}{8}=\frac{32}{35},$$

i.e. the approximate field minimizing the integral J corresponds to $\frac{32}{35}Z$ electrons instead of to Z in the atom.

Problem 178. Screening of K electrons

To determine the screening correction to the binding energy of a K electron by using the Tietz approximation to the Thomas-Fermi model.

Solution. Suppose one of the two K electrons and one unit of the nuclear charge to be removed from the atom of charge Z. The result will be a neutral atom of charge $Z-1$. Then, add again the removed nuclear charge but neglect its influence upon the $Z-1$ remaining electrons. The result (in atomic units) is a charge distribution with the electrostatic potential

$$\Phi(r)=\frac{1}{r}+\frac{Z-1}{r}\varphi(x) \tag{178.1}$$

with $\varphi(x)$ the Thomas-Fermi function to the variable

$$x=\frac{r}{a};\qquad a=0.88534(Z-1)^{-\frac{1}{3}}. \tag{178.2}$$

The potential energy of an electron (charge -1) moving in this potential would be

$$V(r)=-\Phi(r); \tag{178.3}$$

this therefore is the potential energy field in which the removed K electron would move if replaced into the atom.

Let us now apply perturbation theory. Without screening we should have

$$V_0(r)=-\frac{Z}{r};$$

with screening we have

$$V(r) = -\frac{1}{r} + \frac{Z-1}{r}\varphi(x) = V_0(r) + \frac{Z-1}{r}[1-\varphi(x)]. \qquad (178.4)$$

The eigenvalue E_0 and eigenfunction $u_0(r)$ of the K electron without screening are

$$E_0 = -\tfrac{1}{2}Z^2; \quad u_0 = \frac{Z^{\frac{3}{2}}}{\sqrt{\pi}}\,\mathrm{e}^{-Zr}. \qquad (178.5)$$

The first-order energy shift due to screening then becomes

$$\varDelta E_s = \int d\tau\, u_0^* \frac{Z-1}{r}(1-\varphi)u_0$$

or, after inserting u_0,

$$\varDelta E_s = 4Z^3(Z-1)\int_0^\infty dr\, r\,\mathrm{e}^{-2Zr}[1-\varphi(x)]. \qquad (178.6)$$

We now are prepared to introduce the Tietz approximation (Problem 176),

$$\varphi(x) = \frac{1}{(1+\alpha x)^2}; \quad \alpha = 0.53625 \qquad (178.7)$$

To evaluate the integral (178.6) we then use the auxiliary variable

$$t = \beta(1+\alpha x) \qquad (178.8)$$

with

$$\beta = 2Z\frac{a}{\alpha} = 3.302\,Z(Z-1)^{-\frac{1}{3}} \qquad (178.9)$$

thus obtaining

$$\varDelta E_s = Z(Z-1)\mathrm{e}^{\beta}\int_\beta^\infty dt\,\mathrm{e}^{-t}\left(t-\beta-\frac{\beta^2}{t}+\frac{\beta^3}{t^2}\right). \qquad (178.10)$$

The exponential integral occurring here

$$E_1(\beta) = \int_\beta^\infty \frac{dt}{t}\,\mathrm{e}^{-t}, \qquad (178.11)$$

is a well-known function whose asymptotic behaviour at large values of β is described by the semiconvergent series

$$E_1(\beta) = \frac{e^{-\beta}}{\beta}\left\{1 - \frac{1!}{\beta} + \frac{2!}{\beta^2} - \frac{3!}{\beta^3} \pm \cdots\right\}. \tag{178.12}$$

The integral over the last term in (178.10) may be reduced to this function,

$$E_2(\beta) = \int_\beta^\infty \frac{dt}{t^2} e^{-t} = \frac{e^{-\beta}}{\beta} - E_1(\beta). \tag{178.13}$$

Thus we find

$$\Delta E_s = Z(Z-1)\{1+\beta^2 - \beta^2(1+\beta)e^\beta E_1(\beta)\}, \tag{178.14}$$

or by expansion, according to (178.12), for $\beta \gg 1$, the semiconvergent series

$$\Delta E_s = Z(Z-1)\frac{4}{\beta}\left(1 - \frac{9/2}{\beta} + \frac{24}{\beta^2} - \frac{150}{\beta^3} \pm \cdots\right). \tag{178.15}$$

Turning now to the numerical evaluation of the energy shift ΔE_s, we find β from (178.9) indeed to be large (cf. table next page) so that (178.15) is a reasonable approximation. In the column marked $(1-\cdots)$ in the table the values of the bracket in (178.15) are reproduced; the series converges rapidly. The energy shifts, ΔE_s, lie between 26% and 12% of $|E_0|$ if Z varies from 20 to 80. They are therefore corrections only which, however, are not so very small that a second-order perturbation calculation would not change them by several per cent. On the other hand, the Thomas-Fermi model used is too rough to make such a change in the values physically significant.

In x-ray spectroscopy it is customary to describe the energy shift by a screening constant s defined by

$$E = -\tfrac{1}{2}Z^2 + \Delta E_s = -\tfrac{1}{2}(Z-s)^2; \tag{178.16}$$

this definition renders

$$s = Z\left(1 - \sqrt{1 - \frac{\Delta E_s}{|E_0|}}\right). \tag{178.17}$$

The screening constants given in the table are computed according to this formula. Since $\Delta E_s \lesssim \frac{1}{4}|E_0|$, we may roughly expand the radical in (178.17) and write

$$s \simeq Z\frac{\Delta E_s}{2|E_0|} = \frac{\Delta E_s}{Z},$$

which is about proportional to $Z^{\frac{1}{3}}$. If we try a rough formula of this type,

$$s = \sigma Z^{\frac{1}{3}},$$

we obtain the following pairs of values:

$Z=$	20	50	80
$\sigma=$	1.03	1.12	1.15

Z	β	$(1-\cdots)$	ΔE_s	$\lvert E_0 \rvert$	s	ΔE_r	s corr.
20	24.75	0.847	52.2	200	2.81		
30	32.25	0.880	95.2	450	3.36		
40	38.96	0.897	144.3	800	3.79	− 17.1	3.35
50	45.12	0.910	198.0	1250	4.13	− 41.7	3.26
60	50.90	0.921	257.3	1800	4.42	− 86.3	2.94
70	56.35	0.927	318.6	2450	4.71	−160	2.34
80	61.56	0.932	384.0	3200	4.96	−273	1.62

It is, however, useless to go into such detail. One glance at the experimental values shows that our s values are rather good up to about $Z=50$ but that, instead of the predicted slow rise of s with increasing Z beyond this value, the s values do not exceed a maximum of 3.7 and then, first slowly and above $Z=70$ rapidly, begin to fall again. Such a discrepancy at high values of Z clearly must be explained as a relativistic effect. This is, in essence, if only in a rough way, shown in the last two columns of the table. According to relativistic quantum mechanics (cf. Problem 203) the unperturbed K electron energy is lowered by the amount

$$\Delta E_r = -\frac{1}{8} Z^2 \left(\frac{Z}{137}\right)^2;$$

these shifts have to be added to the ΔE_s screening shifts before calculating screening constants as given in the last column. Since

$$\Delta E = \Delta E_s + \Delta E_r$$

becomes, for large values of Z, increasingly smaller than the original ΔE_s, the deviations from the unscreened nuclear field, i.e. the corrected s values, will also become increasingly smaller. This is corroborated by experiment. In a strict sense, of course, relativistic corrections should not be applied only to E_0 but to ΔE_s as well. The results may therefore still be a little rough, but scarcely more so than corresponds to the general application of the Thomas-Fermi field which neglects all special shell structure effects.

V. Non-Stationary Problems

Problem 179. Two-level system with time-independent perturbation

Given an atomic system with only two stationary states $|1\rangle$ and $|2\rangle$ and energies $\hbar\omega_1 < \hbar\omega_2$. At the time $t=0$, the system being in its ground state, a perturbation W not depending upon time is switched on. The probability shall be calculated of finding the system in either state at the time t.

Solution. Let H be the hamiltonian of the unperturbed system with

$$H|1\rangle = \hbar\omega_1|1\rangle; \qquad H|2\rangle = \hbar\omega_2|2\rangle \tag{179.1}$$

defining its two stationary states. Then, the Schrödinger equation with perturbation,

$$-\frac{\hbar}{i}\dot{\psi} = (H+W)\psi, \tag{179.2}$$

is to be solved in terms of the stationary functions:

$$\psi(t) = c_1(t)\,\mathrm{e}^{-i\omega_1 t}|1\rangle + c_2(t)\,\mathrm{e}^{-i\omega_2 t}|2\rangle. \tag{179.3}$$

It must be possible thus to construct the *exact* solution, because $|1\rangle$ and $|2\rangle$ form a complete orthonormal set so that (179.3) is just the expansion of ψ with respect to this set with time-dependent coefficients. The latter have to be determined with initial conditions

$$c_1(0) = 1; \qquad c_2(0) = 0. \tag{179.4}$$

If we put (179.3) into (179.2) and multiply[1] by either $\langle 1|$ or $\langle 2|$, we find two differential equations of the first order for the coefficients:

---

[1] "Multiplication" here means the formation of scalar products in Hilbert space, i.e. matrix elements between a pair of states.

$$-\frac{\hbar}{i}\dot{c}_1 e^{-i\omega_1 t} = \langle 1|W|1\rangle c_1 e^{-i\omega_1 t} + \langle 1|W|2\rangle c_2 e^{-i\omega_2 t},$$
$$-\frac{\hbar}{i}\dot{c}_2 e^{-i\omega_2 t} = \langle 2|W|1\rangle c_1 e^{-i\omega_1 t} + \langle 2|W|2\rangle c_2 e^{-i\omega_2 t}. \tag{179.5}$$

Let us briefly write

$$\langle \mu|W|\nu\rangle = W_{\nu\mu},$$

then, in consequence of the hermiticity of the operator W, the diagonal matrix elements W_{11} and W_{22} are real, whereas the complex off-diagonal elements are conjugate:

$$W_{12} = W_{21}^*.$$

Using the abbreviation

$$\omega_0 = \omega_2 - \omega_1, \tag{179.6}$$

so that $\hbar\omega_0$ is the energy difference between the two levels, Eqs. (179.5) may be written

$$\left.\begin{aligned} \hbar i\dot{c}_1 &= W_{11} c_1 + W_{21} e^{-i\omega_0 t} c_2, \\ \hbar i\dot{c}_2 &= W_{12} e^{i\omega_0 t} c_1 + W_{22} c_2 . \end{aligned}\right\} \tag{179.7}$$

Apparently it is possible to solve these equations by

$$c_1 = A e^{-i\omega t}; \qquad c_2 = B e^{-i(\omega-\omega_0)t}. \tag{179.8}$$

This can immediately be seen when (179.8) is put into (179.7) leading to the linear equations

$$(W_{11} - \hbar\omega) A + W_{21} B = 0,$$
$$W_{12} A + (W_{22} - \hbar\omega + \hbar\omega_0) B = 0.$$

The determinant of these two equations vanishes for the two frequencies

$$\omega_{\mathrm{I,II}} = \frac{W_{11}}{\hbar} + \frac{1}{2}\gamma \pm \sigma \tag{179.9}$$

with

$$\hbar\gamma = W_{22} - W_{11} + \hbar\omega_0,$$
$$\hbar\sigma = \sqrt{\tfrac{1}{4}\gamma^2 + |W_{12}|^2}. \tag{179.10}$$

Further, we obtain

$$B_{\mathrm{I,II}} = \frac{\hbar\omega_{\mathrm{I,II}} - W_{11}}{W_{21}} A_{\mathrm{I,II}}. \tag{179.11}$$

Thence,

$$c_1(t) = A_\mathrm{I} e^{-i\omega_\mathrm{I} t} + A_\mathrm{II} e^{-i\omega_\mathrm{II} t};$$

$$c_2(t) = \frac{1}{W_{21}} e^{i\omega_0 t}\{(\hbar\omega_\mathrm{I} - W_{11}) A_\mathrm{I} e^{-i\omega_\mathrm{I} t} + (\hbar\omega_\mathrm{II} - W_{11}) A_\mathrm{II} e^{-i\omega_\mathrm{II} t}\}.$$

The initial conditions (179.4) permit evaluation of the constants A_I and A_II. After a straightforward computation, we then arrive at the result:

$$c_1(t) = \exp\left[-i\left(\frac{W_{11}}{\hbar} + \frac{1}{2}\gamma\right)t\right]\left\{\cos\sigma t + i\frac{\gamma}{2\sigma}\sin\sigma t\right\}; \tag{179.12a}$$

$$c_2(t) = -i\frac{W_{12}}{\hbar\sigma}\exp\left[-i\left(\frac{W_{11}}{\hbar} + \frac{1}{2}\gamma - \omega_0\right)t\right]\sin\sigma t. \tag{179.12b}$$

The probability, then, of finding the system in the excited state will be

$$|c_2(t)|^2 = \frac{|W_{12}|^2}{\hbar^2\sigma^2}\sin^2\sigma t$$

or, using (179.10),

$$|c_2(t)|^2 = \frac{4|W_{12}|^2}{(\hbar\gamma)^2 + 4|W_{12}|^2}\sin^2\sigma t. \tag{179.13}$$

The probability of finding it in the original ground state again, on the other hand, becomes

$$|c_1(t)|^2 = \cos^2\sigma t + \left(\frac{\gamma}{2\sigma}\right)^2\sin^2\sigma t$$

or

$$|c_1(t)|^2 = 1 - \frac{4|W_{12}|^2}{(\hbar\gamma)^2 + 4|W_{12}|^2}\sin^2\sigma t. \tag{179.14}$$

Note that the sum of (179.13) and (179.14) will be 1. The system is oscillating between the two levels with a time period π/σ.

Problem 180. Periodic perturbation of a two-level system

Given the same two-level system as in the preceding problem. At the time $t=0$, however, let a *periodic* perturbation $W\cos\omega t$ be switched on (e.g. a light wave) with its frequency ω *almost* corresponding to the energy difference $\hbar\omega_0 = \hbar(\omega_2 - \omega_1)$ of the two levels. Again the probability of finding the system in either state after switching off the perturbation at the time t shall be found.

Solution. The Schrödinger equation

$$-\frac{\hbar}{i}\dot{\psi} = [H + W\cos\omega t]\psi \tag{180.1}$$

is to be solved by the function

$$\psi(t) = c_1(t)\,\mathrm{e}^{-i\omega_1 t}|1\rangle + c_2(t)\,\mathrm{e}^{-i\omega_2 t}|2\rangle \tag{180.2}$$

with the initial condition

$$\psi(0) = 1, \text{ or } c_1(0) = 1; \quad c_2(0) = 0. \tag{180.3}$$

Here, $|1\rangle$ and $|2\rangle$ are solutions of the stationary state equations

$$H|1\rangle = \hbar\omega_1|1\rangle; \quad H|2\rangle = \hbar\omega_2|2\rangle, \tag{180.4}$$

and $|1\rangle$ and $|2\rangle$ may be supposed to be orthonormal. By putting (180.2) into (180.1) and by scalar multiplication of the result with either $\langle 1|$ or $\langle 2|$, we arrive at two differential equations for $c_1(t)$ and $c_2(t)$, viz.

$$\begin{aligned} -\frac{\hbar}{i}\dot{c}_1\,\mathrm{e}^{-i\omega_1 t} &= \cos\omega t\{\langle 1|W|1\rangle c_1\,\mathrm{e}^{-i\omega_1 t} + \langle 1|W|2\rangle c_2\,\mathrm{e}^{-i\omega_2 t}\}; \\ -\frac{\hbar}{i}\dot{c}_2\,\mathrm{e}^{-i\omega_2 t} &= \cos\omega t\{\langle 2|W|1\rangle c_1\,\mathrm{e}^{-i\omega_1 t} + \langle 2|W|2\rangle c_2\,\mathrm{e}^{-i\omega_2 t}\}. \end{aligned} \tag{180.5}$$

Let us here introduce the above abbreviation

$$\omega_0 = \omega_2 - \omega_1$$

and write

$$\omega - \omega_0 = \Delta\omega; \tag{180.6}$$

then it is to be supposed that

$$|\Delta\omega| \ll \omega_0. \tag{180.7}$$

Eq. (180.5) thus becomes

$$i\dot{c}_1 = \frac{1}{2\hbar}\{\langle 1|W|1\rangle(\mathrm{e}^{i\omega t} + \mathrm{e}^{-i\omega t})c_1 + \langle 1|W|2\rangle(\mathrm{e}^{i\Delta\omega t} + \mathrm{e}^{-i(\omega+\omega_0)t})c_2\};$$

$$i\dot{c}_2 = \frac{1}{2\hbar}\{\langle 2|W|1\rangle(\mathrm{e}^{i(\omega+\omega_0)t} + \mathrm{e}^{-i\Delta\omega t})c_1 + \langle 2|W|2\rangle(\mathrm{e}^{i\omega t} + \mathrm{e}^{-i\omega t})c_2\}.$$

Here we meet a very pronounced distinction between high-frequency terms with frequencies of the orders ω and 2ω, and low-frequency terms with $\Delta\omega$. Averaging over a time interval $2\pi/\omega$, all high-frequency contributions will cancel. So, if we replace c_1 and c_2 by

$$C_\mu(t) = \frac{1}{2\tau}\int_{t-\tau}^{t+\tau} dt'\,c_\mu(t') \quad \text{with } \tau = \pi/\omega,$$

these averages satisfy the much simpler equations

$$\left.\begin{aligned} i\dot{C}_1 &= \frac{1}{2\hbar}\langle 1|W|2\rangle\, e^{i\Delta\omega t}\, C_2; \\ i\dot{C}_2 &= \frac{1}{2\hbar}\langle 2|W|1\rangle\, e^{-i\Delta\omega t}\, C_1 \end{aligned}\right\} \tag{180.8}$$

where the slowly varying factors $\exp(\pm i\Delta\omega t)$ on the right-hand side are treated as constants in averaging.

The differential equations (180.8) permit exact solution by differentiation of either, and elimination of the other variable and its first derivative by using the original equations (180.8). We thus arrive at

$$\left.\begin{aligned} \ddot{C}_1 - i\Delta\omega\,\dot{C}_1 + \tfrac{1}{4}\Omega^2 C_1 &= 0; \\ \ddot{C}_2 + i\Delta\omega\,\dot{C}_2 + \tfrac{1}{4}\Omega^2 C_2 &= 0 \end{aligned}\right\} \tag{180.9}$$

with

$$\Omega^2 = \frac{1}{\hbar^2}\langle 1|W|2\rangle\langle 2|W|1\rangle = \frac{1}{\hbar^2}|\langle 2|W|1\rangle|^2 . \tag{180.10}$$

Use of the abbreviation

$$R = \sqrt{\Omega^2 + (\Delta\omega)^2} \tag{180.11}$$

leads to the following solutions satisfying the initial conditions (180.3):

$$\begin{aligned} C_1 &= e^{i\frac{\Delta\omega}{2}t}\left\{\cos\frac{Rt}{2} + A\sin\frac{Rt}{2}\right\}; \\ C_2 &= e^{-i\frac{\Delta\omega}{2}t}\, B\sin\frac{Rt}{2} . \end{aligned} \tag{180.12}$$

The remaining two integration constants A and B can be computed by putting (180.12) into the first-order equations (180.8). The result is

$$A = -i\frac{\Delta\omega}{R}; \qquad B = -i\frac{\langle 2|W|1\rangle}{\hbar R} . \tag{180.13}$$

The probability of finding the system, at the time t, in the excited state is now

$$|C_2|^2 = \frac{\Omega}{\Omega^2 + (\Delta\omega)^2}\sin^2\frac{Rt}{2} \tag{180.14}$$

and that of finding it in its ground state again is

$$|C_1|^2 = \cos^2\frac{Rt}{2} + \frac{(\Delta\omega)^2}{\Omega^2 + (\Delta\omega)^2}\sin^2\frac{Rt}{2} . \tag{180.15}$$

According to (180.14), the excitation is a typical resonance process, its probability rapidly decreasing with increasing values of $|\Delta\omega|$. Needless to say, this holds only as long as condition (180.7) remains satisfied. The process is periodically repeated with the frequency R determined by Eq. (180.11), i.e. mainly by the value of the matrix element, so that after a time interval

$$t_n = \frac{2\pi n}{R}, \qquad n=1,2,3,\ldots \tag{180.16}$$

the system will be found in the ground state again. If the periodic perturbation is performed, e.g. by a light wave switched on at time $t=0$ and switched off again at $t=t_n$, no resultant change will have affected the system.

Application. Let the hamiltonian H describe the Zeeman effect in a one-electron S state produced by a magnetic field $\mathscr{H}_0$ in z direction. Then $\hbar\,(\omega_2-\omega_1)$ is the level splitting between the two spin orientations, i.e. $\omega_0=2\mu\mathscr{H}_0/\hbar$ (with $|2\rangle$ the upper, $|1\rangle$ the lower state). Let now the perturbation consist of a periodic magnetic field, $\mathscr{H}'\cos\omega t$, so that

$$W=-\mu(\boldsymbol{\sigma}\cdot\mathscr{H}')\cos\omega t, \qquad \mu=-\frac{e\hbar}{2mc}.$$

If the field $\mathscr{H}'$ is parallel to $\mathscr{H}_0$, the matrix element $\langle 1|W|2\rangle$ vanishes; the states $|1\rangle$ and $|2\rangle$ are then independently perturbed and no transitions are induced. If, on the other hand, $\mathscr{H}'$ is chosen perpendicular to $\mathscr{H}_0$, say in x direction, the diagonal matrix elements of W vanish and we find exactly the case described above with

$$\langle 1|W|2\rangle = \langle 2|W|1\rangle = -\mu\mathscr{H}'.$$

Resonance will occur in this device if $\omega\simeq\omega_0$, as before, and the system will alternate between its two magnetic states. This is the simplest case of *paramagnetic resonance*.

Problem 181. Dirac perturbation method

Let an atomic system have the non-degenerate stationary states ψ_k. Let it be in its ground state ψ_0 at time $t=0$ and then a perturbation be switched on (depending upon, or independent of, time) inducing transitions to other states ψ_l. The probability shall be determined of finding the system, after switching off the perturbation at time t, in a state ψ_l, supposing the perturbation to be small.

Solution. Let the unperturbed states satisfy the Schrödinger equation

$$-\frac{\hbar}{i}\dot{\psi}_k=H\psi_k \quad \text{with} \quad \psi_k=|k\rangle\,\mathrm{e}^{-i\omega_k t}; \qquad E_k=\hbar\omega_k \tag{181.1}$$

and

$$\langle l|k\rangle=\delta_{kl}. \tag{181.2}$$

After switching on the perturbation W, we have the differential equation

$$-\frac{\hbar}{i}\dot{\psi}=(H+W)\psi \tag{181.3}$$

with a wave function ψ which may be expanded into a series

$$\psi=\sum_k a_k(t)\psi_k. \tag{181.4}$$

In consequence of (181.2), it then follows from (181.4) that[2]

$$\sum_k |a_k(t)|^2=1. \tag{181.5}$$

Each $|a_k|^2$ then is the probability of finding the system in the state ψ_k at the time t.

Introducing the sum (181.4) into the differential equation (181.3) we find

$$-\frac{\hbar}{i}\sum_k(\dot{a}_k-i\omega_k a_k)\psi_k=\sum_k(\hbar\omega_k+W)a_k\psi_k$$

or, forming the scalar product in Hilbert space with $\langle l|$ and making use of (181.2),

$$\dot{a}_l=-\frac{i}{\hbar}\sum_k \mathrm{e}^{-i(\omega_k-\omega_l)t}\langle l|W|k\rangle a_k. \tag{181.6}$$

In this equation, so far, nothing has yet been neglected. It corresponds to the fact that the time rate of any state $|l\rangle$ depends upon *all* states of the system combining with $|l\rangle$ under the action of the perturbation. This, of course, is a consequence of (181.5): If one of the coefficients, a_l, is changed, the other coefficients are bound to change as well in order to keep the sum (181.5) constant. (Cf. Problem 179 for a system with two states only.)

If the perturbation is *small*, we may in first approximation insert on the right-hand side of (181.6) the initial values

$$a_k(0)=\delta_{k0}. \tag{181.7}$$

Then, the set of equations (181.6) becomes, for $l\neq 0$,

$$\dot{a}_l=-\frac{i}{\hbar}\mathrm{e}^{-i(\omega_0-\omega_l)t}\langle l|W|0\rangle. \tag{181.8}$$

[2] If a continuous part of the spectrum exists besides the discontinuous part, it can, by the use of a periodicity volume, be transformed into a formally discontinuous spectrum and thus be included into the sums (181.4) and (181.5) without further mathematical difficulties.

(This is a much more specialized behaviour than that of Problem 179, because it neglects backward transitions from $|l\rangle$ to $|0\rangle$ and the like.) Integration of (181.8) yields

$$a_l(t) = -\frac{i}{\hbar}\int_0^t dt \langle l|W|0\rangle \, e^{-i(\omega_0-\omega_l)t}. \tag{181.9}$$

The integral, of course, depends very much on how the perturbation W, and thus its matrix element, depend upon time.

Our approximation is valid only if

$$|\langle l|W|0\rangle|\, t/\hbar \ll 1, \tag{181.10}$$

so that the coefficient $a_l(t)$ remains small throughout. Since it follows from (181.10) that

$$(\omega_l-\omega_0)t \ll \frac{\hbar(\omega_l-\omega_0)}{|\langle l|W|0\rangle|}$$

and the excitation energy in the numerator is usually much larger than the matrix element in the denominator, the exponent in (181.8) or (181.9) may still be quite large so that there occur periodic oscillations of $a_l(t)$ which do not quite agree with the basic idea underlying our first-order perturbation approximation. However, in the following problems we shall show how to eliminate this difficulty.

Problem 182. Periodic perturbation: Resonance

Let the atomic system of the preceding problem be perturbed by a periodic field

$$W(t) = \mathscr{W} e^{-i\omega t} + \mathscr{W}^\dagger e^{i\omega t}. \tag{182.1}$$

Discuss resonance absorption and the effect of a finite frequency width of the irradiated field upon the transitions.

Solution. If (182.1) is put into the general first-order perturbation formula (182.9) and the integration is performed, we get

$$a_l(t) = -\frac{i}{\hbar}\left\{\langle l|\mathscr{W}|0\rangle \frac{e^{i(\omega_l-\omega_0-\omega)t}-1}{i(\omega_l-\omega_0-\omega)} + \langle l|\mathscr{W}^\dagger|0\rangle \frac{e^{i(\omega_l-\omega_0+\omega)t}-1}{i(\omega_l-\omega_0+\omega)}\right\}. \tag{182.2}$$

The excitation energy, $E_{\rm ex}=\hbar(\omega_l-\omega_0)$ being positive, the first term has resonance if $\hbar\omega=E_{\rm ex}$ whereas the second term never shows resonance. Thus, if Bohr's frequency condition

$$\omega=\omega_l-\omega_0 \tag{182.3}$$

holds, the system can absorb energy from the alternating field applied:

$$|a_l(t)|^2=\frac{4|\langle l|\mathscr{W}|0\rangle|^2}{\hbar^2}\cdot\frac{\sin^2\frac{1}{2}(\omega_l-\omega_0-\omega)t}{(\omega_l-\omega_0-\omega)^2}. \tag{182.4}$$

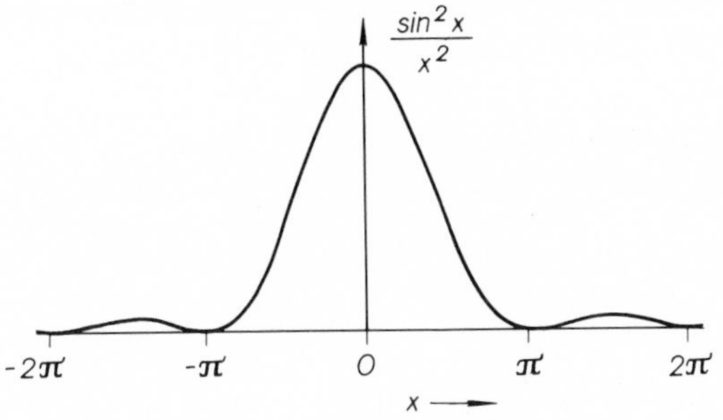

Fig. 70. The natural line shape $\sin^2 x/x^2$

This formula should still be corrected for the finite frequency width of the irradiated field. Let $\rho(\omega)\,d\omega$ be its intensity between ω and $\omega+d\omega$, then we have

$$|a_l(t)|^2=\int d\omega\;\rho(\omega)\cdot 4|\langle l|\mathscr{W}|0\rangle|^2\,\frac{\sin^2\frac{1}{2}(\omega_l-\omega_0-\omega)t}{\hbar^2(\omega_l-\omega_0-\omega)^2} \tag{182.5}$$

or, with

$$\tfrac{1}{2}[\omega-(\omega_l-\omega_0)]\,t=x$$

as integration variable,

$$|a_l(t)|^2=2t\int dx\,\rho\,(\omega_l-\omega_0+2x/t)\left|\langle l|\frac{\mathscr{W}}{\hbar}|0\rangle\right|^2\cdot\frac{\sin^2 x}{x^2}.$$

Here, the last factor, $\sin^2 x/x^2$, has a pronounced maximum at $x=0$ whence it decreases rapidly on both sides (Fig. 70) so that $|x|<\pi$ brings the main contribution to the integral

$$\int dx\frac{\sin^2 x}{x^2}=\pi.$$

Within this range of x values we have $|2x/t|<2\pi/t$ or, since condition (181.10) of the preceding problem must hold, i.e. since

$$|\langle l|\mathscr{W}|0\rangle| \ll \frac{\hbar}{t},$$

and since this matrix element will usually be very small compared with the excitation energy, we find that the argument of ρ may be replaced simply by $\omega_l-\omega_0$. A similar argument obtains for the matrix element which too may be treated as a constant, independent of x, so that we arrive at

$$|a_l(t)|^2 = 2\pi t \left|\langle l|\frac{\mathscr{W}}{\hbar}|0\rangle\right|^2 \rho(\omega_l-\omega_0). \tag{182.6}$$

The probability of finding the system in any state $|l\rangle$ thus increases in proportion to the time. Therefore we may reasonably define a *transition probability*

$$P_l = \frac{1}{t}|a_l(t)|^2 \tag{182.7}$$

independent of time which becomes

$$P_l=2\pi\left|\langle l|\frac{\mathscr{W}}{\hbar}|0\rangle\right|^2 \rho(\omega_l-\omega_0). \tag{182.8}$$

NB. The last result shows a close similarity to the Golden Rule to be discussed in Problem 183. It should, however, be borne in mind that the Golden Rule describes summation over close-lying *final* states, whereas we have introduced a summation over a continuum of *initial* field properties. It has not been shown here that this summation must necessarily be performed in the probabilities, Eq. (182.5), and not in the amplitude formula (182.2).

Problem 183. Golden Rule for scattering

Let a beam of particles with initial momentum $\boldsymbol{p}_i=\hbar\boldsymbol{k}_i$ be elastically scattered by a potential $W(r)$ into states of final momentum $\boldsymbol{p}_f=\hbar\boldsymbol{k}_f$ within the solid angle element $d\Omega_f$. The differential cross section $d\sigma/d\Omega_f$ shall be derived by the Dirac perturbation method.

Solution. We may gather from (181.9) that, in the first Dirac approximation,

$$a_f(t)= -\frac{i}{\hbar}\int_0^t dt\langle f|W|i\rangle e^{-i(\omega_i-\omega_f)t}. \tag{183.1}$$

This may be integrated, if the matrix element is supposed not to depend upon time, so that we arrive at the basic formula

$$|a_f(t)|^2 = \frac{4|\langle f|W|i\rangle|^2}{\hbar^2(\omega_i-\omega_f)^2}\sin^2\frac{(\omega_i-\omega_f)t}{2}. \tag{183.2}$$

Now both the initial and final states are lying in the continuous spectrum. Using a normalization volume V, the respective wave functions are

$$|i\rangle = V^{-\frac{1}{2}}\,e^{i\boldsymbol{k}_i\cdot\boldsymbol{r}}; \qquad \langle f| = V^{-\frac{1}{2}}\,e^{-i\boldsymbol{k}_f\cdot\boldsymbol{r}}. \tag{183.3}$$

There are, even for finite volume V, a great many final states in the vicinity of $\langle f|$ and, in the limit $V\to\infty$, there will be an infinite number of them in an infinitesimal surrounding. To ask what is the probability of *one* final state $\langle f|$ with sharp $\boldsymbol{k}_f$ thus becomes meaningless; we may only ask with what probability a certain interval will be reached.

Let $\rho_f(E_f)dE_f$ levels be lying in the interval dE_f at the final energy E_f with their momenta within the solid angle element $d\Omega_f$, then the transition probability per unit time to this angular interval becomes

$$dT = \frac{1}{t}\int dE_f\,\rho_f(E_f)|a_f(t)|^2. \tag{183.4}$$

This definition is reasonable only because this expression does not depend on time and the integral goes over a very narrow energy region.

If now we introduce the variable

$$x = \tfrac{1}{2}(\omega_f-\omega_i)t$$

and put $E_f=\hbar\omega_f$ so that we get

$$dE_f = \frac{2\hbar}{t}dx$$

we find, according to (183.2) and (183.4),

$$dT = \frac{2}{\hbar}\int dx\,\rho_f(E_f)\frac{\sin^2 x}{x^2}|\langle f|W|i\rangle|^2$$

where the considerations of the preceding problem will again hold for the integral, so that

$$dT = \frac{2\pi}{\hbar}\rho_f(E)|\langle f|W|i\rangle|^2 \tag{183.5}$$

since integration over an extremely large interval in x about $x=0$ corresponds to a very narrow one only in energy about the resonance energy $E_f=E_i$. The differential notation, dT, is appropriate because of the infinitesimal interval $d\Omega_f$ still contained in ρ_f (which itself might perhaps better be written $d\rho_f$). Eq. (183.5) is called the *Golden Rule*.

This transition probability still depends in an obvious way upon the normalization volume V and the initial velocity $v_0=\hbar k_i/m$ of the particles hitting the obstacle,

$$dT=\frac{v_0}{V}\,d\sigma\,, \tag{183.6}$$

where $d\sigma$ is independent of V and therefore a quantity of physical significance. It has the dimension of cm^2 and is identical with the differential cross section thus to be found from the Golden Rule expression (183.5):

$$d\sigma=\frac{2\pi}{\hbar}\,\rho_f(E)\,\frac{V}{v_0}\,|\langle f|W|i\rangle|^2\,. \tag{183.7}$$

Here we still have to evaluate the final state density ρ_f and the matrix element.

The final state density may be derived from the fact that one state (if the particles have no spin) falls in an element d^3p of momentum space of the amount $(2\pi\hbar)^3/V$. Therefore, in an arbitrary element d^3p there will lie

$$\frac{d^3p\,V}{8\pi^3\hbar^3}$$

states. With

$$d^3p=p^2\,dp\,d\Omega=mp\,dE\,d\Omega$$

we have

$$\rho_f\,dE_f=\frac{mp_f\,dE_f\,d\Omega_f\,V}{8\pi^3\hbar^3}\,, \tag{183.8}$$

so that

$$\frac{d\sigma}{d\Omega_f}=\left(\frac{mV}{2\pi\hbar^2}\right)^2\frac{k_f}{k_i}|\langle f|W|i\rangle|^2\,d\Omega_f \tag{183.9}$$

where we may omit the factor $k_f/k_i=1$ for elastic scattering.

The matrix element, for a potential $W(r)$ of central symmetry, formed with the plane waves (183.3) runs

$$\langle f|W|i\rangle=\frac{1}{V}\int d^3x\,e^{-i\boldsymbol{K}\cdot\boldsymbol{r}}\,W(r) \tag{183.10}$$

with $\boldsymbol{K}=\boldsymbol{k}_f-\boldsymbol{k}_i$ the momentum transfer vector (in units of $\hbar$). The integration over the polar angles leads to

$$\langle f|W|i\rangle = \frac{4\pi}{V}\int_0^\infty dr\, r^2\, W(r)\frac{\sin Kr}{Kr} \tag{183.11}$$

so that finally we arrive at the cross section formula

$$\frac{d\sigma}{d\Omega_f} = \left|\frac{2m}{\hbar^2}\int_0^\infty dr\, r^2\, W(r)\frac{\sin Kr}{Kr}\right|^2, \tag{183.12}$$

i.e. the result of the first Born approximation (cf. Problem 105). This of course, was only to be expected since we treated the scattering potential as a perturbation already in our starting equation (183.1) and consequently used plane waves to describe the initial and final states.

Problem 184. Born scattering in momentum space

To derive the differential scattering cross section in momentum space by a time-dependent perturbation method in the first approximation. Let the perturbation be switched on at $t=0$ and be constant thereafter.

Solution. According to Problem 14 the time-dependent Schrödinger equation

$$-\frac{\hbar}{i}\dot{\psi} = -\frac{\hbar^2}{2m}\nabla^2\psi + V(\boldsymbol{r})\psi \tag{184.1}$$

corresponds to the integro-differential equation in momentum space,

$$-\frac{\hbar}{i}\dot{f}(\boldsymbol{k},t) = \frac{\hbar^2}{2m}k^2 f(\boldsymbol{k},t) + \int d^3k'\, W(\boldsymbol{k}-\boldsymbol{k}')\, f(\boldsymbol{k}',t) \tag{184.2}$$

with $f(\boldsymbol{k},t)$ the Fourier transform of $\psi(\boldsymbol{r},t)$ and $W(\boldsymbol{k})$ of $V(\boldsymbol{r})$ in the normalization used in Problem 14. Eq. (184.2) may be somewhat simplified by writing f in the form

$$f(\boldsymbol{k},t)=v(\boldsymbol{k},t)\mathrm{e}^{-i\omega t}; \qquad \omega = \frac{\hbar}{2m}k^2. \tag{184.3}$$

We then find

$$\frac{\partial v(\boldsymbol{k},t)}{\partial t} = -\frac{i}{\hbar}\int d^3k'\, W(\boldsymbol{k}-\boldsymbol{k}')\mathrm{e}^{i(\omega-\omega')t}\, v(\boldsymbol{k}',t). \tag{184.4}$$

To solve (184.4) in first approximation we replace v under the integral by the unperturbed function

$$v_0(\mathbf{k},t)=C(2\pi)^{\frac{3}{2}}\,\delta(\mathbf{k}-\mathbf{k}_0), \tag{184.5}$$

i.e. by the Fourier transform of the plane wave

$$\psi_0(\mathbf{r},t)=C\,\mathrm{e}^{i\mathbf{k}_0\cdot\mathbf{r}}. \tag{184.6}$$

Eq. (184.4) then simply becomes

$$\frac{\partial v(\mathbf{k},t)}{\partial t}=-\frac{i}{\hbar}\,C(2\pi)^{\frac{3}{2}}\,W(\mathbf{k}-\mathbf{k}_0)\,\mathrm{e}^{i(\omega-\omega_0)t}$$

and yields the integral

$$v(\mathbf{k},t)=-\frac{C}{\hbar}(2\pi)^{\frac{3}{2}}\,W(\mathbf{k}-\mathbf{k}_0)\frac{\mathrm{e}^{i(\omega-\omega_0)t}-1}{\omega-\omega_0}. \tag{184.7}$$

The probability of finding a particle of momentum $\mathbf{k}$ within d^3k at the time t is then (cf. Problem 15)

$$|v(\mathbf{k},t)|^2\,d^3k=\frac{|C|^2(2\pi)^3}{\hbar^2}\,|W(\mathbf{k}-\mathbf{k}_0)|^2\,t^2\,\frac{\sin^2 x}{x^2}\,k^2\,dk\,d\Omega$$

with the abbreviation

$$x=\tfrac{1}{2}(\omega-\omega_0)\,t.$$

This expression has still to be integrated over the energy resonance (cf. Problem 183); since

$$k\,dk=\frac{2m}{\hbar t}\,dx$$

and

$$\int_{-\infty}^{+\infty} dx\,\frac{\sin^2 x}{x^2}=\pi,$$

we find an expression linear in t, and the transition probability, defined by

$$dT=\frac{1}{t}\,d\Omega\int dk\,k^2\,|v(\mathbf{k},t)|^2, \tag{184.8}$$

becomes

$$dT=\frac{|C|^2(2\pi)^4\,m k}{\hbar^3}\,|W(\mathbf{k}-\mathbf{k}_0)|^2\,d\Omega. \tag{184.9}$$

The differential cross section is defined by putting

$$dT = |C|^2 v_0 d\sigma = |C|^2 \frac{\hbar k}{m} d\sigma; \tag{184.10}$$

therefore

$$d\sigma = \frac{(2\pi)^4 m^2}{\hbar^4} |W(\boldsymbol{k}-\boldsymbol{k}_0)|^2 d\Omega. \tag{184.11}$$

This general formula may still be simplified for a central-force potential $V(r)$ which permits integration over the polar angles in its Fourier transform:

$$W(\boldsymbol{k}-\boldsymbol{k}_0) = \frac{1}{(2\pi)^3} \int d^3x \, e^{i\boldsymbol{k}\cdot\boldsymbol{r}} V(r) = \frac{4\pi}{(2\pi)^3} \int_0^\infty dr \, r^2 V(r) \frac{\sin Kr}{Kr} \tag{184.12}$$

with $\boldsymbol{K} = \boldsymbol{k} - \boldsymbol{k}_0$ the momentum transfer (in units of $\hbar$), i.e. with

$$K = 2k_0 \sin\frac{\vartheta}{2} \tag{184.13}$$

where ϑ is the angle of deflection. Putting (184.12) into (184.11) we finally arrive at the well-known result of Born,

$$\frac{d\sigma}{d\Omega} = \left\{ \frac{2m}{\hbar^2} \int_0^\infty dr \, r^2 V(r) \frac{\sin Kr}{Kr} \right\}^2. \tag{184.14}$$

NB. The transition from the momentum function f to v is, generally speaking, the transition from Schrödinger to interaction representation as v would no longer depend upon time if there were no interaction $W(\boldsymbol{k})$.—The cross section formula (184.14) has also been derived in Problem 105.—It should further be noted that the Fourier transform $W(\boldsymbol{k}-\boldsymbol{k}_0)$ is—except for a normalization factor—the matrix element of the potential $V(\boldsymbol{r})$ in ordinary space:

$$(2\pi)^3 W(\boldsymbol{k}-\boldsymbol{k}_0) = \int d^3k' \, e^{-i\boldsymbol{k}\cdot\boldsymbol{r}} V(\boldsymbol{r}) e^{i\boldsymbol{k}_0\cdot\boldsymbol{r}} = \langle \boldsymbol{k} | V | \boldsymbol{k}_0 \rangle.$$

This permits e.g. to write the cross section formula (184.11) as well in the form

$$d\sigma = \left| \frac{m}{2\pi\hbar^2} \langle \boldsymbol{k} | V | \boldsymbol{k}_0 \rangle \right|^2 d\Omega. \tag{184.15}$$

Problem 185. Coulomb excitation of an atom

Let an electron of velocity v pass an alkali atom at a distance b ("collision parameter"); let its velocity be large compared to the velocity of the valence electron in the atom. Its Coulomb interaction with the latter may originate an excitation, the cross section of which shall be determined by a perturbation procedure.

Solution. The position of the atomic electron is shown in Fig. 71, its position $\boldsymbol{r}$ relative to the atomic core having coordinates x, y, z, as indicated. Its interaction with the electron passing at the point P is

$$V(t) = \frac{e^2}{R(t)} \tag{185.1}$$

with

$$R^2 = (vt - x)^2 + y^2 + (b - z)^2 . \tag{185.2}$$

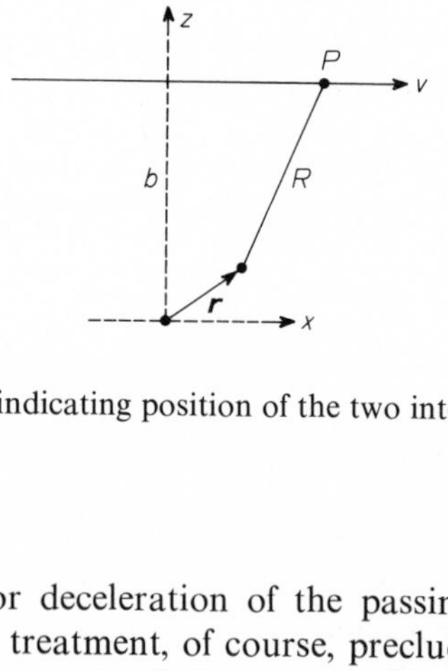

Fig. 71. Notations indicating position of the two interacting particles

Here any deflection or deceleration of the passing electron has been neglected. Its classical treatment, of course, precludes effects of overlap such as exchange phenomena. It is a reasonable approximation for a wavelength small compared to the distance b and to the atomic dimensions. Let us further assume the collision parameter b to be very large compared to the atomic dimensions so that we may expand

$$\frac{1}{R} = \frac{1}{\sqrt{(vt)^2 + b^2}} \left\{ 1 + \frac{vtx + bz}{(vt)^2 + b^2} \right\} . \tag{185.2'}$$

The matrix element between the ground state (subscript: 0) and an excited state (k) then becomes, in this approximation,

$$\langle k|V(t)|0\rangle = \frac{e^2}{[(vt)^2+b^2]^{\frac{3}{2}}}\{vt\langle k|x|0\rangle + b\langle k|z|0\rangle\}, \qquad (185.3)$$

the first (static) term in (185.2′), only contributing to the diagonal elements $\langle 0|V|0\rangle$ of elastic scattering.

If the unperturbed wave functions, $|k\rangle$, satisfy a Schrödinger equation

$$H|k\rangle = \hbar\omega_k|k\rangle$$

the perturbed wave function may be written

$$\psi = e^{-i\omega_0 t}|0\rangle + \sum_k{}' a_k(t)e^{-i\omega_k t}|k\rangle \qquad (185.4)$$

with (cf. Problem 181)

$$a_k(t) = -\frac{i}{\hbar}\int_{-\infty}^{t} dt\, e^{i(\omega_k-\omega_0)t}\langle k|V(t)|0\rangle. \qquad (185.5)$$

The probability of finding the atom in the excited state $|k\rangle$ after passage of the colliding electron is then

$$p_k = |a_k(\infty)|^2, \qquad (185.6)$$

and the excitation cross section of the state $|k\rangle$ is obtained by integration over the collision parameter:

$$\sigma_k = 2\pi\int_0^{\infty} db\, b|a_k(\infty)|^2. \qquad (185.7)$$

Thus the problem is reduced mainly to the calculation of $a_k(\infty)$ with the matrix element (185.3):

$$a_k(\infty) = -\frac{i}{\hbar}e^2\int_{-\infty}^{+\infty} dt\frac{e^{i(\omega_k-\omega_0)t}}{[v^2t^2+b^2]^{\frac{3}{2}}}\{vt\langle k|x|0\rangle + b\langle k|z|0\rangle\}.$$

With the abbreviations

$$\beta = \frac{(\omega_k-\omega_0)b}{v}; \qquad s = \frac{v}{b}t \qquad (185.8)$$

this may be written

$$a_k(\infty) = -\frac{ie^2}{\hbar v b}\left\{\langle k|x|0\rangle \int_{-\infty}^{+\infty} ds \frac{s e^{i\beta s}}{(1+s^2)^{\frac{3}{2}}} + \langle k|z|0\rangle \int_{-\infty}^{+\infty} ds \frac{e^{i\beta s}}{(1+s^2)^{\frac{3}{2}}}\right\}. \quad (185.9)$$

The two integrals can be determined using the integral representation of the modified Hankel function

$$\int_0^{\infty} ds \frac{\cos\beta s}{(1+s^2)^{\frac{3}{2}}} = \beta K_1(\beta). \quad (185.10)$$

This integral may equally well be written

$$\frac{1}{2}\int_0^{\infty} \frac{ds\, e^{i\beta s}}{(1+s^2)^{\frac{3}{2}}} + \frac{1}{2}\int_0^{\infty} \frac{ds\, e^{-i\beta s}}{(1+s^2)^{\frac{3}{2}}} = \frac{1}{2}\int_{-\infty}^{+\infty} \frac{ds\, e^{i\beta s}}{(1+s^2)^{\frac{3}{2}}}$$

so that

$$\int_{-\infty}^{+\infty} ds \frac{e^{i\beta s}}{(1+s^2)^{\frac{3}{2}}} = 2\beta K_1(\beta). \quad (185.11)$$

The derivative with respect to β is

$$i\int_{-\infty}^{+\infty} ds \frac{s e^{i\beta s}}{(1+s^2)^{\frac{3}{2}}} = 2\frac{d}{d\beta}[\beta K_1(\beta)] = -2\beta K_0(\beta). \quad (185.12)$$

We therefore obtain

$$a_k(\infty) = -\frac{e^2}{\hbar v}\cdot\frac{2}{b}\{\langle k|x|0\rangle(-\beta K_0(\beta)) + i\langle k|z|0\rangle \beta K_1(\beta)\}. \quad (185.13)$$

If the atoms are not oriented, in the statistical average

$$\overline{\langle k|x|0\rangle} = \overline{\langle k|z|0\rangle} = 0; \quad \overline{|\langle k|x|0\rangle|^2} = \overline{|\langle k|z|0\rangle|^2} \quad (185.14)$$

so that

$$\overline{|a_k(\infty)|^2} = \left(\frac{e^2}{\hbar v}\right)^2 \cdot \frac{4}{b^2}\overline{|\langle k|x|0\rangle|^2}\,\beta^2[K_0^2(\beta) + K_1^2(\beta)]. \quad (185.15)$$

According to (185.7) and (185.8) this leads to the excitation cross section of the state $|k\rangle$:

$$\sigma_k = 8\pi\left(\frac{e^2}{\hbar v}\right)^2 \overline{|\langle k|x|0\rangle|^2}\int_0^{\infty} d\beta\, \beta[K_0^2(\beta) + K_1^2(\beta)]. \quad (185.16)$$

The last integral can be evaluated for a finite lower bound

$$\int_{\beta}^{\infty} d\beta\,\beta\,[K_0^2(\beta)+K_1^2(\beta)]=\beta\,K_0(\beta)\,K_1(\beta). \tag{185.17}$$

For small values of β, the limiting values to the functions are

$$\beta\,K_1(\beta)\to 1; \qquad K_0(\beta)\to C+\log\frac{2}{\beta} \tag{185.18}$$

with $C=0.5772\ldots$ the Euler constant. The integral therefore diverges at small values of β or, according to (185.8), of the collision parameter b. This divergence is caused by the expansion (185.2′) of the interaction which only holds if the distance r of the atomic electron from the atomic nucleus is small compared to the collision parameter b. The divergence can to some extent be remedied by a cut-off at $b=r_0$, with r_0 something like the atomic radius or, using (185.8), at

$$\beta_{\min}=\frac{r_0(\omega_k-\omega_0)}{v}. \tag{185.19}$$

Since v shall be large compared to the atomic electron velocity, $\beta_{\min}\ll 1$ so that the formulae (185.18) obtain. We then arrive at the cross section formula

$$\sigma_k=8\pi\left(\frac{e^2}{\hbar v}\right)^2 \overline{|\langle k|x|0\rangle|^2}\left\{\log\frac{2v}{r_0(\omega_k-\omega_0)}-C\right\}. \tag{185.20}$$

No very exact knowledge of the cut-off radius is needed, since the logarithm varies rather slowly with its argument.

Literature. The method is an abridged form of that used in the theory of nuclear Coulomb excitation, cf. Alder, K., Winther, W.: Dan. Mat.-Fys. Medd. **29**, 19 (1955).

Problem 186. Photoeffect

A light wave of linear polarization ($\mathscr{E}\parallel x$, $\mathscr{H}\parallel y$) propagating in z direction falls on a hydrogen atom with the electron in its ground state. What is the angular distribution of photoelectrons emitted? What is the differential cross section of photo-emission? Retardation effects shall be neglected and the final state be approximated by a plane wave.

Solution. The light wave may be described by a vector potential $\boldsymbol{A}$ with

$$A_x=\frac{c}{\omega}\mathscr{E}_0\cos\left[\omega\left(t-\frac{z}{c}\right)+\delta\right]; \quad A_y=0; \quad A_z=0, \tag{186.1}$$

then the following field strength components can be derived:

$$\mathscr{E}_x = -\frac{1}{c}\dot{A}_x = \mathscr{E}_0 \sin\left[\omega\left(t-\frac{z}{c}\right)+\delta\right];$$

$$\mathscr{H}_y = \frac{\partial A_x}{\partial z} = \mathscr{E}_0 \sin\left[\omega\left(t-\frac{z}{c}\right)+\delta\right],$$

all other components vanishing. The average Poynting vector in z direction is

$$\overline{S} = \frac{c}{4\pi}\overline{\mathscr{E}_x \mathscr{H}_y} = \frac{c}{8\pi}\mathscr{E}_0^2$$

so that there are

$$n = \frac{c\,\mathscr{E}_0^2}{8\pi\hbar\omega} \tag{186.2}$$

incident photons per cm² and sec.

The interaction energy between light and electron is, according to Problem 125,

$$W = -\frac{e\hbar}{mc}i(\boldsymbol{A}\cdot\nabla) = \mathsf{W}\,\mathrm{e}^{-i\omega t} + \mathsf{W}^{\dagger}\,\mathrm{e}^{i\omega t} \tag{186.3}$$

where

$$\mathsf{W} = -\frac{e\hbar}{2m\omega}\mathscr{E}_0\, i\,\mathrm{e}^{-i\delta}\frac{\partial}{\partial x}. \tag{186.4}$$

Here the retardation factor, $\exp\left(i\frac{\omega z}{c}\right)$ has been taken $=1$.

We then may apply the method developped in Problem 182. Only the interaction term W leads to a resonance denominator $\omega_f-\omega_i-\omega$, thus satisfying energy conservation. Putting

$$x = \tfrac{1}{2}(\omega_f-\omega_i-\omega)t$$

we get

$$|a_f(t)|^2 = \frac{4}{\hbar^2}|\langle f|\mathsf{W}|i\rangle|^2\frac{\sin^2 x}{x^2}$$

and thence the transition probability P_f, from the initial state $|i\rangle$ to the final state $|f\rangle$,

$$P_f = \frac{2\pi}{\hbar}\rho_f|\langle f|\mathsf{W}|i\rangle|^2. \tag{186.5}$$

Here ρ_f, the final state density in the energy scale refers to the outgoing electron and is, according to (183.8),

$$\rho_f = \frac{mV}{8\pi^3 \hbar^2} k_f \, d\Omega_f \tag{186.6}$$

if V is the normalization volume and $\hbar k_f$ the momentum of the photoelectron. The differential cross section for photo-emission into the solid angle element $d\Omega_f$ is P_f/n so that, gathering expressions from (186.2) and (186.4–6), we arrive at the general formula

$$d\sigma = \frac{e^2 V}{2\pi m c \omega} k_f \, d\Omega_f \left| \langle f | \frac{\partial}{\partial x} | i \rangle \right|^2 . \tag{186.7}$$

We now are dealing with wave functions in a potential field of central symmetry, with the ground state $|i\rangle$ independent of polar angles. Then,

$$\frac{\partial}{\partial x} |i\rangle = \frac{d|i\rangle}{dr} \sin\vartheta \cos\varphi \tag{186.8}$$

is proportional to a spherical harmonic of the first order so that the matrix element must necessarily vanish if the electron is not emitted in a P state.

Let the final state be approximately described by a plane wave,

$$|f\rangle = V^{-\frac{1}{2}} e^{i k_f r \cos\gamma} = \frac{1}{kr} \sum_{l=0}^{\infty} (2l+1) i^l j_l(k_f r) P_l(\cos\gamma) \tag{186.9}$$

where γ is the angle between the directions of $\boldsymbol{k}_f$ with polar angles Θ, Φ and $\boldsymbol{r}$ with angles ϑ and φ. Then only the term $l=1$ (P term) can contribute to the matrix element so that

$$\langle f | \frac{\partial}{\partial x} | i \rangle = \frac{3i}{\sqrt{V}} \int_0^{\infty} dr \, r^2 \frac{j_1(kr)}{kr} \frac{d|i\rangle}{dr} \oint d\Omega \cos\gamma \sin\vartheta \cos\varphi .$$

Because of

$$\cos\gamma = \cos\vartheta \cos\Theta + \sin\vartheta \sin\Theta (\cos\varphi \cos\Phi + \sin\varphi \sin\Phi)$$

the angular integral yields

$$\sin\Theta \cos\Phi \oint d\Omega \sin^2\vartheta \cos^2\varphi = \frac{4\pi}{3} \sin\Theta \cos\Phi$$

so that we arrive at

$$\langle f|\frac{\partial}{\partial r}|i\rangle = \frac{4\pi i}{\sqrt{V}}\sin\Theta\cos\Phi\int_0^\infty dr\, r^2\frac{j_1(k_f r)}{k_f r}\frac{d|i\rangle}{dr}$$

and, according to (186.7),

$$\frac{d\sigma}{d\Omega_f} = \frac{8\pi e^2}{mc\omega k_f}\left\{\int_0^\infty dr\, r\, j_1(k_f r)\frac{d|i\rangle}{dr}\right\}^2 \sin^2\Theta\cos^2\Phi. \qquad (186.10)$$

Here the radial function j_1 stemming from the plane-wave approximation should be replaced by a more correct expression for quantitative calculations. The angular distribution of the photo-electrons, however, is correct and in complete agreement with the classical expectation, since $\sin^2\Theta\cos^2\Phi$ has a maximum in the direction x of the electrical field strength.

NB. With

$$|i\rangle = \pi^{-\frac{1}{2}} a^{-\frac{3}{2}} e^{-\frac{r}{a}}; \qquad a = \frac{\hbar^2}{(Z-s)me^2}$$

for a K shell electron (screening constant s, cf. Problem 178) we get for the integral in (186.10)

$$J = \int_0^\infty dr\, r\, j_1(k_f r)\frac{d|i\rangle}{dr} = -\frac{1}{k_f^2 a^2\sqrt{\pi a}}\int_0^\infty dx\, x\left(\frac{\sin x}{x} - \cos x\right) e^{-\frac{x}{k_f a}}$$

where $x = k_f r$. This integral can be evaluated in an elementary way and yields

$$J = -\frac{2}{\sqrt{\pi a}} \cdot \frac{k_f^2 a^2}{(1+k_f^2 a^2)^2}.$$

This formula can only hold for $k_f a \gg 1$, because otherwise the plane-wave approximation would be quite insufficient; so we may write

$$J \simeq -\frac{2}{\sqrt{\pi a}}(k_f a)^{-2}$$

and

$$\frac{d\sigma}{d\Omega_f} = \frac{32 e^2}{mca^5} \cdot \frac{1}{\omega k_f^5}$$

or, with

$$\frac{\hbar^2 k_f^2}{2m} = \hbar\omega_f,$$

finally,

$$\frac{d\sigma}{d\Omega_f} \simeq 8(Z-s)^5 \sin^2\Theta\cos^2\Phi \cdot \frac{e^2}{\hbar c} \cdot \frac{m^2 e^{10}}{\hbar^7} \cdot \frac{1}{\omega\omega_f^2 k_f}.$$

This formula shows the main features of more exact calculations: rapid increase with $Z-s$, rapid decrease with growing quantum energy $\hbar\omega$, roughly proportional to $\omega^{-3.5}$, order of magnitude in atomic units $e^2/\hbar c$, and correct angular distribution.

Literature. Stobbe, M.: Ann. Physik **7**, 661 (1930). — *Retardation effects in hydrogen:* Sommerfeld, A., Schur, G., Ann. Physik **4**, 409 (1930). — *Relativistic treatment:* Sauter, F., Ann. Physik **11**, 454 (1931).

187. Dispersion of light. Oscillator strengths

A light wave as defined in the preceding problem (but take $\delta=0$) interacts with an atom. The induced polarization shall be calculated from which the oscillator strengths may be derived. Only one electron shall be taken into account, and all matrix elements, neglecting retardation, reduced to matrix elements of electrical dipole moments.

Solution. In the Dirac notation of Problem 181 the atomic wave function under the action of the light wave may be written

$$\psi=\sum_l a_l(t)|l\rangle\, e^{-i\omega_l t}$$

if $|l\rangle$ is a state of the unperturbed atom. Using (182.2) for the coefficient $a_l(t)$ and omitting all switch-on effects of the light wave, we arrive at

$$\psi=|0\rangle\, e^{-i\omega_0 t}-\frac{1}{\hbar}\sum_l\left\{\langle l|\mathrm{W}|0\rangle\frac{e^{i(\omega_l-\omega_0-\omega)t}}{\omega_l-\omega_0-\omega}+\langle l|\mathrm{W}^\dagger|0\rangle\frac{e^{i(\omega_l-\omega_0+\omega)t}}{\omega_l-\omega_0+\omega}\right\}|l\rangle\, e^{-i\omega_l t}.$$

Here $|0\rangle$ denotes the atomic ground state and $|l\rangle$ any excited state so that $\omega_l-\omega_0>0$ and only the first term in the sum shows resonance. Neglecting the other term we then may start from the wave function

$$\psi=\left\{|0\rangle-\frac{1}{\hbar}e^{-i\omega t}\sum_l\frac{\langle l|\mathrm{W}|0\rangle}{\omega_l-\omega_0-\omega}|l\rangle\right\}e^{-i\omega_0 t}. \qquad (187.1)$$

Now we know that for the optical properties the induced dipole moment $\boldsymbol{p}_{\mathrm{ind}}$ plays the essential role. It follows from (187.1) according to

$$\boldsymbol{p}_{\mathrm{ind}}=-e\left\{\int d^3x\,\psi^*\boldsymbol{r}\,\psi-\langle 0|\boldsymbol{r}|0\rangle\right\}. \qquad (187.2)$$

Neglecting second-order corrections we thus get

$$\boldsymbol{p}_{\mathrm{ind}}=\frac{e}{\hbar}\sum_l\frac{\langle 0|\boldsymbol{r}|l\rangle\langle l|\mathrm{W}|0\rangle\, e^{-i\omega t}+\langle l|\boldsymbol{r}|0\rangle\langle l|\mathrm{W}|0\rangle^*\, e^{i\omega t}}{\omega_l-\omega_0-\omega}. \qquad (187.3)$$

This expression can be much simplified if we replace the matrix elements $\langle l|\mathsf{W}|0\rangle$ of the interaction

$$\mathsf{W} = -\frac{e\hbar}{2m\omega}\mathscr{E}_0 i \frac{\partial}{\partial x} \tag{187.4}$$

by matrix elements of the atomic moment $\boldsymbol{p}$, in field direction,

$$\langle l|p_x|0\rangle = -e\langle l|x|0\rangle. \tag{187.5}$$

This can be accomplished by using the relation

$$\langle l|\frac{\partial}{\partial x}|k\rangle = \frac{m}{\hbar}(\omega_k-\omega_l)\langle l|x|k\rangle \tag{187.6}$$

holding between any pair of states $|l\rangle$ and $|k\rangle$.

Eq. (187.6) may be derived from the two Schrödinger equations

$$\left\{-\frac{\hbar^2}{2m}\nabla^2+V\right\}|k\rangle = \hbar\omega_k|k\rangle; \quad \left\{-\frac{\hbar^2}{2m}\nabla^2+V\right\}\langle l| = \hbar\omega_l\langle l|$$

from which there follows, by multiplying by x, forming matrix elements and subtracting both equations in order to eliminate the term with V:

$$-\frac{\hbar^2}{2m}\{\langle l|x\nabla^2|k\rangle - \langle k|x\nabla^2|l\rangle^*\} = \hbar(\omega_k-\omega_l)\langle l|x|k\rangle. \tag{187.6'}$$

Now, according to

$$\int d^3x(vx)\nabla^2 u = -\int d^3x\nabla(vx)\cdot\nabla u = -\int d^3x\left(x\nabla v\cdot\nabla u + v\frac{\partial u}{\partial x}\right),$$

the curly bracket on the left-hand side may be reshaped into

$$-\langle l|\frac{\partial}{\partial x}|k\rangle + \langle k|\frac{\partial}{\partial x}|l\rangle^* = -2\langle l|\frac{\partial}{\partial x}|k\rangle,$$

proportional to the left-hand side of (187.6) whereas on the right-hand side of (187.6′) we already have the dipole matrix element wanted.

Using (187.6), we may write the induced dipole moment (187.3) in the simpler form

$$\boldsymbol{p}_{\text{ind}} = -\frac{\mathscr{E}_0}{2i\hbar}\sum_l \frac{\langle 0|\boldsymbol{p}|l\rangle\langle l|p_x|0\rangle e^{-i\omega t} - \langle l|\boldsymbol{p}|0\rangle\langle l|p_x|0\rangle^* e^{i\omega t}}{\omega_l-\omega_0-\omega}. \tag{187.7}$$

In a statistical distribution of atoms with dipole moments $\boldsymbol{p}$ oriented in all directions, the y and z components of $\boldsymbol{p}_{\text{ind}}$ will cancel and an induced dipole moment remain only in the x direction, i.e. parallel to the electrical field applied. In the first term of (187.7) we further use the hermiticity of $\langle 0|\boldsymbol{p}|l\rangle = \langle l|\boldsymbol{p}|0\rangle^*$ and then the statistical average relation

$$\overline{|\langle l|p_x|0\rangle|^2} = \tfrac{1}{3}|\langle l|\boldsymbol{p}|0\rangle|^2$$

where the expression on the right-hand side needs no further averaging, being independent of the atom's orientation. Thus we arrive at the result

$$\boldsymbol{p}_{\text{ind}} = \frac{\mathscr{E}_0}{3\hbar} \sum_l \frac{|\langle l|\boldsymbol{p}|0\rangle|^2}{\omega_l - \omega_0 - \omega} \sin\omega t .$$

Since $\mathscr{E}_0 \sin\omega t = \mathscr{E}$ is the instantaneous value of the field strength we may define atomic polarizability α in the usual way by $\boldsymbol{p}_{\text{ind}} = \alpha\mathscr{E}$ so that

$$\alpha = \frac{1}{3} \sum_l \frac{|\langle l|\boldsymbol{p}|0\rangle|^2}{\hbar(\omega_l - \omega_0 - \omega)} . \tag{187.8}$$

In classical optics, the index of refraction, n, is derived from the formula

$$\frac{n^2-1}{n^2+2} = \frac{4\pi}{3} N\alpha \tag{187.9}$$

with N the number of atoms per unit volume. (This expression is called the *refraction*.) The classical polarizability is then evaluated as a sum over all the electrons contributing (subscript λ) and is written in the form

$$\frac{n^2-1}{n^2+2} = \frac{4\pi}{3} N \frac{e^2}{m} \sum_\lambda \frac{f_\lambda}{\omega_\lambda^2 - \omega^2} \tag{187.10}$$

where ω_λ is the eigenfrequency of the λ'th electron and f_λ, the so-called *oscillator strength*, gives the number of electrons per atom in a state of eigenfrequency ω_λ, in this classical picture. That the oscillator strengths turned out experimentally not to be integers raised the first doubts in this classical picture.

The quantum theoretical formula (187.8) gives formally a very similar result. We may write,

$$\alpha = \frac{1}{3} \sum_l \frac{(\omega_l - \omega_0 + \omega)|\langle l|\boldsymbol{p}|0\rangle|^2}{\hbar[(\omega_l - \omega_0)^2 - \omega^2]} \simeq \frac{2\omega}{3\hbar} \sum_l \frac{|\langle l|\boldsymbol{p}|0\rangle|^2}{(\omega_l - \omega_0)^2 - \omega^2} ;$$

then (187.9) leads to the refraction

$$\frac{n^2-1}{n^2+2} = \frac{4\pi}{3} N \frac{e^2}{m} \sum_l \frac{f_l}{(\omega_l - \omega_0)^2 - \omega^2} \tag{187.11}$$

with the oscillator strengths

$$f_l = \frac{2m\omega}{3\hbar e^2} |\langle l|\boldsymbol{p}|0\rangle|^2 . \tag{187.12}$$

The formal similarity of the quantum theoretical result (187.11) with the classical one, (187.10), however, is deceptive. In Eq. (187.11) the sum does not run over electrons but over excited states so that summation over a multitude of terms is necessary even for our one-electron problem. The eigenfrequencies ω_λ are replaced by frequency differences, $\omega_l - \omega_0$. Finally, the oscillator strengths f_l no longer mean numbers of electrons but are rather involved intensity constants to be computed from dipole transition matrix elements according to (187.12). It is therefore no longer surprising that these numbers turn out not to be integers.

Problem 188. Spin flip in a magnetic resonance device

Let a particle of spin $\frac{1}{2}\hbar$ and of magnetic moment μ pass in y direction through an homogeneous magnetic field $\mathscr{H}_0$ parallel to the z axis. The particle spin in this field will be oriented in either $+z$ or $-z$ direction. Let us assume it to point in positive z direction. When passing the point $y=0$ at the time $t=0$, the particle enters an additional homogeneous field $\mathscr{H}'$ parallel to the x axis. It leaves this auxiliary field at $y=l$ at $t=t_0$. What is the probability of a spin flip during this time interval?

Solution. In the Schrödinger equation

$$-\frac{\hbar}{i}\dot{\psi} = -\frac{\hbar^2}{2m}\nabla^2\psi - \boldsymbol{\mu}\cdot\mathscr{H}\psi \tag{188.1}$$

the last term is the interaction energy of the magnetic field $\mathscr{H}$ with the magnetic moment $\boldsymbol{\mu}$ of the particle. The latter is defined as the vector operator

$$\boldsymbol{\mu} = \mu\boldsymbol{\sigma} \tag{188.2}$$

with $\boldsymbol{\sigma}$ the spin vector whose three components are the Pauli matrices (cf. Problem 129).

For $t<0$, only the field $\mathscr{H}_0 \| z$ is acting on the particle; the solution of (188.1) then is

$$\psi = e^{i(ky-\omega t)}\begin{pmatrix}1\\0\end{pmatrix} \tag{188.3}$$

and the energy of the particle

$$\hbar\omega = \frac{\hbar^2 k^2}{2m} - \mu\mathscr{H}_0. \tag{188.4}$$

If now, at $t=0$, the field $\mathcal{H}'\|x$ is switched on, the state of the particle undergoes a change so that its wave function may be written

$$\psi = e^{i(ky-\omega_0 t)}\left\{a(t)\begin{pmatrix}1\\0\end{pmatrix}+b(t)\begin{pmatrix}0\\1\end{pmatrix}\right\} = e^{i(ky-\omega_0 t)}\begin{pmatrix}a(t)\\b(t)\end{pmatrix} \tag{188.5}$$

with the abbreviation

$$\hbar\omega = \frac{\hbar^2 k^2}{2m}. \tag{188.6}$$

If path curvature by the Lorentz force acting perpendicular to the y direction may be neglected, it is safe to assume that the momentum $\hbar k$ in y direction is unchanged throughout.

Putting (188.5) in (188.1) we have to be a little cautious about the magnetic interaction:

$$(\boldsymbol{\mu}\cdot\mathcal{H}) = \mu(\mathcal{H}_0\sigma_z+\mathcal{H}'\sigma_x) = \mu\begin{pmatrix}\mathcal{H}_0 & \mathcal{H}'\\ \mathcal{H}' & -\mathcal{H}_0\end{pmatrix}$$

and therefore

$$(\boldsymbol{\mu}\cdot\mathcal{H})\begin{pmatrix}a\\b\end{pmatrix} = \mu\begin{pmatrix}\mathcal{H}_0 a+\mathcal{H}' b\\ \mathcal{H}' a-\mathcal{H}_0 b\end{pmatrix}.$$

Separating the two components of the Schrödinger equation, we thus arrive at

$$\left.\begin{aligned} -\frac{\hbar}{i}\dot{a} &= -\mu(\mathcal{H}_0 a+\mathcal{H}' b);\\ -\frac{\hbar}{i}\dot{b} &= -\mu(\mathcal{H}' a-\mathcal{H}_0 b).\end{aligned}\right\} \tag{188.7}$$

To solve these equations we set

$$a(t) = A e^{i\omega' t}; \qquad b(t) = B e^{i\omega' t};$$

then (188.7) becomes the algebraic set of linear equations

$$(\mu\mathcal{H}_0-\hbar\omega')A+\mu\mathcal{H}' B=0;$$
$$\mu\mathcal{H}' A-(\mu\mathcal{H}_0+\hbar\omega')B=0.$$

The determinant vanishes for two different values of ω',

$$\omega' = \pm\frac{\mu}{\hbar}\sqrt{\mathcal{H}_0^2+\mathcal{H}'^2}; \tag{188.8}$$

in what follows we shall write ω' and $-\omega'$ for these two roots. Then the set (188.7) will be solved by

$$a(t)=A_{\mathrm{I}}\,\mathrm{e}^{i\omega' t}+A_{\mathrm{II}}\,\mathrm{e}^{-i\omega' t};$$
$$b(t)=B_{\mathrm{I}}\,\mathrm{e}^{i\omega' t}+B_{\mathrm{II}}\,\mathrm{e}^{-i\omega' t}$$

with

$$B_{\mathrm{I,II}}=\frac{\pm\sqrt{\mathscr{H}_0^2+\mathscr{H}'^2}-\mathscr{H}_0}{\mathscr{H}'}A_{\mathrm{I,II}}\,.$$

There still remain two integration constants, A_{I} and A_{II}, to be determined from the initial conditions,

$$a(0)=1;\qquad b(0)=0. \tag{188.9}$$

This leads to

$$A_{\mathrm{I}}=\frac{\sqrt{\mathscr{H}_0^2+\mathscr{H}'^2}+\mathscr{H}_0}{2\sqrt{\mathscr{H}_0^2+\mathscr{H}'^2}};\qquad A_{\mathrm{II}}=\frac{\sqrt{\mathscr{H}_0^2+\mathscr{H}'^2}-\mathscr{H}_0}{2\sqrt{\mathscr{H}_0^2+\mathscr{H}'^2}}$$

and yields after some straightforward reshaping the formulae

$$\left.\begin{aligned} a(t)&=\cos\omega' t+i\frac{\mathscr{H}_0}{\sqrt{\mathscr{H}_0^2+\mathscr{H}'^2}}\sin\omega' t;\\ b(t)&=i\frac{\mathscr{H}'}{\sqrt{\mathscr{H}_0^2+\mathscr{H}'^2}}\sin\omega' t.\end{aligned}\right\} \tag{188.10}$$

It can easily be checked that

$$|a(t)|^2+|b(t)|^2=1\,.$$

The probability of spin flip, i.e. of finding the particle with spin downward, in the negative z direction, after its leaving the auxiliary field $\mathscr{H}'$ at the time $t=t_0=l/v$ will be, according to (188.10) and (188.8),

$$|b(t_0)|^2=\frac{\mathscr{H}'^2}{\mathscr{H}_0^2+\mathscr{H}'^2}\sin^2\left\{\frac{\mu}{\hbar}\sqrt{\mathscr{H}_0^2+\mathscr{H}'^2}\,t_0\right\}. \tag{188.11}$$

Eq. (188.11) shows that the experimental device may be used to determine the magnetic moment of an atom of spin $\frac{1}{2}\hbar$ (as an alkali atom in its ground state). If atoms are focussed if they do not undergo spin flip, but defocussed if they do, and the magnetic fields are varied during the course of the experiment, the beam intensity will become a minimum,

$$|a(t_0)|^2_{\min} = \frac{\mathcal{H}_0^2}{\mathcal{H}_0^2 + \mathcal{H}'^2}$$

if

$$\frac{\mu}{\hbar}\sqrt{\mathcal{H}_0^2 + \mathcal{H}'^2}\,\frac{l}{v} = \frac{\pi}{2}. \tag{188.12}$$

Such a determination is, of course, possible only if a velocity selection is applied and v is well known. Our representation of the method has necessarily been simplified by neglecting detail like the deflection by the Lorentz force, field inhomogeneities used for focussing, stray fields and —most important—the non-static changes of the magnetic moment by the Zeeman effect. Our problem corresponds rather to the Paschen-Back effect of decoupled moments which, however, may contradict the idea of fields weak enough to allow perpendicular momentum transfer to be neglected.

NB. A particle "at the point $y=0$ at the time $t=0$, etc." should, of course, be described by a wave packet, cf. Problem 17. For the present purpose, however, this is of little avail and has therefore been omitted.

VI. The Relativistic Dirac Equation

Remark. In this chapter we use the fourth coordinate $x_4 = ict$ and Euclidian metric. Greek subscripts (e.g. x_μ) run over $\mu = 1,2,3,4$, Latin subscripts (x_k) over $k = 1,2,3$, only.

Problem 189. Iteration of the Dirac equation

To derive commutation relations of the γ's from the relativistic dispersion law of plane Dirac waves. An irreducible representation of the γ's shall then be given that makes γ_4 diagonal.

Solution. If the Dirac equation for the force-free case,

$$\sum_\mu \gamma_\mu \partial_\mu \psi + \varkappa \psi = 0 \tag{189.1}$$

is to be solved by a plane wave

$$\psi = C\, e^{i(\boldsymbol{k}\boldsymbol{r} - \omega t)} \tag{189.2}$$

the γ's must satisfy the algebraic relation

$$i \sum_\mu k_\mu \gamma_\mu + \varkappa = 0; \qquad k_4 = \frac{\omega}{c}. \tag{189.3}$$

Now the γ's must be independent of the special choice of the k_μ's. The latter can only be eliminated from (189.3) by using the relativistic dispersion law,

$$\sum_\mu k_\mu^2 \equiv k^2 - \frac{\omega^2}{c^2} = -\varkappa^2, \tag{189.4}$$

which can be constructed from (189.3) by iteration:

$$\varkappa^2 = \left(i \sum_\mu k_\mu \gamma_\mu \right)^2 = - \sum_\mu \sum_\nu k_\mu k_\nu \gamma_\mu \gamma_\nu . \tag{189.5}$$

The latter equation becomes identical with (189.4) if, and only if, the double sum is reduced to diagonal terms of appropriate normalization, in fact, if

$$\gamma_\mu \gamma_\nu + \gamma_\nu \gamma_\mu = 2\delta_{\mu\nu}. \tag{189.6}$$

If an analogous procedure is applied directly to the Dirac equation (189.1) without using plane waves, we have

$$\varkappa^2 \psi = \Big(\sum_\mu \gamma_\mu \partial_\mu\Big)^2 \psi = \sum_\mu \sum_\nu \gamma_\mu \gamma_\nu \partial_\mu \partial_\nu .$$

With the anticommutators (189.6) this simply leads to

$$\Box^2 \psi - \varkappa^2 \psi = 0, \tag{189.7}$$

i.e. to the Klein-Gordon equation.

There exist irreducible representations of the γ_μ's by 4×4 matrices. If γ_μ is one such representation, any unitary transformation $U^\dagger \gamma_\mu U$ will produce another set. Therefore, one of the matrices, say γ_4, may always be supposed to be diagonal. As $\gamma_4^2 = 1$, its eigenvalues must be $+1$ and -1. Thus we may start the construction of the matrix set by writing

$$\gamma_k = \begin{pmatrix} \boldsymbol{A}_k & \boldsymbol{B}_k \\ \boldsymbol{C}_k & \boldsymbol{D}_k \end{pmatrix} \quad (k=1,2,3); \quad \gamma_4 = \begin{pmatrix} \boldsymbol{1} & \boldsymbol{0} \\ \boldsymbol{0} & -\boldsymbol{1} \end{pmatrix} \tag{189.8}$$

with all bold face letters meaning 2×2 matrices. From (189.6) we then find

$$\gamma_i \gamma_k + \gamma_k \gamma_i = \begin{pmatrix} \boldsymbol{B}_i \boldsymbol{C}_k + \boldsymbol{B}_k \boldsymbol{C}_i; & \boldsymbol{0} \\ \boldsymbol{0} & \boldsymbol{C}_i \boldsymbol{B}_k + \boldsymbol{C}_k \boldsymbol{B}_i \end{pmatrix} = 2\delta_{ik}; \tag{189.9a}$$

$$\gamma_k \gamma_4 + \gamma_4 \gamma_k = \begin{pmatrix} 2\boldsymbol{A}_k; & \boldsymbol{0} \\ \boldsymbol{0}; & -2\boldsymbol{D}_k \end{pmatrix} = 0; \tag{189.9b}$$

From (189.9b) we get $\boldsymbol{A}_k = 0$, $\boldsymbol{D}_k = 0$ for the three first matrices, and from (189.9a) it follows that

$$\boldsymbol{B}_k \boldsymbol{C}_k = 1; \quad \boldsymbol{B}_i \boldsymbol{C}_k + \boldsymbol{C}_k \boldsymbol{B}_i = 0. \tag{189.10}$$

These are three anticommutation rules for 2×2 matrices allowing reduction to the Pauli matrices, σ_k (cf. Problem 129). If a and b are two numbers, Eqs. (189.10) are satisfied by

$$\boldsymbol{B}_k = a\sigma_k; \quad \boldsymbol{C}_k = b\sigma_k; \quad ab = 1, \tag{189.11}$$

so that any representation

$$\gamma_k = \begin{pmatrix} \boldsymbol{0}; & b^{-1}\sigma_k \\ b\sigma_k; & \boldsymbol{0} \end{pmatrix}; \quad \gamma_4 = \begin{pmatrix} \boldsymbol{1} & \boldsymbol{0} \\ \boldsymbol{0} & -\boldsymbol{1} \end{pmatrix} \tag{189.12}$$

satisfies the commutation rules (189.6). The *standard representation* (often used in the following problems) is obtained by the choice $b=i$; the four matrices then become

$$\gamma_1=\begin{pmatrix}0&0&0&-i\\0&0&-i&0\\0&i&0&0\\i&0&0&0\end{pmatrix};\quad \gamma_2=\begin{pmatrix}0&0&0&-1\\0&0&1&0\\0&1&0&0\\-1&0&0&0\end{pmatrix};$$

$$\gamma_3=\begin{pmatrix}0&0&-i&0\\0&0&0&i\\i&0&0&0\\0&-i&0&0\end{pmatrix};\quad \gamma_4=\begin{pmatrix}1&0&0&0\\0&1&0&0\\0&0&-1&0\\0&0&0&-1\end{pmatrix}. \tag{189.13}$$

NB. With $b=a=1$ one obtains the set of matrices

$$\alpha_k=\begin{pmatrix}0&\sigma_k\\\sigma_k&0\end{pmatrix} \tag{189.14}$$

instead of the three γ_k's. Together with $\gamma_4\equiv\beta$ as above, they obey the same commutation laws. They are connected with the γ's by the relations

$$\gamma_k=-i\beta\alpha_k;\quad \gamma_4=\beta. \tag{189.15}$$

The α's are used to advantage in the Dirac hamiltonian, cf. Problem 200.

Problem 190. Plane Dirac waves of positive energy

To determine in standard representation the spinor amplitudes of plane Dirac waves of positive and negative helicity, but of positive energy only.

Solution. With

$$\psi=C\,e^{i(\boldsymbol{k}\boldsymbol{r}-\omega t)} \tag{190.1}$$

we obtain [cf. Eq. (189.3)] in standard representation the four component equations

$$\begin{aligned}
(k_x-ik_y)C_4+k_zC_3+\left(-\frac{\omega}{c}+\varkappa\right)C_1&=0,\\
(k_x+ik_y)C_3-k_zC_4+\left(-\frac{\omega}{c}+\varkappa\right)C_2&=0,\\
-(k_x-ik_y)C_2-k_zC_1+\left(\frac{\omega}{c}+\varkappa\right)C_3&=0,\\
-(k_x+ik_y)C_1+k_zC_2+\left(\frac{\omega}{c}+\varkappa\right)C_4&=0.
\end{aligned} \tag{190.2}$$

Using the abbreviations

$$k\eta = \frac{\omega}{c} - \varkappa; \qquad \frac{k}{\eta} = \frac{\omega}{c} + \varkappa, \tag{190.3}$$

we have in η a suitable parameter in which to express the most important particle quantities, viz. its momentum

$$p = \hbar k = mc\frac{2\eta}{1-\eta^2}, \tag{190.4a}$$

its kinetic energy

$$E = \hbar\omega = mc^2\frac{1+\eta^2}{1-\eta^2} \tag{190.4b}$$

and its velocity

$$v = c\frac{2\eta}{1+\eta^2}. \tag{190.4c}$$

It is also useful to introduce two polar angles ϑ and φ, determining the direction of the vector $\boldsymbol{k}$,

$$k_x \pm i k_y = k \sin\vartheta\, e^{\pm i\varphi}; \qquad k_z = k\cos\vartheta. \tag{190.5}$$

The equations (190.2) may then be written in the form

$$\begin{aligned}
\sin\vartheta\, e^{-i\varphi} C_4 + \cos\vartheta\, C_3 - \eta\, C_1 &= 0, \\
\sin\vartheta\, e^{+i\varphi} C_3 - \cos\vartheta\, C_4 - \eta\, C_2 &= 0, \\
-\sin\vartheta\, e^{-i\varphi} C_2 - \cos\vartheta\, C_1 + \frac{1}{\eta} C_3 &= 0, \\
-\sin\vartheta\, e^{+i\varphi} C_1 + \cos\vartheta\, C_2 + \frac{1}{\eta} C_4 &= 0.
\end{aligned} \tag{190.6}$$

This is a homogeneous system of linear equations for the four C_μ's. Its determinant vanishes, as may be easily checked, but it can also be factorized in two factors each of which is zero. It is not therefore possible to express all the C_μ's as multiples of one of them, and there remain two C_μ's to be chosen arbitrarily. Therefore we shall proceed in another way.

We first look for eigenfunctions of the helicity operator

$$\mathsf{h} = \frac{1}{k}\sum_j \sigma_j k_j, \tag{190.7}$$

i.e. of the operator "spin component in the direction of $\boldsymbol{k}$". Since, in standard representation, the spin matrices consist merely of doubled Pauli 2×2 matrices s_j,

$$\sigma_j = \begin{pmatrix} s_j & 0 \\ 0 & s_j \end{pmatrix},$$

we find the definition (190.7):

$$\mathsf{h} = \begin{pmatrix} \cos\vartheta; & \sin\vartheta\, e^{-i\varphi}; & 0; & 0 \\ \sin\vartheta\, e^{i\varphi}; & -\cos\vartheta; & 0; & 0 \\ 0; & 0; & \cos\vartheta; & \sin\vartheta\, e^{-i\varphi} \\ 0; & 0; & \sin\vartheta\, e^{i\varphi}; & -\cos\vartheta \end{pmatrix}. \tag{190.8}$$

Let the eigenvalue be h; then the eigenvalue problem $\mathsf{h}C = h\cdot C$ can be decomposed into one pair of equations for C_1 and C_2,

$$\begin{aligned} C_1\cos\vartheta + C_2\sin\vartheta\, e^{-i\varphi} &= hC_1; \\ C_1\sin\vartheta\, e^{i\varphi} - C_2\cos\vartheta &= hC_2 \end{aligned} \tag{190.9}$$

and the same pair for C_3 and C_4. The determinant of (190.9) vanishes if $h=\pm 1$; thus we arrive at two solutions:

$h=+1$ (spin parallel to $\boldsymbol{k}$):

$$C_2 = \tan\frac{\vartheta}{2}\, e^{i\varphi} C_1; \qquad C_4 = \tan\frac{\vartheta}{2}\, e^{i\varphi} C_3 \tag{190.10}$$

and

$h=-1$ (spin antiparallel to $\boldsymbol{k}$):

$$C_2 = -\cot\frac{\vartheta}{2}\, e^{i\varphi} C_1; \qquad C_4 = -\cot\frac{\vartheta}{2}\, e^{i\varphi} C_3. \tag{190.11}$$

C_1 and C_3 may still be chosen arbitrarily.

Let us now put these results into the set of Eqs. (190.6). With the elementary identities

$$\sin\vartheta\tan\frac{\vartheta}{2} + \cos\vartheta = 1; \qquad \sin\vartheta\cot\frac{\vartheta}{2} - \cos\vartheta = 1;$$

$$\sin\vartheta - \cos\vartheta\tan\frac{\vartheta}{2} = \tan\frac{\vartheta}{2}; \qquad \sin\vartheta + \cos\vartheta\cot\frac{\vartheta}{2} = \cot\frac{\vartheta}{2}$$

all four equations (190.6) reduce to

$$C_3 = \eta C_1 \quad \text{for } h = +1 \tag{190.12}$$

and to

$$C_3 = -\eta C_1 \quad \text{for } h = -1, \tag{190.13}$$

respectively. If, therefore, we normalize so that[1]

$$\int_V d^3x\, \psi^\dagger \psi = \int_V d^3x\, C^\dagger C = 1 \tag{190.14}$$

we obtain the following spinor amplitudes:

$$\text{for } h = +1 \qquad C_+ = \frac{1}{\sqrt{V(1+\eta^2)}} \begin{pmatrix} \cos\frac{\vartheta}{2}\, e^{-\frac{i}{2}\varphi} \\ \sin\frac{\vartheta}{2}\, e^{+\frac{i}{2}\varphi} \\ \eta\cos\frac{\vartheta}{2}\, e^{-\frac{i}{2}\varphi} \\ \eta\sin\frac{\vartheta}{2}\, e^{+\frac{i}{2}\varphi} \end{pmatrix} \tag{190.15}$$

and

$$\text{for } h = -1 \qquad C_- = \frac{1}{\sqrt{V(1+\eta^2)}} \begin{pmatrix} \sin\frac{\vartheta}{2}\, e^{-\frac{i}{2}\varphi} \\ -\cos\frac{\vartheta}{2}\, e^{+\frac{i}{2}\varphi} \\ -\eta\sin\frac{\vartheta}{2}\, e^{-\frac{i}{2}\varphi} \\ \eta\cos\frac{\vartheta}{2}\, e^{+\frac{i}{2}\varphi} \end{pmatrix}. \tag{190.16}$$

NB. The unrelativistic case, according to (190.4c), leads to $\eta \ll 1$. The components ψ_3 and ψ_4 of the spinor then may be neglected, and we fall back upon the two-component Pauli spin theory.

---

[1] It should be noted that this normalization is Lorentz-invariant, the integral being proportional to the electric charge total inside the volume V. Another Lorentz-invariant normalization often used is $\bar{\psi}\psi = 1$.

Problem 191. Transformation properties of a spinor

How does a spinor ψ transform under infinitesimal Lorentz transformation?

Solution. An infinitesimal Lorentz transformation is defined by

$$x'_\mu = x_\mu + \sum_\rho \varepsilon_{\mu\rho} x_\rho; \qquad \varepsilon_{\mu\rho} = -\varepsilon_{\rho\mu}; \qquad |\varepsilon_{\mu\rho}| \ll 1. \tag{191.1}$$

Here all ε_{kl} are real, and the three ε_{k4} purely imaginary. The Dirac equation

$$\sum_\mu \gamma_\mu D_\mu \psi + \varkappa \psi = 0 \tag{191.2}$$

shall be transformed into

$$\sum_\mu \gamma_\mu D'_\mu \psi' + \varkappa \psi' = 0 \tag{191.2'}$$

with unchanged coefficients γ_μ and $\varkappa$. The operators D_μ are four-vectors transforming under the same law as the coordinates (191.1):

$$D'_\mu = D_\mu + \sum_\rho \varepsilon_{\mu\rho} D_\rho; \tag{191.3}$$

the transformation formula of ψ may be

$$\psi' = (1+\xi)\psi \tag{191.4}$$

with an infinitesimal ξ linear in the $\varepsilon_{\mu\rho}$'s which is a Clifford number.

We start with Eq. (191.2′) in which we put D'_μ from (191.3) and ψ' from (191.4):

$$\sum_\mu \gamma_\mu \Big(D_\mu + \sum_\rho \varepsilon_{\mu\rho} D_\rho\Big)(1+\xi)\psi + \varkappa(1+\xi)\psi = 0.$$

If we multiply from the left side by $(1-\xi)$, the last term goes over into $\varkappa\psi$, i. e. into the last term of (191.2). The operator ξ therefore must be chosen in such a way as to make

$$\sum_\mu (1-\xi)\gamma_\mu \Big(D_\mu + \sum_\rho \varepsilon_{\mu\rho} D_\rho\Big)(1+\xi)\psi = \sum_\mu \gamma_\mu D_\mu \psi. \tag{191.5}$$

From this equation ξ shall be determined. Neglecting all second-order contributions, we find

$$\sum_\mu (\gamma_\mu \xi - \xi \gamma_\mu) D_\mu \psi + \sum_\mu \sum_\rho \varepsilon_{\mu\rho} \gamma_\mu D_\rho \psi = 0$$

or, exchanging the dummies μ and ρ in the double sum (and writing ν instead of ρ),

$$\sum_\mu \Big\{(\gamma_\mu \xi - \xi \gamma_\mu) - \sum_\nu \varepsilon_{\mu\nu} \gamma_\nu\Big\} D_\mu \psi = 0.$$

Since this relation is supposed to hold for any ψ, each sum term is bound to vanish separately:

$$\gamma_\mu \xi - \xi \gamma_\mu = \sum_\nu \varepsilon_{\mu\nu} \gamma_\nu . \tag{191.6}$$

The only Clifford number linear in the $\varepsilon_{\mu\nu}$'s satisfying these four commutation relations is

$$\xi = \tfrac{1}{4} \sum_\rho \sum_\sigma \varepsilon_{\rho\sigma} \gamma_\rho \gamma_\sigma . \tag{191.7}$$

This can easily be proved by direct evaluation of the commutators (191.6): We have

$$\gamma_\mu \xi - \xi \gamma_\mu = \tfrac{1}{4} \sum_\rho \sum_\sigma \varepsilon_{\rho\sigma} (\gamma_\mu \gamma_\rho \gamma_\sigma - \gamma_\rho \gamma_\sigma \gamma_\mu);$$

here we find

$$\gamma_\rho \gamma_\sigma \gamma_\mu = \gamma_\rho (-\gamma_\mu \gamma_\sigma + 2\delta_{\mu\sigma}) = (\gamma_\mu \gamma_\rho - 2\delta_{\mu\rho}) \gamma_\sigma + 2\gamma_\rho \delta_{\mu\sigma}$$

so that

$$\gamma_\mu \gamma_\rho \gamma_\sigma - \gamma_\rho \gamma_\sigma \gamma_\mu = 2(\gamma_\sigma \delta_{\mu\rho} - \gamma_\rho \delta_{\mu\sigma});$$

therefore

$$\gamma_\mu \xi - \xi \gamma_\mu = \tfrac{1}{2} \sum_\rho \sum_\sigma \varepsilon_{\rho\sigma} (\gamma_\sigma \delta_{\mu\rho} - \gamma_\rho \delta_{\mu\sigma})$$
$$= \tfrac{1}{2} \Big(\sum_\sigma \varepsilon_{\mu\sigma} \gamma_\sigma - \sum_\rho \varepsilon_{\rho\mu} \gamma_\rho \Big) = \sum_\nu \varepsilon_{\mu\nu} \gamma_\nu ,$$

and that was to be proved.

Hence the spinor ψ transforms according to

$$\psi' = \psi + \tfrac{1}{4} \sum_\rho \sum_\sigma \varepsilon_{\rho\sigma} \gamma_\rho \gamma_\sigma \psi . \tag{191.8}$$

Problem 192. Lorentz covariants

Which Lorentz covariants of the form

$$G = \bar{\psi} \Gamma \psi ; \qquad \bar{\psi} = \psi^\dagger \gamma_4 \tag{192.1}$$

can be constructed, Γ being one of the 16 basis elements of the Clifford algebra?

Solution. The 16 basis elements of the algebra can be grouped into five sets as follows:

(1) $1,$

(2) $\gamma_1, \gamma_2, \gamma_3, \gamma_4,$

(3) $\gamma_1\gamma_2, \gamma_1, \gamma_3, \gamma_1\gamma_4, \gamma_2\gamma_3, \gamma_2\gamma_4, \gamma_3\gamma_4,$

(4) $\gamma_2\gamma_3\gamma_4, \gamma_3\gamma_4\gamma_1, \gamma_4\gamma_1\gamma_2, \gamma_1\gamma_2\gamma_3,$

(5) $\gamma_1\gamma_2\gamma_3\gamma_4 .$ (192.2)

Expressions of the form (192.1) will be formed with each of these five sets separately.

Before performing this programme in detail, however, let us investigate the transformation properties of any of the 16 quantities (192.1) under the infinitesimal Lorentz transformation

$$x'_\mu = x_\mu + \sum_\rho \varepsilon_{\mu\rho} x_\rho . \tag{192.3}$$

In the preceding problem it has been shown that ψ under this transformation becomes

$$\psi' = (1+\xi)\psi \tag{192.4a}$$

with

$$\xi = \tfrac{1}{4}\sum_\rho \sum_\sigma \varepsilon_{\rho\sigma}\gamma_\rho\gamma_\sigma ; \qquad \gamma_\mu \xi - \xi\gamma_\mu = \sum_\nu \varepsilon_{\mu\nu}\gamma_\nu . \tag{192.4b}$$

The transform of G is

$$G' = \psi'^\dagger \gamma_4 \Gamma \psi' = \psi^\dagger (1+\xi^\dagger)\gamma_4 \Gamma (1+\xi)\psi .$$

Formally we may write

$$G' = \bar{\psi}\Gamma'\psi ; \qquad \Gamma' = \gamma_4(1+\xi^\dagger)\gamma_4\Gamma(1+\xi). \tag{192.5}$$

This can be further simplified when we know a little more about the hermitian conjugate $\xi^\dagger$. According to Eq. (192.4b), it should obey the commutation relation

$$\xi^\dagger\gamma_4 - \gamma_4\xi^\dagger = \sum_k \varepsilon^*_{4k}\gamma_k$$

since $\gamma_\nu^\dagger = \gamma_\nu$. Now, the rotation angles ε_{4k} are purely imaginary so that $\varepsilon^*_{4k} = -\varepsilon_{4k}$, i. e.

$$\gamma_4\xi^\dagger - \xi^\dagger\gamma_4 = \sum_k \varepsilon_{4k}\gamma_k$$

and

$$\gamma_4\xi^\dagger\gamma_4 = \gamma_4\Big\{\gamma_4\xi^\dagger - \sum_k \varepsilon_{4k}\gamma_k\Big\} = \xi^\dagger - \sum_k \varepsilon_{4k}\gamma_4\gamma_k . \tag{192.6}$$

It further follows from (192.4b) with $\varepsilon^*_{kl} = \varepsilon_{kl}$ (real rotations in 3-space) that

$$\xi = \tfrac{1}{4}\sum_k\sum_l \varepsilon_{kl}\gamma_k\gamma_l + \tfrac{1}{2}\sum_k \varepsilon_{k4}\gamma_k\gamma_4 \tag{192.7a}$$

has the hermitian conjugate

$$\xi^\dagger = -\tfrac{1}{4}\sum_k\sum_l \varepsilon_{kl}\gamma_k\gamma_l + \tfrac{1}{2}\sum_k \varepsilon_{k4}\gamma_k\gamma_4 \tag{192.7b}$$

so that

$$\xi + \xi^\dagger = \sum_k \varepsilon_{k4}\gamma_k\gamma_4 . \tag{192.8}$$

Putting (192.8) into (192.6) we obtain

$$\gamma_4 \xi^\dagger \gamma_4 = -\xi. \tag{192.9}$$

With this last result, the "transformed operator" Γ', Eq. (192.5), may finally be written

$$\Gamma' = \Gamma + (\Gamma \xi - \xi \Gamma). \tag{192.10}$$

We now may easily apply this simple result to the five sets of quantities defined by (192.2), consecutively.

1. With $\Gamma = 1$, Eq. (192.10) gives immediately $\Gamma' = 1$ so that we find

$$G = \bar{\psi}\psi \to G' = \bar{\psi}\psi; \qquad G' = G. \tag{192.11}$$

The quantity G therefore behaves as a *scalar*.

2. With $\Gamma = \gamma_\mu$, Eqs. (192.10) and (192.4b) yield

$$\Gamma' = \gamma_\mu + \sum_\nu \varepsilon_{\mu\nu} \gamma_\nu$$

so that we arrive at the transformation formulae

$$G_\mu = \bar{\psi}\gamma_\mu \psi \to G'_\mu = G_\mu + \sum_\nu \varepsilon_{\mu\nu} G_\nu, \tag{192.12}$$

i. e. the G_μ's transform as the components of a *vector*.

3. It is suitable first to decompose the products $\gamma_\mu \gamma_\nu$ into a symmetrical and an antisymmetrical part,

$$\gamma_\mu \gamma_\nu = \tfrac{1}{2}(\gamma_\mu \gamma_\nu + \gamma_\nu \gamma_\mu) + \tfrac{1}{2}(\gamma_\mu \gamma_\nu - \gamma_\nu \gamma_\mu).$$

The first part reduces to $\delta_{\mu\nu}$, i. e. to the scalar (192.11) multiplied by the unit tensor. New evidence apparently comes only from the antisymmetrical part; so we confine our discussion to the combination

$$\sigma_{\mu\nu} = \tfrac{1}{2}(\gamma_\mu \gamma_\nu - \gamma_\nu \gamma_\mu). \tag{192.13}$$

According to (192.4b) we have

$$\xi \gamma_\mu \gamma_\nu = \Big(\gamma_\mu \xi - \sum_\rho \varepsilon_{\mu\rho} \gamma_\rho\Big)\gamma_\nu = \gamma_\mu \Big(\gamma_\nu \xi - \sum_\rho \varepsilon_{\nu\rho}\gamma_\rho\Big) - \sum_\rho \varepsilon_{\mu\rho}\gamma_\rho \gamma_\nu$$

or

$$\gamma_\mu \gamma_\nu \xi - \xi \gamma_\mu \gamma_\nu = \sum_\rho (\varepsilon_{\nu\rho}\gamma_\mu \gamma_\rho + \varepsilon_{\mu\rho}\gamma_\rho \gamma_\nu)$$

so that the transformation formula runs as follows:

$$G_{\mu\nu} = \bar{\psi}\sigma_{\mu\nu}\psi \to G'_{\mu\nu} = G_{\mu\nu} + \sum_\rho (\varepsilon_{\mu\rho} G_{\rho\nu} + \varepsilon_{\nu\rho} G_{\mu\rho}). \tag{192.14}$$

This is the behaviour of a *tensor* (of rank 2) under infinitesimal rotation.

4. The products of three γ's can be written in a simpler way by introducing the Clifford number

$$\gamma_5 = \gamma_1 \gamma_2 \gamma_3 \gamma_4. \tag{192.15}$$

Then the four products become

$$\gamma_2\gamma_3\gamma_4 = \gamma_1\gamma_5; \quad -\gamma_3\gamma_4\gamma_1 = \gamma_2\gamma_5; \quad \gamma_4\gamma_1\gamma_2 = \gamma_3\gamma_5; \quad -\gamma_1\gamma_2\gamma_3 = \gamma_4\gamma_5.$$

Since γ_5 anticommutes with all four γ_μ,

$$\gamma_\mu\gamma_5 + \gamma_5\gamma_\mu = 0, \tag{192.16}$$

it will commute with ξ so that Eq. (192.10) applied to $\Gamma = \gamma_\mu\gamma_5$ leads to

$$\Gamma' = \gamma_\mu\gamma_5 + (\gamma_\mu\gamma_5\xi - \xi\gamma_\mu\gamma_5) = [\gamma_\mu + (\gamma_\mu\xi - \xi\gamma_\mu)]\gamma_5.$$

Thus we essentially fall back upon case 2 and find the transformation laws of the components of a vector:

$$G_\mu = \bar{\psi}\gamma_\mu\gamma_5\psi \to G'_\mu = G_\mu + \sum_\nu \varepsilon_{\mu\nu} G_\nu. \tag{192.17}$$

Strictly speaking, this is not a (polar) vector but a *pseudovector*, as will be shown in the following problem.

5. According to (192.15) and (192.16) we obtain

$$G = \bar{\psi}\gamma_5\psi \to G' = G, \tag{192.18}$$

i. e. for this last combination there holds the transformation law of a scalar. We shall see in the next problem that the quantity is more correctly classified a *pseudoscalar*.

Problem 193. Parity transformation

How do the five Lorentz covariants of the preceding problem transform under reflection of space coordinates (i. e. under parity transformation)?

Solution. We start by investigating the behaviour of the spinor ψ under the reflection process under consideration. It is defined by the postulate that the Dirac equation

$$\sum_\mu \gamma_\mu D_\mu \psi + \varkappa\psi = 0 \tag{193.1}$$

shall transform into

$$\sum_\mu \gamma_\mu D'_\mu \psi' + \varkappa\psi' = 0 \tag{193.1'}$$

under the parity transformation

$$x'_k = -x_k; \quad x'_4 = x_4. \tag{193.2}$$

The same relations as (193.2) will hold for the operators ∂_μ. A more detailed investigation is necessary of the vector potential. The electric field components $\mathscr{E}_k$ are coupled with the components A_μ by the relations

$$\mathscr{E}_k = -\frac{1}{c}\dot{A}_k - \partial_k \Phi = -i(\partial_4 A_k - \partial_k A_4).$$

Since the electric field is a polar 3-vector, it changes sign with the x_k's under transformation. This leads to

$$A'_k = -A_k; \qquad A'_4 = A_4. \tag{193.3}$$

Thus A_μ undergoes the same transformation as ∂_μ and, in consequence thereof, D_μ transforms in the same way. Therefore, instead of Eq. (1′) we may write

$$-\sum_k \gamma_k D_k \psi' + \gamma_4 D_4 \psi' + \varkappa \psi' = 0.$$

It is immediately seen that with

$$\psi' = \gamma_4 \psi \tag{193.4}$$

this leads back to (193.1), thus determining the parity transformation of a spinor.

Any quantity

$$G = \bar{\psi} \Gamma \psi = \psi^\dagger \gamma_4 \Gamma \psi$$

then transforms into

$$G' = \psi'^\dagger \gamma_4 \Gamma \psi' = \psi^\dagger \Gamma \gamma_4 \psi = \bar{\psi} \gamma_4 \Gamma \gamma_4 \psi$$

so that we may write

$$G = \bar{\psi} \Gamma \psi \to G' = \bar{\psi} \Gamma' \psi; \qquad \Gamma' = \gamma_4 \Gamma \gamma_4. \tag{193.5}$$

Applied to the five covariants of the preceding problem we have

$$(1) \qquad G = \bar{\psi}\psi; \quad \Gamma = 1; \quad \Gamma' = 1; \quad G' = G; \tag{193.6}$$

$$(5) \quad G = \bar{\psi}\gamma_5\psi; \quad \Gamma = \gamma_5; \quad \Gamma' = \gamma_4\gamma_5\gamma_4 = -\gamma_5; \quad G' = -G. \tag{193.7}$$

The behaviour of these two quantities was the same under rotation, but it now is opposite under space reflection, (1) being called a (genuine) *scalar* and (5) a *pseudoscalar*.

$$\begin{aligned}(2) \quad G_\mu = \bar{\psi}\gamma_\mu\psi; \quad \Gamma = \gamma_\mu; \quad & \Gamma'_k = \gamma_4\gamma_k\gamma_4 = -\gamma_k; \quad G'_k = -G_k; \\ & \Gamma'_4 = \gamma_4 \qquad\qquad\qquad\quad G'_4 = +G_4.\end{aligned} \tag{193.8}$$

$$\begin{aligned}(4) \qquad & G_\mu = \bar{\psi}\gamma_\mu\gamma_5\psi; \quad \Gamma = \gamma_\mu\gamma_5; \\ & \Gamma'_k = \gamma_4\gamma_k\gamma_5\gamma_4 = +\gamma_k\gamma_5; \quad G'_k = +G_k; \\ & \Gamma'_4 = \gamma_5\gamma_4 = -\gamma_4\gamma_5; \quad G'_4 = -G_4.\end{aligned} \tag{193.9}$$

Both quantities behave as 4-vectors under rotation, but reversely under space reflection, (2) being called a (polar or genuine) *vector* and (4) an axial or *pseudovector*.

(3)
$$G_{\mu\nu} = \bar{\psi}\sigma_{\mu\nu}\psi; \quad \Gamma = \tfrac{1}{2}(\gamma_\mu\gamma_\nu - \gamma_\nu\gamma_\mu);$$
$$\Gamma'_{kl} = \Gamma_{kl}; \quad G'_{kl} = G_{kl} \qquad (193.10)$$
$$\Gamma'_{k4} = -\Gamma_{k4}; \quad G'_{k4} = -G_{k4}.$$

Since there is only one tensor, no further classification is necessary.

Problem 194. Charge conjugation

To construct from the spinor ψ solving the Dirac equation for a particle of charge e the charge conjugate spinor ψ_c describing the behaviour of a particle of the opposite charge $-e$.

Solution. Let us use the abbreviation $a_\mu = (e/\hbar c)A_\mu$ with A_μ the electromagnetic 4-potential in this problem. Then the Dirac equation for the particle of charge e runs

$$\sum_\mu \gamma_\mu(\partial_\mu - i a_\mu)\psi + \varkappa\psi = 0. \qquad (194.1)$$

The equation to be constructed for the opposite charge then must be

$$\sum_\mu \gamma_\mu(\partial_\mu + i a_\mu)\psi_c + \varkappa\psi_c = 0. \qquad (194.2)$$

The charge conjugate ψ_c which solves the latter equation shall be connected with the solution ψ of (194.1).

The operator $\partial_\mu + i a_\mu$ occurring in (194.2) can be introduced by using the adjoint equation of (194.1),

$$\sum_\mu (\partial_\mu + i a_\mu)\bar{\psi}\gamma_\mu - \varkappa\bar{\psi} = 0; \qquad \bar{\psi} = \psi^\dagger \gamma_4. \qquad (194.3)$$

Transposing the latter again, we find

$$\sum_\mu \tilde{\gamma}_\mu(\partial_\mu + i a_\mu)\tilde{\bar{\psi}} - \varkappa\tilde{\bar{\psi}} = 0$$

where

$$\tilde{\bar{\psi}} = \tilde{\gamma}_4\psi^*. \qquad (194.4)$$

If we multiply the last equation with a Clifford number C,

$$\sum_\mu C\tilde{\gamma}_\mu(\partial_\mu + i a_\mu)\tilde{\gamma}_4\psi^* - \varkappa C\tilde{\gamma}_4\psi^* = 0,$$

the result becomes identical with (194.2) if we choose C so that the two relations

$$-C\tilde{\gamma}_\mu\tilde{\gamma}_4\psi^* = \gamma_\mu\psi_c; \quad C\tilde{\gamma}_4\psi^* = \psi_c \tag{194.5}$$

are satisfied. Here ψ_c may be eliminated in order first to determine C:

$$-C\tilde{\gamma}_\mu\tilde{\gamma}_4\psi^* = \gamma_\mu C\tilde{\gamma}_4\psi^*$$

where we omit $\tilde{\gamma}_4\psi^*$ and are left with the four relations

$$\gamma_\mu C = -C\tilde{\gamma}_\mu \tag{194.6}$$

to determine C.

Since the problem is homogeneous, there is of course always an arbitrary factor in ψ_c. It is reasonable to fix it as far as possible by postulating that charge conjugation shall not alter the normalization,

$$\psi_c^\dagger\psi_c = \psi^\dagger\psi = (\psi^\dagger\psi)^*. \tag{194.7}$$

Now, from (194.5) we have

$$\psi_c^\dagger = \tilde{\psi}\gamma_4^* C^\dagger,$$

hence

$$\psi_c^\dagger\psi_c = \tilde{\psi}\gamma_4^* C^\dagger C\tilde{\gamma}_4\psi^* = (\psi^*)^\dagger(\gamma_4^* C^\dagger C\tilde{\gamma}_4)(\psi^*),$$

and that becomes identical with (194.7) if

$$\gamma_4^* C^\dagger C\tilde{\gamma}_4 = 1$$

or, since $\gamma_4^\dagger = \gamma_4$, $\tilde{\gamma}_4 = \gamma_4^*$,

$$C^\dagger C = 1. \tag{194.8}$$

It follows that C is an unitary operator.

Specializing to standard representation we note that

$$\tilde{\gamma}_1 = -\gamma_1; \quad \tilde{\gamma}_2 = +\gamma_2; \quad \tilde{\gamma}_3 = -\gamma_3; \quad \tilde{\gamma}_4 = +\gamma_4. \tag{194.9}$$

Thence Eq. (194.6) leads to C commuting with γ_1 and γ_3, but anti-commuting with γ_2 and γ_4. This will be performed by

$$C = \gamma_2\gamma_4 \tag{194.10}$$

which is the only one of the 16 basis elements of the Clifford algebra satisfying the four conditions (194.6). It may be noted that from (194.10) we find

$$C^\dagger = -C; \quad C^2 = -1. \tag{194.11}$$

It then follows from (194.5) that the charge conjugate wave function in standard representation is

$$\psi_c = \gamma_2\psi^*. \tag{194.12}$$

Problem 195. Mixed helicity states

A plane Dirac wave runs in z direction. It shall be shown that it is impossible to construct a spinor amplitude which simultaneously makes ψ an eigenfunction of σ_x.

Solution. a) For a plane wave,

$$\psi = C\,e^{i(kz-\omega t)} \tag{195.1}$$

there follows, from the Dirac equation, the algebraic relation

$$\Omega C \equiv \left(ik\gamma_3 - \frac{\omega}{c}\gamma_4 + \varkappa\right) C = 0 \tag{195.2}$$

for the spinor amplitude C. The operator Ω defined by (195.2), however, does not commute with

$$\sigma_x = -i\gamma_2\gamma_3 \tag{195.3}$$

since

$$\sigma_x \Omega = +k\gamma_2 + i\frac{\omega}{c}\gamma_2\gamma_3\gamma_4 - i\varkappa\gamma_2\gamma_3$$

but

$$\Omega\sigma_x = -k\gamma_2 + i\frac{\omega}{c}\gamma_2\gamma_3\gamma_4 - i\varkappa\gamma_2\gamma_3 .$$

Therefore ψ cannot be eigenfunction to both operators.

b) In standard representation Eq. (195.2) would run as follows:

$$\begin{aligned} kC_3 + \left(-\frac{\omega}{c} + \varkappa\right) C_1 &= 0; \\ -kC_4 + \left(-\frac{\omega}{c} + \varkappa\right) C_2 &= 0; \\ -kC_1 + \left(\frac{\omega}{c} + \varkappa\right) C_3 &= 0; \\ kC_2 + \left(\frac{\omega}{c} + \varkappa\right) C_4 &= 0. \end{aligned} \tag{195.2'}$$

With the abbreviations

$$\frac{\omega}{c} - \varkappa = k\eta; \qquad \frac{\omega}{c} + \varkappa = \frac{k}{\eta} \tag{195.3'}$$

this leads to

$$C_3 = \eta C_1; \qquad C_4 = -\eta C_2 . \tag{195.4}$$

On the other hand, the eigenvalue problem

$$\sigma_x C = \lambda \cdot C \tag{195.5}$$

with λ being an eigenvalue demands

$$\sigma_x C = \begin{pmatrix} 0 & 1 & 0 & 0 \\ 1 & 0 & 0 & 0 \\ 0 & 0 & 0 & 1 \\ 0 & 0 & 1 & 0 \end{pmatrix} \begin{pmatrix} C_1 \\ C_2 \\ C_3 \\ C_4 \end{pmatrix} = \begin{pmatrix} C_2 \\ C_1 \\ C_4 \\ C_3 \end{pmatrix} = \begin{pmatrix} \lambda C_1 \\ \lambda C_2 \\ \lambda C_3 \\ \lambda C_4 \end{pmatrix}$$

so that

$$C_2 = \lambda C_1; \quad C_1 = \lambda C_2$$

and

$$C_4 = \lambda C_3; \quad C_3 = \lambda C_4. \tag{195.6}$$

Both pairs of equations can only be satisfied for $\lambda = \pm 1$. Using (195.6) we may eliminate C_2 and C_4 from Eq. (195.4) thus arriving at

$$C_3 = \eta C_1 \quad \text{and} \quad \lambda C_3 = -\eta \lambda C_1,$$

two relations which contradict each other. Therefore the spinor C satisfying (195.4) cannot simultaneously satisfy (195.6) as had to be proved.

NB. In the unrelativistic limit $\eta \to 0$, the amplitudes C_3 and C_4 and hence the second pair of equations (195.6) drop out so that no contradiction remains.

Problem 196. Spin expectation value

The expectation value of σ_x shall be calculated for a superposition of two plane waves in z direction having opposite helicities.

Solution. With the spinor amplitudes [cf. Eqs. (190.15, 16) for $\vartheta = 0$]

$$C_+ = \frac{1}{\sqrt{V(1+\eta^2)}} \begin{pmatrix} 1 \\ 0 \\ \eta \\ 0 \end{pmatrix} \quad \text{and} \quad C_- = \frac{1}{\sqrt{V(1+\eta^2)}} \begin{pmatrix} 0 \\ -1 \\ 0 \\ \eta \end{pmatrix} \tag{196.1}$$

for positive and negative helicities, respectively, we construct the mixed state amplitude,

$$C = C_+ \cos\alpha \, e^{i\beta} + C_- \sin\alpha \, e^{-i\beta} \tag{196.2}$$

of the same normalization,

$$\int d^3x \, C^\dagger C = 1 \tag{196.3}$$

with arbitrary but real constants α and β. The expectation value of σ_x is then defined by

$$\langle\sigma_x\rangle = \int C^\dagger \sigma_x C d^3x. \tag{196.4}$$

From

$$\sqrt{V(1+\eta^2)}\;\sigma_x C_+ = \begin{pmatrix}0\\1\\0\\\eta\end{pmatrix}; \quad \sqrt{V(1+\eta^2)}\;\sigma_x C_- = \begin{pmatrix}-1\\0\\\eta\\0\end{pmatrix}$$

we obtain the products

$$C_+^\dagger \sigma_x C_+ = 0; \quad C_-^\dagger \sigma_x C_+ = -\frac{1}{V}\frac{1-\eta^2}{1+\eta^2};$$

$$C_+^\dagger \sigma_x C_- = -\frac{1}{V}\frac{1-\eta^2}{1+\eta^2}; \quad C_-^\dagger \sigma_x C_- = 0.$$

Then,

$$\langle\sigma_x\rangle = -\cos\alpha\sin\alpha(e^{2i\beta}+e^{-2i\beta})\frac{1-\eta^2}{1+\eta^2}$$

or

$$\langle\sigma_x\rangle = -\sin 2\alpha\cos 2\beta\frac{1-\eta^2}{1+\eta^2}. \tag{196.5}$$

The absolute value of the expectation value of σ_x therefore always turns out to be smaller than 1. In the extreme relativistic case where η approches unity, σ_x becomes very small so that almost complete orientation of the spin parallel or antiparallel to the direction of propagation is obtained. In the unrelativistic limit, on the other hand, where η is very small, polarization perpendicular to the direction of propagation becomes possible with $\beta=0$ and $\alpha=\mp\frac{\pi}{4}$ leading to $\langle\sigma_x\rangle=\pm 1$.

Problem 197. Algebraic properties of a Dirac wave spinor

Given a potential $V(z)$. The wave spinor of a state, with spin in either positive or negative z direction, may not depend upon x and y (one-dimensional problem). As far as possible, the Clifford algebra shall be used without recourse to matrix representations. There will then remain four functions of z satisfying a set of coupled differential equations. They shall finally be expressed by the four component wave functions in standard representation.

Solution. The wave spinor may be written

$$\psi(z,t) = e^{-iEt/\hbar} u(z) \tag{197.1}$$

where the spinor $u(z)$ satisfies the one-dimensional Dirac equation

$$\gamma_3 \frac{du}{dz} + \gamma_4 Q(z) u + \varkappa u = 0; \quad Q(z) = \frac{V(z) - E}{\hbar c}. \tag{197.2}$$

Since this equation is entirely built up within a sub-body of which the Clifford numbers $1, \gamma_3, \gamma_4, \gamma_3\gamma_4$ form the basis, it should be solved by a spinor of the form

$$v(z) = A(z) + B(z)\gamma_3 + C(z)\gamma_4 + D(z)\gamma_3\gamma_4. \tag{197.3}$$

Of course, if v solves the Dirac equation (197.2), so does any spinor

$$u = v\Gamma \tag{197.4}$$

with Γ any constant Clifford number, including elements formed with γ_1 and γ_2. On the other hand, v commutes with the spin operator

$$\sigma_z = -i\gamma_1\gamma_2 \tag{197.5}$$

of which it yet is no eigenspinor. The extension (197.4), however, permits the solution of the Dirac equation to be made an eigenspinor of σ_z. We find

$$\sigma_z u = \sigma_z v \Gamma = v \sigma_z \Gamma.$$

So, if Γ is any eigenspinor of σ_z,

$$\sigma_z \Gamma = \pm \Gamma, \tag{197.6}$$

we arrive at

$$\sigma_z u = \pm u. \tag{197.7}$$

The two eigenvalues $+1$ and -1 are called *helicities* (cf. Problem 190). Now it can easily be seen that

$$\Gamma_+ = 1 - i\gamma_1\gamma_2 = 1 + \sigma_z \tag{197.8a}$$

and

$$\Gamma_- = 1 + i\gamma_1\gamma_2 = 1 - \sigma_z \tag{197.8b}$$

are such eigenspinors with eigenvalues ± 1 of σ_z:

$$\sigma_z \Gamma_\pm = \sigma_z(1 \pm \sigma_z) = \sigma_z \pm 1 = \pm(1 \pm \sigma_z) = \pm \Gamma_\pm.$$

Our argument thus leads to

$$u(z) = v(z)\,(1 \mp i\gamma_1\gamma_2) \tag{197.9}$$

where $v(z)$ still remains to be determined by setting it into the Dirac equation (197.2). A simple calculation along these lines then leads to

$$(B'+QC+\varkappa A)+\gamma_3(A'-QD+\varkappa B)+\gamma_4(D'+QA+\varkappa C) \\ +\gamma_3\gamma_4(C'-QB+\varkappa D)=0 \qquad (197.10)$$

with the prime denoting differentiation. This expression is zero if, and only if, each of the four brackets vanishes. Thus we find the four functions A, B, C, D to satisfy a set of coupled differential equations.

$$B'+QC+\varkappa A=0; \quad A'-QD+\varkappa B=0; \\ D'+QA+\varkappa C=0; \quad C'-QB+\varkappa D=0. \qquad (197.11)$$

These equations become even simpler if combined into two pairs,

$$(B-D)'+(\varkappa-Q)(A-C)=0; \\ (A-C)'+(\varkappa+Q)(B-D)=0 \qquad (197.12\text{a})$$

and

$$(B+D)'+(\varkappa+Q)(A+C)=0; \\ (A+C)'+(\varkappa-Q)(B+D)=0, \qquad (197.12\text{b})$$

so that the first pair of equations, (197.12a), only connects the two functions

$$w_1=\tfrac{1}{2}(B-D); \quad w_3=\tfrac{1}{2}(A-C) \qquad (197.13\text{a})$$

with one another and the other pair, (197.12b), the two functions

$$w_2=\tfrac{1}{2}(A+C); \quad w_4=\tfrac{1}{2}(B+D). \qquad (197.13\text{b})$$

Putting these into $v(z)$, Eq. (197.3), we finally find

$$v(z)=(w_2+w_4\gamma_3)(1+\gamma_4)+(w_3+w_1\gamma_3)(1-\gamma_4). \qquad (197.14)$$

Each of the two terms in (197.14) *separately* satisfies the Dirac equation (197.2) if Eqs. (197.12a, b) are satisfied. Multiplication on the right-hand side of each term with either Γ_+ or Γ_-, Eqs. (197.8a, b), makes it an eigenspinor of σ_z too.

It remains to show how the four functions w_μ are connected with the four component wave functions u_μ of the standard representation. In this matrix description the Dirac equation (197.2) consists of the following component equations:

$$\left.\begin{aligned} -iu_3'+(Q+\varkappa)u_1&=0,\\ iu_4'+(Q+\varkappa)u_2&=0,\\ iu_1'+(\varkappa-Q)u_3&=0,\\ -iu_2'+(\varkappa-Q)u_4&=0. \end{aligned}\right\} \qquad (197.15)$$

Comparing this set with (197.12a, b) we are led to identify

$$u_1 = w_1; \quad u_3 = i w_3; \quad u_2 = i w_2; \quad u_4 = w_4 \tag{197.16}$$

or

$$A = -i(u_2 + u_3); \quad B = u_4 + u_1;$$
$$C = -i(u_2 - u_3); \quad D = u_4 - u_1. \tag{197.17}$$

For $u_2 = u_4 = 0$, the helicity will be $+1$, for $u_1 = u_3 = 0$, it will be -1.

Problem 198. Current in algebraic formulation

To determine the components of the electrical current for the eigenspinor

$$u(z) = (w_3 + w_1 \gamma_3)(1 - \gamma_4)(1 - i\gamma_1 \gamma_2) \tag{198.1}$$

of the preceding problem.

Solution. The components of the electrical four-current are defined by

$$s_\mu = iec\bar{u}\gamma_\mu u; \quad \bar{u} = u^\dagger \gamma_4 \tag{198.2}$$

where, in our example,

$$u^\dagger = (1 - i\gamma_1\gamma_2)(1 - \gamma_4)(w_3^* + w_1^* \gamma_3) \tag{198.3}$$

since the Clifford numbers $i\gamma_1\gamma_2$, γ_4, γ_3 are hermitian operators. Thus we obtain

$$s_\mu = iec(1 - i\gamma_1\gamma_2)(1 - \gamma_4)(w_3^* + w_1^*\gamma_3)\gamma_4\gamma_\mu(w_3 + w_1\gamma_3)(1 - \gamma_4)(1 - i\gamma_1\gamma_2). \tag{198.4}$$

For the components s_1 and s_2 the Clifford number γ_μ may be shifted through two places towards the end of the expression, whereas γ_4 may be shifted one place towards its front:

$$s_{1,2} = iec(1 - i\gamma_1\gamma_2)(\gamma_4 - 1)(w_3^* - w_1^*\gamma_3)(w_3 - w_1\gamma_3)(1 + \gamma_4)\gamma_{1,2}(1 - i\gamma_1\gamma_2).$$

The operator $1 - i\gamma_1\gamma_2$ commutes with γ_3 as well as with γ_4, so that

$$s_{1,2} = iec(\gamma_4 - 1)\{(|w_3|^2 + |w_1|^2)$$
$$- (w_1^* w_3 + w_3^* w_1)\gamma_3\}(1 + \gamma_4)(1 - i\gamma_1\gamma_2)\gamma_{1,2}(1 - i\gamma_1\gamma_2).$$

The last three factors give either, for $\mu = 1$,

$$(1 - i\gamma_1\gamma_2)(\gamma_1 - i\gamma_2) = \gamma_1 + i\gamma_2 - i\gamma_2 - \gamma_1 = 0$$

or, for $\mu = 2$,

$$(1 - i\gamma_1\gamma_2)(\gamma_2 + i\gamma_1) = \gamma_2 - i\gamma_1 + i\gamma_1 - \gamma_2 = 0.$$

Therefore, as was to be expected, no current components exist perpendicular to the z direction.

For s_3 we may write in a similar way,

$$s_3 = iec(\gamma_4 - 1)(w_3^* - w_1^* \gamma_3)(w_3 \gamma_3 + w_1)(1-\gamma_4)(1 - i\gamma_1\gamma_2)^2.$$

The square at the end of this expression gives

$$(1 - i\gamma_1\gamma_2)^2 = 2(1 - i\gamma_1\gamma_2) \tag{198.5}$$

so that we find

$$s_3 = 2iec(\gamma_4 - 1)\{(w_3^* w_1 - w_1^* w_3) + (|w_3|^2 - |w_1|^2)\gamma_3\}(1-\gamma_4)(1 - i\gamma_1\gamma_2).$$

Shifting the front factor $(\gamma_4 - 1)$ one place towards the right, we get

$$s_3 = 2iec\{(w_1^* w_3 - w_3^* w_1)(1-\gamma_4) + (|w_1|^2 - |w_3|^2)\gamma_3(1+\gamma_4)\}(1-\gamma_4)(1 - i\gamma_1\gamma_2).$$

Since

$$(1+\gamma_4)(1-\gamma_4) = 0; \quad (1-\gamma_4)^2 = 2(1-\gamma_4) \tag{198.6}$$

the second term in the curly bracket does not contribute and we arrive at

$$s_3 = 4iec(w_1^* w_3 - w_3^* w_1)(1-\gamma_4)(1 - i\gamma_1\gamma_2). \tag{198.7}$$

Finally, we get in a similar fashion

$$\begin{aligned} s_4 &= iec(1 - i\gamma_1\gamma_2)(1-\gamma_4)(w_3^* + w_1^*\gamma_3)(w_3 + w_1\gamma_3)(1-\gamma_4)(1 - i\gamma_1\gamma_2) \\ &= 2iec(1-\gamma_4)\{(|w_1|^2 + |w_3|^2) + (w_1^* w_3 + w_3^* w_1)\gamma_3\}(1-\gamma_4)(1 - i\gamma_1\gamma_2) \\ &= 4iec(|w_1|^2 + |w_3|^2)(1-\gamma_4)(1 - i\gamma_1\gamma_2). \end{aligned} \tag{198.8}$$

The expressions for s_3 and s_4 are still Clifford numbers but of the same shape. They are to be compared with the normalization expression,

$$\bar{u}u = (1 - i\gamma_1\gamma_2)(1-\gamma_4)(w_3^* + w_1^*\gamma_3)\gamma_4(w_3 + w_1\gamma_3)(1-\gamma_4)(1 - i\gamma_1\gamma_2)$$

which by the same procedure may be brought into the form

$$\bar{u}u = 4(|w_1|^2 - |w_3|^2)(1-\gamma_4)(1 - i\gamma_1\gamma_2). \tag{198.9}$$

Gathering up the results, except for a common factor

$$\Gamma = 4(1-\gamma_4)(1 - i\gamma_1\gamma_2),$$

we just have simple c-number expressions for the current density in z direction,

$$s_3 = iec(w_1^* w_3 - w_3^* w_1)\Gamma, \tag{198.10}$$

for the charge density ρ following from $s_4 = ic\rho$,

$$\rho = e(|w_1|^2 + |w_3|^2)\Gamma \tag{198.11}$$

and for the normalization expression

$$\bar{u}u = (|w_1|^2 - |w_3|^2)\,\Gamma. \tag{198.12}$$

Using the standard components, $u_1 = w_1$ and $u_3 = i w_3$, introduced in the preceding problem, these expression may as well be written

$$s_3 = ec(u_1^* u_3 + u_3^* u_1)\,\Gamma; \quad \rho = e(|u_1|^2 + |u_3|^2)\,\Gamma;$$
$$\bar{u}u = (|u_1|^2 - |u_3|^2)\,\Gamma. \tag{198.13}$$

It should be noted that in standard representation the operator Γ becomes very simple. We have

$$1 - \gamma_4 = \begin{pmatrix} 0 & 0 & 0 & 0 \\ 0 & 0 & 0 & 0 \\ 0 & 0 & 2 & 0 \\ 0 & 0 & 0 & 2 \end{pmatrix}; \quad 1 - i\gamma_1\gamma_2 = 1 + \sigma_3 = \begin{pmatrix} 2 & 0 & 0 & 0 \\ 0 & 0 & 0 & 0 \\ 0 & 0 & 2 & 0 \\ 0 & 0 & 0 & 0 \end{pmatrix};$$

thence we find the product

$$\Gamma = 16 \begin{pmatrix} 0 & 0 & 0 & 0 \\ 0 & 0 & 0 & 0 \\ 0 & 0 & 1 & 0 \\ 0 & 0 & 0 & 0 \end{pmatrix} \tag{198.14}$$

a matrix which, in diagonal form, consists of only one element.

Problem 199. Conduction current and polarization current

a) The electrical current density (particle charge e),

$$s_\nu = iec\bar{\psi}\gamma_\nu\psi; \quad s_k = j_k; \quad s_4 = ic\rho \tag{199.1}$$

shall be shown to satisfy the equation of continuity,

$$\sum_\mu \frac{\partial s_\mu}{\partial x_\mu} = 0 \quad \text{or} \quad \operatorname{div} \boldsymbol{j} + \frac{\partial \rho}{\partial t} = 0. \tag{199.2}$$

b) The vector s_ν shall be decomposed,

$$s_\nu = s_\nu^C + s_\nu^P \tag{199.3}$$

so that the space part of s_ν^C, the *conduction current*, is of the same form as the unrelativistic expression for j_k. The remaining part, s_k^P, is then called the *polarization current*.

Solution. a) In order to prove (199.1) we have to supplement the Dirac equation

$$\sum_\mu \gamma_\mu(\partial_\mu - i a_\mu)\psi + \varkappa\psi = 0; \qquad \partial_\mu = \frac{\partial}{\partial x_\mu}; \qquad a_\mu = \frac{e}{\hbar c} A_\mu \quad (199.4\,\text{a})$$

by an analogous differential equation for $\bar\psi = \psi^\dagger \gamma_4$. The operators

$$D_k = \partial_k - i a_k, \qquad D_4 = \partial_4 - i a_4$$

have the complex conjugates

$$D_k^* = \partial_k + i a_k, \qquad D_4^* = -(\partial_4 + i a_4),$$

since x_k and a_k are real, but x_4 and a_4 imaginary. The conjugate equation of (199.4a),

$$\sum_\mu D_\mu^* \psi^\dagger \gamma_\mu + \varkappa\psi^\dagger = 0$$

or, with $\bar\psi = \psi^\dagger \gamma_4$,

$$-\sum_k D_k^* \bar\psi \gamma_k + D_4^* \bar\psi \gamma_4 + \varkappa\bar\psi = 0$$

may be written

$$\sum_\mu (\partial_\mu + i a_\mu)\bar\psi \gamma_\mu - \varkappa\bar\psi = 0. \qquad (199.4\,\text{b})$$

From (199.4a, b) it then follows by elimination of the mass terms that

$$\sum_\mu \{\bar\psi \gamma_\mu(\partial_\mu - i a_\mu)\psi + (\partial_\mu + i a_\mu)\bar\psi \gamma_\mu \cdot \psi\} = 0.$$

Here the a_μ terms cancel and the rest may be written

$$\sum_\mu \partial_\mu(\bar\psi \gamma_\mu \psi) = 0,$$

in agreement with the equation of continuity (199.2).

b) The unrelativistic expression of the space part of the electrical current is, according to Problem 126,

$$j_k = \frac{i e \hbar}{2m}(\psi\,\partial_k \psi^* - \psi^*\,\partial_k \psi + 2 i a_k \psi^* \psi), \qquad (199.5)$$

i. e. mainly a bilinear combination of wave functions and their space derivatives. In order to give s_ν, Eq. (199.1), a similar form, we may there replace either $\bar\psi$ according to (199.4a) or ψ according to (199.4b) by their first derivatives:

$$s_\nu = \frac{iec}{\varkappa}\sum_\mu (\partial_\mu + i a_\mu)\bar\psi \gamma_\mu \cdot \gamma_\nu \psi = -\frac{iec}{\varkappa}\bar\psi \gamma_\nu \sum_\mu \gamma_\mu(\partial_\mu - i a_\mu)\psi.$$

Symmetrizing by taking half the sum of these two expressions and putting $ec/\varkappa = e\hbar/m$, we find

$$s_\nu = \frac{ie\hbar}{2m}\sum_\mu \left\{ \frac{\partial\bar\psi}{\partial x_\mu}\gamma_\mu\gamma_\nu\psi - \bar\psi\gamma_\nu\gamma_\mu\frac{\partial\psi}{\partial x_\mu} + i a_\mu \bar\psi(\gamma_\mu\gamma_\nu + \gamma_\nu\gamma_\mu)\psi \right\}. \quad (199.6)$$

Using the commutation rules

$$\gamma_\mu\gamma_\nu + \gamma_\nu\gamma_\mu = 2\delta_{\mu\nu}$$

in the second and third terms, we may reshape this into

$$s_\nu = \frac{ie\hbar}{2m}\left\{ \sum_\mu \left(\frac{\partial\bar\psi}{\partial x_\mu}\gamma_\mu\gamma_\nu\psi + \bar\psi\gamma_\mu\gamma_\nu\frac{\partial\psi}{\partial x_\mu} \right) - 2\bar\psi\frac{\partial\psi}{\partial x_\nu} + 2ia_\nu\bar\psi\psi \right\}$$

or, splitting off the sum its diagonal term $\mu = \nu$,

$$s_\nu = \frac{ie\hbar}{2m}\left\{ \frac{\partial\bar\psi}{\partial x_\nu}\psi - \bar\psi\frac{\partial\psi}{\partial x_\nu} + 2ia_\nu\bar\psi\psi \right\} + \frac{ie\hbar}{2m}\sum_\mu{}' \frac{\partial}{\partial x_\mu}(\bar\psi\gamma_\mu\gamma_\nu\psi). \quad (199.7)$$

The first term of this decomposition exactly matches Eq. (199.5) and is therefore the conduction current s_ν^C as defined above. The other term then is the so-called polarization current,

$$s_\nu^P = \frac{ie\hbar}{2m}\sum_\mu{}' \frac{\partial}{\partial x_\mu}(\bar\psi\gamma_\mu\gamma_\nu\psi). \quad (199.8)$$

NB. This decomposition has first been studied by W. Gordon, Z. Physik **50**, 630 (1928). The space part of s_ν^P may be written

$$\boldsymbol{s}^P = \frac{e}{m}\operatorname{curl}(\bar\psi\boldsymbol{S}\psi) - \frac{e\hbar i}{2mc}\frac{\partial}{\partial t}(\bar\psi\boldsymbol{\alpha}\psi)$$

where $S_k = \frac{\hbar}{2}\sigma_k$ are the components of the spin vector (written in 4×4 reducible matrices) and the α_k are the matrices defined at the end of Problem 189. In a plane wave the polarization current vanishes.

Problem 200. Splitting up of Dirac equations into two pairs

Write the Dirac equation in hamiltonian form and split up the resulting four-component equation of standard representation into a pair of two component equations. Pauli matrices will occur in the latter. Show that for rest mass zero (e. g. for a neutrino) two two-component theories are possible.

Solution. The Dirac equation

$$\left.\begin{aligned}&\sum_\mu \gamma_\mu D_\mu \psi + \varkappa \psi = 0; \qquad D_\mu = \partial_\mu - \frac{ie}{\hbar c} A_\mu;\\ &A_4 = i\Phi; \qquad e\Phi = V; \qquad \partial_4 = -\frac{i}{c}\,\partial_t\end{aligned}\right\} \tag{200.1}$$

can be written, distinguishing time and space derivatives,

$$\sum_{n=1}^{3} \gamma_n D_n \psi + \gamma_4 \left(-\frac{i}{c}\,\partial_t + \frac{V}{\hbar c}\right) \psi + \varkappa \psi = 0 .$$

Multiplication from the left by $c\hbar\gamma_4$ renders

$$\hbar c \sum_n \gamma_4 \gamma_n D_n \psi - i\hbar\,\partial_t \psi + V\psi + mc^2 \gamma_4 \psi = 0$$

or

$$-\frac{\hbar}{i}\frac{\partial \psi}{\partial t} = H\psi$$

with

$$H = \hbar c \sum_{n=1}^{3} \gamma_4 \gamma_n \left(\partial_n - \frac{ie}{\hbar c} A_n\right) + V + mc^2 \gamma_4 \tag{200.2}$$

the hamiltonian.

In standard representation we have

$$\gamma_n = \begin{pmatrix} \mathit{0} & -is_n \\ is_n & \mathit{0} \end{pmatrix}; \qquad \gamma_4 = \begin{pmatrix} \mathit{1} & \mathit{0} \\ \mathit{0} & -\mathit{1} \end{pmatrix} \tag{200.3}$$

with s_n the three Pauli matrices and $\mathit{1}$ and $\mathit{0}$ standing, respectively, for the 2×2 matrices of unity and zero. Then,

$$\alpha_n = i\gamma_4\gamma_n = \begin{pmatrix} \mathit{0} & s_n \\ s_n & \mathit{0} \end{pmatrix} \tag{200.4}$$

so that the hamiltonian (200.2) splits up in the form

$$H = \begin{pmatrix} V + mc^2; & -\hbar c i \sum_n s_n D_n \\ -\hbar c i \sum_n s_n D_n; & V - mc^2 \end{pmatrix}. \tag{200.5}$$

If here we introduce the two-component quantities ψ_a and ψ_b so that

$$\psi = \begin{pmatrix} \psi_a \\ \psi_b \end{pmatrix}$$

is the four-component Dirac spinor, the differential equation splits up into the wanted pair of two-component equations, viz.

$$\left.\begin{aligned} -\frac{\hbar}{i}\frac{\partial\psi_a}{\partial t} &= -i\hbar c(\vec{s}\cdot\vec{D})\psi_b+(V+mc^2)\psi_a, \\ -\frac{\hbar}{i}\frac{\partial\psi_b}{\partial t} &= -i\hbar c(\vec{s}\cdot\vec{D})\psi_a+(V-mc^2)\psi_b \end{aligned}\right\} \tag{200.6}$$

or, for a stationary state of positive energy, E,

$$\left.\begin{aligned} (\vec{s}\cdot\vec{D})\psi_b - i\frac{E-V-mc^2}{\hbar c}\psi_a &= 0; \\ (\vec{s}\cdot\vec{D})\psi_a - i\frac{E-V+mc^2}{\hbar c}\psi_b &= 0. \end{aligned}\right\} \tag{200.7}$$

Neutrino theory: If $m=0$, the two equations become identical so that

$$\psi_b = \lambda\psi_a \quad \text{with } \lambda = \pm 1 \tag{200.8}$$

are two possible solutions with ψ_a to be determined from

$$\left\{(\vec{s}\cdot\vec{D}) - \lambda i\frac{E-V}{\hbar c}\right\}\psi_a = 0. \tag{200.9}$$

Since the two systems decouple, there emerge two independent two-component theories of particles of rest-mass zero. It can easily be seen that, in the force-free case, the parameter λ becomes identical with the helicity quantum number. For this purpose we study a plane wave in z direction,

$$\psi_a = C\,\mathrm{e}^{ikz}$$

with C a constant two-component spinor. The first term of (200.9) then becomes

$$s_3\partial_3\psi_a = ik(s_3\psi_a)\mathrm{e}^{ikz}$$

and its second term

$$-\lambda i\frac{E}{\hbar c} = -ik\lambda$$

so that we have

$$s_3 C = \lambda C. \tag{200.10}$$

Therefore λ is the eigenvalue of the spin component (in units of $\hbar/2$) in the direction of propagation ("helicity"). With the Pauli matrix

$$s_3 = \begin{pmatrix} 1 & 0 \\ 0 & -1 \end{pmatrix}$$

Eq. (200.10) then is solved for $\lambda=+1$ by

$$C=\begin{pmatrix}1\\0\end{pmatrix} \quad \text{and} \quad \psi=\begin{pmatrix}C\\C\end{pmatrix}e^{ikz} \tag{200.11a}$$

whereas for the helicity $\lambda=-1$ it is solved by

$$C=\begin{pmatrix}0\\1\end{pmatrix} \quad \text{and} \quad \psi=\begin{pmatrix}C\\-C\end{pmatrix}e^{ikz}. \tag{200.11b}$$

NB. Experience shows that neutrinos always have helicity $h=-1$ so that only the second theory actually describes natural phenomena.

Now, the operator

$$\gamma_5=\gamma_1\gamma_2\gamma_3\gamma_4$$

in standard representation becomes

$$\gamma_5=\begin{pmatrix}0&-1\\-1&0\end{pmatrix}$$

so that

$$1+\gamma_5=\begin{pmatrix}1&-1\\-1&1\end{pmatrix}; \quad 1-\gamma_5=\begin{pmatrix}1&1\\1&1\end{pmatrix}.$$

If these operators act on any $\lambda=+1$ solution of (200.9), i.e. on

$$\psi_+=\begin{pmatrix}\psi_a\\\psi_a\end{pmatrix}$$

they lead to

$$(1+\gamma_5)\psi_+=0; \quad (1-\gamma_5)\psi_+=2\psi_+. \tag{200.12a}$$

Acting on any $\lambda=-1$ solution of (200.9), i.e. on

$$\psi_-=\begin{pmatrix}\psi_a\\-\psi_a\end{pmatrix}$$

they render

$$(1+\gamma_5)\psi_-=2\psi_-; \quad (1-\gamma_5)\psi_-=0. \tag{200.12b}$$

It cannot be decided whether, due to some unknown principle, only ψ_- is realized in nature, or whether the interaction operator producing neutrinos contains a factor $1+\gamma_5$, thus making creation of $\lambda=+1$ neutrinos impossible. It should be noted, however, that $1+\gamma_5$ is an operator without defined parity.

Problem 201. Central forces in Dirac theory

To use the splitting up of the Dirac equations in standard representation into a pair of two-component equations (Problem 200) in order to construct eigenspinors of a central potential field $V(r)$ which are simultaneously eigenspinors of the total angular momentum operators $\boldsymbol{J}^2$ and J_z. The calculations may be restricted to $m_j=+\frac{1}{2}$.

Solution. According to (200.7) we may start from the differential equations

$$\left.\begin{aligned}\sum_{n=1}^{3} s_n \partial_n \psi_b - i\frac{E-V(r)-mc^2}{\hbar c}\psi_a = 0,\\ \sum_{n=1}^{3} s_n \partial_n \psi_a - i\frac{E-V(r)+mc^2}{\hbar c}\psi_b = 0\end{aligned}\right\} \tag{201.1}$$

where s_n $(n=1,2,3)$ denote the three Pauli matrices, and

$$\psi = \begin{pmatrix}\psi_a\\ \psi_b\end{pmatrix} \tag{201.2}$$

is the Dirac spinor, composed of the two-component spinors ψ_a and ψ_b.

The angular momentum operators, deriving from $\boldsymbol{J}=\boldsymbol{L}+\frac{\hbar}{2}\boldsymbol{\sigma}$, have four-component standard representations which do not mix the first two components (ψ_a) with the other two (ψ_b) since the four-component extension of the spin matrices,

$$\sigma_n = \begin{pmatrix} s_n & 0\\ 0 & s_n\end{pmatrix},$$

is diagonal in the Pauli matrices. If therefore ψ_a and ψ_b are two-component eigenspinors of $\boldsymbol{J}^2$ and J_z, so will ψ, Eq. (201.2), be.

Now, in Problem 133, we have already constructed the two-component eigenspinors $u_{j,l}$ of $\boldsymbol{J}^2$ and J_z with quantum numbers $j=l\pm\frac{1}{2}$ and $m_j=+\frac{1}{2}$, viz.

$$u^{\mathrm{I}} = u_{j,j-\frac{1}{2}} = \frac{f_l(r)}{\sqrt{2l+1}}\begin{pmatrix}\sqrt{l+1}\,Y_{l,0}\\ -\sqrt{l}\,Y_{l,1}\end{pmatrix} = \frac{f_{j-\frac{1}{2}}(r)}{\sqrt{2j}}\begin{pmatrix}\sqrt{j+\frac{1}{2}}\,Y_{j-\frac{1}{2},0}\\ -\sqrt{j-\frac{1}{2}}\,Y_{j-\frac{1}{2},1}\end{pmatrix} \tag{201.3a}$$

and

$$u^{\mathrm{II}} = u_{j,j+\frac{1}{2}} = \frac{g_l(r)}{\sqrt{2l+1}}\begin{pmatrix}\sqrt{l}\,Y_{l,0}\\ \sqrt{l+1}\,Y_{l,1}\end{pmatrix} = \frac{g_{j+\frac{1}{2}}(r)}{\sqrt{2(j+1)}}\begin{pmatrix}\sqrt{j+\frac{1}{2}}\,Y_{j+\frac{1}{2},0}\\ \sqrt{j+\frac{3}{2}}\,Y_{j+\frac{1}{2},1}\end{pmatrix}. \tag{201.3b}$$

We shall try to solve our problem by combinations of these two spinors, taking either

$$\psi = \begin{pmatrix}u^{\mathrm{I}}\\ u^{\mathrm{II}}\end{pmatrix} \tag{201.4a}$$

or

$$\psi = \begin{pmatrix}u^{\mathrm{II}}\\ u^{\mathrm{I}}\end{pmatrix}, \tag{201.4b}$$

where the normalization of $f(r)$ and $g(r)$ is still left open and may be different in (201.4a) and (201.4b).

To follow this programme, according to (201.1), we need the expressions

$$S u^{\mathrm{I,II}} = \sum_{n=1}^{3} s_n \partial_n u^{\mathrm{I,II}} = \begin{pmatrix} \partial_z; & \partial_x - i\partial_y \\ \partial_x + i\partial_y; & -\partial_z \end{pmatrix} u^{\mathrm{I,II}}. \tag{201.5}$$

Using the well-known formulae

$$\pm(\partial_x \pm i\partial_y)(F(r)\,Y_{l,m}) = \sqrt{\frac{(l\pm m+2)(l\pm m+1)}{(2l+3)(2l+1)}}\left(F' - \frac{l}{r}F\right) Y_{l+1,m\pm1} - \sqrt{\frac{(l\mp m)(l\mp m-1)}{(2l+1)(2l-1)}}\left(F' + \frac{l+1}{r}F\right) Y_{l-1,m\pm1} \tag{201.6a}$$

and

$$\partial_z(F(r)\,Y_{l,m}) = \sqrt{\frac{(l+m+1)(l-m+1)}{(2l+3)(2l+1)}}\left(F' - \frac{l}{r}F\right) Y_{l+1,m} + \sqrt{\frac{(l+m)(l-m)}{(2l+1)(2l-1)}}\left(F' + \frac{l+1}{r}F\right) Y_{l-1,m} \tag{201.6b}$$

we arrive after some cumbersome but elementary reshaping of the expressions

$$S u^{\mathrm{I}} = \frac{1}{\sqrt{2l+1}} \begin{pmatrix} \sqrt{l+1}\,\partial_z(f_l Y_{l,0}) - \sqrt{l}\,(\partial_x - i\partial_y)(f_l Y_{l,1}) \\ \sqrt{l+1}\,(\partial_x + i\partial_y)(f_l Y_{l,0}) + \sqrt{l}\,\partial_z(f_l Y_{l,1}) \end{pmatrix}$$

and

$$S u^{\mathrm{II}} = \frac{1}{\sqrt{2l+1}} \begin{pmatrix} \sqrt{l}\,\partial_z(g_l Y_{l,0}) + \sqrt{l+1}\,(\partial_x - i\partial_y)(g_l Y_{l,1}) \\ \sqrt{l}\,(\partial_x + i\partial_y)(g_l Y_{l,0}) - \sqrt{l+1}\,\partial_z(g_l Y_{l,1}) \end{pmatrix}$$

at the results

$$S u^{\mathrm{I}} = \frac{1}{\sqrt{2l+3}}\left(f_l' - \frac{l}{r}f_l\right)\begin{pmatrix} \sqrt{l+1}\,Y_{l+1,0} \\ \sqrt{l+2}\,Y_{l+1,1} \end{pmatrix} = \frac{1}{\sqrt{2(j+1)}}\left(f_{j-\frac{1}{2}}' - \frac{j-\frac{1}{2}}{r} f_{j-\frac{1}{2}}\right)\begin{pmatrix} \sqrt{j+\frac{1}{2}}\,Y_{j+\frac{1}{2},0} \\ \sqrt{j+\frac{3}{2}}\,Y_{j+\frac{1}{2},1} \end{pmatrix} \tag{201.7a}$$

for $j = l + \frac{1}{2}$, and

$$S u^{\mathrm{II}} = \frac{1}{\sqrt{2l-1}}\left(g_l' + \frac{l+1}{r}g_l\right)\begin{pmatrix} \sqrt{l}\,Y_{l-1,0} \\ -\sqrt{l-1}\,Y_{l-1,1} \end{pmatrix} = \frac{1}{\sqrt{2j}}\left(g_{j+\frac{1}{2}}' + \frac{j+\frac{3}{2}}{r} g_{j+\frac{1}{2}}\right)\begin{pmatrix} \sqrt{j+\frac{1}{2}}\,Y_{j-\frac{1}{2},0} \\ -\sqrt{j-\frac{1}{2}}\,Y_{j-\frac{1}{2},1} \end{pmatrix} \tag{201.7b}$$

for $j = l - \frac{1}{2}$.

In order now to satisfy Eqs. (201.1), we first try the type of solution (201.4a):

$$S u^{\mathrm{II}} - i\frac{E-V-mc^2}{\hbar c}u^{\mathrm{I}} = 0; \qquad S u^{\mathrm{I}} - i\frac{E-V+mc^2}{\hbar c}u^{\mathrm{II}} = 0. \quad (201.8)$$

Putting u^{I} and u^{II} from (201.3a, b) and Su^{II}, Su^{I} from (201.7a, b) into (201.8), we find (omitting the subscripts of f and g)

$$S u^{\mathrm{II}} - i\frac{E-V-mc^2}{\hbar c}u^{\mathrm{I}}$$

$$= \frac{1}{\sqrt{2j}}\left(g' + \frac{j+\frac{3}{2}}{r}g - i\frac{E-V-mc^2}{\hbar c}f\right)\begin{pmatrix} \sqrt{j+\frac{1}{2}}\, Y_{j-\frac{1}{2},0} \\ -\sqrt{j-\frac{1}{2}}\, Y_{j-\frac{1}{2},1}\end{pmatrix}$$

and

$$S u^{\mathrm{I}} - i\frac{E-V+mc^2}{\hbar c}u^{\mathrm{II}}$$

$$= \frac{1}{\sqrt{2(j+1)}}\left(f' - \frac{j-\frac{1}{2}}{r}f - i\frac{E-V+mc^2}{\hbar c}g\right)\begin{pmatrix} \sqrt{j+\frac{1}{2}}\, Y_{j+\frac{1}{2},0} \\ \sqrt{j+\frac{3}{2}}\, Y_{j+\frac{1}{2},1}\end{pmatrix}.$$

These expressions vanish if the radial functions $f(r)$ and $g(r)$ satisfy the coupled differential equations

$$\left.\begin{aligned} g' + \frac{j+\frac{3}{2}}{r}g - i\frac{E-V(r)-mc^2}{\hbar c}f &= 0, \\ f' - \frac{j-\frac{1}{2}}{r}f - i\frac{E-V(r)+mc^2}{\hbar c}g &= 0. \end{aligned}\right\} \quad (201.9\text{a})$$

The type of solution (201.4b), on the other hand, with u^{I} and u^{II} exchanged in (201.8) leads in the same way to formulae in which the role of the two equations (201.8) and thus the signs of the rest-mass terms are exchanged. Thus we arrive, not at (201.9a) but at the set of differential equations

$$\left.\begin{aligned} g' + \frac{j+\frac{3}{2}}{r}g - i\frac{E-V(r)+mc^2}{\hbar c}f &= 0, \\ f' - \frac{j-\frac{1}{2}}{r}f - i\frac{E-V(r)-mc^2}{\hbar c}g &= 0. \end{aligned}\right\} \quad (201.9\text{b})$$

The differential equations (201.9a) and (201.9b) have still to be solved for any given potential $V(r)$, separately, thus providing the complete solution. By their coupling they determine the relative normalization of the two functions f and g.

It should be noted that, in contrast to the unrelativistic spin theory, contributions of different l values are mixed up in the four-component Dirac spinor so that l is no longer a good quantum number, but j and m_j of course still are.

Problem 202. Kepler problem in Dirac theory

To specialize the solutions of the central-force problem to the potential

$$V(r) = -\frac{Ze^2}{r} \tag{202.1}$$

and to determine the eigenvalues.

Solution. In the preceding problem we have found two sets of solutions for the general central-force problem, the differential equations for whose radial parts had been written up in Eqs. (201.9a) and (201.9b). Let us first discuss the system (201.9a). With the abbreviation

$$\beta = \frac{Ze^2}{\hbar c} = \frac{Z}{137}, \tag{202.2}$$

which, it should be borne in mind, generally is a very small number, and

$$\frac{\mu}{a} = \frac{mc^2 - E}{\hbar c}; \qquad \frac{1}{\mu a} = \frac{mc^2 + E}{\hbar c}$$

or

$$\mu = \sqrt{\frac{mc^2 - E}{mc^2 + E}}; \qquad a = \frac{\hbar c}{\sqrt{(mc^2 - E)(mc^2 + E)}} \tag{202.3}$$

the differential equations (201.9a) may be written for potential (202.1):

$$\left.\begin{aligned} g' + \frac{j+\frac{3}{2}}{r} g + i\left(\frac{\mu}{a} - \frac{\beta}{r}\right) f &= 0; \\ f' - \frac{j-\frac{1}{2}}{r} f - i\left(\frac{1}{\mu a} + \frac{\beta}{r}\right) g &= 0. \end{aligned}\right\} \tag{202.4}$$

These equations are to be solved.

We start by discussing their behaviour for very large and very small values of r. For $r \to \infty$ Eqs. (202.4) become

$$g' + \frac{i\mu}{a} f = 0; \qquad f' - \frac{i}{\mu a} g = 0$$

with normalizable solutions

$$g = C\,e^{-r/a}; \quad f = -C\frac{i}{\mu}e^{-r/a}.$$

There is another set of type $e^{+r/a}$ which we need not discuss. For $r \to 0$, on the other hand, we expect regular solutions of the form

$$g = A r^{s-1}; \quad f = B r^{s-1}.$$

Putting these in (202.4) we get

$$(s-1)A + (j+\tfrac{3}{2})A - i\beta B = 0,$$
$$(s-1)B - (j-\tfrac{1}{2})B - i\beta A = 0.$$

Vanishing of the determinant of this set of linear equations leads to

$$s = \sqrt{(j+\tfrac{1}{2})^2 - \beta^2}. \tag{202.5}$$

Combining these results, it seems reasonable to put

$$\left.\begin{aligned} g &= C r^{s-1} e^{-r/a} G(r); \\ f &= -\frac{i}{\mu} C r^{s-1} e^{-r/a} F(r); \end{aligned}\right\} \tag{202.6}$$

then, from (202.4), there follow the differential equations

$$\left.\begin{aligned} G' + \left(\frac{s+(j+\frac{1}{2})}{r} - \frac{1}{a}\right)G + \left(\frac{1}{a} - \frac{\beta}{\mu r}\right)F = 0, \\ F' + \left(\frac{s-(j+\frac{1}{2})}{r} - \frac{1}{a}\right)F + \left(\frac{1}{a} + \frac{\beta\mu}{r}\right)G = 0. \end{aligned}\right\} \tag{202.7}$$

Adding and substracting these equations, respectively, and putting

$$G + F = v(r); \quad G - F = w(r) \tag{202.8}$$

we arrive at

$$\left.\begin{aligned} v' + \frac{s+p}{r}v &= -(\mathsf{k}+q)\frac{w}{r}, \\ w' + \left(\frac{s-p}{r} - \frac{2}{a}\right)w &= -(\mathsf{k}-q)\frac{v}{r} \end{aligned}\right\} \tag{202.9}$$

with

$$p = \frac{\beta}{2}\left(\mu - \frac{1}{\mu}\right); \quad q = \frac{\beta}{2}\left(\mu + \frac{1}{\mu}\right); \quad \mathsf{k} = j + \frac{1}{2}. \tag{202.10}$$

From the first equation (202.9) we get

$$w = -\frac{1}{\mathsf{k}+q}\{r v' + (s+p)v\}; \quad w' = -\frac{1}{\mathsf{k}+q}(r v'' + (s+p+1)v'\}. \tag{202.11}$$

This we put in the second equation (202.9) which thus becomes a second-order equation for only v, viz.

$$r v'' + \left[(2s+1) - \frac{2}{a} r\right] v' - \frac{2}{a}(s+p) v = 0. \tag{202.12}$$

This is a Kummer's equation; its solution, in arbitrary normalization and regular at the origin, is the confluent series

$$v = {}_1F_1\left(s+p, 2s+1; 2\frac{r}{a}\right). \tag{202.13}$$

From (202.11) we then derive $w(r)$ using the general formula

$$\left(z\frac{d}{dz} + a\right) {}_1F_1(a, c; z) = a\, {}_1F_1(a+1, c; z);$$

the result is

$$w = -\frac{s+p}{\kappa+q}\, {}_1F_1\left(s+p+1, 2s+1; 2\frac{r}{a}\right). \tag{202.14}$$

Putting the expressions (202.13) and (202.14) in (202.8) leads on to G and F and thence, using (202.6), we finally arrive at the radial functions

$$\left.\begin{aligned} g &= \frac{1}{2} C r^{s-1} e^{-r/a} \left\{ {}_1F_1\left(s+p, 2s+1; 2\frac{r}{a}\right) \right. \\ &\qquad \left. - \frac{s+p}{\kappa+q}\, {}_1F_1\left(s+p+1, 2s+1; 2\frac{r}{a}\right) \right\}; \\ f &= -\frac{i}{2\mu} C r^{s-1} e^{-r/a} \left\{ {}_1F_1\left(s+p, 2s+1; 2\frac{r}{a}\right) \right. \\ &\qquad \left. + \frac{s+p}{\kappa+q}\, {}_1F_1\left(s+p+1, 2s+1; 2\frac{r}{a}\right) \right\}. \end{aligned}\right\} \tag{202.15}$$

The two confluent series are asymptotically proportional to $e^{+2r/a}$, thus destroying normalizability unless their first parameters are zero or negative integer:

$$s+p = -n_r; \quad n_r = 0, 1, 2, 3, \ldots \tag{202.16}$$

If $s+p=0$, in the second confluent series we still have $s+p+1=+1$; however, a factor $s+p$ makes this part of Eqs. (202.15) vanish anyway in this particular case, so that (202.16) is the complete eigenvalue condition. Replacing in (202.16) p by (202.10) and taking μ from (202.3), we may introduce the energy and write

$$\frac{1}{2}\beta\left\{\sqrt{\frac{mc^2-E}{mc^2+E}}-\sqrt{\frac{mc^2+E}{mc^2-E}}\right\}=-(n_r+s).$$

This may be resolved for E:

$$E=\frac{mc^2}{\sqrt{1+\dfrac{\beta^2}{(n_r+s)^2}}}, \tag{202.17}$$

thus providing us with the energy level formula wanted.

We have so far not yet considered the second set of radial equations, (201.9b), which follows by replacing μ by $-1/\mu$ in (202.4) and consequently, q by $-q$ and p by $-p$ so that the eigenvalue condition (202.16) is changed into $s-p=-n_r$ which, however, alters nothing whatsoever in the energy level formula. Each energy term therefore is degenerate with two solutions to it.

If $\beta>1$ ($Z>137$) the exponent s, Eq. (202.5), will become imaginary for the ground state, so that the boundary condition at $r=0$ cannot be simply satisfied. For very large Z the potential hole may even become so deep that for the lowest bound state $E<-mc^2$. According to (202.3), a then becomes imaginary and the solutions g and f, Eq. (202.6), no longer decrease exponentially at large values of r. This is a consequence of the electron wave penetrating into the domain of negative energies (Klein's paradox), a phenomenon explained in some detail for a potential step in Problem 207 below (case c).

Problem 203. Hydrogen atom fine structure

For a hydrogen atom, the parameter β of the preceding problem becomes identical with Sommerfeld's fine-structure constant,

$$\alpha=\frac{e^2}{\hbar c}=\frac{1}{137}.$$

This parameter is small enough to justify power expansion of the results. This shall be performed, thus confirming the unrelativistic theory and adding its first relativistic correction.

Solution. The expansion of s, Eq. (202.5), leads to

$$s=\left(j+\frac{1}{2}\right)-\frac{\alpha^2}{2j+1}+O(\alpha^4). \tag{203.1}$$

If this is inserted into the energy formula (202.17), the principal quantum number,

$$n = n_r + j + \tfrac{1}{2}, \tag{203.2}$$

may be introduced with advantage. We then have

$$E = mc^2 \left\{1 + \left[\alpha^2 \Big/ \left(n - \frac{\alpha^2}{2j+1}\right)^2\right]\right\}^{-\frac{1}{2}}$$

or

$$E = mc^2 \left\{1 - \frac{\alpha^2}{2n^2}\left[1 + \frac{\alpha^2}{n^2}\left(\frac{n}{j+\frac{1}{2}} - \frac{3}{4}\right)\right] + O(\alpha^6)\right\}. \tag{203.3}$$

Since

$$mc^2 \alpha^2 = \frac{me^4}{\hbar^2}$$

we arrive at the level formula for hydrogen, in first relativistic approximation:

$$E = mc^2 - \frac{me^4}{2\hbar^2 n^2}\left[1 + \frac{\alpha^2}{n^2}\left(\frac{n}{j+\frac{1}{2}} - \frac{3}{4}\right)\right], \tag{203.4}$$

where the first term is the rest energy, the second term, for $\alpha^2 = 0$, the unrelativistic Balmer term (cf. Problem 67), and the square bracket provides a first relativistic correction of the order of $\alpha^2 = 0{,}532 \times 10^{-4}$ or about $\frac{1}{200}$ percent of the binding energy. Since this correction depends upon j as well as upon n, each unrelativistic level will split up into several fine-structure components.

Next let us expand the length parameter a, Eq. (202.3). This leads to

$$a = \frac{\hbar^2}{me^2} n \left[1 - \frac{\alpha^2}{2n^2}\left(\frac{n}{j+\frac{1}{2}} - 1\right)\right] + O(\alpha^4). \tag{203.5}$$

Without the relativistic correction in the square bracket, this is the well-known Bohr radius of the n-th hydrogen orbital. Since the argument $2r/a$ occurs in the confluent series and $e^{-r/a}$ as a factor, in the radial wave functions (202.15), the atomic size is determined by a in a form quite similar to that in unrelativistic theory (cf. the following problem).

To perform the transition from the relativistic wave functions (202.15) to those of the unrelativistic Schrödinger theory we need the expansions of the parameters μ and q, Eqs. (202.3) and (202.10):

$$\mu = \frac{\alpha}{2n}\left[1 + \frac{\alpha^2}{n^2}\left(\frac{n}{2j+1} - \frac{1}{4}\right)\right]; \tag{203.6}$$

$$q = n\left[1 - \frac{\alpha^2}{n^2}\left(\frac{n}{2j+1} - \frac{1}{2}\right)\right]. \tag{203.7}$$

The factor in front of the second confluent series in (202.15), in unrelativistic approximation, is

$$\frac{s+p}{\mathrm{k}+q} = -\frac{n_r}{j+\frac{1}{2}+n} = -\frac{n_r}{n_r+(2j+1)}, \tag{203.8}$$

i. e. of the order of magnitude 1 (except in the case $n_r=0$ when it vanishes). The factor μ, according to (203.6) being of the order α, therefore makes f about two powers of 10 bigger than g. In unrelativistic approach and arbitrary normalization we entirely neglect g and conclude:

$$\left.\begin{aligned} g &= 0, \\ f &= r^{j-\frac{1}{2}} e^{-\frac{r}{a}} \left\{ {}_1F_1\left(-n_r, 2j+2; 2\frac{r}{a}\right) \right. \\ &\quad \left. - \frac{n_r}{n_r+2j+1} \, {}_1F_1\left(-n_r+1, 2j+2; 2\frac{r}{a}\right) \right\}. \end{aligned}\right\} \tag{203.9}$$

If in the function f we put $j=l+\frac{1}{2}$, it is indeed transformed into the Schrödinger wave function [cf. (67.12)]

$$f_{\mathrm{Schr}} = r^l e^{-\gamma r} \, {}_1F_1(l+1-n, 2l+2; 2\gamma r) \tag{203.9 S}$$

where $\gamma=1/a$. This is readily seen as follows. With $j=l+\frac{1}{2}$, Eq. (203.2) renders for the principal quantum number $n=n_r+l+1$ so that $n_r=n-l-1$ and

$$f = r^l e^{-\gamma r} \left\{ {}_1F_1(l+1-n, 2l+3; 2\gamma r) - \frac{n-l-1}{n+l-1} \, {}_1F_1(l+2-n, 2l+3; 2\gamma r) \right\}.$$

Here we apply the general relation

$$a \, {}_1F_1(a+1, c+1; z) = (a-c) \, {}_1F_1(a, c+1; z) + c \, {}_1F_1(a, c; z)$$

which permits with $a=l+1-n$, $c=2l+2$, $z=2\gamma r$ transformation of the curly bracket

$${}_1F_1(a, c+1; z) + \frac{a}{c-a} \, {}_1F_1(a+1, c+1; z)$$

into

$$\frac{c}{c-a} \, {}_1F_1(a, c; z) = \frac{2l+2}{n+l+1} \, {}_1F_1(l+1-n, 2l+2; 2\gamma r)$$

as required in Eq. (203.9 S).

The second solution needs separate treatment. With μ replaced by $-1/\mu$, we now have f the small and g the big solution. We use, except for a normalization factor, the unrelativistic approach

$$\left.\begin{aligned} g &= r^{j-\frac{1}{2}} e^{-\gamma r} \{{}_1F_1(-n_r, 2j+2; 2\gamma r) - {}_1F_1(1-n_r, 2j+2; 2\gamma r)\}, \\ f &= 0 \end{aligned}\right\} \quad (203.10)$$

where we now have to put $j = l - \frac{1}{2}$:

$$g = r^{l-1} e^{-\gamma r} \{{}_1F_1(-n_r, 2l+1; 2\gamma r) - {}_1F_1(1-n_r, 2l+1; 2\gamma r)\}.$$

To show that this is again the function (203.9 S), we apply the general formulae

$$a\{{}_1F_1(a+1, c-1; z) - {}_1F_1(a, c-1; z)\} = z \frac{d}{dz} {}_1F_1(a, c-1; z)$$

and

$$(c-1) \frac{d}{dz} {}_1F_1(a, c-1; z) = a\, {}_1F_1(a+1, c; z)$$

which indeed perform the desired transformation into

$$g = r^l e^{-\gamma r} {}_1F_1(1-n_r, 2l+2; 2\gamma r)$$

except for a factor $2\gamma/(2l+1)$. Here, according to (203.2), $1-n_r = j + \frac{3}{2} - n$ which, with $j = l - \frac{1}{2}$, again becomes $l+1-n$ as in Eq. (203.9 S).

Hitherto we have treated l as a convenient parameter without referring to its physical significance. In order to check the latter, let us calculate the expectation values of the operator $\boldsymbol{L}^2$ for the two types of solution. Using just the simple relation

$$\boldsymbol{L}^2 Y_{l,m} = \hbar^2 l(l+1) Y_{l,m}$$

we get for both solutions the expectation value

$$\langle \boldsymbol{L}^2 \rangle = \hbar^2 \frac{\int_0^\infty dr\, r^2 \{(j-\frac{1}{2})(j+\frac{1}{2})|f|^2 + (j+\frac{1}{2})(j+\frac{3}{2})|g|^2\}}{\int_0^\infty dr\, r^2 \{|f|^2 + |g|^2\}}. \quad (203.11)$$

For the first solution we may neglect $|g|^2$, which is then of an order α^2 smaller than $|f|^2$; then it follows that

$$\langle \boldsymbol{L}^2 \rangle = (j-\tfrac{1}{2})(j+\tfrac{1}{2})\, \hbar^2 \quad (203.12\text{a})$$

corresponding to $j - \frac{1}{2} = l$. For the second solution, inversely, $|f|^2$ may be neglected and we arrive at

$$\langle \boldsymbol{L}^2 \rangle = (j+\tfrac{1}{2})(j+\tfrac{3}{2})\, \hbar^2 \quad (203.12\text{b})$$

leading to $j+\frac{1}{2}=l$. These are exactly the substitutions used above. In other words, in unrelativistic approximation, when f and g are no longer mixed in the same wave spinor, l again becomes a good quantum number.

Supplement. The eigenspinors are composed of

$$u^{\mathrm{I}}=\frac{1}{\sqrt{2j}}f(r)\begin{pmatrix}\sqrt{j+\frac{1}{2}}\,Y_{j-\frac{1}{2},0}\\ -\sqrt{j-\frac{1}{2}}\,Y_{j-\frac{1}{2},1}\end{pmatrix}$$

and

$$u^{\mathrm{II}}=\frac{1}{\sqrt{2(j+1)}}g(r)\begin{pmatrix}\sqrt{j+\frac{1}{2}}\,Y_{j+\frac{1}{2},0}\\ \sqrt{j+\frac{3}{2}}\,Y_{j+\frac{1}{2},1}\end{pmatrix}$$

in either of the two forms

$$\psi_a=\begin{pmatrix}u^{\mathrm{I}}\\ u^{\mathrm{II}}\end{pmatrix}\quad\text{or}\quad\psi_b=\begin{pmatrix}u^{\mathrm{II}}\\ u^{\mathrm{I}}\end{pmatrix}.$$

We have seen that for the first solution, ψ_a, $g\ll f$, for the other, ψ_b, $f\ll g$ so that they approximately simplify to

$$\psi_a=\begin{pmatrix}u^{\mathrm{I}}\\ 0\end{pmatrix}\quad\text{and}\quad\psi_b=\begin{pmatrix}u^{\mathrm{II}}\\ 0\end{pmatrix}.$$

Thus there remains the unrelativistic two-component spin theory in which now $j=l+\frac{1}{2}$ for ψ_a and $j=l-\frac{1}{2}$ for ψ_b without any mixing of different l-values for the same j.

We further know that the "big" function in ψ_a is

$$f=r^{j-\frac{1}{2}}\,e^{-r/n}\,{}_1F_1(j+\tfrac{1}{2}-n,2j+1;2\gamma r)\quad\text{with } j=l+\tfrac{1}{2}$$

and in ψ_b,

$$g=r^{j+\frac{1}{2}}\,e^{-r/n}\,{}_1F_1(j+\tfrac{3}{2}-n,2j+3;2\gamma r)\quad\text{with } j=l-\tfrac{1}{2}.$$

If, and only if, the first parameter of the confluent series is zero or a negative integer, there exists a normalizable solution. Thus we arrive at the lowest possible states given in the following table. Their radial wave functions, either f or g, are identical with those of the unrelativistic theory treated in problem 67; their two-component character agrees with the unrelativistic spin theory, cf. Problem 133.

j	Solution ψ_a	ψ_b	Spectroscopic symbol for ψ_a	ψ_b
$\frac{1}{2}$	$n\geqq 1$	$n\geqq 2$	$n\,S_{\frac{1}{2}}$	$n\,P_{\frac{1}{2}}$
$\frac{3}{2}$	$n\geqq 2$	$n\geqq 3$	$n\,P_{\frac{3}{2}}$	$n\,D_{\frac{3}{2}}$
$\frac{5}{2}$	$n\geqq 3$	$n\geqq 4$	$n\,D_{\frac{5}{2}}$	$n\,F_{\frac{5}{2}}$

The energy levels, according to (203.4), are, except for the residual energy,

$$E_{n,j}=-E_n^0-\Delta E_{n,j}$$

with (in multiples of the atomic unit $me^4/\hbar^2=27.2$ eV)

$$E_n^0=\frac{1}{2n^2}$$

the unrelativistic Balmer binding energy, and

$$\Delta E_{n,j} = \frac{\alpha^2}{2n^4}\left(\frac{n}{j+\frac{1}{2}} - \frac{3}{4}\right)$$

its relativistic correction. For $n=3$ the energy values and their corrections are tabulated. The table shows that the splitting is larger the lower the level. That is the reason why the red $H\alpha$ line ($n=3 \rightarrow n=2$) appears to be roughly a doublet of

$$(2.08 - 0.42) \times 10^{-6} \times 27.2\,\mathrm{eV}$$

or $0.365\ \mathrm{cm}^{-1}$ splitting.

n	E_n^0	$(2/\alpha^2)\,\Delta E_{n,j}$ for			$10^6 \times \Delta E_{n,j}$ for		
		$j=\frac{1}{2}$	$j=\frac{3}{2}$	$j=\frac{5}{2}$	$j=\frac{1}{2}$	$j=\frac{3}{2}$	$j=\frac{5}{2}$
1	$\frac{1}{2}$	$\frac{1}{4}$	—	—	6.68	—	—
2	$\frac{1}{8}$	$\frac{5}{64}$	$\frac{1}{64}$	—	2.08	0.42	—
3	$\frac{1}{18}$	$\frac{1}{36}$	$\frac{1}{108}$	$\frac{1}{324}$	0.74	0.25	0.08

Problem 204. Radial Kepler solutions at positive kinetic energies

To determine the radial wave functions and their asymptotic behaviour if the electron in the Coulomb field has positive kinetic energy, $E - mc^2 > 0$, at infinite distances from the centre of attraction.

Solution. Since $mc^2 - E < 0$, we replace the relations (202.3) of Problem 202 by the definitions

$$\eta k = \frac{E - mc^2}{\hbar c}; \qquad \frac{k}{\eta} = \frac{E + mc^2}{\hbar c} \tag{204.1a}$$

where k is now the wave number at infinity. This is equivalent with

$$\eta = \sqrt{\frac{E - mc^2}{E + mc^2}}; \qquad k = \frac{\sqrt{(E - mc^2)(E + mc^2)}}{\hbar c}. \tag{204.1b}$$

The differential equations to be solved then run as follows,

$$\left.\begin{aligned} g' + \frac{j+\frac{3}{2}}{r} g + i\left(-\eta k - \frac{\beta}{r}\right) f &= 0, \\ f' - \frac{j-\frac{1}{2}}{r} f - i\left(\frac{k}{\eta} + \frac{\beta}{r}\right) g &= 0. \end{aligned}\right\} \tag{204.2}$$

We solve these equations in full analogy to Problem 202 by putting

$$g = \frac{1}{2}(w + v)\, r^{s-1}\, e^{ikr}; \qquad f = \frac{1}{2\eta}(w - v)\, r^{s-1}\, e^{ikr} \tag{204.3}$$

with

$$s=\sqrt{(j+\tfrac{1}{2})^2-\beta^2}\,. \tag{204.4}$$

After a straightforward calculation, using the abbreviations

$$P=\frac{\beta}{2}\left(\frac{1}{\eta}-\eta\right); \quad Q=\frac{\beta}{2}\left(\frac{1}{\eta}+\eta\right) \tag{204.5}$$

and the variable

$$z=-2ikr \tag{204.6}$$

we arrive at

$$z\frac{d^2w}{dz^2}+[(2s+1)-z]\frac{dw}{dz}-(s-iQ)\,w=0 \tag{204.7}$$

and

$$v=-\frac{1}{j+\frac{1}{2}+iP}\left[z\frac{dw}{dz}+(s-iQ)\,w\right]. \tag{204.8}$$

The solution of the differential equation (204.7), regular at the origin, in arbitrary normalization, is

$$w=C\,{}_1F_1(s-iQ,2s+1;z)\,. \tag{204.9}$$

From (204.8), using the formula

$$\left\{z\frac{d}{dz}+a\right\}{}_1F_1(a,c;z)=a\,{}_1F_1(a+1,c;z)\,,$$

we then find

$$v=-C\frac{s-iQ}{j+\frac{1}{2}+iP}\,{}_1F_1(1+s-iQ,2s+1;z)\,. \tag{204.10}$$

Eqs. (204.9) and (204.10) provide the complete solution of the radial problem in arbitrary normalization.

For positive real values of r, the variable z, Eq. (204.6), is negative imaginary. We therefore may apply without further precaution the asymptotic formula

$${}_1F_1(a,c;z)\to e^{-i\pi a}\frac{\Gamma(c)}{\Gamma(c-a)}z^{-a}+\frac{\Gamma(c)}{\Gamma(a)}e^z z^{a-c} \tag{204.11}$$

thus getting

$$w\to\frac{C\Gamma(2s+1)\,e^{-\frac{\pi Q}{2}}}{(2kr)^s}\left\{\frac{e^{-\frac{i\pi s}{2}+iQ\log 2kr}}{\Gamma(1+s+iQ)}+\frac{i\,e^{-\frac{i\pi s}{2}-iQ\log 2kr}}{\Gamma(s-iQ)}\cdot\frac{e^{-2ikr}}{2kr}\right\}$$

and

$$v \to -\frac{C\,\Gamma(2s+1)\,e^{-\frac{\pi Q}{2}}}{(2kr)^s}\,\frac{s-iQ}{j+\frac{1}{2}+iP}\left\{\frac{e^{-\frac{i\pi s}{2}+iQ\log 2kr}}{i\Gamma(s+iQ)}\cdot\frac{1}{2kr}\right.$$
$$\left.+\frac{e^{\frac{i\pi s}{2}-iQ\log 2kr}}{\Gamma(1+s-iQ)}\,e^{-2ikr}\right\}.$$

Here, the second term in the w bracket and the first term in the v bracket are of an order $1/kr$ smaller than the other terms, respectively, and may thus be neglected. Therefore, using (204.3), we arrive at the results

$$\begin{aligned} rg &\to C_1\,e^{i(kr+Q\log 2kr)} + C_2\,e^{-i(kr+Q\log 2kr)},\\ \eta r f &\to C_1\,e^{i(kr+Q\log 2kr)} - C_2\,e^{-i(kr+Q\log 2kr)} \end{aligned} \tag{204.12}$$

with the complex amplitude constants

$$\begin{aligned} C_1 &= \frac{C\Gamma(2s+1)\,e^{-\frac{\pi Q}{2}}}{2(2k)^s}\cdot\frac{e^{-\frac{i\pi s}{2}}}{\Gamma(1+s+iQ)};\\ C_2 &= -\frac{C\Gamma(2s+1)\,e^{-\frac{\pi Q}{2}}}{2(2k)^s}\,\frac{s-iQ}{j+\frac{1}{2}+iP}\,\frac{e^{\frac{i\pi s}{2}}}{\Gamma(1+s-iQ)}. \end{aligned} \tag{204.13}$$

The constants C_1 and C_2 can, of course, only differ by a phase factor since incoming and outgoing partial waves must have equal amplitudes. Indeed both the factors by which they differ from one another,

$$\frac{s-iQ}{j+\frac{1}{2}+iP} = e^{2i\delta} \quad\text{and}\quad \frac{\Gamma(1+s+iQ)}{\Gamma(1+s-iQ)} = e^{2i\zeta} \tag{204.14}$$

are phase factors. For the second this is obvious; the first one yields

$$\left|\frac{s-iQ}{j+\frac{1}{2}+iP}\right|^2 = \frac{s^2+Q^2}{(j+\frac{1}{2})^2+P^2} = 1$$

because of

$$Q^2-P^2=\beta^2 \quad\text{and}\quad s^2=(j+\tfrac{1}{2})^2-\beta^2.$$

We therefore finally arrive at

$$g=|C_1|\,e^{i\delta}\,2i\,\frac{\sin\xi}{r};\qquad f=|C_1|\,e^{i\delta}\,2\,\frac{\cos\xi}{\eta r} \tag{204.15}$$

with

$$\xi = kr+Q\log 2kr-\frac{\pi s}{2}-\zeta-\delta. \tag{204.16}$$

NB. The result may be compared with the unrelativistic theory where $\eta \ll 1$ so that $f \gg g$. This gives

$$f \propto \frac{1}{r} \sin\left(kr + Q \log 2kr - \frac{\pi}{2}(s+1) - \zeta - \delta\right) \tag{204.17}$$

where, according to (204.5) and (204.1 b), Q may be expressed by E and further by the velocity v at infinity. From the well-known relation

$$E = mc^2 \left(1 - \frac{v^2}{c^2}\right)^{-\frac{1}{2}}$$

one easily deduces

$$Q = \frac{Ze^2}{\hbar v},$$

i.e. Q becomes identical with $\varkappa$, Problem 110. From (204.1 b) one further gathers that $\hbar k$ is the momentum of the electron, at infinity. The first two terms in the argument of (204.17) therefore agree with the classical expressions. The constant phase angles, in the same approximation, follow from

$$s \to j + \tfrac{1}{2} = l + 1 \quad \text{and} \quad P \to Q = \varkappa;$$

they become

$$\frac{\pi}{2}(s+1) + \zeta + \delta \to \frac{\pi}{2}(l+2) + \arg\Gamma(l+2+i\varkappa) - \arg(l+1+i\varkappa)$$

$$= \pi + \frac{\pi}{2} l + \arg\Gamma(l+1+i\varkappa),$$

and as the wave function is defined except for a sign (i.e., for a phase π) we arrive finally at

$$f \propto \frac{1}{r} \sin\{kr + \varkappa \log 2kr - \frac{\pi l}{2} - \arg\Gamma(l+1+i\varkappa)\}$$

in complete agreement with the results of Problem 111, but for an attractive Coulomb potential.

Problem 205. Angular momentum expansion of plane Dirac wave

The angular momentum eigenspinors of Problem 190 shall be used to construct the plane wave, of helicity $h = +1$, in z direction:

$$\psi = C \begin{pmatrix} 1 \\ 0 \\ \eta \\ 0 \end{pmatrix} e^{ikz} = C \begin{pmatrix} 1 \\ 0 \\ \eta \\ 0 \end{pmatrix} \sum_{l=0}^{\infty} \sqrt{4\pi(2l+1)}\, i^l\, Y_{l,0}\,. \tag{205.1}$$

Solution. We start by solving Eqs. (201.9 a, b) in the force-free case. Using the abbreviation η, introduced for plane waves in Problem 190, we have with

$$\frac{E-mc^2}{\hbar c}=k\eta\,;\qquad \frac{E+mc^2}{\hbar c}=\frac{k}{\eta}\,, \tag{205.2}$$

instead of (201.9 a)

$$g'+\frac{j+\frac{3}{2}}{r}g-ik\eta f=0\,; \tag{205.3a}$$

$$f'-\frac{j-\frac{1}{2}}{r}f-i\frac{k}{\eta}g=0\,. \tag{205.3b}$$

From (205.3 b) we get

$$i\frac{k}{\eta}g=f'-\frac{j-\frac{1}{2}}{r}f \tag{205.4}$$

and by differentiation

$$i\frac{k}{\eta}g'=f''-\frac{j-\frac{1}{2}}{r}f'+\frac{j-\frac{1}{2}}{r^2}f\,.$$

Putting these expressions into (205.3 a), a second-order equation for f results:

$$f''+\frac{2}{r}f'+\left[k^2-\frac{(j-\frac{1}{2})(j+\frac{1}{2})}{r^2}\right]f=0 \tag{205.5}$$

with the solution

$$f=\frac{1}{kr}j_{j-\frac{1}{2}}(kr) \tag{205.6}$$

regular at the origin, in arbitrary normalization. Putting then (205.6) into (205.4), we find

$$i\frac{k}{\eta}g=\frac{1}{kr}\left\{j'_{j-\frac{1}{2}}-\frac{j+\frac{1}{2}}{kr}j_{j-\frac{1}{2}}\right\}$$

with the prime denoting differentiation with respect to the argument kr. The general formula

$$j'_l(z)-\frac{l+1}{z}j_l(z)=-j_{l+1}(z)$$

then permits to write

$$g=\frac{i\eta}{kr}j_{j+\frac{1}{2}}(kr)\,. \tag{205.7}$$

Thus the normalization of g is fixed relative to that of f. Eqs. (205.6) and (205.7) allow us to write the full four-component spinor according to (201.4a) for given j and $m_j = +\frac{1}{2}$:

$$\psi_j^{\mathrm{I}} = \frac{1}{kr}\begin{pmatrix} \sqrt{\dfrac{j+\frac{1}{2}}{2j}}\, j_{j-\frac{1}{2}}(kr)\, Y_{j-\frac{1}{2},0} \\ -\sqrt{\dfrac{j-\frac{1}{2}}{2j}}\, j_{j-\frac{1}{2}}(kr)\, Y_{j-\frac{1}{2},1} \\ i\eta\sqrt{\dfrac{j+\frac{1}{2}}{2(j+1)}}\, j_{j+\frac{1}{2}}(kr)\, Y_{j+\frac{1}{2},0} \\ i\eta\sqrt{\dfrac{j+\frac{3}{2}}{2(j+1)}}\, j_{j+\frac{1}{2}}(kr)\, Y_{j+\frac{1}{2},1} \end{pmatrix}. \tag{205.8}$$

The second solution derived in Problem 201 is obtained by replacing η by $1/\eta$ in (205.3a, b) so that f is still given by (205.6), but in (205.7) the factor η has to be put in the denominator instead of in the numerator. From (201.4b) with an arbitrary normalization we get for the second solution,

$$\psi_j^{\mathrm{II}} = \frac{1}{kr}\begin{pmatrix} \dfrac{i}{\eta}\sqrt{\dfrac{j+\frac{1}{2}}{2(j+1)}}\, j_{j+\frac{1}{2}}(kr)\, Y_{j+\frac{1}{2},0} \\ \dfrac{i}{\eta}\sqrt{\dfrac{j+\frac{3}{2}}{2(j+1)}}\, j_{j+\frac{1}{2}}(kr)\, Y_{j+\frac{1}{2},1} \\ \sqrt{\dfrac{j+\frac{1}{2}}{2j}}\, j_{j-\frac{1}{2}}(kr)\, Y_{j-\frac{1}{2},0} \\ -\sqrt{\dfrac{j-\frac{1}{2}}{2j}}\, j_{j-\frac{1}{2}}(kr)\, Y_{j-\frac{1}{2},1} \end{pmatrix}. \tag{205.9}$$

In order to construct a solution like (205.1), we have to write

$$\psi = \sum_j (A_j \psi_j^{\mathrm{I}} + B_j \psi_j^{\mathrm{II}}). \tag{205.10}$$

In this expansion it is, of course, possible to change the dummy j in a different way in different sums. We shall do this by again using the orbital momentum quantum number l so that in all sums over

$$j_{j-\frac{1}{2}}(kr)\, Y_{j-\frac{1}{2},m}$$

(with $m=0$ or 1) we shall put $j=l+\frac{1}{2}$, and in all sums over

$$j_{j+\frac{1}{2}}(kr)\, Y_{j+\frac{1}{2},m}$$

we shall put $j=l-\frac{1}{2}$. We then obtain from (205.8) and (205.9):

$$\psi = \sum_{l=0}^{\infty} \frac{j_l(kr)}{\sqrt{2l+1}\,kr} \begin{pmatrix} \left(A_{l+\frac{1}{2}}\sqrt{l+1} + B_{l-\frac{1}{2}}\frac{i}{\eta}\sqrt{l}\right) Y_{l,0} \\ \left(-A_{l+\frac{1}{2}}\sqrt{l} + B_{l-\frac{1}{2}}\frac{i}{\eta}\sqrt{l+1}\right) Y_{l,1} \\ (i\eta A_{l-\frac{1}{2}}\sqrt{l} + B_{l+\frac{1}{2}}\sqrt{l+1})\, Y_{l,0} \\ (i\eta A_{l-\frac{1}{2}}\sqrt{l+1} - B_{l+\frac{1}{2}}\sqrt{l})\, Y_{l,1} \end{pmatrix}. \tag{205.11}$$

Here it is correct to sum over all $l \geqq 0$. In the terms $j=l+\frac{1}{2}$, this follows directly from $j \geqq \frac{1}{2}$. On the other hand, for $j=l-\frac{1}{2}$ the sums would start with $l=1$. In the second and fourth line, however, the zeroth term vanishes because $Y_{0,1}=0$. There remain $j=l-\frac{1}{2}$ terms in the first and third line with $l=0$, but again these vanish due to the factor $\sqrt{l}$.

In order to make the sum (205.11) identical with (205.1) we need only put

$$\left.\begin{aligned} A_{l+\frac{1}{2}} &= \sqrt{4\pi}\, C i^l \sqrt{l+1}, \\ B_{l+\frac{1}{2}} &= \eta \sqrt{4\pi}\, C i^l \sqrt{l+1} \end{aligned}\right\} \tag{205.12}$$

as can be directly checked.

Problem 206. Scattering by a central force potential

A plane Dirac wave of positive helicity is scattered by a central-force potential. To determine the asymptotic behaviour of the scattered wave if the phases are taken from the solutions of the radial wave equations.

Solution. In Problem 201 it has been shown that there are two sets of radial equations, viz.

$$\left.\begin{aligned} \textit{Set I:}\quad & g_j' + \frac{j+\frac{3}{2}}{r} g_j - ik\eta f_j + iU(r) f_j = 0, \\ & f_j' - \frac{j-\frac{1}{2}}{r} f_j - i\frac{k}{\eta} g_j + iU(r) g_j = 0 \end{aligned}\right\} \tag{206.1 a}$$

with $U(r)=V(r)/\hbar c$. It can easily be seen that, for $U(r)$ decreasing more steeply than $1/r$, the asymptotic solution may be written

$$g_j(r)\to\frac{i\eta}{r}\sin\sigma_j;\qquad f_j(r)\to\frac{1}{r}\cos\sigma_j \tag{206.2a}$$

with

$$\sigma_j=kr-(j+\tfrac{1}{2})\frac{\pi}{2}+\alpha_j. \tag{206.3a}$$

The two functions are mutually shifted by a phase angle $\pi/2$ and their respective amplitudes, given in arbitrary normalization, are coupled in such a way that, as the unrelativistic limit ($\eta\to 0$) is approached, f_j becomes the big and g_j the small wave function. If f_j is chosen to be real, g_j is purely imaginary. The phase α_j is determined by integrating the set (206.1 a) with the boundary conditions $g_j(0)=0$, $f_j(0)=0$. In the classical approach, $j=l+\frac{1}{2}$ so that we get

$$f_j(r)\to\frac{1}{r}\sin\left(kr-l\frac{\pi}{2}+\alpha_{l+\frac{1}{2}}\right);\qquad g_j(r)\to 0$$

at large r.

$$\text{Set } II:\quad \left.\begin{aligned} &g_j'+\frac{j+\frac{3}{2}}{r}g_j-i\frac{k}{\eta}f_j+iU(r)f_j=0,\\ &f_j'-\frac{j-\frac{1}{2}}{r}f_j-ik\eta g_j+iU(r)g_j=0\end{aligned}\right\} \tag{206.1b}$$

with the asymptotic solution

$$g_j\to\frac{i}{\eta r}\sin\tau_j;\qquad f_j\to\frac{1}{r}\cos\tau_j \tag{206.2b}$$

and

$$\tau_j=kr-(j+\tfrac{1}{2})\frac{\pi}{2}+\beta_j \tag{206.3b}$$

with other phase constants β_j than in case I. Since (206.1 b) differs from (206.1 a) by η being replaced by $1/\eta$, in the classical limit there will g_j become the big, and f_j the small wave function. This leads to the identification $j=l-\frac{1}{2}$ in this approach:

$$g_j(r)\to\frac{1}{r}\sin\left(kr-l\frac{\pi}{2}+\beta_{l-\frac{1}{2}}\right);\qquad f_j(r)\to 0.$$

We have further seen in Problem 201 that there exist two solutions ψ_j^{I} and ψ_j^{II} for each value of j belonging to spin in positive z direction, constructed from these two sets. They behave asymptotically as

$$\psi_j^{\mathrm{I}} \to \frac{1}{kr}\begin{pmatrix} \sqrt{\dfrac{j+\frac{1}{2}}{2j}}\cos\sigma_j\, Y_{j-\frac{1}{2},0} \\ -\sqrt{\dfrac{j-\frac{1}{2}}{2j}}\cos\sigma_j\, Y_{j-\frac{1}{2},1} \\ i\eta\sqrt{\dfrac{j+\frac{1}{2}}{2(j+1)}}\sin\sigma_j\, Y_{j+\frac{1}{2},0} \\ i\eta\sqrt{\dfrac{j+\frac{3}{2}}{2(j+1)}}\sin\sigma_j\, Y_{j+\frac{1}{2},1} \end{pmatrix};$$

$$\psi_j^{\mathrm{II}} \to \frac{1}{kr}\begin{pmatrix} \dfrac{i}{\eta}\sqrt{\dfrac{j+\frac{1}{2}}{2(j+1)}}\sin\tau_j\, Y_{j+\frac{1}{2},0} \\ \dfrac{i}{\eta}\sqrt{\dfrac{j+\frac{3}{2}}{2(j+1)}}\sin\tau_j\, Y_{j+\frac{1}{2},1} \\ \sqrt{\dfrac{j+\frac{1}{2}}{2j}}\cos\tau_j\, Y_{j-\frac{1}{2},0} \\ -\sqrt{\dfrac{j-\frac{1}{2}}{2j}}\cos\tau_j\, Y_{j-\frac{1}{2},1} \end{pmatrix}. \tag{206.4}$$

The most general solution is

$$\psi = \sum_j (A_j\psi_j^{\mathrm{I}} + B_j\psi_j^{\mathrm{II}}) \tag{206.5}$$

which, by changing the summation subscript j into $l\pm\frac{1}{2}$ everywhere so that spherical harmonics of the order l emerge in the sum, may be written

$$\psi \to \frac{1}{kr}\sum_{l=0}^{\infty}\frac{1}{\sqrt{2l+1}}$$

$$\begin{pmatrix} \left[\sqrt{l+1}\,A_{l+\frac{1}{2}}\cos\sigma_{l+\frac{1}{2}} + \dfrac{i}{\eta}\sqrt{l}\,B_{l-\frac{1}{2}}\sin\tau_{l-\frac{1}{2}}\right] Y_{l,0} \\ \left[-\sqrt{l}\,A_{l+\frac{1}{2}}\cos\sigma_{l+\frac{1}{2}} + \dfrac{i}{\eta}\sqrt{l+1}\,B_{l-\frac{1}{2}}\sin\tau_{l-\frac{1}{2}}\right] Y_{l,1} \\ \left[i\eta\sqrt{l}\,A_{l-\frac{1}{2}}\sin\sigma_{l-\frac{1}{2}} + \sqrt{l+1}\,B_{l+\frac{1}{2}}\cos\tau_{l+\frac{1}{2}}\right] Y_{l,0} \\ \left[i\eta\sqrt{l+1}\,A_{l-\frac{1}{2}}\sin\sigma_{l-\frac{1}{2}} - \sqrt{l}\,B_{l+\frac{1}{2}}\cos\tau_{l+\frac{1}{2}}\right] Y_{l,1} \end{pmatrix}. \tag{206.6}$$

This has the same structure as the plane wave (205.11, 12) with which it becomes identical if $\alpha_j=0$ and $\beta_j=0$ for all j. Let us write the plane wave

$$\psi^0 = \sum_j (A_j^0 \psi_j^{0\,\mathrm{I}} + B_j^0 \psi_j^{0\,\mathrm{II}}) \tag{206.7}$$

with

$$\left.\begin{aligned} A_{l+\frac{1}{2}}^0 &= \sqrt{4\pi}\, C i^l \sqrt{l+1}\,; \quad B_{l+\frac{1}{2}}^0 = \eta\, A_{l+\frac{1}{2}}^0\,; \\ \sigma_{l-\frac{1}{2}}^0 &= \tau_{l-\frac{1}{2}}^0 = kr - l\frac{\pi}{2}\,. \end{aligned}\right\} \tag{206.8}$$

The boundary conditions of a scattering problem, for $r\to\infty$, are then satisfied if, and only if, the difference

$$\psi_s = \psi - \psi^0 \tag{206.9}$$

contains no incoming spherical wave parts of the form e^{-ikr}/kr but only outgoing waves so that it may be identified with the scattered wave. This, according to (206.6), leads to four coefficient relations, viz.

$$\begin{aligned}
\sqrt{l+1}\, A_{l+\frac{1}{2}}\, \mathrm{e}^{i\left(\frac{\pi}{2}-\alpha_{l+\frac{1}{2}}\right)} + \frac{1}{\eta}\sqrt{l}\, B_{l-\frac{1}{2}}\, \mathrm{e}^{-i\beta_{l-\frac{1}{2}}} &= \sqrt{l+1}\, A_{l+\frac{1}{2}}^0\, \mathrm{e}^{i\frac{\pi}{2}} + \frac{1}{\eta}\sqrt{l}\, B_{l-\frac{1}{2}}^0\,, \\
-\sqrt{l}\, A_{l+\frac{1}{2}}\, \mathrm{e}^{i\left(\frac{\pi}{2}-\alpha_{l+\frac{1}{2}}\right)} + \frac{1}{\eta}\sqrt{l+1}\, B_{l-\frac{1}{2}}\, \mathrm{e}^{-i\beta_{l-\frac{1}{2}}} &= -\sqrt{l}\, A_{l+\frac{1}{2}}^0\, \mathrm{e}^{i\frac{\pi}{2}} + \frac{1}{\eta}\sqrt{l+1}\, B_{l-\frac{1}{2}}^0\,, \\
-\eta\sqrt{l}\, A_{l-\frac{1}{2}}\, \mathrm{e}^{-i\alpha_{l-\frac{1}{2}}} + \sqrt{l+1}\, B_{l+\frac{1}{2}}\, \mathrm{e}^{i\left(\frac{\pi}{2}-\beta_{l+\frac{1}{2}}\right)} &= -\eta\sqrt{l}\, A_{l-\frac{1}{2}}^0 + \sqrt{l+1}\, B_{l+\frac{1}{2}}^0\, \mathrm{e}^{i\frac{\pi}{2}}\,, \\
-\eta\sqrt{l+1}\, A_{l-\frac{1}{2}}\, \mathrm{e}^{-i\alpha_{l-\frac{1}{2}}} - \sqrt{l}\, B_{l+\frac{1}{2}}\, \mathrm{e}^{i\left(\frac{\pi}{2}-\beta_{l+\frac{1}{2}}\right)} &= -\eta\sqrt{l+1}\, A_{l-\frac{1}{2}}^0 - \sqrt{l}\, B_{l+\frac{1}{2}}^0\, \mathrm{e}^{i\frac{\pi}{2}}\,.
\end{aligned} \tag{206.10}$$

These equations are satisfied if, and only if,

$$A_j = A_j^0\, \mathrm{e}^{i\alpha_j}\,; \qquad B_j = B_j^0\, \mathrm{e}^{i\beta_j}\,. \tag{206.11}$$

The scattered wave then can be shown by straightforward calculation to behave asymptotically as

$$\psi_s \to \sqrt{4\pi}\, C \frac{\mathrm{e}^{ikr}}{2ikr} \sum_{l=0}^{\infty} \frac{1}{\sqrt{2l+1}} \begin{pmatrix} [(l+1)(\mathrm{e}^{2i\alpha_{l+\frac{1}{2}}}-1) + l(\mathrm{e}^{2i\beta_{l-\frac{1}{2}}}-1)]\, Y_{l,0} \\ \sqrt{l(l+1)}\,(\mathrm{e}^{2i\beta_{l-\frac{1}{2}}} - \mathrm{e}^{2i\alpha_{l+\frac{1}{2}}})\, Y_{l,1} \\ \eta\,[l(\mathrm{e}^{2i\alpha_{l-\frac{1}{2}}}-1) + (l+1)(\mathrm{e}^{2i\beta_{l+\frac{1}{2}}}-1)]\, Y_{l,0} \\ \eta\sqrt{l(l+1)}\,(\mathrm{e}^{2i\alpha_{l-\frac{1}{2}}} - \mathrm{e}^{2i\beta_{l+\frac{1}{2}}})\, Y_{l,1} \end{pmatrix}. \tag{206.12}$$

Problem 207. Continuous potential step

A plane wave of positive helicity $h=+1$ is falling perpendicularly from negative z upon a potential step described by

$$V(z)=\tfrac{1}{2}V_0\left(1+\tanh\frac{z}{l}\right), \tag{207.1}$$

i.e. with the potential increasing from $V=0$ at $z=-\infty$ to $V=+V_0$ at $z=+\infty$, within a layer around $z=0$ of a thickness of the order l. The coefficient of transmission of the step shall be investigated for the cases of different step heights, viz.

$$\left.\begin{array}{ll}\text{Case a:} & V_0<E-mc^2,\\ \text{Case b:} & E-mc^2<V_0<E+mc^2,\\ \text{Case c:} & E+mc^2<V_0.\end{array}\right\} \tag{207.2}$$

The three cases are sketched in Fig. 72.

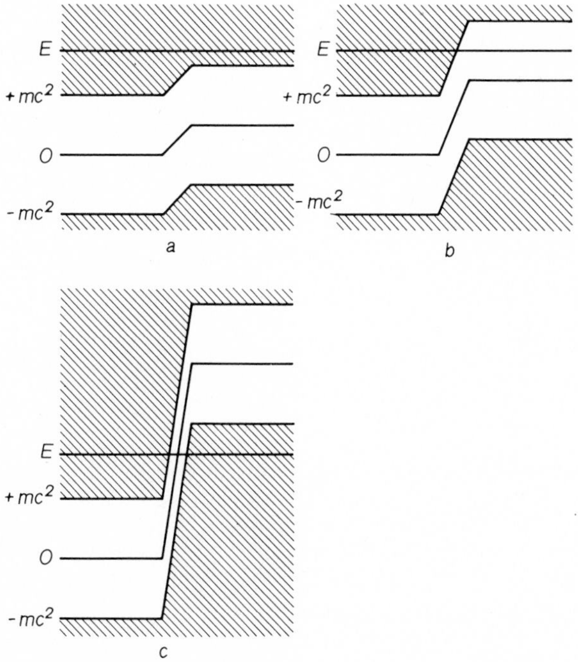

Fig. 72a—c. Three potential steps of different heights. The energy domains in which the particle can move are indicated by hatching

Solution. This is a special case of Problem 197. For helicity $h=+1$, the component functions u_2 and u_4 vanish so that we are left with a set of only two differential equations,

$$\left.\begin{aligned} -i\frac{du_3}{dz}+(\varkappa+Q)\,u_1=0,\\ i\frac{du_1}{dz}+(\varkappa-Q)\,u_3=0 \end{aligned}\right\} \tag{207.3}$$

with

$$Q=\frac{V(z)-E}{\hbar c}; \qquad \varkappa=\frac{mc^2}{\hbar c}. \tag{207.4}$$

Let us introduce, instead of u_1 and u_3, their symmetric and antisymmetric combinations,

$$\varphi_s=u_1+u_3; \qquad \varphi_a=u_1-u_3, \tag{207.5}$$

then, instead of (207.3), we have the equations

$$-\varkappa\varphi_s=i\varphi_a'+Q\varphi_a; \qquad \varkappa\varphi_a=i\varphi_s'-Q\varphi_s \tag{207.6}$$

which permit a simpler elimination of one of the two functions than do the original equations (207.3). We find

$$\varphi_s''+[Q^2-\varkappa^2+iQ']\,\varphi_s=0. \tag{207.7}$$

We have to solve this differential equation with fitting boundary conditions and afterwards to derive φ_a from the second equation (207.6) in order to solve the problem completely.

The differential equation (207.7) can be written with rational coefficient functions if we use the variable

$$x=(1+e^{2z/l})^{-1} \tag{207.8}$$

instead of z with

$$\frac{du}{dz}=-\frac{2}{l}x(1-x)\frac{du}{dx}; \qquad V(z)=V_0(1-x). \tag{207.9}$$

Further, we shall introduce an energy unit $2\hbar c/l$ and use the dimensionless abbreviations

$$\varepsilon=\frac{El}{2\hbar c}; \qquad \varepsilon_0=\frac{lmc^2}{2\hbar c}; \qquad v_0=\frac{lV_0}{2\hbar c}. \tag{207.10}$$

Eq. (207.7) then is transformed into

$$x(1-x)\frac{d}{dx}\left[x(1-x)\frac{d\varphi_s}{dx}\right]+\{[v_0(1-x)-\varepsilon]^2-\varepsilon_0^2+iv_0x(1-x)\}\,\varphi_s=0. \tag{207.11}$$

If we put

$$\varphi_s = x^\nu (1-x)^\mu f(x) \tag{207.12}$$

with

$$\nu^2 = \varepsilon_0^2 - (\varepsilon - v_0)^2 ; \qquad \mu^2 = \varepsilon_0^2 - \varepsilon^2 \tag{207.13}$$

this reduces to the hypergeometrical equation

$$x(1-x) f'' + [(2\nu+1) - (2\nu+2\mu+2)x] f' - (\mu+\nu-i v_0)(\mu+\nu-i v_0+1) f = 0 \tag{207.14}$$

of which we need, as we shall immediately see, only the solution regular at $x=0$,

$$f(x) = {}_2F_1(\mu+\nu-i v_0, \mu+\nu+i v_0+1, 2\nu+1; x). \tag{207.15}$$

In order to see this, we next study the *boundary conditions.* According to (207.8) we have $x=1$ for $z=-\infty$, and $x=0$ for $z=+\infty$. Further, according to (207.10) and (207.13), we have

$$\mu = -i\frac{l}{2}k ; \quad k^2 = \frac{E^2-(mc^2)^2}{(\hbar c)^2} = \left(\frac{p}{\hbar}\right)^2 ; \tag{207.16}$$

hence μ is always an imaginary parameter proportional to the particle momentum $p=\hbar k$ for $V=0$, i.e. on the far left. The hypergeometric series (207.15), in the neighbourhood of $x=1$, may be transformed according to the rule

$$\begin{aligned} {}_2F_1(a,b,c;x) = {} & \frac{\Gamma(c)\Gamma(c-a-b)}{\Gamma(c-a)\Gamma(c-b)}\,{}_2F_1(a,b,a+b-c+1;1-x) \\ & + (1-x)^{c-a-b}\frac{\Gamma(c)\Gamma(a+b-c)}{\Gamma(a)\Gamma(b)}\,{}_2F_1(c-a,c-b,c-a-b+1;1-x) \end{aligned}$$

which, with (207.8) and (207.15), leads for $z\to-\infty$, $x\simeq 1$ to

$$\begin{aligned} \varphi_s \to (1-x)^\mu \Bigg\{ & \frac{\Gamma(2\nu+1)\Gamma(-2\mu)}{\Gamma(\nu-\mu+i v_0+1)\Gamma(\nu-\mu-i v_0)} \\ & + (1-x)^{-2\mu}\frac{\Gamma(2\nu+1)\Gamma(2\mu)}{\Gamma(\nu+\mu-i v_0)\Gamma(\nu+\mu+i v_0+1)} \Bigg\} \end{aligned}$$

or, with

$$1-x = \frac{e^{2z/l}}{1+e^{2z/l}} \to e^{2z/l}$$

and (207.16), to

$$\varphi_s \to A\,e^{ikz} + B\,e^{-ikz}; \tag{207.17}$$

$$A = \frac{\Gamma(2\nu+1)\Gamma(2\mu)}{\Gamma(\nu+\mu-i v_0)\Gamma(\nu+\mu+i v_0+1)}; \quad B = \frac{\Gamma(2\nu+1)\Gamma(-2\mu)}{\Gamma(\nu-\mu-i v_0)\Gamma(\nu-\mu+i v_0+1)} \tag{207.18}$$

where B differs from A just by the sign of μ. The function φ_s therefore is, at large negative z, composed of an incoming wave with amplitude A and a reflected wave with amplitude B, conforming to the physical problem. Hence the special solution (207.15) satisfies the boundary condition at large negative values of z. The same composition of two waves holds for φ_a if we apply the second equation (207.6) to the asymptotic function (207.17),

$$\varphi_a \to A\left(\frac{E}{mc^2} - \frac{k}{\varkappa}\right)e^{ikz} + B\left(\frac{E}{mc^2} + \frac{k}{\varkappa}\right)e^{-ikz}. \tag{207.19}$$

The electrical current density (cf. Problem 198, Eq. (198.13))

$$j_z = ec(u_1^* u_3 + u_3^* u_1) = \tfrac{1}{2}ec(|\varphi_s|^2 - |\varphi_a|^2) \tag{207.20}$$

is then, except for interference terms,

$$j_z = j_{\text{in}} - j_{\text{refl}} \tag{207.21}$$

with the incident current

$$j_{\text{in}} = \tfrac{1}{2}ec|A|^2\left\{1 - \left(\frac{E}{mc^2} - \frac{k}{\varkappa}\right)^2\right\} = ec|A|^2\,\frac{pc(E-pc)}{(mc^2)^2} \tag{207.22}$$

and the reflected current

$$j_{\text{refl}} = \tfrac{1}{2}ec|B|^2\left\{1 - \left(\frac{E}{mc^2} + \frac{k}{\varkappa}\right)^2\right\} = ec|B|^2\,\frac{pc(E+pc)}{(mc^2)^2} \tag{207.23}$$

where the energy formula

$$E = \sqrt{(pc)^2 + (mc^2)^2}$$

is to be remembered.

Let us now pass on to a discussion of the behaviour of the wave function on the right-hand side of the potential step, near $x=0$ or for $z \to +\infty$. It then follows from (207.12) and (207.15) directly that

$$\varphi_s \to x^\nu = e^{-\frac{2}{l}\nu z} \tag{207.24}$$

where, according to (207.10) and (207.13)

$$\nu = -i\frac{l}{2}k'; \quad k'^2 = \frac{(E-V_0)^2 - (mc^2)^2}{(\hbar c)^2}. \tag{207.25}$$

Here we have to distinguish three cases according to Eq. (207.2). If $E-V_0>mc^2$ or $V_0-E>mc^2$ (cases a and c), $k'^2>0$ and k' is real. If, on the other hand, $|E-V_0|<mc^2$ (case b), ν will be real and k' imaginary. In the last case, with $\nu>0$, Eq. (207.24) describes total reflection of the incident wave so that the *coefficient of reflection*,

$$R=j_{\text{refl}}/j_{\text{in}}=\left|\frac{B}{A}\right|^2\frac{E+pc}{E-pc}, \tag{207.26}$$

must become $=1$.

This can easily be derived from A and B, Eq. (207.18), by using the identity $\Gamma(z+1)=z\Gamma(z)$,

$$\frac{B}{A}=\frac{\nu+\mu+i v_0}{\nu-\mu+i v_0}\cdot\frac{\Gamma(\nu+\mu-i v_0)\Gamma(\nu+\mu+i v_0)}{\Gamma(\nu-\mu+i v_0)\Gamma(\nu-\mu-i v_0)}\cdot\frac{\Gamma(-2\mu)}{\Gamma(2\mu)}. \tag{207.27}$$

Since

$$\mu=-i\sigma$$

is always imaginary, the last of the three fractions in (207.27) never contributes to the absolute square $|B/A|^2$. If ν is real (case b), the second factor also is the ratio of two conjugate numbers not contributing. There then remains

$$\left|\frac{B}{A}\right|^2=\frac{\nu^2+(v_0-\sigma)^2}{\nu^2+(v_0+\sigma)^2}=\frac{2v_0(\varepsilon-\sigma)}{2v_0(\varepsilon+\sigma)}=\frac{E-pc}{E+pc},$$

so that (207.26) indeed yields $R=1$.

In cases a and c, on the other hand, ν is imaginary so that a running wave exists on the far right,

$$\varphi_s=e^{ik'z}$$

and according to (207.6),

$$\varphi_a=\left(-\frac{k'}{\varkappa}+\frac{E-V_0}{mc^2}\right)e^{ik'z}=\frac{E'-p'c}{mc^2}e^{ik'z}$$

where $p'=\hbar k'$ is the particle momentum for $z\to+\infty$, and $E'=E-V_0$. The electrical current transmitted is then, according to (207.20),

$$j_{\text{trans}}=\tfrac{1}{2}ec\left\{1-\left(\frac{E'-p'c}{mc^2}\right)^2\right\}=ec\frac{p'c(E'-p'c)}{(mc^2)^2}. \tag{207.28}$$

This yields, with (207.22), a *coefficient of transmission*,

$$T=j_{\text{trans}}/j_{\text{in}}=\frac{1}{|A|^2}\cdot\frac{p'c(E'-p'c)}{pc(E-pc)}. \tag{207.29}$$

In now computing $|A|^2$ from Eq. (207.18) we use, besides $\Gamma(z+1)=z\Gamma(z)$, the general formula

$$|\Gamma(\pm iy)|^2 = \frac{\pi}{y\sinh\pi y} \quad (y \text{ real});$$

we then have, with

$$\mu=-i\sigma, \qquad \nu=-i\sigma',$$

$$1/|A|^2 = \frac{\sigma}{\sigma'}\cdot\frac{\sigma+\sigma'-v_0}{\sigma+\sigma'+v_0}\cdot\frac{\sinh 2\pi\sigma \sinh 2\pi\sigma'}{\sinh\pi(\sigma+\sigma'+v_0)\sinh\pi(\sigma+\sigma'-v_0)}. \tag{207.30}$$

Combining (207.30) with (207.29), there occurs a factor

$$\frac{p'c(E'-p'c)}{pc(E-pc)}\cdot\frac{\sigma}{\sigma'}\cdot\frac{\sigma+\sigma'-v_0}{\sigma+\sigma'+v_0} = \frac{(E'-p'c)(pc+p'c-V_0)}{(E-pc)(pc+p'c+V_0)}$$

which can easily be shown to be $=1$: if we replace V_0 by $E-E'$, the expression may be written

$$\frac{(E'-p'c)(E'+p'c)-(E'-p'c)(E-pc)}{(E-pc)(E+pc)-(E-pc)(E'-p'c)} = 1$$

since $E'^2-(p'c)^2=(mc^2)^2=E^2-(pc)^2$ in the first terms of numerator and denominator. We therefore finally obtain the coefficient of transmission,

$$T = \frac{\sinh 2\pi\sigma \sinh 2\pi\sigma'}{\sinh\pi(\sigma+\sigma'+v_0)\sinh\pi(\sigma+\sigma'-v_0)}. \tag{207.31}$$

The denominator can be suitably reshaped so that the characteristic quantity v_0, proportional to the product of breadth and height of the step and independent of the particle energy, is isolated from the momentum quantities σ and σ':

$$T = \frac{\sinh 2\pi\sigma \sinh 2\pi\sigma'}{\sinh^2\pi(\sigma+\sigma')\cosh^2\pi v_0 - \cosh^2\pi(\sigma+\sigma')\sinh^2\pi v_0}. \tag{207.32}$$

In case a of Eq. (207.2), we still have $\sigma+\sigma'-v_0>0$ or $V_0<(p+p')c$. This is the normal case which also occurs in unrelativistic theory. In case c, on the other hand, we find $\sigma+\sigma'-v_0<0$ and therefore $T<0$. The wave then penetrates, for large positive z, into the domain of negative energies (cf. Fig. 72) where negative electrical current accompanies positive momentum. If $\pi v_0 \gg 1$ or

$$V_0 \gg \frac{2\pi\hbar c}{l},$$

the expression (207.32) becomes

$$T = -4\sinh\frac{\pi l p}{\hbar}\sinh\frac{\pi l p'}{\hbar}\, e^{-\frac{\pi l V_0}{\hbar c}},$$

i.e. the penetrability of the potential step from positive to negative energies rapidly becomes very small with increasing "step size" $V_0 l$. Since, in case c, $V_0 > mc^2$, the exponential in T contributes a factor smaller than

$$e^{-\frac{\pi l m c}{\hbar}} = e^{-\pi l/\lambda}$$

with $\lambda = \hbar/mc$ the Compton wavelength.

Literature. Klein, O.: Z. f. Physik **53**, 157 (1929); Sauter, F.: Z. f. Physik **69**, 742; **73**, 547 (1931).

Problem 208. Plane wave at a potential jump

A plane Dirac wave of arbitrary polarization falls obliquely on a potential jump smaller than its kinetic energy. The laws of reflection and refraction shall be derived, and the state of polarization of the transmitted wave calculated.

Solution. Let the incident wave ψ be described by a wave vector $\boldsymbol{k}$ in the direction defined by polar angles ϑ, φ, the reflected wave ψ' by $\boldsymbol{k}'$ in the direction ϑ', φ', and the transmitted wave ψ'' by $\boldsymbol{k}''$ in the direction ϑ'', φ''. Let further $z=0$ be the refracting surface and ψ and ψ' be defined for $z<0$, ψ'' for $z>0$ (see Fig. 73).

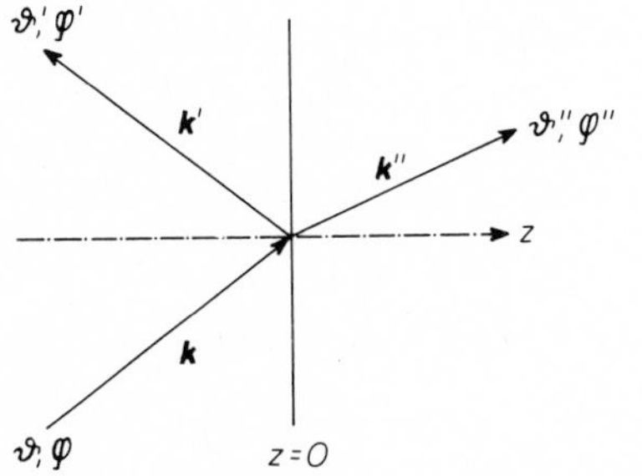

Fig. 73. The three wave vectors at a potential jump

On the surface $z=0$ there must hold for all values of x and y the relation

$$\psi + \psi' = \psi'' \tag{208.1}$$

according to which the three $\boldsymbol{k}$ vectors have the same x and y components:

$$k \sin\vartheta \cos\varphi = k \sin\vartheta' \cos\varphi' = k'' \sin\vartheta'' \cos\varphi'' ;$$

$$k \sin\vartheta \sin\varphi = k \sin\vartheta' \sin\varphi' = k'' \sin\vartheta'' \sin\varphi'' .$$

These relations are satisfied if

$$\varphi=\varphi'=\varphi''; \tag{208.2}$$

$$\vartheta'=\pi-\vartheta; \tag{208.3}$$

$$k\sin\vartheta=k''\sin\vartheta''. \tag{208.4}$$

Eq. (208.2) shows that the three $\boldsymbol{k}$ vectors are lying in the same meridional plane, which we may choose to be the x, z plane, so that the y components of the three vectors vanish along with $\varphi=0$, $\varphi'=0$, $\varphi''=0$. Eq. (208.3) is the law of reflection, and Eq. (208.4) the law of refraction, the index of refraction being $n=k''/k$. These two laws are very much the same as those holding for unrelativistic Schrödinger waves (Problem 45).

Additional information arises for the *polarization*. Using $\varphi=0$ and $\vartheta'=\pi-\vartheta$, the three wave functions in standard representation run as follows:

$$\psi=\frac{e^{i\boldsymbol{k}\boldsymbol{r}}}{\sqrt{V(1+\eta^2)}}\begin{pmatrix}A\cos\frac{\vartheta}{2}+B\sin\frac{\vartheta}{2}\\ A\sin\frac{\vartheta}{2}-B\cos\frac{\vartheta}{2}\\ \eta(A\cos\frac{\vartheta}{2}-B\sin\frac{\vartheta}{2})\\ \eta(A\sin\frac{\vartheta}{2}+B\cos\frac{\vartheta}{2})\end{pmatrix};$$

$$\psi'=\frac{e^{i\boldsymbol{k}'\boldsymbol{r}}}{\sqrt{V(1+\eta^2)}}\begin{pmatrix}C\sin\frac{\vartheta}{2}+D\cos\frac{\vartheta}{2}\\ C\cos\frac{\vartheta}{2}-D\sin\frac{\vartheta}{2}\\ \eta(C\sin\frac{\vartheta}{2}-D\cos\frac{\vartheta}{2})\\ \eta(C\cos\frac{\vartheta}{2}+D\sin\frac{\vartheta}{2})\end{pmatrix};$$

$$\psi''=\frac{e^{i\boldsymbol{k}''\boldsymbol{r}}}{\sqrt{V(1+\eta''^2)}}\begin{pmatrix}E\cos\frac{\vartheta''}{2}+F\sin\frac{\vartheta''}{2}\\ E\sin\frac{\vartheta''}{2}-F\cos\frac{\vartheta''}{2}\\ \eta''(E\cos\frac{\vartheta''}{2}-F\sin\frac{\vartheta''}{2})\\ \eta''(E\sin\frac{\vartheta''}{2}+F\cos\frac{\vartheta''}{2})\end{pmatrix}. \tag{208.5}$$

Here, the first parts of the spinors, proportional to the constants A, C, and E, respectively, belong to waves with helicity $+1$, the parts with B, D, and F to helicity -1.

The boundary condition (208.1) applied to the amplitudes then leads to the following four-component relations:

$$\begin{aligned}A+pB+pC+D&=r(E+qF),\\ pA-B+C-pD&=r(qE-F),\\ A-pB+pC-D&=r\lambda(E-qF),\\ pA+B+C+pD&=r\lambda(qE+F)\end{aligned} \tag{208.6}$$

with the abbreviations

$$p=\tan\frac{\vartheta}{2}; \quad q=\tan\frac{\vartheta''}{2}; \quad r=\sqrt{\frac{1+\eta^2}{1+\eta''^2}\,\frac{\cos\frac{\vartheta''}{2}}{\cos\frac{\vartheta}{2}}}; \quad \lambda=\frac{\eta''}{\eta}. \tag{208.7}$$

By combination we find from Eqs. (208.6)

$$\begin{aligned} A+pC&=\tfrac{1}{2}r[(1+\lambda)E+(1-\lambda)qF],\\ pA+C&=\tfrac{1}{2}r[(1+\lambda)qE-(1-\lambda)F] \end{aligned} \tag{208.8 a}$$

and

$$\begin{aligned} B+pD&=\tfrac{1}{2}r[-(1-\lambda)qE+(1+\lambda)F],\\ pB+D&=\tfrac{1}{2}r[(1-\lambda)E+(1+\lambda)qF]. \end{aligned} \tag{208.8 b}$$

From the first pair of equations we then eliminate C and from the other pair D so that there remain relations connecting the amplitude coefficients E and F of the transmitted wave with A and B of the incident wave, viz.

$$\begin{aligned} (1-p^2)A&=\tfrac{1}{2}r[(1+\lambda)(1-pq)E+(1-\lambda)(p+q)F];\\ (1-p^2)B&=\tfrac{1}{2}r[-(1-\lambda)(p+q)E+(1+\lambda)(1-pq)F]. \end{aligned} \tag{208.9}$$

The expectation value of the helicity (or, briefly, the polarization) of the incident wave is defined by

$$h=\frac{A^2-B^2}{A^2+B^2}=\frac{1-(B/A)^2}{1+(B/A)^2} \tag{208.10}$$

and that of the transmitted wave by

$$h''=\frac{E^2-F^2}{E^2+F^2}=\frac{1-(F/E)^2}{1+(F/E)^2}. \tag{208.11}$$

From (208.9) we have

$$B/A=\frac{-(1-\lambda)(p+q)+(1+\lambda)(1-pq)(F/E)}{(1+\lambda)(1-pq)+(1-\lambda)(p+q)(F/E)}. \tag{208.12}$$

The expression can be much simplified by using the abbreviation

$$u=\frac{1-\lambda}{1+\lambda}\cdot\frac{p+q}{1-pq}=\frac{\eta-\eta''}{\eta+\eta''}\tan\frac{\vartheta+\vartheta''}{2} \tag{208.13}$$

which allows us to write

$$B/A=\frac{(F/E)-u}{1+u(F/E)}. \tag{208.14}$$

If, into this latter expression, h instead of B/A and h'' instead of F/E are introduced according to (208.10) and (208.11), a simple calculation finally leads to

$$h'' = \frac{1-u^2}{1+u^2}\,h - \frac{2u}{1+u^2}\sqrt{1-h^2}\,. \tag{208.15}$$

If the incident wave is completely polarized in either of the two directions ($h = \pm 1$) we find

$$h'' = \pm \frac{1-u^2}{1+u^2}\,,$$

i.e. a partial depolarization will occur at the potential jump which is, however, of second order only in the parameter u. An entirely unpolarized beam ($h=0$), on the other hand, leads to

$$h'' = -\frac{2u}{1+u^2}$$

the potential jump giving rise to at least partial polarization. The latter effect is of first order in u. Since u in reasonable arrangements generally turns out to be rather small (about 0.1), the partial polarization of an unpolarized beam will be of greater interest than the partial depolarization of a beam of well-defined helicity. It should further be remarked that u vanishes for perpendicular incidence, so that grazing incidence will favour the effect.

Problem 209. Reflected intensity at a potential jump

To calculate the reflection coefficient of the plane wave of mixed helicity investigated in the preceding problem, and to derive the law of continuity of the electrical current at the surface.

Solution. The electrical current density

$$j_l = iec\psi^\dagger \gamma_4 \gamma_l \psi = ec\psi^\dagger \alpha_l \psi$$

can be expressed by the spinor components ψ_μ in standard representation as follows:

$$\begin{aligned} j_x &= ec(\psi_1^*\psi_4 + \psi_2^*\psi_3 + \psi_3^*\psi_2 + \psi_4^*\psi_1)\,, \\ j_y &= eci(-\psi_1^*\psi_4 + \psi_2^*\psi_3 - \psi_3^*\psi_2 + \psi_4^*\psi_1)\,, \\ j_z &= ec(\psi_1^*\psi_3 - \psi_2^*\psi_4 + \psi_3^*\psi_1 - \psi_4^*\psi_2)\,. \end{aligned}$$

Since the exponential factors of ψ_μ^* and ψ_ν in a plane wave cancel out, there only remain the spinor amplitudes which, using the Dirac wave functions of Eq. (208.5), are all real according to our choice of $\varphi=0$. Indeed, this leads to $j_y=0$ above. The two other components become for the incident wave

$$j_x=\frac{2ec}{V(1+\eta^2)}\left\{\left(A\cos\frac{\vartheta}{2}+B\sin\frac{\vartheta}{2}\right)\eta\left(A\sin\frac{\vartheta}{2}+B\cos\frac{\vartheta}{2}\right)\right.$$
$$\left.+\left(A\sin\frac{\vartheta}{2}-B\cos\frac{\vartheta}{2}\right)\eta\left(A\cos\frac{\vartheta}{2}-B\sin\frac{\vartheta}{2}\right)\right\};$$

$$j_z=\frac{2ec}{V(1+\eta^2)}\left\{\left(A\cos\frac{\vartheta}{2}+B\sin\frac{\vartheta}{2}\right)\eta\left(A\cos\frac{\vartheta}{2}-B\sin\frac{\vartheta}{2}\right)\right.$$
$$\left.-\left(A\sin\frac{\vartheta}{2}-B\cos\frac{\vartheta}{2}\right)\eta\left(A\sin\frac{\vartheta}{2}+B\cos\frac{\vartheta}{2}\right)\right\}$$

which may be simplified to

$$j_x=\frac{2ec\eta}{V(1+\eta^2)}(A^2+B^2)\sin\vartheta\,;\quad j_z=\frac{2ec\eta}{V(1+\eta^2)}(A^2+B^2)\cos\vartheta\,. \tag{209.1}$$

By an analogous procedure we arrive at

$$j'_x=\frac{2ec\eta}{V(1+\eta^2)}(C^2+D^2)\sin\vartheta\,;\; j'_z=-\frac{2ec\eta}{V(1+\eta^2)}(C^2+D^2)\cos\vartheta \tag{209.2}$$

for the reflected[2], and at

$$j''_x=\frac{2ec\eta''}{V(1+\eta''^2)}(E^2+F^2)\sin\vartheta''\,;\quad j''_z=\frac{2ec\eta''}{V(1+\eta''^2)}(E^2+F^2)\cos\vartheta'' \tag{209.3}$$

for the transmitted wave.

In order to calculate the three currents we first express the amplitude constants C, D, E, F by those, A and B, of the incident wave by solving the linear system (208.6) or (208.8a, b). The result of this rather cumbersome but elementary computation is

$$C=\frac{-\alpha A+\beta B}{\Delta};\qquad D=-\frac{\beta A+\alpha B}{\Delta}; \tag{209.4}$$

$$E=\frac{\rho A+\sigma B}{2r\lambda};\qquad F=\frac{-\sigma A+\rho B}{2r\lambda} \tag{209.5}$$

---

[2] Interference terms of incident and reflected waves may be omitted for the present purpose since $\int d^3x\, e^{i(\boldsymbol{k}-\boldsymbol{k}')\boldsymbol{r}}=0$. When dealing only with local dependence of densities, these interferences are of interest, cf. Problem 23.

with the abbreviations

$$\begin{aligned}
\alpha &= 2(\lambda^2+1)p(1+q^2)-4\lambda q(1+p^2);\\
\beta &= (\lambda^2-1)(1-p^2)(1+q^2);\\
\Delta &= (\lambda+1)^2(1-pq)^2+(\lambda-1)^2(p+q)^2;\\
\rho\Delta &= (\lambda+1)(\Delta-\alpha p)-(\lambda-1)\beta=4\lambda(\lambda+1)(1-pq)(1-p^2);\\
\sigma\Delta &= (\lambda-1)(p\Delta-\alpha)+(\lambda+1)\beta p=4\lambda(\lambda-1)(p+q)(1-p^2).
\end{aligned} \tag{209.6}$$

From Eqs. (209.4) and (209.5) it further follows that

$$\begin{aligned}
\Delta^2\cdot(C^2+D^2)&=(\alpha^2+\beta^2)(A^2+B^2);\\
4r^2\lambda^2(E^2+F^2)&=(\rho^2+\sigma^2)(A^2+B^2).
\end{aligned} \tag{209.7}$$

If the expressions (209.6) are put into (209.7) we find by elementary though lengthy calculation,

$$\Delta\cdot(\rho^2+\sigma^2)=16\lambda^2(1-p^2)^2 \tag{209.8}$$

and

$$\Delta^2-(\alpha^2+\beta^2)=\Delta\cdot 4\lambda(1-p^2)(1-q^2). \tag{209.9}$$

It is now easy to express the reflected and transmitted currents by the incident one. We find for the z components perpendicular to the potential jump surface,

$$j_z'=-\frac{C^2+D^2}{A^2+B^2}j_z=-\frac{\alpha^2+\beta^2}{\Delta^2}j_z \tag{209.10}$$

and

$$\begin{aligned}
j_z''&=\frac{\eta''}{1+\eta''^2}\cdot\frac{1+\eta^2}{\eta}\cdot\frac{\cos\vartheta''}{\cos\vartheta}\cdot\frac{E^2+F^2}{A^2+B^2}j_z\\
&=\frac{\eta''}{\eta}\cdot\frac{1+\eta^2}{1+\eta''^2}\cdot\frac{\cos\vartheta''}{\cos\vartheta}\cdot\frac{\rho^2+\sigma^2}{4r^2\lambda^2}j_z.
\end{aligned}$$

In the last formula we introduce the definitions (208.7) which render

$$\frac{\eta''}{\eta}\cdot\frac{1+\eta^2}{1+\eta''^2}\cdot\frac{\cos\vartheta''}{\cos\vartheta}\cdot\frac{1}{r^2}=\lambda\frac{1-q^2}{1-p^2};$$

hence

$$j_z''=\frac{1-q^2}{1-p^2}\cdot\frac{\rho^2+\sigma^2}{4\lambda}j_z. \tag{209.11}$$

Eqs. (209.10) and (209.11) when combined with (209.8) and (209.9) then yield

$$j_z' = -\left\{1 - \frac{4\lambda(1-p^2)(1-q^2)}{\Delta}\right\} j_z\,; \quad j_z'' = \frac{4\lambda(1-p^2)(1-q^2)}{\Delta} j_z \tag{209.12}$$

with Δ defined in (209.6). From (209.12) there follows immediately the equation of continuity,

$$j_z + j_z' = j_z''\,. \tag{209.13}$$

The quantity

$$R = 1 - \frac{4\lambda(1-p^2)(1-q^2)}{\Delta} \tag{209.14}$$

is the coefficient of reflection. At normal incidence ($p=0$, $q=0$) it simply becomes

$$R = \left(\frac{\lambda-1}{\lambda+1}\right)^2. \tag{209.15}$$

VII. Radiation Theory

Problem 210. Quantization of Schrödinger field

The quantization of a force-free Schrödinger wave field into particles obeying either Bose or Fermi statistics shall be discussed using suitable expressions for energy, momentum and electric charge of the field.

Solution. Let us start by treating the force-free Schrödinger equation as a classical wave field, ψ being simply a scalar function of space coordinates and time. Then, in the usual normalization, we have the following integral expressions (cf. Problems 3 and 5): the total field energy is

$$W = -\frac{\hbar^2}{2m}\int d^3x\,\psi^*\,\nabla^2\psi\,, \tag{210.1}$$

the total momentum of the field is

$$\boldsymbol{P} = \frac{\hbar}{i}\int d^3x\,\psi^*\,\nabla\psi\,, \tag{210.2}$$

and the total electric charge of the field is

$$Q = e\int d^3x\,\psi^*\,\psi\,. \tag{210.3}$$

Here m and e are phenomenological constants not yet explained as particle properties since, so far, the field does not yet consist of particles.

We further know that a Schrödinger field must satisfy two conjugate wave equations,

$$-\frac{\hbar}{i}\,\frac{\partial\psi}{\partial t} = -\frac{\hbar^2}{2m}\nabla^2\psi\,;\qquad \frac{\hbar}{i}\,\frac{\partial\psi^*}{\partial t} = -\frac{\hbar^2}{2m}\nabla^2\psi^*\,. \tag{210.4}$$

They may be solved by plane waves which we normalize within an arbitrary periodicity cube of volume $\mathscr{V}$; the complete solution then may be written in the form

$$\psi(\boldsymbol{r},t) = \mathscr{V}^{-\frac{1}{2}}\sum_{\boldsymbol{k}} c_{\boldsymbol{k}}\,\mathrm{e}^{i(\boldsymbol{k}\boldsymbol{r}-\omega t)} \tag{210.5}$$

with the law of dispersion following from (210.4),

$$\hbar\omega = \frac{\hbar^2 k^2}{2m}. \tag{210.6}$$

If we put (210.5) in the expressions (210.1) to (210.3) and use the orthonormality relation of the plane waves,

$$\frac{1}{\mathscr{V}} \int_{\mathscr{V}} d^3x \, e^{i(\boldsymbol{k}-\boldsymbol{k}')\boldsymbol{r}} = \delta_{\boldsymbol{k},\boldsymbol{k}'}$$

we get

$$W = \frac{\hbar^2}{2m} \sum_{\boldsymbol{k}} k^2 c_{\boldsymbol{k}}^* c_{\boldsymbol{k}}; \quad \boldsymbol{P} = \sum_{\boldsymbol{k}} \hbar \boldsymbol{k} c_{\boldsymbol{k}}^* c_{\boldsymbol{k}}; \quad Q = e \sum_{\boldsymbol{k}} c_{\boldsymbol{k}}^* c_{\boldsymbol{k}}. \tag{210.7}$$

This classical theory now we shall quantize. The wave function ψ shall be replaced by an operator ψ which operates on Hilbert vectors χ of particle numbers. The same then holds for the coefficients $c_{\boldsymbol{k}}$ of the Fourier series (210.5). They have to be replaced by such operators $c_{\boldsymbol{k}}$ and their hermitian conjugates $c_{\boldsymbol{k}}^\dagger$ as to make the eigenvalues of $c_{\boldsymbol{k}}^\dagger c_{\boldsymbol{k}}$ integers, viz.

$$\begin{array}{lll} \text{either} & N_{\boldsymbol{k}} = 0, 1, 2, 3, \ldots & \text{in the Bose case} \\ \text{or} & N_{\boldsymbol{k}} = 0, 1 \text{ only} & \text{in the Fermi case}. \end{array} \tag{210.8}$$

If this is done, the three expressions (210.7), too, become operators with the eigenvalues

$$W = \sum_{\boldsymbol{k}} N_{\boldsymbol{k}} E_k; \quad E_k = \frac{\hbar^2 k^2}{2m}; \tag{210.9a}$$

$$\boldsymbol{P} = \sum_{\boldsymbol{k}} N_{\boldsymbol{k}} \boldsymbol{p}_{\boldsymbol{k}}; \quad \boldsymbol{p}_{\boldsymbol{k}} = \hbar \boldsymbol{k}; \tag{210.9b}$$

$$Q = e \sum_{\boldsymbol{k}} N_{\boldsymbol{k}}, \tag{210.9c}$$

describing a system of particles without interactions of which $N_{\boldsymbol{k}}$ are in the state $\boldsymbol{k}$ and have each the energy E_k, the momentum $\boldsymbol{p}_{\boldsymbol{k}}$ and the charge e.

Quantization leading to the required eigenvalues (210.8) is performed if the coefficients satisfy the following commutation rules[1,2]:

$$\begin{array}{ll} & [c_{\boldsymbol{k}}; c_{\boldsymbol{k}'}^\dagger]_- \equiv c_{\boldsymbol{k}} c_{\boldsymbol{k}'}^\dagger - c_{\boldsymbol{k}'}^\dagger c_{\boldsymbol{k}} = \delta_{\boldsymbol{k}\boldsymbol{k}'} \text{ in the Bose case} \\ \text{or} & [c_{\boldsymbol{k}}; c_{\boldsymbol{k}'}^\dagger]_+ \equiv c_{\boldsymbol{k}} c_{\boldsymbol{k}'}^\dagger + c_{\boldsymbol{k}'}^\dagger c_{\boldsymbol{k}} = \delta_{\boldsymbol{k}\boldsymbol{k}'} \text{ in the Fermi case}. \end{array} \tag{210.10}$$

---

[1] The commutator notation $[a, b] = ab - ba$ used in this chapter differs by a factor $i/\hbar$ from the one used in chapter I.

[2] For the Bose case, it has been shown in Problem 31 that the eigenvalues (210.8) are the consequence of the commutation rule (210.10). The same method may be applied in the Fermi case.

From these relations the commutator of the wave functions can easily be constructed:

$$[\psi(\boldsymbol{r},t);\psi^\dagger(\boldsymbol{r}',t)]_\pm = \frac{1}{\mathscr{V}}\sum_{\boldsymbol{k}}\sum_{\boldsymbol{k}'}[c_{\boldsymbol{k}};c^\dagger_{\boldsymbol{k}'}]_\pm\, e^{i(\boldsymbol{k}\boldsymbol{r}-\boldsymbol{k}'\boldsymbol{r}')-i(\omega-\omega')t}$$

$$= \frac{1}{\mathscr{V}}\sum_{\boldsymbol{k}} e^{i\boldsymbol{k}(\boldsymbol{r}-\boldsymbol{r}')} = \delta(\boldsymbol{r}-\boldsymbol{r}')\,. \tag{210.11}$$

This commutation relation apparently holds for both commutator signs.

Problem 211. Scattering in Born approximation

The quantized Schrödinger theory shall be applied to the elastic scattering of a particle in a central force potential, $V(r)$.

Solution. Let the force-free quantized field of the preceding problem be disturbed by the potential $V(r)$; then the hamiltonian W of the field has to be supplemented by the perturbation energy,

$$W' = \int d^3x\, \psi^\dagger V \psi\,. \tag{211.1}$$

If we stick to the first approximation, we may put the plane-wave decomposition (210.5) of the preceding problem into W' for ψ and $\psi^\dagger$, thus getting

$$W' = \frac{1}{\mathscr{V}}\sum_{\boldsymbol{k}}\sum_{\boldsymbol{k}'} c^\dagger_{\boldsymbol{k}'} c_{\boldsymbol{k}} \int d^3x\, V(r)\, e^{i(\boldsymbol{k}-\boldsymbol{k}')\boldsymbol{r}}\, e^{i(\omega'-\omega)t}\,. \tag{211.2}$$

The integrals in (211.2) are well known from the Born approximation (cf. Problem 105); with the abbreviation

$$\boldsymbol{k}-\boldsymbol{k}' = \boldsymbol{K} \tag{211.3}$$

we again arrive at the expression (note that its dimension is erg·cm^3)

$$\langle \boldsymbol{k}'|V|\boldsymbol{k}\rangle = \int d^3x\, V(r)\, e^{i\boldsymbol{K}\boldsymbol{r}} = 4\pi\int_0^\infty dr\, r^2\, V(r)\frac{\sin Kr}{Kr} \tag{211.4}$$

and may briefly write

$$W' = \frac{1}{\mathscr{V}}\sum_{\boldsymbol{k}}\sum_{\boldsymbol{k}'} c^\dagger_{\boldsymbol{k}'} c_{\boldsymbol{k}} \langle \boldsymbol{k}'|V|\boldsymbol{k}\rangle\, e^{i(\omega'-\omega)t}\,. \tag{211.5}$$

We now turn to describing the scattering process. If in the initial state there is just one particle in the state $\boldsymbol{k}_0$ and all other one-particle states are empty, so that the initial Hilbert vector is

$$\chi_i = |0 \ldots 1_{\boldsymbol{k}_0} \ldots 0_{\boldsymbol{k}_f} \ldots\rangle, \tag{211.6a}$$

and if the final state is defined by there being just one particle in the state $\boldsymbol{k}_f$,

$$\chi_f = |0 \ldots 0_{\boldsymbol{k}_0} \ldots 1_{\boldsymbol{k}_f} \ldots\rangle, \tag{211.6b}$$

then we need the matrix element

$$\langle \chi_f | W' | \chi_i \rangle \tag{211.7}$$

to determine the transition probability between these two states.

Using the general rules

$$c_{\boldsymbol{k}} |0_{\boldsymbol{k}}\rangle = 0; \quad c_{\boldsymbol{k}} |1_{\boldsymbol{k}}\rangle = |0_{\boldsymbol{k}}\rangle$$

we find

$$c_{\boldsymbol{k}} \chi_i = \delta_{\boldsymbol{k}\boldsymbol{k}_0} |0_{\boldsymbol{k}_0}\rangle,$$

i.e. all terms of the sum over $\boldsymbol{k}$ in Eq. (211.5) vanish when W' is applied to χ_i except the term $\boldsymbol{k} = \boldsymbol{k}_0$ and here the operator $c_{\boldsymbol{k}_0}$ exactly annihilates the initial particle, thus generating the state vector of vacuum, $|0\rangle$. If now $c_{\boldsymbol{k}'}^{\dagger}$ is applied, all sum terms will contribute, viz.

$$W' |1_{\boldsymbol{k}_0}\rangle = \frac{1}{\mathscr{V}} \sum_{\boldsymbol{k}'} \langle \boldsymbol{k}' | V | \boldsymbol{k}_0 \rangle \, \mathrm{e}^{i(\omega' - \omega_0)t} |1_{\boldsymbol{k}'}\rangle .$$

The matrix element (211.7) then again leaves only one term of this sum, in consequence of the orthogonality,

$$\langle \chi_f | 1_{\boldsymbol{k}'} \rangle = \langle 1_{\boldsymbol{k}_f} | 1_{\boldsymbol{k}'} \rangle = \delta_{\boldsymbol{k}_f \boldsymbol{k}'},$$

therefore

$$\langle \chi_f | W' | \chi_i \rangle = \frac{1}{\mathscr{V}} \langle \boldsymbol{k}_f | V | \boldsymbol{k}_0 \rangle \, \mathrm{e}^{i(\omega' - \omega_0)t} . \tag{211.8}$$

From the matrix element we now finally proceed to the differential cross section using the Golden Rule,

$$d\sigma = \frac{2\pi}{\hbar} \rho_f \, |\langle \chi_f | W' | \chi_i \rangle|^2 \cdot \frac{\mathscr{V}}{v} \tag{211.9}$$

with

$$\rho_f = \frac{\mathscr{V} p^2 \, dp \, d\Omega}{(2\pi\hbar)^3 \, dE} = \frac{m^2 v \mathscr{V}}{8\pi^3 \hbar^3} d\Omega \tag{211.10}$$

where all the quantities refer to the final state. Putting (211.10) in (211.9) and using the matrix element formula (211.8), the (large but otherwise arbitrary) periodicity volume $\mathscr{V}$ cancels out and we arrive at the result,

$$\frac{d\sigma}{d\Omega}=\left(\frac{m}{2\pi\hbar^2}\right)^2 |\langle \boldsymbol{k}_f| V |\boldsymbol{k}_0\rangle|^2 . \tag{211.11}$$

With the expression (211.4), this completely agrees with Born's first approximation formula, cf. Problems 105 and 184.

Problem 212. Quantization of classical radiation field

A classical Maxwell field in vacuo, underlying the condition of periodicity within a cube $\mathscr{V}=L^3$, shall be quantized into photons, using the classical expressions of field energy and momentum.

Solution. The classical radiation field is described by a vector potential $\boldsymbol{A}$ satisfying the differential equations

$$\Box^2 \boldsymbol{A}=0 \quad \text{and} \quad \operatorname{div}\boldsymbol{A}=0 \tag{212.1}$$

in the usual gauge leading to transverse waves, The physical meaning of these equations can either be explained by coupling $\boldsymbol{A}$ with the field strengths,

$$\mathscr{E} = -\frac{1}{c}\dot{\boldsymbol{A}}\,; \quad \mathscr{H}=\operatorname{rot}\boldsymbol{A}\,, \tag{212.2}$$

or by the mechanical expressions of total field energy,

$$W=\frac{1}{8\pi}\int d^3x(\mathscr{E}^2+\mathscr{H}^2)=\frac{1}{8\pi}\int d^3x\left\{\frac{1}{c^2}\dot{\boldsymbol{A}}^2+(\operatorname{rot}\boldsymbol{A})^2\right\}, \tag{212.3}$$

and of total field momentum,

$$\boldsymbol{P}=\frac{1}{4\pi c}\int d^3x(\mathscr{E}\times\mathscr{H})=-\frac{1}{4\pi c^2}\int d^3x(\dot{\boldsymbol{A}}\times\operatorname{rot}\boldsymbol{A})\,. \tag{212.4}$$

The differential equations (212.1) are solved by plane waves in the usual standard form,

$$\boldsymbol{A} = \sqrt{\frac{4\pi c^2}{\mathscr{V}}}\sum_{\boldsymbol{k}\lambda}\boldsymbol{u}_{\boldsymbol{k}\lambda}(q_{\boldsymbol{k}\lambda}\,e^{i(\boldsymbol{k}\boldsymbol{r}-\omega t)}+q^*_{\boldsymbol{k}\lambda}\,e^{-i(\boldsymbol{k}\boldsymbol{r}-\omega t)}) \tag{212.5}$$

with $\lambda=1,2$ denoting the two states of transverse polarization and $\boldsymbol{u}_{\boldsymbol{k}\lambda}$ a unit vector. There hold three orthogonality relations,

$$(\boldsymbol{u}_{\boldsymbol{k}1}\cdot\boldsymbol{k})=0\,; \quad (\boldsymbol{u}_{\boldsymbol{k}2}\cdot\boldsymbol{k})=0\,; \quad (\boldsymbol{u}_{\boldsymbol{k}1}\cdot\boldsymbol{u}_{\boldsymbol{k}2})=0\,. \tag{212.6}$$

The normalization factor in front of the summation sign in (212.5) is arbitrary and merely conventional. The $\boldsymbol{k}$ vectors are restricted by the periodicity within the cube $\mathscr{V}=L^3$ to

$$\boldsymbol{k}=\frac{2\pi}{L}\boldsymbol{n} \tag{212.7}$$

with integer components $n_i=0, \pm 1, \pm 2, \dots$ The frequency ω is connected with k through the law of dispersion,

$$\omega = k c\,. \tag{212.8}$$

Since each term in the sum (212.5) consists of two complex conjugate members, the vector potential $\boldsymbol{A}$ is a real function of $\boldsymbol{r}$ and t, as necessary in the classical Maxwell theory.

Putting (212.5) in the energy integral (212.3), we get

$$W=\frac{c^2}{2\mathscr{V}}\sum_{\boldsymbol{k}\lambda}\sum_{\boldsymbol{k}'\lambda'}\int d^3x\left\{-\frac{\omega\omega'}{c}\boldsymbol{u}_{\boldsymbol{k}'\lambda'}\boldsymbol{u}_{\boldsymbol{k}\lambda}-(\boldsymbol{k}'\times\boldsymbol{u}_{\boldsymbol{k}'\lambda'})(\boldsymbol{k}\times\boldsymbol{u}_{\boldsymbol{k}\lambda})\right\}$$
$$\times\left\{q_{\boldsymbol{k}'\lambda'}\,\mathrm{e}^{i(\boldsymbol{k}'\boldsymbol{r}-\omega' t)}-q^*_{\boldsymbol{k}'\lambda'}\,\mathrm{e}^{-i(\boldsymbol{k}'\boldsymbol{r}-\omega' t)}\right\}$$
$$\times\left\{q_{\boldsymbol{k}\lambda}\,\mathrm{e}^{i(\boldsymbol{k}\boldsymbol{r}-\omega t)}-q^*_{\boldsymbol{k}\lambda}\,\mathrm{e}^{-i(\boldsymbol{k}\boldsymbol{r}-\omega t)}\right\}.$$

Multiplication of the two last brackets and integration leaves either $\boldsymbol{k}'=-\boldsymbol{k}$ in the products of type qq and q^*q^*, or $\boldsymbol{k}'=\boldsymbol{k}$ for the types qq^* and q^*q. The first bracket then becomes, when use is yet made of (212.6),

$$-\frac{\omega^2}{c}-(\boldsymbol{k}'\cdot\boldsymbol{k})=\begin{cases}0 & \text{if } \boldsymbol{k}'=-\boldsymbol{k},\\ -2\dfrac{\omega^2}{c^2}\delta_{\lambda\lambda'} & \text{if } \boldsymbol{k}'=+\boldsymbol{k}.\end{cases}$$

Hence, there remains

$$W=\sum_{\boldsymbol{k}\lambda}\omega^2(q_{\boldsymbol{k}\lambda}q^*_{\boldsymbol{k}\lambda}+q^*_{\boldsymbol{k}\lambda}q_{\boldsymbol{k}\lambda})\,. \tag{212.9}$$

For the momentum, Eq. (212.4), a similar calculation leads to

$$\boldsymbol{P}=\sum_{\boldsymbol{k}\lambda}\omega\,\boldsymbol{k}(q_{\boldsymbol{k}\lambda}q^*_{\boldsymbol{k}\lambda}+q^*_{\boldsymbol{k}\lambda}q_{\boldsymbol{k}\lambda})\,. \tag{212.10}$$

Now we are ready to proceed to quantizing the classical radiation field by replacing the amplitudes $q_{\boldsymbol{k}\lambda}$ and $q^*_{\boldsymbol{k}\lambda}$ by operators and their hermitian conjugates. Written in the form

$$q_{\boldsymbol{k}\lambda}=C_k b_{\boldsymbol{k}\lambda} \quad\text{and}\quad q^\dagger_{\boldsymbol{k}\lambda}=C_k b^\dagger_{\boldsymbol{k}\lambda} \tag{212.11}$$

with a real normalization factor C_k, the expressions (212.9) and (212.10) turn into

$$W=\sum_{\boldsymbol{k}\lambda} \omega^2 C_k^2 (b_{\boldsymbol{k}\lambda} b_{\boldsymbol{k}\lambda}^\dagger + b_{\boldsymbol{k}\lambda}^\dagger b_{\boldsymbol{k}\lambda});$$

and (212.12)

$$\boldsymbol{P}=\sum_{\boldsymbol{k}\lambda} \omega \boldsymbol{k} C_k^2 (b_{\boldsymbol{k}\lambda} b_{\boldsymbol{k}\lambda}^\dagger + b_{\boldsymbol{k}\lambda}^\dagger b_{\boldsymbol{k}\lambda}).$$

With the operators $b_{\boldsymbol{k}\lambda}$ and $b_{\boldsymbol{k}\lambda}^\dagger$ chosen to satisfy the commutation rules

$$b_{\boldsymbol{k}\lambda} b_{\boldsymbol{k}'\lambda'}^\dagger - b_{\boldsymbol{k}'\lambda'}^\dagger b_{\boldsymbol{k}\lambda} = \delta_{\boldsymbol{k}\boldsymbol{k}'} \delta_{\lambda\lambda'}, \tag{212.13}$$

and all other combinations commuting, the eigenvalues of the operators $b_{\boldsymbol{k}\lambda}^\dagger b_{\boldsymbol{k}\lambda}$ become integers, $N_{\boldsymbol{k}\lambda}$, including zero,

$$N_{\boldsymbol{k}\lambda}=0,1,2,3,\ldots \tag{212.14}$$

and those of $b_{\boldsymbol{k}\lambda} b_{\boldsymbol{k}\lambda}^\dagger$ become $N_{\boldsymbol{k}\lambda}+1$. (Cf. Problem 31). If we further put

$$C_k = \sqrt{\frac{\hbar}{2\omega}}, \tag{212.15}$$

the operators (212.12) become

$$W=\sum_{\boldsymbol{k}\lambda} \frac{\hbar\omega}{2} (b_{\boldsymbol{k}\lambda} b_{\boldsymbol{k}\lambda}^\dagger + b_{\boldsymbol{k}\lambda}^\dagger b_{\boldsymbol{k}\lambda})$$

and (212.16)

$$\boldsymbol{P}=\sum_{\boldsymbol{k}\lambda} \hbar \boldsymbol{k} (b_{\boldsymbol{k}\lambda} b_{\boldsymbol{k}\lambda}^\dagger + b_{\boldsymbol{k}\lambda}^\dagger b_{\boldsymbol{k}\lambda})$$

with eigenvalues

$$W=\sum_{\boldsymbol{k}\lambda} \hbar\omega (N_{\boldsymbol{k}\lambda}+\tfrac{1}{2}); \qquad \boldsymbol{P}=\sum_{\boldsymbol{k}\lambda} \hbar\boldsymbol{k} (N_{\boldsymbol{k}\lambda}+\tfrac{1}{2}). \tag{212.17}$$

This permits interpretation of $N_{\boldsymbol{k}\lambda}$ as the number of photons in a state defined by $\boldsymbol{k}$ and λ, each of the photons having the energy $\hbar\omega$ and the momentum in the wave propagation direction $\hbar k = \hbar\omega/c$.

There occurs a zero-point energy, i.e. an energy of the vacuum,

$$W_0=\sum_{\boldsymbol{k}\lambda} \frac{\hbar\omega}{2}. \tag{212.18}$$

In spite of its being infinitely large, it has no serious physical consequences and may be normalized away by using

$$W-W_0=\sum_{\boldsymbol{k}\lambda} \hbar\omega N_{\boldsymbol{k}\lambda}, \tag{212.19}$$

i.e. the difference of the actual state from vacuum. In the momentum the zero-point contribution vanishes anyway, because in the sum pairs of terms with opposite $\boldsymbol{k}$ vectors cancel each other out.

The vector potential $\boldsymbol{A}$ then becomes an operator, creating or annihilating photons. It may be gathered from Eqs. (212.5), (212.11) and (212.15) that we shall write in quantum theory:

$$\boldsymbol{A} = \sum_{\boldsymbol{k}\lambda} \sqrt{\frac{2\pi c\hbar}{\mathscr{V} k}} \{b_{\boldsymbol{k}\lambda} e^{i(\boldsymbol{kr}-\omega t)} + b^{\dagger}_{\boldsymbol{k}\lambda} e^{-i(\boldsymbol{kr}-\omega t)}\} \boldsymbol{u}_{\boldsymbol{k}\lambda}. \qquad (212.20)$$

Problem 213. Emission probability of a photon

The probability that an electron from an excited state in a central potential field $V(r)$ will jump to a lower level and emit a photon shall be determined. Retardation effects may be neglected.

Solution. The interaction of matter (the electron) with radiation is given by the classical Maxwell expression,

$$H' = \frac{1}{c}\int d^3x (\boldsymbol{A}\cdot\boldsymbol{j}) \qquad (213.1)$$

with $\boldsymbol{A}$ the vector potential and $\boldsymbol{j}$ the electrical current density of matter. The first has been translated into the theory of quantized fields in Problem 212[3],

$$\boldsymbol{A} = \sum_{\boldsymbol{k}\lambda} \sqrt{\frac{2\pi\hbar c}{k\mathscr{V}}} \boldsymbol{u}^{(\lambda)}_{\boldsymbol{k}} (b_{\boldsymbol{k}\lambda} e^{i\boldsymbol{kr}} + b^{\dagger}_{\boldsymbol{k}\lambda} e^{-i\boldsymbol{kr}}). \qquad (213.2)$$

The current can be taken from the quantized Schrödinger field,

$$\psi = \sum_n c_n u_n(\boldsymbol{r}); \qquad -\frac{\hbar^2}{2m}\nabla^2 u_n + V u_n = E_n u_n \qquad (213.3)$$

using the formula

$$\boldsymbol{j} = -\frac{e\hbar}{2mi}(\psi^{\dagger}\nabla\psi - \nabla\psi^{\dagger}\cdot\psi)$$

(with the electron charge $-e$), hence

$$\boldsymbol{j} = -\frac{e\hbar}{2mi}\sum_n\sum_{n'} (u^*_{n'}\nabla u_n - u_n \nabla u^*_{n'})\, c^{\dagger}_{n'} c_n. \qquad (213.4)$$

---

[3] Omission of the time factors of (212.20) and (210.5) in (213.2) and (213.3) corresponds to the transition from the Schrödinger to the Heisenberg picture.

The functions $u_n(\boldsymbol{r})$ and $u_{n'}^*(\boldsymbol{r})$ are one-particle wave functions according to (213.3), the subscript n (or n') being a comprehensive notation for a set of three quantum numbers. The c_n and $c_{n'}^\dagger$ are the operators used in Problem 210 and may satisfy either of the commutation rules

$$\begin{aligned} c_n c_{n'}^\dagger \pm c_{n'}^\dagger c_n &= \delta_{nn'}, \\ c_n c_{n'} \pm c_{n'} c_n &= 0. \end{aligned} \tag{213.5}$$

Spontaneous emission of a photon is a process in which the electron jumps from an initial state n_i to a final state n_f or, in the wording of quantized wave theory, an electron in the initial state n_i is annihilated and an electron in the final state n_f created instead. At the same time a photon in a state $(\boldsymbol{k}, \lambda)$ is created. Such a process will be originated by a term with the operator product

$$b_{\boldsymbol{k}\lambda}^\dagger c_{n_f}^\dagger c_{n_i}$$

in the hamiltonian. If we put (213.2) and (213.4) in (213.1) such a term will indeed occur; we write it in the form

$$\langle f|H'|i\rangle\, b_{\boldsymbol{k}\lambda}^\dagger c_{n_f}^\dagger c_{n_i} \tag{213.6a}$$

with

$$\langle f|H'|i\rangle = \frac{1}{c}\int d^3x \sqrt{\frac{2\pi\hbar c}{k\mathscr{V}}}\,\frac{e\hbar}{2m}\, i\, \mathrm{e}^{-i\boldsymbol{k}\boldsymbol{r}}\, \boldsymbol{u}_{\boldsymbol{k}}^{(\lambda)}\cdot(u_{n_f}^*\nabla u_{n_i} - u_{n_i}\nabla u_{n_f}^*) \tag{213.6b}$$

the matrix element of H' between initial and final state. The transition probability then follows from the Golden Rule (cf. Problem 183),

$$P = \frac{2\pi}{\hbar}\,\rho_f\,|\langle f|H'|i\rangle|^2. \tag{213.7}$$

The density ρ_f of final states in the energy scale is completely determined by the photon:

$$\rho_f = \frac{k^2\, dk\, d\Omega_{\boldsymbol{k}}\, \mathscr{V}}{(2\pi)^3\hbar c\, dk} = \frac{\mathscr{V}}{8\pi^3\hbar c}\, k^2\, d\Omega_{\boldsymbol{k}} \tag{213.8}$$

with $d\Omega_{\boldsymbol{k}}$ the solid angle element into which the photon is emitted.

There remains the evaluation of the integral in (213.6b),

$$I = \int d^3x\, \mathrm{e}^{-i\boldsymbol{k}\boldsymbol{r}}\,(u_{n_f}^*\nabla u_{n_i} - u_{n_i}\nabla u_{n_f}^*).$$

Retardation may be neglected if the wavelength of the emitted light is large compared with the atomic integration domain, i.e., if in the integrand

$e^{-i\boldsymbol{k}\boldsymbol{r}}=1$ is a good approximation. Of the two terms of the integral, the second then may be partially integrated, thus yielding

$$I=2\int d^3x\, u^*_{n_f}\,\nabla u_{n_i}\,.$$

From the Schrödinger equations for u_{n_i} and $u^*_{n_f}$ an identity may be derived (cf. Problem 187),

$$I=\frac{2m}{\hbar^2}(E_i-E_f)\int d^3x\, u^*_{n_f}\,\boldsymbol{r}\,u_{n_i}$$

and from the conservation of energy there follows

$$E_i-E_f=\hbar\omega$$

so that the matrix element (213.6b) becomes

$$\langle f|H'|i\rangle=\frac{e}{c}\,i\sqrt{\frac{2\pi\hbar c}{k\mathscr{V}}}\,\omega\,\langle f|\boldsymbol{u}_{\boldsymbol{k}}^{(\lambda)}\cdot\boldsymbol{r}\,|i\rangle \tag{213.9}$$

with

$$\langle f|\boldsymbol{u}_{\boldsymbol{k}}^{(\lambda)}\cdot\boldsymbol{r}\,|i\rangle=\int d^3x\, u^*_{n_f}(\boldsymbol{u}_{\boldsymbol{k}}^{(\lambda)}\cdot\boldsymbol{r})\,u_{n_i}\,. \tag{213.10}$$

By gathering the expressions (213.8) and (213.9) into (213.7), we then finally obtain

$$P_{\boldsymbol{k},\lambda}=\frac{d\Omega_{\boldsymbol{k}}}{2\pi}\,\frac{e^2}{\hbar c}\,\frac{\omega^3}{c^2}\,|\langle f|\boldsymbol{u}_{\boldsymbol{k}}^{(\lambda)}\cdot\boldsymbol{r}\,|i\rangle|^2 \tag{213.11a}$$

or, if we introduce $\nu=\omega/2\pi$ instead of ω, in the usual notation,

$$P_{\boldsymbol{k},\lambda}=\frac{e^2}{\hbar c}\,\frac{4\pi^2\nu^3}{c^2}\,d\Omega_{\boldsymbol{k}}\,|\langle f|\boldsymbol{u}_{\boldsymbol{k}}^{(\lambda)}\cdot\boldsymbol{r}\,|i\rangle|^2\,. \tag{213.11b}$$

The remaining matrix element in this formula may still be decomposed into two factors according to

$$\langle f|\boldsymbol{u}_{\boldsymbol{k}}^{(\lambda)}\cdot\boldsymbol{r}\,|i\rangle=\boldsymbol{u}_{\boldsymbol{k}}^{(\lambda)}\cdot\boldsymbol{r}_{if} \tag{213.12}$$

with the first factor depending only upon the direction and polarization of the photon emitted, and the second only upon atomic parameters.

Problem 214. Angular distribution of radiation

The final formulae of the preceding problem shall be used to investigate the angular distribution of photons emitted in a $P\to S$ transition of one electron.

Solution. Let Θ and Φ be polar angles defining the direction of $\boldsymbol{k}$. We then may define two polarization states, one with $\boldsymbol{u}_{\boldsymbol{k}}^{(1)}$ in the meridional plane, the other with $\boldsymbol{u}_{\boldsymbol{k}}^{(2)}$ perpendicular to it. The two unit vectors then have the components

$$u_x^{(1)}=\cos\Theta\cos\Phi\,;\quad u_y^{(1)}=\cos\Theta\sin\Phi\,;\quad u_z^{(1)}=-\sin\Theta \tag{214.1a}$$

and

$$u_x^{(2)}=-\sin\Phi\,;\quad u_y^{(2)}=\cos\Phi\,;\quad u_z^{(2)}=0\,. \tag{214.1b}$$

To construct the components of $\boldsymbol{r}_{if}$ we first write the three components x, y, z of $\boldsymbol{r}$ in terms of spherical harmonics:

$$\left.\begin{aligned} x&=r\sqrt{\frac{2\pi}{3}}(Y_{1,1}+Y_{1,-1})\,;\\ y&=-ir\sqrt{\frac{2\pi}{3}}(Y_{1,1}-Y_{1,-1})\,;\\ z&=r\sqrt{\frac{4\pi}{3}}\,Y_{1,0}\,. \end{aligned}\right\} \tag{214.2}$$

Of these three expressions we then form matrix elements between the two electrons states,

$$|i\rangle=v(r)\,Y_{1,m}(\vartheta,\varphi)\,;\qquad |f\rangle=u(r)\,Y_{0,0}(\vartheta,\varphi)\,. \tag{214.3}$$

It should be noted that $Y_{0,0}=(4\pi)^{-\frac{1}{2}}$ is a constant. Thence we get,

$$\left.\begin{aligned} x_{if}&=\frac{1}{\sqrt{6}}R(\delta_{m,1}+\delta_{m,-1}),\\ y_{if}&=-\frac{i}{\sqrt{6}}R(\delta_{m,1}-\delta_{m,-1}),\\ z_{if}&=\frac{1}{\sqrt{3}}R\,\delta_{m,0} \end{aligned}\right\} \tag{214.4}$$

with the abbreviation

$$R=\int_0^\infty dr\,r^3\,u^*(r)\,v(r)=\sqrt{3}\,|\boldsymbol{r}_{if}| \tag{214.5}$$

for the radial integral.

The probability of emission of a photon of polarization λ in the solid angle element $d\Omega_{\boldsymbol{k}}$ is, according to (213.11a) of the preceding problem,

$$P_{\boldsymbol{k},\lambda} = \frac{e^2\omega^3}{\hbar c^3}\frac{d\Omega_{\boldsymbol{k}}}{2\pi}|\boldsymbol{u}_{\boldsymbol{k}}^{(\lambda)}\cdot\boldsymbol{r}_{if}|^2. \tag{214.6}$$

With (214.1a, b) for $\boldsymbol{u}_{\boldsymbol{k}}^{(\lambda)}$ and (214.4) for $\boldsymbol{r}_{if}$ we find for the scalar product $\boldsymbol{u}_{\boldsymbol{k}}^{(\lambda)}\cdot\boldsymbol{r}_{if}$ in the case $\lambda=1$

$$\left.\begin{aligned} &\text{for } m=+1: && \frac{R}{\sqrt{6}}\cos\Theta\, e^{-i\Phi},\\ &\text{for } m=0: && -\frac{R}{\sqrt{3}}\sin\Theta,\\ &\text{for } m=-1: && \frac{R}{\sqrt{6}}\cos\Theta\, e^{i\Phi} \end{aligned}\right\} \tag{214.7}$$

and in the case $\lambda=2$

$$\left.\begin{aligned} &\text{for } m=+1: && -i\frac{R}{\sqrt{6}}e^{-i\Phi},\\ &\text{for } m=0: && 0,\\ &\text{for } m=-1: && +i\frac{R}{\sqrt{6}}e^{i\Phi}. \end{aligned}\right\} \tag{214.8}$$

Thus, if we write

$$P_{k,\lambda} = \frac{e^2\omega^3}{\hbar c^3}\frac{R^2}{6\pi}d\Omega_{\boldsymbol{k}}\cdot D_{\boldsymbol{k}\lambda}(\Theta,\Phi), \tag{214.9}$$

the directional factor $D_{\boldsymbol{k}\lambda}$ will be given by the expressions assembled in the accompanying table. In the case $\lambda=2$, all directions of photon

Directional factors $D_{\boldsymbol{k}\lambda}$ for

m	$\lambda=1$	$\lambda=2$
$+1$	$\frac{1}{2}\cos^2\Theta$	$\frac{1}{2}$
0	$\sin^2\Theta$	0
-1	$\frac{1}{2}\cos^2\Theta$	$\frac{1}{2}$

emission have equal probabilities; only from the initial state $m=0$ are no photons of this polarization emitted at all. The P state with $m=0$, only decaying therefore under emission of a photon in polarization state $\lambda=1$, has an angular distribution as $\sin^2\Theta$, the main direction of emission for $m=0$ thus falling in the equatorial plane $\Theta=90°$. For $\lambda=1$ and $m=\pm 1$, however, the distribution becomes proportional to $\cos^2\Theta$ and the emission occurs preferably in the directions $\Theta=0°$ and $\Theta=180°$.

Problem 215. Transition probability

What is the transition probability of an electron from a higher P to a lower S state by emission of a photon of any direction or polarization whatsoever? As an example, the mean lifetime of the $2P$ state in a hydrogen atom shall be computed.

Solution. In the preceding problem, the differential emission probability for the photon going into the solid angle element $d\Omega_{\boldsymbol{k}}$ in the direction Θ, Φ was computed for $m=+1, 0, -1$ and both states of polarization. Gathering these formulae and performing summation over both polarization states, we obtain

$$\sum_\lambda P_{\boldsymbol{k}\lambda} = \frac{e^2\omega^3}{\hbar c^3}\cdot\frac{R^2}{6\pi}\,d\Omega_{\boldsymbol{k}}\cdot\sin^2\Theta \qquad \text{for } m=0; \tag{215.1a}$$

$$\sum_\lambda P_{\boldsymbol{k}\lambda} = \frac{e^2\omega^3}{\hbar c^3}\cdot\frac{R^2}{6\pi}\,d\Omega_{\boldsymbol{k}}\cdot\frac{1}{2}(1+\cos^2\Theta) \qquad \text{for } m=\pm 1. \tag{215.1b}$$

Integration of these expressions over all directions of the photon then yields the same transition probability from P to S state for all three initial values of m, viz.

$$P = \frac{e^2\omega^3}{\hbar c^3}\cdot\frac{R^2}{6\pi}\cdot\frac{8\pi}{3} = \frac{4}{9}\,\frac{e^2\omega^3}{\hbar c^3}R^2. \tag{215.2}$$

Since R has been defined in some detail in the preceding problem, we may now immediately turn to the example. From the $2P$ state of a hydrogen atom transitions are possible only to the $1S$ ground state. The eigenfunctions of the two states are

$$|i\rangle = \tfrac{1}{12}\sqrt{6}\,r\,e^{-\frac{r}{2}}\,Y_{1,m} \tag{215.3a}$$

and

$$|f\rangle = 2e^{-r}\,Y_{0,0} \tag{215.3b}$$

in atomic units (unit length: $\hbar^2/me^2$). It then follows from (214.5) that

$$R = \int_0^\infty dr\,r^3\cdot\frac{\sqrt{6}}{12}\,r\,e^{-\frac{r}{2}}\cdot 2e^{-r} = \frac{1}{\sqrt{6}}\left(\frac{2}{3}\right)^5\cdot 24; \qquad R^2 = \frac{2^{15}}{3^9}$$

or, if we make the notation independent of units again,

$$R^2 = \frac{2^{15}}{3^9} \frac{\hbar^4}{m^2 e^4}. \tag{215.4}$$

The light frequency ω, in our example, follows from the hydrogen level formula

$$\omega = \frac{3}{8} \frac{m e^4}{\hbar^3}. \tag{215.5}$$

Putting R^2 from (215.4) and ω from (215.5) in (215.2) and suitably arranging the factors, we arrive at the result

$$P = \left(\frac{2}{3}\right)^8 \left(\frac{e^2}{\hbar c}\right)^3 \frac{m e^4}{\hbar^2}. \tag{215.6}$$

The reciprocal value of this transition probability is the mean lifetime τ of the $2P$ state, since there are no other ways of decaying in this special case. Hence,

$$\tau = \left(\frac{3}{2}\right)^8 \left(\frac{\hbar c}{e^2}\right)^3 \frac{\hbar^3}{m e^4}. \tag{215.7}$$

The quantity

$$\frac{\hbar^3}{m e^4} = 2.4187 \times 10^{-17}\ \text{sec} \tag{215.8}$$

is a suitable time unit for lifetimes of excited electron states. The factor

$$\frac{\hbar c}{e^2} = 137.0373$$

is the reciprocal fine-structure constant. The numerical value of the mean lifetime of the hydrogen $2P$ state then becomes

$$\tau = 1.5953 \times 10^{-9}\ \text{sec.}$$

Problem 216. Selection rules for dipole radiation

It has been shown in Problem 213 that the transition probabilities between one-electron states in an atom depend upon the matrix elements of the electrical dipole moment, if the photon wavelength is large compared to atomic dimensions. Selection rules shall be derived from this fact. What can be concluded for the normal Zeeman effect?

Solution. The probability that a photon is emitted into a solid angle element $d\Omega_{\boldsymbol{k}}$ in the direction $\boldsymbol{k}$ (polar angles Θ, Φ) in a state λ of polarization is, according to (213.11b),

$$P_{\boldsymbol{k},\lambda} = \frac{e^2}{\hbar c} \cdot \frac{4\pi^2 \nu^3}{c^2} d\Omega_{\boldsymbol{k}} |\langle f | \boldsymbol{u}_{\boldsymbol{k}}^{(\lambda)} \cdot \boldsymbol{r} | i \rangle|^2 \tag{216.1}$$

in dipole approximation. Here $\boldsymbol{u}_{\boldsymbol{k}}^{(\lambda)}$ is a unit vector in the direction of polarization, viz. according to (214.1a, b), either with the components

$$u_x^{(1)} = \cos\Theta \cos\Phi; \quad u_y^{(1)} = \cos\Theta \sin\Phi; \quad u_z^{(1)} = -\sin\Theta \tag{216.2a}$$

or

$$u_x^{(2)} = -\sin\Phi; \quad u_y^{(2)} = \cos\Phi; \quad u_z^{(2)} = 0. \tag{216.2b}$$

The two atomic states have wave functions

$$|i\rangle = \varphi_i(r)\, Y_{l,m}(\vartheta,\varphi); \qquad \langle f| = \varphi_f(r)\, Y^*_{l',m'}(\vartheta,\varphi)$$

so that the matrix element of $\boldsymbol{r}$ has the components

$$\left.\begin{aligned} \langle f | x \pm i y | i \rangle &= \int_0^\infty dr\, r^3 \varphi_i \varphi_f \oint d\Omega\, Y^*_{l',m'} \sin\vartheta\, e^{\pm i\varphi}\, Y_{l,m}, \\ \langle f | z | i \rangle &= \int_0^\infty dr\, r^3 \varphi_i \varphi_f \oint d\Omega\, Y^*_{l',m'} \cos\vartheta\, Y_{l,m}. \end{aligned}\right\} \tag{216.3}$$

We may now use the relations

$$\left.\begin{aligned} \sin\vartheta\, e^{\pm i\varphi}\, Y_{l,m} &= \pm A_{l+1,\pm m+1}\, Y_{l+1,m\pm 1} \mp A_{l,\mp m}\, Y_{l-1,m\pm 1}, \\ \cos\vartheta\, Y_{l,m} &= B_{l+1,m}\, Y_{l+1,m} + B_{l,m}\, Y_{l-1,m} \end{aligned}\right\} \tag{216.4}$$

with coefficients

$$A_{l,m} = \sqrt{\frac{(l+m)(l+m-1)}{(2l+1)(2l-1)}}; \qquad B_{l,m} = \sqrt{\frac{(l+m)(l-m)}{(2l+1)(2l-1)}} \tag{216.5}$$

which permit angular integrations in (216.3),

$$\begin{aligned} \langle f | x \pm i y | i \rangle = R_{if} \{ &\pm A_{l+1,\pm m+1}\, \delta_{l',l+1}\, \delta_{m',m\pm 1} \\ &\mp A_{l,\mp m}\, \delta_{l',l-1}\, \delta_{m',m\pm 1} \}, \end{aligned} \tag{216.6a}$$

$$\langle f | z | i \rangle = R_{if} \{ B_{l+1,m}\, \delta_{l',l+1} + B_{l,m}\, \delta_{l',l-1} \}\, \delta_{m',m} \tag{216.6b}$$

where the abbreviation R_{if} stands for the radial part,

$$R_{if} = \int_0^\infty dr\, r^3 \varphi_i(r)\, \varphi_f(r). \tag{216.7}$$

These matrix elements vanish if not $l' = l \pm 1$, thus providing us with the first basic selection rule for dipole transitions. Further, it is seen that for $m'=m\pm 1$ only the matrix element of $x\pm iy$, and for $m'=m$ of z, do not vanish. Any other changes of the quantum numbers l and m cannot occur in dipole transitions.

Combining these selection rules with the polarization vectors (216.2a, b), we may compute the matrix elements

$$\langle f|\boldsymbol{u}_{\boldsymbol{k}}^{(\lambda)}\cdot\boldsymbol{r}|i\rangle$$

occurring in (216.1), for the two polarization states $\lambda=1$ and $\lambda=2$. The results are summarized in the accompanying table.

m'	λ	$l'=l+1$	$l'=l-1$
$m+1$	1	$\frac{1}{2}\cos\Theta\, e^{-i\Phi} R_{if} A_{l+1,m+1}$	$-\frac{1}{2}\cos\Theta\, e^{-i\Phi} R_{if} A_{l,-m}$
	2	$-\frac{i}{2}e^{-i\Phi} R_{if} A_{l+1,m+1}$	$\frac{i}{2}e^{-i\Phi} R_{if} A_{l,-m}$
m	1	$-\sin\Theta\, R_{if} B_{l+1,m}$	$-\sin\Theta\, R_{if} B_{l,m}$
	2	0	0
$m-1$	1	$-\frac{1}{2}\cos\Theta\, e^{i\Phi} R_{if} A_{l+1,-m+1}$	$\frac{1}{2}\cos\Theta\, e^{i\Phi} R_{if} A_{l,m}$
	2	$-\frac{i}{2}e^{i\Phi} R_{if} A_{l+1,-m+1}$	$\frac{i}{2}e^{i\Phi} R_{if} A_{l,m}$

If the radiating atoms are not oriented in space, observation gives only intensity averages over all directions. The orientation, though, which is performed under Zeeman effect conditions gives more infor-

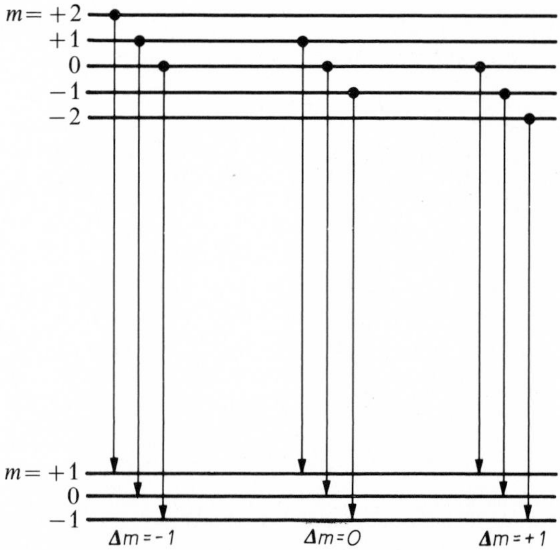

Fig. 74. Zeeman transitions $D\rightarrow P$. There are three lines of different polarization according to $\Delta m=+1, 0, -1$

mation. The polar axis $\Theta=0$ then coincides with the direction of the magnetic field. If we observe light emission in field direction, therefore, the matrix elements for $m'=m$ will vanish, so that only the lines $m'=m+1$ and $m'=m-1$ are observed. In any direction perpendicular to the field, on the other hand, we have $\cos\Theta=0$ so that $m'=m\pm 1$ lines appear in polarization state 2, but $m'=m$ in state 1, only.

As an example let us take the Zeeman transition from a D state (5 components) to a P state (3 components). If the frequency of the line emitted in absence of the magnetic field is ω_0, we then find three possible frequencies in the magnetic field (Fig. 74), viz.

$$\left.\begin{aligned} \omega_1 &= \omega_0+\omega_L \quad \text{for } m'=m-1,\\ \omega_0 & \quad \text{for } m'=m,\\ \omega_{-1} &= \omega_0-\omega_L \quad \text{for } m'=m+1 \end{aligned}\right\} \text{ with } \omega_L=\frac{e\mathcal{H}}{2mc}.$$

Observing in field direction ($\Theta=0$), we find the middle line (ω_0) does not occur and there is a doublet of frequencies $\omega_0+\omega_L$ and $\omega_0-\omega_L$. In perpendicular observation all three lines occur, forming the full Zeeman triplet, though in different states of polarization.

Problem 217. Intensities of Lyman lines

To compare the intensities of emission of the two first Lyman lines of atomic hydrogen, Lyα and Lyβ.

Solution. We are concerned with the two transitions

$$\text{Ly}\,\alpha\text{: } 2p\to 1s \quad\text{and}\quad \text{Ly}\,\beta\text{: } 3p\to 1s.$$

The emission probability, integrated over all directions and summed up over both polarizations, is

$$P=\frac{4}{3}\frac{e^2\omega^3}{\hbar c^3}\,|\langle f|\boldsymbol{r}|i\rangle|^2; \tag{217.1}$$

the intensity of a spectral line (energy per second) is proportional to $\omega\cdot P$ so that for the two lines under consideration we have the intensity ratio

$$\frac{I_\alpha}{I_\beta}=\left(\frac{E_\alpha}{E_\beta}\right)^4\left|\frac{\langle 1s|\boldsymbol{r}|2p\rangle}{\langle 1s|\boldsymbol{r}|3p\rangle}\right|^2. \tag{217.2}$$

Here the energy differences in atomic units are

$$E_\alpha=\tfrac{1}{2}-\tfrac{1}{8}=\tfrac{3}{8} \quad\text{and}\quad E_\beta=\tfrac{1}{2}-\tfrac{1}{18}=\tfrac{4}{9}. \tag{217.3}$$

We still have to calculate the two matrix elements.

From Problem 67 we know the wave function of the final state to be

$$|1s\rangle = \frac{1}{\sqrt{\pi}} e^{-r}, \tag{217.4a}$$

whereas for the initial state we have either in the Lyα case

$$|2p\rangle = \frac{1}{4\sqrt{2\pi}} r e^{-\frac{1}{2}r} \cos\vartheta \tag{217.4b}$$

or in the Lyβ case

$$|3p\rangle = \frac{4}{27\sqrt{2\pi}} \left(r - \frac{1}{6} r^2\right) e^{-\frac{1}{3}r} \cos\vartheta. \tag{217.4c}$$

Here we have arbitrarily used the p states with $m=0$, a choice which does not restrict generality as long as no directional effects are discussed. Since the vector $\boldsymbol{r}$ has components

$$x \pm iy = r \sin\vartheta\, e^{\pm i\varphi} \quad \text{and} \quad z = r\cos\vartheta,$$

it is immediately seen that the matrix elements of $x \pm iy$ vanish in consequence of the integration over φ. Therefore only $\langle f|z|i\rangle$ remains to be computed. We obtain,

$$\langle 1s|z|2p\rangle = \frac{1}{4\pi\sqrt{2}} \oint d\Omega \cos^2\vartheta \int_0^\infty dr\, r^4 e^{-\frac{3}{2}r}$$

and

$$\langle 1s|z|3p\rangle = \frac{4}{27\pi\sqrt{2}} \oint d\Omega \cos^2\vartheta \int_0^\infty dr\, r^3 \left(r - \frac{1}{6} r^2\right) e^{-\frac{4}{3}r}.$$

Elementary evaluation leads to

$$\langle 1s|z|2p\rangle = \frac{1}{\sqrt{2}} \cdot \frac{256}{243}; \qquad \langle 1s|z|3p\rangle = \frac{1}{\sqrt{2}} \cdot \frac{27}{64}. \tag{217.5}$$

Thus, Eq. (217.2) with (217.3) and (217.5) yields the final result

$$\frac{I_\alpha}{I_\beta} = \left(\frac{27}{32}\right)^4 \cdot \left(\frac{256}{243} \cdot \frac{64}{27}\right)^2 = 0.510 \times 6.23 \tag{217.6}$$

or

$$I_\alpha / I_\beta = 3.18.$$

Literature. Radial matrix elements for other pairs of hydrogen states are given by Bethe, H. A., Salpeter, E. E., in: *Encyclopedia of Physics*, vol. **35** (1956), cf. their section 63, and especially table 13.

Problem 218. Compton effect

The scattering of a photon by a free electron at rest shall be investigated in the unrelativistic frame.

Solution. In the presence of a radiation field, the electric current density of an electron field is described by[4]

$$\boldsymbol{j} = -\frac{e\hbar}{2mi}(\psi^\dagger \nabla\psi - \nabla\psi^\dagger \cdot \psi) - \frac{e^2}{mc}\boldsymbol{A}\psi^\dagger\psi = \boldsymbol{j}' + \boldsymbol{j}''. \tag{218.1}$$

The interaction energy of the two fields ψ and $\boldsymbol{A}$ is

$$W = \frac{1}{c}\int d^3x\, \boldsymbol{j}\cdot\boldsymbol{A} = W' + W''. \tag{218.2}$$

If for ψ the quantized Schrödinger field,

$$\psi = \frac{1}{\sqrt{\mathscr{V}}}\sum_{\boldsymbol{q}} c_{\boldsymbol{q}} e^{i\boldsymbol{q}\boldsymbol{r}} \tag{218.3}$$

and for $\boldsymbol{A}$ the quantized radiation field

$$\boldsymbol{A} = \sum_{\boldsymbol{k}\lambda}\sqrt{\frac{2\pi\hbar c}{\mathscr{V}k}}\, \boldsymbol{u}_{\boldsymbol{k}}^{(\lambda)}(b_{\boldsymbol{k}\lambda}e^{i\boldsymbol{k}\boldsymbol{r}} + b_{\boldsymbol{k}\lambda}^\dagger e^{-i\boldsymbol{k}\boldsymbol{r}}) \tag{218.4}$$

are put in (218.2), it can easily be seen that W'' (arising from $\boldsymbol{j}''$) will contribute to scattering in first-order, whereas W' (arising from $\boldsymbol{j}'$) does so only in second-order approximation. We shall therefore in what follows confine our attention to the term

$$W'' = -\frac{e^2}{mc^2}\int d^3x\, \boldsymbol{A}^2\psi^\dagger\psi. \tag{218.5}$$

This interaction term can easily be understood from the viewpoint of the classical picture in which the electric field strength $\mathscr{E} = -\frac{1}{c}\dot{\boldsymbol{A}}$ of the incident light wave plucks at the electron according to its equation of motion,

$$m\ddot{\boldsymbol{r}} = -e\mathscr{E} = \frac{e}{c}\dot{\boldsymbol{A}}; \quad \text{hence } \dot{\boldsymbol{r}} = \frac{e}{mc}\boldsymbol{A}.$$

This originates an induced current density $\boldsymbol{j}'' = \rho\dot{\boldsymbol{r}} = \frac{e}{mc}\rho\boldsymbol{A}$, if ρ denotes the

[4] In the radiation problems previously dealt with, the last supplement term in (218.1) would not have contributed in first order.

charge density, and the interaction between the current and the radiation field is, according to Maxwell's theory,

$$W'' = \frac{1}{c} \int d^3x\, \boldsymbol{j}'' \cdot \boldsymbol{A} = \frac{e}{mc^2} \int d^3x\, \rho\, \boldsymbol{A}^2 .$$

If here we set $\rho = -e\psi^\dagger \psi$, Eq. (218.5) results.

The Compton scattering process is, in first order, described by a hamiltonian term with the operator combination

$$c^\dagger_{\boldsymbol{q}'} c_{\boldsymbol{q}} b^\dagger_{\boldsymbol{k}'\lambda'} b_{\boldsymbol{k}\lambda} \tag{218.6}$$

if a photon in the state $(\boldsymbol{k}, \lambda)$ and an electron of momentum $\hbar \boldsymbol{q}$ of the initial state are both annihilated and replaced by the newly created photon in the state $(\boldsymbol{k}', \lambda')$ and electron of momentum $\hbar \boldsymbol{q}'$.

If we pick out of (218.5) the factor of the operator product (218.6), we have the matrix element,

$$\langle f | W'' | i \rangle = -\frac{e^2}{mc^2} \int d^3x \frac{2\pi\hbar c}{\mathscr{V}^2 \sqrt{kk'}} (\boldsymbol{u}_{\boldsymbol{k}}^{(\lambda)} \cdot \boldsymbol{u}_{\boldsymbol{k}'}^{(\lambda')}) e^{i(\boldsymbol{k}+\boldsymbol{q}-\boldsymbol{k}'-\boldsymbol{q}')\boldsymbol{r}} . \tag{218.7}$$

The integral in (218.7) vanishes unless

$$\boldsymbol{k} + \boldsymbol{q} = \boldsymbol{k}' + \boldsymbol{q}', \tag{218.8}$$

i. e. unless the law of momentum conservation holds for the process. If it holds, (218.7) becomes

$$\langle f | W'' | i \rangle = -\frac{2\pi e^2 \hbar}{mc\mathscr{V}} \frac{(\boldsymbol{u}_{\boldsymbol{k}}^{(\lambda)} \cdot \boldsymbol{u}_{\boldsymbol{k}'}^{(\lambda')})}{\sqrt{kk'}} . \tag{218.9}$$

For determining the cross section we apply the Golden Rule,

$$d\sigma(\boldsymbol{k}', \lambda') = \frac{2\pi}{\hbar} \frac{\mathscr{V}}{c} \rho_f |\langle f | W'' | i \rangle|^2 \tag{218.10}$$

where the final density of states follows from

$$\rho_f = \frac{k'^2\, dk'\, d\Omega'\, \mathscr{V}}{8\pi^3\, dE_f} \tag{218.11}$$

and the final energy is

$$E_f = \hbar c k' + \frac{\hbar^2}{2m} \boldsymbol{q}'^2 = \hbar c \left\{ k' + \frac{1}{2\varkappa} (\boldsymbol{k}' - \boldsymbol{k} - \boldsymbol{q})^2 \right\} \tag{218.12}$$

with $\varkappa = mc/\hbar$.

Let us now deal with polarization. In Fig. 75 the directions $\boldsymbol{k}$ and $\boldsymbol{k}'$ of the photon before and after the collision are lying in the plane of drawing. The vectors $\boldsymbol{u}_{\boldsymbol{k}}^{(1)}$ and $\boldsymbol{u}_{\boldsymbol{k}'}^{(1)}$ both lie in this plane, $\boldsymbol{u}_{\boldsymbol{k}}^{(2)}$ and $\boldsymbol{u}_{\boldsymbol{k}'}^{(2)}$ (not drawn) are perpendicular to it. The scalar products in (218.9) can then be read directly from the figure:

$$(\boldsymbol{u}_{\boldsymbol{k}}^{(1)} \cdot \boldsymbol{u}_{\boldsymbol{k}'}^{(1)}) = \cos\vartheta; \qquad (\boldsymbol{u}_{\boldsymbol{k}}^{(1)} \cdot \boldsymbol{u}_{\boldsymbol{k}'}^{(2)}) = 0; \tag{218.13a}$$

$$(\boldsymbol{u}_{\boldsymbol{k}}^{(2)} \cdot \boldsymbol{u}_{\boldsymbol{k}'}^{(1)}) = 0; \qquad (\boldsymbol{u}_{\boldsymbol{k}}^{(2)} \cdot \boldsymbol{u}_{\boldsymbol{k}'}^{(2)}) = 1. \tag{218.13b}$$

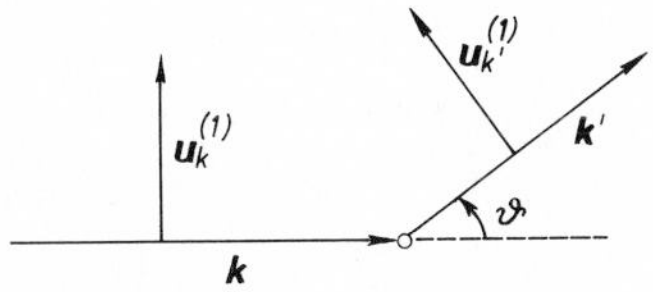

Fig. 75. Compton effect. Definition of polarization vectors in the initial state ($\boldsymbol{k}$) and final state ($\boldsymbol{k}'$) of the photon. Vectors $\boldsymbol{u}_{\boldsymbol{k}}^{(1)}$ and $\boldsymbol{u}_{\boldsymbol{k}'}^{(1)}$ in the $\boldsymbol{k}$, $\boldsymbol{k}'$ plane, vectors $\boldsymbol{u}_{\boldsymbol{k}}^{(2)}$ and $\boldsymbol{u}_{\boldsymbol{k}'}^{(2)}$ (not drawn) perpendicular to it

These relations show that there can only be transitions with both polarization vectors, before and after the collision, lying in the $\boldsymbol{k}, \boldsymbol{k}'$ plane, or with both perpendicular to it. In the first case the transition probability becomes proportional to $\cos^2\vartheta$, in the second case it is independent of the scattering angle. If the incident light is unpolarized we have to average over λ and to sum up over the final states λ' thus getting

$$\sum_{\lambda'} \overline{(\boldsymbol{u}_{\boldsymbol{k}}^{(\lambda)} \cdot \boldsymbol{u}_{\boldsymbol{k}'}^{(\lambda')})^2} = \tfrac{1}{2}(1+\cos^2\vartheta). \tag{218.14}$$

For the further discussion, let us suppose the electron to be at rest in the initial state,

$$\boldsymbol{q} = 0. \tag{218.15}$$

Then the law of energy conservation reads, according to (218.12),

$$\hbar c \left\{ k' + \frac{1}{2\varkappa}(\boldsymbol{k}'-\boldsymbol{k})^2 \right\} = \hbar c k. \tag{218.16}$$

This is a quadratic equation for the determination of k' because

$$(\boldsymbol{k}'-\boldsymbol{k})^2 = k'^2 + k^2 - 2kk'\cos\vartheta.$$

Its solution is

$$k' = k\cos\vartheta - \varkappa + \sqrt{\varkappa^2 + 2\varkappa k(1-\cos\vartheta) - k^2\sin^2\vartheta}. \tag{218.17}$$

To evaluate ρ_f, Eq. (218.11), we derive from (218.16)

$$\frac{dE_f}{dk'} = \hbar c \left\{ 1 + \frac{1}{\varkappa}(k' - k\cos\vartheta) \right\};$$

therefore,

$$\rho_f = \frac{\mathscr{V}}{8\pi^3 \hbar c} \cdot \frac{k'^2}{1 + \frac{1}{\varkappa}(k' - k\cos\vartheta)} d\Omega', \tag{218.18}$$

and the differential scattering cross section, according to (218.10), (218.9), (218.14) and (218.18) becomes

$$d\sigma = \left(\frac{e^2}{mc^2}\right)^2 \frac{k'^2}{1 + \frac{1}{\varkappa}(k' - k\cos\vartheta)} \frac{\frac{1}{2}(1 + \cos^2\vartheta)}{kk'} d\Omega', \tag{218.19}$$

where k' still may be replaced by the full expression (218.17).

So far, these are all rigorous unrelativistic formulae. They can, of course, be used only as long as the electron does not receive a kinetic energy comparable with mc^2:

$$E_{\text{kin}} = \hbar c(k - k') \ll mc^2 \quad \text{or} \quad k - k' \ll \varkappa .$$

It is therefore reasonable to expand (218.17) and (218.19) in powers of $k/\varkappa$:

$$\frac{k'}{\varkappa} = \frac{k}{\varkappa} - \frac{k^2}{\varkappa^2}(1 - \cos\vartheta) + \cdots$$

and

$$d\sigma = \frac{1}{2}\left(\frac{e^2}{mc^2}\right)^2 \left\{ 1 - 2\frac{k}{\varkappa}(1 - \cos\vartheta) \right\} (1 + \cos^2\vartheta) d\Omega'. \tag{218.20}$$

The second-order contribution from W', Eq. (218.2) vanishes for $q=0$. In the relativistic treatment, the Dirac expression (199.1) for the current density is generally used which only in second order can originate Compton transitions. The expression, however, may be split up according to Problem 199, so that the relativistic treatment may be performed in complete analogy to the unrelativistic method of the present Problem.

The total cross section then follows by elementary integration of (218.20) over all directions and may be written

$$\sigma = \frac{8\pi}{3}\left(\frac{e^2}{mc^2}\right)^2 \frac{1}{1 + 2\frac{k}{\varkappa}} . \tag{218.21}$$

It is well known from classical electrodynamics that the so-called Thomson cross section,

$$\sigma_0 = \frac{8\pi}{3}\left(\frac{e^2}{mc^2}\right)^2 = 6.652 \times 10^{-25}\ \mathrm{cm}^2, \tag{218.22}$$

represents the long-wavelength limit to our problem. The additional factor in (218.21), lowering the cross section with increasing photon energy ($k/\varkappa = \hbar\omega/mc^2$), is a first quantum theoretical correction, sufficient as long as $k/\varkappa \ll 1$, or as long as the wavelength is still large compared with the Compton wavelength $1/\varkappa = \hbar/mc$. (For $\hbar\omega = mc^2 = 0.51$ MeV or $k = \varkappa$, the wavelength is $\lambda = 2\pi\hbar/mc$.)

NB 1. The second-order contribution from W', Eq. (218.2) vanishes for $q = 0$. In the relativistic treatment, usually the current density expression (199.1) is used, which can only in second order originate Compton transitions. If this expression is split up according to Problem 199, we may formulate the relativistic treatment in complete analogy to the unrelativistic one of the present problem.

NB 2. At higher photon energies the electron field has to be treated by the Dirac theory. The result then is the Klein-Nishina formula instead of (218.21). That our approximation can be used in rather a wide domain of energies may be seen from the following figures. For $k/\varkappa = 0.2$, Eq. (218.21) gives $\sigma/\sigma_0 = 0.714$ whereas the rigorous Klein-Nishina formula leads to 0.737. At $k/\varkappa = 1$ the two values are $\sigma/\sigma_0 = 0.333$ from (218.21) and 0.431 (Kl.-N.). The real values of cross sections decrease much more slowly with increasing energy than those of our approximation, e.g. at $k/\varkappa = 1000$ we find $\sigma/\sigma_0 = 0.0050$ instead of the exact value 0.0215.

Problem 219. Bremsstrahlung

In an unrelativistic treatment, the production of an x ray photon by an electron passing a heavy nucleus may be dealt with as a second-order process in which the nucleus is simply described by its electrostatic field and its mass is supposed to be infinitely large. The bremsstrahlung spectrum shall be calculated in this approximation.

Solution. Reproduced in Fig. 76 are the two simplest possible graphs of this process. In the initial state there is the nucleus at rest and an electron of momentum $\hbar\boldsymbol{q}$, in the final state the infinitely heavy nucleus is still at rest, the electron has a smaller momentum $\hbar\boldsymbol{q}'$ and a photon $(\boldsymbol{k}, \lambda)$ has been created. This creation process, together with simple Rutherford scattering of the electron at the nucleus, gives rise to the two vertices in one or the other order of succession. In consequence of the infinite mass of the nucleus, not energy but momentum may be transferred to it ($M=\infty$, p finite, $p^2/2M=0$, $v=0$) so that between the initial and final states of the other particles energy conservation will still hold, but conservation of momentum will not.

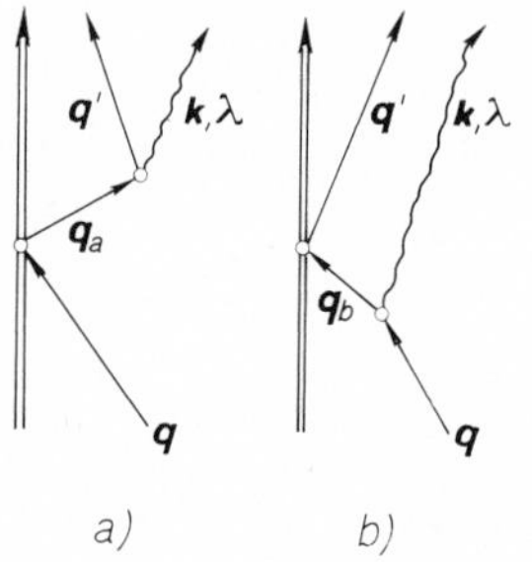

Fig. 76a and b. Lowest Feynman graphs for bremsstrahlung in unrelativistic approach. Double line nucleus, single line electron, wavy line photon

The perturbation energy consists of two terms,

$$H' = H_1 + H_2 \tag{219.1}$$

viz.

$$H_1 = -Ze^2 \int d^3x \frac{1}{r} \psi^\dagger \psi, \tag{219.2}$$

the Coulomb interaction of the nucleus (charge Ze) and electron (charge density $\rho = -e\psi^\dagger\psi$), and

$$H_2 = \frac{1}{c}\int d^3x\,(\boldsymbol{A}\cdot\boldsymbol{j}); \quad \boldsymbol{j} = -\frac{e\hbar}{2mi}(\psi^\dagger\nabla\psi - \nabla\psi^\dagger\cdot\psi), \tag{219.3}$$

the radiation interaction. The field operators are

$$\boldsymbol{A} = \sum_{\boldsymbol{k},\lambda} \sqrt{\frac{2\pi\hbar c}{k\mathscr{V}}}\, \boldsymbol{u}_{\boldsymbol{k}}^{(\lambda)} (b_{\boldsymbol{k}\lambda} e^{i\boldsymbol{k}\boldsymbol{r}} + b_{\boldsymbol{k}\lambda}^\dagger e^{-i\boldsymbol{k}\boldsymbol{r}}) \tag{219.4}$$

and

$$\psi = \frac{1}{\sqrt{\mathscr{V}}}\sum_\nu c_\nu e^{i\boldsymbol{q}_\nu \boldsymbol{r}}; \quad \nabla\psi = \frac{i}{\sqrt{\mathscr{V}}}\sum_\nu \boldsymbol{q}_\nu c_\nu e^{i\boldsymbol{q}_\nu\boldsymbol{r}}. \tag{219.5}$$

Putting (219.4) and (219.5) in (219.2) and (219.3) and performing the space integrations, we arrive at

$$H_1 = -\frac{Ze^2}{\mathscr{V}} \sum_{\mu\nu} c_\mu^\dagger c_\nu \frac{4\pi}{|\boldsymbol{q}_\mu - \boldsymbol{q}_\nu|^2} \tag{219.6}$$

and

$$H_2 = -\frac{e\hbar}{2mc}\sum_{\boldsymbol{k},\lambda}\sum_{\mu,\nu}\sqrt{\frac{2\pi\hbar c}{k\mathscr{V}}}\,\boldsymbol{u}_{\boldsymbol{k}\lambda}\cdot(\boldsymbol{q}_\mu+\boldsymbol{q}_\nu)c_\mu^\dagger c_\nu\{b_{\boldsymbol{k}\lambda}\delta_{\boldsymbol{k},\boldsymbol{q}_\mu-\boldsymbol{q}_\nu}+b_{\boldsymbol{k}\lambda}^\dagger\delta_{\boldsymbol{k},\boldsymbol{q}_\nu-\boldsymbol{q}_\mu}\}. \tag{219.7}$$

For graph a in Fig. 76, we have to look up the factors of $c_{\boldsymbol{q}}c_{\boldsymbol{q}_a}^\dagger$ in H_1 and of $c_{\boldsymbol{q}_a}c_{\boldsymbol{q}'}^\dagger b_{\boldsymbol{k}\lambda}^\dagger$ in H_2 to obtain the matrix elements

$$\langle\boldsymbol{q}_a|H_1|\boldsymbol{q}\rangle = -\frac{4\pi Ze^2}{\mathscr{V}}\cdot\frac{1}{|\boldsymbol{q}-\boldsymbol{q}_a|^2} \tag{219.8a}$$

and

$$\langle\boldsymbol{q}',\boldsymbol{k}\lambda|H_2|\boldsymbol{q}_a\rangle = -\frac{e\hbar}{2mc}\sqrt{\frac{2\pi\hbar c}{k\mathscr{V}}}\,\boldsymbol{u}_{\boldsymbol{k}\lambda}\cdot(\boldsymbol{q}_a+\boldsymbol{q}'). \tag{219.8b}$$

No law of conservation holds at the first vertex but, at the second, conservation of momentum yields

$$\boldsymbol{k}+\boldsymbol{q}'=\boldsymbol{q}_a \tag{219.9a}$$

so that, with $\boldsymbol{u}_{\boldsymbol{k}\lambda}$ perpendicular to $\boldsymbol{k}$,

$$\boldsymbol{u}_{\boldsymbol{k}\lambda}\cdot(\boldsymbol{q}_a+\boldsymbol{q}')=2\boldsymbol{u}_{\boldsymbol{k}\lambda}\cdot\boldsymbol{q}'. \tag{219.10a}$$

For the graph b in Fig. 76 we want the factors of $c_{\boldsymbol{q}}c_{\boldsymbol{q}_b}^\dagger b_{\boldsymbol{k}\lambda}^\dagger$ in H_2 and of $c_{\boldsymbol{q}_b}c_{\boldsymbol{q}'}^\dagger$ in H_1, viz.

$$\langle\boldsymbol{q}_b,\boldsymbol{k}\lambda|H_2|\boldsymbol{q}\rangle = -\frac{e\hbar}{2mc}\sqrt{\frac{2\pi\hbar c}{k\mathscr{V}}}\,\boldsymbol{u}_{\boldsymbol{k}\lambda}\cdot(\boldsymbol{q}+\boldsymbol{q}_b) \tag{219.11a}$$

and

$$\langle\boldsymbol{q}'|H_1|\boldsymbol{q}_b\rangle = -\frac{4\pi Ze^2}{\mathscr{V}}\cdot\frac{1}{|\boldsymbol{q}'-\boldsymbol{q}_b|^2}. \tag{219.11b}$$

In this case, momentum conservation holds at the first vertex,

$$\boldsymbol{k}-\boldsymbol{q}+\boldsymbol{q}_b=0 \tag{219.9b}$$

and therefore

$$\boldsymbol{u}_{\boldsymbol{k}\lambda}\cdot(\boldsymbol{q}+\boldsymbol{q}_b)=2\boldsymbol{u}_{\boldsymbol{k}\lambda}\cdot\boldsymbol{q}. \tag{219.10b}$$

The energy of the initial state,

$$E_i = \frac{\hbar^2 q^2}{2m} \tag{219.12}$$

must be equal to the energy of the final state,

$$E_f = \frac{\hbar^2 q'^2}{2m} + \hbar c k, \tag{219.13}$$

hence

$$q^2 = q'^2 + 2\varkappa k; \qquad \varkappa = \frac{mc}{\hbar}. \tag{219.14}$$

Using (219.9a, b) we find for the two intermediate states

$$E_a = \frac{\hbar^2 q_a^2}{2m} = \frac{\hbar^2}{2m}(\boldsymbol{q}' + \boldsymbol{k})^2; \tag{219.15}$$

$$E_b = \frac{\hbar^2 q_b^2}{2m} + \hbar c k = \frac{\hbar^2}{2m}\{(\boldsymbol{q} - \boldsymbol{k})^2 + 2\varkappa k\}. \tag{219.16}$$

With these abbreviations we now may write the second-order matrix element in the form

$$\langle f|H'|i\rangle = \frac{\langle f|H_2|a\rangle\langle a|H_1|i\rangle}{E_i - E_a} + \frac{\langle f|H_1|b\rangle\langle b|H_2|i\rangle}{E_i - E_b},$$

which, collecting the matrix elements from (219.8a, b) and (219.11a, b) and the expressions for $\boldsymbol{q}_a$ and $\boldsymbol{q}_b$, may be written in more detail

$$\langle f|H'|i\rangle = \frac{4\pi Z e^2}{\mathscr{V}} \cdot \frac{e\hbar}{mc}\sqrt{\frac{2\pi\hbar c}{\mathscr{V}k}}\,\frac{1}{|\boldsymbol{q} - \boldsymbol{q}' - \boldsymbol{k}|^2}\left\{\frac{(\boldsymbol{u}_{\boldsymbol{k}\lambda}\cdot\boldsymbol{q}')}{E_i - E_a} + \frac{(\boldsymbol{u}_{\boldsymbol{k}\lambda}\cdot\boldsymbol{q})}{E_i - E_b}\right\}. \tag{219.17}$$

To find the cross section, we must apply the Golden Rule and therefore need the final state density ρ_f. This is a little difficult to determine because, in consequence of non-conservation of momentum, the two final particles are emitted in independent directions. For one particle (1) we know that

$$\rho_1 = \frac{d^3 p_1 \mathscr{V}}{h^3 dE_1} = \frac{p_1^2}{v_1}\frac{\mathscr{V}}{h^3}d\Omega_1.$$

The other particle (2), for which ρ_2 must be a similar expression, is bound to lie within an interval dE_2 whose width and position are already determined by energy conservation when the interval for the first particle is given. Hence,

$$\rho_f = \rho_1 \rho_2 \, dE_2$$

or, in our special case, if 1 is the unrelativistic electron ($\boldsymbol{q}'$) and 2 the photon ($\boldsymbol{k}$), so that

$$\frac{p_1^2}{v_1} = m\hbar q'; \qquad \frac{p_2^2}{v_2} = \frac{1}{c}\hbar^2 k^2; \qquad dE_2 = \hbar c\, dk,$$

we have

$$\rho_f = \frac{m\hbar q'}{8\pi^3\hbar^3}\mathscr{V}\,d\Omega' \cdot \frac{\hbar^2 k^2}{8\pi^3\hbar^3 c}\mathscr{V}\,d\Omega_k \cdot \hbar c\,dk. \tag{219.18}$$

From the general formula

$$d\sigma = \frac{\mathscr{V}}{v_1}\cdot\frac{2\pi}{\hbar}\rho_f |\langle f|H|i\rangle|^2,$$

giving the differential cross section of the electron being scattered into $d\Omega'$ and the photon of polarization λ falling in the interval dk and within the solid-angle element $d\Omega_{\boldsymbol{k}}$, we then find by inserting (219.17) and (219.18):

$$d\sigma = \frac{e^2}{\hbar c}\cdot\frac{Z^2 e^4}{\pi^2}\cdot\frac{q' k}{q}\cdot\frac{1}{|\boldsymbol{q}-\boldsymbol{q}'-\boldsymbol{k}|^4}\left\{\frac{\boldsymbol{u}_{k\lambda}\cdot\boldsymbol{q}'}{E_i-E_a}+\frac{\boldsymbol{u}_{k\lambda}\cdot\boldsymbol{q}}{E_i-E_b}\right\}^2 d\Omega_k\,d\Omega'\,dk. \tag{219.19}$$

There remains the problem of determining the photon energy spectrum, whatever the directions of both particles or the polarization of the photon emitted. This means integration over the directions and summation over λ. The integration procedures in such problems are often rather laborious. In the present case, however, they become very simple, as shall now be shown.

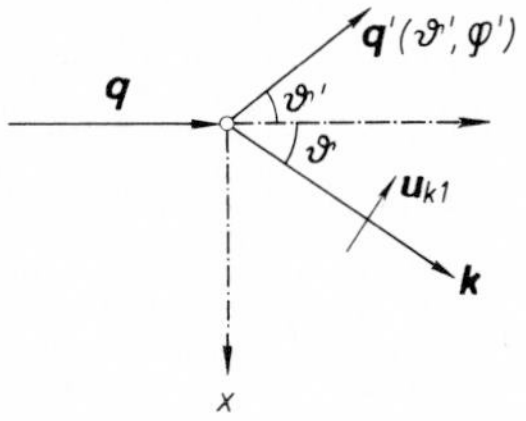

Fig. 77. Notations for bremsstrahlung

In Fig. 77 the three momenta have been drawn in a coordinate system. They are not coplanar, i.e. if the xz plane is chosen so that $\boldsymbol{q}$ and $\boldsymbol{k}$ fall in this plane, $\boldsymbol{q}'$ will have a y component. The components of the three momenta thus become

$$\boldsymbol{q} = q(0,0,1);$$
$$\boldsymbol{k} = k(\sin\vartheta, 0, \cos\vartheta);$$
$$\boldsymbol{q}' = q'(\sin\vartheta'\cos\varphi', \sin\vartheta'\sin\varphi', \cos\vartheta')$$

and of the two polarization vectors

$$\boldsymbol{u}_{\boldsymbol{k}1} = (-\cos\vartheta, 0, \sin\vartheta); \qquad \boldsymbol{u}_{\boldsymbol{k}2} = (0, 1, 0).$$

For further evaluation of the cross section, we now make the tyipically unrelativistic approximation to neglect the photon momentum compared to that of the electron, since from (219.14) it may be concluded that $k \ll q$ and $k \ll q'$ because $\varkappa$ is large. This allows us to simplify the energy denominators in (219.19). Making use of Eqs. (219.12), (219.15) and (219.16), we get

$$E_i - E_a = \frac{\hbar^2}{2m}(q^2 - q'^2 - 2\boldsymbol{q}'\cdot\boldsymbol{k} - k^2);$$

$$E_i - E_b = \frac{\hbar^2}{2m}(-2\varkappa k + 2\boldsymbol{q}\cdot\boldsymbol{k} - k^2).$$

In both expressions the two last terms then may be neglected, and $2\varkappa k$ replaced by $q^2 - q'^2$ according to (219.14) so that

$$E_i - E_a \simeq \frac{\hbar^2}{2m}(q^2 - q'^2) \simeq -(E_i - E_b). \tag{219.20}$$

The two energy denominators in (219.19) thus becoming opposite and equal, we may simply subtract the two numerators for either $\lambda = 1$ or $\lambda = 2$:

$$\begin{aligned} \boldsymbol{u}_{\boldsymbol{k}1}\cdot\boldsymbol{q}' - \boldsymbol{u}_{\boldsymbol{k}1}\cdot\boldsymbol{q} &= q'(-\cos\vartheta\sin\vartheta'\cos\varphi' + \sin\vartheta\cos\vartheta') - q\sin\vartheta; \\ \boldsymbol{u}_{\boldsymbol{k}2}\cdot\boldsymbol{q}' - \boldsymbol{u}_{\boldsymbol{k}2}\cdot\boldsymbol{q} &= q'\sin\vartheta'\sin\varphi'. \end{aligned} \tag{219.21}$$

To perform summation over λ we then square and add these two expressions.

Finally, the Rutherford denominator in (219.19) may be written in the same approximation

$$(\boldsymbol{q} - \boldsymbol{q}' - \boldsymbol{k})^4 \simeq (\boldsymbol{q} - \boldsymbol{q}')^4 = (q^2 + q'^2 - 2qq'\cos\vartheta')^2. \tag{219.22}$$

Assembling all these factors, it is seen at once that the angles ϑ and φ' occur in the sum of the squares of (219.21) only, so that integration over these angles can be performed separately in these terms by a straightforward elementary calculation:

$$\int_0^{2\pi} d\varphi' \oint d\Omega_k \sum_\lambda (\boldsymbol{u}_{\boldsymbol{k}\lambda}\cdot\boldsymbol{q}' - \boldsymbol{u}_{\boldsymbol{k}\lambda}\cdot\boldsymbol{q})^2 = \frac{16\pi^2}{3}(q^2 + q'^2 - 2qq'\cos\vartheta'). \tag{219.23}$$

Fortunately we find here the same bracket in the numerator of the cross section that according to (219.22) occurs twice in its denominator. Gathering up all these factors, we then arrive at

$$d\sigma(k) = Z^2\left(\frac{e^2}{\hbar c}\right)^3 \cdot \frac{16}{3}\,\frac{q'}{q}\,\frac{dk}{k}\int\limits_{-1}^{+1}\frac{d\cos\vartheta'}{q^2+q'^2-2qq'\cos\vartheta'}$$

where $d\sigma(k)$ stands for the partial cross section of a photon created in the energy interval dk, whatever else may have happened. The integral permits elementary evaluation,

$$\int\limits_{-1}^{+1}\frac{d\cos\vartheta'}{q^2+q'^2-2qq'\cos\vartheta'} = \frac{1}{qq'}\log\frac{q+q'}{q-q'},$$

leaving us finally with the result

$$d\sigma(k) = \frac{16}{3}Z^2\left(\frac{e^2}{\hbar c}\right)^3\left(\frac{1}{q^2}\log\frac{q+q'}{q-q'}\right)\frac{dk}{k}. \tag{219.24}$$

Using (219.14), the momenta may be eliminated from this formula and q' be expressed by the energies E of the incident electron, and $E_k = \hbar c k$ of the outgoing photon:

$$\log\frac{q+q'}{q-q'} = \log\frac{(q+q')^2}{q^2-q'^2} = \log\left\{\frac{(\sqrt{E}+\sqrt{E-E_k})^2}{E_k}\right\}.$$

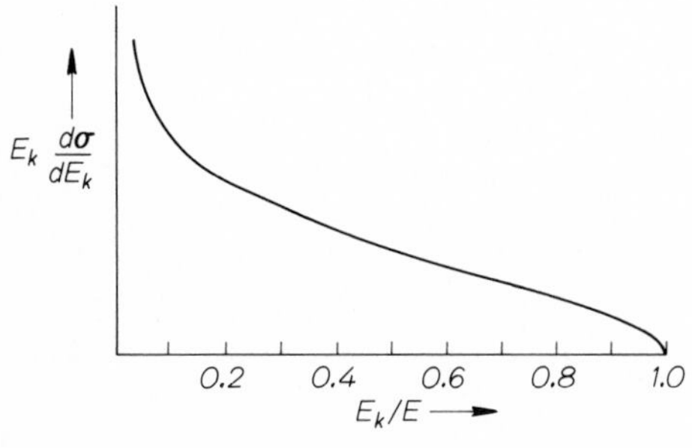

Fig. 78. Bremsstrahlung intensity distribution in unrelativistic theory. The logarithmic divergence at $E_k=0$ does not occur if screening of the Coulomb field is taken into account

This bremsstrahlung spectrum has been shown in Fig. 78. It shows a remarkable singularity for the production of photons of very small energy, the so-called infrared divergence.

Literature. For relativistic treatment and for screening problems cf. Heitler, W.: Quantum Theory of Radiation, 3rd ed., Oxford 1954, pp. 242—256.

Mathematical Appendix

Coordinate systems

We start with rectangular coordinates x, y, z and list the transformation formulae for some frequently occurring curvilinear coordinate systems. We add the formulae of transformation for the distance from the coordinate centre,

$$r = \sqrt{x^2+y^2+z^2}$$

and of the Laplacian,

$$\nabla^2 = \frac{\partial^2}{\partial x^2} + \frac{\partial^2}{\partial y^2} + \frac{\partial^2}{\partial z^2}.$$

a) Spherical polar coordinates. Let the z axis be the polar axis and denote by ϑ the angle between the vector $\boldsymbol{r}$ and this axis, and by φ the azimuth angle about this axis (cf. Fig. 33 on p. 145, vol. I).

$$x = r \sin\vartheta \cos\varphi; \quad y = r \sin\vartheta \sin\varphi; \quad z = r\cos\vartheta;$$

$$\nabla^2 u = \frac{\partial^2 u}{\partial r^2} + \frac{2}{r}\frac{\partial u}{\partial r} + \frac{1}{r^2}\left\{\frac{1}{\sin\vartheta}\frac{\partial}{\partial\vartheta}\left(\sin\vartheta\frac{\partial u}{\partial\vartheta}\right) + \frac{1}{\sin^2\vartheta}\frac{\partial^2 u}{\partial\varphi^2}\right\}.$$

b) Circular cylindrical coordinates. Let the z axis be the common axis of circular cylinders of radii ρ=constans. The angle φ again shall be the azimuth angle about the z axis, and ρ, φ, z be chosen as coordinates.

$$x = \rho\cos\varphi; \quad y = \rho\sin\varphi; \quad z = z; \quad r = \sqrt{\rho^2+z^2};$$

$$\nabla^2 u = \frac{\partial^2 u}{\partial \rho^2} + \frac{1}{\rho}\frac{\partial u}{\partial \rho} + \frac{1}{\rho^2}\frac{\partial^2 u}{\partial\varphi^2} + \frac{\partial^2 u}{\partial z^2}.$$

c) Parabolic coordinates. Let the z axis be the common axis of rotation for two sets of paraboloids ξ=constans and η=constans, all having their foci in the coordinate centre ($z=0$) and opened the one towards the positive, the other towards the negative z axis. Again φ

shall be the azimuth angle about the z axis. There are two customary choices then of the coordinates ξ, η, φ.

First system. $x=\sqrt{\xi\eta}\cos\varphi;\quad y=\sqrt{\xi\eta}\sin\varphi;\quad z=\frac{1}{2}(\xi-\eta);$
$r=\frac{1}{2}(\xi+\eta);\quad \xi=r+z;\quad \eta=r-z;\quad \rho=\sqrt{\xi\eta};$

$$\nabla^2 u=\frac{4}{\xi+\eta}\left\{\frac{\partial}{\partial\xi}\left(\xi\frac{\partial u}{\partial\xi}\right)+\frac{\partial}{\partial\eta}\left(\eta\frac{\partial u}{\partial\eta}\right)+\frac{\xi+\eta}{4\xi\eta}\frac{\partial^2 u}{\partial\varphi^2}\right\}.$$

Second system. $x=\xi\eta\cos\varphi;\quad y=\xi\eta\sin\varphi;\quad z=\frac{1}{2}(\xi^2-\eta^2);$
$r=\frac{1}{2}(\xi^2+\eta^2);\quad \xi^2=r+z;\quad \eta^2=r-z;\quad \rho=\xi\eta;$

$$\nabla^2 u=\frac{1}{\xi^2+\eta^2}\left\{\frac{1}{\xi}\frac{\partial}{\partial\xi}\left(\xi\frac{\partial u}{\partial\xi}\right)+\frac{1}{\eta}\frac{\partial}{\partial\eta}\left(\eta\frac{\partial u}{\partial\eta}\right)+\left(\frac{1}{\xi^2}+\frac{1}{\eta^2}\right)\frac{\partial^2 u}{\partial\varphi^2}\right\}.$$

d) Ellipsoidal coordinates. Two points lying on the z axis at $z=\pm c$ shall be chosen as common foci to a set of prolate ellipsoids of rotation described by ξ=constans. There exists a set of two-sheet hyperboloids of rotation orthogonally intersecting them and having the same foci, η=constans. Again φ shall be the azimuth angle about the z axis. Let r_1 and r_2 be the distances from the two foci $z=-c$ and $z=+c$, respectively.

$$x=c\sqrt{(\xi^2-1)(1-\eta^2)}\cos\varphi;\quad y=c\sqrt{(\xi^2-1)(1-\eta^2)}\sin\varphi;\quad z=c\,\xi\eta;$$

$$r_1=c(\xi+\eta);\quad r_2=c(\xi-\eta);\quad \xi=\frac{1}{2c}(r_1+r_2);\quad \eta=\frac{1}{2c}(r_1-r_2);$$

$$r=c\sqrt{\xi^2+\eta^2-1};\quad \rho=c\sqrt{(\xi^2-1)(1-\eta^2)}.$$

Domains of values: $1\le\xi<\infty;\ -1\le\eta\le+1;\ 0\le\varphi<2\pi.$

$$\nabla^2 u=\frac{1}{c^2(\xi^2-\eta^2)}\left\{\frac{\partial}{\partial\xi}(\xi^2-1)\frac{\partial u}{\partial\xi}+\frac{\partial}{\partial\eta}(1-\eta^2)\frac{\partial u}{\partial\eta}+\frac{\xi^2-\eta^2}{(1-\eta^2)(\xi^2-1)}\frac{\partial^2 u}{\partial\varphi^2}\right\}.$$

Γ function

The *Γ* function is a generalization of the *factorial*, the latter being defined only for positive integers by

$$n!=1\cdot 2\cdot 3\ldots n \tag{1}$$

and having the special property

$$(n+1)!=(n+1)\,n!. \tag{2}$$

It can equally well be defined by the *Euler integral*,

$$n! = \int_0^\infty dt\, e^{-t} t^n \tag{3}$$

which supplements the value $0!=1$ which would otherwise be meaningless.

The generalizations of (2) and (3) for any complex number $z=x+iy$ are

$$\Gamma(z+1)=z\,\Gamma(z); \qquad \Gamma(n+1)=n! \tag{4}$$

and

$$\Gamma(z) = \int_0^\infty dt\, e^{-t} t^{z-1} \quad (\text{if } \operatorname{Re} z>0). \tag{5}$$

This function is meromorphic and has poles along the negative real axis at $z=-n$ $(n=0,1,2,\ldots)$ with residues $(-1)^n/n!$.

Special values.

$$\Gamma(1)=0!=1; \qquad \Gamma(n)=(n-1)!; \tag{6}$$

$$\Gamma(\tfrac{1}{2})=\sqrt{\pi}; \qquad \Gamma(\tfrac{3}{2})=(\tfrac{1}{2})!=\tfrac{1}{2}\sqrt{\pi}; \tag{7}$$

$$\Gamma(n+\tfrac{3}{2})=(n+\tfrac{1}{2})! = \frac{(2n+1)!}{2^{2n+1}\,n!}\sqrt{\pi}. \tag{8}$$

Relations between functions of different arguments.

$$\Gamma(z)\,\Gamma(1-z) = \frac{\pi}{\sin \pi z}; \tag{9}$$

$$\Gamma(2z) = \frac{1}{\sqrt{\pi}}\, 2^{2z-1}\,\Gamma(z)\,\Gamma(z+\tfrac{1}{2}). \tag{10}$$

Infinite series or product expansions. To compute the complex number

$$\Gamma(x+iy)=\xi e^{i\eta} \tag{11a}$$

we may use the expansions

$$\xi=\Gamma(x)\prod_{n=0}^{\infty}\left\{1+\frac{y^2}{(x+n)^2}\right\}^{-\frac{1}{2}} \tag{11b}$$

and

$$\eta=y\left\{-C+\sum_{n=1}^{\infty}\left(\frac{1}{n}-\frac{1}{y}\tan^{-1}\frac{y}{x+n-1}\right)\right\} \tag{11c}$$

with

$$C = \int_0^\infty dt\, e^{-t} \log\frac{1}{t} = 0.577215\ldots \tag{11d}$$

the Euler constant. For $x=1$ we get

$$\xi^2 = |\Gamma(1+iy)|^2 = \frac{\pi y}{\sinh \pi y}. \tag{12}$$

Asymptotic behaviour. For $|z| \gg 1$ and $|\arg z| < \pi$ (i.e. excluding the negative real axis with its poles) we may use *Stirling's formula*

$$\log \Gamma(z) = \left(z - \frac{1}{2}\right) \log z - z + \frac{1}{2} \log 2\pi + O\left(\frac{1}{z}\right) \tag{13}$$

or

$$\Gamma(z) \simeq \sqrt{\frac{2\pi}{z}}\, e^{z(\log z - 1)}. \tag{14}$$

The formula is often used for

$$z! = \Gamma(z+1) = z\,\Gamma(z) \simeq \sqrt{2\pi z}\, e^{z(\log z - 1)} \left\{1 + \frac{1}{12z} + \frac{1}{288z^2} + \cdots\right\}. \tag{15}$$

The last series is semiconvergent. Putting it $=1$, we get the following comparison:

n	$n!$	$\sqrt{2\pi n}\, e^{n(\log n - 1)}$
0	1	0
1	1	0.925
2	2	1.920
3	6	5.836
4	24	23.506
5	120	118.01

Bessel functions

The differential equation

$$u'' + \frac{1}{z} u' + \left(1 - \frac{\nu^2}{z^2}\right) u = 0 \tag{1}$$

is solved by

$$u = A\, J_\nu(z) + B\, N_\nu(z) \tag{2}$$

or by

$$u = C_1 H_\nu^{(1)}(z) + C_2 H_\nu^{(2)}(z). \tag{3}$$

The function J_ν is called *Bessel function* (in the proper sense), the function N_ν *Neumann function.* If ν is not an integer, the definition

$$N_\nu(z) = \frac{1}{\sin\pi\nu}(\cos\pi\nu\, J_\nu(z) - J_{-\nu}(z)) \tag{4}$$

may be used, otherwise (i.e. for $\nu=n$ with $n=0, \pm 1, \pm 2, \ldots$),

$$J_{-n}(z) = (-1)^n J_n(z) \tag{5}$$

is no longer linearly independent of J_n. The function $N_n(z)$ still may then be defined by its asymptotics (see below).

The Bessel function J_ν can be defined by its power series

$$J_\nu(z) = \left(\frac{z}{2}\right)^\nu \sum_{n=0}^{\infty} \frac{(-1)^n}{n!\,\Gamma(\nu+n+1)}\left(\frac{z}{2}\right)^{2n}, \tag{6}$$

converging in the whole z plane cut along the negative real axis, since $z=0$ is a branch point.

The functions $H_\nu^{(1)}$ and $H_\nu^{(2)}$ are called *Hankel functions* of the first and second kind. They are defined by

$$H_\nu^{(1)}(z) = J_\nu(z) + i N_\nu(z); \qquad H_\nu^{(2)}(z) = J_\nu(z) - i N_\nu(z). \tag{7}$$

The fundamental system (2) of solutions has real values for real z; its Wronskian is $2/(\pi z)$. The Wronskian of (3) is $-4i/(\pi z)$. If ν is not integer, J_ν and $J_{-\nu}$ form a third fundamental system of solutions with the Wronskian $-2\sin\pi\nu/(\pi z)$.

Recurrence relations. For each one of the four types of functions defined by (2) and (3) there hold the relations,

$$u_{\nu-1} + u_{\nu+1} = \frac{2\nu}{z} u_\nu; \qquad u_{\nu-1} - u_{\nu+1} = 2u'_\nu \tag{8a}$$

or

$$u_{\nu+1} = \frac{\nu}{z} u_\nu - u'_\nu; \qquad u_{\nu+1} = \frac{2\nu}{z} u_\nu - u_{\nu-1}. \tag{8b}$$

Asymptotic behaviour. With the abbreviation

$$\zeta = z - \frac{\pi}{2}(\nu + \tfrac{1}{2}) \tag{9a}$$

for $|z| \gg 1 + |\nu|$ and $|\arg z| < \pi$, i.e. for large values of $|z|$ in the z plane cut along the negative real axis, there hold the asymptotic formulae

$$\left.\begin{aligned} J_\nu(z) &\to \sqrt{\frac{2}{\pi z}}\cos\zeta; & N_\nu(z) &\to \sqrt{\frac{2}{\pi z}}\sin\zeta; \\ H_\nu^{(1)}(z) &\to \sqrt{\frac{2}{\pi z}}\,e^{i\zeta}; & H_\nu^{(2)}(z) &\to \sqrt{\frac{2}{\pi z}}\,e^{-i\zeta}. \end{aligned}\right\} \tag{9b}$$

Functions of the type $z^{-\frac{1}{2}} H_\nu^{(1,2)}(z)$ for real z describe outgoing and incoming spherical waves, respectively.

Modified Bessel functions. The functions

$$I_\nu(x) = i^{-\nu} J_\nu(i x) \tag{10}$$

and

$$K_\nu(x) = \frac{i\pi}{2} i^\nu H_\nu^{(1)}(i x) \tag{11a}$$

or, if ν is not an integer,

$$K_\nu(x) = \frac{\pi}{2 \sin \nu\pi} [I_{-\nu}(x) - I_\nu(x)] \tag{11b}$$

have real values for real positive x. The last function is of special interest because of its asymptotic behaviour for large x:

$$K_\nu(x) \to \sqrt{\frac{\pi}{2x}}\, e^{-x}. \tag{12}$$

More formulae for K_0 and K_1 are given in Problem 185. — In Problem 99 the differential equation

$$u'' - g^2 x^{-n} u = 0 \tag{13a}$$

has been solved by

$$u = \sqrt{x}\, K_{\frac{1}{2\lambda}}\left(\frac{g}{\lambda} x^{-\lambda}\right) \quad \text{with } \lambda = \frac{n-2}{2}. \tag{13b}$$

In some optical diffraction problems the function

$$\mathrm{Ai}(x) = \frac{1}{\pi} \sqrt{\frac{x}{3}}\, K_{\frac{1}{3}}(\tfrac{2}{3} x^{\frac{3}{2}}) \tag{14a}$$

plays a role. It is called the *Airy function*. Its analytic continuation to negative x values yields

$$\mathrm{Ai}(-x) = \tfrac{1}{3}\sqrt{x}\{J_{\frac{1}{3}}(\tfrac{2}{3} x^{\frac{3}{2}}) + J_{-\frac{1}{3}}(\tfrac{2}{3} x^{\frac{3}{2}})\}. \tag{14b}$$

In this book, the Airy function has been used in Problem 40 where it is shown in Fig. 28. It might also be used with advantage in Problem 117 where, however, regress to the functions J_ν and I_ν with $\nu = \pm\frac{1}{3}$ has been preferred.

Spherical Bessel functions. The index values $\nu = l + \frac{1}{2}$ with integer $l = 0, 1, 2, \ldots$ play an important role, because they occur in factorizing the solutions of the wave equation in spherical polar coordinates. It is usual to introduce the four standard types[1]

---

[1] In the literature the functions here denoted by j_l etc are often denoted by $\hat{j}_l$ etc with $j_l = \frac{1}{z}\hat{j}_l$. This has the advantage that $h_l^{(1,2)}(z) = \frac{1}{z}\hat{h}_l^{(1,2)}(z)$ then become outgoing and incoming spherical waves.

$$j_l(z) = \sqrt{\frac{\pi z}{2}}\, J_{l+\frac{1}{2}}(z); \tag{15}$$

$$n_l(z) = \sqrt{\frac{\pi z}{2}}\, N_{l+\frac{1}{2}}(z) = (-1)^{l+1}\sqrt{\frac{\pi z}{2}}\, J_{-(l+\frac{1}{2})}(z); \tag{16}$$

$$h_l^{(1)}(z) = j_l(z) + i\, n_l(z); \quad h_l^{(2)}(z) = j_l(z) - i\, n_l(z). \tag{17}$$

They are solutions of the differential equation

$$u'' + \left(1 - \frac{(l+\frac{1}{2})^2}{z^2}\right) u = 0 \tag{18}$$

and have very simple asymptotics:

$$\begin{aligned} &j_l(z) \to \sin\left(z - \frac{l\pi}{2}\right); \qquad && n_l(z) \to -\cos\left(z - \frac{l\pi}{2}\right); \\ &h_l^{(1)} \to i^{-(l+1)} e^{iz}; && h_l^{(2)}(z) \to i^{l+1} e^{-iz}. \end{aligned} \tag{19}$$

The most important Wronskians are

$$j_l n_l' - n_l j_l' = 1; \quad h_l^{(1)} h_l^{(2)\prime} - h_l^{(2)} h_l^{(1)\prime} = -2i. \tag{20}$$

For $|z| \ll l + \frac{1}{2}$ we have

$$j_l(z) \simeq \frac{2^l\, l!}{(2l+1)!}\, z^{l+1}; \quad n_l(z) = -\frac{(2l)!}{2^l\, l!}\, z^{-l}. \tag{21}$$

Recurrence relations. For each one of the four types of functions defined by (15)–(17) there hold the relations

$$u_{l+1} = \frac{2l+1}{z}\, u_l - u_{l-1}; \quad u_{l+1} = \frac{l+1}{z}\, u_l - u_l'. \tag{22}$$

The first of these relations may also be used for constructing functions with negative l.

The spherical Bessel functions are elementary functions. The simplest of them are assembled in the following survey.

$$j_0 = \sin z; \qquad n_0 = -\cos z;$$

$$j_1 = \frac{\sin z}{z} - \cos z; \qquad n_1 = -\frac{\cos z}{z} - \sin z;$$

$$j_2 = \left(\frac{3}{z^2} - 1\right)\sin z - \frac{3}{z}\cos z; \qquad n_2 = -\left(\frac{3}{z^2} - 1\right)\cos z - \frac{3}{z}\sin z$$

and

$$h_0^{(1)} = -i\,e^{iz}; \qquad h_0^{(2)} = i\,e^{-iz};$$

$$h_1^{(1)} = \left(-\frac{i}{z} - 1\right)e^{iz}; \qquad h_1^{(2)} = \left(\frac{i}{z} - 1\right)e^{-iz};$$

$$h_2^{(1)} = \left(-\frac{3i}{z^2} - \frac{3}{z} + i\right)e^{iz}; \qquad h_2^{(2)} = \left(\frac{3i}{z^2} - \frac{3}{z} - i\right)e^{-iz}.$$

Legendre functions

The differential equation

$$(1-z^2)u'' - 2zu' + \nu(\nu+1)u = 0 \tag{1}$$

is of the hypergeometric type with the singularities at $z=\pm 1$ and ∞. The complete solution can be written

$$u_\nu = A\,{}_2F_1\left(-\nu,\,\nu+1,\,1;\,\frac{1-z}{2}\right)$$

$$+B z^{-\nu-1}\,{}_2F_1\left(\frac{\nu}{2}+1,\,\frac{\nu}{2}+\frac{1}{2},\,\nu+\frac{3}{2};\,\frac{1}{z^2}\right), \tag{2}$$

where the factor of A is called a Legendre function of the first kind, $P_\nu(z)$, and the factor of B is (save of a normalization factor) a Legendre function of the second kind, $Q_\nu(z)$.

If ν is an integer, $l=0,1,2,\ldots$, the function of the first kind becomes a polynomial. For $z=x$ with real x and $|x|\le 1$ or $x=\cos\vartheta$, Legendre polynomials have a simple geometrical meaning being connected with spherical harmonics according to

$$P_l(\cos\vartheta) = \sqrt{\frac{4\pi}{2l+1}}\,Y_{l,0}. \tag{3}$$

Properties of Legendre polynomials. The polynomials form an orthogonal set:

$$\int_{-1}^{+1} dx\,P_l(x)\,P_{l'}(x) = \frac{1}{l+\frac{1}{2}}\,\delta_{ll'}. \tag{4}$$

The first polynomials are

$$P_0(x)=1; \quad P_1(x)=x; \quad P_2(x)=\tfrac{3}{2}x^2-\tfrac{1}{2};$$
$$P_3(x)=\tfrac{5}{2}x^3-\tfrac{3}{2}x; \quad P_4(x)=\tfrac{35}{8}x^4-\tfrac{15}{4}x^2+\tfrac{3}{8}. \tag{5}$$

They are either even or odd functions of x, according to whether the index l is even or odd, so that

$$P_l(-x)=(-1)^l P_l(x). \tag{6}$$

Polynomials of higher order than $l=1$ may be derived from the recurrence relation

$$(l+1)P_{l+1}(x)+l P_{l-1}(x)=(2l+1)x P_l(x). \tag{7}$$

The derivatives are connected with the polynomials by

$$(1-x^2)P_l'=l(P_{l-1}-x P_l)=(l+1)(x P_l-P_{l+1}) \tag{8a}$$

whence there follows

$$(2l+1)P_l=P_{l+1}'-P_{l-1}'. \tag{8b}$$

At $x=\pm 1$ we have

$$P_l(\pm 1)=(-1)^l; \tag{9a}$$

the n'th derivative at $x=\pm 1$ becomes

$$\frac{d^n P_l(\pm 1)}{dx^n}=(\mp 1)^{l+n}\frac{(l+n)!}{2^n n!(l-n)!}. \tag{9b}$$

The definition of the polynomials by (2) may be written, with $x=\cos\vartheta$,

$$P_l(\cos\vartheta)={}_2F_1\left(l+1,\ -l,\ 1;\ \sin^2\frac{\vartheta}{2}\right)=\sum_{n=0}^{l}(-1)^n\frac{(l+n)!}{n!^2(l-n)!}\sin^{2n}\frac{\vartheta}{2}$$

$$=1-\frac{l(l+1)}{1!^2}\sin^2\frac{\vartheta}{2}+\frac{(l-1)l(l+1)(l+2)}{2!^2}\sin^4\frac{\vartheta}{2}\ \ldots \tag{10}$$

If $l\gg 1$ and $\left|\sin\frac{\vartheta}{2}\right|$ of the order of $1/l$, this series simplifies to the Bessel series so that we get

$$P_l(\cos\vartheta)\simeq J_0\left((2l+1)\sin\frac{\vartheta}{2}\right), \tag{11}$$

with an error of the order of $1/l^2$. The zeros of P_l are still given rather accurately even up to large angles ϑ. For the example $l=10$ the approximation has been represented in Fig. 55 (p. 275 of vol. I).

For geometrical relations, cf. spherical harmonics.

Legendre functions of the first kind. Expansions. If ν is not an integer, the expansion (10) is no longer finite so that $P_\nu(x)$ becomes a transcendental with singularities at $x = \pm 1$. It may then be expanded into a series of Legendre polynomials,

$$P_\nu(x) = \frac{\sin \pi \nu}{\pi} \sum_{n=0}^{\infty} \frac{(-1)^n (2n+1)}{(\nu - n)(\nu + n + 1)} P_n(x). \tag{12}$$

This is a special case of the general type of expansion

$$f(x) = \sum_{n=0}^{\infty} (2n+1) f_n P_n(x) \tag{13a}$$

with coefficients

$$f_n = \tfrac{1}{2} \int_{-1}^{+1} dx\, f(x) P_n(x) \tag{13b}$$

which is always possible since the Legendre polynomials form a *complete* orthogonal set.

The following examples of such expansions (with $|x| < 1$) are often useful:

$$e^{ixy} = \frac{1}{y} \sum_{l=0}^{\infty} (2l+1) i^l j_l(y) P_l(x); \tag{14}$$

$$\frac{\sin(y\sqrt{2(1-x)})}{y\sqrt{2(1-x)}} = \sum_{l=0}^{\infty} (2l+1) j_l^2(y) P_l(x). \tag{15}$$

The series (14) and (15) are generally used with $y = kr$ and $x = \cos \vartheta$ so that $y\sqrt{2(1-x)} = 2kr \sin \frac{\vartheta}{2}$. They hold for all real values of y. Another important example may even be used to define the Legendre polynomials:

$$\frac{1}{\sqrt{1 - 2xy + y^2}} = \sum_{n=0}^{\infty} y^n P_n(x) \quad \text{if } |y| < 1. \tag{16}$$

Legendre functions of the second kind. The expansion

$$\frac{1}{z - x} = \sum_{n=0}^{\infty} (2n+1) Q_n(z) P_n(x) \tag{16a}$$

where z is any complex number, except real values between -1 and $+1$, leads to

$$Q_n(z) = \frac{1}{2} \int_{-1}^{+1} \frac{dx\, P_n(x)}{z - x}. \tag{16b}$$

The functions $Q_n(z)$ are called Legendre functions of the second kind. From the symmetry of (16a) in z and x, it follows at once that they satisfy the differential equation (1) of the polynomials $P_n(x)$. They have, however, branch points at $z=\pm 1$ with logarithmic singularities. The simplest of these functions are

$$Q_0(z) = \frac{1}{2}\log\frac{z+1}{z-1}; \qquad Q_1(z) = P_1(z)Q_0(z) - 1;$$
$$Q_2(z) = P_2(z)Q_0(z) - \frac{3}{2}z. \tag{17}$$

The higher Q_n's may be determined from the recurrence relation (7) which holds for the Q's as well as for the P's. Their general form is

$$Q_n(z) = P_n(z)Q_0(z) - W_{n-1}(z) \tag{18}$$

with W_{n-1} a polynomial of degree $n-1$, either even or odd, according to whether the index n is even or odd.

Spherical harmonics

In factorizing the solutions of the wave equation in spherical polar coordinates defined by

$$x = r\sin\vartheta\cos\varphi; \qquad y = r\sin\vartheta\sin\varphi; \qquad z = r\cos\vartheta \tag{1}$$

with z the polar axis, the angular part of the solution satisfies the differential equation

$$\frac{1}{\sin\vartheta}\frac{\partial}{\partial\vartheta}\left(\sin\vartheta\frac{\partial u}{\partial\vartheta}\right) + \frac{1}{\sin^2\vartheta}\frac{\partial^2 u}{\partial\varphi^2} + l(l+1)u = 0, \tag{2}$$

where the separation parameter l is integer, $l=0,1,2,\ldots$ By further factorization,

$$u = \Theta(\vartheta)\,e^{im\varphi} \tag{3a}$$

with $m=0,\pm 1,\pm 2,\ldots$ we get

$$\frac{1}{\sin\vartheta}\frac{d}{d\vartheta}\left(\sin\vartheta\frac{d\Theta}{d\vartheta}\right) + \left[l(l+1) - \frac{m^2}{\sin^2\vartheta}\right]\Theta = 0 \tag{3b}$$

or, using

$$t = \frac{z}{r} = \cos\vartheta$$

as variable,

$$(1-t^2)\frac{d^2\Theta}{dt^2} - 2t\frac{d\Theta}{dt} + \left[l(l+1) - \frac{m^2}{1-t^2}\right]\Theta = 0. \tag{3c}$$

This differential equation is a slight generalization of the one of the Legendre polynomial $P_l(t)$ into which it passes over for $m=0$, but it remains of hypergeometric type for $m \neq 0$. Its only regular solution, in arbitrary normalization, may be written

$$\Theta_{l,m} = (1-t^2)^{\frac{m}{2}} \frac{d^m P_l(t)}{dt^m} \tag{3d}$$

for $m \geq 0$, and $\Theta_{l,-m}$ turns out to be the same function. Since the polynomial P_l cannot be differentiated more than l times without vanishing, there exist $2l+1$ regular solutions with integer $|m| \leq l$ for every value of l.

We still have a free hand concerning normalization. We adopt its most customary form, which best reflects the geometrical meaning of these solutions regular on the entire surface of the unit sphere,

$$Y_{l,m}(\vartheta, \varphi) = \frac{1}{\sqrt{2\pi}} \mathscr{P}_l^m(\vartheta)\, e^{im\varphi} \tag{4}$$

with

$$\oint d\Omega\, |Y_{l,m}|^2 = 1 \tag{5}$$

or

$$\int_0^\pi d\vartheta \sin\vartheta\, |\mathscr{P}_l^m(\vartheta)|^2 = 1. \tag{6}$$

Eq. (3d) then has to be normalized according to

$$\mathscr{P}_l^m = \sqrt{\frac{2l+1}{2}\frac{(l-m)!}{(l+m)!}}\,(1-t^2)^{\frac{m}{2}} \frac{d^m P_l(t)}{dt^m} \tag{7}$$

for $m \geq 0$. To include negative values of m, we define

$$\mathscr{P}_l^{-m}(\vartheta) = (-1)^m \mathscr{P}_l^m(\vartheta). \tag{8}$$

In this standard form the spherical harmonics $Y_{l,m}$ up to $l=3$ are tabulated on p. 174 of vol. I. For $m=0$ we have the useful relations

$$\mathscr{P}_l^0 = \sqrt{l+\tfrac{1}{2}}\, P_l \quad \text{and} \quad Y_{l,0} = \sqrt{\frac{2l+1}{4\pi}}\, P_l. \tag{9}$$

Recurrence relations. There exist several important relations, connecting spherical harmonics with such of neighbouring values of l and m, viz.

$$\sin\vartheta\, e^{i\varphi}\, Y_{l,m} = a_{l,m}\, Y_{l+1,m+1} - a_{l-1,-m-1}\, Y_{l-1,m+1}; \tag{10a}$$

$$\sin\vartheta\, e^{-i\varphi}\, Y_{l,m} = -a_{l,-m}\, Y_{l+1,m-1} + a_{l-1,m-1}\, Y_{l-1,m-1}; \tag{10b}$$

$$\cos\vartheta\, Y_{l,m} = b_{l,m}\, Y_{l+1,m} + b_{l-1,m}\, Y_{l-1,m} \tag{11}$$

with the abbreviations

$$a_{l,m} = \sqrt{\frac{(l+m+1)(l+m+2)}{(2l+1)(2l+3)}}; \qquad b_{l,m} = \sqrt{\frac{(l+m+1)(l-m+1)}{(2l+1)(2l+3)}}. \tag{12}$$

By repeated application of the relations (10) and (11) higher powers of $\sin\vartheta$ and $\cos\vartheta$ may be multiplied by $Y_{l,m}$ so that indices differing by more than ± 1 from l and m may appear on the right-hand side.

Derivatives. If the operators

$$r\left(\frac{\partial}{\partial x} \pm i\frac{\partial}{\partial y}\right) = e^{\pm i\varphi}\left\{\sin\vartheta\, r\frac{\partial}{\partial r} + \cos\vartheta\frac{\partial}{\partial\vartheta} \pm \frac{i}{\sin\vartheta}\frac{\partial}{\partial\varphi}\right\}$$

and

$$r\frac{\partial}{\partial z} = \cos\vartheta\, r\frac{\partial}{\partial r} - \sin\vartheta\frac{\partial}{\partial\vartheta} \tag{13}$$

are applied to a spherical harmonic, the following results are obtained:

$$r\left(\frac{\partial}{\partial x} \pm i\frac{\partial}{\partial y}\right) Y_{l,m} = \mp l\, a_{l,\pm m}\, Y_{l+1,m\pm 1} \mp (l+1)\, a_{l-1,\mp m-1}\, Y_{l-1,m\pm 1};$$

$$r\frac{\partial}{\partial z} Y_{l,m} = -b_{l,m}\, Y_{l+1,m} + (l+1)\, b_{l-1,m}\, Y_{l-1,m} \tag{14}$$

where $a_{l,m}$ and $b_{l,m}$ are defined by (12).

In the theory of angular momentum, the three hermitian operators L_x, L_y, L_z with

$$L_z = -i\left(x\frac{\partial}{\partial y} - y\frac{\partial}{\partial x}\right) \tag{15}$$

etc. cyclic, play a large role. In spherical polar coordinates we may write

$$L_\pm = L_x \pm i L_y = i e^{\pm i\varphi}\left(\mp i\frac{\partial}{\partial\vartheta} + \cot\vartheta\frac{\partial}{\partial\varphi}\right) \tag{16a}$$

and

$$L_z = -i\frac{\partial}{\partial\varphi}. \tag{16b}$$

Applied to a spherical harmonic, L_+ and L_- turn out to be shift operators which raise or lower the index m by 1:

$$L_+ Y_{l,m} = -\sqrt{l(l+1)-m(m+1)}\, Y_{l,m+1}; \tag{17a}$$

$$L_- Y_{l,m} = -\sqrt{l(l+1)-m(m-1)}\, Y_{l,m-1}. \tag{17b}$$

On the other hand, $Y_{l,m}$ is an eigenfunction of L_z:

$$L_z Y_{l,m} = m\, Y_{l,m}. \tag{18}$$

The same holds for the second-order operator

$$L^2 = L_x^2 + L_y^2 + L_z^2 = \tfrac{1}{2}(L_+ L_- + L_- L_+) + L_z^2, \tag{19}$$

viz.

$$L^2 Y_{l,m} = l(l+1)\, Y_{l,m}. \tag{20}$$

There further hold the relations

$$L_- L_+ Y_{l,m} = [l(l+1)-m(m+1)]\, Y_{l,m}; \tag{21a}$$

$$L_+ L_- Y_{l,m} = [l(l+1)-m(m-1)]\, Y_{l,m}. \tag{21b}$$

The operators L_i here defined are identical with the angular momentum component operators used in the text of this book, except for a factor $\hbar$.

Orthogonality and expansion. The spherical harmonics satisfy the orthonormality relations

$$\oint d\Omega\, Y^*_{l',m'} Y_{l,m} = \delta_{ll'}\, \delta_{mm'}. \tag{22}$$

They form a complete set of functions in the sense that any regular function on the surface of the unit sphere may be expanded into a series

$$f(\vartheta,\varphi) = \sum_{l=0}^{\infty} \sum_{m=-l}^{+l} f_{l,m}\, Y_{l,m}(\vartheta,\varphi) \tag{23a}$$

with coefficients

$$f_{l,m} = \oint d\Omega\, Y^*_{l,m}(\vartheta,\varphi)\, f(\vartheta,\varphi). \tag{23b}$$

A few important expansions are the following ones.

1. Expansion of the Legendre polynomial $P_l(\cos\gamma)$ where γ is the angle between the directions to the points (ϑ,φ) and (ϑ',φ') on the unit sphere:

$$P_l(\cos\gamma) = \frac{4\pi}{2l+1} \sum_{m=-l}^{+l} Y^*_{l,m}(\vartheta',\varphi')\, Y_{l,m}(\vartheta,\varphi). \tag{24}$$

This formula is equivalent to a transformation by rotating the polar axis through an angle γ. It mixes only spherical harmonics of the same order l. It is also called the addition theorem.

2. Plane wave. If the plane wave propagates along the polar axis, only Legendre polynomials occur:

$$e^{ikz} = e^{ikr\cos\vartheta} = \frac{1}{kr}\sum_{l=0}^{\infty}\sqrt{4\pi(2l+1)}\, i^l j_l(kr)\, Y_{l,0}(\vartheta). \tag{25}$$

This formula has been proved in extenso in Problem 81. Incidentally, it leads by inversion to an integral representation of the spherical Bessel functions:

$$j_l(z) = i^{-l}\frac{z}{2}\int_{-1}^{+1} dt\, e^{izt} P_l(t). \tag{26}$$

If the plane wave runs in the direction of a vector $\boldsymbol{k}$ with polar angles Θ, Φ, the expansion may be generalized to

$$e^{i\boldsymbol{k}\cdot\boldsymbol{r}} = \frac{4\pi}{kr}\sum_{l=0}^{\infty} i^l j_l(kr)\, Y^*_{l,m}(\Theta,\Phi)\, Y_{l,m}(\vartheta,\varphi). \tag{27}$$

3. Spherical wave, to be used as Green's function of the wave equation. If $\boldsymbol{r}$ and $\boldsymbol{r}'$ are vectors the directions of which are defined by polar angles ϑ, φ and ϑ', φ', and γ is the angle between them, then

$$\frac{e^{ik|\boldsymbol{r}-\boldsymbol{r}'|}}{4\pi|\boldsymbol{r}-\boldsymbol{r}'|} = \frac{1}{rr'}\sum_{l=0}^{\infty}\sqrt{\frac{2l+1}{4\pi}}\, \Gamma_l(r,r')\, Y_{l,0}(\cos\gamma) \tag{28a}$$

with

$$\Gamma_l(r,r') = \begin{cases} \dfrac{i}{k} j_l(kr) h_l^{(1)}(kr') & \text{for } r<r', \\ \dfrac{i}{k} j_l(kr') h_l^{(1)}(kr) & \text{for } r>r'. \end{cases} \tag{28b}$$

In the limit $k \to 0$ this yields the well-known formula

$$\frac{1}{|\boldsymbol{r}-\boldsymbol{r}'|} = \begin{cases} \dfrac{1}{r'}\sum\limits_{l=0}^{\infty}\left(\dfrac{r}{r'}\right)^l P_l(\cos\gamma) & \text{for } r<r', \\ \dfrac{1}{r}\sum\limits_{l=0}^{\infty}\left(\dfrac{r'}{r}\right)^l P_l(\cos\gamma) & \text{for } r>r'. \end{cases} \tag{29}$$

The hypergeometric series

The differential equation

$$z(1-z)v'' + \{(c-2\lambda)-(a+b+1-2\lambda-2\mu)z\}v' + \left\{\frac{\lambda(\lambda-c+1)}{z} + \frac{\mu(\mu-a-b-c)}{1-z} - [(\lambda+\mu)(\lambda+\mu-a-b)+ab]\right\}v = 0 \tag{1}$$

has regular singularities only at $z=0,1,\infty$. Putting

$$v(z)=z^{\lambda}(1-z)^{\mu}u(z) \tag{2}$$

it can be brought into the standard form

$$z(1-z)u''+[c-(a+b+1)z]u'-ab\,u=0 \tag{3}$$

which is called the hypergeometric or Gaussian differential equation. For all values of its three parameters a, b, c, except $c=-n$ with $n=0,1,2,\ldots$, it has a solution which is regular and does not vanish at $z=0$ which, when normalized according to $u(0)=1$, is called the hypergeometric series ${}_2F_1(a,b,c;z)$. Solving the differential equation (3) by series expansion at the origin we find[2]

$$\begin{aligned}{}_2F_1(a,b,c;z) = 1 &+ \frac{ab}{c}\frac{z}{1!} + \frac{a(a+1)b(b+1)}{c(c+1)}\frac{z^2}{2!} \\ &+ \frac{a(a+1)(a+2)b(b+1)(b+2)}{c(c+1)(c+2)}\frac{z^3}{3!} + \cdots\end{aligned} \tag{4a}$$

or

$${}_2F_1(a,b,c;z) = \frac{\Gamma(c)}{\Gamma(a)\Gamma(b)} \sum_{n=0}^{\infty} \frac{\Gamma(a+n)\Gamma(b+n)}{\Gamma(c+n)\,n!} z^n . \tag{4b}$$

This function is invariant with respect to exchanging the parameters a and b. If $a=-n$ or $b=-n$ $(n=0,1,2,\ldots)$ it becomes a polynomial of degree n, called a Jacobi polynomial according to the definition

$$J_n(p,q;z) = {}_2F_1(-n,p+n,q;z).$$

These polynomials form an orthogonal set according to

$$\int_0^1 dz\, z^{q-1}(1-z)^{p-q} J_m J_n = 0 \quad \text{for } m \neq n.$$

The hypergeometric series does not exist for $c=-n$; in that case however the limiting process

$$\lim_{c\to -n} \frac{{}_2F_1(a,b,c;z)}{\Gamma(c)} \tag{5}$$

$$= \frac{\Gamma(a+n+1)\Gamma(b+n+1)}{\Gamma(a)\Gamma(b)} \frac{z^{n+1}}{(n+1)!}\, {}_2F_1(a+n+1,b+n+1,n+2;z)$$

leads to a solution of the differential equation (3).

---

[2] The notation ${}_2F_1$ was introduced by Pochhammer who generalized the hypergeometric series to ${}_nF_m$ with $n+m$ parameters of which products of n appear in the numerator and of m in the denominator of the series in the same way as 2, resp. 1 in Eq. (4a). In our context we need only ${}_2F_1$ and the confluent series ${}_1F_1$.

If none of the three parameters a, b, c are either zero or negative integers, the series (4a, b) converges absolutely for $|z|<1$. Its analytic continuation beyond this circle can be made unique by a branch cut extending from $z=1$ to $z=\infty$.

The following formulae may serve for its continuation beyond $|z|<1$:

$$\begin{aligned} {}_2F_1(a,b,c;z) &= \frac{\Gamma(c)\Gamma(c-a-b)}{\Gamma(c-a)\Gamma(c-b)}\,{}_2F_1(a,b,a+b-c+1;1-z) \\ &+ \frac{\Gamma(c)\Gamma(a+b-c)}{\Gamma(a)\Gamma(b)}(1-z)^{c-a-b}\,{}_2F_1(c-a,c-b,c-a-b+1;1-z) \end{aligned} \tag{6}$$

and

$$\begin{aligned} {}_2F_1(a,b,c;z) &= \frac{\Gamma(c)\Gamma(b-a)}{\Gamma(b)\Gamma(c-a)}(-z)^{-a}\,{}_2F_1\left(a,a-c+1,a-b+1;\frac{1}{z}\right) \\ &+ \frac{\Gamma(c)\Gamma(a-b)}{\Gamma(a)\Gamma(c-b)}(-z)^{-b}\,{}_2F_1\left(b,b-c+1,b-a+1;\frac{1}{z}\right). \end{aligned} \tag{7}$$

The last formula determines the asymptotic behaviour of ${}_2F_1$ for $z\to\infty$:

$${}_2F_1(a,b,c;z) \to \frac{\Gamma(c)\Gamma(b-a)}{\Gamma(b)\Gamma(c-a)}(-z)^{-a} + \frac{\Gamma(c)\Gamma(a-b)}{\Gamma(a)\Gamma(c-b)}(-z)^{-b}. \tag{8}$$

The general solution of the hypergeometric differential equation for $|z|<1$ is

$$u = C_1\,{}_2F_1(a,b,c;z) + C_2 z^{1-c}\,{}_2F_1(a+1-c,b+1-c,2-c;z). \tag{9}$$

Only for integer $c=0,\pm1,\pm2,\ldots$ the two special solutions used in (9) become identical, as is easily seen by applying Eq. (5). The second solution then has a logarithmic singularity at $z=0$.

In the following, the most important formulae for the practical use of the hypergeometric series have been collected.

$$\begin{aligned} {}_2F_1(a,b,c;z) &= \frac{a(c-b)}{c(a-b)}\,{}_2F_1(a+1,b,c+1;z) + \frac{b(c-a)}{c(b-a)}\,{}_2F_1(a,b+1,c+1;z) \\ &= \frac{c-a}{c}\,{}_2F_1(a,b,c+1;z) + \frac{a}{c}\,{}_2F_1(a+1,b,c+1;z), \end{aligned}$$

$$z\,{}_2F_1(a,b,c;z)=\frac{c-1}{a-b}\left[{}_2F_1(a-1,b,c-1;z)-{}_2F_1(a,b-1,c-1;z)\right]$$

$$=\frac{c-1}{a-1}\left[{}_2F_1(a-1,b,c-1;z)-{}_2F_1(a-1,b-1,c-1;z)\right]$$

$$=\frac{c-a}{a-b}\,{}_2F_1(a-1,b,c;z)+\frac{c-b}{b-a}\,{}_2F_1(a,b-1,c;z)+{}_2F_1(a,b,c;z),$$

$$(1-z)\,{}_2F_1(a,b,c;z)=\frac{c-1}{a-1}\,{}_2F_1(a-1,b-1,c-1;z)+\frac{a-c}{a-1}\,{}_2F_1(a-1,b,c;z)\,.$$

Derivatives:

$$\frac{d}{dz}\,{}_2F_1(a,b,c;z)=\frac{ab}{c}\,{}_2F_1(a+1,b+1,c+1;z),$$

$$z(1-z)\frac{d}{dz}\,{}_2F_1(a,b,c;z)=\frac{a(c-a)}{a-b}\,{}_2F_1(a-1,b,c;z)+\frac{b(c-b)}{b-a}\,{}_2F_1(a,b-1,c;z)$$

$$+\frac{b(c-b)-a(c-a)}{a-b}\,{}_2F_1(a,b,c;z).$$

The confluent series

If, in the hypergeometric differential equation, we perform the limiting process $b\to\infty$, $z=x/b$, we get Kummer's differential equation,

$$\frac{d^2u}{dx^2}+(c-x)\frac{du}{dx}-au=0. \qquad (1)$$

The singularity at $z=1$ has been shifted to $x=\infty$, so that in the complex x plane there is a regular singularity still at $x=0$ but an irregular singularity at $x=\infty$ caused by the confluence of the two singularities at $z=1$ and $z=\infty$. Hence the name of the solution.

The general solution of (1) is

$$u=C_1\,{}_1F_1(a,c;x)+C_2\,x^{1-c}\,{}_1F_1(a-c+1,2-c;x) \qquad (2)$$

with the so-called confluent series ${}_1F_1$ being defined by

$${}_1F_1(a,c;z)=1+\frac{a}{c}\,\frac{z}{1!}+\frac{a(a+1)}{c(c+1)}\,\frac{z^2}{2!}+\frac{a(a+1)(a+2)}{c(c+1)(c+2)}\,\frac{z^3}{3!}+\cdots \qquad (3\text{a})$$

or

$${}_1F_1(a,c;z)=\frac{\Gamma(c)}{\Gamma(a)}\sum_{n=0}^{\infty}\frac{\Gamma(a+n)}{\Gamma(c+n)}\,\frac{z^n}{n!}\,. \qquad (3\text{b})$$

This series converges absolutely in the whole z plane. To make it unique, a branch cut has to be made from $z=0$ to $z=\infty$ which, in the standard notation used in this book, runs along the positive imaginary axis.

Only for $c=-n$, $n=0,1,2,\ldots$ is the series (3a, b) not defined, but in this case

$$\lim_{c\to -n}\frac{{}_1F_1(a,c;z)}{\Gamma(c)}=\frac{\Gamma(a+n+1)}{\Gamma(a)}\,\frac{z^{n+1}}{(n+1)!}\,{}_1F_1(a+n+1,n+2;z) \tag{4}$$

is a solution of (1) (with z instead of x).

The asymptotic behaviour of the confluent series for $|z|\to\infty$ is given by

$${}_1F_1(a,c;z)\to \mathrm{e}^{-i\pi a}\frac{\Gamma(c)}{\Gamma(c-a)}z^{-a}+\frac{\Gamma(c)}{\Gamma(a)}\mathrm{e}^{z}z^{a-c}. \tag{5}$$

This formula does not hold for $a=-n$, $n=0,1,2,\ldots$ where, according to (3), the function ${}_1F_1(-n,c;z)$ becomes a polynomial of degree n. Of special interest among these are the Laguerre polynomials

$$L_n^{(m)}(z)=\frac{(n+m)!}{n!\,m!}\,{}_1F_1(-n,m+1;z) \tag{6}$$

and the Hermite polynomials (cf. Problem 30)

$$\left.\begin{aligned} H_{2n}(z)&=(-1)^n\frac{(2n)!}{n!}\,{}_1F_1(-n,\tfrac{1}{2};z^2),\\ H_{2n+1}(z)&=(-1)^n\frac{(2n+1)!}{n!}\,2z\,{}_1F_1(-n,\tfrac{3}{2};z^2).\end{aligned}\right\} \tag{7}$$

Finally, we again list some important formulae for the confluent series:

$$\begin{aligned}{}_1F_1(a,c;z)&=\frac{a}{c}\,{}_1F_1(a+1,c+1;z)+\frac{c-a}{c}\,{}_1F_1(a,c+1;z)\\ &=\mathrm{e}^{z}\,{}_1F_1(c-a,c;-z);\end{aligned}$$

$$z\frac{d}{dz}\,{}_1F_1(a,c;z)=a\{{}_1F_1(a+1,c;z)-{}_1F_1(a,c;z)\};$$

$$\frac{d}{dz}\,{}_1F_1(a,c;z)=\frac{a}{c}\,{}_1F_1(a+1,c+1;z).$$

Some functions defined by integrals

Error integral and related functions. The error integral is defined by

$$\operatorname{erfc} z = \int_z^\infty dt\, e^{-t^2};$$

as an alternative also the definition

$$\operatorname{erf} z = \int_0^z dt\, e^{-t^2} = z\, {}_1F_1(\tfrac{1}{2}, \tfrac{3}{2}; -z^2)$$

is often used. Both integrals are connected by the complete integral

$$\operatorname{erfc} z + \operatorname{erf} z = \int_0^\infty dt\, e^{-t^2} = \tfrac{1}{2}\sqrt{\pi}.$$

There holds the power expansion

$$\operatorname{erf} z = z - \frac{z^3}{1!\,3} + \frac{z^5}{2!\,5} - \cdots;$$

the function $\operatorname{erfc} z$ can for real positive argument $z \gg 1$ be represented by the semiconvergent series

$$\operatorname{erfc} z \to \frac{e^{-z^2}}{2z} \sum_{n=0}^{\infty} (-1)^n \frac{(2n)!}{2^{2n} n!} z^{-2n} = \frac{e^{-z^2}}{2z}\left(1 - \frac{1}{2z^2} + \frac{3}{4z^4} \cdots\right).$$

From the identity

$$F(z,\beta) = \int_z^\infty dt\, e^{-\beta t^2} = \beta^{-\frac{1}{2}} \operatorname{erfc}(\sqrt{\beta}\, z)$$

there follows by differentiation,

$$\frac{\partial F}{\partial \beta} = -\int_z^\infty dt\, t^2 e^{-\beta t^2} = -\frac{1}{2}\beta^{-\frac{3}{2}} \operatorname{erfc}(\sqrt{\beta}\, z) - \frac{z}{2\beta} e^{-\beta z^2}$$

so that, with $\beta = 1$, we get the reduction formula

$$\int_z^\infty dt\, t^2 e^{-t^2} = \frac{1}{2} \operatorname{erfc} z + \frac{z}{2} e^{-z^2}.$$

In this manner, by repeated differentiations, all integrals of the form

$$\int_z^\infty dt\, t^{2n} e^{-t^2}$$

may be finally reduced to the error function. For $z=0$, this leads to the special formula for the complete integrals,

$$\int_0^\infty dt\, t^{2n} e^{-t^2} = \sqrt{\pi}\, \frac{(2n-1)!}{2^{2n}(n-1)!}.$$

With $t^2 = x$, this may be written as an Euler integral so that

$$\int_0^\infty dt\, t^{2n} e^{-t^2} = \tfrac{1}{2}\Gamma(n+\tfrac{1}{2}).$$

Exponential integral. This function is defined by

$$\mathrm{Ei}\, z = \int_{-\infty}^{z} \frac{dt}{t} e^{t}$$

and is of special interest for negative real values $z = -x$ where we write

$$\mathrm{E}_1(x) = -\mathrm{Ei}(-x) = \int_x^\infty \frac{dt}{t} e^{-t}.$$

There holds the power series

$$\mathrm{E}_1(x) = -C + \log\frac{1}{x} + x - \frac{x^2}{2!2} + \frac{x^3}{3!3} - \cdots$$

where

$$C = \int_0^\infty dt\, e^{-t} \log\frac{1}{t} = 0.577\,215\ldots$$

is the Euler constant. For $x \gg 1$, there holds the semiconvergent series

$$\mathrm{E}_1(x) = \frac{e^{-x}}{x}\left(1 - \frac{1!}{x} + \frac{2!}{x^2} - \cdots\right).$$

The exponential integral can be generalized to

$$\mathrm{E}_n(x) = \int_x^\infty \frac{dt}{t^n} e^{-t}.$$

These integrals may, by partial integration, be reduced to $\mathrm{E}_1(x)$ according to

$$\mathrm{E}_n(x) = \frac{1}{n-1}\left[x^{n-1} e^{-x} - \mathrm{E}_{n-1}(x)\right].$$

Index

for both volumes

Numbers refer to problems, not to pages. Volume I comprises the Problems 1—128, Volume II the Problems 129—219. The letter A refers to the Mathematical Appendix at the end of Volume II.